Omics for Environmental Engineering and Microbiology Systems

Bioremediation using microbes is a sustainable technology for biodegradation of target compounds, and an omics approach gives more clarity on these microbial communities. This book provides insights into the complex behavior of microbial communities and identifies enzymes/metabolites and their degradation pathways. It describes the application of microbes and their derivatives for the bioremediation of potentially toxic and novel compounds. It highlights the existing technologies along with industrial practices and real-life case studies.

Features:

- Includes recent research and development in the areas of omics and microbial bioremediation.
- Covers the broad environmental pollution control approaches such as metagenomics, metabolomics, fluxomics, bioremediation, and biodegradation of industrial wastes.
- Reviews metagenomics and waste management, and recycling for environmental cleanup.
- Describes the metagenomic methodologies and best practices, from sample collection to data analysis for taxonomies.
- Explores various microbial degradation pathways and detoxification mechanisms for organic and inorganic contaminants of wastewater with their gene expression.

This book is aimed at graduate students and researchers in environmental engineering, soil remediation, hazardous waste management, environmental modeling, and wastewater treatment.

Omics for Environmental Engineering and Microbiology Systems

Edited by
Vineet Kumar
Vinod Kumar Garg
Sunil Kumar
Jayanta Kumar Biswas

CRC Press
Taylor & Francis Group
Boca Raton London New York

CRC Press is an imprint of the
Taylor & Francis Group, an **Informa** business

Designed cover image: Shutterstock

First edition published 2023
by CRC Press
6000 Broken Sound Parkway NW, Suite 300, Boca Raton, FL 33487-2742

and by CRC Press
4 Park Square, Milton Park, Abingdon, Oxon, OX14 4RN

CRC Press is an imprint of Taylor & Francis Group, LLC

ISBN: 978-1-032-16283-6 (hbk)
ISBN: 978-1-032-16285-0 (pbk)
ISBN: 978-1-003-24788-3 (ebk)

DOI: 10.1201/9781003247883

Typeset in Times
by codeMantra

*This book is truly dedicated to my family for their abundant
support, patience, understanding, endless support,
love, and educating us to date, without them this book
would not have been possible.* —**Vineet Kumar**

*This book is dedicated to my children and my wife. Their patience
and understanding have given me the time and inspiration
to research and edit this project.* —**Vinod Kumar Garg**

*Dedicated to our students and to our teachers, and mentors,
from whom I continue to learn, and to my family for their
support, blessings, motivation, and love.* —**Sunil Kumar**

*Dedicated to my family especially my wife without whose support
this book would not have been possible.* —**Jayanta Kumar Biswas**

Contents

Preface

The concentrations of inorganic, organic, and organometallic contaminants in different environmental matrices (soil, water, and air) are increasing at an unprecedented rate and scale mainly due to diverse anthropogenic activities. Sustainable development entails the promotion and development of innovative and environment friendly green technologies for the cleanup of contaminants from the polluted environment. In the last few decades, innovative and environmentally friendly technologies, such as bioremediation using potent organisms, particularly the unseen intelligence of bacteria, yeast, plants, fungi, algae, and actinomycetes, acting in parallel or sequentially, to degrade, neutralize, remove, and/or transform contaminants from polluted environment have gained worldwide attention. It is based on the acceleration of the natural metabolic process of microorganisms that have the capacity to enzymatically attack the toxic contaminants and degrade or mineralize them into to carbon dioxide, water, microbial biomass, inorganic salts, and other by-products (metabolites), which are less toxic than the parental compounds. Bioremediation potentiality of microbes mainly relies on the enzymes produced by them and takes part in the metabolic pathways. Bioremediation technology can be applied *in situ* or *ex situ*, depending on the site in which they will be applied for cleanup. *In situ* method is performed at the original site of contamination, whereas *ex situ* treatment refers to the removal of soil or water from the original contaminated sites to subsequent treatment. In the last few decades, an array of bioremediation techniques such as bioaugmentation, biostimulation, natural attenuation, biosorption, bioleaching, composting, phytoremediation, and microbe-assisted phytoremediation have been developed, with very high public acceptance. Since bioremediation and these techniques not only represent an emerging (green) technology, but also present a great advantage of being cost-effective, sustainable, and promising technologies when compared to the conventional methods, the book, *Omics for Environmental Engineering and Microbiology Systems*, presents an up-to-date and comprehensive collection of chapters contributed by leading experts in relevant fields of bioremediation research and applications working in the top institutions around the globe.

The post-omics approaches, such as genomics, metagenomics, meta-transcriptomics, metaproteomics, proteogenomics, metabolomics, and metafluxomics provide a nexus among the genetics and functional resemblances between various microbial communities, hastening the process of bioremediation of pollutants. In recent years, the application of "omics" technologies in biodegradation and bioremediation research has generated a plethora of new information providing a greater understanding of the key pathways and new insights into the adaptability of organisms in stress environment and how the microbes degrade/transform or metabolize a huge range of contaminants including persistent organic pollutants. Omics approaches, such as genomics, metagenomics, metabolomics, transcriptomics, proteomics, and fluxomics give more insights into the microbial communities inhabiting a particular environmental niche. Chapters 1, 2, 3, 4, 9, and 12 present the emergence, current status, challenges, and future prospects of the "brave new world" of omics in their new avatars (genomics, proteomics, metagenomics, and other meta-omics) for environmental remediation.

The microbial world exhibits far greater metabolic versatility than is found in the macro-organismic world. The metabolic diversity of microorganisms increases resiliency to environmental perturbations. This calls for bioprospecting of microbial species, populations, and communities endowed with the efficiency of remediating contaminants from the environment and promoting sustainable agriculture, which is illustrated in Chapters 11, 14, 15, and 21. Chapters 8, 10, and 23 portray a rosy picture of genetic engineering as an incredibly important tool for the trade of editing the genetic text of microbes or generating tailor-made microbes having magical power for environmental decontamination.

Quorum sensing or bacterial cell-to-cell signaling is involved in sensing, monitoring, and adapting to their external environment as well as counting their local population numbers and formation of a physiologically integrated community of bacterial cells enclosed in a matrix called biofilm, which represents a perfect theater of activity and presents a huge potential for the remediation of environmental contaminants. Chapters 5, 7, and 24 decode the codes of quorum sensing and furnish fascinating facets of biofilms and their potential environmental applications. Wastewater is teeming with microbes that can be employed as engineers' engineer for wastewater treatment. The primary role of microbes is as biocatalysts of bio-geochemical changes mediating kinetically inhibited, but thermodynamically favorable reactions. Chapters 6 and 17 provide a synoptic view of the integrated omics approaches toward understanding and improving remediation of contaminants from wastewater.

Microorganisms present in the contaminated environment are highly adaptable to toxic environments and play a crucial role in the degradation, detoxification, and/or removal of toxic pollutants by their successful metabolic processes, but until recently, it has been very hard to determine their exact mechanisms in the removal of chemical pollutants. As Mother Nature's janitors, the microbes inclusive of their ubiquitous presence and unimaginable efficiency render critical service in cleaning up contaminants in the environment. The narrative of bioremediation as a product of microbial tolerance and wisdom is elucidated by Biswas et al. in Chapter 16 entitled, "Omics reflection on the bacterial escape from the toxic trap of metal(loid)s: Cracking the code of contaminants stress, resistance repertoire, and remediation", and by Khanam et al. in Chapter 20 entitled, "Metal(loid)-microbe interactions: Trading on tolerance and transformation for environmental remediation". Microbial populations are capable of rapid evolutionary responses to environmental perturbations due to mechanisms that generate relatively high mutation frequencies and a capacity for horizontal gene transfer (HGT) crossing huge phylogenetic boundaries. Chapter 22 gives an insight into the significance of HGT among pathogens bacteria in acquiring antibiotic resistance. Chapter 25 presents promises of artificial intelligence in waste(water) management.

As editors, we have sincerely attempted to provide a comprehensive account and state-of-the-art of omics approaches to solving biochemical facets of the jigsaw puzzle of biodegradation and bioremediation of contaminated environmental matrices. The book provides insights into omics technologies in a more comprehensive manner, which enables the analysis of microbial behavior at the community level under different environmental stresses during the degradation and detoxification of environmental contaminants and ecological restoration. This book provides information on the developed omics approaches, upgradation of existing technologies, advent of newer technologies, and focus areas for further research and developments. It also highlights the opportunities in the existing technologies along with industrial practices and real-life case studies. All chapters include fundamentals of the processes involved, which will offer up-to-date and comprehensive ideas on omics, and knowhow of appropriate bioremediation technology for environmental remediation to the broad-spectrum and cross-disciplinary audience of academic institutions and research institutes, comprising students, researchers, and scientists in the fields of chemistry, biology, environmental biotechnology, environmental microbiology, biochemistry, environmental science, chemical and environmental engineering, and other related ones. We also hope that the book will also be very useful for engineering consultants, industrial waste managers, and regulators representing government and non-government/private organizations.

Acknowledgments

This book, *Omics for Environmental Engineering and Microbiology Systems*, is the result of dedicated efforts of numerous individuals, many of whom deserve special mention: First, we, the editors, would like to acknowledge and appreciate the efforts of all the authors who responded to our request and shared their knowledge with us in the form of manuscript containing the recent and updated information on the topic and made this primer a reality.

We are also very grateful to many anonymous reviewers who took the time to critically review individual manuscripts of this book, and their useful comments have been gratefully received, which improved the book manuscripts.

We would like to thank our colleagues who took the time to read over individual manuscripts of this book and those who reviewed the entire manuscript, from whose insights and suggestions we have benefited greatly and borrowed freely and, in some cases, spared us from the embarrassment of seeing our mistakes perpetuated in print.

We are grateful to those publishers and individuals who have granted permission to authors to reproduce illustrations and/or tables.

Dr. Vineet Kumar expresses his sincere thanks to Dr. Sushil Kumar Shahi, Associate Professor in the Department of Botany at Guru Ghasidas Vishwavidyalaya (GGV, A Central University), Chhattisgarh, India, for providing a fantastic facility in his laboratory during the compilation of book manuscripts. We also record our warmest appreciation to Ms. Sakshi Agrawal and Mr. Chandan Das, Ph.D. Scholar in the Department of Botany at GGV, Chhattisgarh, India, who helped us compilation of the manuscripts till submission to publication. Many thanks are due to our staff, students, colleagues, and mentors who have all contributed in small and big ways to this effort.

We would also like to thank CRC Press (Taylor & Francis Group) for giving us the opportunity to accomplish this book project and share the knowledge with the academic and scientific fraternity. We are particularly indebted to Dr. Gagandeep Singh, Senior Publisher (Engineering) at CRC Press (Taylor & Francis Group India Pvt. Ltd.), for the execution of the publishing agreement, encouragement, support, valuable suggestions, and unconditional support till the submission of manuscripts for production.

We are grateful to Ms. Aditi Mittal, in her role as Editorial Assistant, Engineering at CRC Press (Taylor & Francis Group India Pvt. Ltd.), for constant critical advice and invaluable support throughout the project.

Many thanks are due to Cathy Hurren, Senior Production Editor (Business, Economics, Environment and Sustainability) at CRC Press (Taylor & Francis Group, Florida), who coordinated the entire project. She systematically managed the book's production schedule and progress; without her prodding, this book would never have been completed. We have been fortunate to continue to work again with Cathy Hurren, the production manager.

We wish to warmly and gratefully acknowledge the Project Manager, Sathya Devi, at Codemantra who helped directly or indirectly in the accomplishment of this work and who helped to make *Omics for Environmental Engineering and Microbiology Systems*, First Edition, a reality, many of whom deserve special mention.

The editors are extremely grateful to Sathya Devi and her team for transforming literally more than 350 of pages of text and art manuscript into the superb learning tool you have in front of you. The team of Taylor & Francis Publishing Group has played a great role throughout, always helpful and supportive.

Last but not the least, the editors would like to acknowledge their family members who provide the encouragement, inspiration, endurance, and moral support that make everything possible. Any success that we have achieved or will achieve in the future would not be possible without the love and moral support of our beloved families.

Finally, we would like to apologize in advance for any errors that may occur in the text and express our heartfelt embarrassment.

We should be pleased to receive any comments on the content and style of *Omics for Environmental Engineering and Microbiology Systems*, from students, professionals, environmentalists, and policymakers, all of which will be given serious consideration for inclusion in any further editions.

Vineet Kumar
Vinod Kumar Garg
Sunil Kumar
Jayanta Kumar Biswas

Editors

Dr. Vineet Kumar is presently working as an Assistant Professor in the School of Engineering & Sciences at GD Goenka University, Gurugram, Haryana, India. Prior to joining at GD Goenka University, Dr. Kumar served in various reputed institutions in India like CSIR-National Environmental Engineering Research Institute (NEERI), Maharashtra, India; Guru Ghasidas Vishwavidyalaya (A Central University), Chhatisgarh, India; Jawaharlal Nehru University, Delhi; Dr. Shakuntala Misra National Rehabilitation University, Uttar Pradesh, India, etc. He received his M.Sc. (2008) and M.Phil. (2012) in Microbiology at Ch. Charan Singh University, Meerut, India. Subsequently, he earned his Ph.D. (2018) in Environmental Microbiology from Babasaheb Bhimrao Ambedkar (A Central) University, Lucknow, India. He was awarded a Rajiv Gandhi National Fellowship by the University Grants Commission, India, for his work on "Distillery Wastewater Treatment" in 2012. His research interest includes bioremediation, phytoremediation, metagenomics, wastewater treatment, environmental monitoring, waste management, bioenergy, and biofuel production. Currently, his research mainly focuses on the development of integrated and sustainable treatment techniques that can help in minimizing or eliminating hazardous waste in the environment. He has published more than 32 research and review articles in reputed international journals. In addition, he is the author of four proceeding papers, 45 book chapters, and four scientific magazine articles. Moreover, he has published two authored and 15 edited books on different aspects of phytoremediation, bioremediation, wastewater treatment, omics, genomics, and metagenomics from CRC Press; Elsevier, the USA; Springer Nature; and Wiley. His recently published book is Recent Advances in Distillery Waste Management for Environmental Safety (CRC Press; Taylor & Francis Group, the USA). He has presented several papers relevant to his research areas at national and international conferences. He is an active member of numerous scientific societies including the Microbiology Society (UK), the Indian Science Congress Association (India), the Association of Microbiologists of India (India), etc. Dr. Kumar has been serving as a guest editor and reviewer in many prestigious international journals in his research area. As part of his interest in teaching biology, he is the founder of the Society for Green Environment, India (www.sgeindia.org).

Vinod Kumar Garg is presently working as a Professor in the Department of Environmental Science and Technology at Central University of Punjab, Bathinda, Punjab, India. Previously, he had worked at Guru Jambheshwar University of Science and Technology and CCS Haryana Agricultural University, Hisar, India, in different capacities. Prof. Garg is a well-rounded researcher with more than 30 years of experience in leading, supervising, and undertaking research in the broad field of solid and hazardous waste management. He and his research group are working on water and wastewater pollution monitoring and abatement, solid waste management, pesticide degradation, and heavy metal and radionuclide detoxification from the contaminated environment. He has published more than 200 research and review articles, 22 proceedings, and six editorials in peer-reviewed journals of international and national journals of repute with more than citations 15,000 and h-index 65. In addition, he has published two books and 12 book chapters and completed 10 sponsored research projects as PI funded by various agencies and departments. He was awarded "Thomson Reuters Research Excellence – India Citation Awards 2012". He is a member of the editorial board and reviewing committee of several journals and professional societies. He has guided 24 Ph.D. and 80 M.Sc. students for research work. He is an active member of various scientific societies and organizations, including the Indian Science Congress, Kolkata; Biotech Research Society of India; National Solid Waste Association of India (NSWAI), Mumbai; and Indian Nuclear Society. He was elected a Fellow of the Biotech Research Society of India. He has served on the editorial board of various scientific journals in his research areas.

Sunil Kumar is a well-rounded researcher with more than 20 years of experience in leading, supervising, and undertaking research in the broad field of environmental engineering and science with a focus on solid and hazardous waste management. His primary area of expertise is solid waste management (municipal solid waste, electronic waste, biomedical waste, etc.) over a wide range of environmental topics, including contaminated sites, EIA, and wastewater treatment. He has contributed extensively to these fields and has a citation of 8825, h-index of 44, and i10-index of 168 (Google Scholar). His contributions since inception at CSIR-National Environmental Engineering Research Institute (NEERI), India, in 2000 include 300 refereed journal publications, 5 books, and 40 book chapters, 10 edited volumes, and numerous project reports to various governmental and private, local, and international academic/research bodies. He is the Associate Editor of peer-reviewed journals of international repute, i.e., *Environmental Chemistry Letter, International Journal of Environmental Science & Technology, ASCE, and Journal of Hazardous, Toxic and Radioactive Waste*. He also serves as an editorial board member in Bioresource Technology, Elsevier. He has completed more than 22 research projects as PI with 17 (seven awarded) Ph.D. and 17 M. Phil/M. Tech theses/dissertations under his supervision.

The list of his collaborations is long and includes key Indian universities such as IIT Kharagpur, IIT Delhi, and IIT Mumbai and prestigious regional institutes such as Asian Institute of Technology (AIT) and Kasetsart University, Bangkok, Hong Baptist University, universities in the USA (Columbia, Texas A&M), University of Calgary, Canada, and universities in Europe (UN University Dresden, and University of Uppsala, Sweden). He has contributed immensely to the advancement of environmental engineering/science fields in India and internationally as an expert committee member for revision of Solid Waste Management Rules, NGT members for Solid Waste Rules, organizing workshops/conferences, and delivering invited speeches at both Indian and international venues. Dr. Kumar has achieved recognition and was awarded Outstanding Scientist in 2011 and 2016 at CSIR-NEERI for his Scientific Excellence in the field of Research & Development in Solid Waste Management. Dr. Kumar was also awarded the most prestigious award Alexander von Humboldt-Stiftung Jean-Paul-Str.12 D-53173 Bonn, Germany, as a Senior Researcher for developing a Global Network and Excellence for more advanced research and technology innovation.

Jayanta Kumar Biswas is a Professor at the Department of Ecological Studies and International Centre for Ecological Engineering, University of Kalyani, India. He obtained M.Sc. in Zoology, M.Phil. in Ecology, and Ph.D. for his work on ecotechnological management of aquatic systems. His spectrum of research interest and expertise spans the following focal areas: ecotoxicology and bioremediation of toxic metal(loid)s in soil and water; ecological engineering and ecotechnological remediation of water and soil contaminants; environmental microbiology; nanobiotechnology; wastewater treatment and resource recovery; and environmental sustainability. A consistent rank holder all through his academic career, Professor Biswas received many scholarships, awards, and fellowships, including Fellow, West Bengal Academy of Science and Technology (FAScT), Fellow, National Environmentalists Association (FNEA), Zoological Society of India (FZSI), Zoological Society, Kolkata (FZS), Outstanding Reviewer Award (Chemosphere (Elsevier) & Environmental Geochemistry and Health (Springer). He is credited with publishing 6 books and 160 original research papers in reputed international journals, with 2335 citations, cumulative IF >550. He is serving as associate editor, advisory/editorial board member of several international journals of repute, namely *Science of the Total Environment* (Elsevier), *Heliyon* (Cell Press/Elsevier), *Environmental Chemistry Letters* (Springer), Ambio (Springer), *Environmental Technology & Innovation* (Elsevier), *Environmental Geochemistry and Health* (Springer), *Environmental Management* (Springer), *Ecotoxicology* (Springer), *Archives of Environmental Contamination and Toxicology* (Springer), *Environmental Chemistry and Ecotoxicology* (Elsevier), *Current Pollution Reports* (Springer), *Sustainable Environment* (Taylor & Francis), and *Energy & Environment* (SAGE). He has visited several countries with government travel fellowships, to present research papers and chair scientific sessions in international conferences.

Contributors

Richard Andi Solórzano Acosta
Escuela de Ingeniería Ambiental
Universidad César Vallejo
Víctor Larco Herrera, Peru

Nahid Akhtar
School of Bioengineering and
    Biosciences
Lovely Professional University
Phagwara, India

Balram Ambade
Department of Chemistry
National Institute of Technology
Jamshedpur, India

Tanuj Kumar Ambwani
Department of Veterinary Physiology
    and Biochemistry
G.B. Pant University of Agriculture &
    Technology
Pantnagar, India

Vandana Anand
Division of Microbial Technology
CSIR-National Botanical Research
    Institute
Lucknow, India
and
Academy of Scientific and Innovative
    Research, AcSIR
Ghaziabad, India

Wilgince Apollon
Departamento de Ingeniería Agrícola y
    de los Alimentos
Universidad Autónoma de Nuevo León
General Escobedo, México

Madhumita Barooah
Department of Agricultural
    Biotechnology and DBT-North
    East Centre for Agricultural
    Biotechnology
Assam Agricultural University
Jorhat, India

Xiomara Gisela Mendoza Beingolea
Faculty of Pharmacy and Biochemistry
Universidad Nacional Mayor de San
    Marcos
Lima, Perú

Archisman Bhunia
Institute's Innovation Council
University of Engineering &
    Management
Kolkata, India

Jayanta Kumar Biswas
Department of Ecological Studies
and
International Centre for Ecological
    Engineering
University of Kalyani
Nadia, India

Nivedita Chatterjee
Institute's Innovation Council
University of Engineering &
    Management
Kolkata, India

Chirag Chopra
School of Bioengineering and
    Biosciences
Lovely Professional University
Phagwara, India

Jayashankar Das
Valnizen Healthcare
Vile Parle, Mumbai, India

Sanchita Das
Wastewater Technology Division
CSIR-National Environmental
    Engineering Research Institute
    (CSIR-NEERI)
Nagpur, India

Sushma Dave
Department of chemistry
Jodhpur Institute of Engineering and
    Technology
Jodhpur, India

Kulal Deekshitha
Department of Chemical Engineering
National Institute of Technology
    Karnataka Surathkal
Mangalore, India

Daljeet Singh Dhanjal
School of Bioengineering and
    Biosciences
Lovely Professional University
Phagwara, India

Varsha Dharmesh
Division of Microbial Technology
CSIR-National Botanical Research
    Institute
Lucknow, India
and
Academy of Scientific and Innovative
    Research, AcSIR
Ghaziabad, India

Fakhr-un-Nisa
Department of Chemistry
The Women University Multan
Multan, Pakistan

Héctor Flores-Breceda
Departamento de Ingeniería Agrícola y
    de los Alimentos
Universidad Autónoma de Nuevo León
General Escobedo, México

Ayushman Gadnayak
Centre for Biotechnology
School of Pharmaceutical Sciences,
    IMS and SUM Hospital, Siksha
    'O' Anusandhan (Deemed to be
    University)
Bhubaneswar, India

Julián Gamboa-Delgado
Facultad de Ciencias Biológicas,
    Laboratorio de Maricultura
Universidad Autónoma de Nuevo León
General Escobedo, México

Celestino García-Gómez
Facultad de Agronomía
Universidad Autónoma de Nuevo León
General Escobedo, México

Azaruddin V. Gohil
School of Science
P. P. Savani University
Surat, India

Juan Florencio Gómez-Leyva
Laboratorio de Biología Molecular
TecNM-Instituto Tecnológico de
    Tlajomulco (ITTJ),
Tlajomulco de Zúñiga, México

Abhineet Goyal
School of Bioengineering and
    Biosciences
Lovely Professional University
Phagwara, India

Dibya Jyoti Hazarika
Department of Agricultural
    Biotechnology and DBT-North
    East Centre for Agricultural
    Biotechnology
Assam Agricultural University
Jorhat, India

Edwin Hualpa-Cutipa
Faculty of Pharmacy and Biochemistry,
    Biotechnology and Omics in Life
    Sciences Research Group
Universidad Nacional Mayor de San
    Marcos
Lima, Perú

Rao Muhammad Shahbaz Mahboob
Department of Plant Pathology
University of Agriculture Faisalabad
    (UAF)
Faisalabad, Pakistan

Ingrid Maldonado Jimenez
Escuela de Posgrado
Universidad Nacional del Altiplano de
    Puno
Puno, Perú

Sathish-Kumar Kamaraj
Laboratorio de Medio Ambiente
    Sostenible
TecNM-Instituto Tecnológico El Llano
    Aguascalientes (ITEL)
Aguascalientes, México

Nensi K. Thumar
School of Science
P P Savani University,
Surat, India

Jasvinder Kaur
Division of Microbial Technology
CSIR-National Botanical Research
    Institute
Lucknow, India
and
Department of Botany
Kumaun University
Nainital, India

Mahrukh Khan
Department of Chemistry
The Women University Multan
Multan, Pakistan

Rubina Khanam
Department of Soil Science
ICAR-National Rice Research Institute
Cuttack, India

Ashwani Kumar
Department of Botany
Metagenomics and Secretomics
    Research Laboratory, Dr. Harisingh
    Gour University (Central University)
Sagar, India

Manish Kumar
Department of Botany
Guru Ghasidas Vishwavidyalaya
Bilaspur, India

Vineet Kumar
Department of Basic and Applied
    Sciences
School of Engineering and Sciences
G D Goenka University, Sohna Road
Gurugram, India

Vandana Kumari
Department of Soil Science
Dr. Rajendra Prasad Central
    Agricultural University
Samastipur, India

Pedda Ghouse Peera Sheikh Kulsum
Department of Soil Science
C V Raman Global University
Bhubaneswar, India

Daniela Landa-Acuña
Department of Biology
National Agrarian University La Molina
   (UNALM)
and
Facultad de Ingeniería
Universidad Privada del Norte
Lima, Perú

Ranjan Laik
Department of Soil Science
Dr. Rajendra Prasad Central
   Agricultural University
Pusa, India

Antonio Leija-Tristan
Department of Ecology
Universidad Autonoma de Nuevo Leon
Barragán, Mexico

Joshua Lerner
Department of Ecology and
   Conservation Biology
Texas A&M AgriLife
College Station, Texas

Alejandro Isabel Luna-Maldonado
Facultad de Agronomía
Universidad Autónoma de Nuevo León
General Escobedo, México

Akshita Maheshwari
Division of Microbial Technology
CSIR-National Botanical Research
   Institute
Lucknow, India

Rao Muhammad Mahtab Mahboob
Department of Computer Science
University of Agriculture Faisalabad
   (UAF)
Faisalabad, Pakistan
and
Department of Computer Science
Institute of Management and Applied
   Sciences,
Khanewal, Pakistan

Julia Mariana Márquez-Reyes
Facultad de Agronomía
Universidad Autónoma de Nuevo León
General Escobedo, México

Arti Mishra
Amity Institute of Microbial
   Technology
Amity University Uttar Pradesh
Noida, India

Gerardo Méndez-Zamora
Departamento de Ingeniería Agrícola y
   de los Alimentos
Universidad Autónoma de Nuevo León
General Escobedo, México

Monojit Mondal
Department of Ecological Studies
University of Kalyani
Nadia, India

Raj Mukhopadhyay
Division of Irrigation and Drainage
   Engineering
ICAR-Central Soil Salinity Research
   Institute
Karnal, India

Sara Musaddiq
Department of Chemistry
The Women University Multan
Multan, Pakistan

Kiran Mustafa
Department of Chemistry
The Women University Multan
Multan, Pakistan
and
Higher Education Department
Govt. Graduate College for Women
Khanewal, Pakistan

Juan Nápoles-Armenta
Facultad de Agronomía
Universidad Autónoma de Nuevo León
General Escobedo, México

Kumar Narayan
Department of Human Molecular
    Genetics
Jiwaji University
Gwalior, India

Debabrata Nath
Department of Soil Science
Dr. Rajendra Prasad Central
    Agricultural University
Samastipur, India

Ananya Nayak
Centre for Biotechnology, SPS
IMS and SUM Hospital, Siksha
    'O' Anusandhan (Deemed to be
    University)
Bhubaneswar, India

Sukdeb Pal
Wastewater Technology Division
CSIR-National Environmental
    Engineering Research Institute
Nagpur, India
and
Academy of Scientific and Innovative
    Research (AcSIR)
Ghaziabad, India

Kripa Pancholi
Department of Microbiology
Parul Institute of Applied Science, Parul
    University
Vadodara, India

Niharika Pandey
Department of Biotechnology
M. B. Gov. P. G. College
Haldwani, India

Hiren K. Patel
School of Science
P. P. Savani University
Surat, India

Priyank D. Patel
School of Science
P. P. Savani University
Surat, India

Pritam Patil
Department of Biotechnology and
    Medical Engineering
National Institute of Technology
Rourkela, India

Shalini Porwal
Amity Institute of Microbial
    Technology
Amity University Uttar Pradesh
Noida, India

Bhupendra Pushkar
Department of Biotechnology
University of Mumbai
Mumbai, India

Aman Raj
Department of Botany
Metagenomics and Secretomics
    Research Laboratory, Dr. Harisingh
    Gour University (Central University)
Sagar, India

Humberto Rodríguez-Fuentes
Facultad de Agronomía
Universidad Autónoma de Nuevo León
General Escobedo, México

Julio Reynaldo Ruiz Quiroz
Faculty of Pharmacy and Biochemistry
Universidad Nacional Mayor de San
    Marcos
Lima, Perú

Pushpa Ruwali
Department of Biotechnology
M. B. Gov. P. G. College
Haldwani, India

Maheswata Sahoo
Centre for Biotechnology
IMS and SUM Hospital, Siksha
    'O' Anusandhan (Deemed to be
    University)
Bhubaneswar, India

Swayamprabha Sahoo
Centre for Biotechnology
IMS and SUM Hospital, Siksha
    'O' Anusandhan (Deemed to be
    University)
Bhubaneswar, India

Menaka Devi Salam
Amity Institute of Microbial
    Technology
Amity University
Noida, India

María Elena Salazar Salvatierra
Faculty of Pharmacy and Biochemistry
Universidad Nacional Mayor de San
    Marcos
Lima, Perú

Angana Sarkar
Department of Biotechnology and
    Medical Engineering
National Institute of Technology
Rourkela, India

Anupama Shrivastav
Department of Microbiology
Parul University
Vadodara, India

Rangabhashiyam Selvasembian
Department of Biotechnology
SASTRA Deemed University
Thanjavur, India

Abhilasha Singh
Institute's Innovation Council
University of Engineering &
    Management
Kolkata, India

Pallavi Singh
Division of Microbial Technology
CSIR-National Botanical Research
    Institute
Lucknow, India

Rahul Vikram Singh
Department of Biological Science
Academy of Scientific and Innovative
    Research (AcSIR)
Ghaziabad, India

Reena Singh
School of Bioengineering and
    Biosciences
Lovely Professional University
Phagwara, India

Asmeeta Sircar
Institute's Innovation Council
University of Engineering &
    Management
Kolkata, India

Pooja Sevak
Department of Biotechnology
University of Mumbai
Mumbai, India

Neha Sharma
Amity Institute of Microbial
    Technology
Amity University
Noida, India

Kartikeya Shukla
Amity Institute of Environmental
    Sciences
Amity University
Noida, India

Smriti Shukla
Amity Institute of Environmental
    Toxicology, Safety and Management
Amity University
Noida, India

Fatima Estefanía Soto-Zamora
Facultad de Ciencias Biológicas,
    Laboratorio de Maricultura
Universidad Autónoma de Nuevo León
General Escobedo, México

Sonal Srivastava
Division of Microbial Technology
CSIR-National Botanical Research
    Institute
Lucknow, India
and
Academy of Scientific and Innovative
    Research, AcSIR
Ghaziabad, India

Suchi Srivastava
Division of Microbial Technology
CSIR-National Botanical Research
    Institute
Lucknow, India
and
Academy of Scientific and Innovative
    Research, AcSIR
Ghaziabad, India

Nency K. Thummar
P. P. Savani University
Surat, India

Milton Torres-Ceron
Department of Ecology and
    Conservation Biology
Texas A&M AgriLife
College Station, Texas

Ajit Varma
Amity Institute of Microbial
    Technology
Amity University Uttar Pradesh
Noida, India

Satish Kumar Verma
Department of Botany
Banaras Hindu University (BHRU)
Varanasi, India

Juan Antonio Vidales-Contreras
Facultad de Agronomía
Universidad Autónoma de Nuevo León
General Escobedo, México

Shetty K. Vidya
Department of Chemical Engineering
National Institute of Technology
    Karnataka Surathkal
Mangalore, India

Meththika Vithanage
Faculty of Applied Sciences
University of Sri Jayewardenepura
Nugegoda, Sri Lanka

Atif Khurshid Wani
School of Bioengineering and
  Biosciences
Lovely Professional University
Phagwara, India

Evelin Yana-Neira
Ciencia, Tecnología y Medio Ambiente,
  Escuela de Posgrado
Universidad Nacional del Altiplano de
  Puno
Puno, Perú

Isabel Navarro Zabarburú
Faculty of Pharmacy and Biochemistry
Universidad Nacional Mayor de San
  Marcos
Lima, Perú

1 Omics to Field Bioremediation
Current Status, Challenges, and Future Opportunities

Pritam Patil and Angana Sarkar
National Institute of Technology

CONTENTS

1.1 INTRODUCTION

Evolution of humans has enhanced the standard of living of society due to developments in technology. These developments have led to vast industrialization across the globe. This has made humans to achieve many milestones that were supposed to be highly difficult. Along with this development, many new difficulties have raised. Pollution is the biggest difficulty that faces human being today. The vast industrialization caused contamination of soil and water around it. New agricultural practices have led to soil deterioration. Plastic, which was thought to be the greatest discovery, has now become the greatest problem. The main reason behind pollution is the lack

DOI: 10.1201/9781003247883-1

of tools and technology to tackle it. Contamination of soil and water is also a growing concern among these issues. In a survey done in India, it was found that the soil
of Chennai and air in Delhi are the most contaminated and affected by a contaminant, namely polychlorinated biphenyls (PCBs) (Web ref. 1). The level of contamination of soil in India is twice the amount found globally. Remediation is a way to
reduce this contamination; it involves the removal of contaminants from the source
through different ways such as chemical and physical removal. In physical removal,
nets or sieves are used to remove large particles that can be seen by naked eyes. On
the contrary, different chemicals are used in chemical removal, which react with
contaminants and neutralize them. But these are conventional ways and have their
own challenges; for example, physical methods involve high cost and require huge
assemblies for large-scale applications. Some contaminants do not react with chemicals used for remediation; for example plastic does not react with any chemicals. So
nowadays bioremediation is used to remove such contaminants (Kumar et al., 2018).
It is a method in which biological entities such as microorganisms and plants are used
to remove contaminants from different sources. Microbial bioremediation has shown
the highest efficiency as compared to other methods. Along with these methods,
omics approaches have aided the bioremediation by making it more cost-effective,
productive, less time-consuming, and efficient (Kumar et al., 2020).

1.1.1 WHAT ARE OMICS?

According to the Medical Dictionary, the definition of omics is the study of a large
amount of data representing an entire set of molecules such as genes, proteins, and
metabolites. Also, it refers to the quantitative measurement of global sets of molecules in a bio-sample. And in this, high-throughput techniques along with biostat-
ics and analysis tools are used. So, the main concept behind these studies is that a
complex system can be understood efficiently when it is considered as a whole. These
approaches are used to conduct a study in a non-biased and non-targeted manner so
that the efficiency can be increased. These technologies adopt a holistic approach to
study a system and these can also be considered as high-dimensional biology and the
incorporation of these all systems is called systems biology. These studies generate
hypotheses by collecting and analyzing the data, and the hypotheses that are generated from the data are further tested (Horgan and Kenny, 2011). Many different technologies come under this classification, such as next-generation sequencing (NGS)
and mass spectrometry.

1.1.2 DIFFERENT TACTICS OF OMICS STUDIES

There are four main approaches of omics studies, namely genomics, transcriptomics,
proteomics, and metabolomics (Figure 1.1). Each approach is specific to a set of
molecules that it studies.

a. Genomics

The genomics approach studies the genome. It involves sequencing, mapping, and structural analysis of the whole genome of an organism or sample.

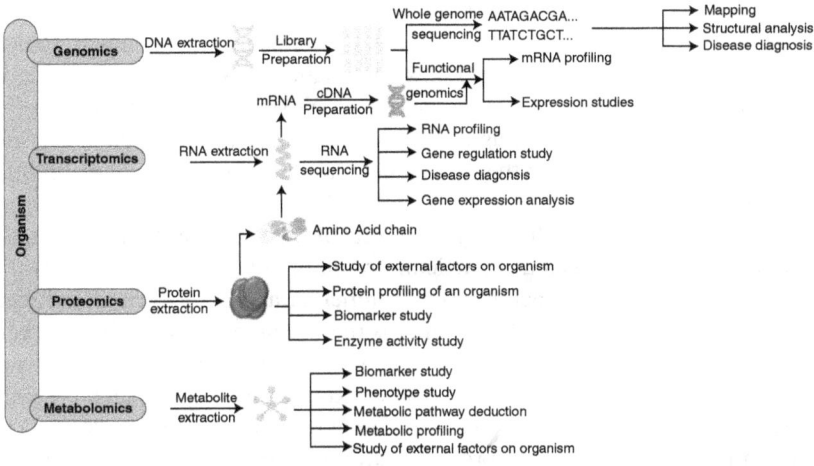

FIGURE 1.1 Approaches of omics study.

Due to advancements in technology, this approach has become more efficient. It differs from genetics, as genetics is the study of heredity, i.e., how a
gene or phenotype is carried from one generation to the next generation. So,
it focuses on a limited number of genes rather than considering the whole
genomes (Vailati-Riboni et al., 2017). Due to the surge in the fields of NGS,
genomics, and genomic data analysis, genomics has tremendously developed.
The basic process of the genomics study includes the collection of samples;
the extraction of the whole genome, its sequencing, and library preparation;
and then the analysis of the data that have been retrieved (Kumar et al., 2021).
All these steps use many tools that have been developed for many decades.
The collected data are stored in libraries so that they are available to all who
wish to pursue research in this field (Misra et al., 2018). The breakthrough
in the field of genomics was the Human Genome Project (HGP), which gave
insights into the human genome and its various aspects. It also gave insights
into various gene-related diseases from one individual to other. Genomics
can be divided into various types according to the function as metagenomics (study of sample from the environment), epigenomics (factors that affect
gene expression without changing the sequence of the genome), structural
genomics (characterizing the physical nature of genome), functional genomics (characterizing the function of genomes), etc. (Vlaanderen et al., 2009).

b. Transcriptomics

Some say this is a subtype of genomics as tools. The process of study is
almost the same, but the major difference is in the set of molecules that it
studies. While genomics is the study of the whole genome, transcriptomics
is a study of the product of genomes, i.e., RNAs. They are considered in the
same category because both are nucleic acid studies. So, sometimes this is
considered under functional genomics. So, transcriptomics studies the whole
set of mRNAs found in the sample or organism. It is also considered gene

expression profiling as mRNAs are the product of gene expression. It studies the level of gene expression in a sample or organism at that moment. Hence, these studies are important to know the relation between the difference in profiling and its effect on phenotype. These are very variable studies as gene expression changes according to various factors such as the type of cell and environmental factors, so they are difficult to reproduce (Celis et al., 2000).

c. Proteomics

It can be defined as the study of all sets of expressed protein in a sample, individual, or organism at a particular point or environmental condition. It represents the relation between environmental and other factors and their effect on gene expression. Proteomics is important in the field of biomarkers as proteins are most commonly affected by any environmental stimulus or any disease condition. And also, protein expression is easy to identify compared to nucleic acid studies due to different reasons (Rifai et al., 2006). The main tools for the study of proteome are mass spectrophotometry, protein microarrays, etc. It also provides insights into the role of a particular protein in a system. Many different approaches can be utilized for this study, such as the quantitative approach in which the concentration of proteins or a particular protein in a sample at different conditions or moments is determined. This study will give information on the rate of production and degradation of proteins inside the cell. Hence, if a protein is present in large amounts, then it can be considered that it is a very important protein for the cell (Wingren and Borrebaeck, 2006). Another approach is a qualitative one, in which the presence of different proteins can be found out in a wide range of samples and thus comparative studies can be done (Sellers and Yates, 2003).

d. Metabolomics

As the name suggests, metabolomics is the study of small metabolites such as lipids and vitamins in a sample or organism. Metabolites represent the energy transfer intermediates in the cell and in between different cells. This study mainly comprises the metabolite profiling in a sample. Metabolic phenotypes are the result of different processes that a cell undergoes due to the relationship between the genome and the factors affecting it. This study has not been developed yet like other studies as qualitative and quantitative study of metabolites is challenging due to perturbations in the metabolites profile due to changes in environmental conditions (Claudino et al., 2007). But it has some advantages over other systems as the metabolite is the closest to the phenotype a system exhibits, which can be studied easily if any changes happen. Also, it is the last product to be formed, and the changes in metabolome can be related to the changes in the genome quite comfortably. And also due to the large number of molecules to be studied, it is the most complex of all other omics (Horgan and Kenny, 2011).

1.1.3 APPLICATIONS OF OMICS IN VARIOUS FIELDS

Omics have been of great importance in the study of various fields such as pharmaceuticals, agriculture, environment, health care, and biomedicine. Due to their flexibility and the reproducibility of results, these are becoming popular in research areas. These are also based on tools such as microarrays, 2D gel electrophoresis and 2DLC/MS, which make them less time-consuming and more efficient, and also with the ability to produce large-scale data and with the help of computer-based modeling, they can theoretically build models that can explain how biological processes take place. In the field of pharmaceuticals, omics can be applied to every step to find drug targets and personalize medicines. Due to omics, we came to know that around 30,000–40,000 genes are present in our cell and thus same number of proteins from which almost every protein can be a drug target. Many tools are being used to construct a huge amount of data, such as nuclear magnetic resonance (NMR), RNAi, gene transfection, and gene knockout modeling. These data are used to build a database that can be used by researchers across the world (Yan et al., 2015). In the field of human observational studies, omics have come up as an important tool for making biomarker discoveries for a large set of biomolecules. But their applications in the field of occupational environmental health are limited due to some issues; for example, the interpretation of data is a bit tedious. Till now, these have been used in the investigation of the effects of benzene and arsenic on human health. For example, epigenomics is used in studies of the disease. Epigenomics is focused on two main modifications of the DNA, which are histone modifications and methylation. Hypermethylation is an indicator of gene silencing in case of any disease- or tissue-specific expression, etc. (Jirtle and Skinner, 2007). In the case of environmental epidemiology, omics are used for the construction or studies of various biomarkers that can be used to study the effects of various environmental hazards. These biomarkers may include metabolites (metabolomics), DNA (genomics), and proteins (proteomics).

1.2 BIOREMEDIATION

Bioremediation means the removal or the reversal of the damage. In a large context, it is applied to the environment. Hence, it can be said that remediation means stopping or reversing the process of environmental damage by any means. Bioremediation means stopping this damage by using any biological entity or any biological means, as biological approaches are most of the time harmless and efficient. Conventionally, remediation was done by landfilling or by using any chemical resources to stop the damage. But these processes are not efficient and most importantly not safe. And they do not destroy the pollutants. Hence, new approaches have been developed, which aim at either the destruction of pollutants, or conversion into harmless products. These approaches include incineration and chemical decomposition, but these methods also have drawbacks. For example, incineration produces ash and harmful gases and chemicals may be released into the environment. Hence, bioremediation becomes an option to completely remove pollution utilizing any biological entity or process (Vidali, 2001).

1.2.1 WHAT IS BIOREMEDIATION AND ITS NEED

It is defined as the process in which waste (organic or inorganic) is degraded into a harmless substance or is reduced to the level below which it cannot be harmful to the environment utilizing any biological entity, especially microorganisms. It was first limited to only microorganisms, but due to advancements in the field, various biological entities are used now. Nowadays, we use bacteria, fungi, and plants (phytoremediation) to convert harmful substances into harmless ones (Chandra and Kumar, 2018). This is achieved by microbes and plants using the waste as a raw material for their growth. The conversion of waste into metabolites and then the use of these metabolites for growth occurs through many different metabolic processes. It is achieved in the environment by the action of many different microbes altogether and not by any single microbe; hence, naturally, it takes place only at places where microbial and plant growth is feasible. Nowadays, in vitro bioremediation is gaining importance. The waste will be first treated by microbes and then disposed of in the environment, or the waste will be collected from the environment, brought to the sites for degradation, and then converted into harmless or industrially important compounds (Malik et al., 2016). The main advantage of bioremediation is that, once a pollutant is metabolized by microbes, its products are often utilized by other microbes; thus, the cycle goes on and the pollution reduces (Coelho et al., 2015).

Similar to other different ways of remediation, bioremediation also has different challenges. For example, we cannot predict whether a pollutant can be degraded by a microbe. Many pollutants such as chlorinated compounds are resistant to microbial attacks; hence, the degradation will be very slow or will not take place at all. It is dependent on many different factors that are related to the growth of microbes, such as pH, temperature, nutrient availability, type of pollutant, and degradation process (aerobic or anaerobic). Hence, to predict the rate of bioremediation, all these factors are to be taken into consideration; hence, it is very difficult to predict (Vidali, 2001).

1.2.2 MEANS OF BIOREMEDIATION

There are three main ways of bioremediation (Figure 1.2). Each way has its advantages and disadvantages. *Ex situ* bioremediation means the removal of waste away from the site. In this method, the waste or waste-contaminated soil or water is taken to a different bioremediation site and the process is carried out. It contains different ways such as landfarming, composting, and bio-piles. In landfarming, the contaminated soil is drawn out, and then it is spread over an already prepared land, and then farming is done so that plants or microbes will digest the waste. It is a very cheap form of bioremediation as there is no maintenance cost involved. In composting, soil or mostly organic waste is taken and mixed with microbes, where microbes degrade the organic waste into compost which can further be used as fertilizer for crops. A bio-pile is a hybrid method of both landfarming and composting, as it contains a pile in which the contaminated soil or waste is kept. Then engineered microbes are introduced into it. This is specially used for the surface degradation of petroleum products that contaminate the soil. It prevents these hydrocarbons from leaching into the soil (Maletić et al., 2013).

In situ bioremediation means degradation of waste in its place. It is used widely as it does not require excavation and transport of contaminated soil. Thus, it is a very low-cost way of bioremediation. It involves approaches such as bioventing, bioaugmentation, and biosparging. In bioventing, as the name suggests, vents are made to

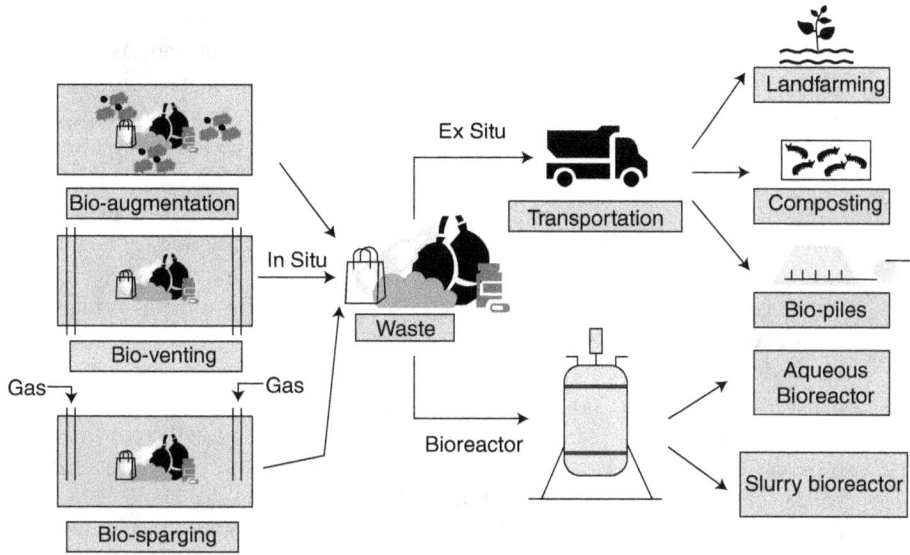

FIGURE 1.2   Techniques of bioremediation.

facilitate the flow of air (oxygen) deep into the soil to boost the growth of indigenous microbes. Wells are prepared deep inside the soil, and the flow rate is maintained in such a way that it will only boost the growth and does not release contaminants into the atmosphere (Höhener and Ponsin, 2014). Bioaugmentation involves the introduction of exogenous microbes into the contaminated sites to increase the efficacy of the process. But it depends on how well the exogenous microbes compete with the indigenous ones to carry out the process (Anh et al., 2021). Biosparging involves sparging of high-pressure air into the groundwater so that microbes can grow and then degrade the waste in a contaminated site. It facilitates the mixing of contamination and soil and hence increases the efficiency.

Bioreactors can be called a subtype of ex situ bioremediation. In this process, waste-contaminated soil is brought to the decontamination site and the degradation process takes place inside a bioreactor. Its advantage is that microbes are kept under well-maintained process parameters, so the growth of microbes is boosted and thus the efficiency of bioremediation. But the main disadvantage is that it is very costly as it requires electricity to run the bioreactor, involves maintenance costs, includes transport cost of waste, etc. (Harekrushna and Kumar, 2012).

1.2.3   PHYTOREMEDIATION

This is another different approach of bioremediation that has recently been developed. In this approach, plants are used instead of microbes for the degradation of contaminants (Kumar and Chandra, 2018). Plants through their roots take the contaminants and convert them into metabolites that they require and utilize them for their growth (Chandra et al., 2018a, b). This is a very innovative approach as it has a dual role in saving the environment by reducing pollution and, as the plant grows, by utilizing carbon dioxide from the atmosphere and thus reducing the risk of global warming. It is divided into different types based on their fate after going into the plant, pollutant type, i.e., inorganic or organic, etc. The following are some approaches of phytoremediation:

1. **Phytoextraction**: This method is also called phytoaccumulation or rhizoac-cumulation as plants accumulate pollutants into their different parts such as shoot, root, or leaves. Also, it can be utilized to produce a mass of plants (Meagher, 2000; Chandra and Kumar, 2017).
2. **Phytodegradation**: This is also called rhizodegradation. In this method, the contaminant is broken down into smaller parts by the action of different enzymes secreted by plant roots and microbes associated with root rhizo-spheres. The symbiotic relationship between plants and microbes becomes very useful in this method (San et al., 2013).
3. **Phytostabilization**: In this method, the pollutant is restricted from moving or leaching into the soil or water by producing a mass of plants inside the contaminant (Meagher, 2000).
4. **Phytotransformation**: This involves the uptake of contaminants and con-version into substances that are not harmful (Kuiper et al., 2004).

1.2.4   LIMITATIONS OR DISADVANTAGES OF BIOREMEDIATION

- Some pollutants that cannot be degraded by bioremediation are present as microbes or plants cannot utilize them, for example chlorinated compounds, plastic, and radionuclides.
- Sometimes, microbes can produce toxic by-products or metabolites during the process of bioremediation, which will add to the contamination.
- There should be a small-scale study for every contaminated site as a single process cannot be used for every contaminant and its site. So, every process has to be tailor-made for each site considering its various factors. Hence, this becomes very hectic and time-consuming.
- In phytoremediation, it takes a very long period for plant growth and ulti-mately for remediation.
- It is very difficult to predict the nature of large-scale plants considering the small- and pilot-scale studies (Tyagi and Kumar, 2021).

1.3   OMICS AND BIOREMEDIATION: CURRENT STATUS

Omics have gained much importance in the field of bioremediation also. They are mostly used to study the contamination sites, their microbial diversity, and the mech-anism behind the degradation. And since these tools are being used, bioremediation has also become more efficient and cost-effective. Conventionally, bioremediation was studied on a cultural basis to find the interaction between pollutants and microbes and also between different microbes from the same site. But, since it takes long time for microbial growth and study, molecular tools such as omics have become an alternative solution that requires very little time compared to others. Also, they can be used for modeling any bioremediation process so that we can predict different aspects of the process, such as rate, efficacy, and cost (Czaplicki and Gunsch, 2016). The following are some examples of different omics in bioremediation.

1.3.1 GENOMICS AND METAGENOMICS IN BIOREMEDIATION

In bioremediation, genomics is used to study the functional genes responsible for the degradation of a particular pollutant. Also, it is used to identify the microbes and their lineage (phylogenic) by using techniques such as 16S rRNA or 18S rRNA. These data are also uploaded to different databases and used by researchers worldwide. NGS is also used to sequence the whole genome of microbes and utilize that data to identify some very useful genes. From the metagenomic data, researchers can find insights into the microbial community in a particular contamination site and can study which microbes are very sensitive, which microbes have the potential to degrade the pollutant, etc. (Kumar and Chandra, 2020). This is performed by extracting the nucleic acids (DNA and RNA) from the collected samples. Further these nucleic acids are studied for detection of functional gene and further to identify the microorganisms expressing them (Desai et al., 2010). These samples are used for sequencing the 16S rRNA genes to identify the microbe. Fingerprinting is another tool from genomics used for the study, and it involves many different approaches such as amplified fragment length polymorphism (AFLP), amplified ribosomal DNA restriction analysis (ARDRA), automated ribosomal intergenic spacer analysis (ARISA), length heterogeneity PCR (LH-PCR), single-strand conformation polymorphism (SSCP), denaturing gradient gel electrophoresis (DGGE)/temperature gradient gel electrophoresis (TGGE), terminal restriction fragment length polymorphism (T-RFLP), and randomly amplified polymorphic DNA analysis (RAPD). These all techniques are used for profiling the microbes and for finding out the key genes or key microbes that are crucial for bioremediation (Nocker et al., 2007). For example, Zhao et al. (2020) described and used a technique called DNA assembler-assisted pathway assembly to prepare a strain that can mineralize the γ-hexachlorocyclohexane (HCH) and 1,2,3-trichloropropane, which are found in pesticides and are harmful to the non-target organisms, so they need to be degraded (Table 1.1). Dumont et al. (2006) showed that DNA stable isotope probing (DNA-SIP) in a metagenomic approach can be beneficial for the identification of methanotrophic microbes from the soil (Table 1.1).

1.3.2 PROTEOMICS IN BIOREMEDIATION

The proteomic study gives insights into the change in the concentration of different proteins or a particular protein according to the environmental conditions or exposure. This can be very useful in determining the potential of a microbe to degrade any pollutant present in the sample. Nowadays, proteomic studies are directly applied to the microbial community (meta-proteomics) to study their functional role in a given ecosystem. Many different tools are used in proteomics, such as 2D gel electrophoresis, mass spectrometry, matrix-assisted laser desorption ionization

TABLE 1.1

Examples of Applications of Genomics in Bioremediation

Pollutant/Contaminant	Microorganisms	Techniques Used	Reference
Petroleum products	*Achromobacter* sp. HZ01	Genome sequencing using paired-end sequencing strategy	Hong et al. (2017)
Metal reduction, Methane oxidation	Consortium	FISH, pyrosequencing, paired-end sequencing, 16s rRNA	Haroon et al. (2013)
Sodium benzoate, chloroalkane, selenite reduction	*Haloferax* sp.	Shotgun sequencing, 16s rRNA	Naik et al. (2021)
Xenobiotics, dioxin	*Pseudomonas* BUN14	GC-FID, phylogenetic comparison	Mahjoubi et al. (2021)
γ-**H**exachlorocyclohexane, 1,2,3-trichloropropane	*Pseudomonas putida* KT2440	DNA assembler-assisted pathway assembly	Zhao et al. (2020)
Direct Black G dye	*Anoxybacillus* sp.	qRT-PCR, BLAST	Chen et al. (2021)
Uranium	*Rhodanobacter, Rhodocyclaceae*	16S rRNA microarray	Green et al. (2012)
Methane, methanol	Methanotrophs	DNA-SIP, 16s rRNA, DGGE	Dumont et al. (2006)

time-of-flight mass spectrometry (MALDI-ToF-MS), and liquid chromatography linked to MS via electrospray ionization source (Table 1.2). A proteomic study on aromatic hydrocarbon catabolic pathways of *Pseudomonas putida* KT2440 strain using 2DE/MS and cleavable isotope-coded affinity tag analysis found that there were more than 80 unique proteins responsible for the degradation of vanillin, benzoate, and p-hydroxybenzoate (Kim et al., 2004) (Table 1.2). In a study on *Providencia rettgeri*, it was seen that it can degrade selenite, which was proved by using liquid chromatography-mass spectrometry and parallel reaction managing (Huang et al., 2021).

1.3.3 Transcriptomics in Bioremediation

This focuses on the functional insights into the microbe by studying their mRNA profiles. The basic steps involve the extraction of mRNA and then the production of cDNA and its sequencing to find out the functional gene for that specific mRNA. Nowadays, this approach is used for the total community of microbes in a sample from contaminated soil or water and this approach is called meta-transcriptomics. Tools used for this study are DNA microarrays, RT-PCR, cDNA sequencers, etc. Jennings et al. (2009), in their study, used the DNA microarrays technique to identify the gene that is expressed more in the presence of cis-dichloroethane (cDCE) (Table 1.3). Lu et al. described the heavy metal utilization by the microalgae *Auxenochlorella protothecoides*. They specifically showed results for cadmium utilization by given

TABLE 1.2

Examples of Application of Proteomics in Bioremediation

Pollutant/Contaminant	Microorganisms	Techniques Used	Reference
High molecular weight polycyclic aromatic hydrocarbons	*Mycobacterium vanbaalenii* PYR-1	2D gel electrophoresis (2DE)	Kim et al. (2004)
Aromatic hydrocarbons	*Pseudomonas putida* KT2440	2DE/MS	Kim et al. (2006)
2,4-**Dichlorophenoxyacetic** acid and chlorobenzene	Consortium	2DE, LC-ESI-MS	Benndorf et al. (2007)
Activated sludge	Consortium	2DE, Q-ToF/MS, MALDI-ToF/MS	Wilmes et al. (2008)
Uranium	*Geobacter* sp.	LC-MS	Wilkins et al. (2009)
Triphenyl phosphate	*Pycnoporus sanguineus*	UPLC, LC-MS	Feng et al. (2021)
Selenite	*Providencia rettgeri*	LC-MS, iTRAQ, PRM	Huang et al. (2021)

TABLE 1.3

Examples of Application of Transcriptomics in the Field of Bioremediation

Pollutant/Contaminant	Microorganisms	Techniques Used	Reference
cis-Dichloroethane	*Polaromonas* sp. JS666	DNA microarrays	Jennings et al. (2009)
Toluene	*Cladophialophora immunda*	RT-PCR, sequencing	Blasi et al. (2017)
Polycyclic aromatic hydrocarbon	*Dentipellis* sp. KUC8613	RNA sequencing (RNA-Seq), comparative analysis of sequence	Park et al. (2019)
Polychlorinated biphenyls	*Cladosporium* sp. TM138-S3	RNA-Seq, phylogenetic analysis	Nikolaivits et al. (2021)
Methyl parathion	*Burkholderia cenocepacia*	RNA-Seq, RNA mapping	Ortiz-Hernández et al. (2021)
Cadmium	*Auxenochlorella prototothecoides*	RNA-Seq, RSEM	Lu et al. (2019)

microalgae. They first extracted the RNA and then sequencing was performed and then library construction was done. After mapping was done, the quantification of gene expression was carried out, to have an idea of expression levels with different concentrations of cadmium (Lu et al., 2019) (Table 1.3).

1.3.4 Metabolomics in Bioremediation

In metabolism, we study the entire metabolite profile of a microbe. Microorganisms produce many different metabolites in different concentrations in response to environmental

TABLE 1.4

Examples of the Application of Metabolomics in Bioremediation

Pollutant/Contaminant	Microorganisms	Techniques Used	Reference
Phenanthrene	*Sinorhizobium* sp. C4	GC-MS	Keum et al. (2009)
Different carbon sources	*Shewanella* sp.	GC-MS, genetic algorithms	Tang et al. (2009)
2-Chlorotoluene	*Pseudomonas* sp.	GC-MS	Haro and Lorenzo (2001)
Hydrocarbon	Consortium	GC-MS	Parisi et al. (2009)
2-Fluoromuconate	*Rhodococcus opacus* 1cp	NMR	Solyanikova et al. (2003)
Fluorobenzoate	*Sphingomonas* sp.	19F-NMR	Hidde Boersma et al. (2004)
Metyrapone	*Mycobacterium*	H-NMR	Combourieu et al. (2004)

stresses. This study can be done to understand the expression of certain genes through the production of metabolites. This study also aims at the quantification of these metabolites by using many different analytical and separation techniques. The metabolomic study encompasses different tools such as NMR, capillary electrophoresis-based mass spectroscopy, high-pressure liquid chromatography (HPLC), direct injection mass spectrometry (DIMS), liquid chromatography (LC), and Fourier transformed infrared spectroscopy (FTIR) (Mapelli et al., 2009). The new approach with metabolomics now is studies that focus on real-time flux analysis of metabolites in a cell over a time period. This is called fluxomics. However, there has not been much work done in the field of metabolomics used for bioremediation purposes. A study was done on a microbial consortium for hydrocarbon degradation in anaerobic conditions. It showed that hydrocarbons from a petroleum-contaminated site were efficiently degraded by the consortium (Parisi et al., 2009) (Table 1.4).

1.4   FUTURE PROSPECTS

Although currently omics are used widely in many different fields, there are still many challenges faced, such as the complexity of data searching, lack of techniques to make the process easy, and tedious protocol. Omics have made many bioremediation strategies applicable in the field. Some things, which were assumed to be very difficult, are now becoming too easy due to applications of omics. A huge amount of data come from the omics studies, which have to be maintained and monitored properly, so there is still a need for better algorithms and bioinformatics tools to achieve the same. Also, there is a need for new tools in metabolomic and proteomic studies, as they are not that much developed compared to other omics approaches. There is a need for the application of the functional approach in omics studies so that we can focus mainly on the useful data. In this way, the study can become easy and more productive.

1.5   CONCLUSIONS

Omics tools have been proved as efficient and cost-effective methods to be used in the field of bioremediation. Many difficulties faced due to the use of conventional approaches have been solved; for example, the need of studying every organism separately is solved by metagenomics. Functional genomics has made the process less time-consuming and focused. Proteomics and metabolomics explained how environmental factors or stimuli affect an organism. Currently, researchers are facing many challenges while using these tools as this is a relatively new area compared to conventional approaches. But efforts are being made to make this a viable and productive approach for bioremediation.

REFERENCES

Web Reference 1: - https://www.downtoearth.org.in/news/environment/chennai-s-soil-and-delhi-s-air-most-contaminated-due-to-pcb-concentration-study-57217.

Anh, H.T.H., E. Shahsavari, N.J. Bott, and A.S. Ball. "Bioaugmentation of seafood processing wastewater enhances the removal of inorganic nitrogen and chemical oxygen demand." *Aquaculture* 542(2021): 736818.

Benndorf, D., G.U. Balcke, H. Harms, and M. Von Bergen. "Functional metaproteome analysis of protein extracts from contaminated soil and groundwater." *The ISME Journal* 1, no. 3 (2007): 224–234.

Blasi, B., H. Tafer, C. Kustor, C. Poyntner, K. Lopandic, and K. Sterflinger. "Genomic and transcriptomic analysis of the toluene degrading black yeast *Cladophialophora immunda*." *Scientific Reports* 7, no. 1 (2017): 1–13.

Celis, J.E., M. Kruhøffer, I. Gromova, C. Frederiksen, M. Østergaard, T. Thykjaer, P. Gromov et al. "Gene expression profiling: Monitoring transcription and translation products using DNA microarrays and proteomics." *FEBS Letters* 480, no. 1 (2000): 2–16.

Chandra, R., N.K. Dubey, and V. Kumar. *Phytoremediation of Environmental Pollutants.* Boca Raton, FL: CRC Press, (2018a) Doi: 10.1201/9781315161549.

Chandra, R., and V. Kumar. Phytoextraction of heavy metals by potential native plants and their microscopic observation of root growing on stabilised distillery sludge as a prospective tool for In-situ phytoremediation of industrial waste. *Environmental Science and Pollution Research* 24(2017): 2605–2619 Doi: 10.1007/s11356-016-8022-1.

Chandra, R., and V. Kumar. Phytoremediation: A green sustainable technology for industrial waste management. In: Chandra, R., N. Dubey, and V. Kumar (eds.) *Phytoremediation of Environmental Pollutants.* Boca Raton, FL: CRC Press, (2018) Doi: 10.1201/9781315161549-1.

Chandra, R., V. Kumar, and K. Singh. Hyperaccumulator versus nonhyperaccumulator plants for environmental waste management. In: Chandra, R., N. Dubey, and V. Kumar (eds.) *Phytoremediation of Environmental Pollutants.* Boca Raton, FL: CRC Press, (2018b). Doi: 10.1201/9781315161549-1.

Chen, G., X. An, H. Li, F. Lai, E. Yuan, X. Xia, and Q. Zhang. "Detoxification of azo dye direct black G by thermophilic *Anoxybacillus* sp. PDR2 and its application potential in bioremediation." *Ecotoxicology and Environmental Safety* 214(2021): 112084.

Claudino, W.M., A. Quattrone, L. Biganzoli, M. Pestrin, I. Bertini, and A.D. Leo. "Metabolomics: Available results, current research projects in breast cancer, and future applications." *Journal of Clinical Oncology* 25, no. 19 (2007): 2840–2846.

Coelho, L.M., H.C. Rezende, L.M. Coelho, P.A. de Sousa, D.F. Melo, and N.M. Coelho. "Bioremediation of polluted waters using microorganisms." *Advances in Bioremediation of Wastewater and Polluted Soil* 10(2015): 60770.

Combourieu, B., P. Besse, M. Sancelme, E. Maser, and A.-M. Delort. "Evidence of metyrapone reduction by two mycobacterium strains shown by 1 H NMR." *Biodegradation* 15, no. 2 (2004): 125–132.

Czaplicki, L.M., and C.K. Gunsch. "Reflection on molecular approaches influencing state-of-the-art bioremediation design: Culturing to microbial community fingerprinting to omics." *Journal of Environmental Engineering* 142, no. 10 (2016): 03116002.

Desai, C., H. Pathak, and D. Madamwar. "Advances in molecular and "-omics" technologies to gauge microbial communities and bioremediation at xenobiotic/anthropogen contaminated sites." *Bioresource Technology* 101, no. 6 (2010): 1558–1569.

Dumont, M.G., S.M. Radajewski, C.B. Miguez, I.R. McDonald, and J.C. Murrell. "Identification of a complete methane monooxygenase operon from soil by combining stable isotope probing and metagenomic analysis." *Environmental Microbiology* 8, no. 7 (2006): 1240–1250.

Feng, M., J. Zhou, X. Yu, H. Wang, Y. Guo, and W. Mao. "Bioremediation of triphenyl phosphate by *Pycnoporus sanguineus*: Metabolic pathway, proteomic mechanism and biotoxicity assessment." *Journal of Hazardous Materials* 417(2021): 125983.

Green, S.J., O. Prakash, P. Jasrotia, W.A. Overholt, E. Cardenas, D. Hubbard, J.M. Tiedje et al. "Denitrifying bacteria from the genus Rhodanobacter dominate bacterial communities in the highly contaminated subsurface of a nuclear legacy waste site." *Applied and Environmental Microbiology* 78, no. 4 (2012): 1039–1047.

Harekrushna, S., and D.C. Kumar. "A review on: Bioremediation." *International Journal of Research in Chemistry and Environment* 2, no. 1 (2012): 13–21.

Haro, M.-A., and V. de Lorenzo. "Metabolic engineering of bacteria for environmental applications: Construction of *Pseudomonas* strains for biodegradation of 2-chlorotoluene." *Journal of Biotechnology* 85, no. 2 (2001): 103–113.

Haroon, M.F., S. Hu, Y. Shi, M. Imelfort, J. Keller, P. Hugenholtz, Z. Yuan, and G.W. Tyson. "Anaerobic oxidation of methane coupled to nitrate reduction in a novel archaeal lineage." *Nature* 500, no. 7464 (2013): 567–570.

Hidde Boersma, F.G., W. Colin McRoberts, S.L. Cobb, and C.D. Murphy. "A 19F NMR study of fluorobenzoate biodegradation by *Sphingomonas* sp. HB-1." *FEMS Microbiology Letters* 237, no. 2 (2004): 355–361.

Höhener, P., and V. Ponsin. "In situ vadose zone bioremediation." *Current Opinion in Biotechnology* 27 (2014): 1–7.

Hong, Y.-H., C.-C. Ye, Q.-Z. Zhou, X.-Y. Wu, J.-P. Yuan, J. Peng, H. Deng, and J.-H. Wang. "Genome sequencing reveals the potential of *Achromobacter* sp. HZ01 for bioremediation." *Frontiers in Microbiology* 8(2017): 1507.

Horgan, R.P., and L.C. Kenny. "'Omic' technologies: Genomics, transcriptomics, proteomics and metabolomics." *The Obstetrician & Gynaecologist* 13, no. 3 (2011): 189–195.

Huang, S.W., Y. Wang, C. Tang, H.L. Jia, and L. Wu. "Speeding up selenite bioremediation using the highly selenite-tolerant strain *Providencia rettgeri* HF16-A novel mechanism of selenite reduction based on proteomic analysis." *Journal of Hazardous Materials* 406(2021): 124690.

Jennings, L.K., M.M.G. Chartrand, G. Lacrampe-Couloume, B.S. Lollar, J.C. Spain, and J.M. Gossett. "Proteomic and transcriptomic analyses reveal genes upregulated by cis-dichloroethene in *Polaromonas* sp. strain JS666." *Applied and Environmental Microbiology* 75, no. 11 (2009): 3733–3744.

Jirtle, R.L., and M.K. Skinner. "Environmental epigenomics and disease susceptibility." *Nature Reviews Genetics* 8, no. 4 (2007): 253–262.

Keum, Y.S., J.S. Seo, Q.X. Li, and J.H. Kim. "Comparative metabolomic analysis of *Sinorhizobium* sp. C4 during the degradation of phenanthrene." *Applied Microbiology and Biotechnology* 80, no. 5 (2008): 863–872.

Kim, S.-J., R.C. Jones, C.-J. Cha, O. Kweon, R.D. Edmondson, and C.E. Cerniglia. "Identification of proteins induced by polycyclic aromatic hydrocarbon in *Mycobacterium vanbaalenii* PYR-1 using two-dimensional polyacrylamide gel electrophoresis and de novo sequencing methods." *Proteomics* 4, no. 12 (2004): 3899–3908.

Kim, Y.H., K. Cho, S.-H. Yun, J.Y. Kim, K.-H. Kwon, J.S. Yoo, and S. Il Kim. "Analysis of aromatic catabolic pathways in *Pseudomonas putida* KT 2440 using a combined proteomic approach: 2-DE/MS and cleavable isotope-coded affinity tag analysis." *Proteomics* 6, no. 4 (2006): 1301–1318.

Kuiper, I., E.L. Lagendijk, G.V. Bloemberg, and B.J.J. Lugtenberg. "Rhizoremediation: A beneficial plant-microbe interaction." *Molecular Plant-Microbe Interactions* 17, no. 1 (2004): 6–15.

Kumar, V., and R. Chandra. Bacterial assisted phytoremediation of industrial waste pollutants and eco-restoration. In R. Chandra, N.K. Dubey, and V. Kumar. (eds.) *Phytoremediation of Environmental Pollutants*. Boca Raton, FL: CRC Press (2018).

Kumar, V., and R. Chandra. Metagenomics analysis of rhizospheric bacterial communities of *Saccharum arundinaceum* growing on organometallic sludge of sugarcane molasses-based distillery. *3 Biotech* 10, no. 7 (2020): 316. Doi: 10.1007/s13205-020-02310-5.

Kumar, V., S.K. Shahi, and S. Singh. Bioremediation: An eco-sustainable approach for restoration of contaminated sites. In: Singh J., D. Sharma, G. Kumar, and N. Sharma. (Eds.) *Microbial Bioprospecting for Sustainable Development*. Springer: Singapore (2018), Doi: 10.1007/978-981-13-0053-0_6.

Kumar, V., K. Singh, M.P. Shah, A.K. Singh, A. Kumar, and Y. Kumar. Application of omics technologies for microbial community structure and function analysis in contaminated environment. In Shah, M.P., A. Sarkar, and S. Mandal. (Ed.), *Wastewater Treatment: Cutting Edge Molecular Tools, Techniques & Applied Aspects in Waste Water Treatment*. Elsevier (2021), Doi: 10.1016/B978-0-12-821925-6.00013-7.

Kumar, V., I.S. Thakur, A.K. Singh, and M.P. Shah. Application of metagenomics in remediation of contaminated sites and environmental restoration. In: Shah, M., S. Rodriguez-Couto, and S.S. Sengor. (Eds.), *Emerging Technologies in Environmental Bioremediation*. Elsevier (2020), Doi: 10.1016/B978-0-12-819860-5.00008-0.

Lu, J.J., Y.L. Ma, G.L. Xing, W.L. Li, X.X. Kong, J.Y. Li, L.J. Wang, H.L. Yuan, and J.S. Yang. "Revelation of microalgae's lipid production and resistance mechanism to ultra-high Cd stress by integrated transcriptome and physiochemical analyses." *Environmental Pollution* 250(2019): 186–195.

Mahjoubi, M., H. Aliyu, M. Neifar, S. Cappello, H. Chouchane, Y. Souissi, A.S. Masmoudi, D.A. Cowan, and A. Cherif. "Genomic characterization of a polyvalent hydrocarbonoclastic bacterium *Pseudomonas* sp. strain BUN14." *Scientific Reports* 11, no. 1 (2021): 1–13.

Maletić, S., B. Dalmacija, and S. Rončević. "Petroleum hydrocarbon biodegradability in soil– Implications for bioremediation." *Hydrocarbon* 1, (2013): 43.

Malik, S., S.A.L. Andrade, M.H. Mirjalili, R. R. Arroo, M. Bonfill, and P. Mazzafera. "Biotechnological approaches for bioremediation: *In vitro* hairy root culture." In *Transgenesis and Secondary Metabolism*, vol. 1. Springer International Publishing, (2016).

Mapelli, V., L. Olsson, and J. Nielsen. "Metabolic footprinting in microbiology: Methods and applications in functional genomics and biotechnology." *Trends in Biotechnology* 26, no. 9 (2008): 490–497.

Meagher, R.B. "Phytoremediation of toxic elemental and organic pollutants." *Current Opinion in Plant Biology* 3, no. 2 (2000): 153–162.

Misra, B.B., C. Langefeld, M. Olivier, and L.A. Cox. "Integrated omics: Tools, advances and future approaches." *Journal of Molecular Endocrinology* 62, no. 1 (2019): R21–R45.

Naik, M.M., M. Imran, D.C. Vaigankar, S.Y. Mujawar, A.D. Malik, and S.K. Gaonkar. "Genome guided bioprospecting of extremely halophilic *Haloferax* sp. AS1 for CAZymes, bioremediation and study metabolic versatility." *Proceedings of the National Academy of Sciences, India Section B: Biological Sciences*, India (2021): 1–9.

Nikolaivits, E., R. Siaperas, A. Agrafiotis, J. Ouazzani, A. Magoulas, A. Gioti, and E. Topakas. "Functional and transcriptomic investigation of laccase activity in the presence of PCB29 identifies two novel enzymes and the multicopper oxidase repertoire of a marine-derived fungus." *Science of the Total Environment* 775(2021): 145818.

Nocker, A., Burr, M., and Camper, A. K. Genotypic microbial community profiling: a critical technical review. *Microbial Ecology* 54(2) (2007): 276–289.

Ortiz-Hernández, M.L., Y. Gama-Martínez, M. Fernández-López, M.L. Castrejón-Godínez, S. Encarnación, E. Tovar-Sánchez, E. Salazar, A. Rodríguez, and P. Mussali-Galante. "Transcriptomic analysis of *Burkholderia cenocepacia* CEIB S5-2 during methyl parathion degradation." *Environmental Science and Pollution Research* 28(31) (2021): 1–18.

Parisi, V.A., G.R. Brubaker, M.J. Zenker, R.C. Prince, L.M. Gieg, M.L.B. Da Silva, P.JJ Alvarez, and Joseph M. Suflita. "Field metabolomics and laboratory assessments of anaerobic intrinsic bioremediation of hydrocarbons at a petroleum-contaminated site." *Microbial Biotechnology* 2, no. 2 (2009): 202–212.

Park, H., B. Min, Y. Jang, J. Kim, A. Lipzen, A. Sharma, B. Andreopoulos et al. "Comprehensive genomic and transcriptomic analysis of polycyclic aromatic hydrocarbon degradation by a mycoremediation fungus, *Dentipellis* sp. KUC8613." *Applied Microbiology and Biotechnology* 103, no. 19 (2019): 8145–8155.

Rifai, N., M.A. Gillette, and S.A. Carr. "Protein biomarker discovery and validation: The long and uncertain path to clinical utility." *Nature Biotechnology* 24, no. 8 (2006): 971–983.

San, M., A.P. Ravanel, and M. Raveton. "A comparative study on the uptake and translocation of organochlorines by *Phragmites australis*." *Journal of Hazardous Materials* 244 (2013): 60–69.

Sellers, T.A., and J.R. Yates. "Review of proteomics with applications to genetic epidemiology." *Genetic Epidemiology: The Official Publication of the International Genetic Epidemiology Society* 24, no. 2 (2003): 83–98.

Solyanikova, I.P., O.V. Moiseeva, S. Boeren, M.G. Boersma, M.P. Kolomytseva, J. Vervoort, I.M.C.M. Rietjens, L.A. Golovleva, and W.J.H. van Berkel. "Conversion of 2-fluoromuconate to cis-dienelactone by purified enzymes of *Rhodococcus opacus* 1cp." *Applied and Environmental Microbiology* 69, no. 9 (2003): 5636–5642.

Tang, Y.J., H.G. Martin, P.S. Dehal, A. Deutschbauer, X. Llora, A. Meadows, A. Arkin, and J.D. Keasling. "Metabolic flux analysis of *Shewanella* spp. reveals evolutionary robustness in central carbon metabolism." *Biotechnology and Bioengineering* 102, no. 4 (2009): 1161–1169.

Tyagi, B., and N. Kumar. "Bioremediation: Principles and applications in environmental management." In *Bioremediation for Environmental Sustainability*, pp. 3–28. Elsevier, (2021).

Vailati-Riboni, M., V. Palombo, and J.J. Loor. "What are omics sciences?" In *Periparturient Diseases of Dairy Cows*, pp. 1–7. Springer, Cham, (2017).

Vidali, M. "Bioremediation: An overview." *Pure and Applied Chemistry* 73, no. 7 (2001): 1163–1172.

Vlaanderen, J., L.E. Moore, M.T. Smith, Q. Lan, L. Zhang, C.F. Skibola, N. Rothman, and R. Vermeulen. "Application of OMICS technologies in occupational and environmental health research; current status and projections." *Occupational and Environmental Medicine* 67, no. 2 (2010): 136–143.

Wilkins, M.J., N.C. VerBerkmoes, K.H. Williams, S.J. Callister, P.J. Mouser, H. Elifantz, A.L. N'Guessan et al. "Proteogenomic monitoring of Geobacter physiology during stimulated uranium bioremediation." *Applied and Environmental Microbiology* 75, no. 20 (2009): 6591–6599.

Wilmes, P., M. Wexler, and P.L. Bond. "Metaproteomics provides functional insight into activated sludge wastewater treatment." *PLoS One* 3, no. 3 (2008): e1778.

Wingren, C., and C.A.K. Borrebaeck. "Antibody microarrays: Current status and key technological advances." *Omics: A Journal of Integrative Biology* 10, no. 3 (2006): 411–427.

Yan, S.-K., R.-H. Liu, H.-Z. Jin, X.-R. Liu, J. Ye, L. Shan, and W.-D. Zhang. ""Omics" in pharmaceutical research: Overview, applications, challenges, and future perspectives." *Chinese Journal of Natural Medicines* 13, no. 1 (2015): 3–21.

Zhao, Y., Y. Che, F. Zhang, J. Wang, W. Gao, T. Zhang, and C. Yang. "Development of an efficient pathway construction strategy for rapid evolution of the biodegradation capacity of *Pseudomonas putida* KT2440 and its application in bioremediation." *Science of the Total Environment* 761 (2021): 143239.

2 Role of Genomics, Metagenomics, and Other Meta-Omics Approaches for Expunging the Environmental Contaminants by Bioremediation

Atif Khurshid Wani, Daljeet Singh Dhanjal,
Nahid Akhtar, Chirag Chopra,
Abhineet Goyal, and Reena Singh
Lovely Professional University

CONTENTS

DOI: 10.1201/9781003247883-2

2.1   INTRODUCTION

Pollution continues to toxify diverse ecosystems and deteriorate human health. The mismanagement and mishandling of industrial effluents, household wastes, insecticides, herbicides, etc., are damaging the already deteriorating ecosystems (Kumar et al., 2020a; Akhtar and Mannan, 2020b; Dutta et al., 2021). The statistics by WHO about the impact of pollution is also perplexing. Around the world, 2.2 billion people are without safe water services, while 144 million use contaminated water. It is supposed that half of the world's population might have to live in water-scarce regions by 2025 (Akhtar and Mannan, 2020a, b). In the late 20th century, it was estimated that more than 20 million hectares of land were polluted. This continues to engulf more land with rapidly growing anthropogenic activities such as industrialization and urbanization (Manisalidis et al., 2020). It impacts the economy by putting the burden of a 5% GDP loss on developing nations (Reddy and Behera, 2006), accounting for more than 10% of direct and indirect deaths (Landrigan et al., 2018). Thus, it is vital to frame strategies to fight various environmental contaminants because of their prolonged persistence in nature. Several physical and chemical methods have been adopted to eliminate pollution (Abdulsalam et al., 2012). However, they are not extensively used because of their high cost, low efficacy, and harmful by-product synthesis (Khan et al., 2019; Kumar et al., 2021a).

Bioremediation is a practical, economic, and eco-friendly strategy to combat various pollutants (Olawale et al., 2020; Bhat et al., 2018). This ensures the restoration and prevention of further pollution expansion in the actual natural setup (Yerushalmi et al., 2003; Kumar et al., 2018). The bioremediation approach to fighting pollution comprises bioaugmentation and bio-stimulation. The former involves exogenous microorganism inoculation into the contaminated site, whereas the latter depends more on the native microorganisms (Ali et al., 2020; Awasthi et al., 2020). Tremendous progress has been made in exploring the environmental remediation approaches, but inadequate information about beneficial microorganisms' growth factors and metabolism limits their execution. However, the metagenomics approach has emerged as a novel concept for understanding the effective microorganisms, functional genes, bioactive molecules, and enzymes (Table 2.2) in a particular environmental sample. There are diverse projects of metagenomics-mediated bioremediation that have been set up in terrestrial and marine ecosystems (Figure 2.1) (Mani and

Kumar, 2014; Marco, 2008). These projects include bioremediation processes mediated by enzymes such as oxygenases, peroxidases, and hydrolases (Table 2.2) derived from microbial sources using genomic, metagenomic, and meta-transcriptomic approaches (Figure 2.3). The rapid analysis of microbes and their associated products has become possible after the advent of metabolomics, metafluxomics (Figure 2.3), and computational approaches (Table 2.3). Although certain limitations are associated with metagenomics-led bioremediation approaches, they are likely to be mitigated with more advancements in molecular biology techniques.

This chapter aims to discuss bioremediation, different molecular techniques, and enzyme catalysis involved in bioremediation. Moreover, it also enlightens us about genomics, metagenomics, and different *meta-omics* approaches to eradicating various environmental contaminants through bioremediation.

2.2 UNDERSTANDING BIOREMEDIATION

Microorganisms can eliminate pollutants from the contaminated environment by a series of enzymatic processes (Coelho et al., 2015). Many microorganisms execute bioremediation reactions (Table 2.1) (Akhtar and Mannan, 2020b). The oxidation of toxic organic contaminants to by-products of non-toxic nature is a common bioremediation process (Lovley, 1995). Oxygen is a well-studied electron acceptor for numerous microbial metabolic processes and extensive degradation of contaminants such as xenobiotics, heavy metals (Figure 2.1), and pesticides (El Mamouni et al., 2002). The xenobiotic plastic polyethylene terephthalate is among the most abundant pollutants found on the earth. Although it is chemically stable with higher microbial degradation refractory, many microorganisms have shown PET hydrolase property and enable the usage of the derived monomers, which include terephthalate and ethylene glycol (Carr et al., 2020). Although a diverse microbial community can carry out the process of degradation, the species of *Pseudomonas* and some related organisms have been thoroughly investigated due to their contaminant degradation ability (Tables 2.1 and 2.2) (Ojewumi et al., 2018). Several toxified environments, water body sediments, aquifers, and submerged soils, are very much anoxic (oxygen deficient); thus, many organisms oxidize the pollutants taking advantage of other electron acceptors such as sulphate nitrate and iron oxides (Coates et al., 1998). These electron acceptors are used depending on their availability and competition among specific respiratory microbes for electron donors.

A typical example is the ability of Fe(III) to act as an electron acceptor to oxidize organic substances beneath the surface and amplify the Fe(III) availability for stimulating the microbial anaerobic degradation contaminants that are chemically organic (Wang et al., 2020). In marine environments, sulfate is a crucial oxidizing agent for the anaerobic degradation of pollutants. This is attributed to the marine environments because of the higher sulfur concentrations in seawater, which can accelerate the pollutant degradation process in aquifers. *Desulfobacterium* and *Desulfobacula* species include the sulfate-reducing microorganisms that can carry out hydrocarbons' oxidation with sulfate acting as an oxidizing agent (Mahjoubi et al., 2018; Zhang et al., 2015). Many contaminants act as potential oxidizing agents rather than reducing agents in the bioremediation process. One familiar example of bioremediation of this type is the reductive dechlorination process, which involves chlorine atom replacement by microorganisms that sequentially results in ethene from PCE (tetrachloroethene) to TCE (trichloroethene) to DCE (dichloroethene) to vinyl chloride (Xiao et al., 2020). Many *Dehalococcoides* species carry out dehalogenation by catalyzing this process in polluted subsurfaces (Sims et al., 1990). Some microorganisms also reduce dangerous inorganic contaminants such as percolate and nitrate and convert them into non-dangerous innocuous products (Evans and Trute, 2006). Metals, a class of pollutants that serve as potential oxidizing agents in microbial respiration, do not harm them but alter the solubility (Rehan and Alsohim, 2019).

A typical example is the reduction of oxidized and soluble uranium (U_{vi}) (Table 2.1) to the insoluble uranium (U_{iv}) by *Geobacter* species using uranium as an oxidizing agent, thus resulting in the growth stimulation of *Geobacter* in polluted uranium environments that ultimately precipitates uranium from polluted groundwater and in this way averts its further spread (Newsome et al., 2014). Bioremediation for detoxification or contaminant removal is intrinsic and occurs without human interference. Thus, intrinsic bioremediation is the least dangerous as it does not pose health risks because of its localized execution (Kumar and Chandra, 2020b). However, the bioremediation process rate is usually too slow to meet the required demand of degradation processes. The speeding up of the process involves manipulating microorganisms that would ardently advance the activity of microorganisms involved in degradation processes (Donati et al., 2019). The engineered bioremediation approaches include electron donor and electron acceptor addition for microbial growth stimulation and metabolic vivification, the addition of growth- and activity-limiting nutrients of microorganisms, and manipulation of bioremediation-associated genes (Hlihor et al., 2017).

2.3 MOLECULAR BIOLOGY OF MICROORGANISMS-MEDIATED BIOREMEDIATION

Multiple factors influence the process of bioremediation, such as nutrient availability (Atagana et al., 2003), chemical nature of pollutants (Vidali, 2001), moisture content (Cho et al., 2000), temperature (Delille et al., 2004), and microbial diversity (Leal et al., 2017). In studying the bioremediation process, most laboratory investigations are carried out by incubating the environment's contaminated samples to evaluate the contaminant immobilization or degradation rates (Azubuike et al., 2016). These studies give an idea about the microorganisms' metabolic potential but provide minor information about the microorganisms involved in bioremediation (Sharma, 2021). Bioremediation investigations through pure cultures are essential for molecular analysis and the microorganism's physiological status (Perumbakkam et al., 2006). Genes that code for catabolic pathways are chromosomal or extrachromosomal (transposon, plasmid, or conjugative element) (Santero and Díaz, 2020). It is crucial to understand the genomics of catabolism in bioremediation to understand the degradation potential, physiology, and metabolism of the microorganisms. The molecular analysis was done by Imperato et al. on *Pseudomonas* strains that degrade naphthalene (Table 2.1) showed the presence of genes that cause the degradation of pollutants such as

TABLE 2.1
Microorganisms that Execute Remediation Processes Against Different Contaminants

Microorganisms	Compound Class	Compound/Element Name	References
Candida viswanathii	Hydrocarbons	Phenanthrene and benzopyrene	Hesham et al. (2012)
Coprinellus radians	Hydrocarbons	Dibenzofurans and methylnapthalenes	Aranda et al. (2010)
Bacillus safensis	Metals	Cadmium	Priyalaxmi et al. (2014)
Saccharomyces cerevisiae	Heavy metals	Nickel, mercury, and lead	Infante et al. (2014)
Aspergillus niger, Candida krusei, Candida glabrata	Crude oil	Petroleum	Burghal et al. (2016)
Penicillium ochrochloron	Dyes	Malachite green	Shedbalkar and Jadhav (2011)
Pseudomonas putida, Arthrobacter sp.	Pesticides	Malathion and Ridomil MZ 68	Hussaini et al. (2013); Pérez et al. (2016)
Microbacterium profundi	Metals	Iron	Wu et al. (2015)
Klebsiella oxytoca, Bacillus macerans	Textile effluents and vat dyes	Anthraquinone and acids	Adebajo et al. (2017), Benkhaya et al. (2018)
Gloeophyllum striatum	Hydrocarbons	Striatum pyrene, lignin peroxidase, and dibenzothiophene	Yadav et al. (2011)
Rhodococcus sp.	Xenobiotics	Dioxins	Peng et al. (2013)
Dehalococcoides sp.	Halogenated compounds	Atrazine and vinyl chloride	Xiao et al. (2020)
Mycobacterium PYR-1	Polycyclic hydrocarbons	Pyrene	Kim et al. (2010)
Cunninghamella elegans	Heavy metals	Cobalt, chromium, copper, and lead	De Oliveira Franco et al. (2004)
Penicillium chrysogenum	Hydrocarbons	Toluene, ethylbenzene, and triphenylmethane	Pereira et al. (2014)
Alcaligenes odorans, Bacillus subtilis	Hydrocarbons	Phenol	Singh et al. (2013)
Pseudomonas aeruginosa	Metals	Copper, nickel, and chromium	Sankar et al. (2011)
Cladophora sp., Spirogyra sp.	Metals	Lead and copper	Lee and Chang (2011)
Ganoderma lucidum	Metals	Argon	Loukidou et al. (2003)
Hydrodictyon, Rhizoclonium, and Oedogonium species	Metals	Vanadium and arsenic	Saunders et al. (2012)
Arthrobacter sp.	Pesticides	Endosulfate	Kafilzadeh et al. (2015)
Pseudomonas putida	Aromatic hydrocarbons	Naphthalene	Dutta et al. (2019)
Trametes trogii	Industrial dyes	Malachite green	Yan et al. (2014)

BTEX (benzene, toluene, ethylbenzene and xylene), terephthalate, anthranilate, and alkanes (Table 2.1) (2019). These isolated strains from contaminated oil contain a set of plasmids, but only one plasmid from one of the strains bears the complete *nah* gene set.

On the other hand, other strains' nah genes are associated with transposable sequences of DNA in integrated chromosomal form. *Sphingopyxis lindanitolerans* genomic analysis carried out by Kaminski et al. revealed two plasmids having *lin* genes linked with a transposable element (insertion sequences) and are responsible for the biodegradation pathway of lindane (2019). The multiple strain genomic analysis (MSGA) of the biphenyl-contaminated sites revealed a higher presence of *bph* genes linked with conjugative elements. It showed molecular dissemination among bacteria through lateral transfer (Hirose et al., 2019).

Evaluating critical genes involved in bioremediation can provide more insights into the other microbial processes. Thus, it is vital for an effective bioremediation process that the involved genes express efficiently in the substrate's presence to halt the wasteful and excessive synthesis of catabolic enzymes (Bartholameuz et al., 2020). The independently evolving regulatory systems of the catabolic genes adjust to the different molecules to be metabolized to prevent the gratuitous induction mediated by a molecule that cannot be metabolized (Floriano et al., 2019). Thus, the higher degree of pollutant degradation is directly associated with the higher mRNA concentrations (Nouaille et al., 2017). A typical example is the positive correlation of the levels of mRNA concentrations with the *nah*A gene involved in aerobic naphthalene degradation (Table 2.1) (Dutta et al., 2019). There are factors other than the regulatory gene transcription that efficiently control the process of bioremediation. For example, the reduction of mercury from $Hg^{(II)}$ (soluble) to $Hg^{(0)}$ (volatile) is a potent way to remove mercury from water. The expressed *mer*A gene concentration that causes the reduction of $Hg^{(II)}$ was found to be highest in mercury-contaminated water with a higher rate of $Hg^{(II)}$. However, the merA concentration was not always proportional to $Hg^{(II)}$ reduction rate (Table 2.1) (Naguib et al., 2018).

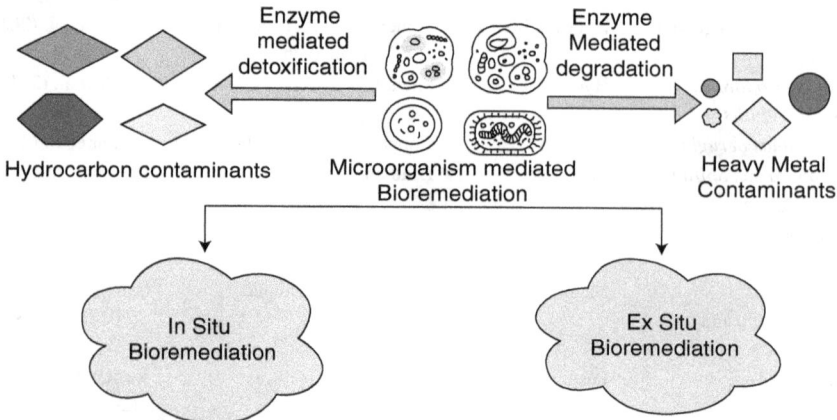

FIGURE 2.1 Microorganism-mediated degradation/detoxification of heavy metals and hydrocarbons through *in situ* and *ex-situ* approaches.

2.4 MICROBIAL ENZYME CATALYSIS FOR BIOREMEDIATION

Enzymes degrade recalcitrant contaminants (Table 2.2) by lowering the activation energy, thereby ensuring the substrates' easy and fast disintegration. Enzymes have gained momentum in the bioremediation process because of the diverse mode of action against different contaminants (Tables 2.1 and 2.2). The use of enzymes has mitigated the conventional slow degradation processes (Camacho-Pérez et al., 2012). The eco-friendly mode of action supported by the smaller size makes microbial origin enzymes a potent anti-contaminant agent with vigorous activity, mobility, effectiveness, and harmlessness (Chandra et al., 2017). Sharma et al. reported in depth the enzymatic pathways for the degradation of various pollutants (2018).

2.4.1 OXYGENASES FOR BIOREMEDIATION

Oxygenases are part of the oxidoreductase enzyme group that transfers molecular oxygen for oxidizing their substrates using NADH/FAD/NADPH as a co-substrate (Rao et al., 2010). Based on the oxygen atoms utilized for oxygenation, these can either be monooxygenases or dioxygenases. These are crucial for the catabolism to increase their water solubility or reactivity or to ensure aromatic ring cleavage. The halogenated compounds include diversified environmental contaminants in fungicides, plasticizers, herbicides, insecticides, and chemical synthesis intermediates. Using a specific oxygenase ensures the easy and effective degradation of pollutants (Hazen, 2010). The molecular oxygen induction into the substrate is mediated by monooxygenases and is further categorized into P450- and flavin-dependent monooxygenases based on a cofactor's presence (Roper and Grogan, 2016). The biocatalytic action of monooxygenases in bioremediation is attributed to their stereoselectivity and regioselectivity. The chemical reactions such as dehalogenation, ammonification, desulfurization, denitrification, and cleavage, important for bioremediation, are catalyzed by monooxygenases (Arora, 2010). Methane monooxygenase is a widely studied and evaluated biocatalyst that mediates the degradation of alkanes, alkenes, and ethers. The dioxygenase enzyme, isolated from soil bacteria, serves as a potent biocatalyst for degrading heterocyclic aromatic compounds (Table 2.2) (dos Santos et al., 2015). OxDBase is a database developed explicitly for biodegradative oxygenases. Till now, this database has compiled the information of almost 250 monooxygenases and dioxygenases and can be freely accessed at www.imtech.res.in/raghava/oxdbase/.

2.4.2 OXIDOREDUCTASES FOR BIOREMEDIATION

Various microorganisms utilize the mechanism of removal of toxic compounds by oxidative reactions-catalyzed oxidoreductase enzymes to detoxify the environment (Table 2.2). Microorganisms gain energy from the cascade of enzyme-mediated metabolic reactions to cleave the chemical bonds and ensure electron transfer from a donor (reduced form) to the acceptor, thus oxidizing harmful contaminants to non-toxic compounds (Alneyadi et al., 2018; Barber et al., 2020). The oxidoreductases detoxify the xenobiotic (phenolic and anilinic) compounds and cooperate in phenolic humification formed from lignin decomposition in the soil environment through polymerization copolymerization reactions helpful in the degradation of dyes (Martínez et al., 2017). Many bacterial biocatalysts reduce metals of radioactive nature from soluble oxidized to insoluble reduced forms by taking up the organic compound-associated electrons and utilizing the radioactive metal as an oxidizing agent (Mani and Kumar, 2014; Kumar, 2018). The chlorinated compounds of phenolic nature generated from pulp and paper industries are among the harmful effluent wastes that can be partially degraded during the bleaching process (Bosso and Cristinzio, 2014). In many fungal species, the presence of extracellular biocatalysts such as peroxidase, manganese, lignin peroxidase, and laccase helps remove phenolic compounds from polluted sites. These biocatalysts are synthesized by the mycelium of fungal species in the vicinity and thus act on soil contaminants effectively (Alneyadi et al., 2018).

2.4.3 PEROXIDASES FOR BIOREMEDIATION

Peroxidases are a diverse group of ubiquitous enzymes that mediate lignin oxidation by utilizing H_2O_2 (hydrogen peroxide). These peroxidases have been grouped based on the heme group's presence or absence (Pande et al., 2019). Group I consists of intracellular peroxidases that include ascorbate peroxidase, cytochrome peroxidase, and catalase peroxidases. Group II is a category of secretory biosynthetic fungal enzymes that include manganese peroxidase and lignin peroxidase from *Coprinus cinereus* and *Phanerochaete chrysosporium*. Group II peroxidases are well known for degrading wood lignin (Darwesh et al., 2019). Peroxidases are produced from different sources, including animals, plants, and microbes. Microbial sources of peroxidases include *Bacillus subtilis, Candida krusei, Bacillus sphaericus*, *P. chrysosporium,* and *Thermobifida fusca*. Several microbial peroxidases have been used to degrade the xenobiotic dyes, which are commonly challenging to degrade (Bansal and Kanwar, 2013; Zainith et al., 2020). The removal of acid orange (azo dye) and chromate by *Brevibacterium casei* have been studied in detail. *Pleurotus ostreatus*, known for the production of an extracellular peroxidase, has been reported as having decolourization properties against Remazol blue and other groups such as polymeric dyes, heterocyclic azo dyes, and triarylmethane (Bansal and Kanwar, 2013).

2.4.4 HYDROLASES FOR BIOREMEDIATION

The disruption of chemical bonding by hydrolytic action of enzymes reduces the toxicity of the contaminants. This is crucial for degrading organophosphate, xenobiotic, and carbamate compounds (dos Santos Aguilar and Sato, 2018). Heptachlor and DDT insecticides are stable in aerobic conditions but are readily degraded by anaerobic microbes by the hydrolytic action (Table 2.2). The hydrolases are ubiquitous and stereoselective. The extracellular hydrolytic enzymes, including lipases, amylases, xylanases, DNases, and pullulanases, have extensively been used in different fields as feed additives, food additives, biomedical agents, and chemical biocatalysts (Kasana, 2010). *Candida rugose*-derived hydrolytic lipase is effective in triolein hydrolysis. Lipase acts by breaking the triolein ester bonds and results in monoolein, diolein, and glycerol. Lipase activity is considered an essential indicator parameter during the degradation of hydrocarbons in contaminated soils (Karigar and Rao, 2011). Cellulase hydrolytic enzymes, which are effective in crystalline cellulose's biodegradation, have been obtained from microbial sources. The cellulases of acidic nature are biosynthesized by *Humicola* and *Trichoderma* fungi, while cellulases of alkaline nature are synthesized by *Bacillus* strains (Table 2.2). Cellulases effectively detoxify microfibrils and remove ink in the paper industry, helping in brewing and ethanol production. On the other hand, protease hydrolytic enzymes cause the proteinaceous breakdown of biological and industrial contaminants, which enter the environment. Owing to the diverse occurrence of proteases, they are obtained from different sources, including animals, plants, and microbes. Among the microorganisms, *Bacillus* sp. and *Aspergillus* sp. are potential sources of hydrolases (Table 2.2) (Razzaq et al., 2019).

2.5 EXPLORING GENOMES FOR BIOREMEDIATION

Molecular approaches are commonly used monitoring tools for studying the potency of microbial remediation processes in diverse contaminated environments (Figure 2.2) (Lovley, 2003). The conventional culturable and non-culturable molecular biology techniques have been evaluated for identifying the microbial structure and composition at the polluted sites (Azubuike et al., 2016). The conventional culturable methods include randomly amplified polymorphic DNA (RAPD), amplified ribosomal DNA restriction analysis (ARDRA), automated ribosomal intergenic spacer analysis (RISA), and polymerase chain reaction-length heterogeneity (PCR-LH) (Bharagava et al., 2018; Kumar et al., 2020b). Depending on in situ methods, the initial bioremediation studies by culture are monitored by colony-forming units (CFUs). For example, the sensitivity of *Pseudomonas* sp. toward toluene was analyzed by CFU analysis after subjecting them to solvent shock. The CFU analysis continues to be among the preferred and robust methods for analyzing and monitoring the microbial niche (Pandey et al., 2009). Genome sequencing is particularly beneficial in understanding the relevant organisms for bioremediation. For example, the molecular studies of *Geobacter* species have indicated its importance in detoxifying the metal and organic pollutants in subsurface-contaminated sites. The genomic studies of *Geobacter* species have identified the flagella-encoding genes that were earlier considered non-motile. The advancements in molecular studies revealed the production of flagella in *Geobacter metallireducens* only in the presence of Mn(IV) and Fe(III) oxides (Mouser et al., 2009).

Naturally occurring metabolic activities are the basics for exploring genomes concerning their associated applications (Dangi et al., 2019; Kumar et al., 2021b). Thus, it is essential to get an insight into the genes and their functions to understand their role in degrading some harmful contaminants (Santero and Díaz, 2020). Most microbes (Tables 2.1 and 2. 2) possess enzymes (Table 2.2) that mediate the bioremediation pathways, utilizing contaminants as substrates and converting them to less toxic compounds. The microbes act as communities against a specific pollutant, and the activity is monitored by a set of biomarkers and biosensors (Abatenh et al., 2017; Jaiswal and Shukla, 2020). The analysis of microbial communities can be done

TABLE 2.2

Microbial Enzyme-Mediated Bioremediation Processes

Microorganism(s)	Enzyme(s)	Contaminant(s)	Application(s)	Reference(s)
Desulfitobacterium hafniense	Cytochromes and oxidoreductases	Phenols and chlorinated solvents	Metal reduction and dehalogenation	Kim et al. (2012)
Agrobacterium radiobacter	Hydrolase	Organophosphorous	Hydrolysis and higher substrate affinity	Blatchford et al. (2012), Bosso and Cristinzio (2014)
Loligo vulgaris	Diisopropylfluorophosphatase	Organophosphates	Biosensing, bioremediation, and immobilization	Jain et al. (2019)
Pseudomonas putida, P. aeruginosa	Catechol dioxygenase	Aromatic organic compounds	Clean up oil-contaminated soils	Mesarch et al. (2000), Rodríguez-Salazar et al. (2020)
Serratia marcescens	Catalase	Hydrocarbons	Degradation of crude oil-polluted environments	Kaushal et al. (2018), Zeng et al. (2010)
Bacillus subtilis, Candida krusei, Schizophyllum sp.	Manganese peroxidase	Dyes	Decolourization	Bansal and Kanwar (2013), Yao et al. (2013)
Brevundimonas diminuta	Phosphotriesterase	Organophosphorous insecticides	Hydrolysis	Santillan et al. (2016)
Bacillus stearothermophilus, B. coagulans	Lipase	Hydrocarbons (plastics)	Thermolabile, industrial and solid waste degradation	Kanmani et al. (n.d.)
Shewanella oneidensis	Cytochrome oxidases	Metals (uranium and chromium)	Redox reaction catalysis	Heidelberg et al. (2001), Kouzuma et al. (2015)
Bacillus sp.	Protease	Leather	Proteolysis	Razzaq et al. (2019)
Aspergillus sp.	Lipase, tannase, and protease	Chromium, lead, cadmium, etc	Fat hydrolysis, reduction, and proteolysis.	Deshmukh et al. (2016)
Doratomyces nanus, Phoma eupyrena, Aspergillus niger, and Myceliophthora	Lignin peroxidase	Azo, heterocyclic, and remazol dyes	Decolouration and lignolytic activity	Deshmukh et al. (2016), Gulzar et al. (2017)
Trichoderma and Humicola	Cellulase	Cellulose	Decolouration	Karigar and Rao (2011)
Dentipellis sp.	Dehydrogenase	Aromatic hydrocarbons	Oxidation	Akhtar and Mannan (2020b)
Phanerochaete chrysosporium	Cytochrome monooxygenase	Hydrocarbons, xenobiotics, and fatty acids	Oxidation-reduction	Alneyadi et al. (2018)

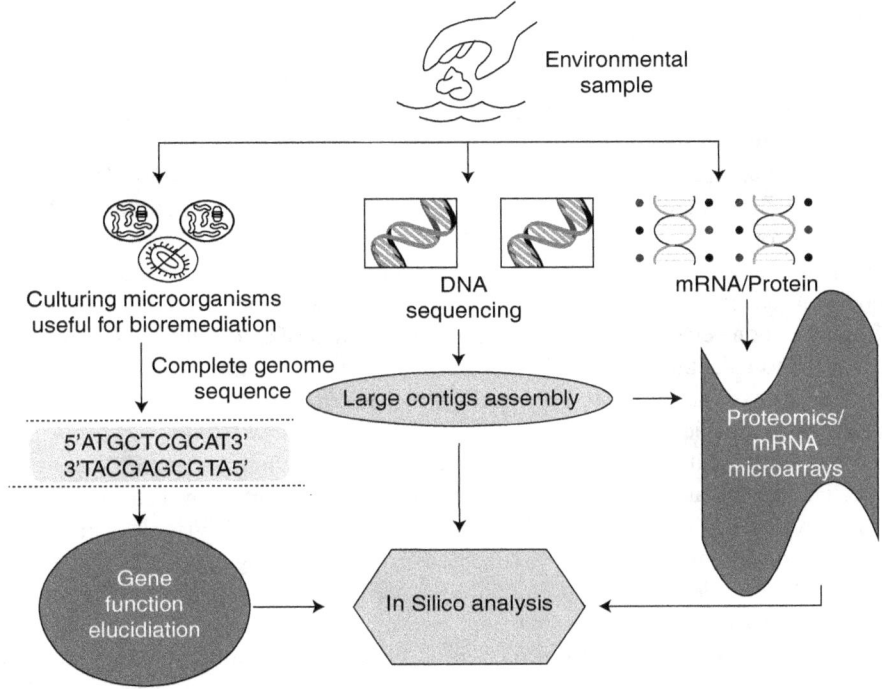

FIGURE 2.2 Genomic and metagenomic approaches for understanding the role of microbes in the contaminated sites.

by a set of molecular tools such as gel electrophoresis (denaturing gradient and temperature gradient) and RFLP (terminal) (Chikere, 2013). These techniques exploit the base-pairing differences between G-C and A-T, thus electrophorese (acrylamide) the amplified DNA fragments. The fragments with higher GC content stay stable for extended periods because of the extra hydrogen bond until they encounter the denaturing agent of higher concentration. The denatured DNA fragments move slowly in acrylamide gel compared to double-stranded fragments. Thus, fragments of different sequences can be easily separated. The same methodology has been employed in the case of terminal RFLP, in which the fluorescently labelled 5′-end is exposed to amplification followed by the restriction digestion and then finally separation by either acrylamide or capillary electrophoresis (Lovley, 2003; Płaza et al., 2001; Rhee et al., 2004).

2.6 METAGENOMICS AND *META-OMICS* APPROACHES: A BOON TO BIOREMEDIATION

The shortcomings of culture-dependent approaches have led to the increased demand and importance of culture-independent approaches (Phulpoto et al., 2020). This approach has efficiently been employed to understand genome-driven processes, such as bioremediation, after advancing methodologies that involve direct microbial genomic DNA isolation from the environment (Costa et al., 2020). The isolated DNA can subsequently be utilized in DNA-RNA hybridization, DNA-DNA hybridization, and selective biomarker gene amplification. The metagenomics approach has been endorsed to isolate, identify, and evaluate the specific microbial genes involved in bioremediation processes such as accumulation, resistance, filtration, cleavage, detoxification, and metabolism to reduce environmental contaminants (Jaiswal and Shukla, 2020; Phulpoto et al., 2020; Kumar and Chandra, 2020a). Duong et al. successfully isolated the XMcr-6 strain of *Bacillus cereus* from the contaminated chromium site. Based on the 16S rRNA and biochemical attributes, the *Bacillus* strain showed a reduction potential under aerobic conditions of about 100 mg/L Cr(VI) in 48 hours (Duong et al., 2013). The soil and groundwater of Oak Ridge (FRC area), the USA, due to its acidic nature, high calcium content, high nitrate concentration, and uranium contamination, posed an excellent bioremediation challenge. However, several studies identified *Acidovorax, Geobacter, Geothrix,* and *Desulfosporosinus* spp. to cause detoxification by denitrification and dehalogenation processes. Wang et al. reported that the association of *MtrC* and *OmcA* ensures efficient heavy metal (Cr and Fe) reduction (Hemme et al., 2010; Mani and Kumar, 2014; Vrionis et al., 2005).

The metagenomics approach has been utilized in the fungal analysis of the sediments in Zuari inlets (Mandovi), Goa, India. Various fungi such as *Saccharomyces, Dothideomycetes, Agaricomycetes,* and *Sordariomycetes* have been studied. The study hypothesized the importance of these fungi in decontaminating the environment and suggested their further utilization in bioremediation processes (Haldar and Nazareth, 2019). Li et al. conducted similar research on the mangrove forest sediments, highlighting the importance of microorganisms in methanogenesis and nitrogen fixation (2019). Naphthalene, a ubiquitous contaminant, is a major anaerobic environment pollutant found challenging to degrade and detoxify. However, metabolomics and metagenomics play a crucial role in isolating, understanding, and evaluating naphthalene-degrading microorganisms. One of the studies reported *Desulfuromonadales* as a potential naphthalene-degrading microorganism (Toth et al., 2018). The presence of vinyl chloride, a carcinogen, has also been challenging earlier. The metagenomics approach has helped identify several *Proteobacteria, Bacteroides,* and *Actinobacteria* as potential sources for vinyl chloride degradation (Liu et al., 2018). The ammonia-oxidizing archaea (AOA) known for nitrogen bioremediation have been studied using a metagenomics approach. The AOA genomes isolated from the estuary of the Pearl River were compared, and the comparative analysis of SCMI *Nitrosopumilus maritimus* strains and SPOT01 *Nitrosomatrinus catalina* strains showed genomic similarity with each other (Zou et al., 2019). A metagenomic analysis indicated contaminant-degrading genes that can be beneficial for eutrophic environments. The combined studies of metabolomics, metagenomics and meta-transcriptomics have been employed to evaluate and examine the microorganisms involved in the biodegradation of bisphenol A. The *Pseudomonas* and *Sphingomonas* species co-cultured demonstrated an increased biodegradation potential of bisphenol A compared to the separate culture of *Sphingomonas* species (Yu et al., 2019). The metagenomics approach also showed the microbial community's presence in the degradation of toxic hydrocarbons (tarballs), which includes *Halomonas* and *Marinobacter* (Fernandes et al., 2019). Metagenomics combined with meta-transcriptomics (Figure 2.3) studies have been extended in studying the bioaugmentation effect while removing dichloroethylene and trichloroethylene from contaminated sites. Math et al. reported the presence of 3,5,6 trichloro-2-pyridinol-encoding gene, screened by using a metagenomic library, utilized for chlorpyrifos degradation, which is harmful to both animals and humans (Watahiki et al., 2019).

Metagenomics and other novel molecular biology techniques have diversified microbiological studies by focusing on the applications associated with microbial communities and their gene products. Metagenomics (Figure 2.2), supported by the next-generation sequencing, has enabled the evaluation and investigation of the microbiomes of different environments such as the human gut (Almeida et al., 2019), sea sediments (Marshall et al., 2018), marine water (Langridge, 2009), hydrothermal springs (Saxena et al., 2017), and soils (Daniel, 2005), each studies providing more significant insights into the function, structure, evolution, and adaptations of the microbial communities. Bioremediation driven by metagenomics has increased the degree of degradation of various xenobiotics either by *in situ* or *ex situ* treatment (Figure 2.1) (Haiser and Turnbaugh, 2013). Metagenomics can be essential in tracking air, water, and soil pollution by using specific genes as biomarkers (Bibby et al., 2019; Hong et al., 2020). Biosurfactants, amphiphilic compounds, are known for hydrocarbon emulsification in water and promote their degradation. Thus, biosurfactant-synthesizing microbial strains were employed to degrade hydrocarbons of oil spills in marine water (Thies et al., 2016). Besides the above-listed bioremediation benefits of metagenomics, several bacterial species have also been evaluated from nitrifying sludge, which includes *Betaproteobacteria, Bacteroidetes*, and *Alphaproteobacteria* (Ciesielski et al., 2018). There has been an enormous increase in the upregulation of contaminant-degrading genes revealed by meta-transcriptomics. The diversity of genes encoding hydroxylating dioxygenases (HDs) in several bacterial communities found in mangrove and oilfield sediments has been reported using metagenomics and other associated molecular approaches. The identified bacteria with HDs diversity include *Polymorphum gilvum, Burkholderia, Mycobacterium, Rhodococcus*, and *Pseudomonas* (de Sousa et al., 2017). The molecular techniques led by metagenomics have also been applied in investigating cholesterol degradation. The report suggested the possible role of C25 dehydrogenase as a cholesterol degradation indicator (Wei et al., 2018). Metagenomics-driven microbial studies also demonstrated the five phyla involved in the degradation of antimony and arsenic, such as *Nitrospirae, Firmicutes, Actinobacteria, Tenericutes* and *Gemmatimonadetes* (Luo et al., 2014).

2.6.1 META-TRANSCRIPTOMICS

Transcriptome, a complete mRNA set plus a noncoding transcript of RNA, acts as a connecting link (Figure 2.3) between genomics and proteomics, thereby regulating gene expression (Milward et al., 2016). Since metagenomics provides little clarification about the gene activity, the need for a more advanced approach is obvious. Meta-transcriptomics development is a gateway to predicting and exploring the functional aspects of microorganisms (Figure 2.3). This is of enormous interest

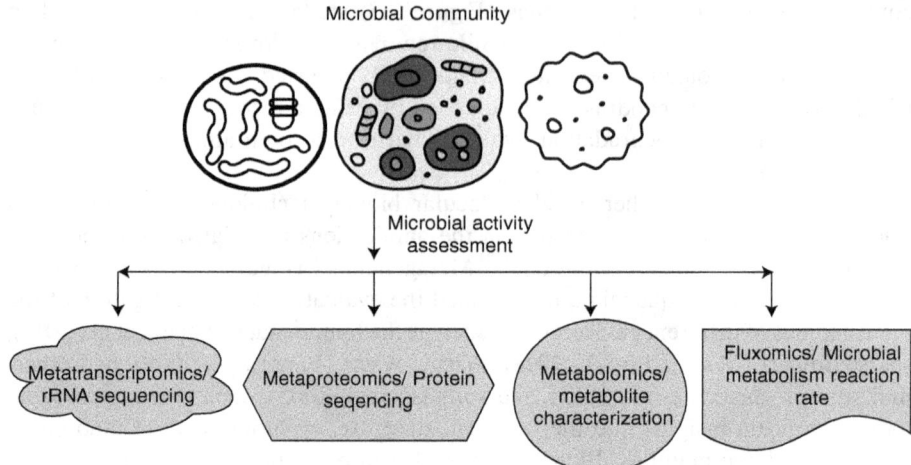

FIGURE 2.3 Microbial community analysis by different metagenomics approaches (Zuñiga et al., 2017).

concerning the bioremediation processes. This approach has been applied to a herbicide degrading the *tdfA* gene, coding for dioxygenase and ultimately quantified efficiently (Nicolaisen et al., 2008). The genome-centric meta-transcriptomics investigation reported that the extracellular electron transfer activity is performed by nine *Pelobacter/Geobacter* microbes depending upon their surface potential and preferable substrates. These redox reactions, coupled with the microbial communities' gene regulation mechanisms, play a crucial role in environmental remediation by microbes (Ishii et al., 2018).

2.6.2 META-PROTEOMICS

Proteomics, the study of proteins biosynthesized in a cell(s)/organism(s), has enhanced the microbial community-mediated environmental remediation by ensuring rapid and relevant protein identification (Basu and Stolz, 2014). The potential to translate the transcribed genomic data into proteins elucidates the proteome understanding, thereby mediating the microbial response against the different contaminants (López-Barea and Gómez-Ariza, 2006). Because of its more significant potential and efficiency than conventional genomics, meta-proteomics helps to analyze microbial communities functionally. The catabolic enzymes that play a crucial role in microbe-mediated degradation pathways are driven by mass spectroscopy (MS) and electrophoretic techniques (Mueller and Pan, 2013). For example, Lacerda et al. (2007) used two-dimensional (2D) gel electrophoresis and MS to evaluate the identified microbial proteins exposed to water contaminated with cadmium. The proteomic studies have led to novel gene identifications during the toluene and ethylbenzene anaerobic degradation (Kühner et al., 2005). Research by Konopka and Wilkins reported the identification of *Geobacter* spp. They used meta-proteomics

(Figure 2.3) in carbon stimulation (Konopka and Wilkins, 2012). Peroxidase enzyme was identified from PYR-1 *Mycobacterium* sp. strain synthesized subject to pyrene exposure. This was followed by reasonable induction and increased expression after further phenanthrene, dibenzothiophene, and pyrene exposure (Wang et al., 2000). In another research, NG80-2, a thermophilic strain, was isolated from an oil reservoir, and the monooxygenase *ladA* gene function was studied. This reported the alkane-degradation potential of the NG80-2 strain, which is among the significant environmental contaminants (Feng et al., 2007).

2.6.3 METABOLOMICS

Besides conventional genomics and advanced meta-transcriptomics/proteomics (Figure 2.3), metabolomics, a metabolite profile study, is among the recent omics approaches that are helping the expansion of microbial studies (Figure 2.3) (Villas-Bôas and Bruheim, 2007). This has also helped to develop bioremediation strategies, which have already been made easy by metagenomics approaches (Figure 2.2) (Bonifay et al., 2016). A typical example is *Sinorhizobium* sp. metabolome analysis during the biodegradation of phenanthrene (Keum et al., 2008). The metabolome evaluation of *Bacillus* sp. was performed during the mesotrione (herbicide) degradation (Batisson et al., 2009). There have been several diversified approaches that have been expanded for metabolomics and thereby applied in microbial bioremediation. These strategies include mineralization assessment utilizing metabolic footprinting, metabolic engineering to enhance biodegradation, and biomolecular connectivity analysis (Malla et al., 2018). One of the researchers used the FT-IR approach for biochemical monitoring in phenol-resistant bacteria (Tables 2.1 and 2.2) (Wharfe et al., 2010). Metabolomics aims to utilize the existing analytical tools for microbial metabolite evaluation (Yang et al., 2019). This includes the NMR analysis of organic pollutants in soil bioremediation (Boersma et al., 2001). The degradation of demeton-S-methyl (pesticide) by *Corynebacterium glutamicum* was studied by Girbal et al. and analyzed by NMR (2000). Moreover, the bioremediation kinetics can be directly studied by ^{1}H-NMR and 2D-NMR spectra analysis (Emwas et al., 2019).

2.6.4 METAFLUXOMICS

Omics analysis, in general, has diversified the knowledge about the physiology and genetics of the microbial communities (Franzosa et al., 2015). The insight into *Pseudomonas* sp.'s molecular background, involved in the cyanide and trinitrotoluene degradation, has been studied on a large scale (Pascual et al., 2015). Metabolic flux evaluation is used in calculating fluxes from stable isotopes acquired during the isotope labelling experiments (Antoniewicz et al., 2007). Isotopic tracers and metabolomics have been utilized to find the flux processes that drive the essential microbial functions, such as methanogenesis and syntrophic oxidation of acetate (Mosbæk et al., 2016). Although metabolite flux analysis is utilized in co-cultures for flux measurements, it is problematic in communities because metabolite pools are not assigned easily to the individual cells. The possible reaction number in the microbiome is much higher than in the organism individually (Gebreselassie and Antoniewicz, 2015). Since fluxes are the final representatives of the cellular outcome regulation at all stages, metafluxomics development is essential in advancing and demonstrating microbiome engineering (Nielsen, 2003).

2.7   COMPUTATIONAL TOOLS FOR METAGENOMICS-MEDIATED BIOREMEDIATION

The genomics, metagenomics, and other omics approaches discussed above boost our analytical understanding of various microorganisms' degradation abilities and enzymes. However, to predict the organisms functioning in hybrid environments, it is of utmost importance to have advanced metabolism models to understand multiple reactions occurring in the microbial cell(s). The computational tools (Table 2.3) related to cell metabolism make it possible. Bioinformatics holds prime importance in metagenomic bioremediation because it allows easy and productive analysis of metagenomics data. One of the recent studies has manifested the importance of bioinformatics by identifying and characterizing the soil microbes by metagenomic approaches. This study used publicly available metagenomes and combined some bioinformatics tools to prospect the functional and phylogenetic characteristics of the soil microbes (Huson et al., 2007). Mukherjee et al. used bioinformatics tools and a metagenomics approach to study the bacterial characteristics in response to hydrocarbon pollution in different contaminated sites. They analyzed more than 50 16S rRNA sets from 12 polluted sites to decode the metagenomic features of resident microbial communities (2017). This section provides more insights into the standard computational datasets utilized in metagenomics data analysis with a possible role in understanding the molecular mechanism of bioremediation processes.

2.7.1   MEGAN: MEtaGenome ANalyzer

MEGAN, an all-inclusive toolbox, is a bioinformatics tool used for significant metagenomics data analysis. The DNA contigs (or reads) are correlated with the available sequence databases using BLAST (Huson, 2015). This computational tool's utilization thus helps explore customized taxonomy and enables functional analysis by SEED or KEGG. MEGAN, an easy and stand-alone program, needs only a BLAST file as input data. Thus, MEGAN performs functional and taxonomical analysis of the metagenomic data and analyzes the meta-transcriptomic data. The functional analysis is done by mapping contigs to the KEGG, COG, and SEED classifications, while in taxonomic analysis, the reads are placed in the NCBI. This further ensures a higher degree of comparison among the multiple samples by clustering methods and principal coordinate analysis (PCoA) (Huson, 2015; Huson et al., 2007, 2018). The program is written in Java and works in most operating systems (Windows, Linux, Unix, and Mac OS X). It was developed by Daniel Houson and his co-workers (Huson et al., 2007) and is downloadable at www-ab.informatik.uni-tuebingen.de/software/megan.

TABLE 2.3
Other Computational Tools Useful for Metagenomics and Other Omics Approaches

Computational Tool	Feature(s)/Application(s)	Links
MetaVelvet	Assembly of metagenomic contigs	http://metavelvet.dna.bio.keio.ac.jp/
METAGENassist	Comparative metagenomic analysis for phenotype mapping	www.metagenassist.ca
metagenomeSeq	Offers bioconductor for the relative analysis of 16S rRNA in meta-profiling	http://bioconductor.org/packages/release/bioc/html/metagenomeSeq.html
MetaClusterTA	Binning of sequence contigs and reads for taxonomic analysis and annotation	http://i.cs.hku.hk/~alse/MetaCluster
MetaPhlAn2	Advanced profiling of metagenomes and strain identification	http://segatalab.cibio.unitn.it/tools/metaphlan2/
MEGAN-CE	Open access tool for microbiome sequence analysis	https://github.com/danielhusonmegan-ce
MEGAHIT	Single server assembly of large metagenomic datasets	https://hku-bal.github.io/megabox
MetaBAT	Binning bulk metagenomes	https://bitbucket.org/berkeleylab/metabat
MetaBAT 2	Elimination of manual setup parameter for the novel binning algorithm	https://bitbucket.org/berkeleylab/metabat
MetaSort	Sorting mini-metagenomes based on a single-cell sequence approach	https://sourceforge.net/projects/metasort/
MetaPath	Evaluation of metabolic pathways in microbial microcosm samples	http://metapath.cbcb.umd.edu/
MetaGeneMark	Metagenome exon prediction using a heuristic model	http://exon.gatech.edu/index.html
METAREP	Metagenomic data analysis and comparison	www.jcvi.org/metarep
MetaQUAST	Genome assembly and alignment, a modified QUAST tool	http://bioinf.spbau.ru/metaquast

2.7.2 MG-RAST: Metagenomic Rapid Annotations using Subsystems Technology

MG-RAST is an automated open-source tool for the easy phylogenetic and reliable, functional analysis of the metagenomes. This is among the broad metagenomic data repositories and is used in comparing sequences with nucleotide and amino acid databases. Thus, MG-RAST offers comparative analysis, quality control, archiving services, and annotation of amplicon sequences and metagenomics with other computational tools (Kumavath and Deverapalli, 2013). Besides analyzing metagenomic data, it also supports the sequence processing of meta-transcriptomes and amplicons. The pipeline is divided into five levels: data hygiene, feature extraction, feature annotation, profile generation, and data loading (Keegan et al., 2016). The pipeline is applied in Perl by utilizing multiple open-source tools such as SEED, NCBI BLAST, Sun Grid Engine, and SQLite. The users can upload raw sequences in FASTA format, leading to sequence normalization, processing, and finally, the automatic generation of summaries (Kisand et al., 2012). Jung et al. used MG-RAST for the metagenomic functional analysis of bacterial communities in hydrocarbon-contaminated sites for understanding bioremediation. The calculated alpha diversity by MG-RAST showed that, in the control treatment, the 10^{-2}-inoculated samples contained higher OTUs than the 10^{-5} inoculated samples(Jung et al., 2016). This tool was created by Argonne National Laboratory of Chicago University and can be accessed at http:// metagenomics.anl.gov/.

2.7.3   SMASHCOMMUNITY: SIMPLE METAGENOMICS ANALYSIS SHELL FOR MICROBIAL COMMUNITIES

There is an exponential increase in the metagenomic data for computational analysis (Singh et al., 2009). MEGAN and MG-RAST carry out the phylogenetic and functional analyses of the metagenomic datasets. However, they only count the contigs or reads that map the known genes and do not evaluate quantitative aspects. They are not developed for assembling the reads of metagenomic studies. Therefore, they cannot identify multi-domain genes and operons in complex metagenomes (Arumugam et al., 2010). SmashCommunity, a stand-alone pipeline, allows annotation and analysis of metagenomic data to quantitatively estimate metagenomic phylogeny and the associated function composition, correlate several metagenomes, and construct visual models of these analyses (Arumugam et al., 2010; Letunic and Bork, 2007). It shares design principles with the SmashCell documented in Perl with well-defined inter-modular interfaces, a database installed locally, and a modular architecture (Harrington et al., 2010). The exercises such as gene prediction or sequence assembly in the metagenomic analysis are executed as modules (Muller et al., 2009). The exercises on platforms other than the SmashCommunity can be introduced into the SmashCommunity using GFF and ACE files (Arumugam et al., 2010). The SmashCommunity documentation and source code can be accessed at www.bork.embl.de/software/smash.

2.7.4   IMG/M: INTEGRATED MICROBIAL GENOMES AND MICROBIOMES

The IMG/M bioinformatics tool contains bacterial, archaeal, viral, and eukaryotic genomes; metagenomes from microbial microcosms; and single-cell genomes. The aim of IMG/M is to assist in the analysis, distribution, and annotation of microbiome datasets and microbial genomes. DOE's Joint Genome Institute (JGI) generates data in sequences collected from public sequence repositories or submitted by scientists individually (Chen et al., 2017). NCBI serves as the significant genomic data source for the IMG. The data files have to pass the submission system of IMG and its annotation pipeline before being integrated into the databank of IMG (Markowitz et al., 2010). IMG has advanced several folds since its inception, providing a related evaluation of the genomic and metagenomic data. The Expert Review (ER) version

I, e IMG ER platform supports the scientific community for curation and annotation of their genome and metagenome of interest (Chen et al., 2017; Markowitz et al., 2010). Another version known as IMG-HMP (Human Microbiome Project) is solely dedicated to the genome analysis related to human microbial microcosm (Chen et al., 2017). On the other hand, IMG-ABC and IMG-VR systems are meant to analyze bacterial and viral genomes (Paez-Espino et al., 2017, 2019). Detailed information about the IMG/M system is available at http://img.jgi.doe.gov/.

2.8 DRAWBACKS AND WAY FORWARD IN METAGENOMICS-MEDIATED BIOREMEDIATION

The metagenomics approach has given a deeper insight into prokaryotic diversity and its associated advantages. However, some complications curtail the efficacy of the metagenomics pipeline. The experimental metagenomics network starts with the sample collection, and consequently, its storage is quite challenging in several cases (Laudadio et al., 2019). The studies of the deep-sea sediments and samples of extreme environments are problematic both in the sample collection process and storage setup (Bang et al., 2018). An important aspect of complete sequencing in metagenomics is the abundant input DNA; however, this is difficult when dealing with extinct representative (paleogenomics) samples from paleontological or archaeological sites, museum artefacts, and ice cores (Hofreiter, 2008). Although metagenomics ensures the identification and characterization of the microbial microcosms through genomic studies, identifying a large number of genes in a primordial nucleic acid is a laborious task. The culture-based genomic studies help examine and evaluate a single microorganism. Still, while studying the environmental samples, the diverse genomic content of unknown species needs precise identification, evaluation, and sorting. However, in metagenomics, the species present in abundance are represented more than those microorganisms present in less number in a particular environment (Wooley and Ye, 2009). However, this problem can be resolved by multiplying reads per nucleotide (coverage depth) and deep DNA sequencing. This will ensure the representation of the lesser found microorganisms in the metagenome datasets (Xu and Zhao, 2018). Besides, sequencing long stable genomes are difficult with the existing sequencing tools; thus, eDNA is cleaved into smaller fragments before the metagenomics sequence analysis is carried out. Therefore, the short reads generated require reassembly before using computational tools, but reassembly into contigs is challenging with the existing tools for metagenome analysis (Amarasinghe et al., 2020). The limited reference sequence availability for comparative genomic analysis in metagenomics databases for certain environment-specific microorganisms such as airborne and human gut is also a persistent challenge in advancing meta-omics studies (Behzad et al., 2015; Marcelino et al., 2020).

Besides some existing experimental drawbacks, developments in metagenomics continue to transform our outlook on microbial diversity. With the availability of high-performance sequence tools, there is a possibility for understanding the nature of metamorphosis induced by microorganisms. The benefits of eDNA analysis already transpiring microbial niche studies, biomonitoring, sustainable conservation (bioremediation), and surveillance tools can be made more accurate and efficient with further progress. Considering the existing meta-omics research about bioremediation, there is a possibility of mining the data to lay out the absolute microbe-mediated bioremediation mechanism pathways. There is an acute need for standardizing the protocols for different stages of metagenomics studies, such as sample collection, analysis, evaluation, repositioning, sequencing, and transmission. The accurate and novel biomarker identification and characterization to apply a proper bioremediation operation are likely to get attention in future metagenomics research.

2.9 CONCLUSIONS

Environmental contaminants such as hydrocarbons, heavy metals, xenobiotics, insecticides, and herbicides are burgeoning in abundant quantities. There is a dire need to look for possible biodegradation approaches to detoxify these recalcitrant pollutants since chemical treatments pose a challenge because of their eco-unfriendly nature. Based on several studies done in the past, microorganisms have been reported to have various contaminant degradation properties through a series of enzymatic processes. Metagenomics and several meta-omics approaches act as a gateway to the exploration of genes involved in microbial remediation. Different enzyme classes such as hydrolases, oxidoreductases, and proteases derived from diverse microbial communities have been listed with degradation properties against certain specific contaminants. This review provides a deep insight into the role of molecular biology, metagenomics, metabolomics, meta-transcriptomics, and metafluxomics techniques in fighting against harmful environmental pollutants. Advanced computational tools have also been developed to speed up microbial genetic analysis to understand the underlying mechanism of bioremediation. This review further highlights the drawbacks associated with metagenomics, but at the same time, it gives a glimpse of possible future trends in the field of meta-omics and bioremediation.

REFERENCES

Abatenh, E., Gizaw, B., Tsegaye, Z., & Wassie, M. (2017). The role of microorganisms in bioremediation- A review. *Open Journal of Environmental Biology*, 2(1), 038–046. Doi: 10.17352/ojeb.000007.

Abdulsalam, S., Adefila, S. S., Bugaje, I. M., & Ibrahim, S. (2012). Bioremediation of soil contaminated with used motor oil in a closed system. *Journal of Bioremediation and Biodegradation*, 03(12), 3–9. Doi: 10.4172/2155-6199.1000172.

Adebajo, S., Balogun, S., & Akintokun, A. (2017). Decolourization of vat dyes by bacterial isolates recovered from local textile mills in Southwest, Nigeria. *Microbiology Research Journal International*, 18(1), 1–8. Doi: 10.9734/mrji/2017/29656.

Akhtar, N., & Mannan, M. A. (2020a). Mycoremediation: An unexplored gold mine. In *New and Future Developments in Microbial Biotechnology and Bioengineering* (pp. 11–24). Elsevier. Doi: 10.1016/b978-0-12-821007-9.00002-4.

Akhtar, N., & Mannan, M. A. (2020b). Mycoremediation: Expunging environmental pollutants. In *Biotechnology Reports* (Vol. 26). Elsevier B.V. Doi: 10.1016/j.btre.2020.e00452.

Ali, N., Dashti, N., Khanafer, M., Al-Awadhi, H., & Radwan, S. (2020). Bioremediation of soils saturated with spilled crude oil. *Scientific Reports*, 10(1), 1–9. Doi: 10.1038/s41598-019-57224-x.

Almeida, A., Mitchell, A. L., Boland, M., Forster, S. C., Gloor, G. B., Tarkowska, A., Lawley, T. D., & Finn, R. D. (2019). A new genomic blueprint of the human gut microbiota. *Nature*, *568*(7753), 499–504. Doi: 10.1038/s41586-019-0965-1.

Alneyadi, A. H., Rauf, M. A., & Ashraf, S. S. (2018). Oxidoreductases for the remediation of organic pollutants in water–a critical review. *Critical Reviews in Biotechnology*, *38*(7), 971–988. Doi: 10.1080/07388551.2017.1423275.

Amarasinghe, S. L., Su, S., Dong, X., Zappia, L., Ritchie, M. E., & Gouil, Q. (2020). Opportunities and challenges in long-read sequencing data analysis. *Genome Biology*, *21*(1), 1–16. Doi: 10.1186/s13059-020-1935-5.

Antoniewicz, M. R., Kelleher, J. K., & Stephanopoulos, G. (2007). Elementary metabolite units (EMU): A novel framework for modeling isotopic distributions. *Metabolic Engineering*, *9*(1), 68–86. Doi: 10.1016/j.ymben.2006.09.001.

Aranda, E., Ullrich, R., & Hofrichter, M. (2010). Conversion of polycyclic aromatic hydrocarbons, methyl naphthalenes and dibenzofuran by two fungal peroxygenases. *Biodegradation*, *21*(2), 267–281. Doi: 10.1007/s10532-009-9299-2.

Arora, P. K. (2010). Application of monooxygenases in dehalogenation, desulphurization, denitrification and hydroxylation of aromatic compounds. *Journal of Bioremediation & Biodegradation*, *01*(03), 1–8. Doi: 10.4172/2155-6199.1000112.

Arumugam, M., Harrington, E. D., Foerstner, K. U., Raes, J., & Bork, P. (2010). Smashcommunity: A metagenomic annotation and analysis tool: Fig. 1. *Bioinformatics*, *26*(23), 2977–2978. Doi: 10.1093/bioinformatics/btq536.

Atagana, H. I., Haynes, R. J., & Wallis, F. M. (2003). Optimization of soil physical and chemical conditions for the bioremediation of creosote-contaminated soil. *Biodegradation*, *14*(4), 297–307. Doi: 10.1023/A:1024730722751.

Awasthi, M. K., Ravindran, B., Sarsaiya, S., Chen, H., Wainaina, S., Singh, E., Liu, T., Kumar, S., Pandey, A., Singh, L., & Zhang, Z. (2020). Metagenomics for taxonomy profiling: Tools and approaches. *Bioengineered*, *11*(1), 356–374. Doi: 10.1080/21655979.2020.1736238.

Azubuike, C. C., Chikere, C. B., & Okpokwasili, G. C. (2016). Bioremediation techniques–classification based on site of application: Principles, advantages, limitations and prospects. *World Journal of Microbiology and Biotechnology*, *32*(11), 1–18. Doi: 10.1007/s11274-016-2137-x.

Bang, C., Dagan, T., Deines, P., Dubilier, N., Duschl, W. J., Fraune, S., Hentschel, U., Hirt, H., Hülter, N., Lachnit, T., Picazo, D., Pita, L., Pogoreutz, C., Rädecker, N., Saad, M. M., Schmitz, R. A., Schulenburg, H., Voolstra, C. R., Weiland-Bräuer, N., … Bosch, T. C. G. (2018). Metaorganisms in extreme environments: Do microbes play a role in organismal adaptation? *Zoology*, *127*, 1–19. Doi: 10.1016/j.zool.2018.02.004.

Bansal, N., & Kanwar, S. S. (2013). Peroxidase(s) in environment protection. *The Scientific World Journal*, *2013*. Doi: 10.1155/2013/714639.

Barber, E. A., Liu, Z., & Smith, S. R. (2020). Organic contaminant biodegradation by oxidoreductase enzymes in wastewater treatment. *Microorganisms*, *8*(1). Doi: 10.3390/microorganisms8010122.

Bartholameuz, E. M., Hettiaratchi, J. P. A., Steele, M. & Kumar, S. (2020). Reaction kinetic analysis of manganese peroxidase augmented aerobic waste degradation. *Journal of Hazardous, Toxic, and Radioactive Waste*, *24*(4), 04020043.

Basu, P., & Stolz, J. F. (2014). Application of proteomics in bioremediation. In P. Chovanec (Ed.), *Microbial Metal and Metalloid Metabolism* (pp. 247–P2). ASM Press. Doi: 10.1128/9781555817190.ch13.

Batisson, I., Crouzet, O., Besse-Hoggan, P., Sancelme, M., Mangot, J. F., Mallet, C., & Bohatier, J. (2009). Isolation and characterization of mesotrione-degrading *Bacillus* sp. from soil. *Environmental Pollution*, *157*(4), 1195–1201. Doi: 10.1016/j.envpol.2008.12.009.

Behzad, H., Gojobori, T., & Mineta, K. (2015). Challenges and opportunities of airborne metagenomics. *Genome Biology and Evolution*, *7*(5), 1216–1226. Doi: 10.1093/gbe/evv064.

Benkhaya, S., El Harfi, S., & El Harfi, A. (2017). Classifications, Properties and Applications of Textile Dyes: A Review. *Applied Journal of Environmental Engineering Science*, 3(3), 311–320.

Bharagava, R. N., Purchase, D., Saxena, G., & Mulla, S. I. (2018). Applications of metagenomics in microbial bioremediation of pollutants: From genomics to environmental cleanup. In *Microbial Diversity in the Genomic Era* (pp. 459–477). Elsevier. Doi: 10.1016/B978-0-12-814849-5.00026-5.

Bhat, S. A., Singh, S., Singh, J., Kumar, S., & Vig, A. P. (2018). Bioremediation and detoxification of industrial wastes by earthworms: Vermicompost as powerful crop nutrient in sustainable agriculture. *Bioresource Technology*, 252, 172–179. Doi: 10.1016/j.biortech.2018.01.003.

Bibby, K., Crank, K., Greaves, J., Li, X., Wu, Z., Hamza, I. A., & Stachler, E. (2019). Metagenomics and the development of viral water quality tools. *NPJ Clean Water*, 2(1), 1–13. Doi: 10.1038/s41545-019-0032-3.

Blatchford, P. A., Scott, C., French, N., & Rehm, B. H. A. (2012). Immobilization of organophosphohydrolase *OpdA* from agrobacterium radiobacter by overproduction at the surface of polyester inclusions inside engineered *Escherichia coli*. *Biotechnology and Bioengineering*, 109(5), 1101–1108. Doi: 10.1002/bit.24402.

Boersma, M. G., Solyanikova, I. P., Van Berkel, W., Vervoort, J., Golovleva, L. A., & Rietjens, I. (2001). F NMR metabolomics for the elucidation of microbial degradation pathways of fluorophenols. *Journal of Industrial Microbiology & Biotechnology*, 26. https://academic.oup.com/jimb/article/26/1-2/22/6015282.

Bonifay, V., Aydin, E., Aktas, D. F., Sunner, J., & Suflita, J. M. (2016). *Metabolic Profiling and Metabolomic Procedures for Investigating the Biodegradation of Hydrocarbons* (pp. 111–161). Doi: 10.1007/8623_2016_225.

Bosso, L., & Cristinzio, G. (2014). A comprehensive overview of bacteria and fungi used for pentachlorophenol biodegradation. *Reviews in Environmental Science and Biotechnology*, 13(4), 387–427. Doi: 10.1007/s11157-014-9342-6.

Burghal, A. A., Abu-Mejdad, N. M. J. A., & Al-Tamimi, W. H. (2016). Engineering and technology (A high impact factor. *International Journal of Innovative Research in Science*, 5(2). Doi: 10.15680/IJIRSET.2016.0502068.

Camacho-Pérez, B., Ríos-Leal, E., Rinderknecht-Seijas, N., & Poggi-Varaldo, H. M. (2012). Enzymes involved in the biodegradation of hexachlorocyclohexane: A mini review. *Journal of Environmental Management*, 95(SUPPL.) Doi: 10.1016/j.jenvman.2011.06.047.

Carr, C. M., Clarke, D. J., & Dobson, A. D. W. (2020). Microbial polyethylene terephthalate hydrolases: Current and future perspectives. *Frontiers in Microbiology*, 11, 2825. Doi: 10.3389/fmicb.2020.571265.

Chandra, R., Kumar, V., & Yadav, S. (2017). Extremophilic ligninolytic enzymes. In: Sani R., Krishnaraj R. (eds) *Extremophilic Enzymatic Processing of Lignocellulosic Feedstocks to Bioenergy*. Springer, Cham. Doi: 10.1007/978-3-319-54684-1_8.

Chen, I. M. A., Markowitz, V. M., Chu, K., Palaniappan, K., Szeto, E., Pillay, M., Ratner, A., Huang, J., Andersen, E., Huntemann, M., Varghese, N., Hadjithomas, M., Tennessen, K., Nielsen, T., Ivanova, N. N., & Kyrpides, N. C. (2017). IMG/M: Integrated genome and metagenome comparative data analysis system. *Nucleic Acids Research*, 45(D1), D507–D516. Doi: 10.1093/nar/gkw929.

Chikere, C. (2013). Application of molecular microbiology techniques in bioremediation of hydrocarbons and other pollutants. *British Biotechnology Journal*, 3(1), 90–115. Doi: 10.9734/bbj/2013/2389.

Cho, Y. G., Rhee, S. K., & Lee, S. T. (2000). Effect of soil moisture on bioremediation of chlorophenol-contaminated soil. *Biotechnology Letters*, 22(11), 915–919. Doi: 10.1023/A:1005612232079.

Ciesielski, S., Czerwionka, K., Sobotka, D., Dulski, T., & Makinia, J. (2018). The metagenomic approach to characterization of the microbial community shift during the long-term cultivation of anammox-enriched granular sludge. *Journal of Applied Genetics*, *59*(1), 109–117. Doi: 10.1007/s13353-017-0418-1.

Coates, J. D., Bruce, R. A., & Haddock, J. D. (1998). Anoxic bioremediation of hydrocarbons. *Nature*, *396*(6713), 730. Doi: 10.1038/25470.

Coelho, L. M., Rezende, H. C., Coelho, L. M., de Sousa, P. A. R., Melo, D. F. O., & Coelho, N. M. M. (2015). Bioremediation of polluted waters using microorganisms. *Advances in Bioremediation of Wastewater and Polluted Soil*. Doi: 10.5772/60770.

Costa, O. Y. A., Costa, O. Y. A., De Hollander, M., Pijl, A., Liu, B., & Kuramae, E. E. (2020). Cultivation-independent and cultivation-dependent metagenomes reveal genetic and enzymatic potential of microbial community involved in the degradation of a complex microbial polymer. *Microbiome*, *8*(1), 76. Doi: 10.1186/s40168-020-00836-7.

Dangi, A. K., Sharma, B., Hill, R. T., & Shukla, P. (2019). Bioremediation through microbes: Systems biology and metabolic engineering approach. *Critical Reviews in Biotechnology*, *39*(1), 79–98. Doi: 10.1080/07388551.2018.1500997.

Daniel, R. (2005). The metagenomics of soil. *Nature Reviews Microbiology*, *3*(6), 470–478. Doi: 10.1038/nrmicro1160.

Darwesh, O. M., Matter, I. A., & Eida, M. F. (2019). Development of peroxidase enzyme immobilized magnetic nanoparticles for bioremediation of textile wastewater dye. *Journal of Environmental Chemical Engineering*, *7*(1), 102805. Doi: 10.1016/j.jece.2018.11.049.

De Oliveira Franco, L., Maia, R. D. C. C., Porto, A. L. F., Messias, A. S., Fukushima, K., & De Campos-Takaki, G. M. (2004). Heavy metal biosorption by chitin and chitosan isolated from *Cunninghamella elegans* (IFM 46109). *Brazilian Journal of Microbiology*, *35*(3), 243–247. Doi: 10.1590/s1517-83822004000200013.

de Sousa, S. T. P., Cabral, L., Lacerda Júnior, G. V., & Oliveira, V. M. (2017). Diversity of aromatic hydroxylating dioxygenase genes in mangrove microbiome and their biogeographic patterns across global sites. *MicrobiologyOpen*, *6*(4), 490. Doi: 10.1002/mbo3.490.

Delille, D., Coulon, F., & Pelletier, E. (2004). Effects of temperature warming during a bioremediation study of natural and nutrient-amended hydrocarbon-contaminated sub-Antarctic soils. *Cold Regions Science and Technology*, *40*(1–2), 61–70. Doi: 10.1016/j.coldregions.2004.05.005.

Deshmukh, R., Khardenavis, A. A., & Purohit, H. J. (2016). Diverse metabolic capacities of fungi for bioremediation. *Indian Journal of Microbiology*, *56*(3), 247–264. Doi: 10.1007/s12088-016-0584-6.

Donati, E. R., Sani, R. K., Goh, K. M., & Chan, K.-G. (2019). Editorial: Recent advances in bioremediation/biodegradation by extreme microorganisms. *Frontiers in Microbiology*, *10*(Aug), 1851. Doi: 10.3389/fmicb.2019.01851.

dos Santos, D. F. K., Istvan, P., Noronha, E. F., Quirino, B. F., & Krüger, R. H. (2015). New dioxygenase from metagenomic library from Brazilian soil: Insights into antibiotic resistance and bioremediation. *Biotechnology Letters*, *37*(9), 1809–1817. Doi: 10.1007/s10529-015-1861-x.

dos Santos Aguilar, J. G., & Sato, H. H. (2018). Microbial proteases: Production and application in obtaining protein hydrolysates. *Food Research International*, *103*, 253–262. Elsevier Ltd. Doi: 10.1016/j.foodres.2017.10.044.

Duong, T. T. T., Verma, S. L., Penfold, C., & Marschner, P. (2013). Nutrient release from composts into the surrounding soil. *Geoderma*, *195–196*, 42–47. Doi: 10.1016/j.geoderma.2012.11.010.

Dutta, D., Arya, S., & Kumar, S. (2021). Industrial wastewater treatment: Current trends, bottlenecks, and best practices. *Chemosphere*, *285*, 131245. Doi: 10.1016/j.chemosphere.2021.131245.

Dutta, K., Shityakov, S., Khalifa, I., Ballav, S., Jana, D., Manna, T., Karmakar, M., Raul, P., Guchhait, K. C., & Ghosh, C. (2019). Enhanced biodegradation of naphthalene by *Pseudomonas* sp. consortium immobilized in calcium alginate beads. *bioRxiv*, 631135. Doi: 10.1101/631135.

El Mamouni, R., Jacquet, R., Gerin, P., & Agathos, S. N. (2002). Influence of electron donors and acceptors on the bioremediation of soil contaminated with trichloroethene and nickel: Laboratory- and pilot-scale study. *Water Science and Technology : A Journal of the International Association on Water Pollution Research*, 45(10), 49–54.

Emwas, A. H., Roy, R., McKay, R. T., Tenori, L., Saccenti, E., Nagana Gowda, G. A., Raftery, D., Alahmari, F., Jaremko, L., Jaremko, M., & Wishart, D. S. (2019). Nmr spectroscopy for metabolomics research. *Metabolites*, 9(7). Doi: 10.3390/metabo9070123.

Evans, P. J., & Trute, M. M. (2006). *In situ* bioremediation of nitrate and perchlorate in vadose zone soil for groundwater protection using gaseous electron donor injection technology. *Water Environment Research*, 78(13), 2436–2446. Doi: 10.2175/106143006x123076.

Feng, L., Wang, W., Cheng, J., Ren, Y., Zhao, G., Gao, C., Tang, Y., Liu, X., Han, W., Peng, X., Liu, R., & Wang, L. (2007). Genome and proteome of long-chain alkane degrading *Geobacillus thermodenitrificans* NG80–2 isolated from a deep-subsurface oil reservoir. *Proceedings of the National Academy of Sciences of the United States of America*, 104(13), 5602–5607. Doi: 10.1073/pnas.0609650104.

Fernandes, C., Kankonkar, H., Meena, R. M., Menezes, G., Shenoy, B. D., & Khandeparker, R. (2019). Metagenomic analysis of tarball-associated bacteria from Goa, India. *Marine Pollution Bulletin*, 141, 398–403. Doi: 10.1016/j.marpolbul.2019.02.040.

Floriano, B., Santero, E., & Reyes-Ramírez, F. (2019). Biodegradation of tetralin: Genomics, gene function and regulation. *Genes*, 10(5). MDPI AG. Doi: 10.3390/genes10050339.

Franzosa, E. A., Hsu, T., Sirota-Madi, A., Shafquat, A., Abu-Ali, G., Morgan, X. C., & Huttenhower, C. (2015). Sequencing and beyond: Integrating molecular "omics" for microbial community profiling. *Nature Reviews Microbiology*, 13(6), 360–372. Doi: 10.1038/nrmicro3451.

Gebreselassie, N. A., & Antoniewicz, M. R. (2015). 13C-metabolic flux analysis of co-cultures: A novel approach. *Metabolic Engineering*, 31, 132–139. Doi: 10.1016/j.ymben.2015.07.005.

Girbal, L., Hilaire, D., Leduc, S., Delery, L., Rols, J. L., & Lindley, N. D. (2000). Reductive cleavage of demeton-S-methyl by *Corynebacterium glutamicum* in cometabolism on more readily metabolizable substrates. *Applied and Environmental Microbiology*, 66(-3), 1202–1204. Doi: 10.1128/AEM.66.3.1202-1204.2000.

Gulzar, T., Huma, T., Jalal, F., Iqbal, S., Abrar, S., Kiran, S., Nosheen, S., Hussain, W., & Rafique, M. A. (2017). Bioremediation of synthetic and industrial effluents by *Aspergillus niger* isolated from contaminated soil following a sequential strategy. *Molecules (Basel, Switzerland)*, 22(12), 2244. Doi: 10.3390/molecules22122244.

Haiser, H. J., & Turnbaugh, P. J. (2013). Developing a metagenomic view of xenobiotic metabolism. *Pharmacological Research*, 69(1), 21–31. Doi: 10.1016/j.phrs.2012.07.009.

Haldar, S., & Nazareth, S. W. (2019). Diversity of fungi from mangrove sediments of Goa, India, obtained by metagenomic analysis using Illumina sequencing. *3 Biotech*, 9(5), 1–5. Doi: 10.1007/s13205-019-1698-4.

Harrington, E. D., Arumugam, M., Raes, J., Bork, P., & Relman, D. A. (2010). SmashCell: A software framework for the analysis of single-cell amplified genome sequences. *Bioinformatics*, 26(23), 2979–2980. Doi: 10.1093/bioinformatics/btq564.

Hazen, T. C. (2010). Cometabolic bioremediation. In *Handbook of Hydrocarbon and Lipid Microbiology* (pp. 2505–2514). Springer, Berlin Heidelberg. Doi: 10.1007/978-3-540-77587-4_185.

Heidelberg, J. F., Paulsen, I. T., Nelson, K. E., Gaidos, E. J., Nelson, W. C., Read, T. D., Eisen, J. A., Seshadri, R., Ward, N., Methe, B., Clayton, R. A., Meyer, T., Tsapin, A.,

Scott, J., Beanan, M., Brinkac, L., Daugherty, S., Deboy, R. T., Dodson, R. J., … Fraser, C. M. (2001). Genome sequence of the dissimilatory metal ion-reducing bacterium *Shewanella oneidensis*. *Nature Biotechnology* (10), 5251. Doi: 10.1038/nbt749.

Hemme, C. L., Deng, Y., Gentry, T. J., Fields, M. W., Wu, L., Barua, S., Barry, K., Tringe, S. G., Watson, D. B., He, Z., Hazen, T. C., Tiedje, J. M., Rubin, E. M., & Zhou, J. (2010). Metagenomic insights into evolution of a heavy metal-contaminated groundwater microbial community. *ISME Journal*, 4(5), 660–672. Doi: 10.1038/ismej.2009.154.

Hesham, A. E. L., Khan, S., Tao, Y., Li, D., Zhang, Y., & Yang, M. (2012). Biodegradation of high molecular weight PAHs using isolated yeast mixtures: Application of metagenomic methods for community structure analyses. *Environmental Science and Pollution Research*, 19(8), 3568–3578. Doi: 10.1007/s11356-012-0919-8.

Hirose, J., Fujihara, H., Watanabe, T., Kimura, N., Suenaga, H., Futagami, T., Goto, M., Suyama, A., & Furukawa, K. (2019). Biphenyl/PCB degrading bph genes of ten bacterial strains isolated from biphenyl-contaminated soil in Kitakyushu, Japan: Comparative and dynamic features as integrative conjugative elements (ICEs). *Genes*, 10(5). Doi: 10.3390/genes10050404.

Hlihor, R. M., Gavrilescu, M., Tavares, T., Favier, L., & Olivieri, G. (2017). Bioremediation: An overview on current practices, advances, and new perspectives in environmental pollution treatment. *BioMed Research International*, 2017. Doi: 10.1155/2017/6327610.

Hofreiter, M. (2008). Palaeogenomics. *Comptes Rendus - Palevol*, 7(2–3), 113–124. Doi: 10.1016/j.crpv.2007.12.005.

Hong, P. Y., Mantilla-Calderon, D., & Wang, C. (2020). Metagenomics as a tool to monitor reclaimed-water quality. *Applied and Environmental Microbiology*, 86(16), 1–15. Doi: 10.1128/AEM.00724-20.

Huson, D. H. (2015). Metagenome analyzer (MEGAN): Metagenomic expert resource. In *Encyclopedia of Metagenomics* (pp. 383–389). Springer US. Doi: 10.1007/978-1-4899-7478-5_4.

Huson, D. H., Albrecht, B., Bağci, C., Bessarab, I., Górska, A., Jolic, D., & Williams, R. B. H. (2018). MEGAN-LR: New algorithms allow accurate binning and easy interactive exploration of metagenomic long reads and contigs. *Biology Direct*, 13(1). Doi: 10.1186/s13062-018-0208-7.

Huson, D. H., Auch, A. F., Qi, J., & Schuster, S. C. (2007). Megan analysis of metagenomic data. *Genome Research*, 17(3), 377–386. Doi: 10.1101/gr.5969107.

Hussaini, S. Z., Shaker, M., & Asef Iqbal, M. (2013). Isolation of bacterial for degradation of selected pesticides. *Advances in Bioresearch*, 4(3), 82–85.

Imperato, V., Portillo-Estrada, M., McAmmond, B. M., Douwen, Y., Van Hamme, J. D., Gawronski, S. W., Vangronsveld, J., & Thijs, S. (2019). Genomic diversity of two hydrocarbon-degrading and plant growth-promoting *Pseudomonas* species isolated from the oil field of Bóbrka (Poland). *Genes*, 10(6), 443. Doi: 10.3390/genes10060443.

Infante, C. J., De Arco, D. R., & Angulo, E. M. (2014). Removal of lead, mercury and nickel using the yeast *Saccharomyces cerevisiae* Remoción de plomo, mercurio y níquel utilizando la levadura *Saccharomyces cerevisiae*. *Rev.MVZ Córdoba*, 19(2), 4141–4149.

Ishii, S., Suzuki, S., Tenney, A., Nealson, K. H., & Bretschger, O. (2018). Comparative metatranscriptomics reveals extracellular electron transfer pathways conferring microbial adaptivity to surface redox potential changes. *ISME Journal*, 12(12), 2844–2863. Doi: 10.1038/s41396-018-0238-2.

Jain, M., Yadav, P., Joshi, A., & Kodgire, P. (2019). Advances in detection of hazardous organophosphorus compounds using organophosphorus hydrolase based biosensors. *Critical Reviews in Toxicology*, 49(5), 387–410. Doi: 10.1080/10408444.2019.1626800.

Jaiswal, S., & Shukla, P. (2020). Alternative strategies for microbial remediation of pollutants via synthetic biology. *Frontiers in Microbiology*, 11, 808. Doi: 10.3389/fmicb.2020.00808.

Jung, J., Philippot, L., & Park, W. (2016). Metagenomic and functional analyses of the consequences of reduction of bacterial diversity on soil functions and bioremediation in diesel-contaminated microcosms. *Scientific Reports, 6*(1), 1–10. Doi: 10.1038/srep23012.

Kafilzadeh, F., Ebrahimnezhad, M., & Tahery, Y. (2015). Isolation and identification of endosulfan-degrading bacteria and evaluation of their bioremediation in Kor river, Iran. *Osong Public Health and Research Perspectives, 6*(1), 39–46. Doi: 10.1016/j.phrp.2014.12.003.

Kaminski, M. A., Sobczak, A., Dziembowski, A., & Lipinski, L. (2019). Genomic analysis of γ-hexachlorocyclohexane-degrading *Sphingopyxis lindanitolerans* WS5A3p strain in the context of the pangenome of *Sphingopyxis. Genes, 10*(9). Doi: 10.3390/genes10090688.

Kanmani, P., Aravind, J, & Kumaresan, K. (n.d.). *An Insight into Microbial Lipases and Their Environmental Facet.* Doi: 10.1007/s13762-014-0605-0.

Karigar, C. S., & Rao, S. S. (2011). Role of microbial enzymes in the bioremediation of pollutants: A review. *Enzyme Research, 2011*(1). Doi: 10.4061/2011/805187.

Kasana, R. C. (2010). Proteases from psychrotrophs: An overview. *Critical Reviews in Microbiology, 36*(2), 134–145. Doi: 10.3109/10408410903485525.

Kaushal, J., Mehandia, S., Singh, G., Raina, A., & Arya, S. K. (2018). Catalase enzyme: Application in bioremediation and food industry. *Biocatalysis and Agricultural Biotechnology, 16*, 192–199. Elsevier Ltd. Doi: 10.1016/j.bcab.2018.07.035.

Keegan, K. P., Glass, E. M., & Meyer, F. (2016). MG-RAST, a metagenomics service for analysis of microbial community structure and function. *Methods in Molecular Biology, 1399*, 207–233. Doi: 10.1007/978-1-4939-3369-3_13.

Keum, Y. S., Seo, J. S., Li, Q. X., & Kim, J. H. (2008). Comparative metabolomic analysis of *Sinorhizobium* sp. C4 during the degradation of phenanthrene. *Applied Microbiology and Biotechnology, 80*(5), 863–872. Doi: 10.1007/s00253-008-1581-4.

Khan, I., Aftab, M., Shakir, S. U., Ali, M., Qayyum, S., Rehman, M. U., Haleem, K. S., & Touseef, I. (2019). Mycoremediation of heavy metal (Cd and Cr)–polluted soil through indigenous metallotolerant fungal isolates. *Environmental Monitoring and Assessment, 191*(9), 1–11. Doi: 10.1007/s10661-019-7769-5.

Kim, S. H., Harzman, C., Davis, J. K., Hutcheson, R., Broderick, J. B., Marsh, T. L., & Tiedje, J. M. (2012). Genome sequence of *Desulfitobacterium hafniense* DCB-2, a gram-positive anaerobe capable of dehalogenation and metal reduction. *BMC Microbiology, 12*, 21. Doi: 10.1186/1471-2180-12-21.

Kim, S.-J., Kweon, O., & Cerniglia, C. E. (2010). Degradation of polycyclic aromatic hydrocarbons by *Mycobacterium* strains. In *Handbook of Hydrocarbon and Lipid Microbiology* (pp. 1865–1879). Springer Berlin Heidelberg. Doi: 10.1007/978-3-540-77587-4_136.

Kisand, V., Valente, A., Lahm, A., Tanet, G., & Lettieri, T. (2012). Phylogenetic and functional metagenomic profiling for assessing microbial biodiversity in environmental monitoring. *PLoS One, 7*(8), e43630. Doi: 10.1371/journal.pone.0043630.

Konopka, A., & Wilkins, M. J. (2012). Application of meta-transcriptomics and -proteomics to analysis of in situ physiological state. *Frontiers in Microbiology, 3*(May). Doi: 10.3389/fmicb.2012.00184.

Kouzuma, A., Kasai, T., Hirose, A., & Watanabe, K. (2015). Catabolic and regulatory systems in *Shewanella oneidensis* MR-1 involved in electricity generation in microbial fuel cells. *Frontiers in Microbiology, 6*(Jun), 609. Doi: 10.3389/fmicb.2015.00609.

Kühner, S., Wöhlbrand, L., Fritz, I., Wruck, W., Hultschig, C., Hufnagel, P., Kube, M., Reinhardt, R., & Rabus, R. (2005). Substrate-dependent regulation of anaerobic degradation pathways for toluene and ethylbenzene in a denitrifying bacterium, strain EbN1. *Journal of Bacteriology, 187*(4), 1493–1503. Doi: 10.1128/JB.187.4.1493-1503.2005.

Kumar, V. (2018). Mechanism of microbial heavy metal accumulation from polluted environment and bioremediation. In: Sharma, D., & Saharan, B.S. (eds.), *Microbial Fuel Factories.* Boca Raton, FL: CRC Press.

Kumar, V., & Chandra, R. (2020a). Bioremediation of melanoidins containing distillery waste for environmental safety. In: Bharagava, R., & Saxena, G. (eds.), *Bioremediation of Industrial Waste for Environmental Safety*. Springer, Singapore. Doi: 10.1007/978-981-13-3426-9_20.

Kumar, V., & Chandra, R. (2020b). Metagenomics analysis of rhizospheric bacterial communities of *Saccharum arundinaceum* growing on organometallic sludge of sugarcane molasses-based distillery. *3 Biotech 10*(7), 316. Doi: 10.1007/s13205-020-02310-5.

Kumar, V., Singh, K., & Shah, M. P. (2021a). Advanced oxidation processes for complex wastewater treatment. In: Shah, M. P. (ed.), *Advance Oxidation Process for Industrial Effluent Treatment*. Elsevier. Doi: 10.1016/B978-0-12-821011-6.00001-3.

Kumar, V., Shahi, S. K., & Singh, S. (2018). Bioremediation: An eco-sustainable approach for restoration of contaminated sites. In: Singh, J., Sharma, D., Kumar, G., & Sharma N. (eds.), *Microbial Bioprospecting For Sustainable Development*. Springer, Singapore. Doi: 10.1007/978-981-13-0053-0_6.

Kumar, V., Singh, K., Shah, M. P., Singh, A. K, Kumar, A., & Kumar, Y. (2021b). Application of omics technologies for microbial community structure and function analysis in contaminated environment. In Shah, M. P., Sarkar, A., & Mandal, S., (eds.), *Wastewater Treatment: Cutting Edge Molecular Tools, Techniques & Applied Aspects*. Elsevier. Doi: 10.1016/B978-0-12-821925-6.00013-7.

Kumar, V., Thakur, I. S., & Shah, M. P. (2020a). Bioremediation approaches for pulp and paper industry wastewater treatment: Recent advances and challenges. In: Shah, M. P. (ed.), *Microbial Bioremediation & Biodegradation*. Springer, Singapore. Doi: 10.1007/978-981-15-1812-6_1.

Kumar, V., Thakur, I. S., Singh, A. K., & Shah, M. P. (2020b). Application of metagenomics in remediation of contaminated sites and environmental restoration. In: Shah, M., Rodriguez-Couto, S., & Sengor, S. S. (eds.), *Emerging Technologies in Environmental Bioremediation*. Elsevier. Doi: 10.1016/B978-0-12-819860-5.00008-0.

Kumavath, R. N., & Deverapalli, P. (2013). Scientific swift in bioremediation: An overview. In *Applied Bioremediation - Active and Passive Approaches*. InTech. Doi: 10.5772/56409.

Lacerda, C. M. R., Choe, L. H., & Reardon, K. F. (2007). Metaproteomic analysis of a bacterial community response to cadmium exposure. *Journal of Proteome Research*, 6(3), 1145–1152. Doi: 10.1021/pr060477v.

Landrigan, P. J., Fuller, R., Acosta, N. J. R., Adeyi, O., Arnold, R., Basu, N., Baldé, A. B., Bertollini, R., Bose-O'Reilly, S., Boufford, J. I., Breysse, P. N., Chiles, T., Mahidol, C., Coll-Seck, A. M., Cropper, M. L., Fobil, J., Fuster, V., Greenstone, M., Haines, A., ... Zhong, M. (2018). The Lancet commission on pollution and health. *The Lancet, 391* (10119), 462–512. Lancet Publishing Group. Doi: 10.1016/S0140-6736(17)32345-0.

Langridge, G. (2009). Testing the water: Marine metagenomics. *Nature Reviews Microbiology*, 7(8), 552–552. Doi: 10.1038/nrmicro2188.

Laudadio, I., Fulci, V., Stronati, L., & Carissimi, C. (2019). Next-generation metagenomics: Methodological challenges and opportunities. *OMICS: A Journal of Integrative Biology*, 23(7), 327–333. Doi: 10.1089/omi.2019.0073.

Leal, A. J., Rodrigues, E. M., Leal, P. L., Júlio, A. D. L., Fernandes, R. de C. R., Borges, A. C., & Tótola, M. R. (2017). Changes in the microbial community during bioremediation of gasoline-contaminated soil. *Brazilian Journal of Microbiology*, 48(2), 342–351. Doi: 10.1016/j.bjm.2016.10.018.

Lee, Y. C., & Chang, S. P. (2011). The biosorption of heavy metals from aqueous solution by *Spirogyra* and *Cladophora* filamentous macroalgae. *Bioresource Technology*, 102(9), 5297–5304. Doi: 10.1016/j.biortech.2010.12.103.

Letunic, I., & Bork, P. (2007). Interactive Tree Of Life (iTOL): An online tool for phylogenetic tree display and annotation. *Bioinformatics*, 23(1), 127–128. Doi: 10.1093/bioinformatics/btl529.

Li, Y., Zheng, L., Zhang, Y., Liu, H., & Jing, H. (2019). Comparative metagenomics study reveals pollution induced changes of microbial genes in mangrove sediments. *Scientific Reports*, *9*(1), 1–11. Doi: 10.1038/s41598-019-42260-4.

Liu, X., Wu, Y., Wilson, F. P., Yu, K., Lintner, C., Cupples, A. M., & Mattes, T. E. (2018). Integrated methodological approach reveals microbial diversity and functions in aerobic groundwater microcosms adapted to vinyl chloride. *FEMS Microbiology Ecology*, *94*(9). Doi: 10.1093/femsec/fiy124.

López-Barea, J., & Gómez-Ariza, J. L. (2006). Environmental proteomics and metallomics. *Proteomics*, *6*(S1), S51–S62. Doi: 10.1002/pmic.200500374.

Loukidou, M. X., Matis, K. A., Zouboulis, A. I., & Liakopoulou-Kyriakidou, M. (2003). Removal of As(V) from wastewaters by chemically modified fungal biomass. *Water Research*, *37*(18), 4544–4552. Doi: 10.1016/S0043-1354(03)00415-9.

Lovley, D. R. (1995). Bioremediation of organic and metal contaminants with dissimilatory metal reduction. *Journal of Industrial Microbiology*, *14*(2), 85–93. Doi: 10.1007/BF01569889.

Lovley, D. R. (2003). Cleaning up with genomics: Applying molecular biology to bioremediation. *Nature Reviews Microbiology*, *1*(1), 35–44. Doi: 10.1038/nrmicro731.

Luo, J., Bai, Y., Liang, J., & Qu, J. (2014). Metagenomic approach reveals variation of microbes with arsenic and antimony metabolism genes from highly contaminated soil. *PLoS One*, *9*(10), 108185. Doi: 10.1371/journal.pone.0108185.

Mahjoubi, M., Cappello, S., Souissi, Y., Jaouani, A., & Cherif, A. (2018). Microbial bioremediation of petroleum hydrocarbon– contaminated marine environments. *Recent Insights in Petroleum Science and Engineering*. Doi: 10.5772/intechopen.72207.

Malla, M. A., Dubey, A., Yadav, S., Kumar, A., Hashem, A., & Abd-Allah, E. F. (2018). Understanding and designing the strategies for the microbe-mediated remediation of environmental contaminants using omics approaches. *Frontiers in Microbiology*, *9*(Jun), 1132. Doi: 10.3389/fmicb.2018.01132.

Mani, D., & Kumar, C. (2014). Biotechnological advances in bioremediation of heavy metals contaminated ecosystems: An overview with special reference to phytoremediation. *International Journal of Environmental Science and Technology*, *11*(3), 843–872. Doi: 10.1007/s13762-013-0299-8.

Manisalidis, I., Stavropoulou, E., Stavropoulos, A., & Bezirtzoglou, E. (2020). Environmental and health impacts of air pollution: A review. *Frontiers in Public Health*, *8*, 14. Doi: 10.3389/fpubh.2020.00014.

Marcelino, V. R., Holmes, E. C., & Sorrell, T. C. (2020). The use of taxon-specific reference databases compromises metagenomic classification. *BMC Genomics*, *21*(1), 184. Doi: 10.1186/s12864-020-6592-2.

Marco, D. (2008). Metagenomics and the niche concept. *Theory in Biosciences*, *127*(3), 241–247. Doi: 10.1007/s12064-008-0028-x.

Markowitz, V. M., Chen, I.-M. A., Palaniappan, K., Chu, K., Szeto, E., Grechkin, Y., Ratner, A., Anderson, I., Lykidis, A., Mavromatis, K., Ivanova, N. N., & Kyrpides, N. C. (2010). The integrated microbial genomes system: An expanding comparative analysis resource. *Nucleic Acids Research*, *38*, D382–D390. Doi: 10.1093/nar/gkp887.

Marshall, I. P. G., Karst, S. M., Nielsen, P. H., & Jørgensen, B. B. (2018). Metagenomes from deep Baltic sea sediments reveal how past and present environmental conditions determine microbial community composition. *Marine Genomics*, *37*, 58–68. Doi: 10.1016/j.margen.2017.08.004.

Martínez, A. T., Ruiz-Dueñas, F. J., Camarero, S., Serrano, A., Linde, D., Lund, H., Vind, J., Tovborg, M., Herold-Majumdar, O. M., Hofrichter, M., Liers, C., Ullrich, R., Scheibner, K., Sannia, G., Piscitelli, A., Pezzella, C., Sener, M. E., Kılıç, S., van Berkel, W. J. H., … Alcalde, M. (2017). Oxidoreductases on their way to industrial biotransformations. *Biotechnology Advances*, *35*(6), 815–831. Doi: 10.1016/j.biotechadv.2017.06.003.

Mesarch, M. B., Nakatsu, C. H., & Nies, L. (2000). Development of catechol 2,3-dioxygenase-specific primers for monitoring bioremediation by competitive quantitative PCR. *Applied and Environmental Microbiology, 66*(2), 678–683. Doi: 10.1128/AEM.66.2.678-683.2000.

Milward, E. A., Shahandeh, A., Heidari, M., Johnstone, D. M., Daneshi, N., & Hondermarck, H. (2016). Transcriptomics. *Encyclopedia of Cell Biology, 4*, 160–165. Doi: 10.1016/B978-0-12-394447-4.40029-5.

Mosbæk, F., Kjeldal, H., Mulat, D. G., Albertsen, M., Ward, A. J., Feilberg, A., & Nielsen, J. L. (2016). Identification of syntrophic acetate-oxidizing bacteria in anaerobic digesters by combined protein-based stable isotope probing and metagenomics. *ISME Journal, 10*(10), 2405–2418. Doi: 10.1038/ismej.2016.39.

Mouser, P. J., Holmes, D. E., Perpetua, L. A., DiDonato, R., Postier, B., Liu, A., & Lovley, D. R. (2009). Quantifying expression of *Geobacter* spp. oxidative stress genes in pure culture and during in situ uranium bioremediation. *ISME Journal, 3*(4), 454–465. Doi: 10.1038/ismej.2008.126.

Mueller, R. S., & Pan, C. (2013). Sample handling and mass spectrometry for microbial metaproteomic analyses. *Methods in Enzymology, 531*, 289–303. Doi: 10.1016/B978-0-12-407863-5.00015-0.

Mukherjee, A., Chettri, B., Langpoklakpam, J. S., Basak, P., Prasad, A., Mukherjee, A. K., Bhattacharyya, M., Singh, A. K., & Chattopadhyay, D. (2017). Bioinformatic approaches including predictive metagenomic profiling reveal characteristics of bacterial response to petroleum hydrocarbon contamination in diverse environments. *Scientific Reports, 7*(1), 1108. Doi: 10.1038/s41598-017-01126-3.

Muller, J., Szklarczyk, D., Julien, P., Letunic, I., Roth, A., Kuhn, M., Powell, S., Von Mering, C., Doerks, T., Jensen, L. J., & Bork, P. (2009). EggNOG v2.0: Extending the evolutionary genealogy of genes with enhanced non-supervised orthologous groups, species and functional annotations. *Nucleic Acids Research, 38*(SUPPL.1). Doi: 10.1093/nar/gkp951.

Naguib, M. M., El-Gendy, A. O., & Khairalla, A. S. (2018). Microbial diversity of mer operon genes and their potential rules in mercury bioremediation and resistance. *The Open Biotechnology Journal, 12*(1), 56–77. Doi: 10.2174/1874070701812010056.

Newsome, L., Morris, K., & Lloyd, J. R. (2014). The biogeochemistry and bioremediation of uranium and other priority radionuclides. *Chemical Geology, 363*, 164–184. Elsevier. Doi: 10.1016/j.chemgeo.2013.10.034.

Nicolaisen, M. H., Bælum, J., Jacobsen, C. S., & Sørensen, J. (2008). Transcription dynamics of the functional tfdA gene during MCPA herbicide degradation by *Cupriavidus necator* AEO106 (pRO101) in agricultural soil. *Environmental Microbiology, 10*(3), 571–579. Doi: 10.1111/j.1462-2920.2007.01476.x.

Nielsen, J. (2003). It is all about metabolic fluxes. *Journal of Bacteriology, 185*(24), 7031–7035. Doi: 10.1128/JB.185.24.7031-7035.2003.

Nouaille, S., Mondeil, S., Finoux, A. L., Moulis, C., Girbal, L., & Cocaign-Bousquet, M. (2017). The stability of an mRNA is influenced by its concentration: A potential physical mechanism to regulate gene expression. *Nucleic Acids Research, 45*(20), 11711–11724. Doi: 10.1093/nar/gkx781.

Ojewumi, M. E., Okeniyi, J. O., Ikotun, J. O., Okeniyi, E. T., Ejemen, V. A., & Popoola, A. P. I. (2018). Bioremediation: Data on *Pseudomonas aeruginosa* effects on the bioremediation of crude oil polluted soil. *Data in Brief, 19*, 101–113. Doi: 10.1016/j.dib.2018.04.102.

Olawale, O., Obayomi, K. S., Dahunsi, S. O., & Folarin, O. (2020). Bioremediation of artificially contaminated soil with petroleum using animal waste: Cow and poultry dung. *Cogent Engineering, 7*(1), 1721409. Doi: 10.1080/23311916.2020.1721409.

Paez-Espino, D., Chen, I.-M. A., Palaniappan, K., Ratner, A., Chu, K., Szeto, E., Pillay, M., Huang, J., Markowitz, V. M., Nielsen, T., Huntemann, M., K Reddy, T. B., Pavlopoulos, G. A., Sullivan, M. B., Campbell, B. J., Chen, F., McMahon, K., Hallam, S. J., Denef,

V., ... Kyrpides, N. C. (2017). IMG/VR: A database of cultured and uncultured DNA viruses and retroviruses. *Nucleic Acids Research, 45*(D1), D457–D465. Doi: 10.1093/nar/gkw1030.

Paez-Espino, D., Roux, S., Chen, I.-M. A., Palaniappan, K., Ratner, A., Chu, K., Huntemann, M., Reddy, T. B. K., Pons, J. C., Llabrés, M., Eloe-Fadrosh, E. A., Ivanova, N. N., & Kyrpides, N. C. (2019). IMG/VR v.2.0: An integrated data management and analysis system for cultivated and environmental viral genomes. *Nucleic Acids Research, 47* (D1), D678–D686. Doi: 10.1093/nar/gky1127.

Pande, V., Pandey, S. C., Joshi, T., Sati, D., Gangola, S., Kumar, S., & Samant, M. (2019). Biodegradation of toxic dyes: A comparative study of enzyme action in a microbial system. In *Smart Bioremediation Technologies: Microbial Enzymes* (pp. 255–287). Elsevier. Doi: 10.1016/B978-0-12-818307-6.00014-7.

Pandey, J., Chauhan, A., & Jain, R. K. (2009). Integrative approaches for assessing the ecological sustainability of in situ bioremediation. *FEMS Microbiology Reviews, 33*(2), 324–375. Doi: 10.1111/j.1574-6976.2008.00133.x.

Pascual, J., Udaondo, Z., Molina, L., Segura, A., Esteve-Núñez, A., Caballero, A., Duque, E., Ramos, J. L., & van Dillewijn, P. (2015). Draft genome sequence *of Pseudomonas putida* JLR11, a facultative anaerobic 2,4,6-trinitrotoluene biotransforming bacterium. *Genome Announcements, 3*(5). Doi: 10.1128/genomeA.00904-15.

Peng, P., Yang, H., Jia, R., & Li, L. (2013). Biodegradation of dioxin by a newly isolated *Rhodococcus* sp. with the involvement of self-transmissible plasmids. *Applied Microbiology and Biotechnology, 97*(12), 5585–5595. Doi: 10.1007/s00253-012-4363-y.

Pereira, P., Enguita, F. J., Ferreira, J., & Leitão, A. L. (2014). DNA damage induced by hydroquinone can be prevented by fungal detoxification. *Toxicology Reports, 1*, 1096–1105. Doi: 10.1016/j.toxrep.2014.10.024.

Pérez, M., Rueda, O. D., Bangeppagari, M., Johana, J. Z., Ríos, D., Rueda, B. B., Sikandar, I. M., & Naga, R. M. (2016). Evaluation of various pesticides-degrading pure bacterial cultures isolated from pesticide-contaminated soils in Ecuador. *African Journal of Biotechnology, 15*(40), 2224–2233. Doi: 10.5897/ajb2016.15418.

Perumbakkam, S., Hess, T. F., & Crawford, R. L. (2006). A bioremediation approach using natural transformation in pure-culture and mixed-population biofilms. *Biodegradation, 17*(6), 545–557. Doi: 10.1007/s10532-005-9025-7.

Phulpoto, A. H., Maitlo, M. A., & Kanhar, N. A. (2020). Culture-dependent to culture-independent approaches for the bioremediation of paints: A review. *International Journal of Environmental Science and Technology, 18*(3), 241–262. Doi: 10.1007/s13762-020-02801-1.

Płaza, G., Ulfig, K., Hazen, T. C., & Brigmon, R. L. (2001). Use of molecular techniques in bioremediation. *Acta Microbiologica Polonica, 50*(3–4), 205–218.

Priyalaxmi, R., Murugan, A., Raja, P., & Raj, K. D. (2014). Bioremediation of cadmium by *Bacillus safensis* (JX126862), a marine bacterium isolated from mangrove sediments. *International Journal of Current Microbiology and Applied Sciences, 3*(12). http://www.ijcmas.com.

Rao, M. A., Scelza, R., Scotti, R., & Gianfreda, L. (2010). Role of enzymes in the remediation of polluted environments. *Journal of Soil Science and Plant Nutrition, 10*(3), 333–353. Doi: 10.4067/S0718-95162010000100008.

Razzaq, A., Shamsi, S., Ali, A., Ali, Q., Sajjad, M., Malik, A., & Ashraf, M. (2019). Microbial proteases applications. *Frontiers in Bioengineering and Biotechnology, 7*(Jun), 110. Doi: 10.3389/fbioe.2019.00110.

Reddy, V. R., & Behera, B. (2006). Impact of water pollution on rural communities: An economic analysis. *Ecological Economics, 58*(3), 520–537. Doi: 10.1016/j.ecolecon.2005.07.025.

Rehan, M., & Alsohim, S. (2019). Bioremediation of heavy metals. *Environmental Chemistry and Recent Pollution Control Approaches*. Doi: 10.5772/intechopen.88339.

Rhee, S. K., Liu, X., Wu, L., Chong, S. C., Wan, X., & Zhou, J. (2004). Detection of genes involved in biodegradation and biotransformation in microbial communities by using 50-mer oligonucleotide microarrays. *Applied and Environmental Microbiology*, *70*(7), 4303–4317. Doi: 10.1128/AEM.70.7.4303-4317.2004.

Rodríguez-Salazar, J., Almeida-Juarez, A. G., Ornelas-Ocampo, K., Millán-López, S., Raga-Carbajal, E., Rodríguez-Mejía, J. L., Muriel-Millán, L. F., Godoy-Lozano, E. E., Rivera-Gómez, N., Rudiño-Piñera, E., & Pardo-López, L. (2020). Characterization of a novel functional trimeric catechol 1,2-dioxygenase from a *Pseudomonas stutzeri* isolated from the gulf of Mexico. *Frontiers in Microbiology*, *11*, 1100. Doi: 10.3389/fmicb.2020.01100.

Roper, L., & Grogan, G. (2016). Biocatalysis for organic chemists: Hydroxylations. In *Organic Synthesis Using Biocatalysis* (pp. 213–241). Elsevier Inc. Doi: 10.1016/B978-0-12-411 518-7.00008-1.

Sankar, N. S., Mrinal, B., Dipak, P., & Saidur, R. (2011). Biodegradation potential of bacterial isolates from tannery effluent with special reference to hexavalent chromium. *Journal of Biotechnology, Bioinformatics and Bioengineering*, *1*(3), 381–386.

Santero, E., & Díaz, E. (2020). Special issue: Genetics of biodegradation and bioremediation. *Genes*, *11*(4), 441. Doi: 10.3390/genes11040441.

Santillan, J. Y., Dettorre, L. A., Lewkowicz, E. S., & Iribarren, A. M. (2016). New and highly active microbial phosphotriesterase sources. *FEMS Microbiology Letters*, *363*(24), fnw276. Doi: 10.1093/femsle/fnw276.

Saunders, R. J., Paul, N. A., Hu, Y., & de Nys, R. (2012). Sustainable sources of biomass for bioremediation of heavy metals in waste water derived from coal-fired power generation. *PLoS One*, *7*(5). Doi: 10.1371/journal.pone.0036470.

Saxena, R., Dhakan, D. B., Mittal, P., Waiker, P., Chowdhury, A., Ghatak, A., & Sharma, V. K. (2017). Metagenomic analysis of hot springs in central India reveals hydrocarbon degrading thermophiles and pathways essential for survival in extreme environments. *Frontiers in Microbiology*, *7*(Jan), 2123. Doi: 10.3389/fmicb.2016.02123.

Sharma, I. (2021). Bioremediation techniques for polluted environment: Concept, advantages, limitations, and prospects. *Trace Metals in the Environment - New Approaches and Recent Advances*. Doi: 10.5772/intechopen.90453.

Sharma, B., Dangi, A. K., & Shukla, P. (2018). Contemporary enzyme based technologies for bioremediation: A review. *Journal of Environmental Management*, *210*, 10–22. Doi: 10.1016/j.jenvman.2017.12.075.

Shedbalkar, U., & Jadhav, J. P. (2011). Detoxification of malachite green and textile industrial effluent by *Penicillium ochrochloron*. *Biotechnology and Bioprocess Engineering*, *16*(1), 196–204. Doi: 10.1007/s12257-010-0069-0.

Sims, J. L., Suflita, J. M., & Russell, H. H. (1990). Reductive dehalogenation: A subsurface bioremediation process. *Remediation Journal*, *1*(1), 75–93. Doi: 10.1002/rem.3440010109.

Singh, A., Kumar, V., & Jn, S. (2013). Assessment of bioremediation of oil and phenol contents in refinery waste water via bacterial consortium. *Journal of Petroleum and Environmental Biotechnology*, *4*, 3. Doi: 10.4172/2157-7463.1000145.

Singh, A. H., Doerks, T., Letunic, I., Raes, J., & Bork, P. (2009). Discovering functional novelty in metagenomes: Examples from light-mediated processes. *Journal of Bacteriology*, *91*(1), 32–41. Doi: 10.1128/JB.01084-08.

Thies, S., Rausch, S. C., Kovacic, F., Schmidt-Thaler, A., Wilhelm, S., Rosenau, F., Daniel, R., Streit, W., Pietruszka, J., & Jaeger, K.-E. (2016). Metagenomic discovery of novel enzymes and biosurfactants in a slaughterhouse biofilm microbial community. *Scientific Reports*, *6*(1), 27035. Doi: 10.1038/srep27035.

Toth, C., Berdugo-Clavijo, C., O'Farrell, C., Jones, G., Sheremet, A., Dunfield, P., & Gieg, L. (2018). Stable isotope and metagenomic profiling of a methanogenic naphthalene-degrading enrichment culture. *Microorganisms*, *6*(3), 65. Doi: 10.3390/microorganisms 6030065.

Vidali, M. (2001). Bioremediation. An overview. *Pure and Applied Chemistry*, *73*(7), 1163–1172. Doi: 10.1351/pac200173071163.

Villas-Bôas, S. G., & Bruheim, P. (2007). The potential of metabolomics tools in bioremediation studies. *OMICS A Journal of Integrative Biology*, *11*(3), 305–313. OMICS. Doi: 10.1089/omi.2007.0005.

Vrionis, H. A., Anderson, R. T., Ortiz-Bernad, I., O'Neill, K. R., Resch, C. T., Peacock, A. D., Dayvault, R., White, D. C., Long, P. E., & Lovley, D. R. (2005). Microbiological and geochemical heterogeneity in an *in situ* uranium bioremediation field site. *Applied and Environmental Microbiology*, *71*(10), 6308–6318. Doi: 10.1128/AEM.71.10.6308-6 318.2005.

Wang, R. F., Wennerstrom, D., Cao, W. W., Khan, A. A., & Cerniglia, C. E. (2000). Cloning, expression, and characterization of the katG gene, encoding catalase-peroxidase, from the polycyclic aromatic hydrocarbon-degrading bacterium *Mycobacterium* sp. strain PYR-1. *Applied and Environmental Microbiology*, *66*(10), 4300–4304. Doi: 10.1128/A EM.66.10.4300-4304.2000.

Wang, X., Aulenta, F., Puig, S., Esteve-Núñez, A., He, Y., Mu, Y., & Rabaey, K. (2020). Microbial electrochemistry for bioremediation. *Environmental Science and Ecotechnology*, *1*, 100013. Doi: 10.1016/j.ese.2020.100013.

Watahiki, S., Kimura, N., Yamazoe, A., Miura, T., Sekiguchi, Y., Noda, N., Matsukura, S., Kasai, D., Takahata, Y., Nojiri, H., & Fukuda, M. (2019). Ecological impact assessment of a bioaugmentation site on remediation of chlorinated ethylenes by multi-omics analysis. *Journal of General and Applied Microbiology*, *65*(5), 225–233. Doi: 10.2323/jgam.2018.10.003,

Wei, S. T.-S., Wu, Y.-W., Lee, T.-H., Huang, Y.-S., Yang, C.-Y., Chen, Y.-L., & Chiang, Y.-R. (2018). Microbial functional responses to cholesterol catabolism in denitrifying sludge. *MSystems*, *3*(5), 113–131. Doi: 10.1128/msystems.00113-18.

Wharfe, E. S., Jarvis, R. M., Winder, C. L., Whiteley, A. S., & Goodacre, R. (2010). Fourier transform infrared spectroscopy as a metabolite fingerprinting tool for monitoring the phenotypic changes in complex bacterial communities capable of degrading phenol. *Environmental Microbiology*, *12*(12), 3253–3263. Doi: 10.1111/j.1462-2920.2010.02300.x.

Wooley, J. C., & Ye, Y. (2009). Metagenomics: Facts and artifacts, and computational challenges*. *Journal of Computer Science and Technology*, *25*(1), 71–81. Doi: 10.1007/ s11390-010-9306-4.

Wu, Y. H., Zhou, P., Cheng, H., Wang, C. S., Wu, M., & Xu, X. W. (2015). Draft genome sequence of *Microbacterium profundi* Shh49T, an actinobacterium isolated from deep-sea sediment of a polymetallic nodule environment. *Genome Announcements*, *3*(3). Doi: 10.1128/genomeA.00642-15.

Xiao, Z., Jiang, W., Chen, D., & Xu, Y. (2020). Bioremediation of typical chlorinated hydrocarbons by microbial reductive dechlorination and its key players: A review. *Ecotoxicology and Environmental Safety*, *202*, 110925. Doi: 10.1016/j.ecoenv.2020.110925.

Xu, Y., & Zhao, F. (2018). Single-cell metagenomics: Challenges and applications. *Protein & Cell*, *9*(5), 501–510. Doi: 10.1007/s13238-018-0544-5.

Yadav, M., Singh, S. K., Sharma, J. K., & Yadav, K. D. S. (2011). Oxidation of polyaromatic hydrocarbons in systems containing water miscible organic solvents by the lignin peroxidase of *Gleophyllum striatum* MTCC-1117. *Environmental Technology*, *32*(11), 1287–1294. Doi: 10.1080/09593330.2010.535177.

Yan, J., Niu, J., Chen, D., Chen, Y., & Irbis, C. (2014). Screening of trametes strains for efficient decolorization of malachite green at high temperatures and ionic concentrations.

International Biodeterioration and Biodegradation, *87*, 109–115. Doi: 10.1016/j.ibiod. 2013.11.009.

Yang, L., Li, Y., Su, F., & Li, H. (2019). A study of the microbial metabolomics analysis of subsurface wastewater infiltration system. *RSC Advances*, *9*(68), 39674–39683. Doi: 10.1039/c9ra05290a.

Yao, J., Jia, R., Zheng, L., & Wang, B. (2013). Rapid decolorization of azo dyes by crude manganese peroxidase from *Schizophyllum* sp. F17 in solid-state fermentation. *Biotechnology and Bioprocess Engineering*, *18*(5), 868–877. Doi: 10.1007/s12257-013-0357-6.

Yerushalmi, L., Rocheleau, S., Cimpoia, R., Sarrazin, M., Sunahara, G., Peisajovich, A., Leclair, G., & Guiot, S. R. (2003). Enhanced biodegradation of petroleum hydrocarbons in contaminated soil. *Bioremediation Journal*, *7*(1), 37–51. Doi: 10.1080/713914241-274.

Yu, K., Yi, S., Li, B., Guo, F., Peng, X., Wang, Z., Wu, Y., Alvarez-Cohen, L., & Zhang, T. (2019). An integrated meta-omics approach reveals substrates involved in synergistic interactions in a bisphenol A (BPA)-degrading microbial community. *Microbiome*, *7*(1), 16. Doi: 10.1186/s40168-019-0634-5.

Zainith, S., Chowdhary, P., Mani, S., & Mishra, S. (2020). Microbial ligninolytic enzymes and their role in bioremediation. *Microorganisms for Sustainable Environment and Health*, 179–203. Doi: 10.1016/b978-0-12-819001-2.00009-7.

Zeng, H. W., Cai, Y. J., Liao, X. R., Qian, S. L., Zhang, F., & Zhang, D. B. (2010). Optimization of catalase production and purification and characterization of a novel cold-adapted Cat-2 from mesophilic bacterium *Serratia marcescens* SYBC-01. *Annals of Microbiology*, *60*(4), 701–708. Doi: 10.1007/s13213-010-0116-2.

Zhang, Z., Lo, I. M. C., & Yan, D. Y. S. (2015). An integrated bioremediation process for petroleum hydrocarbons removal and odor mitigation from contaminated marine sediment. *Water Research*, *83*, 21–30. Doi: 10.1016/j.watres.2015.06.022.

Zou, D., Li, Y., Kao, S. J., Liu, H., & Li, M. (2019). Genomic adaptation to eutrophication of ammonia-oxidizing archaea in the Pearl River estuary. *Environmental Microbiology*, *21*(7), 2320–2332. Doi: 10.1111/1462-2920.14613.

Zuñiga, C., Zaramela, L., & Zengler, K. (2017). Elucidation of complexity and prediction of interactions in microbial communities. *Microbial Biotechnology*, *10*(6), 1500–1522. Doi: 10.1111/1751-7915.12855.

3 Functional Metagenomics
A Methodological Approach to View the Microbial World: A Review

Kripa Pancholi and Anupama Shrivastav
Parul University

CONTENTS

DOI: 10.1201/9781003247883-3

3.1 INTRODUCTION

The first life as a living organism is microorganisms. They evolve with respect to the changing atmosphere and the components of the environment and show both active and inactive interactions using 16S rRNA (Oyewusi et al. 2021). Microorganisms carry similar properties as well as novel properties such as recycling the organic matter and catalysing the chemical processes in the environment. Thus, it's inevitable to ignore the interaction between environmental components with respect to the environment they are present in. Hence, metagenomics is a cosmic area study. The total genetic material of organisms along with species is known through metagenomics, although 99% of microbial species are "uncultivated" only because of the different environmental conditions (Sharma et al. 2021). Metagenomic analyses reveal the complexity of microbes and show non-targeted microbial contaminants in environmental samples (Awasthi et al. 2020). The screening of those environmental libraries was done by using plasmids, cosmids, fosmids, etc. This method is probably replaced by shotgun metagenomics using a specific region or random parts of DNA (Krüger et al. 2020). Meanwhile, a complete genomic study also known as sequence-based analysis is an approach where the whole genomic sequences of microbial communities are studied without having any pre-existing information. But there remains a limitation; metagenomic libraries have the least frequency of organisms having the same desired functional information. To resolve this problem, the first essential step is to enhance the genome from target populations. One of the effective enhancing methods is stable isotope probing (SIP), which divides the microbial communities into functional categories (Barnett and Buckley 2020). Later, the screening of those libraries is performed through agar plate screening, microarray-based screening, microtiter plate screening, fluorescence-activated cell sorting (FACS), and microfluid-based screening (García-Moyano et al. 2021). The metagenomics is thus altogether a collaboration between molecular biologists, geneticists, microbiologists, plant biologists, veterinarians, animal biology, genetic engineering, clinical traits, marine biology, and other professionals related to the health of the environment (Laudadio et al. 2019). Thus, functional metagenomics helps in challenging the future aspects regarding (1) importance in laboratory-based and field-based investigations, (2) biochemical and biogeochemical processes, (3) validation of metabolic models of marine ecosystem, and (4) studying more prevalent systems and to deeply interpret information of microbes and their hypothesis (Grossart et al. 2020).

3.2 FUNCTIONAL METAGENOMICS

The global increase in population results in an increase in the productivity of medical care facilities and medicines to fight specific diseases, which has exponentially increased the utility of nutritional products. The chemical, pharmaceutical, and nutraceutical industries cannot sufficiently fulfil the demand. Therefore, the booming area of biotechnology and microbiology industries uses microorganisms to produce antibiotics, enzymes, and other organic and bioactive compounds. Metagenomics along with other biotechnological processes favours the industrial production of chemicals, nutraceuticals, biofuels, pharmaceuticals, etc., which plays a major role in bringing back the bioactive molecules and enzymes that were earlier not utilized (Kumar et al. 2021). Metagenomic libraries help in screening such enzymes and metabolites depending on their activities. In functional metagenomics, the cultural mix is extracted from the environment; the DNA comes up with gene libraries through shotgun sequencing, where *E. coli* is used as the cloning host (Snipen et al. 2021). Now when the clones are formed, they are treated with antibiotics having enough concentration to kill the host except for the host having a resistance gene and are classified taxonomically at the end. The only purpose of taxonomic classification is to find the exact species of origin from the environment so that the researchers can investigate the diversities in the microbial community. The classification method allows the study of the relationship between host stages or environmental changes and the species diversity of microbial communities (Awasthi et al. 2020). The functional metagenomics along with sequence-based screening overcomes the limitation of small size, which does not provide enough information regarding resistance gene and the initial host organism. Here, clones growing on antibiotics were selected along with small DNA libraries from metagenomic DNA, which proves that genetic elements have the resistance gene. The functional metagenomics is the only metagenomic screening that allows the isolation of antibiotic resistance genes on the condition of expression of gene resistance. The *E. coli* is the most common cloning host, and when it is intrinsically resistant to antibiotics, functional screening cannot be shown proving the false negatives. The selection of expression host becomes mandatory for metagenomics for the expression of the gene; *E. coli* is sufficient for promoter recognition and translation initiation, but some strains do not efficiently express due to various reasons. During special conditions such as high temperature or active conditions, this issue becomes worst for protein expression, so alternative host expression is done where thermophilic bacterium such as *Thermus thermophilus* is used. It helps in the detection of thermozymes (DeCastro et al. 2016).

3.2.1 ISOLATION OF MOBILE GENETIC ELEMENTS USING FUNCTIONAL METAGENOMICS

The metagenomics is originated from the conventional microbial genome. The conventional method shows that neither pure culture nor isolation and culturing is mandatory for sequencing. The techniques included here isolate antibiotic resistance genes, but the isolation depends on the gene transfer by the specific genetic element. There are different types of mobile genetic elements attenuated into three general classes which differentiate as distinct metagenomic approaches.

3.2.2 CONSOLIDATION OF GENETIC ELEMENTS WITHOUT INDEPENDENT CONJUGAL TRANSFER

Mobile genetic elements allow the mobility of DNA, and when the mobility of DNA chunks is intracellular, they are called transposons. Transposons are DNA sequences that move from one location to another location in the genome. Such a transposable DNA sequence is called transposons. This intracellular mobility is gained through transformation or transduction. The transposons are able to transpose the genome, but are not able to undergo conjugational transfer. Perhaps, transposons of the conjugative elements can be transferred to a whole new host with the help of transformation and this transposable element can be trapped through different methods (Partridge et al. 2018). The vector selected here should have the target site of transposons in a bunch of strains. The activation of the silent gene and the inactivation of the lethal gene results in a change of phenotype that helps in the detection of transposition, and after the transposition, the vector having this element is isolated. DNA sequencing and functional analysis along with transposons trapping allow the isolation of new mobile functional elements having antibiotic resistance genes. This method is used in metagenomics by forming libraries in host or vector. And later, it is transformed to an appropriate host through transposons vector along with screening to deactivate the target.

3.2.3 CONSOLIDATION OF GENETIC ELEMENTS PROFICIENT IN CONCILIATING THEIR OWN TRANSFER

The genetic material that is required to transfer from one host to another through conjugation is found in integrative and conjugative elements (ICE), previously known as conjugative transposons. They have chromosomally integrated mobile elements transferred through horizontal gene transfer. ICE is a widely spread unit that is responsible for the integration, excision, conjugative transfer, and maintenance of a new host genome. Integration is done through the recombination of direct repeats found in the host and ICE. The integrase is involved in excision and forms the intermediate. This intermediate is in a circular manner that is available for conjugative transfer (Botelho and Schulenburg 2021). The replication for this kind of element is done through integration into host replication to survive within them. This happens because of the recombination at a specific site called an integrase. The ICE differs from plasmids in characteristics and evolutionary dynamics.

1. ICE show a feature that is a combination of transposons and prophages; that is, it functions for integration and excision (Botelho and Schulenburg 2021).
2. Conjugative plasmids are not frequently transferred between distant taxa (Cury et al. 2018).
3. An ICE carries dual life with both horizontal gene transfer and vertical transmission.
4. Segregational loss does not affect ICE although it affects plasmids during cell division.

ICE have the same integration site, resulting in strong competition for limited integration sites. Integration can occur at a single attachment site or can be random (Baranowski et al. 2018). An ICE once transferred to the new recipient cell integrates into the chromosome or evolves into an extrachromosomal element (Oliveira et al. 2017). A single-stranded DNA is then transferred to a new host, and the double-stranded elements through DNA polymerase are regenerated. The donor cell maintains the copy of ICE for the reintegration of ICE in chromosomes (Botelho and Schulenburg 2021). The ICE encodes the toxin-antitoxin and restriction-modification that triggers post-segregational damage to daughter cells (Koonin et al. 2020).

3.2.4 Mobilizable Consolidative Elements

Another method for the isolation of mobile genetic elements is through integrative mobilizable elements (IME). IME can be transferred through conjugation. Comparatively, IME are difficult to detect than ICE. Although like ICE they also maintain the health of the host through antibiotic resistance, the antibiotic resistance is shown through enzyme tyrosine recombinase or serine recombinase. The DNA transposons excise DNA and integrate it into plasmids or chromosomes. Unrelated conjugative elements take over to the DNA transport system to promote their own transfer. But to mobilize through ICE, they need to promote their own transfer origin. The nicking of DNA at a sequence called the origin of transfer (oriT) initiates the conjugative transfer. This initiation is done by a protein called relaxase. The relaxase integrates with the coupling protein and binds covalently to DNA, resulting in one of the DNA strands to the active transport system. And the conjugative element encodes that transport system. The Firmicutes characterized till now are majorly encoding their own relaxes belonging to the canonical family (Coluzzi et al. 2018). After the experiment with ICE and IME, the conclusion came that IME that integrate the origin of transfer of both ICE and IME are widely found in streptococci and other Firmicutes. They vehicle antibiotic resistance genes and also offer other helpful traits to bacterial hosts, such as protection and repair. The IME_oriT is the first element that shows the use of conjugative elements as integration host and helper. However, some IME can be harmful to the host (Libante et al. 2020). Here the transfer is mediated by a conjugative element present in a cell other than the element that does not function for conjugative transfer. It becomes mandatory to integrate and promote their own transfer origin into the host replicon to survive just like ICE. For this, the identical methods described above are used for the isolation of such elements from metagenomics.

3.3 TYPES OF SCREENING STRATEGIES

Functional screening is a field for the identification of novel biomolecules that can be used in biotechnology and medicine. Most of the screenings of metagenomic libraries depend on the formation of clones that help in the identification of enzyme activities (Bekele et al. 2021). Some criteria have become important for the screening of metagenomic libraries such as; libraries should have an adequate amount of genes, the size of libraries, the vector and the host correlation, using a host which can express the target gene, the gene expressed in a particular surrogate host, etc. The following are the examples of functional screening used for obtaining the biocatalysts related to metagenomics. This shows the high-throughput screening (HTS) through which we are able to screen 1 billion clones (Sarnaik et al. 2020).

3.3.1 Agar Plate Screening Method

It is the oldest method of screening, and it is used as a state-of-the-art hydrolytic enzyme screening methodology. The novel enzymes such as lipases, cellulases, esterases, proteases, lactases, glycosylases, dehalogenases, and nitrilases work under unfavourable or adverse conditions, and the identification of such enzymes can be done through the functional metagenomic method of agar plate screening (Bekele et al. 2021). The agar plate screening strategies detect non-specific esterase assays (Molitor et al. 2020). The gene that is responsible for the resistance of toxic elements is highlighted through agar plate screening. Such toxic elements can be antibiotics, extreme pH or salt concentrations, or heavy metals. Based on the production of a chromophore or fluorophore, 103–106 clones per day are formed. The method has successfully isolated an enormous amount of typical enzymes from the environment (Bekele et al. 2021).

3.3.2 Microarray-Based Screening

The concept behind DNA microarray-based screening is to know the metagenomic libraries for the selected genes. Unlike agar plates, DNA microarray could be a sequence-based screening method of the metagenomic library (Bekele et al. 2021). The protocol includes the identification of biological photoreceptors supported by already sequenced and annotated genomes. The microarray approach is used for the comparative study of the already existing DNA sequence of the identified gene to the novel DNA sequence. The utilization of microarrays helps in the characterization of many clones of libraries. This format is mentioned as a metagenome microarray (MGA) (Singh et al. 2021). However, it has some limitations, viz. high hybridization efficiency and the presence of target gene from already known protein family that reduces the chances of forming new proteins (Bekele et al. 2021). The microarray is now also used for transformation and adaption for performing screening in biomass valorization (Singh et al. 2021).

3.3.3 Microtiter Plate Screening

The microtiter plate screening includes incubation in microwells where the bacterial culture and enzyme substrate are processed under incubation. Protein library screening using microtiter plate assays is a traditional and straightforward high-throughput approach. The desired protein and its properties are measured through microtiter plates using spectrophotometry or fluorometry. The visible signals like fluorescence are emitted, with the phenomenon of substrate conversion in microtiter plates. It is further used to identify enzyme expressing colonies (Bekele et al. 2021).

3.3.4 Fluorescence-Activated Cell Sorting

Fluorescence-activated cell sorting (FACS) is highly used for screening enzyme libraries because of its ability to analyse 108 mutants per day. FACS authorizes the identification of the biological activity enclosed by a single cell depending on size, shape, and fluorescence. FACS has many advantages (Neun et al. 2020):

1. Single events are deposited into different containers quickly and accurately.
2. Interference of cells during the assembly of laminar flow fluidics of FACS is prevented.
3. The small volume of each droplet results in minimal contamination.

FACS is able to couple a number of different screening methods due to its cell assembly capacities such as droplet classification and reporter-based screening. Recently, this system has assimilated lasers with multiple wavelength capabilities to screen thousands to billions of clones per second or per day, respectively (Bekele et al. 2021).

3.3.5 MICROFLUIDICS-BASED SCREENING

Microfluidics-based screening platform is combined with high throughput sequencing (HTS) technology, which gives dominance over other methods due to its suitability for cell-based assay, low cost of analysis, and simpler operating picolitre volumes of liquids. The rapid examination of various chemicals and biochemicals, genetic or pharmacological tests, etc., is done via this method. Out of thousands of microdroplets, a single droplet has function in the reaction chamber and these microdroplets are produced with the highest speed known. The reaction takes place later and is measured in picolitre volume of droplets, which include the cells, enzymes, substrates, and products. These droplets are distinguished depending on the fluorescence or colour of the product. This can be a limitation because detection is mostly bound to the fluorescence or colour of the product. However, detection methods such as mass spectrometry, nuclear magnetic resonance (NMR), and the colourimetric assay can be used in the future (Bekele et al. 2021). When microfluidics along with FACS works, it shows the highest screening of metagenomic libraries.

3.4 FUNCTIONAL METAGENOMIC LIBRARIES SCREENING

The identification of a particular gene depends on many factors such as the size of the target gene and gene expression availability of metagenome linked with one another, and the screening of such gene libraries is summarized further. Here, we present an overview of screening and show the technical issue that leads to the hit rate problem. One of the advanced issues is the partial expression of foreign genes into *E. coli*. Generally, the common surrogate host is *E. coli* for forming the libraries and so the gene deficiency occurs for recombination and restriction that help in cloning different or modified DNA into *E. coli* (Rizzo and Giudice 2020). Selecting the vector highly depends on the length of inserts, and for this, many types of strains are available as highly competent cells. For smaller fragments, say for example 10 kb, plasmids are used, and for larger fragments, cosmids or fosmids are used independently of the length of inserts, which shows that it depends on the size of clones. Among all the vectors, the plasmids have the strongest borne promoters and high copy number, although they do not improve the hit rate (the number of positives per total number of clones screened) significantly. For example, plasmid and fosmid vectors both clone the same enzyme from the same metagenome, but hit rates were quite similar between both procedures (Tansirichaiya et al. 2021) although the plasmid vector can be orientated twice for increasing the chances of transcribing the clone-inserted fragment which also succeeds the increase in the hit rate. The provenance of the metagenome has an impact on the hit rate. The target gene in the ecosystem is enriched through natural or artificial contamination (Bhatt et al. 2021). The enzymatic activities can be known through agar plates where positive clones are identified through

clear zone through visual screening. But when the metagenomic libraries are cultivated on plates, then the screening can be identified through colour. Enzyme activities are usually expanded with substrates. There can be a low hit rate for a common
reason of imperceptible signals, although the screening can be done without using a
special device. The other substitutional approaches such as cell lysates for screening,
where the library of cells is grown in 96-well plates having lysates formed by the
chemical or physical procedure (Kumar et al. 2020), help in improving the sensitivity. The survival of the host is linked to target activities for obtaining high throughput; this method is used for toxic compounds such as antibiotics or heavy metals
(Xing et al. 2020), which helps in the screening of resistance genes. Some hosts lack
the target gene, and for identifying and selecting the essential gene of such host, this
method can be helpful. And another approach beneficial is reporter assay. Green
fluorescent protein, β-galactosidase (Dhanjal et al. 2020), and tetracycline resistance
gene (Berglund et al. 2020) are sensitive reporter genes expressed when the biological event is attached to the reporter gene (Figure 3.1).

3.5 FUNCTIONAL METAGENOMICS: METHODOLOGICAL APPROACHES TO STUDY MICROBIAL WORLD

3.5.1 SEQUENCING DEPENDING ON STRATEGIES

The quantitative and qualitative species should have the purest form of metagenomic
DNA, which is isolated from the environmental samples, and helps in the initiation of metagenomic analyses (Bharti and Grimm 2021). The isolated metagenomic
DNA is sequenced with the help of tools of bioinformatics. The microbial ribosomal
RNAs (rRNAs) gene is sequenced through the break from culture-dependent to
culture-independent theories for the analysis of environmental samples. The bacterial 16S rRNA gene has a highly conserved primer-binding site within it. And
the identification of bacterial samples can be done through this species-specific signature sequence. This technology favours controlling the phylogenetic relationship
between unculturable bacteria and quantifies its microbial consistency. The function
of bacteria in a particular given environment can be obtained through 16S rRNA
sequencing (Chelliah et al. 2021). The database usually includes information of both
the culturable and unculturable bacteria, i.e. the cultured bacteria through functional
protein and the uncultured bacteria through function assigned through previous studies. The phylogenetic analysis identifies the function of particular species through
functional taxonomic category, and once the identification is done, the comparison
with previously mentioned families or closely related families is made. This process
is applied to certainly most of the various bacterial species across a sample, and
consequently, the community roles are predicted for the microbes lodged within the
sample without the necessity for shotgun sequencing. Out of the available databases,
the 16S rRNA sequencing data are used to characterize the metagenomic sample.
Phylogenetic Investigation of Communities by Reconstruction of Unobserved States
(PICRUSt) (Douglas et al. 2020) works for the prediction of functional properties
of microorganisms by taking an enormous amount of individual species. The 16S
rDNA metabarcoding and single genomics works in combination for the functional

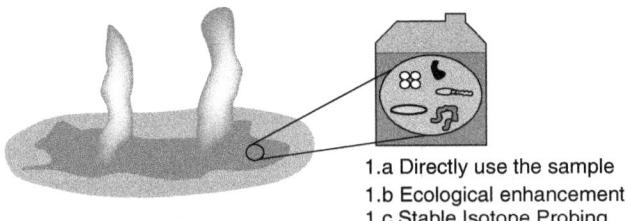

1.a Directly use the sample
1.b Ecological enhancement
1.c Stable Isotope Probing

1. Sample collection from natural or artifical environement (e.g., hot spring)

2. Cell lysis and metagenomic   Optional: Whole Genome Amplification with φ
DNA extraction   29 polymerase if not enough DNA is extracted

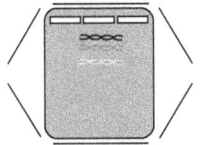

If extracted DNA is too big for the
vector, mechanical shearing should be
performed

3. Size selection of DNA fragments and clean-up

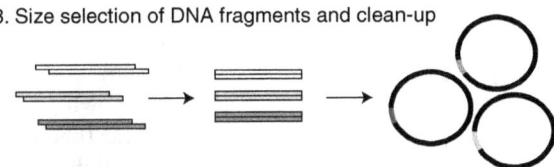

4. DNA fragments are blunt-ended and ligated to a suitable vector

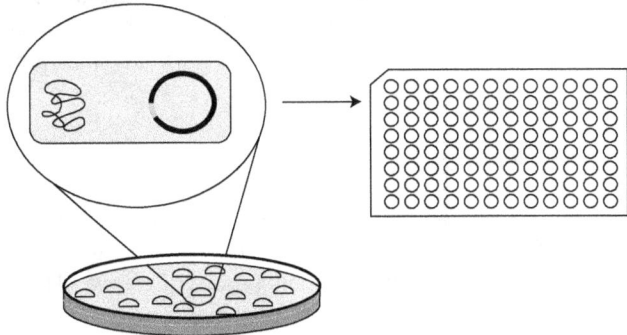

5. DNA is cloned in the selected host, individual colonies are picked and
transferred to individual wells in a microtiter plate

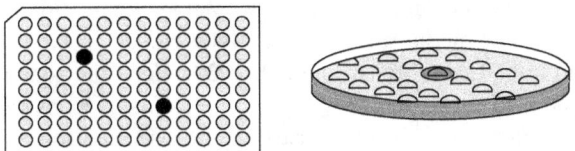

6. Functional screenings are performed at the temperature of the original
environment to maximize the number of hits. If the host is unable to
grow at that temperature a cell lysate may be employed.

FIGURE 3.1   Functional metagenomic libraries screening.

differentiation of microbial difference, and the eukaryotic communities are also studied through these strategies via eukaryotic primers homologous to the bacteria (Wang et al. 2020). The internal transcribed spacer (ITS), which is targeted as a universal DNA marker in fungi, has the noncoding DNA found between the small and large subunits of eukaryotic rRNA, which is targeted as a universal DNA marker in fungi. Next-generation sequencing helps in the amplification of nucleic acid (Kruppa et al. 2018), which is independent of couple sequencing that helps in detecting environmental virome within the epidemic area and diagnosis. Additionally, viruses have auxiliary metabolic genes (AMG) liked by metabolic cells (Martinez-Hernandez et al. 2019). The total metagenomic DNA is sequenced from the environmental DNA through shotgun sequencing where the functional potential of microbial population is known and no prediction is done considering 16S rRNA data. The hindgut microbiota is examined by photoreceptors such as rhodopsin, which were not described earlier, and studied through large-scale metagenomic studies involving shotgun sequencing. Perhaps, *Nasutitermes* species of wood-victualing termite show enormous microbial community and express novel genes involved in cellulose and xylan hydrolysis (Fan et al. 2021). Shotgun metagenomics sequencing also helps in analysing the functional capacity of the healthy human skin microbiome. The Basic Local Alignment Search Tool (BLAST) identifies the gene of interest via random sequencing. Thus, random sequencing can identify the homology of uncultured microorganisms and has the ability to identify the presence of already known genes; also, it provides advantages such as a new variant encoding the particular function indulged to an extreme environment. Furthermore, it is also helpful for studying population dynamics and genomic evolution and the past, present, and future of bacterial DNA in functional metagenomics (Arning and Wilson 2020). So the arrangements and necessity of functions throughout the community are known. Perhaps, it also carries unfavourable limitations or drawbacks.

3.5.2 FUNCTIONAL DEPENDED STRATEGIES

First, the environmental DNA sample is segregated and purified and, then, it is cloned into a suitable vector and inserted into the suitable host (mostly *E. coli*) and its examination is done through functional phenotype sequencing. The phenotype approach detects the expression of phenotypic characters through their metagenomic libraries. Screening helps in revealing the identical clones, i.e. with the same phenotype via screening of multiple clones. This is done through extracellular detection from the host cells having functional protein. Metagenomic clones are usually grown on specific indicator media such as haemolytic clones on blood agar (Islam and Sarma 2021) and lipolytic clones (Kaur et al. 2020). Here, the zone of inhibition reveals the antimicrobial agent produced by active clone; this happens because of the presence of indicator microorganism. Libraries can be screened too respecting the selection perspective, but the clones that have the activity of interest can be consulted through DNA inserts that can grow or survive within. The clone is the sole carbon source, so it carries the ability to metabolize the given subtract, to resist antimicrobial agents, or to grow in the presence of lethal concentrated heavy metals (Kumar et al. 2020). However, the SIGEX (substrate-induced gene expression) screening can be an alternative to recognize novel genes (Azeem et al. 2021). The green fluorescent protein (GFP) is encoded by a reporter gene that combines with environmental DNA. And SIGEX is combined with FACS for selecting GFP-expressing clone. The self-ligated plasmid clones are eliminated despite the restriction of applications of this method. SIGEX is efficient to identify the substrate-induced gene. The product-induced gene expression (PIGEX) is the developed porter system for the screening of metagenomic libraries for enzymatic function. The transcriptional activator sensitive to the product of the specified reaction is used for the system of GFP gene inserts. The clone when exposed to the appropriate substrate possesses the activity of interest whose product activates the transcription of regulator and GFP causing fluorescence that detects the positive clone (Tripathi and Nailwal 2020). One of the phenotype-based functional metagenomic approaches experiences various complications where potential resolutions are currently formulated to identify useful genes or proteins. Transcription, translation, protein folding, and secretion from the surrogate host should be achieved before the screening begins. The presence of genes within the metagenomic library is detected using the screening method itself because the probability of identifying metagenomic clones with desired activity is low (Ma et al. 2021). However, HTS can improve the probability of obtaining these clones by allowing maximum clones screened simultaneously. The obstacles remarked usually under proper circumstances can be overcome overlooking the clones. The methodological approach is a relating approach to express DNA fragments isolated from microorganisms. Due to the presence of DNA regulatory elements placed on vectors, the foreign DNA is transcribed and translated, but still, the chaperones required for protein collapse in the original species of *E. coli* may be absent (Boutin and Dalpke 2020). To overcome this limitation regarding host, the substitution of surrogate host expression can be more efficient to express environmental DNA. For instance, in an activity-based screening library, the *E. coli* host is used as a heterologous host and the metagenomic library is expressed in *Ralstonia metallidurans*. The result showed the activity of two clones, out of which one showed antimicrobial activities and another expressed carotenoid gene cluster, i.e. yellow-coloured clone. Depending on the host, the clones show different metabolic abilities; that is, clones show different metabolic abilities in *R. metallidurans* and *E. coli*. The hosts that cannot be expressed in standard *E. coli* can be identified by the heterologous host. And with this success, the study to compare Proteobacteria as a host was carried out for the same cosmid libraries (Piscotta et al. 2021). The host has the ability to express the libraries, and these libraries screen antimicrobial activities and altered colony morphology. The experimental host is chosen by selecting bacterial species through the dwelling of soil. The heterologous host with the least overlaps is selected for the recovery of active clones. The following study shows the important use of the broad-host-range vector for overcoming the barriers related to hosting expression (Islam and Sarma 2021). A novel proteinaceous antimicrobial agent is identified through activity-based screening active against *Bacillus cereus*. The DNA fragment of such activity was only active in *B. subtilis* and not in *E. coli*. This helps in knowing the multi-host expression system. Other cons of heterologous expression are that DNA fragments can be too short to carry the functional gene operon, so the vectors that denote large DNA inserts are essential (Taton et al. 2020). To overcome the methodological limitations,

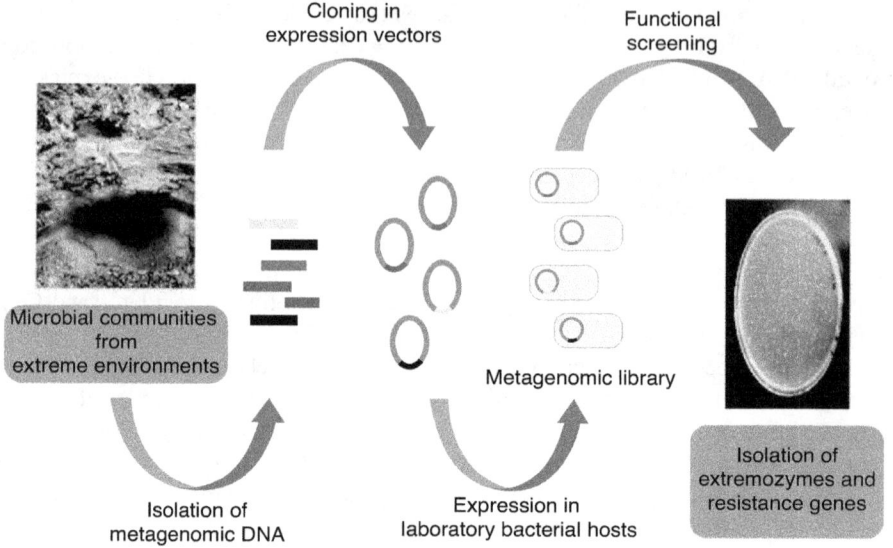

FIGURE 3.2   Diagrammatic representation of functional metagenomic strategies.

shuttle vectors that can corporate in more than one heterologous host and large insert vectors that accommodate gene operon are used (Figure 3.2).

3.6   APPLICATIONS OF FUNCTIONAL METAGENOMICS

Functional metagenomics usually involves multiple steps from the extraction of DNA to the screening and sequencing of libraries. For extraction, various physical, chemical, and biological activities are used and sequence-based screening or function-based screening as mentioned above is used. Hence, the overall identification of environmental samples is done through microbial and functional metagenomics. The environmental sample studies have disclosed the novel enzymes with the potential application of metagenomics. Starting from the pharmaceutical industry to the food industry, the application of enzymes plays a diverse role, which is discovered by metagenomic analysis. The soil habitats are a diverse microbial reservoir. These microbial reservoirs produce genes with novel enzymes such as amylases, β-glucosidases, lipases, oxidases, esterases, and reductases (Sethupathy et al. 2021). Out of them, amylases, the commercially employed industrial enzymes, have various set of enzymes that catalyse the breakdown of starch into short chains of glucose monomers. They are employed in the paper, food, and textile industries, in the production of detergents and sugar, in the biodegradation of alkanes, and also in the synthesis of gold nanoparticles (Hug et al. 2020). The β-glucosidases are another category of industrially used enzymes produced through metagenomics that cleaves glycosidic bonds of carbohydrates that yield glycosyl residues, glycosides, and oligosaccharides. They are used for ethanol production using fermentative bacteria and used in food industries to reduce the bitterness of citrus juices and enhance flavour

(Ariaeenejad et al. 2020). The lipases are another group of hydrolase enzymes that hydrolyse lipid. They are used in various food industries to hydrolyse fats and oils; in dairy industries to remove fats from milk products; in detergent industries to remove tough stains; in sewage systems to degrade lipid impurities; and sometimes in cosmetic industries as surfactants and aromatic compounds (Verma et al. 2021). Similarly, oxidases, esterases, and reductases are enzymes that show industrial applications, too. When the samples during the niche are metagenomically detected, the outbreaks of epidemics can be prevented and this proves to be another great application of metagenomics. The mortality over the past few years has increased, metagenomics profile the resistor of environments of different regions in the planet (Sukhum et al. 2019). The individual microbiome assisted by metagenomic analysis detects all pathogens present irrespective of time and place. The next-generation technologies and sequencing can be of high diagnostic potential. The metagenomics detect and diagnose viral pathogens, but due to the high rate of mutation, it becomes difficult to diagnose. The complication of viruses was more, and culturing them in the laboratory is impossible and hinders the diagnosis and monitoring. Perhaps, metagenomics has overcome all limitations as it considers the characterization of whole DNA. The early detection and monitoring of plant viral pathogens is now possible because of the agriculture metagenomics pipelines. Hence, the compiled study of metagenomic data discloses the profound details regarding the epidemiology of plant viral diseases. Also, metagenomics help in better understanding the viruses that are quite different from standard plant viruses. For instance, the herbivores and rodents are identified through metagenomic profiling of environments as vectors of plant pathogenic virus (Datta et al. 2020). Another application of metagenomics includes environmental monitoring and bioremediation using microbial metagenomics.

3.7   CONCLUSIONS AND FUTURE OBJECTIVES

The enormous diversity and differences have largely been known by metagenomic grants. Functional metagenomics signifies the microbial world and progress in research communities. The indulgence of metagenomics favours the understanding of and benefits from unculturable microbes and their novel enzymes and bioactive sources. As a comparatively new technology, functional metagenomics faces challenges that have yet to be overcome, but the promises have been fulfilled in the history of functional metagenomics. The appropriate solutions and further optimization regarding the commitments of functional metagenomic have been proven. The future of functional metagenomics stays in improving the screening and selection techniques altogether with faster and cheaper rates that will lead to higher expansion technologies in food or pharmaceutical industries. From the uncultured microorganisms, the novel products used at the industrial level can be isolated using functional metagenomics. The retained formulated products and industrially relevant products are authorised and known through identification of proteins and bioactive agents. More or less the identification of proteins and bioactive exaggerate the quantities to industrially relevant products and retain formulated final product and lastly sustain their firmness during shipping and storage.

ABBREVIATIONS

AMG	Auxiliary metabolic genes
BLAST	Basic Local Alignment Search Tool
DDE	DNA transposons
FACS	Fluorescence-activated cell sorting
GFP	Green fluorescent protein
ICE	Integrative and conjugative elements
IME	Integrative mobilizable elements
ITS	Internal transcribed spacer
MGA	Metagenome microarray
NMR	Nuclear magnetic resonance
oriT	Origin of transfer
PICRUSt	Phylogenetic Investigation of Communities by Reconstruction of Unobserved States
PIGEX	Product-induced gene expression
rRNAs	Ribosomal RNAs
SIGEX	Substrate-induced gene expression
SIP	Stable isotope probing

REFERENCES

Ariaeenejad, S., Nooshi-Nedamani, S., Rahban, M., Kavousi, K., Pirbalooti, A.G., Mirghaderi, S., Mohammadi, M., Mirzaei, M. and Salekdeh, G.H., 2020. A novel high glucose-tolerant β-Glucosidase: Targeted computational approach for metagenomic screening. *Frontiers in Bioengineering and Biotechnology*, 8, 813.

Arning, N. and Wilson, D.J., 2020. The past, present and future of ancient bacterial DNA. *Microbial Genomics*, 6(7), 3–13.

Awasthi, M.K., Ravindran, B., Sarsaiya, S., Chen, H., Wainaina, S., Singh, E., Liu, T., Kumar, S., Pandey, A., Singh, L. and Zhang, Z., 2020. Metagenomics for taxonomy profiling: Tools and approaches. *Bioengineered*, 11(1), 356–374.

Azeem, M., Soundari, P.G., Ali, A., Tahir, M.I., Imran, M., Bashir, S., Irfan, M., Li, G., Zhu, Y.G. and Zhang, Z., 2021. Soil metaphenomics: A step forward in metagenomics. *Archives of Agronomy and Soil Science*, 68(8), 1–19.

Baranowski, E., Dordet-Frisoni, E., Sagné, E., Hygonenq, M.C., Pretre, G., Claverol, S., Fernandez, L., Nouvel, L.X. and Citti, C., 2018. The integrative conjugative element (ICE) of mycoplasma agalactiae: Key elements involved in horizontal dissemination and influence of Coresident ICEs. *MBio*, 9(4), e00873–18.

Barnett, S.E. and Buckley, D.H., 2020. Simulating metagenomic stable isotope probing datasets with MetaSIPSim. *BMC Bioinformatics*, 21(1), 1–17.

Bekele, W., Zegeye, A., Simachew, A. and Assefa, G., 2021. Functional Metagenomics from the Rumen Environment—A Review. *Advances in Bioscience and Biotechnology*, 12(5), 125–141.

Berglund, F., Böhm, M.E., Martinsson, A., Ebmeyer, S., Österlund, T., Johnning, A., Larsson, D.J. and Kristiansson, E., 2020. Comprehensive screening of genomic and metagenomic data reveals a large diversity of tetracycline resistance genes. *Microbial Genomics*, 6(11), 5–6.

Bharti, R. and Grimm, D.G., 2021. Current challenges and best-practice protocols for microbiome analysis. *Briefings in Bioinformatics*, 22(1), 178–193.

Bhatt, P., Bhatt, K., Sharma, A., Zhang, W., Mishra, S. and Chen, S., 2021. Biotechnological basis of microbial consortia for the removal of pesticides from the environment. *Critical Reviews in Biotechnology*, *41*(3), 317–338.

Botelho, J. and Schulenburg, H., 2021. The role of integrative and conjugative elements in antibiotic resistance evolution. *Trends in Microbiology*, *29*(1), 8–18.7.

Boutin, S. and Dalpke, A.H., 2020. The microbiome: A reservoir to discover new antimicrobials agents. *Current Topics in Medicinal Chemistry*, *20*(14), 1291–1299.

Chelliah, R., Daliri, E.B.M., Elahi, F., Khan, I., Wei, S., Yeon, S.J., Saravanakumar, K., Madar, I.H., Miskeen, S., Sultan, G. and Arockianathan, M., 2021. Impact of sequencing and bioinformatics tools in food microbiology. In Devarajan Thangadurai, Leo M. L. Nollet, Saher Islam, and Jeyabalan Sangeetha (Eds.), *Sequencing Technologies in Microbial Food Safety and Quality,* 1st ed., 217. CRC Press: Boca Raton, USA.

Coluzzi, C., Guédon, G., Devignes, M.D., Ambroset, C., Loux, V., Lacroix, T., Payot, S. and Leblond-Bourget, N., 2017. A glimpse into the world of integrative and mobilizable elements in streptococci reveals an unexpected diversity and novel families of mobilization proteins. *Frontiers in Microbiology*, *8*, 443.

Cury, J., Oliveira, P.H., de la Cruz, F. and Rocha, E.P., 2018. Host range and genetic plasticity explain the coexistence of integrative and extrachromosomal mobile genetic elements. *Molecular Biology and Evolution*, *35*(9), 2230–2239.

Datta, S., Rajnish, K.N., Samuel, M.S., Pugazlendhi, A. and Selvarajan, E., 2020. Metagenomic applications in microbial diversity, bioremediation, pollution monitoring, enzyme and drug discovery. A review. *Environmental Chemistry Letters*, *18*(4), 1229–1241.

DeCastro, M.E., Rodríguez-Belmonte, E. and González-Siso, M.I., 2016. Metagenomics of thermophiles with a focus on discovery of novel thermozymes. *Frontiers in Microbiology*, *7*, 1521.

Dhanjal, D.S., Chopra, R.S. and Chopra, C., 2020. Metagenomics and enzymes: The novelty perspective. In Reena Singh Chopra, Chirag Chopra, and Neeta Raj Sharma (Eds.), *Metagenomics: Techniques, Applications, Challenges and Opportunities* (pp. 109–131). Springer, Singapore.

Douglas, G.M., Maffei, V.J., Zaneveld, J.R., Yurgel, S.N., Brown, J.R., Taylor, C.M., Huttenhower, C. and Langille, M.G., 2020. PICRUSt2 for prediction of metagenome functions. *Nature Biotechnology*, *38*(6), 685–688.

Fan, M.Z., Wang, W., Cheng, L., Chen, J., Fan, W. and Wang, M., 2021. Metagenomic discovery and characterization of multi-functional and monomodular processive endoglucanases as biocatalysts. *Applied Sciences*, *11*(11), 5150.

García-Moyano, A., Diaz, Y., Navarro, J., Almendral, D., Puntervoll, P., Ferrer, M. and Bjerga, G.E.K., 2021. Two-step functional screen on multiple proteinaceous substrates reveals temperature-robust proteases with a broad-substrate range. *Applied Microbiology and Biotechnology*, *105*(8), 3195–3209.

Grossart, H.P., Massana, R., McMahon, K.D. and Walsh, D.A., 2020. Linking metagenomics to aquatic microbial ecology and biogeochemical cycles. *Limnology and Oceanography*, *65*, S2–S20.

Hug, J.J., Krug, D. and Müller, R., 2020. Bacteria as genetically programmable producers of bioactive natural products. *Nature Reviews Chemistry*, *4*(4), 172–193.

Islam, N.F. and Sarma, H., 2021. Metagenomics approach for selection of biosurfactant producing bacteria from oil contaminated soil: An insight into its technology. In Hemen Sarma and Majeti Narasimha Vara Prasad (Eds.), *Biosurfactants for a Sustainable Future: Production and Applications in the Environment and Biomedicine,* 43–58.

Kaur, R., Kumar, R., Verma, S., Kumar, A., Rajesh, C. and Sharma, P.K., 2020. Structural and functional insights about unique extremophilic bacterial lipolytic enzyme from metagenome source. *International Journal of Biological Macromolecules*, *152*, 593–604.

Koonin, E.V., Makarova, K.S., Wolf, Y.I. and Krupovic, M., 2020. Evolutionary entanglement of mobile genetic elements and host defence systems: Guns for hire. *Nature Reviews Genetics*, *21*(2), 19–131.

Krüger, A., Schäfers, C., Busch, P. and Antranikian, G., 2020. Digitalization in microbiology–Paving the path to sustainable circular bioeconomy. *New Biotechnology*, *59*, 88–96.

Kruppa, J., Jo, W.K., van der Vries, E., Ludlow, M., Osterhaus, A., Baumgaertner, W. and Jung, K., 2018. Virus detection in high-throughput sequencing data without a reference genome of the host. *Infection, Genetics and Evolution*, *66*, 180–187.

Kumar, L., Satyam, R. and Bharadvaja, N., 2021. Metagenomics: A powerful lens viewing the microbial world. In Maulin Shah and Susana Rodriguez-Couto (Eds.), *Wastewater Treatment Reactors* (pp. 185–218). Elsevier.

Kumar, R., Kumar, D., Pandya, L., Pandit, P.R., Patel, Z., Bhairappanavar, S. and Das, J., 2020. Gene-targeted metagenomics approach for the degradation of organic pollutants. In *Emerging Technologies in Environmental Bioremediation* (pp. 257–273). Elsevier.

Kumar Awasthi, M., Ravindran, B., Sarsaiya, S., Chen, H., Wainaina, S., Singh, E., Liu, T., Kumar, S., Pandey, A., Singh, L. and Zhang, Z., 2020. Metagenomics for taxonomy profiling: Tools and approaches. *Bioengineered*, *11*(1), pp.356–374.

Laudadio, I., Fulci, V., Stronati, L. and Carissimi, C., 2019. Next-generation metagenomics: Methodological challenges and opportunities. *Omics: A Journal of Integrative Biology*, *23*(7), 327–333.

Libante, V., Sarica, N., Mohamad Ali, A., Gapp, C., Oussalah, A., Guédon, G., Leblond-Bourget, N. and Payot, S., 2020. Mobilization of IMEs integrated in the oriT of ICEs involves their own relaxase belonging to the Rep-trans family of proteins. *Genes*, *11*(9), 1004.

Ma, F., Guo, T., Zhang, Y., Bai, X., Li, C., Lu, Z., Deng, X., Li, D., Kurabayashi, K. and Yang, G.Y., 2021. An ultrahigh-throughput screening platform based on flow cytometric droplet sorting for mining novel enzymes from metagenomic libraries. *Environmental Microbiology*, *23*(2), 996–1008.

Martinez-Hernandez, F., Fornas, Ò., Gomez, M.L., Garcia-Heredia, I., Maestre-Carballa, L., López-Pérez, M., Haro-Moreno, J.M., Rodriguez-Valera, F. and Martinez-Garcia, M., 2019. Single-cell genomics uncover Pelagibacter as the putative host of the extremely abundant uncultured 37-F6 viral population in the ocean. *The ISME Journal*, *13*(1), 232–236.

Molitor, R., Bollinger, A., Kubicki, S., Loeschcke, A., Jaeger, K.E. and Thies, S., 2020. Agar plate-based screening methods for the identification of polyester hydrolysis by Pseudomonas species. *Microbial Biotechnology*, *13*(1), 274–284.

Neun, S., Zurek, P.J., Kaminski, T.S. and Hollfelder, F., 2020. Ultrahigh throughput screening for enzyme function in droplets. *Methods Enzymology*, *643*, 317–343.

Oliveira, P.H., Touchon, M., Cury, J. and Rocha, E.P., 2017. The chromosomal organization of horizontal gene transfer in bacteria. *Nature Communications*, *8*(1), 1–11.

Oyewusi, H.A., Abdul Wahab, R., Edbeib, M.F., Mohamad, M.A.N., Abdul Hamid, A.A., Kaya, Y. and Huyop, F., 2021. Functional profiling of bacterial communities in Lake Tuz using 16S rRNA gene sequences. *Biotechnology & Biotechnological Equipment*, *35*(1), 1–10.

Partridge, S.R., Kwong, S.M., Firth, N. and Jensen, S.O., 2018. Mobile genetic elements associated with antimicrobial resistance. *Clinical Microbiology Reviews*, *31*(4), e00088–e00081.

Piscotta, F.J., Whitfield, S.T., Nakashige, T.G., Estrela, A.B., Ali, T. and Brady, S.F., 2021. Multiplexed functional metagenomic analysis of the infant microbiome identifies effectors of NF-κB, autophagy, and cellular redox state. *Cell Reports*, *36*(12), 109746.

Rizzo, C. and Lo Giudice, A., 2020. The variety and inscrutability of polar environments as a resource of biotechnologically relevant molecules. *Microorganisms*, *8*(9), 1422.

Sarnaik, A., Liu, A., Nielsen, D. and Varman, A.M., 2020. High-throughput screening for efficient microbial biotechnology. *Current opinion in Biotechnology, 64*, 141–150.

Sethupathy, S., Morales, G.M., Li, Y., Wang, Y., Jiang, J., Sun, J. and Zhu, D., 2021. Harnessing microbial wealth for lignocellulose biomass valorization through secretomics: A review. *Biotechnology for Biofuels, 14*(1), 1–31.

Sharma, P., Tripathi, S. and Chandra, R., 2021. Metagenomic analysis for profiling of microbial communities and tolerance in metal-polluted pulp and paper industry wastewater. *Bioresource Technology, 324*, 124681.

Singh, J., Gupta, M., Singh, K.K., Kumar, A., Yadav, D., Wenjing, W. and Singh, P.K., 2021. Advancement in bioinformatics and microarray-based technologies for genome sequence analysis and its application in bioremediation of soil and water pollutants. In A. Kumar, V.K. Singh, P. Singh, V.K. Mishra (Eds.), *Microbe Mediated Remediation of Environmental Contaminants* (pp. 209–225). Woodhead Publishing.

Snipen, L., Angell, I.L., Rognes, T. and Rudi, K., 2021. Reduced metagenome sequencing for strain-resolution taxonomic profiles. *Microbiome, 9*(1), 1–19.

Sukhum, K.V., Diorio-Toth, L., Dantas, G., 2019, September. Genomic and metagenomic approaches for predictive surveillance of emerging pathogens and antibiotic resistance. *Clinical Pharmacology & Therapeutics, 106*(3), 512–524.

Tansirichaiya, S., Reynolds, L.J. and Roberts, A.P., 2021. Functional metagenomic screening for antimicrobial resistance in the oral microbiome. In *The Oral Microbiome* (Vol. 2327, pp. 31–50). Humana, New York, NY.

Taton, A., Ecker, A., Diaz, B., Moss, N.A., Anderson, B., Reher, R., Leão, T.F., Simkovsky, R., Dorrestein, P.C., Gerwick, L. and Gerwick, W.H., 2020. Heterologous expression of cryptomaldamide in a cyanobacterial host. *ACS Synthetic Biology, 9*(12), 3364–3376.

Tripathi, L.K. and Nailwal, T.K., 2020. Metagenomics: Applications of functional and structural approaches and meta-omics. In Surajit De Mandal and Pankaj Bhatt (Eds.), *Recent Advancements in Microbial Diversity* (pp. 471–505). Academic Press.

Verma, S., Meghwanshi, G.K. and Kumar, R., 2021. Current perspectives for microbial lipases from extremophiles and metagenomics. *Biochimie, 182*, 23–36.

Wang, S., Yan, Z., Wang, P., Zheng, X. and Fan, J., 2020. Comparative metagenomics reveals the microbial diversity and metabolic potentials in the sediments and surrounding seawaters of Qinhuangdao mariculture area. *PloS One, 15*(6), e0234128.

Xing, C., Chen, J., Zheng, X., Chen, L., Chen, M., Wang, L. and Li, X., 2020. Functional metagenomic exploration identifies novel prokaryotic copper resistance genes from the soil microbiome. *Metallomics, 12*(3), 387–395.

4 Analysis of Emerging Microbial Contaminants through Next-Generation Sequencing (NGS)

*Ayushman Gadnayak, Swayamprabha Sahoo,
Ananya Nayak, and Maheswata Sahoo*
Siksha 'O' Anusandhan (Deemed to be University)

Sushma Dave
Jodhpur Institute of Engineering and Technology

Jayashankar Das
Valnizen Healthcare

CONTENTS

4.1   INTRODUCTION

Next-generation sequencing (NGS) technology is associated with an advanced technology that has been responsible for sequencing the whole genome of an organism. This technology is responsible for formulating criteria for analyzing the genetic variation present within a DNA or RNA sequence. The process of NGS technology in sequencing thousands of smaller fragments of genes in a sample is related to the analysis of the contaminated particles. Microbial contamination is one of the potential elements found through the processing of NGS technology to sequence

DOI: 10.1201/9781003247883-4

the genetic material of a particular sample. The high-throughput NGS technology can efficiently analyze the microscopic contaminants of microorganisms from the sample. The study sheds light on the prospects of NGS on analyzing the emerging microbial contaminants through addressing the advantages.

4.2 NEXT-GENERATION SEQUENCING IN THE CLINICAL ASPECT

Next-generation sequencing or NGS technology involves high throughput, speed, and scalability for the advancement of the massively parallel sequencing process. In the clinical aspect, biological samples have been analyzed through the NGS technology followed by identifying genetic variations. The NGS technology is responsible for identifying the multiple or single mode of variation resulting within the DNA or RNA sequences related to the disease condition (Koitzsch et al. 2017). According to the clinical requirements, the NGS technology has provided the opportunity to identify the nucleotide sequences of the entire genome as well as the specific sequence of the gene. The prospect of analyzing the targeted sequence of the gene has formulated the clinical advantages of identifying the defects resulting in a genetic disorder (Brandhagen et al. 2020). The clinical laboratories have been provided with the opportunity to sequence the entire genome or a target region in rapid mode to address the complication.

Figure 4.1 depicts the fundamental steps involved in the NGS technology associated with the clinical pathology laboratories. In this case, the clinical laboratory is involved in the collection of samples from the targeted area, which further derived into DNA fragments. The NGS technology has provided the opportunity to sequence millions of base pairs of DNA or RNA sequences in a limited period of time. A prompt technology for DNA, RNA, and methylation sequencing within the considered clinical sample has been formulated with the NGS technology (Slatko, Gardner, and Ausubel 2018). In the case of a clinical sample, the NGS has developed a process to isolate the targeted DNA (50–20 bp) by cell lysis procedure. Furthermore, the fragmented DNA sequence has been cleaned from the cell lysed to promote the NGS procedure.

Figure 4.2 describes the individual steps related to the clinical sample preparation for the NGS technology to sequence the targeted genome fragment. Moreover, the NGS technology has provided evidence for analyzing the epigenetic factors related to the biological sample. For example, DNA methylation has been identified through the NGS technology, which can be responsible for several human diseases such as autoimmune diseases, cancer, Alzheimer's disease and Parkinson's disorder (Motro and Moran-Gilad 2017). According to the clinical prospect, disease-causing factors at the genome level have been identified and evaluated by the NGS technology based on the requirements of preventive measures. The development of pathological support and clinical measures has been derived through the genome level study of the targeted sequence. It is required to understand that the NGS technology has been involved in the development of clinical support by analyzing defects within the genetic material (Bacher et al. 2018). Meanwhile, it has been observed that the NGS technology within the clinical area provides evidence for disease prognosis and therapeutic decision development.

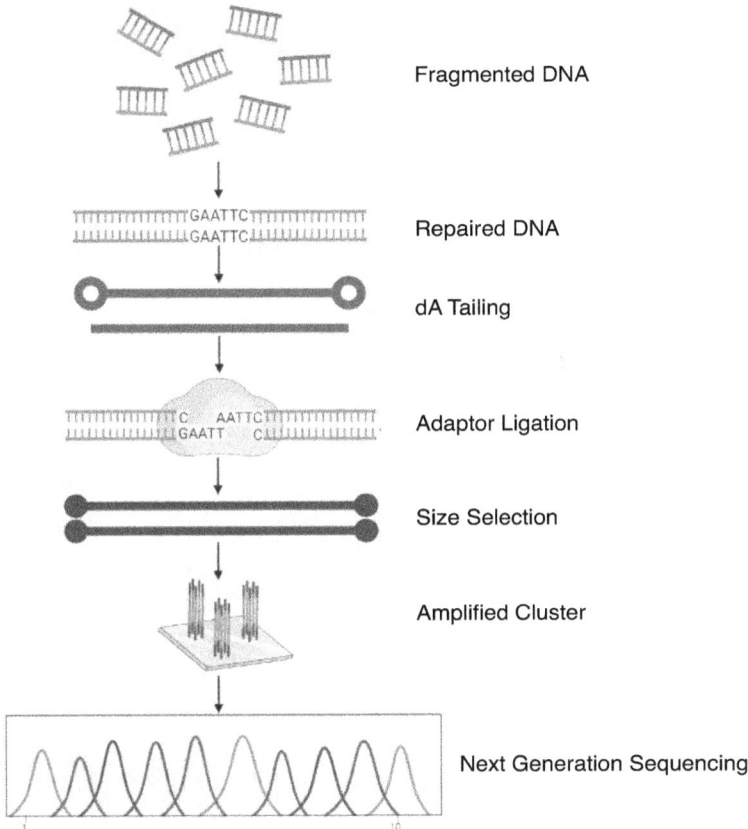

Fragmented DNA

Repaired DNA

dA Tailing

Adaptor Ligation

Size Selection

Amplified Cluster

Next Generation Sequencing

FIGURE 4.1 Basic steps involved in the next-generation sequencing or NGS technology.

4.3 WORKING PRINCIPLE OF NGS TECHNOLOGY

The NGS technology has been involved in the whole-genome sequencing including fragments of the targeted nucleotide sequence to identify the variation in the genome sequences. The working principle of the NGS technology has been associated with the derivation of effective steps such as *preparation, sequencing,* and *analysis.* Library preparation is one of the important and basic parts related to the progression of NGS technology. Primarily, the NGS technology has been involved in the isolation and purification of the genetic material or the targeted sequences from the sample (Yang et al. 2017; Awasthi et al. 2020). The working principle of the NGS technology has been associated with the effective progression of the specific steps related to sequencing (Kumar and Chandra 2020). The negative constituents such as inhibitors within the considered sample have been removed during the DNA extraction.

Figure 4.3 helps in the demonstration of the working principle and individual steps followed by the NGS technology. It is important to understand that the RNA sequencing process through the NGS has been involved in the conversion of RNA

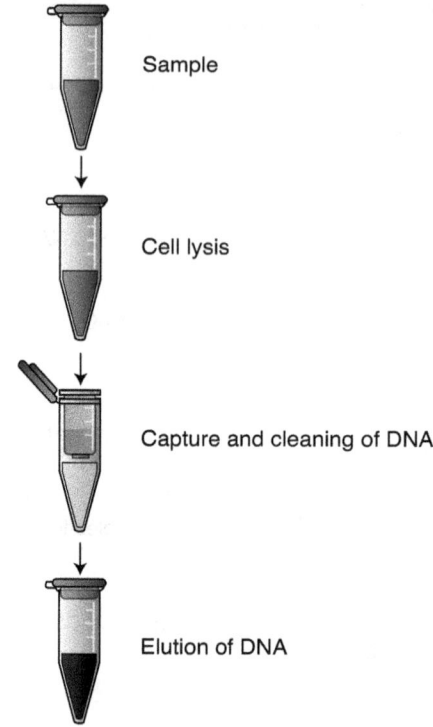

FIGURE 4.2   Sample preparation for clinical aspects involved in the NGS technology.

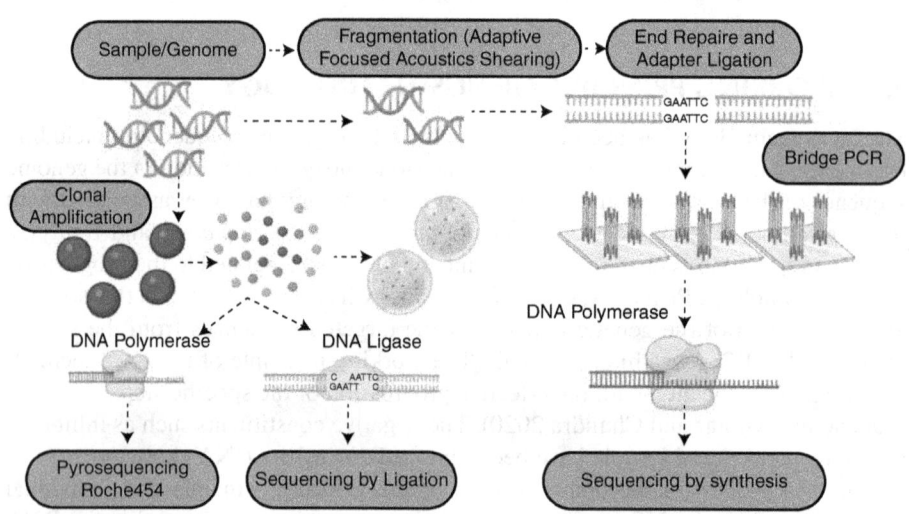

FIGURE 4.3   Working principle of the NGS technology and sequencing of genetic material.

TABLE 4.1

Advantages and Limitations in the Working Principle of Short- and Long-Read Sequencing of NGS

	Short-Read Sequencing	Long-Read Sequencing
Advantage	Short fragment sequencing through high fidelity	Sequence longer regions of genome, including homologous genomic region, repetitive sequences, and whole RNA transcript
Disadvantage	Unable to proceed with repetitive regions and structural variants including homologous regions of genome	Chances of error in scalability and base allocation, including lower read accuracy

into cDNA through reverse transcription technology. The extracted nucleic acid or the fragment of genetic material has been processed through a UV spectrophotometer for the quality assessment of the sample. Library preparation

for the NGS is a critical part of the overall procedure. It has been observed that library preparation is developed through the comparative genome related to the sequencing process. The library preparation step is associated with the development of the ligation procedure of adapters to the end of the DNA or RNA sequence of the sample. The adapters ligated to the end of the DNA or RNA sequence have been involved with the complementary sequences that promote the compatibility of the target sequence to the flow cell (Giannopoulou et al. 2019). Multiplexing or the pooling strategy of DNA or RNA sequencing has been observed through the NGS technology, which has promoted the ligation of a unique index sequence to the library. In the next step, sequencing of the sample is administered through the transferring onto a flow cell or sequencer followed by the cluster generation of fragments. The NGS process provides the opportunity for the short and long runs of the sequences (Table 4.1).

The NGS technology has been involved in the fluorescent tagging in the single-stranded DNA copies, which have been developed through the cluster generation process in sequencing. The targeted nucleotide has been tagged with fluorescent tags and reversible terminator to block the addition of other bases to the fragment of the genome. In the last step, data analysis has been associated with the identification of variation through the unique index and fluorescent tags within the sample (Figure 4.4).

4.4 NGS TECHNOLOGY FOR ANALYZING MICROBIAL CONTAMINANTS

Biological components are involved in the huge emergence of the microbial contaminants and microbial population development. The NGS technology is a potential system to address the presence of the microbial contaminants within the desired sample. Except for the clinical pathology, the NGS technology has emerged into the

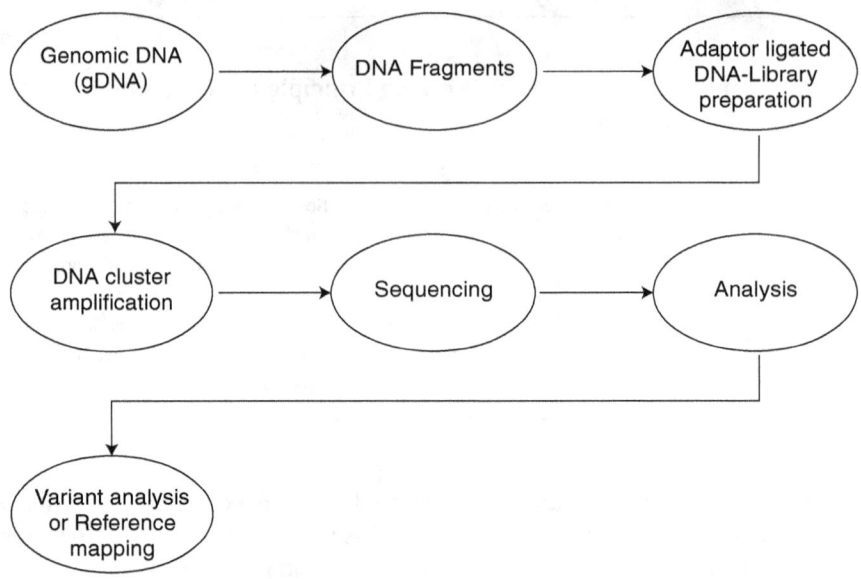

FIGURE 4.4 Working principle or the workflow in the NGS technology.

field of ecology, which helps to detect the presence of pathogens. The NGS is responsible for analyzing the microbial contamination through the processes of pathogen
dissemination and contaminant biodegradation of the environmental contaminants
involved in the emergence of microbes (Kumar et al. 2020, 2021). Ecological contexts of the microbial contamination have been involved with the NGS technology
and sequencing of the samples. Small subunit (SSU) rRNA hypervariable regions
have been identified through the derivation of the NGS technology to evaluate the
analysis of the microbial contamination (Salk et al. 2018). Bioindicators present in
the biological sample have been identified through the analysis of microbial processes that have been considered through the NGS technology. Genetic capabilities
related to the microbial contaminants have been determined through the derivation
of the microbial prospect of the NGS technology. The NGS is used in combination
with amplicon sequencing, meta-transcriptomics analysis, and metagenomics related
to microbial contamination. The population of microorganisms as well as the microbial communities within a sample can be detected through the NGS technology. The
application of the NGS technology related to the identification of microbial contamination has been responsible for deriving the biological risk developed through the
chemical and biological reactions developed by the microorganisms. The NGS helps
to identify the adverse effects of the microbial contamination, such as water quality
assessment. Waterborne microorganism and the risk of disease development can be
analyzed by checking the water quality through the NGS technology.

Figure 4.5 describes the microbial diversity analyzed through the specific steps
associated with the NGS technology. Metagenomics are related to the functional area
of microbial diversity and genetic material studies related to the microorganisms in

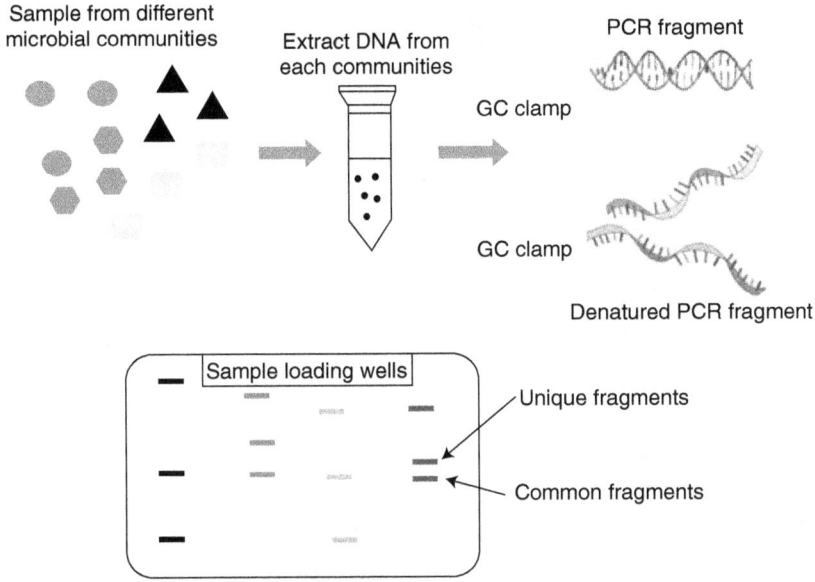

FIGURE 4.5 NGS technology used to analyze microbial community.

the environment (Kumar and Chandra 2020). The NGS technology has provided the opportunity to sequence the entire community or the population of microbes associated with the biological sample. Culture-dependent modalities are involved in the identification of chemical as well as biochemical reactions performed by the microbial population that has been formulated with the NGS technology. In some cases, microorganisms that carried out the biodegradation process have been analyzed based on the quality of the sample through the NGS technology and relative steps. According to the NGS technology, microorganisms can be detected based on the culture-based enumeration and standard cultivation-based methods such as the viable or non-culturable (VBNC) state of the organism (Greay et al. 2018). The standard method of the NGS technology has been involved in the quantification of the microbial population, including culture-based enumeration and VBNC technique to estimate microbial contamination.

Figure 4.6 indicates the involvement of the various processes of the NGS technology related to the identification of the microbial population or contamination related to a specific sample. Genetic material extraction from the considered sample has been the preliminary steps for the NGS technology to identify the presence of host pathogen related to disease development. Molecular analysis through the NGS technology has progressed through the 16S rRNA for bacteria and the 18S rRNA gene for Eukarya related to the small subunit ribosomal RNA (SSU rRNA) gene (Panek et al. 2018). The structural orientation followed by the variable or hypervariable regions is analyzed through the NGS through genetic material analysis for microbial contamination. In this condition, the NGS has employed PCR to amplify the target fragment of the genetic material from SSU rRNA and the functional gene

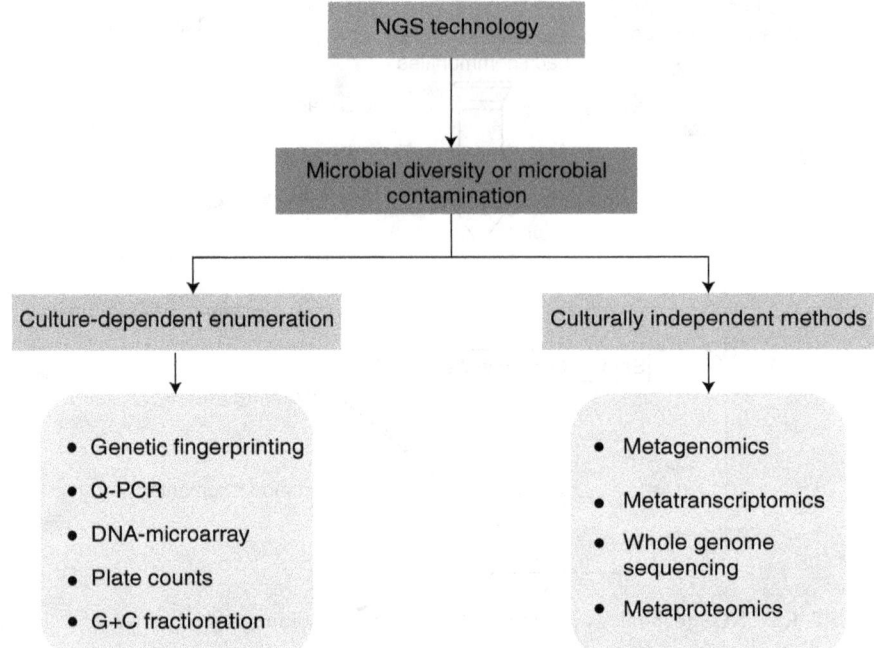

FIGURE 4.6   NGS technology associated with the microbial diversity analysis.

of the microorganism. The microorganism population based on the functionality and gene target has been detected through the molecular analysis strategy of the NGS technology (Degois et al. 2017). It has been found that the NGS technology provides opportunity for the direct detection of the microbial population, quantification of the target, and tracking of the functional activity of the microbial population.

4.5   ADVANTAGES OF ANALYZING MICROBIAL CONTAMINANT THROUGH NGS

Microbial contaminants are present in biological samples such as food and water, and the clinical samples are related to the development of disease. NGS technology application to evaluate and analyze the microbial contaminants are responsible for eliminating the risk of disease development (Schlaberg et al. 2017). Accumulation of whole-genome sequencing related to the NGS technology has been related to the identification of variants of pathogen of interest. Moreover, the NGS technology has a higher accuracy in identifying multiple communities involved in the biological sample. The advantageous aspect of the NGS technology has been the development of pathogen identification as well as the diagnosis process of microbial contamination. The advantages of microbial contamination detection have been associated with the development of quantification of the genetic material (Miller and Chiu

2018). Diversity of the microbial population has been followed by the derivation of genome sequencing through the NGS technology. In some cases, the NGS has provided the authority to the human-derived RNA sequencing to identify the prospects of microbial contamination identification (Brenner et al. 2018). Several environmental resources provided the condition for the contamination developed among the human population related to the identification through the NGS technology. In the case of clinical samples, the NGS technology has provided the authority to the sequence-based microbial signature to analyze the functionality of the pathogenic microorganism.

Table 4.2 addresses the advantages as well as the challenges related to the microbial contamination analysis through the NGS technology. The association of fusiform bacteria related to the development of pathological advantages has been found out through the NGS technology (Park and Nakai 2021). Meanwhile, it has been found that the large volume of the sample has been potentially analyzed through the fast turnover of the NGS technology. The entire genome coverage has been determined through the process of NGS related to the microbial population derivation. One of the important aspects of the diagnostic approaches has been carried out through the NGS technology through comparative analysis followed by data identification prospects. However, NGS has the potential to analyze 1000–100,000 microbial contamination reads by RNA-Seq among one million host reads (Park et al. 2019). The NGS technology has provided an opportunity to distinguish the intra- and inter-species sequences related to the microbial community or population (Figure 4.7).

TABLE 4.2

NGS Technology-Based Advantages Related to the Microbial Contamination Analysis

	Amplicon	Metagenome	Metatranscriptomics	Virome	Culturome
Advantage	Requirement of low biomass, quick analysis applicable for the sample contaminated by host DNA	Analysis of uncultured microbial genome, potential functionality, taxonomic level identification	Analysis of live microbial community through the identification of microbial activity	Potential to identify the various patterns of the genome such as DNA or RNA viruses	Potentially support the microbial isolation related to the development of high-throughput results
Limitation	False interpretation in low biomass and limited gene-level analysis	Host-level contamination, costly, and time-consuming	Contamination in host mRNA and rRNA, complications related to the sample collection and analysis	Difficulty in genome-level analysis and host-level contamination followed by cost-related issue	High risk of contamination by the environment and host, cost-effective

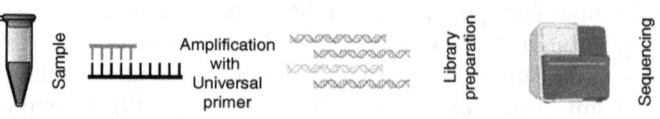

FIGURE 4.7   Advantages of NGS technology in high scalability of sample handling.

TABLE 4.3
NGS Technology Application in Various Fields Related to Microbial Contamination

NGS Application	Human Health	Public Health	Food Industry	Environment
Infectious pathogen	Pathogens related to the disease development are identified through NGS technology	Analysis of the microorganism outbreak within human microbiome	Contamination of microorganism related to food industry has been analyzed	Ecological contamination through pathogens causing disease has been identified
Risk factors	Individual risk of genetic variation through microbial contamination has been defined	Microbial community or population affecting public health has been analyzed through genetic-level analysis	Characteristics of the genome variants of microbial population have been identified through the NGS technology	Environmental sample contamination by biological sample has been identified through the NGS technology
Microbial population	Microbial population affecting human health can be diagnosed through NGS	Microbial community contaminated among the human population can be controlled by analyzing genetic pattern through the NGS technology	Food components contaminated by microbial population can be defined through the NGS technology to eliminate risk	Microbial population emerging within the environmental structure can be analyzed through the NGS technology

4.6   APPLICATIONS OF NGS IN MICROBIAL CONTAMINANT ANALYSIS RELATED TO HUMAN HEALTH

Human health has been supported by the prospect of analyzing microbial contaminants from the biological sample by NGS technology. The application of NGS technology has been associated with the derivation of the effective system of sequencing the entire genome within a short period of time (Hynes et al. 2017) (Table 4.3).

NGS technology has been responsible for promoting diagnostic tools for controlling various disorders. It has been observed that the NGS technology has provided an opportunity to identify the specific region or area that has been affected by microbial

TABLE 4.4

Steps and Outcomes Related to the 16S rRNA and Total Microbiome DNA Sequencing in NGS Technology Application

16S rRNA Sequencing	Total Microbiome DNA Sequencing
• PCR amplification	• Whole-genome sequencing of the microbiome
• Sequencing of the specific 16S rRNA segment	• Host DNA sequencing
• Specific oligonucleotides have been sequenced	• Comparative analysis of the microbial genome and reference genome
• Identification of the microbial species	• Identification of variations in the species of the microbial community

contamination. It is important to understand that the identification of the genetic variation is responsible for the derivation of preventive measures by controlling infection (Dulanto, Chiang, and Dekker 2020). Different strategies of sequencing in NGS technology, such as 16S rRNA and total microbiome DNA sequencing, are responsible for promoting health condition development through specifically analyzing microorganism communities (Table 4.4).

In various fields such as ecology, food industry, clinical pathology, and human health, the NGS technology has been developed efficiently through the analysis of the microbial population. Identification of population as well as species of the microbial community on a clinical sample provides an opportunity to enhance the clinical support and future treatment procedure (Ung et al. 2020). The development of microbial community can be controlled through the NGS technology-based analysis for genetic variations related to an organism or a system.

4.7 CONCLUSIONS

The study delivers the knowledge of the NGS technology in a high-throughput variation to analyze the genetic variation present within a biological sample. The NGS technology has the potential to identify the molecular level change or alterations within the genetic material. The overall working principle of the NGS technology has been responsible for formulating the DNA and RNA sequencing. Microbial contaminants within a biological sample can be effectively analyzed and evaluated through the NGS. The analysis of microbial contamination has been effectively imposing advanced technologies through NGS. The NGS technology has provided the opportunity to identify the presence of emerging microbial contamination, which is helpful to promote diagnostic processes. The aspect of NGS technology is effective to identify the area of the genetic characteristics of the contaminated microbe within a sample, which helps to identify the preventive measure. Human disease and diagnostic processes have been formulated through the execution of the NGS technology. The application of NGS technology has been responsible for developing advanced sequencing processes to study the complications related to the microbial contamination as well as genetic material.

REFERENCES

Alsglobal, 2019. *EnviroMail™ 127- Bacterial Diversity Profiling in NGS*. Available at: https://www.alsglobal.com/%2Fen-in%2Fnews%2Farticles%2F2019%2F09%2Fenviro mail-127-bacterial-diversity-profiling-in-ngs [Accessed on 15 September 2021].

Awasthi, M.K., Ravindran, B., Sarsaiya, S., Chen, H., Wainaina, S., Singh, E., Liu, T., Kumar, S., Pandey, A., Singh, L. and Zhang, Z., 2020. Metagenomics for taxonomy profiling: Tools and approaches. *Bioengineered*, *11*(1), 356–374. Doi: 10.1080/21655979.2020.1736238.

Bacher, U., Shumilov, E., Flach, J., Porret, N., Joncourt, R., Wiedemann, G., Fiedler, M., Novak, U., Amstutz, U. and Pabst, T., 2018. Challenges in the introduction of next-generation sequencing (NGS) for diagnostics of myeloid malignancies into clinical routine use. *Blood Cancer Journal*, *8*(11), 1–10.

Brandhagen, M.D., Just, R.S. and Irwin, J.A., 2020. Validation of NGS for mitochondrial DNA casework at the FBI Laboratory. *Forensic Science International: Genetics*, *44*, 102151.

Brenner, T., Decker, S.O., Grumaz, S., Stevens, P., Bruckner, T., Schmoch, T., Pletz, M.W., Bracht, H., Hofer, S., Marx, G. and Weigand, M.A., 2018. Next-generation sequencing diagnostics of bacteremia in sepsis (Next GeneSiS-Trial): Study protocol of a prospective, observational, noninterventional, multicenter, clinical trial. *Medicine*, *97*(6), e9868.

Degois, J., Clerc, F., Simon, X., Bontemps, C., Leblond, P. and Duquenne, P., 2017. First metagenomic survey of the microbial diversity in bioaerosols emitted in waste sorting plants. *Annals of Work Exposures and Health*, *61*(9), 1076–1086.

Dulanto Chiang, A. and Dekker, J.P., 2020. From the pipeline to the bedside: Advances and challenges in clinical metagenomics. *The Journal of Infectious Diseases*, *221*, S331–S340.

Giannopoulou, E., Katsila, T., Mitropoulou, C., Tsermpini, E.E. and Patrinos, G.P., 2019. Integrating next-generation sequencing in the clinical pharmacogenomics workflow. *Frontiers in Pharmacology*, *10*, 384.

Greay, T.L., Gofton, A.W., Paparini, A., Ryan, U.M., Oskam, C.L. and Irwin, P.J., 2018. Recent insights into the tick microbiome gained through next-generation sequencing. *Parasites & Vectors*, *11*(1), 1–14.

Hynes, S.O., Pang, B., James, J.A., Maxwell, P. and Salto-Tellez, M., 2017. Tissue-based next generation sequencing: Application in a universal healthcare system. *British Journal of Cancer*, *116*(5), 553–560.

Kumar, V. and Chandra, R., 2020. Metagenomics analysis of rhizospheric bacterial communities of *Saccharum arundinaceum* growing on organometallic sludge of sugarcane molasses-based distillery. *3 Biotech*, 10(7), 316. Doi: 10.1007/s13205-020-02310-5.

Kumar, V., Singh, K., Shah, M.P., Singh, A.K, Kumar, A. and Kumar, Y., 2021. Application of omics technologies for microbial community structure and function analysis in contaminated environment. In Shah, M.P., Sarkar, A. and Mandal, S., (Ed.), *Wastewater Treatment: Cutting Edge Molecular Tools, Techniques & Applied Aspects in Waste Water Treatment*. Elsevier. Doi: 10.1016/B978-0-12-821925-6.00013-7.

Kumar, V., Thakur, I.S., Singh, A.K. and Shah, M.P., 2020. Application of metagenomics in remediation of contaminated sites and environmental restoration. In: Shah, M., Rodriguez-Couto, S. and Sengor, S.S., (Eds.), *Emerging Technologies in Environmental Bioremediation*. Elsevier. Doi: 10.1016/B978-0-12-819860-5.00008-0.

Koitzsch, U., Heydt, C., Attig, H., Immerschitt, I., Merkelbach-Bruse, S., Fammartino, A., Büttner, R.H., Kong, Y. and Odenthal, M., 2017. Use of the GeneReader NGS system in a clinical pathology laboratory: A comparative study. *Journal of Clinical Pathology*, *70*(8), pp.725–728.

Miller, S. and Chiu, C., 2018. Metagenomic next-generation sequencing for pathogen detection and identification. In: Tang, Y-W. and Stratton, C.W. *Advanced Techniques in Diagnostic Microbiology* (pp. 617–632). Springer, Cham.

Motro, Y. and Moran-Gilad, J., 2017. Next-generation sequencing applications in clinical bacteriology. *Biomolecular Detection and Quantification, 14*, 1–6.

Panek, M., Paljetak, H.Č., Barešić, A., Perić, M., Matijašić, M., Lojkić, I., Bender, D.V., Krznarić, Ž. and Verbanac, D., 2018. Methodology challenges in studying human gut microbiota–effects of collection, storage, DNA extraction and next generation sequencing technologies. *Scientific Reports, 8*(1), 1–13.

Park, S.J. and Nakai, K., 2021. OpenContami: A web-based application for detecting microbial contaminants in next-generation sequencing data. *Bioinformatics,* 12;37(18), 3021–3022.

Park, S.J., Onizuka, S., Seki, M., Suzuki, Y., Iwata, T. and Nakai, K., 2019. A systematic sequencing-based approach for microbial contaminant detection and functional inference. *BMC Biology, 17*(1), 1–15.

Salk, J.J., Schmitt, M.W. and Loeb, L.A., 2018. Enhancing the accuracy of next-generation sequencing for detecting rare and subclonal mutations. *Nature Reviews Genetics, 19*(5), 269–285.

Schlaberg, R., Chiu, C.Y., Miller, S., Procop, G.W., Weinstock, G., Professional Practice Committee and Committee on Laboratory Practices of the American Society for Microbiology and Microbiology Resource Committee of the College of American Pathologists, 2017. Validation of metagenomic next-generation sequencing tests for universal pathogen detection. *Archives of Pathology and Laboratory Medicine, 141*(6), 776–786.

Slatko, B.E., Gardner, A.F. and Ausubel, F.M., 2018. Overview of next-generation sequencing technologies. *Current Protocols in Molecular Biology, 122*(1), e59.

Ung, L., Bispo, P.J., Doan, T., Van Gelder, R.N., Gilmore, M.S., Lietman, T., Margolis, T.P., Zegans, M.E., Lee, C.S. and Chodosh, J., 2020. Clinical metagenomics for infectious corneal ulcers: Rags to riches? *The Ocular Surface, 18*(1), 1–12.

Yang, X., Chu, Y., Zhang, R., Han, Y., Zhang, L., Fu, Y., Li, D., Peng, R., Li, D., Ding, J. and Li, Z., 2017. Technical validation of a next-generation sequencing assay for detecting clinically relevant levels of breast cancer–related single-nucleotide variants and copy number variants using simulated cell-free DNA. *The Journal of Molecular Diagnostics, 19*(4), 525–536.

5 Quorum Sensing
A Potential Mechanism Toward Microbial Activity and Biofilm Formation

Archisman Bhunia
University of Engineering & Management

Kumar Narayan
Jiwaji University

Abhilasha Singh, Asmeeta Sircar, and Nivedita Chatterjee
University of Engineering & Management

CONTENTS

DOI: 10.1201/9781003247883-5

5.1   INTRODUCTION

Since the 17th century, which marked the first appearance of microscopic organisms into the limelight of science, they have been under intense investigation for decades to characterize their diverging phenomenal behaviors. Years of microbial research has characterized these microorganisms to be freely suspending, planktonic cells that rely on the availability of essentials in their microenvironment. This dependent nature along with coordinated behavior enhances their adaptability for better proliferation. It was until 1978 when the advancement in microbial research introduced a breakthrough theory on biofilm process (Donlan and Costerton, 2002). The theory focuses upon the presence of bacteria within a matrix-enclosed, three-dimensional adhesive mesh, forming a sessile microcolony upon an external surface. It also states a dramatic change in the phenotypic behavior of the cells different from its planktonic counterparts. Although the theory came predominantly from the analytical evidence of biofilm being present mostly in the aquatic ecosystem, exceptions in the investigation were found in the deep groundwater and abyssal oceans (Costerton et al., 1995).

This theory of existence of the bacteria within the ecosystem in a consortium revealed the presence of inter- and/or intracellular communication within planktons. Their formation is a result of cascade of signaling in response to the density gradient. This cell-to-cell communication within the microenvironment is a result of the release of small clusters of signal molecules, i.e., autoinducers (AIs), by the microbes themselves, which respond in accordance with the changing density. This phenomenon further triggers a cascade of responses that regulate the expression of the operon, resulting in a behavioral change (Caicedo et al., 2017). Investigation reports over the past years have concluded quorum sensing to be primarily a population density-dependent signal transduction pathway triggering a network of operon responding to fluctuating microbial population density (Bassler, 1999; Miller and Bassler, 2001; Abisado et al., 2018; Mukherjee and Bassler, 2019).

These signaling molecules are derivatives of several chemical classes and are majorly categorized into (1) fatty acid derivatives and (2) short peptides and amino acids (Whitehead et al., 2001). These signaling molecules identified within the microbes are N-acyl homoserine lactones (AHLs) in gram-negative bacteria, post-transcriptionally modified autoinducing oligopeptides (AIPs) in gram-positive bacteria, and a class of 4,5-dihydroxy-2,3-pentonedione derivatives (called autoinducer-2, i.e., AI-2) in gram-negative and gram-positive bacteria (Li and Nair, 2012). Another type of autoinducer, AI-3, was identified from the isolates of enterohemorrhagic *Escherichia coli* (EHEC) O157:H7 spent media (Sperandio et al., 2003). However, its presence in gram-positive bacteria is not yet documented.

These inducer molecules upon interaction with the operon (R/I system) results in the expression of a diverse array of downstream cellular processes. Such a behavior includes the regulation in biofilm formation, attachment, virulence, motility, conjugation, bioluminescence, competence, and formation of resistance to antibiotics (Pena et al., 2019; Bramhachari et al., 2018; Popham and Stevens, 2006; Zhao et al., 2020). All this switching of the physiological activities, complying with the changing density, helps the microbe to survive and sustain in the stressed microenvironment against all challenges. Aside from the biofilm development and upregulation of activities, quorum sensing is also responsible for the dispersal of bacterial surface molecules (Rossmann et al., 2015). This dispersal of microbes in biofilms and/or planktons helps in eliminating unwanted or decayed cells, or often extending their microcolony into a new environmental niche.

Hence, the dramatic phenomenon of quorum sensing that undergoes in the microbial ecosystem can be simplified in a nutshell as follows:

i. In contrast to population density, the bacterial population releases signaling molecules called autoinducers (AIs).
ii. AIs are sensed by receptors present on the membrane or in the bacterial cytoplasm.
iii. Signaling response triggers the AI synthesis and gene expression, resulting in the cooperative behavioral outcome of the biofilm. This autoinduction loop apparently endorses synchrony within the population.

This chapter will explore various effects, mechanisms, and control measures on how quorum sensing and biofilm genesis interrelate within the microorganisms.

5.2 MECHANISM OF QUORUM SENSING

Biofilm formation has a unique significance of its own in diverse perspectives. But, the formation of this consortium occurs only upon signaling, preferably to say cell-to-cell communication, and is a density-dependent phenomenon (Miller and Bassler, 2001). This phenomenon is regulated by a cascade of signaling mechanisms involving primarily two types of signaling molecules: Acyl homoserine lactone (AHL)s in gram-negative bacteria; AIPs in gram-positive bacteria (Dong et al., 2007). Alongside these signaling molecules, the microbes also associate various signal-regulating protein systems, and a diverse sequence of operons in response to the external stimuli (Papenfort and Bassler, 2016; Liu et al., 2019). As the cascade of signaling is a density-dependent mechanism, it is also influenced by the surrounding physical environment of the microorganism. These factors such as pH, temperature, hydrodynamics, and available nutrient concentration in the microenvironment coming in contact with the surface receptors intensify the accumulation of inducers required to trigger the genes for the expression of various microbial characteristics (Asri et al., 2019; Sehar and Naz, 2016; Prabhu et al., 2019).

5.2.1 Quorum Sensing in Gram-Negative Bacteria

A completely different pathway implying the quorum sensing mechanism has evolved in gram-negative bacteria that in most cases comprise AHLs (also known as autoinducer-1, i.e., AI-1) (Prabhu et al., 2019), whose biosynthesis is regulated by the concentration of S-adenosylmethionine (SAM) within the system (Papenfort and Bassler, 2016). A general illustration of the quorum sensing mechanism with the gram-negative bacteria can be simplified as follows: At high cell density (HCD), the autoinducers synthesized by AHL synthase diffuse into the inner cell membrane and couple with various receptors forming an AHL-receptor complex, which regulates the transcriptional mechanism of gene expression associated with quorum sensing (Figure 5.1) (Subramani and Jayaprakashvel, 2019).

Other than AHL, acting as a signaling molecule, gram-negative bacteria also synthesize other signaling molecules. For example, 2-heptyl-3-hydroxy-4-quinolone is synthesized in *Pseudomonas aeruginosa*; butyrolactone in *Pseudomonas aureofaciens*; 3-hydroxy-palmitic acid methyl ester in *Ralstonia solanacearum*; and diketopiperazines (DKPs) in *Enterobacter agglomerans*, *Pseudomonas alcaligenes*, and *Citrobacter freundii* (Pena et al., 2019).

5.2.2 Quorum Sensing in Gram-Positive Bacteria

In contrast to gram-negative bacteria, gram-positive bacteria are commonly known to use segmented layers of signal transduction for cell-cell communication. This simple model of the signaling system comprises two components: histidine kinase

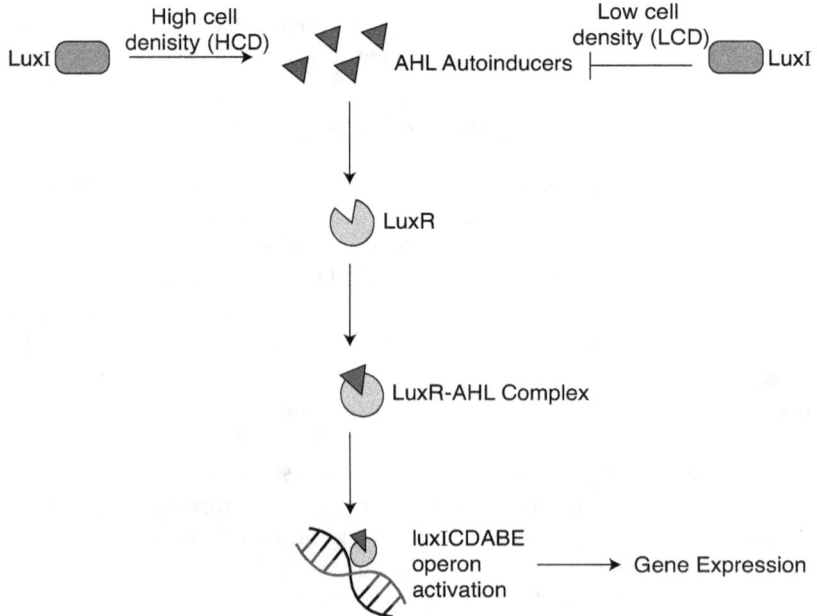

FIGURE 5.1   Quorum sensing circuit in gram-negative bacteria.

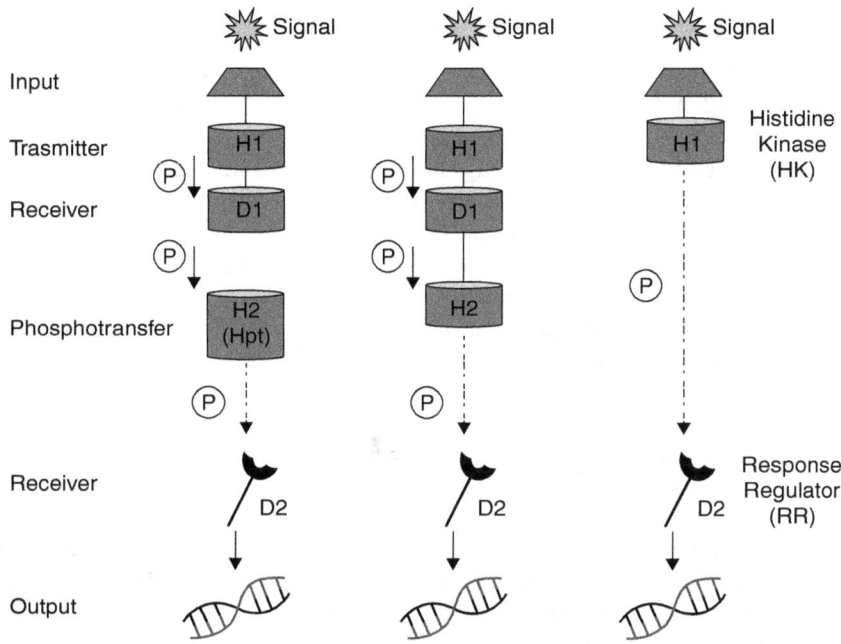

FIGURE 5.2 Quorum sensing circuit in gram-positive bacteria (hybrid, unorthodox, and classical).

(HK) and response regulator (RR), and is known as two-component signal transduction system (TCS) (Liu et al., 2018). Their report also highlights the types of two-/three-/multi-component signal transduction systems, such as classical, unorthodox, and hybrid (Figure 5.2). The mechanism works on the principle of phosphoryl group transfer domains. The HK receives signal molecules as a result of altered external stimuli, and autophosphorylation occurs within the conserved domains of HK along a sequence. At the same time, the cascade of signal triggers the transfer of phosphoryl group to the N-terminal of the aspartic residue in the RR (Liu et al., 2018). This further initiates the operon mechanism for gene expressions. Unlike the AHL signaling molecules in gram-negative bacteria, small sequences of post-translationally modified oligopeptides called autoinducing peptides (AIPs) act as the signal generators (Monnet and Gardan, 2015).

A well-studied model of *Staphylococcus aureus* shows a quorum sensing architecture encoded by accessory gene regulator (*agr*) locus (Sifri, 2008). This locus comprises transcripts such as RNAII and RNAIII. The RNAII encodes for the genes *agr*A, *agr*B, *agr*C, and *agr*D that are involved in the synthesis and sensation of the AIPs to initiate the quorum sensing circuit. RNAIII regulates the production of exoproteins that are secreted in response to the varying cell concentration outside (Sifri, 2008). This entire regulatory mechanism involving the AIP synthesis, as in the AHL-QS system of gram-negative bacteria, is directly influenced by the altering concentration of cells in the microenvironment, leading to the activation and/or inhibition of quorum sensing circuit.

5.2.3 Quorum Sensing in Cross-Species

Apart from the distinct quorum sensing inducers, another type of complex inducer is present in both gram-negative and gram-positive species, i.e., autoinducer-2 (AI-2) (Schauder et al, 2001; Xavier and Bassler, 2003). This AI-2, a furanosyl borate diester, is synthesized as a result of cleavage of the LuxS substrate S-adenosylmethionine and responds to cell density-dependent inter-species communication within the microbial niches (Subramani and Jayaprakashvel, 2019). A diverse inter-species cell-cell communication has been mentioned by Bassler (2002), which is regulated by AI-2. Some of these notable species mentioned are *Vibrio harveyi, Shigella flexneri, Escherichia coli, Clostridium perfringens, Salmonella typhimurium, Borrelia burgdorferi*, and *Actinobacillus actinomycetemcomitans*.

5.3 INFLUENCE OF QUORUM SENSING ON BIOFILM FORMATION AND SIGNAL TRANSDUCTION PATHWAY

Biofilm is a three-dimensional structure comprising of consortia of innumerable planktonic cell colonies coming in close vicinity and adhering to a solid surface. Their structural configuration is primarily influenced by the surrounding fluid dynamics, resulting in the formation of mushroom-shaped sculpture in stagnant water and filamentous structure in dynamic flow (Asri et al., 2019). Aside from the fluid dynamics influencing the structural configuration, the pH and the temperature of the surrounding microenvironment equally regulate the biofilm formation. Although studies have shown varying range of pH and temperature in this prospect, Sehar and Naz (2016) in their work concluded a pH of 7 and a temperature of 40°C to be optimum conditional values for biofilm formation in most of the species. Although the biofilm formation mechanism has been under controversies, many researchers have highlighted the factors of changing gene expression within the plankton colonies as a result of varying extracellular signaling due to the change in the ionic concentration, secondary metabolite formation and nutrients present, fluid forces, electrostatic interactions, and surface charge and topography (López et al., 2010; Sehar and Naz, 2016; Renner and Weibel, 2011).

5.3.1 Biofilm Formation

The consortia of planktonic cells forming biofilms by coming in close vicinity are a result of certain stages of interactions of various components such as extracellular polymeric substance (EPS), lipids, eDNA, enzymes, extracellular proteins, water channels, and debris of cell lysis (Figure 5.3) (Rabin et al., 2015).

Although conflicts are present on the theory involving the mechanism and environmental conditions inducing the biofilm formation, the factors primarily focused upon include nitrogen and oxygen levels, the availability of nutrients, pH, changing temperature, and the extent of desiccation (Vu et al., 2009). Abed et al. (2012) in their study highlighted the generalized mechanism (Figure 5.4) of the formation of the consortium structure in the microenvironment in a simpler way.

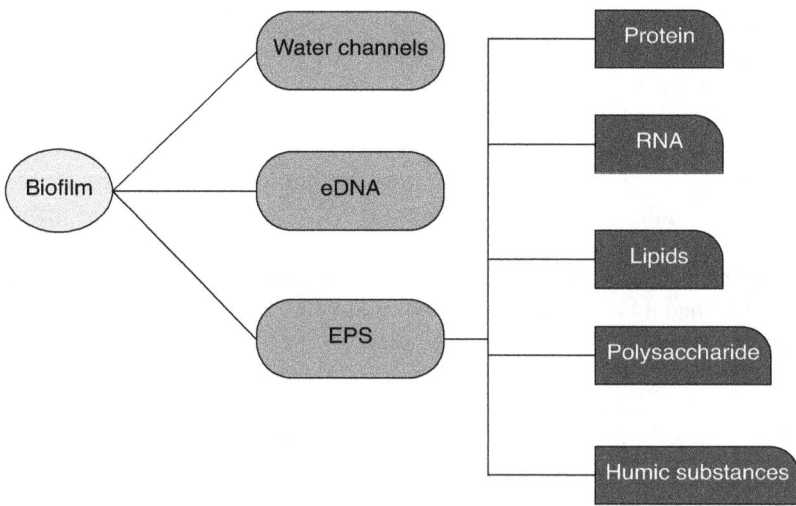

FIGURE 5.3 Components of biofilm.

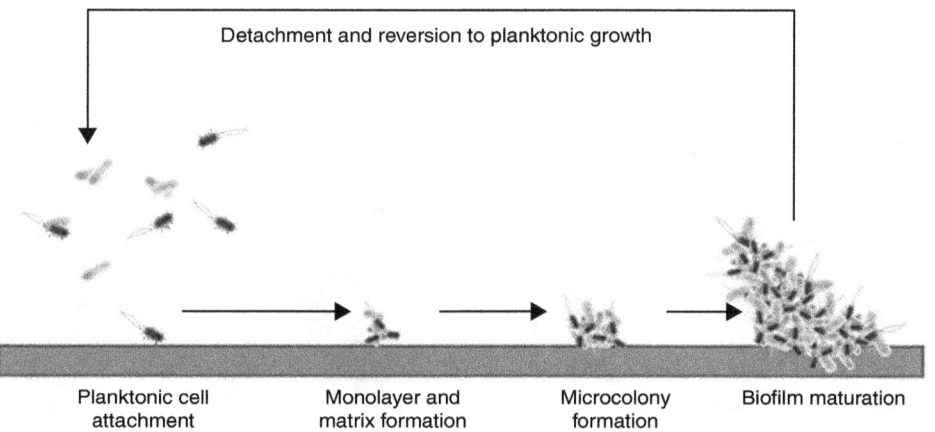

FIGURE 5.4 Mechanism of biofilm formation.

a. The planktonic and/or eukaryotic cells coming in close vicinity under the varying environmental influences of changing pH, chemical constituents, temperature, and the available nutrients regulate their motion in a directed fashion toward an irreversible attachment to a conditioned surface. The attachment is usually reversible in nature and depends upon the binding interactions between the surface and the cells.

b. Followed by the initial attachment occurs an alteration in the intracellular signaling that causes the synthesis of adhesive proteins such as curli or fimbriae, clustering the interacting cells and coating with gel-like matrix called extracellular polymeric substance (EPS). Although conflicts are there upon the mechanism of regulation of EPS synthesis, researchers concluded its purpose of formation as a protective covering and conservation of the three-dimensional structure of the biofilm. The irreversible adhesion also marks the beginning of the maturation and proliferation of the consortia.

c. With the biofilm colony being matured, the overall complexity of the consortia increases as the planktons begin to proliferate and interact with the immediate molecules. The maturation involves alteration in the gene expression and quorum sensing, optimizing the biofilm against challenging environmental conditions for their survival.

d. Following maturation, the interactions too weaken with time resulting in the detachment of the biofilm from the surface as well as among themselves. The process is further fueled by the varying fluid dynamics that costs with the disseverment and displacement of the planktonic cells from the biofilm.

Abed et al. (2012) also highlighted the visualization of the microbial attachment in *Streptococcus crista* CR$_3$, *Candida albicans*, and *Staphylococcus aureus* biofilms that are influenced by various physicochemical properties (temperature, pH, topography, etc.) with the help of scanning electron microscopy (SEM) and environmental SEM (ESEM).

Since the bacteria and other microbes within the consortium bear multidimensional variations, the reason behind the regulation of the scenario in biofilm formation has yet been in dilemma and not well understood. But researchers apprehend the intracellular signaling or the cell-cell communication, well known as quorum sensing (QS), to be a vital regulatory mechanism to synchronize the formation using extracellular QS molecules referred to as autoinducers (AIs) (Papernfort et al., 2017; Subramani and Jayaprakashvel, 2019).

5.3.2 SIGNAL TRANSDUCTION PATHWAY

The concepts of quorum sensing and signal transduction are interrelated and came into the light since 1970s from the density-dependent expression within microbes (Bassler, 1999). The change in the microenvironment results in the synthesis of auto-inducing molecules within the biofilm, hence forming a cascade of signal transduction that regulates the biofilm density formation by altering the principal regulators of EPS synthesis. This signal transduction is termed as TCS, which predominates the mode of response to external stimuli (Subramani and Jayaprakashvel, 2019; Liu et al., 2018). Such a regulatory mechanism is economical for the formation of biofilm and comprises receptor HK and RRs. Liu et al. (2018) portrayed the network of classification of the TCS on the basis of domains transferring the phosphoryl group. But the mechanism can simply be summarized as follows: The HK senses specific signals from external stimuli and autophosphorylates the conserved histidine within its domain. Eventually, a phosphate is also transferred to the N-terminal aspartic residue of the RR domain (Figure 5.5). This two-component signal transduction system is also responsible for various social behaviors of the biofilm colony.

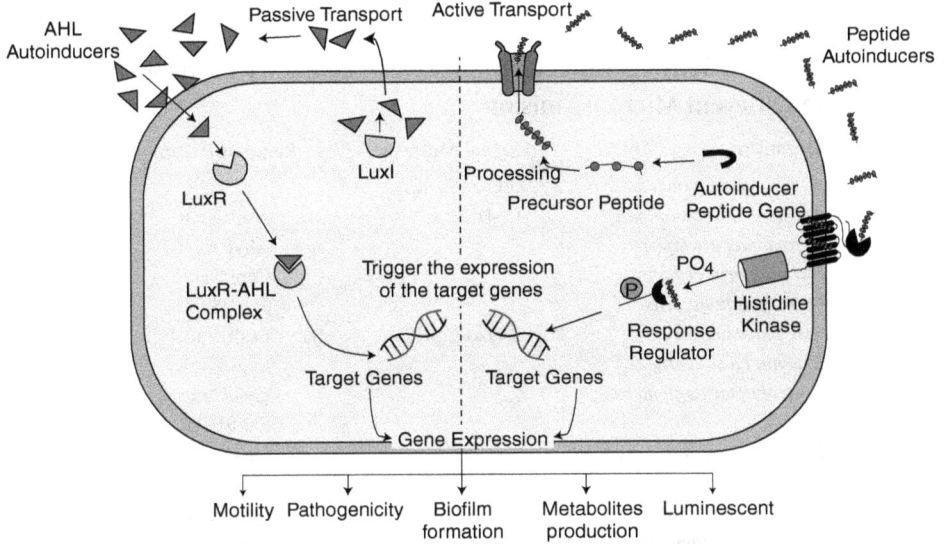

FIGURE 5.5 General signal transduction pathway (LHS: gram-negative; RHS: gram-positive).

This QS process involving the signal transduction pathways responds to two principle autoinducers (AIs): AI-1 and AI-2. The AI-1 is responsible for coordinating the intra-species communion within the biofilm, while the AI-2 regulates the inter-species interaction in the surrounding (Vu et al., 2009). Some other inducers that trigger the QS system are AHL, *Pseudomonas* quinolone signal (PQS), indole, diffusible signal factor (DSF), and hydroxyl-palmitic acid methyl ester (PAME) (Pena et al., 2019). Few other types of quorum sensing molecules are tabulated in Table 5.1 (de Kievit and Iglewski, 2000).

This cascade of regulatory protein activation as a response to the signal molecules, formed due to different external stimuli caused by the change in the density gradient, results in the alteration of behavioral expression within the biofilm community.

5.4 BACTERIAL SOCIAL BEHAVIOR AND QS IN SWITCHING THE EPS PRODUCTION

The notation of bacteria as social and multicellular organisms gained currency in the report of Lyon (2015). The planktons cluster to form the biofilm that is responsible for various behavioral interactions, both inter- and intracellular and with the surrounding environment. Studies reveal the well-studied species such as *E. coli*, *M. xanthus*, and *P. aeruginosa* have been subjected to cognitive parameters even in consortia supporting their survival against all odds. Lyon (2015) and Liu et al. (2018) added the point highlighting the two-component, three-component, and/or multi-component signal transduction system and the extracytoplasmic factors (ECFs) persisting within the bacteria (gram-negative and gram-positive), which regulate

TABLE 5.1

Quorum Sensing Molecules and Their Regulatory Proteins in Different Microorganisms

Organism	Signal Molecule	Regulatory Protein
Agrobacterium tumefaciens	3-Oxo-C_8-HSL	TraI/TraR
Aeromonas salmonicida	C_4-HSL	AsaI/AsaR
Aeromonas hydrophila		AhyI/AhyR
Burkholderia cepacia		CepI/CepR
Serratia liquefaciens		SwrI/SwrR
Erwinia stewartii	3-Oxo-C_6-HSL	EsaI/EsaR
Erwinia carotovora		ExpI/ExpR
Enterobacter agglomerans		CarI/CarR
		EagI/EagR
Pseudomonas aeruginosa	3-Oxo-C_{12}-HSL	LasI/LasR
	C_4-HSL	RhlI/RhlR
Pseudomonas aureofaciens	C_6-HSL	PhzI/PhzR
Rhizobium leguminosarum		RhiI/RhiR
Yersinia enterocolitica		YenI/YenR
Chromobacterium violaceum		CviI/CviR
Rhodobacter sphaeroides	7-cis-C_{14}-HSL	CerI/CerR
Vibrio anguillarum	3-Oxo-C_{10}-HSL	VanI/VanR
Vibrio fischeri	3-Oxo-C_6-HSL	Luxi/LuxR
Vibrio harveyi	3-Hydroxy-C_4-HSL	LuxLM/LuxN
Yersinia pseudotuberculosis	C_8-HSL	YesI/YesR

their genetic transcription behavior in all aspects. Other than regulatory systems, three types of "network architecture," i.e., serial, parallel, and antagonistic, have also been recognized for the operationality of the QS systems (Waters and Bassler, 2005).

By contrast, these regulatory systems are first triggered by AIs that are produced in response to the signals generated due to external interaction. The significance of this mechanism is from the standpoint of its twofold mode of action. First, the signal complexion within the cells increases due to the rise in the concentration of the interacting molecules. Second, the interaction results in the synthesis of inducers that act as proxies initiating the transduction mechanism to regulate the behavior in response (Smith, 2000). This diverse alteration in the behavioral response is observed often in the form of biofilm density regulation for survival against environmental challenges; anchoring of the planktons; motility; inter- and intra-communication and mutual cooperation; bioluminescence; signaling complexity; altered metabolism; alteration in the degree of virulence and cytotoxicity; competence upon interactions; resistance to antibiotics and organic compounds expressing toxicity; and even the detachment, dispersion, and controlled death (apoptosis) of the planktonic cells (Lyon, 2015; Abisado et al., 2018; de Kievit and Iglewski, 2000; Pena et al., 2019; Rutherford and Bassler, 2012).

The diverse expression of behavior by the sub-colonies of the biofilm revealed an interesting insight into the presence of motile cells, sporulation cells, and matrix

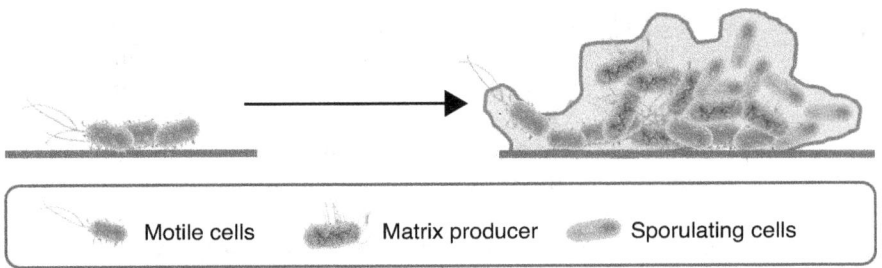

Motile cells   Matrix producer   Sporulating cells

FIGURE 5.6 Type of cells present in the biofilm.

producers (Figure 5.6) that switch the mechanism of cell density formation. The anatomy shows the presence of motile cells at the bottom, facilitating anchorage, the matrix cells lie all over regulating the EPS synthesis, and the sporulating cells lying at the outer surface for dispersion (Romero, 2013).

These cooperative behaviors of the microbial colony are also influenced by the cell density due to its degree of EPS formation. However, the QS system of regulation diverges in different biofilm niches making the definition of QS controversial. For example, *P. aeruginosa* involves LasI/LasR and RhlI/RhlR QS systems for enhancing the degree of end virulence and biofilm formation; LsrR/LsrK system in *E. coli* improves the biofilm architecture; AfeI/AfeR is a QS system in *Acidithiobacillus ferrooxidans* that regulates the AHL level in low-phosphate medium, hence regulating the formation of EPS and biofilm attachment (Vu et al., 2009). Aside from these, exposure to large surface area also enhances the AIs to trigger the genes regulating the matrix-forming cells for EPS synthesis and fimbriae formations to arrest planktons and enhance biofilm density modification till the permissible threshold signal accumulation, before the silencing genes are activated for the same response.

5.5 QUORUM SENSING IN MICROBIAL COLONY-WIDE FUNCTIONS

Quorum sensing, or the cell-to-cell interaction that allows the bacterial species to transfer information of varying cell density and adapt to the organic phenomenon, regulates a broad spectrum of bacterial colony-wide functions accordingly. These functions involve a chain of interrelated mechanisms and are energetically expensive to the microbial community and bring about an impact on its associated environments. Despite a contrast in the regulatory components and molecular mechanisms, the quorum sensing system works on three basic principles (Rutherford and Bassler, 2012) and are as follows:

a. In the first place, the members help to transfer AIs, known as the signaling molecules that are synthesized as a result of altered concentration of the surrounding. At low cell density (LCD), AIs diffuse away, but are present at concentrations below the threshold required for detection. Similarly, at HCD, the increasing production of AIs results in a high concentration gradient resulting in easy detection of AIs.

b. These signaling molecules or AIs are detected by various receptors that are present within the cytoplasm or on the outer surface of the bacterial membrane.
c. Additionally, along with activating the genes for necessary expressions for cooperative behaviors, AIs also enable the activation of AI-producing genes within the cytoplasm. This feed-forward autoinduction loop gradually promotes synchrony within the population.

Alongside the involvement of autoinducer AHL in gram-negative bacteria, the gram-positive bacteria involve segmented pathway framework for quorum sensing. The segment/component pathway involves AIPs that are synthesized ribosomally as initial pro-peptide structures and are later post-translationally modified to functional AIP molecules. Two key elements: HK at the membrane and the intracellular response regulatory (RR) receptors, are involved in this component pathway. Thus, it is commonly recognized as the two-component signal transduction pathway. It was initially identified in *Lactococcus lactis* and *Streptococcus pneumoniae*, and later in different strains of gram-positive bacteria. This pathway regulation is mediated through phosphorylation, which in turn leads to signal amalgamation in kinases that are controlled by environmental cues (Liu et al., 2018). Another pathway involved is a self-signaling pathway where the ribosomally synthesized and post-translationally modified AIPs are synthesized by SecA-dependent systems and activated under favorable parameters of modification. The key difference with the two-component pathway is that the AIPs, on reaching the threshold density gradient, are influxed inside the cell cytoplasm via an oligopeptide transporter system (Rutherford and Bassler, 2012). This regulation within the cascade of gene expression plays a crucial role in numerous colony-wide functions such as bioluminescence, conjugation, competence, sporulation, virulence, and biofilm formation.

5.5.1 Sporulation

Sporulation is a natural phenomenon observed in microbial colonies when their environmental circumstances demeaned and the consortium is deprived of supplements. As observed earlier in *Bacillus subtilis* microcolony, a skewed cell division occurs ensuring the formation of circumstantially unsusceptible spores (Miller and Bassler, 2001). The sporulation sensor kinase, KinA, is one of the primary kinases in *B. subtilis*, which responds to the nutritional shortage by phosphorylation of the primary regulator Spo0A~P. This in turn induces the cascade mechanism resulting in sporulation. KinA mainly comprises three well-defined PAS, namely PAS-A, PAS-B, and PAS-C, collectively recognized as PAC sensor domains. PAS domains help to oversee the intensity of light exposure, redox potential, oxygen concentration, small ligands, and their intracellular energy levels. Thus, KinA with the assistance of those PAS sensor domains senses the undulating concentration of protoplasmic agglomerates, similar to the key metabolites that induce sporulation (López et al., 2010).

Sporulation occurs when the cells' concentration gradient is low, even when the *B. subtilis* cells starved. Regulating the mechanism of sporulation at a high concentration requires several extracellular or environmental stimuli that will further coordinate the quorum sensing mechanism in various aspects. Sporulation process is irreversible, and the peptides regulating the mechanism of sporulation are ComX and CSF (competence and sporulation factor). The later peptide in *B. subtilis*, also known as CSF, is a four-peptide bonded structure and is called pentapeptide. The C-terminus of the precursor peptide, also known as PhrC, comprises five amino acids that are cleaved to form CSF signal molecules. It assimilated extracellularly as a function of increased concentration gradient. However, the role of CSF, as a signaling molecule, is intracellular in nature (Miller and Bassler, 2001).

5.5.2 BIOLUMINESCENCE

The regulatory gene expression in contrast to the changing density gradient was first observed in the form of luminescence in marine bacterium *Vibrio fischeri* in the early 1970s. Since then, it has been accepted as a model subject for exploring gene regulatory circuits in gram-negative Proteobacteria. In case of *V. fischeri*, quorum sensing is known to control the property of bioluminescence or the ability of the bacteria to produce light (Popham and Stevens, 2005). The two important genes involved in this phenomenon of regulatory stratagem are the following: *lux*I, which is responsible for encoding an AI synthase LuxI, and *lux*R, which encodes an AI-dependent activator gene LuxR; both are essential for luminescence. The LuxI synthesizes autoinducer molecules called N-acyl homoserine lactone (AHL) (3-oxo-hexanoyl-homoserine lactone). This AI accumulation on outstretching beyond a critical threshold concentration initiates the formation of LuxR-AHL complexes, resulting in the activation of the lux operon within the bacterial cytoplasm (Subramani and Jayaprakashvel, 2019).

This phenomenon of bioluminescence by host organism is used to attract prey or to avoid predation. An essential component in the expression of bacterial bioluminescence is molecular oxygen, without which luminous bacteria cannot emit light. The availability of molecular oxygen at various concentrations surrounding the external cellular environment is the key stimuli to initiate the gene expression. While luminescent bacterial species are mostly found in deep marine environments forming various microcolonies to accommodate their existence under challenging habitats, a few of them are also found in saline water, freshwater, and terrestrial environments (Popham and Stevens, 2005). Taxonomically stating, luminous bacteria are sub-species of the six genera in three Gammaproteobacteria families known as Vibrionaceae, Enterobacteriaceae, and Shewanellaceae.

5.5.3 VIRULENCE

Virulence is in general defined as the degree of toxicity or the injury-inducing capacity of a microorganism. Virulence and pathogenicity are often used interchangeably. The virulence of bacteria describes their capability of producing an exotoxin and/or endotoxin (Zipfel et al., 2006). *Staphylococcus aureus* is known to uses the Agr quorum sensing circuit to induce inflammation as an effort to enhance the degree of nutritional uptake, causing a manifestation of food poisoning. Fengycin, a cyclic lipopeptide produced from *Bacillus* sp., possesses an amphipathic structure and is known to exhibit antifungal properties. This chemical often helps to disrupt the virulence caused by *S. aureus* by striving for *S. aureus* autoinducer-binding sites, alleviating bacterial consortium formation and thereby poisoning. The autoinducing peptide, also known as AIP, is a precursor manufactured from a compound named *agr*D. AgrB, another AIP transporter, synthesizes the precursor to the developed or fully matured AIP and systematically transports it outside the cell membrane. AIPs can be recognized by a two-component signal transduction pathway, which involves a membrane-bound histidine kinase AgrC and a response regulator AgrA. On phosphorylation, the stimuli from AgrA activate the P2 and P3 promoters, which encode the *agr* operon (called RNAII) and the RNAIII regulatory RNA, respectively. The RNAIII post-transcriptionally triggers the virulence factor synthesis and subdues the expression of repressor of toxins (Rot), thus leading to further activation of virulence factors along the cascade (Rutherford and Bassler, 2012).

P. aeruginosa is another such bacterial species superintended for critical nosocomial septicemia, even potentially fatal infections in an immuno-compromised person, and long-standing infections in patients affected by cystic fibrosis. Its infection widely relies on the host possessing a weakened immune response. The primary virulence factors of *P. aeruginosa* include elastase, phospholipase C, protease A, exotoxins and cytotoxins, flagella, pili, pigment production, and quorum sensing regulatory system proteins that contribute to its infection-causing capability (Zipfel et al., 2006). *P. aeruginosa* mainly harbors three QS systems: two LuxI/LuxR-type QS circuits that function in cascade controlling the articulation of virulence factors. Furthermore, the third system is a non-LuxI/LuxR-type system known as the PQS system. These mechanisms are necessary for the survival and proliferation of these microorganisms in the host system.

5.5.4 COMPETENCE

Back in 1928, *Streptococcus pneumoniae* was the first of its kind bacterial species where the phenomenon of genetic transformation was identified. This phenomenon is exhibited only in the cells potential to collect the exogenous DNA from its adjoining microenvironment. *S. pneumoniae* is a bacterium found to achieve its zenith of competence impulsively. This accession to prowess relies on several complicated physiological actions, of which many are regulated by the gene cascades called quorum sensing system (Ziemichód and Skotarczak, 2017). In *S. pneumoniae*, these signaling molecules comprise oligopeptides, known as competence stimulating peptide (CSP). These oligopeptides are made of 17 amino acids and are derived from the precursor peptide, known as ComC. The oligopeptide CSP is diffused outside the cell membrane via transporter protein, ATP-binding cassette (ABC). ComD kinase is a sensor protein that detects the extracellularly cumulated autoinducer CSP. At an elevated concentration of CSP in the surrounding, ComD undergoes autophosphorylation by the mode of phosphorylation and dephosphorylation pathway. It channelizes the signal to the final acceptor, ComE. Simultaneous phosphorylation of ComE protein initiates the transcription of the *com*X gene encoding an alternative sigma factor ComX, which is responsible for the regulation of genes involved in the expression of competence (Ziemichód and Skotarczak, 2017). This system also facilitates the assemblage of exogenous DNA in *S. pneumoniae*. Com system has also been observed in *B. subtilis*, where the receptor in charge of regulation forms homodimers to cohere with inverted repeats of DNA motif. This also helps the *B. subtilis* cells in precisely responding to external factors and appropriate adaptation to changing environmental conditions.

5.6 EFFECTS OF QUORUM SENSING ON MICROENVIRONMENT

The colossal communication within the microbial community has engaged a cooperative approach to survive in antagonistic environments (Kumar et al., 2020, 2021). A bacterium residing within a biofilm generally exhibits resistance to antibiotics and biocides, which also shelters them from environmentally stressed circumstances and invasions by the host immune response. Moreover, the cells lying at close proximity aid in the horizontal gene transfer and exchange of metabolic by-products inside the biofilm. QS promotes the growth of related strains of bacteria referred to as autoinduction. It simultaneously inhibits the proliferation and maturation of other bacterial/ fungal strains, and other organisms competing for the same ecological niche. QS gene end products regulate a diverse range of transcriptional mechanisms in bacterium, both *in vitro* and *in vivo*. QS systems in organisms contribute to growth potential, biofilm formation, sporulation, antibiotic resistance expression, virulence expression, autolysis, oxidative stress tolerance, metabolic activity, motility, DNA transfer, etc. (Jiang et al., 2019). Biofilms are an omnipresent form of bacterial community. Their formation is an essential requirement adopted for growth, proliferation, and maturation of the microbes, a mean to stretch the existence of the microbial colony. The formation is accompanied with the production of EPS, which leads to foundational switching in the bacterial growth and overall gene expression. The formation of biofilm noticeably reduces the susceptibility of bacteria to antibacterial agents and radiation. Specific QS signaling blockage (quorum quenching) is usually taken into account as a promising measure to avert the formation of biofilms in majority of the pathogens, hence escalating the sensitivity of pathogens to antibacterial agents and surpassing the bactericidal effect of antibiotics (Jiang et al., 2019). But if the signal production is nutrient-finite, then the nutrient-inadequate interior of a biofilm cannot donate to quorum sensing, which restricts the aptness of bacteria to appraise their own population and behave appropriately (Narla et al., 2021). The production of virulence factors helps microbe evade the barriers of immune response and cause pathogenic impacts. A gram-negative bacterium, *P. aeruginosa*, is proficient in existing within a diverse range of environmental conditions. This organism, being an opportunistic pathogen, is known to remain associated with nosocomial infections and is now among the foremost causes of death in severe respiratory infections (Antunes et al., 2010).

5.7 REGULATORY MECHANISM MAINTAINING THE BIOFILM ACTIVITY

A biofilm, as we discussed before, is nothing but a conglomeration of microbes that are usually adhered to a surface and confined in an EPS matrix. The initiation of a biofilm development is a crucial mechanism of microbial resistance, which gradually leads to an uncontrolled pathogenic infection formation among the faunas (Zhou et al., 2020). Not just this, it even allows bacterial species to be a major threat in several different situations, such as in food manufacturing and packaging scale, during industrial effluent treatment, and even in metallurgy (Zhou et al., 2020). Quorum sensing or QS is a microbial transmission process (cell-to-cell) regulated by the diffusible signal inducers called autoinducers (AIs) that are dependent upon the varying population concentration gradient of the microbes. Usually, AI stands for artificial intelligence, and bacterial communication is nothing very different except these networks are very alive and the information transmission is mediated through these autoinducers as they directly mediate the population of these bacteria, thereby showing that signal transmission and QS are directly dependent on the population density of bacteria. They use quorum sensing to regulate diverse sets of functional arrangements, including their magnitude of toxicity and colony formation (Solano et al., 2014).

Therefore, arbitration with quorum sensing by the use of QS-inhibiting agents comprise certain QS inhibitors (QSIs). The quorum quenching (QQ) enzymes bring down or completely quell the colony formation of pathogenic microbes. This process appears to be a favorable way to proceed toward controlling contamination by bacteria. Here, we would find out about the processes that are accountable for QS-regulated biofilm formation and QQ agents that prevent microbial biofilm accumulation and colony-wide activities, and the trending applications of quorum quenchers in wide areas to draw a conclusion upon how regulation of biofilm actually works (Zhou et al., 2020). These quorum sensing circuits are highly complex and also vary within different bacterial species (Solano et al., 2014). This is the reason the pathway to biofilm formation and its regulation by QS circuit is yet difficult to be counted in a generalized way. Depending on the types of engaged AIs, the QS systems can actually be divided diversely into few sub-categories, that is to say, the AHL systems and the AIP systems (these two systems were actually categorized as AI-1 system previously) and finally the AI-2 system and the AI-3 system. The AHL system was primarily identified in gram-negative bacteria. Here, the signaling molecule engaged into the activity is usually AHLs. However, the autoinducing peptides (also known as AIPs) that are involved are employed in the AIP system I, which is found predominantly in gram-positive bacteria (Irie and Parsek, 2008).

There is a vast collection of literature regarding these studies, but hardly any report about AI-2 and AI-3 systems, even though these two systems have been identified to endure in both gram-positive and gram-negative bacteria. Evidence states their involvement in signal transmission activities that are performed inter-specially (Surette and Bassler, 1998). AI-2 signaling molecules are included in a class of furanosyl borate diesters whose precursor is known to be (S)-4,5-dihydroxypentane-2,3-dione also familiar as DPD. In contrast, the AI-3 signaling molecules have not so long ago been recognized to be a type of pyrethroids that are very similar to natural pyrethrins (Kim et al., 2020). The procedure of how the QS regulates the formation of consortium has been depicted for gram-positive and gram-negative bacteria distinctively due to their structural and functional differences.

5.7.1 REGULATION OF QUORUM SENSING IN GRAM-POSITIVE BACTERIA

In the instance of gram-positive bacteria, the autoinducers occupied in the quorum sensing are known as AIPs and the constraints of biofilm formation by the AIP-mediated QS network are a quintessential criterion. Bacteria usually produce a small number of oligopeptides in their cell cytoplasm, and these oligopeptides are classified into mature AIP molecules through post-translational alteration. These signaling molecules are then transported outside the cells during the transmission of information on an intercellular basis (Sturme et al., 2002). No sooner the concentration of AIP hits the threshold limit, it starts binding to the extracellular segments of HK, which is a transmembrane receptor found confined upon the surface of cell membrane, leading to kinase activation. This kinase activation is simultaneously accompanied by a sequence of phosphorylation of response regulatory (RR) factors along the downstream chain. This results in the mediation of the featuring of genes responsible for the formation of biofilm (Zhou et al., 2020) as well as other colony-wide behaviors.

Let us take an example of bacteria, such as *Staphylococcus aureus*. It has five genes, *agr*A, *agr*B, *agr*C, *agr*D, and *hld* that are found in the *agr* operon of the *agr*-AIP system (Painter et al., 2014). The signaling AIP molecule is modified from its precursor peptide AgrD, while a transmembrane protein, AgrB, is accountable for the transformation of AgrD to a mature AIP molecule and the AIP, hence derived, is transported out of the cell membrane. Now, if the extracellular AIP concentration outreaches its limiting value, the AIP, here, will bind to a cell surface receptor domain of AgrC, which is an integral transmembrane protein that functions as a HK, causing the activation of kinase. The AgrC activation, successively, initiates the phosphorylation of the AgrA, the downstream component of RR. After this, the phosphorylated AgrA goes on to bind to the intergenic DNA lying between two promoters P2 and P3, thus activating the promoter transcription process this way. The *agr* operon has been known to perform an integral role during the phase of planktonic dispersion, which is the final stage of biofilm lifecycle. A mutant form of Agr causes the formation of a thicker biofilm in comparison with the wild type, but the increasing biofilm thickness has been accredited to the incapability of the planktons to disengage from the mature consortium, neither to proliferate nor to perish (Painter et al., 2014). There are presently a number of studies that have documented evidence about the modulation of biofilm formation by the AI-2 system in gram-positive bacteria. There are also studies that show that the absence of *lux*S, the gene encoding the AI-2 synthase, actually encourages the *rbf* transcription. This *rbf* is a positive regulator, regulating the evolution of biofilms, resulting in an increased degree of biofilm proliferation and elevation in the manufacturing of polysaccharide intercellular adhesion (PIA) in *S. aureus* (Ma et al., 2017). Nevertheless, a contrasting report, showing the unavailability of *lux*S has reduced the possibility of biofilm formation, was also observed in *Enterococcus faecalis* and *Streptococcus suis*. Nonetheless, the AI-2-mediated system is known to be involved in the "regulation" of biofilm inception in gram-positive bacteria, but its coordinating pathways have not been distinctively categorized yet.

5.7.2 REGULATION OF QUORUM SENSING IN GRAM-NEGATIVE BACTERIA

The quorum sensing system's signaling molecule in gram-negative bacteria is known as AHL. The AHL-mediated QS system was originally observed in a bacteria named *Vibrio fischeri*, a bioluminous marine bacterium species. In fact, the effluxed autoinducer called the N-acyl homoserine lactone (HSL) is known to interact with a regulator, known as LuxR, and this in turn initiates the transcription of *lux* operon at a condition of higher density gradient. This system has turned out to be the basic framework for exploring the quorum sensing in several bacterial species (Miyamoto and Meighen, 2000). In many bacterial species, the AHL signal, N-(3-oxohexanoyl)-L-homoserine lactone (OHHL), is synthesized biologically by the AI synthase named LuxI, and the OHHL molecule obtained as end product disperses/spreads out of the plankton. If the assemblage of OHHL outdoes the critical limit with the elevating cell density, these OHHL molecules bind to LuxR, an OHHL receptor as well as a DNA-binding transcriptional initiator, which activates the expression of genes corresponding to biofilm genesis. In present-day works, an elaborative exploration has shown the regulatory methodology as a classic model that depicts the modulation of biofilm development by the AHL systems as majority, if not all gram-negative bacteria (Zhou et al., 2020).

P. aeruginosa, for example, comprises two AI synthase genes in it, *las*I and *rhl*I, where they share outstanding homologous similarities in sequence to *lux*I of the *V. fischeri* species (Lee and Zhang, 2015). These signaling molecules, N-(3-oxo-dodecanoyl)-L-homoserine lactone (OdDHL) as well as N-butyryl-L-homoserine lactone (BHL), are independently processed by LasI and RhlI, respectively. After reaching the concentration threshold magnitude, these two AHL signaling molecules end up binding to their receptors, LasR and RhlR, respectively, to in turn activate the exhibition of regulatory genes that relate to virulence and biofilm formation of the bacterium. Amid these AHL systems, the *rhl* operon has also been known to partake in the mediation of swarming motility that is seen in the initial phases of biofilm foundation (Khan et al., 2020), and in the biosynthesis of toxic factors and precursors, such as rhamnolipid and pyocyanin (Dusane et al., 2010). On the contrary, the *las* system is known to constrict genes that encode elastase, alkaline protease, endotoxin A, and other genes involved in biofilm formation. In addition to all this, in *P. aeruginosa*, two distinct AHL-regulated systems are also involved, the *pqs* system and the *iqs* system. Both the systems function in a way that is indistinguishable from the *rhl* and *las* operons, even though their AIs, PQS (2-heptyl3-hydroxy-4-quinolone) and IQS (2-(2-hydroxyphenyl)-thiazole-4-carb aldehyde), are chemically distinct molecules from AHLs (Lee and Zhang, 2015). Moreover, the *pqs* system has also been documented in different studies to be connected with the synthesis of extracellular DNA, which is crucial for the creation of microbial consortium (Allesen-Holm et al., 2006). In short, the four AHL systems that are in the premise, i.e., *las*, *rhl*, *iqs*, and *pqs*, cross-convoluted to form a complex network of quorum sensing, which synchronizes the formation of biofilm in bacterial species such as *P. aeruginosa*.

The AHL system is even engaged in the mediation of biofilm genesis in *Escherichia coli* or *E. coli*. But there are some differences that distinguish their AHL system from those in *P. aeruginosa*. Not just the receptor gene, which for *E. coli* is *sdi*A, which is actually homologous to *lux*R in findings, but also the AHL synthase gene that is homologous to *lux*I is nonexistent in this bacterium (Walters and Sperandio, 2006). Therefore, it has been conjectured that the receptor SdiA may also generate a response to the AHLs, synthesized by other microbes, which are used to control biofilm-associated gene expression. Studies show that at higher exogenous AHL concentration, there is an elevation of EPS production in *E. coli* and, with that, the adhesion of bacterial cells increases (Heindl et al., 2014), which therefore validates the fact that bacteria can use the signaling inducers of analogous species for biofilm formation of their own species. Presently in studies, the AI-2 signaling pathway has been found to influence biofilm proliferation in a variety of gram-negative bacterial species such as *Helicobacter pylori*, *E. coli*, *Vibrio parahaemolyticus*, *Vibrio cholerae*, and also *Pseudomonas aeruginosa* (Guo et al., 2018). Having said that, the regulatory mechanisms involving the AI-2 system in *Vibrio cholerae* and *E. coli* have been recognized under different studies, where the AI-2 signaling molecules are channelized into the cells by an ABC transporter protein in *E. coli* the moment the concentration of extracellular signaling inducers reaches a critical limit (Li et al., 2007). Due to this phenomenon, LsrK kinase phosphorylates the end signaling molecule and simultaneously causes its binding to the transcriptional regulator, i.e., LsrR, hence activating the expression of corresponding genes by that mean (Li et al., 2007). In case of *V. cholerae*, the AI-2 inducer is actually perceived through a receptor complex, which is known as LuxPQ. When these signaling molecules' concentration outreaches the threshold value, the activity of LuxQ as a kinase is thereby converted into phosphatase, which goes on and dephosphorylates the downstream control proteins, which is known as LuxO, resulting in the fabrication of a transcriptional regulatory protein recognized as the HapR, by inhibition of the transcriptional genes that relate to biofilm development in the bacterium species (Hammer and Bassler, 2007). AI-3 signaling molecules have been beheld connected to the development of flagellum and adhesin in *E. coli*, but their control procedures for the development of biofilm still remain unclarified.

5.7.3 MECHANISM OF QUORUM QUENCHING IN SUBDUING BIOFILM FORMATION

The phenomenon of interrupting the signal transduction pathway within the cytoplasm to perish certain gene cascade expressing a microbial conduct is known as quorum quenching. The mechanism of quorum quenching agents to check the bacterial biofilm formation has been simplified as follows:

a. It is a process of inhibiting AIs synthesis that initiates the quorum sensing pathway.
b. It causes degradation and/or inactivation of autoinducers using AHL-lactonases, oxidoreductases, antibodies, and other consecutive events.
c. The interference with the signal receptors is caused by the use of AI antagonists.
d. There is an interference with the RR, which in turn disturbs the signaling cascade process.

TABLE 5.2

Quorum Quenching Molecules with Target Bacteria and Effects

Quorum Quenching Molecules	Effects of the Molecule	Target Bacteria
Sitagliptin	Interacts with the LasR receptors and in turn significantly reduces the formation of biofilm.	*P. aeruginosa*
Molecularly imprinted polymers (MIPs)	They capture OdDHL, thus ceasing QS and inhibiting biofilm formation.	
Boronic acid derivate SM23	Decreases 3-oxo-C_{12}-HSL and C_4-HSL and therefore inhibits biofilm formation.	
NADP-dependent reductase BpiB09	Reduces the production of pyocyanin, thereby decreasing motility and ultimately inhibiting biofilm formation.	
N-phenyl-4-(3-phenylthioureido)-benzenesulfonamide	These molecules allosterically modify AI-3 receptor QseC, thereby impeding the expression of virulence and inhibiting the formation of biofilm.	*E. coli*
Fructose furoic acid	These molecules compete with C_8-HSL to downregulate its target specificity during expression and phenotypical characters related to biofilm.	
Acyl-HSL analog J8-C8	Binds to TofI, thus resulting in disturbance during C_8-HSL synthesis, which then affects biofilm formation.	*B. glumae*
AHL-lactonase AiiA	Degrades AHLs to prevent the formation of biofilms and virulence factor production.	*V. cholerae*
Anti-autoinducer monoclonal antibody AP4-24H11	Seizes the autoinducing peptide 4 (AIP-4) and therefore inhibits the formation of biofilm.	*S. aureus*

e. There is a reduction in the extracellular AI accumulation by the inhibition of the AIs efflux, which in turn inhibits cell-to-cell signaling (Zhou et al., 2020).

Some of the known quorum quenching agents are sitagliptin, furanones, molecular imprinted polymers (MIPs), boronic acid derivate SM23, 3-hydroxy-2-methyl-4(1H)-quinolone 2,4-dioxygenase, etc. Few other quenching molecules and their target bacteria are tabulated in Table 5.2 (Zhou et al., 2020).

5.8   QUORUM SENSING IN DISPERSAL AND UPREGULATION OF BIOFILM SURFACTANT MOLECULES

A biofilm can be explicated as a multicellular community of multitudinal microbes that are securely adhered to a surface and are encapsulated extracellularly by a biological matrix of extravasated polymers (Costerton et al., 1995). Biofilm formation initiates from individual members of bacterial species coming in close proximity, leading to a surface contact and reversible attachment, followed by accretion of microcolony, maturation of biofilm, and then ultimately dispersal (Heindl et al., 2014). Biofilms can be formed on an extensive variety of surfaces that include even living tissues. These three-dimensional structures and the methods that cause them are of immense curiosity as they are ubiquitous within the bacterial world and have intense impacts on the society in terms of industrial, medical, and agricultural contexts. The physiological characteristics of bacteria inside a microcolony are very distinctive from the same kind of cells persisting in free-living, planktonic state. This can be clearly understood by the observation that biofilms often manifest dramatic opposition to antimicrobial agents that are both chemical (i.e., antibiotics and disinfectants) and biological (i.e., viruses and protists) in nature. Manipulating the formation of biofilm is therefore quite challenging and needs a strategy of consequential investigation. The inceptive steps of surface clamping that lead to endmost formation of a biofilm are a pronounced target as the control of this phase in the process could be used to restrict the development of biofilms before they have been formed, or even go as far as promoting the formation of biofilm for advantageous operations.

The incorporation techniques of pathogens targeting the host tissues overlap with the methods that pilot the biofilm formation. For many pathogens, biofilm formation is a principal and/or indispensable component of disease advancement. In addition to that, the existence of facultative pathogens in environmental water bodies (static), such as the waterborne pathogens, can be considerably amplified within biofilms, thereby directly influencing the ecology of diseases (Heindl et al., 2014).

Agrobacterium tumefaciens, a pathogen mostly observed in plants, is distinctly proficient in surface fouling and formation of biofilm upon the biotics, and even on abiotic surfaces. Most of the data recorded till date, in studies, are performed in the laboratory environment using the nopaline type of the strain *A. tumefaciens* C58; further investigations on the diversity of *Agrobacterium* species have even disclosed similarity with their tendency of biofilm formation (Abarca-Grau et al., 2011). It has been acknowledged, in the literature, that many a time this interaction between the elucidated adhesion and colony formation methods and/or techniques as well as the ecological interactivity of the bacterium within the soil's vicinity of different roots has persisted to be experimentally upheld, and much of the pertinent environmental pretext for bacterial species such as *A. tumefaciens* has been left crudely understood.

5.8.1 Upregulation of Virulence

The study of motile-to-sessile alteration is irradiation within itself. For observing and apprehending the microbial evolution and establishment, it is essential to consider the function that the transition plays a key role in the pathogenic lifecycle of bacteria. The virulence of bacteria such as *A. tumefaciens* can be regulated by their Ti plasmid, a fraction of which, known as T-DNA, is substituted into the host plant's cells and then finally amalgamated into the target genome, resulting in the development of a tumor structure. A captious part of the bacterial–plant reciprocity is the adjunct of a bacterial genome to a targeted plant genome, accompanied by the translocation process of the T-DNA via type IV secretion apparatus that encompasses the microbial cell wall and dispenses access to the cytoplasm of plant cells (Heindl et al., 2014). Even though the attachment to plant tissue usually steers toward the formation of biofilm, reports concluded that the biofilm formation in a laboratory environment is not an outcome of T-DNA transfer. Howsoever, the transfer of T-DNA is strikingly incompetent and the associated bacterial cells may even be subjected to the plant defense counter (Zipfel et al., 2006). In case of innate infections, densely populated bacterial cells inside a biofilm, which accumulate at the prospective infection site, can assist to overpower the obstructions and encourage the ultimate feasibility of the transfer of T-DNA successfully. Although a congested population of bacteria may not be obligatory for virulence as such, their presence is required for the preservation of pTi and conjugal circulation or diffusion within the colony of bacteria such as *A. tumefaciens* that are confederated with affected plants (Zipfel et al., 2006).

Biofilms that are found within plant tumors usually impart an optimal condition for pTi conjugation, guaranteeing the conservation of the plasmid – and the competency for septicemia – among the colonies of bacterial species that are similar to *A. tumefaciens*. There even exists a supplemental relationship between the formation of biofilm and virulence. For example, in low-phosphate conditions, as in the rhizosphere, both formation rate of biofilm and virulence gene expression are intensified (Danhorn et al., 2007). Now, the two-component architecture such as PhoR/PhoB, which is potential to phosphate sensing, also arbitrates an aggrandized adhesion phenotype, whereas the pTi-encoded architecture VirA/VirG intercedes the virulence response (Danhorn et al., 2007). These controlling frameworks function side by side to let bacterial cells initiate the attachment to floral components and reveal their toxicity genes in a well opportune fashion.

5.8.2 Dispersal of Members of the Colony

The ultimate "step" within the life cycle of a microbial colony is the dispersal of the planktons of the consortium away from the location of adhesion and into the surrounding ecosystem. The tendency to discourage the development of biofilm, segregate the EPS matrix, or instigate the active biofilm dispersal is ecologically and economically pertinent. There are several regulators (initiating substances) that initiate the biofilm dispersal in various microbes, including the processes of quorum quenching (QQ), formation of active molecules such as nitric oxide, and secretion of matrix-decaying exoenzymes such as nucleases and glycoside hydrolase dispersin (McDougald et al., 2012). The D-enantiomer form of amino acids is also incriminated for the process of biofilm dispersal, even though the phenomena may occur due to ancillary effects on protein synthesis (Leiman et al., 2013). The withdrawal of motile innate planktons from the affixed mature consortium upon septation may perform like an interrelated facet of biofilm establishment.

Even though the detachment of independent planktons from a fully developed biofilm has been found occurring at certain points in the lifespan of major multicellular groups, there are a few investigation reports or studies that arise the excitement in Rhizobiaceae, including *A. tumefaciens*. During the scattering of *R. leguminosarum* planktons on abiotic surfaces, it was found that the coordination and process of detachment, as well as the relevance to surface interconnection with the host, have remained undefined. In bacterial species such as *A. tumefaciens*, the introduction of cell-free supernatant of *P. aeruginosa* culture triggers the detachment and dispersal of the planktons of *A. tumefaciens*, although the active component secreted by *P. aeruginosa* has not yet been identified (Leiman et al., 2013). These studies have indicated that the phenomena of detachment and dispersal are elemental phases of the normal evolutionary program in bacterial species.

5.9 CONCLUSIONS

The microbial planktons occupying diverse niche have broadly adopted the quorum sensing system for optimizing and improving their coordinate communal behavior in all aspects. It has developed the phenomenon of manipulating the behavior in response to the presence of surrounding elements, hence acting as a multicellular unit. This dramatic mechanism made the scientific community curious to dive deep into investigating the degree of responses. In this context, researchers used various analytical techniques such as thin-layer chromatography (TLC), liquid chromatography (LC), gas chromatography (GC), capillary electrophoresis, and high-pressure liquid chromatography-tandem mass spectrometry (HPLC-MS/MS) to identify short-, medium-, and long-chain autoinducers synthesized by the isolates (Verbeke et al., 2017). This utilization of the analytical techniques accompanied with the keen domain of computational and mathematical modeling has resulted further in a better understanding of the quorum sensing circuits and their mechanism (Pérez-Velázquez et al., 2016; Gilbert et al., 2019). With this intense contribution, researchers can now configure a quite promising model of the inter- and/or intracellular communication and gene expression within the microbial community.

REFERENCES

Abarca-Grau, A.M., Penyalver, R., Lopez, M.M., & Marco-Noales, E. (2011). Pathogenic and non-pathogenic *Agrobacterium tumefaciens*, *A. rhizogenes*, and *A. vitis* strains form biofilms on abiotic as well as on root surfaces. *Plant Pathology*, 60(3), 416–425. Doi: 10.1111/j.1365–3059.2010.02385.x.

Abed, S.E., Ibnsouda, S.K., Latrache, H., & Hamadi, F. (2012). Scanning Electron Microscopy (SEM) and Environmental SEM: Suitable tools for study of adhesion stage and biofilm formation. In: Viacheslav Kazmiruk (ed). *Scanning Electron Microscopy*. DOI: 10.5772/34990.

Abisado, R.G., Benomar, S., Klaus, J.R., Dandekar, A.A., & Chandler, J.R. (2018). Bacterial quorum sensing and microbial community interactions. *mBio*, 9(3), e02331–e02317. Doi: 10.1128/mBio.02331-17.

Allesen-Holm, M., Barken, K.B., Yang, L., Klausen, M., Webb, J.S., ... Tolker-Nielsen, T. (2006). A characterization of DNA release in *Pseudomonas aeruginosa* cultures and biofilms. *Molecular Microbiology*, 59(4), 1114–1128. Doi: 10.1111/j.1365-2958.2005.0 5008.x.

Antunes, L.C.M., Ferreira, R.B.R., Buckner, M.M.C., & Finlay, B.B. (2010). Quorum sensing in bacterial virulence. *Microbiology*, 156(8). Doi: 10.1099/mic.0.038794-0.

Asri, M., Elabed, S., Ibnsouda Koraichi, S., & El Ghachtouli, N. (2019). Biofilm-Based Systems for Industrial Wastewater Treatment. In: Hussain, C. (ed) *Handbook of Environmental Materials Management*. Springer, Cham. Doi: 10.1007/978-3-319-73645-7_137.

Bassler, B.L. (1999). How bacteria talk to each other: Regulation of gene expression by quo-rum sensing. *Current Opinion in Microbiology*, 2(6), 582–587. Doi: 10.1016/s1369-5274 (99)00025-9.

Bassler, B.L. (2002). Small talk. Cell-to-cell communication in bacteria. *Cell*, 109(4), 421–424. Doi: 10.1016/s0092-8674(02)00749-3.

Bramhachari, P.V., Yugandhar, N.M., Prathyusha, A.M.V.N., Sheela, G.M., Naravula, J., & Venkateswarlu, N. (2018). Quorum sensing regulated swarming motility and migra-tory behavior in bacteria. In P.V. Bramhachari (Eds.), *Implication of Quorum Sensing System in Biofilm Formation and Virulence* (pp. 49–66). Singapore: Springer. Doi: 10.1007/978-981-13-2429-1_5.

Caicedo, J.C., Villamizar, S., & Ferro, J.A. (2017). Quorum sensing, its role in virulence and symptomatology in bacterial citrus canker. *Citrus Pathology*. Doi: 10.5772/66721.

Costerton, J.W., Lewandowski, Z., Caldwell, D.E., Korber, D.R., & Lappin-Scott, H.M. (1995). Microbial biofilms. *Annual Review of Microbiology*, 49, 711–745. Doi: 10.1146/annurev. mi.49.100195.003431.

Danhorn, T., Merritt, P.M., and Fuqua, C. (2007). Motility and chemotaxis in *Agrobacterium tumefaciens* surface attachment and biofilm formation. *Journal of Bacteriology*, 189, 8005–8014. doi: 10.1128/JB.00566-07

de Kievit, T.R., & Iglewski, B.H. (2000). Bacterial quorum sensing in pathogenic relationships. *Infection and Immunity*, 68(9), 4839–4849. Doi: 10.1128/IAI.68.9.4839-4849.2000.

Dong, Y., Wang, L., & Zhang, L. (2007). Quorum-quenching microbial infections: Mechanism and implications. *Philosophical transactions of the Royal Society of London. Series B, Biological Sciences*, 362(1483), 1201–1211. Doi: 10.1098/rstb.2007.2045.

Donlan, R.M., & Costerton, J.W. (2002). Biofilms: Survival mechanisms of clinically relevant microorganisms. *Clinical Microbiology Reviews*, 15(2), 167–193. Doi: 10.1128/CMR.15.2.167-193.2002.

Dusane, D.H., Zinjarde, S.S., Venugopalan, V.P., McLean, R J., Weber, M.M., & Rahman, P.K. (2010). Quorum sensing: Implications on rhamnolipid biosurfactant production. *Biotechnology and Genetic Engineering Reviews*, 27, 159–184. Doi: 10.1080/02648725. 2010.10648149.

Gilbert, D., Heiner, M., Ghanbar, L., & Chodak, J. (2019). Spatial quorum sensing modelling using coloured hybrid Petri nets and simulative model checking. *BMC Bioinformatics*, 20(173). Doi: 10.1186/s12859-019-2690-z.

Guo, M., Fang, Z., Sun, L., Sun, D., Wang, Y., … Gooneratne, R. (2018). Regulation of thermostable direct hemolysin and biofilm formation of *Vibrio parahaemolyticus* by quorum-sensing genes *luxM* and *luxS*. *Current Microbiology*, 75(9), 1190–1197. Doi: 10.1007/s00284-018-1508-y.

Hammer, B.K., & Bassler, B.L. (2003). Quorum sensing controls biofilm formation in *Vibrio cholerae*. *Molecular Microbiology*, 50(1), 101–104. Doi: 10.1046/j.1365-2958.2003.03 688.x

Heindl, J.E., Wang, Y., Heckel, B.C., Mohari, B., Feirer, N., & Fuqua, C. (2014). Mechanisms and regulation of surface interactions and biofilm formation in agrobacterium. *Frontiers in Plant Science*. Doi: 10.3389/fpls.2014.00176.

Irie, Y., & Parsek, M. R. (2008). Quorum sensing and microbial biofilms. *Current Topics in Microbiology and Immunology*, 322, 67–84. Doi: 10.1007/978-3-540-75418-3_4.

Jiang, Q., Chen, J., Yang, C., Yin, Y., & Yao, K. (2019). Quorum sensing: A prospective therapeutic target for bacterial diseases. *BioMed Research International*. Doi: 10.1155/2019/2015978.

Khan, F., Pham, D.T.N., Oloketuyi, S.F., & Kim, Y.M. (2020). Regulation and control-ling the motility properties of *Pseudomonas aeruginosa*. *Applied Microbiology and Biotechnology*, 104(1), 33–49. Doi: 10.1007/s00253-019-10201-w.

Kim, C.S., Gatsios, A., Cuesta, S., Lam, Y.C., Wei, Z., ... Crawford, J.M. (2020). Characterization of autoinducer-3 structure and biosynthesis in *E. coli*. *ACS Central Science*, 6(2), 197–206. Doi: 10.1021/acscentsci.9b01076.

Kumar, V., Singh, K., Shah, M.P., Singh, A.K, Kumar, A., & Kumar, Y. (2021). Application of omics technologies for microbial community structure and function analysis in contaminated environment. In Shah, M.P., Sarkar, A., & Mandal, S., (Eds.), *Wastewater Treatment: Cutting Edge Molecular Tools, Techniques & Applied Aspects in Waste Water Treatment*. Elsevier. Doi: 10.1016/B978-0-12-821925-6.00013-7.

Kumar, V., Thakur, I.S., Singh, A.K., & Shah, M.P. (2020). Application of metagenomics in remediation of contaminated sites and environmental restoration. In: Shah, M., Rodriguez-Couto, S., & Sengor, S.S. (Eds.), *Emerging Technologies in Environmental Bioremediation*. Elsevier. Doi: 10.1016/B978-0-12-819860-5.00008-0.

Lee, J., & Zhang, L. (2015). The hierarchy quorum sensing network in *Pseudomonas aeruginosa*. *Protein & Cell*, 6, 26–41. Doi: 10.1007/s13238-014-0100-x.

Leiman, S.A., May, J.M., Lebar, M.D., Kahne, D., Kolter, R., & Losick, R. (2013). D-Amino acids indirectly inhibit biofilm formation in *Bacillus subtilis* by interfering with protein synthesis. *Journal of Bacteriology*, 195(23), 5391–5395. Doi: 10.1128/JB.00975-13.

Li, J., Attila, C., Wang, L., Wood, T.K., Valdes, J.J., & Bentley, W.E. (2007). Quorum sensing in *Escherichia coli* is signaled by AI-2/LsrR: Effects on small RNA and biofilm architecture. *Journal of Bacteriology*, 189(16), 6011–6020. Doi: 10.1128/Jb.00014-17.

Li, Z., & Nair, S.K. (2012). Quorum sensing: How bacteria can coordinate actively and synchronize their response to external signals? *Protein Science*, 21(10), 1403–1417. Doi: 10.1002/pro.2132.

Liu, C., Sun, D., Zhu, J., & Liu, W. (2018). Two-component signal transduction systems: A major strategy for connecting input stimuli to biofilm formation. *Frontiers in Microbiology*, 9, 3279. Doi: 10.3389/fmicb.2018.03279.

López, D., Vlamakis, H., & Kolter, R. (2010). Biofilms. *Cold Spring Harbor Perspectives in Biology*. Doi: 10.1101/cshperspect.a000398.

Lyon, P. (2015). The cognitive cell: Bacterial behavior reconsidered. *Frontiers in Microbiology*. Doi: 10.3389/fmicb.2015.00264.

Ma, R., Qiu, S., Jiang, Q., Sun, H., Xue, T., ... Sun, B. (2017). AI-2 quorum sensing negatively regulates rbf expression and biofilm formation in *Staphylococcus aureus*. *International Journal of Medical Microbiology*, 307(4–5), 257–267. Doi: 10.1016/j.ijmm.2017.03.003.

McDougald, D., Rice, S.A., Barraud, N., Steinberg, P.D., & Kjelleberg, S. (2012). Should we stay or should we go: Mechanisms and ecological consequences for biofilm dispersal. *Nature Reviews Microbiology*, 10, 39–50. Doi: 10.1038/Nrmicro2695.

Miller, M.B., & Bassler, B.L. (2001). Quorum sensing in bacteria. *Annual Review of Microbiology*, 55, 165–199. Doi: 10.1146/annurev.micro.55.1.165.

Miyamoto, C.M., Lin, Y.H., & Meighen, E.A. (2000). Control of bioluminescence in *Vibrio fischeri* by the LuxO signal response regulator. *Molecular Microbiology*, 36(3), 594–607. Doi: 10.1046/j.1365-2958.2000.01875.x.

Monnet, V., & Gardan, R. (2015). Quorum-sensing regulators in Gram-positive bacteria: 'cherchez le peptide'. *Molecular Microbiology*, 97(2), 181–184. Doi: 10.1111/mmi.13060.

Mukherjee, S., & Bassler, B.L. (2019). Bacterial quorum sensing in complex and dynamically changing environments. *Natural Review Microbiology*, 17(6), 371–382. Doi: 10.1038/s41579-019-0186-5.

Narla, A.V., Borenstein, D.B., & Wingreen, N.S. (2021). A biophysical limit for quorum sensing in biofilms. *PNAS*, 118 (21), e2022818118. Doi: 10.1073/pnas.2022818118.

Painter, K.L., Krishna, A., Wigneshwararaj, S., & Edwards, A.M. (2014). What role does the quorum-sensing accessory gene regulator system play during *Staphylococcus aureus* bacteremia? *Trends in Microbiology*, 22(12), 676–685. Doi: 10.1016/j.tim.2014.09.002.

Papenfort, K., & Bassler, B.L. (2016). Quorum sensing signal-responses systems in gram-negative bacteria. *Nature Reviews Microbiology*, 14(9), 576–588. Doi: 10.1038/nrmicro. 2016.89.

Papernfort, K., Silpe, J.E., Schramma, K.R., Cong, J., Seyedsayamdost, M.R., & Bassler, B.L. (2017). A vibrio cholerae autoinducer-receptor pair that controls biofilm formation. *Nature Chemical Biology*, 13, 551–557. Doi: 10.1038/nchembio.2336.

Pena, R.T., Blasco, L., Ambroa, A., González-Pedrajo, B., Fernández-García, L., ... Tomás, M. (2019). Relationship between quorum sensing and secretion systems. *Frontiers in Microbiology*. Doi: 10.3389/fmicb.2019.01100.

Pérez-Velázquez, J., Gölgeli, M., & García-Contreras, R. (2016). Mathematical modelling of bacterial quorum sensing: A review. *Bulletin of Mathematical Biology*, 78, 1585–1639. Doi: 10.1007/s11538-016-0160-6.

Popham, D.L., & Stevens, A.M. (2005). Bacterial quorum sensing and bioluminescence. *Association for Biology Laboratory Education*, 27, 201–215.

Prabhu, M., Naik, M., & Manerikar, V. Quorum sensing-controlled gene expression systems in gram-positive and gram-negative bacteria. In P.V. Bramhachari (Ed.), *Implication of Quorum Sensing and Biofilm Formation in Medicine, Agriculture and Food Industry* (pp. 21–37). Singapore: Springer. Doi: 10.1007/978-981-32-9409-7.

Rabin, N., Zheng, Y., Opoku-Temeng, C., Du, Y., Bonsu, E., & Sintim, H. (2015). Biofilm formation mechanisms and targets for developing antibiofilm agents. *Future Medicinal Chemistry*, 7(4). Doi: 10.4155/fmc.15.6.

Renner, L.D., & Weibel, D.B. (2011). Physicochemical regulation of biofilm formation. *MRS Bulletin*, 36(5), 347–355. Doi: 10.1557/mrs.2011.65.

Romero, D. (2013). Bacterial determinants of the social behavior of *Bacillus subtilis*. *Research in Microbiology*, 164, 788–798. Doi: 10.1016/j.resmic.2013.06.004.

Rossmann, F.S., Racek, T., Wobser, D., Puchalka, J., Rabener, E.M., ... Huebner, J. (2015). Phage-mediated dispersal of biofilm and distribution of bacterial virulence genes is induced by quorum sensing. *PLOS Pathogens*, 11(2). e1004653. Doi: 10.1371/journal.ppat.1004653.

Rutherford, S.T., & Bassler, B.L. (2012). Bacterial quorum sensing: Its role in virulence and possibilities for its control. *Cold Spring Harbor Perspectives of Medicine*, 2(11), a012427. Doi: 10.1101/cshperspect.a012427.

Schauder, S., Shokat, K., Surette, M.G., & Bassler, B.L. (2001). The LuxS family of bacterial autoinducers: Biosynthesis of a novel quorum-sensing signal molecule. *Molecular Microbiology*, 41(2), 463–476. Doi: 10.1046/j.1365-2958.2001.02532.x.

Sehar, S. & Naz, I. (2016). Role of the biofilms in wastewater treatment. *Microbial Biofilms-Importance and Applications*. Doi: 10.5772/63499.

Sifri, C.D. (2008). Quorum sensing: Bacteria talk sense. *Clinical Infectious Diseases*, 47(8), 1070–1076. Doi: 10.1086/592072.

Smith, J.M. (2000). The concept of information in biology. *Philosophy of Science*, 67(2). Doi: 10.1086/392768.

Solano, C., Echeverz, M., & Lasa, I. (2014). Biofilm dispersion and quorum sensing. *Current Opinion in Microbiology*, 18, 96–104. Doi: 10.1016/j.mib.2014.02.008.

Sperandio, V., Torres, A.G., Jarvis, B., Nataro, J.P., & Kaper, J.B. (2003). Bacteria-host communication: The language of hormones. *PNAS*, 100(15), 8951–8956. Doi: 10.1073/pnas.153710010.

Sturme, M.H., Kleerebezem, M., Nakayama, J., Akkermans, A.D., Vaugha, E.E., & de Vos, W.M. (2002). Cell to cell communication by autoinducing peptides in gram-positive bacteria. *Antonie Van Leeuwenhoek*, 81, 233–243. Doi: 10.1023/a:1020522919555.

Subramani, R., & Jayaprakashvel, M. (2019). Bacterial quorum sensing: Biofilm formation, survival behaviour and antibiotic resistance. In P.V. Bramhachari (Eds.), *Implication of Quorum Sensing and Biofilm Formation in Medicine, Agriculture and Food Industry* (pp. 21–37). Singapore: Springer. Doi: 10.1007/978-981-32-9409-7.

Surette, M.G., & Bassler, B.L. (1998). Quorum sensing in *Escherichia coli* and *Salmonella typhimurium*. *Proceedings of the National Academy of Sciences*, 95(12), 7046–7050. Doi: 10.1073/pnas.95.12.7046.

Verbeke, F., De Craemmer, S., Debunne, N., Janssens, Y., Wynendaele, E., ... De Spiegeller, B. (2017). Peptides as quorum sensing molecules: Measurement techniques and obtained levels *in vitro* and *in vivo*. *Frontiers in Neuroscience*, 11(183). Doi: 10.3389/fnins.2017.00183.

Vu, B., Chen, M., Crawford, R.J., & Ivanova, E.P. (2009). Bacterial extracellular polysaccharides involved in biofilm formation. *Molecules*, 14(7), 2535–2554. Doi: 10.3390/molecules14072535.

Walters, M., & Sperandio, V. (2006). Quorum sensing in *Escherichia coli* and *Salmonella*. *International Journal of Medical Microbiology*, 296(2–3), 125–131. Doi: 10.1016/j.ijmm.2006.01.041.

Waters, C.M., & Bassler, B.L. (2005). Quorum sensing: Cell-to-cell communication in bacteria. *Annual Review of Cell and Developmental Biology*, 21, 319–346. Doi: 10.1146/annurev.cellbio.21.012704.131001.

Whitehead, N.A., Barnard, A.M.L., Slater, H., Simpson, N.J.L., & Salmond, G.P.C. (2001). Quorum-sensing in gram-negative bacteria. *FEMS Microbiology Reviews*, 25(4), 365–404. Doi: 10.1111/j.1574-6976.2001.tb00583.x.

Xavier, K.B., & Bassler, B.L. (2003). LuxS quorum sensing: More than just a numbers game. *Current Opinion in Microbiology*, 6(2), 191–197. Doi: 10.1016/S1369-5274(03)00028-6.

Zhao, X., Yu, Z., & Ding, T. (2020). Quorum-sensing regulation of antimicrobial resistance in bacteria. *Microorganisms*, 8(3), 425. Doi: 10.3390/microorganisms8030425.

Zhou, L., Zhang, Y., Ge., Y., Zhu, X., & Pan, J. (2020). Regulatory mechanisms and promising applications of quorum sensing-inhibiting agents in control of bacterial biofilm formation. *Frontiers in Microbiology*. Doi: 10.3389/fmicb.2020.589640.

Ziemichód, A., & Skotarczak, B. (2017). QS – systems communication of gram-positive bacterial cells. *Acta Biologica*, 24, 51–56. Doi: 10.18276/ab.2017.24-06.

Zipfel, C., Kunze, G., Chinchilla, D., Caniard, A., Jones, J.D., ... Felix, G. (2006). Perception of the bacterial PAMP EF-Tu by the receptor EFR restricts *Agrobacterium*-mediated transformation. *Cell*, 125(4), 749–760. Doi: 10.1016/j.cell.2006.03.037.

6 Integrated Omics Approaches to Understand and Improve Wastewater Remediation

Aman Raj and Ashwani Kumar
Dr. Harisingh Gour University (Central University)

CONTENTS

6.1   INTRODUCTION

Water is of paramount significance in our ordinary routine; as a result, the necessity to enhance and protect its quality is growing all the time. Our precious water supplies are being contaminated by both point and non-point sources. The expeditious industrialization and development have revamped human life, but has overshadowed the negative impacts that it causes to the environment (Chandran et al., 2020). Non-degradable pollutants have emerged due to industrialization, population growth, and changing lifestyles, endangering the environment and human health. The toxic compounds released from the chemical industries, such as pesticides, cosmetics, pharmaceutical wastes, petroleum, and oil, have burdened the water bodies.

DOI: 10.1201/9781003247883-6

Moreover, rapid industrialization and population growth have led to a decrease in the land area leading to water bodies becoming hotspots of waste dumps (Crini and Lichtfouse, 2019; Mohan et al., 2021; Johnson, 2019). Wastewater management is essential for the health of humans and the environment's preservation. Alternative wastewater management techniques are needed, especially in countries that are developing where wastewater output exceeds traditional treatment capacity (Sraw et al., 2018; Kumar et al., 2020a, 2021a). To clean up polluted water bodies, different physical and chemical approaches are used, but most of these methods are expensive, time-consuming, and labour-intensive and ultimately fail to remove micro-contaminants and restore the actual biodiversity of the contaminated water bodies, even after such a long and tedious process (Gupta et al., 2012). Microbial bioremediation solves the above concern as it is the most economical and sustainable method to clean up the wastewater bodies utilizing the potential of a diverse array of indigenous microbes present in the contaminated water bodies (Giri et al., 2021; Agrawal, Kumar, and Shahi, 2021). Microbes can adapt to severe environmental circumstances and are nutritionally adaptable. They also have a variety of intracellular and extracellular enzymes that use hazardous contaminants to transform them into carbon and energy (Chandran et al., 2020; Kumar and Chandra, 2018). They also go through a fast genetic mutation that allows them to gain new metabolic pathways for xenobiotic degradation (Igiri et al., 2018). However, despite its great role in naturally cleaning polluted sites, it has some limitations as the action spectrum is not wide and is time-consuming. Recent advancements in omics biology, such as metagenomics, proteomics, transcriptomics, metabolomics, and fluxomics, have made it possible to comprehend the diversity and operability of microbial systems living in polluted sites, as well as their molecular and biochemical mechanisms and interactions with the environment (Jaiswal, Singh, and Shukla, 2019; Pant et al., 2020; Kumar and Chandra, 2020). Omics technologies have led to identifying a wide variety of microbes along with their genes and proteins that play an essential role in cleaning up contaminated sites (Aguiar-Pulido et al., 2016; Kumar et al., 2021b). The advancement of next-generation sequencing (NGS) technology has aided in the removal of all "Achilles' heels" in the bioremediation process. It is now feasible to view the major pathways involved in the biodegradation of hazardous chemicals and displayed by environmentally beneficial microorganisms due to the advancement of NGS technology (Beale et al., 2017).

Pollutant degradation results in the production of less complex molecules with different chemical characteristics, resulting in a shift in the pH of water bodies, which hinders the growth of certain microorganisms while also promoting the growth of others (Mishra et al., 2019). Metagenomics is being utilized to learn more about the genetic structure of wastewater microorganisms and to find genes involved in bioremediation in a variety of microbial species (Chandra and Kumar, 2017a, Kumar and Chandra, 2020). The microbial communities that exist in the polluted matrix and the potential of specific microbes that can be synergistic or antagonistic entirely influence the process of bioremediation (Kumar et al., 2020b). The microbial diversity and genes found in the effluents are revealed by aligning the reads from the metagenomic library. Silva et al. (2012) discovered a group of microorganisms belonging to the genera *Pseudomonas*, *Diaphorobacter*, and *Comamonas* that contained ben-zoate, biphenyl compounds, naphthalene, phenol, toluene, and biphenyl compounds degrading genes in petroleum refinery effluent (Yu and Zhang, 2012). This means that a wide range of new genes could be present, some of which could help in biore-mediation by degrading other organic molecules (Garrido-Sanz et al., 2019).

Regrettably, metagenomics has the drawback of being unable to determine whether the genes identified can express or not and to overcome these limitations. Metagenomics is integrated with meta-transcriptomics that deals with the mRNA sequencing of the microbial communities expressing whether the change in envi-ronmental condition leads to up- or downregulation of the genes in the microbial environment (Malik et al., 2021). But the unstable nature of mRNA leads to lim-ited usage of meta-transcriptomics and promotes the usage of metaproteomics that identifies the expressed proteins in the microbial communities existing in wastewater conditions and is further employed with metabolomics and fluxomics to understand the complexity and interaction of proteins and enzymes involved in degradation with the surrounding environment and to understand the pathway involved in degradation along with secondary metabolites formed during the microbial degradation process (Suneja and Sharma, 2020). To better understand the molecular mechanism involved in wastewater bioremediation and to uncover the genes involved in degradation, the use of multi-omics tools seems to be the most recommended approach. This chapter is centred on the role of diverse "omics" tools in the remediation of wastewater bodies.

6.2   WASTEWATER AND WASTEWATER TREATMENT PLANTS (WWTPs)

A combination of liquid- or water-carrying trash removed from houses, organiza-tions, business and industrial facilities, hospitals, groundwater, surface water, and stormwater is referred to as wastewater (Jamaly, Giwa, and Hasan, 2015). Wastewater includes a combination of chemical and biological contaminants, depending on its type and source (Wacławek et al., 2017; Kumar et al., 2021c). It usually contains high levels of pollutants that deplete oxygen, pathogenic or disease-causing organisms, organic matter, nitrogenous compounds, inorganic chemicals, minerals, and sand deposits (Jamaly, Giwa, and Hasan, 2015). It could also include toxic substances (Shah and Manan, 2020; Chandra and Kumar, 2017a). Wastewater reclamation and reuse is a significant problem, and scientists are looking for low-cost, effective solu-tions. The treatment of wastewater has three different purposes: water conservation, wastewater reuse, and recycling (Chrzanowski et al., 2011). Unit operations and pro-cedures are now used to deliver primary, secondary, and tertiary treatments. Physical and chemical pre-purification activities are part of primary treatment, whereas bio-logical sewage treatment is part of secondary treatment (Crini and Lichtfouse, 2019). In tertiary treatment methods, wastewater (processed by primary and secondary processes) is converted into high-quality water that may be utilized for a variety of applications, including consumption, commercial, agricultural, and pharmacologi-cal supplies. Primary treatment methods include screening, membrane processing, sedimentation, centrifugation, homogenization, and agitation. When water is highly polluted, these techniques are usually used (Gupta et al., 2012).

In contrast, secondary water treatment, also known as biological treatment, involves the use of microbes to remove soluble and insoluble impurities. The biological treatment of wastewater includes aerobic and anaerobic digestion. A reactor circulates water while a high concentration of microorganisms is maintained. The organic matter is converted into water, carbon dioxide, and ammonia gas by microbes, most of which are bacterial and fungal strains (Kumar et al., 2020c; Suneja and Sharma, 2020). The organic matter is sometimes transformed into various compounds such as alcohol, glucose, and nitrate. In addition, the bacteria degrade inorganic stuff which is hazardous. At this stage, toxic organics and inorganics should be removed from the effluent. There are a variety of aerobic reactor designs in use today, ranging from the widely used activated sludge reactor to more contemporary systems with improved slurry retention, such as biofilm reactor design, membrane reactor design, and aerobic activated sludge reactor designs (Sponza and Gok, 2011). Anaerobic processes convert organic matter to carbon dioxide and methane through a reaction chain that comprises the phases of acidogenesis, lysis, methanogenesis, and acetogenesis (Rodríguez et al., 2015). The activated sludge system reactor is one of the most commonly utilized anaerobic digesters for wastewater treatment (Meena et al., 2021).

Other popular types of anaerobic reactors include the internal circulation reactor, stable granular bed reactor, extended granular sludge blanket reactor, and anaerobic biofilm reactor (Sponza and Gok, 2011). The integration of anoxic, aerobic, and anaerobic processes permits the revocation of not only carbon compounds, but also phosphorous and nitrogen from sewage water, which is critical as the discharge of these nutrients into the ecosystem is one of the main causes of eutrophication in surface waters (Fito, Tefera, and Van Hulle, 2019). Tertiary desalination technologies are critical in the sewage water treatment approach since they are utilized to produce safe drinking water. Distillation, crystallization, sublimation, Soxhlet extraction, oxidization, dissolution, electrolysis, ion exchange, electrodialysis, reverse osmosis, and adsorption are some of the processes utilized for this. Despite such a long, time-taking, and tedious process, we are able to get water with a purity level of 95%–98% (Suneja and Sharma, 2020). Figure 6.1 shows the schematic representation of a WWTP.

Recently, the employment of microorganisms to remediate wastewater has generated some impressive outcomes (Narayanasamy et al., 2015a; Mishra et al., 2019). For economical and efficient wastewater treatment facilities, a thorough understanding of the microbiomes that aid in the conversion of organic and inorganic compounds is required. A consortium of microorganisms has been employed to eliminate toxic contaminants from wastewater in varying circumstances (Bhatt et al., 2021). Although biological wastewater treatment is not well developed due to microorganisms' variable nature and behaviours under various environmental conditions, current research in this area has yielded considerable percentages of pollutants removed from wastewater. The use of multi-omics techniques (Figure 6.2) can enhance the biological treatment process as it enables the culture-independent study of an entire microbial population that would otherwise go unnoticed (Meena et al., 2021; Plewniak et al., 2018; Jeffries et al., 2018).

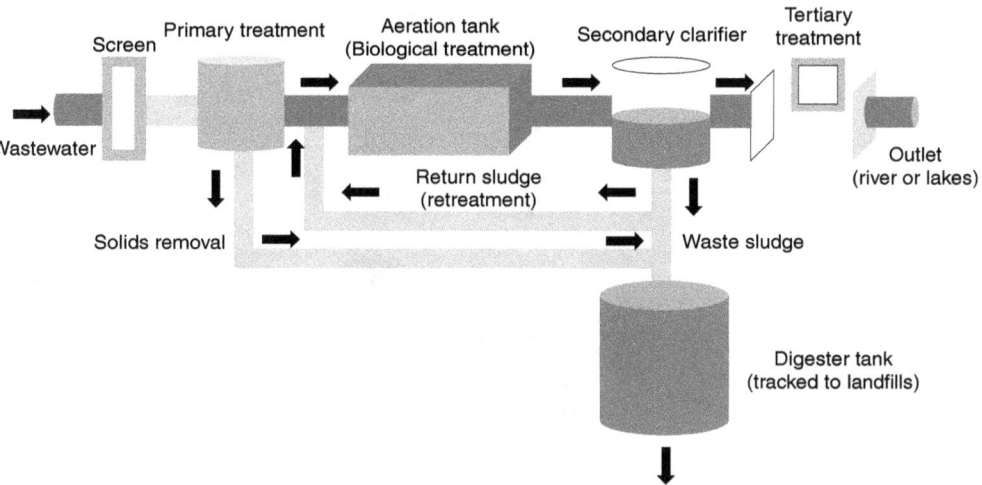

Screen | Primary treatment | Aeration tank (Biological treatment) | Secondary clarifier | Tertiary treatment

Wastewater

Return sludge (retreatment)

Solids removal

Waste sludge

Outlet (river or lakes)

Digester tank (tracked to landfills)

FIGURE 6.1 Schematic representation of a wastewater treatment plant.

6.3 A GLANCE AT THE MAJOR POLLUTANTS OF WASTEWATER AND THE NEED FOR REMEDIATION

Water resources are continually contaminated by effluents discharged by urban and industrial operations, putting a strain on water quality management. Chemicals in unfiltered or poorly treated sewage effluents are hazardous to plants and animals, including humans, and have serious environmental implications. The major contributors of wastewater generation are as follows:

1. **Plant nutrients**: Domestic sewage includes nitrogen and phosphorus compounds, which are two fundamental nutrients required for plant development. Excessive nitrates and phosphates in lakes can cause algae to develop quickly. Eutrophication is a process that accelerates the natural ageing of lakes caused by algal blooms, which are frequently triggered by sewage discharges (Akpor et al., 2014). In untreated wastewater, nitrogen is mostly found as ammonia and organic nitrogen, whereas phosphorus is mostly found as the soluble orthophosphate ion, oxidized phosphate, or different oxygen/phosphorous combinations (Jin et al., 2014). The presence of algal blooms in water has been demonstrated to induce a non-linear decline in water clarity. Another effect of eutrophication is an increase in the amount of chlorine required for water purification, which may increase the risk of cancer (Crini and Lichtfouse, 2019).

2. **Hydrocarbons**: Almost all kinds of life are harmed by industrial effluents, and crude oil pollution is ubiquitous owing to its widespread use, associated disposal procedures, and unintentional spills. Hydrocarbon pollutants found in wastewater effluents have been related to many significant environmental and health effects (Fang et al., 2018). Even though petroleum is an

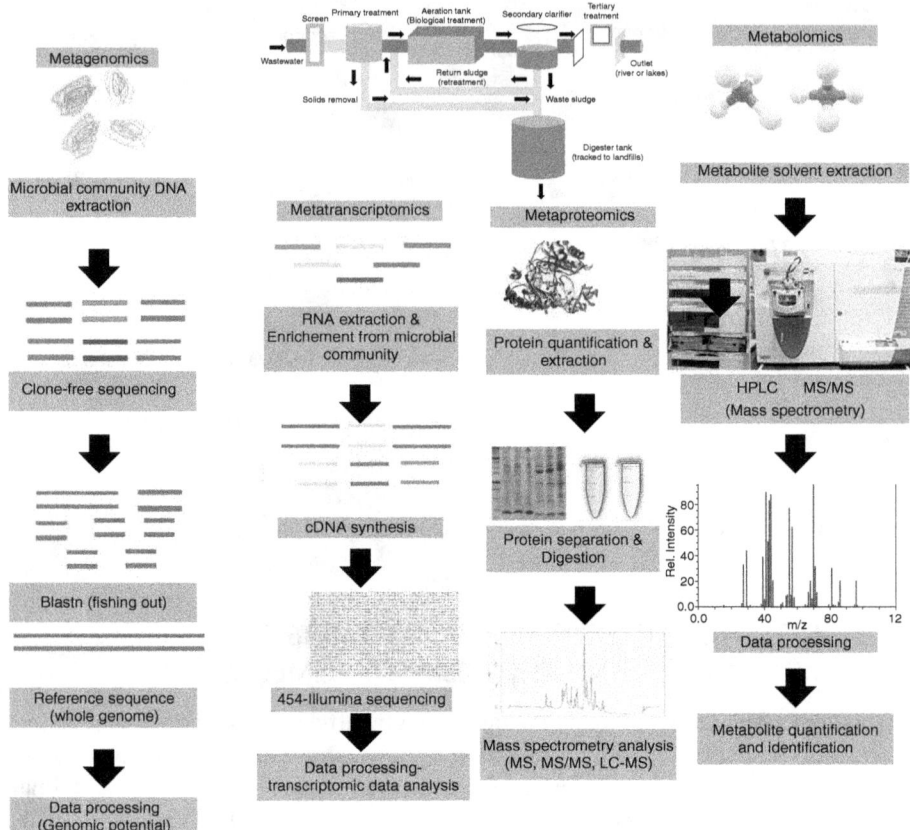

FIGURE 6.2   Integrated omics in wastewater remediation.

essential source of energy and raw material for the chemical industry, it can cause serious problems when it comes in contact with receiving streams, including threats to fisheries, wildlife habitats, the health of humans, and the ruination of ecosystem balance, which may take years if not decades to recover. Exposing rats to hydrocarbons produces reproductive cytotoxicity that is restricted to the compartment of dividing spermatogonia, according to research (Johnson, 2019).

3. **Heavy metals and pesticides**: Industrial activities, petroleum pollution, and agricultural and effluent discharge are the major anthropogenic sources of heavy metals in wastewaters (Akpor et al., 2014; Chandra and Kumar, 2017b; Hemmat-Jou et al., 2018; Bhatt et al., 2019; Kumar, 2021). Although several heavy metals, such as zinc, copper, and iron, are considered necessary in the aquatic environment due to their participation in numerous metabolic processes, they become harmful when present in excessive quantities. The

majority of heavy metals are poisonous and carcinogenic, posing a severe threat to human health and the fauna and flora of receiving water bodies. A variety of heavy metals, including cadmium, copper, fluorine, nickel, and arsenic, are hazardous even in extremely low quantities. Heavy metals are thought to have the ability to bind to proteins, altering their structure and inactivating them, which can lead to health problems (Igiri et al., 2018). The agriculture sector is the major contributor to pesticides in wastewater. Pesticides and their metabolites are persistent, accumulating in the environment as part of soil humus or leaching into groundwater, as well as being absorbed by plant roots or entering aquatic bodies through surface run-off, and so entering the food chain. Pesticides are found to be highly harmful to aquatic life forms in several studies. Aliesterases and acetylcholinesterase activity are known to be inhibited by pesticides, as well as many developmental, histopathological, neurological, and endocrine disturbances. Pesticides are mutagenic and genotoxic to the majority of aquatic life forms, according to a few reports. When compared to mature amphibians, pesticides are particularly harmful to larval life forms (Giri et al., 2021).

4. **Suspended solids**: Suspended particles are another significant feature of sewage. The amount of sludge generated in a treatment plant is proportional to the amount of total suspended particles in the sewage. Industrial and storm sewage may have greater suspended particles concentrations than household sewage. The effectiveness of the treatment process is determined by how effectively a treatment plant eliminates suspended particles and lowers BOD (biological oxygen demand) (Rahman et al., 2017).

5. **Microbes**: The majority of waterborne bacteria that induce human illness are considered to originate in the faeces of sick humans or animals. Per gallon of sewage, there are millions of bacteria. The majority are coliform bacteria from the human intestine, although other microorganisms are also likely to be present in domestic sewage. Coliforms are utilized as sewage contamination indicators. A high coliform level typically implies that sewage contamination has occurred recently (Akpor et al., 2014; Miao and Liu, 2018).

6. **Pharmaceuticals and illicit drugs**: Greater medical requirements, and therefore increased pharmaceutical use, have prompted pharmaceutical businesses to expand output, resulting in the discharge of more pharmaceutical industrial and hospital waste into aquatic environments (Mohan et al., 2021). Roughly 90% of oral medicines are eliminated from the human body and so end up in aquatic systems, contributing to the high incidence of pharmaceuticals and illegal drugs as contaminants in the aquatic system (Gupta et al., 2018). Lipid regulators, antibiotics, antiepileptic drugs, and NSAIDs (non-steroidal anti-inflammatory drugs) are four types of medicines that are often detected in water systems. Pharmaceuticals enter water systems primarily through two routes: improper removal into water treatment systems and scheduled excretion following human ingestion (Tiwari et al., 2016).

Although medicines are used to promote human health by treating and preventing different diseases as well as improving health conditions, their

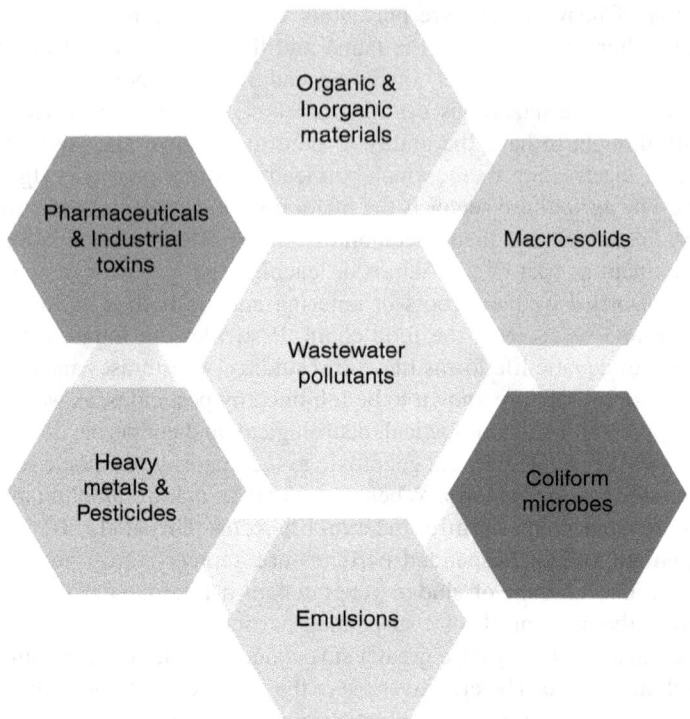

FIGURE 6.3   Major contributors of wastewater.

widespread usage and consumption may pollute water to the point where it can no longer be utilized for a specific purpose. Drug pollution in water systems has the potential to cause the proliferation of antibiotic-resistant microorganisms (Shi, Leong, and Ng, 2017; Fang et al., 2018; Yadav, 2019; Lamilla et al., 2021).

Apart from these major contributors of wastewater pollutants, there exist several other pollutants such as organic wastes, endocrine disruptors, transuranic elements, detergents, and foaming agents, and dyes from tanning and food industries that pollute the water bodies and cause severe threat to the pristine ecosystem (Figure 6.3) (Akpor et al., 2014).

Efficacious treatment discharge into receiving water bodies is intended to stop the harmful effects of untreated and inadequately treated sewage effluents. Wastewater treatment includes being able to enhance the quality of wastewater (Rosso, Muday, and Curran, 2018).

Several techniques have been used to remove contaminants from wastewater. Physical, chemical, and microbiological remediation are all examples of remedial procedures (Gupta et al., 2012). Although these treatment procedures are essential in wastewater remediation, they have faults that demand the employment of a combination of techniques for remediation in certain situations. The use of integrated omics for wastewater bioremediation appears to alleviate the aforementioned problem (Lira et al., 2020; Chu et al., 2018; Yang et al., 2014).

6.4 BIOREMEDIATION: A BRIEF INTRODUCTION

Bioremediation is a waste management technique that employs microorganisms' natural capacity to remove pollutants from industrial wastes, while microbial degradation entails the breakdown of extremely hazardous complex organic contaminants into less toxic forms (Malla et al., 2018a; Kumar et al., 2018). It entails decontamination and mineralization, in which waste is sequentially converted into basic components such as CO_2, water, and methane (Bharagava et al., 2018). However, if the contaminants are persistent, biodegradation may take several phases involving various enzymes or microbial communities. The capacity of microorganisms to break down, eliminate, or convert contaminants relies on their metabolic potential, which is also influenced by the pollutants' bioavailability and accessibility (Gaytán et al., 2020). *In situ* (trash that will be processed on-site) or *ex-situ* (trash that will be processed elsewhere) remediation can be employed. *In situ* bioremediation is generally preferred over *ex-situ* bioremediation due to the high cost of digging contaminated regions to be cleaned.

The following steps of *in situ* remediation are included: bioattenuation (the biological process of breakdown that can be scrutinized by a decline in pollutants intensity over time), bio-stimulation (deliberate stimulation of pollutants breakdown by adding water, essential nutrients, and electron donors or acceptors), and bioaugmentation (adding microbes with potency for biodegradation under suitable laboratory settings) (Bharagava et al., 2019). In industrial effluents, microbes are critical for the breakdown and elimination of organic and inorganic contaminants (Bharagava et al., 2018). However, bioremediation of pollutants may occur at a sluggish rate due to the local microbial community's inability to deteriorate chemicals of considerable significance. To tackle this, external or non-native microorganisms, as well as their enzymes, could be introduced into the system to improve biodegradation. However, because a non-native microorganism brought into an ecosystem may survive in a contaminated system and persist long after the pollutant has been removed, this approach has the potential to be pervasive, harming the environment (Techtmann and Hazen, 2016). As microorganisms are the major factors of bioremediation, changes in their diversity and behaviour might affect the ultimate destiny of pollutants in the environment; it's critical to comprehend the microbial communities involved in pollutant bioremediation.

Microbial bioremediation techniques rely on a diverse group of microbes native to the contaminated area and have enormous metabolic capability. Isolating and purifying such native microorganisms allows researchers to learn more about microbial bioactive molecules and degradation processes. The technique for accessing the microbial world remains a mystery since the bulk of microbes are uncultivable under standard conditions (Malla et al., 2019). Only a small percentage of microorganisms from various environmental samples are easily culturable, making them unavailable for fundamental investigation (Raj et al., 2021; Bursle and Robson, 2016).

Metagenomics and other omics methods are currently being utilized to comprehend the microbiome diversity and activity during bioremediation in a polluted environment. Metagenomics techniques, such as NGS, can also provide reliable information on the major proteins and genes that catalyse the breakdown and detoxification of contaminants in the environment (Beale et al., 2017; Kumar and Dubey, 2020). Researchers have solved the issues concerned with unculturable bacteria, possibly by introducing NGS approaches and *in silico* analysis. These innovations have aided in gathering extensive biological data on microbes, their cellular processes, and biomolecules such as enzymes involved in bioremediation (Malla et al., 2018a; Misra et al., 2019; Pandey et al., 2019).

6.5 OMICS APPROACHES IN WASTEWATER REMEDIATION

Almost 99% of the microorganisms residing in any environment are impossible to culture due to the lack of suitable media or varied environmental conditions. Due to this reason, the majority of microbes remain unexplored (Gutierrez et al., 2018; Chandran, Meena, and Sharma, 2020; Suneja and Sharma, 2020). The development of NGS technologies such as genomics, metagenomics, transcriptomics, metaproteomics, metabolomics, and fluxomics has removed all the barriers faced in the culturable method and has helped explore the uncultivable microbiome of the wastewater treatment plants (Malla et al., 2018b; Narayanasamy et al., 2015b; Rawat and Rangarajan, 2019). Each omics platform's high-throughput data are akin to a signalling cascade in a cell, in which the previous protein alters the action of downstream molecules. Similarly, each set of data gathered through an omics tool is a progression of the last. Omics tools complement each other by assisting in the correlation of DNA sequences with mRNA, protein, and metabolite accumulation, resulting in a clearer and more accurate picture of the biological processes involved in microbial-based wastewater treatment (Rodríguez et al., 2020; Suneja and Sharma, 2020). Table 6.1 shows various NGS technologies.

6.5.1 GENOMICS

Genomics is a new branch of research that integrates molecular biology, recombinant DNA technology, and bioinformatics to unravel the active microbial population in wastewater bioremediation (Awasthi et al., 2020; Plewniak et al., 2018). In a nutshell, genomics is the study of an organism's entire genome. A genomic investigation has revealed essential details on microbial resilience in the polluted environment and the degradation process. As per phylogenetic studies, the adaptation potential of microbes in contaminated settings is carried out inside the genome itself through chemotaxis, nutrient uptake, thermo-sensing, and two-component bacterial systems followed by detoxification on the completion. To date, a wide variety of microbes have been identified and described for the bioremediation of wastewater bodies (Kim et al., 2013; Otsuka, Fukui, and Ozaki, 2009; Staley et al., 2013) Table 6.2. Pesticides, pharmaceutical wastes, and industrial dyes are among the leading toxicants and, through surface run-off and leaching, reach the water bodies, thus contaminating them (Mohan et al., 2021; Sraw et al., 2018; Johnson, 2019). In the case of pesticide

TABLE 6.1
Description of Next-Generation Sequencing Technologies for Omics Data Analysis

S. No.	NGS Technology	Principle	Applications
1.	Pyrosequencing	The detection of pyrophosphate released following nucleotide amalgamation in the newly synthesized DNA strand is the basis for sequencing by synthesis.	Microbial identification and whole-genome sequencing
2.	Nanopore sequencing	The physical changes that occur when DNA sequences pass through nanometre-sized holes under the effect of an electric domain are measured.	Protein, DNA, and RNA analysis
3.	Ion torrent semiconductor sequencing	It works by detecting hydrogen ions produced during DNA polymerization.	Exome, transcriptome, and small RNA sequencing were used to examine microbial diversity in complex environments
4.	ABI SOLiD sequencing	The ligation of DNA fragments is used in this sequencing technology.	Whole-genome sequencing, targeted sequencing, epigenome analysis, and transcriptome analysis
5.	Illumina/Solexa genome sequencing	It's based on dye terminators that can be reversed, which allow single bases to be identified as they're inserted into DNA strands.	To find isoforms, novel transcripts, and gene fusions, researchers use exome sequencing, whole-genome sequencing, and gene expression studies
6.	Helicos HeliScope	Using a very sensitive fluorescence detection technology, a single-molecule sequencing platform has been developed.	Transcript counting and re-sequencing
7.	Pacific Biosciences SMRT sequencing	Sequencing and real-time detection of integrated fluorescently tagged nucleotides using a synthesis technique.	Whole-genome sequencing, targeted sequencing, RNA sequencing, and epigenetic studies

bioremediation, genomics and metagenomics techniques have identified many genes in various microbes that code for enzymes involved in the bioremediation of pesticides. The cloning of several genes in bacteria and the heterologous synthesis of their proteins has allowed researchers to gain a thorough knowledge of their catalytic properties and capability for use in the bioremediation process. The eco-functional and genomic studies of the petroleum-degrading *Franconibacter pulveris* strain DJ34 have been discovered to contain genes involved in the breakdown of different petroleum hydrocarbons that are the major constituent of wastewater bodies.

TABLE 6.2

List of Microbes Isolated from Wastewater Treatment Plants Using Metagenomics Approach and Their Role in Wastewater Remediation

Microbes	Degrading Compound in Wastewater	References
Rhodocyclales	Phosphate removal	Hesselsoe et al. (2009), Wang et al. (2020)
Nitrospira, Nitrosococcus, Nitrosomonas	Nitrate removal	Tiwari et al. (2016)
Thiothrix species	Sulphur	de Graaff, van Loosdrecht, and Pronk (2020), Rubio-Rincón et al. (2017)
Haliscomenobacter hydrossis	Sludge bulking and nitrogen elimination	Tiwari et al. (2016)
Zoogloea	Pharmaceutical degradation	Kang et al. (2018), Gao et al. (2018)
Pseudomonas psychrophila	Sulphamethoxazole removal	Wang and Wang (2018)
Methanomicrobia, Thermoplasmata, Methanobacteria	Anaerobic digestion of metabolites into methane	Shi et al. (2019)
Streptomyces MIUG 4.89	Clofibric acid	Popa Ungureanu, Favier, and Bahrim (2016)
Pseudomonas aeruginosa	Ketoprofen	Ismail et al. (2016)
Trametes versicolor	Ibuprofen	Tiwari et al. (2016)
Paecilomyces lilacinus NH1	Cadmium	Malik et al. (2021)
Rhodococcus species IN306	Polychlorinated biphenyls (PCBs)	Xing et al. (2020)
Pseudomonas sp. CB-3 and *Comamonas* sp. CD-2	PCBs	Xing et al. (2020)
Mycolicibacterium frederiksberg	Petroleum hydrocarbons	Hong et al. (2017)
Alphaproteobacterium, Rhizobium sp. NT-26	Arsenic	Malik et al. (2021)
Trichoderma atroviride	Dichlorvos pesticide	Malik et al. (2021)
Pseudomonas strain CE22	Cephalexin antibiotic	Lin et al. (2015)
Actinoplanes species	Diclofenac	Tiwari et al. (2016), Steliga et al. (2020)
Cunninghamella echinulata	Etonogestrel	Baydoun et al. (2016)
Klebsiella pneumoniae	Degradation of nitrogenous compounds	Padhi et al. (2013)
Achromobacter sp. HZ01	Petroleum hydrocarbons	Hong et al. (2017)
Rhodococcus erythropolis CD 167	Hydrocarbons	Pacwa-Płociniczak et al. (2019)
Thauera sp. M9	4-Chloroaniline	Kumar et al. (2020)
Thauera sp. strain DO	p-Chlorocresol	Ha and Nguyen (2020)
Geobacter sp.	Metal reduction	Clark et al. (2021)

Moreover, they synthesize biosurfactants that enhance the solubility of hydrophobic oil compounds. The biodegradation of aromatic compounds by *Burkholderia* species and *Pseudoxanthomonas spadix* has also been studied using comparative and functional genomics (Pang et al., 2020).

6.5.2 METAGENOMICS

Metagenomics studies a community's genome after extracting DNA straight from the environmental niche, with or without involving an intermediary cloning step (Techtmann and Hazen, 2016). Handelsman and his colleagues originally proposed the idea of metagenomics in 1998; nevertheless, the first evidence of metagenomics comes from Pace and his associates in the year 1985, who were the first to undertake phylogenetic studies of environmental microbial populations (Handelsman, 2004; Araújo et al., 2020). Metagenomics proves to be one of the best tools to explore the microbial diversity of a particular niche. Metagenomics is a new field of genomics that employs a variety of techniques to typify microbiomes in environmental samples and to decipher the genomes of uncultured microbiota, revealing the diversification of taxonomic groups and phylogenetically relevant genes, hydrolytic genes, and entire operons (Bharagava et al., 2019; Uhlik et al., 2013). Function- and sequence-driven approaches are two types of metagenomics approaches. Function-driven metagenomics is building a metagenomic library by copying community DNA in a given host and then assessing the library for a specific activity (Ngara and Zhang, 2018; Datta et al., 2020). After sequencing the positive clones, novel genes that encode for the action could be discovered.

On the other hand, sequence-driven metagenomics entails sequencing the entire metagenomic library while using sequence databases to identify functional genes (Thies et al., 2016). With the introduction of NGS technologies with lower prices and higher sequence efficiency, the diverse microbial population may now be explored for its members in the community and their particular role as per the depth of sequencing (Bashiardes, Zilberman-Schapira and Elinav, 2016). Microbial variability, particular markers for phosphorus removal and other nutrient pathways, and the functions of the entire metagenome, which was originally applied to samples from laboratory-scale reactors, are now routinely surveyed in a variety of full-scale wastewater treatment plants (Wang et al., 2014; Meena et al., 2021). The community-based approach and genome-centric approach are two sorts of studies that can be used to comprehend the community's full functional potential. The whole metagenome of a wastewater treatment plant sample is sequenced and compared to sequence databases in a community-based method to better understand the microbial community's activities and maybe mechanisms behind a few of them (Chandran, Meena, and Sharma, 2020). The genome-based approach, after DNA sequencing, includes genome assembly of wastewater microorganisms using metagenomic sequences, when appropriate genome coverage is available. When the community consists of a few dominant populations, it is generally possible to obtain entire genomes and assess their metabolism in wastewater and contributions to treatment (Miao and Liu, 2018). High-throughput sequencing and metagenomic technologies, such as NGS methods, can be used to monitor the variety of microbial diversity and nitrogen physiological activities in

WWTPs (Silva et al., 2013; Kumar et al., 2021). Denitrifiers that contribute the most to nitrogen removal in WWTPs are primarily members of the genera *Dechloromonas*, *Thauera*, *Paracoccus*, *Hyphomicrobium*, *Denitratisoma*, *Comamonas*, and *Azoarcus*, and the family Comamonadaceae, according to metagenomic analyses. *Thauera* has been found in wastewater treatment bioreactors and is competent in denitrification and organic compound biodegradation (Silva et al., 2012; Jiang et al., 2012; Khan et al., 2002). The synthesis and screening of metagenomic libraries to discover genes involved in bioremediation has been described in many publications. The ecological and physiological abilities of microbial diversity associated with enhanced biological phosphate removal systems by *Candidatus Accumulibacter phosphatis*, a prominent polyphosphate-accumulating organism (PAO) that belongs to an unclassified type of Betaproteobacteria and is commonly found in sewage treatment plants, were studied using metagenomic libraries (Rodríguez et al., 2015).

6.5.3 META-TRANSCRIPTOMICS

Meta-transcriptomics is indeed the study of the whole transcriptional profile of a sample's microbial diversity, including total RNA recovery with or without cloning. Meta-transcriptomics provides an overview of the expression of genes in a particular sample at a certain time and under particular conditions by collecting total mRNA (Aguiar-Pulido et al., 2016; Maroli et al., 2018). It's also known as gene expression profiling since it explains how genes are up- or downregulated in different conditions in microbial communities. For transcriptome study, microarray and sequencing methods are used. Microarrays are used to assess gene expression, whereas RNA sequencing is used to detect the quantity of RNA in a sample using NGS (Chandran, Meena, and Sharma, 2020). The RNA microarray system employs pre-designed probes, making it a more cost-effective, powerful, and improved method for investigating protein expression (Maroli et al., 2018; Chandran, Meena, and Sharma, 2020). A powerful transcriptomics technique, DNA microarray, allows researchers to monitor and evaluate the transcriptional activity of every gene in a microorganism. Researchers have been able to define and analyse the transcripts of genes that are regulated by changes in the environment owing to whole-genome DNA microarrays (Xu et al., 2018). It's been used to investigate physiologic and catabolic transcriptional variations, study the metabolism of microbes from diverse environments, and discover novel microorganisms (Chandran, Meena, and Sharma, 2020). Using advances in NGS technology and bioinformatics tools, this strategy identifies dynamic biochemical pathways without baseline sequencing data and quantifies gene expression without PCR bias (Parada, Needham, and Fuhrman, 2016). It also contains data on the regulation of gene expression in response to various environmental conditions (Bashiardes, Zilberman-Schapira and Elinav, 2016). Changes in DNA and cDNA profile were emphasized in one of the meta-transcriptomics investigations done on the activated sludge-based wastewater treatment facility, showing the relevance of transcriptomics research. It disclosed functional genes involved in the activated sludge treatment process (Wang et al., 2014). Even though that nitrification genes were in short supply, they were substantially expressed.

On the other hand, the extremely abundant denitrification genes had relatively low transcriptional activity, suggesting that nitrification was the major active mechanism of nitrogen metabolism (Yu and Zhang, 2012). Treatment actions controlled by the heterogeneity and interaction of the microbial community may be thoroughly described using a combination of meta-transcriptomics and other omics techniques. In a first effort to characterize the anammox granules, a combination of genome-centred metagenomics and the microbial transcriptomics showed the function of anaerobic microorganisms and their interactions with anammox (anaerobic ammonia oxidation) microbes (Lawson et al., 2017; Rodríguez et al., 2015). Although heterotrophic microbes lack numerous amino acid biosynthesis mechanisms, they may consume peptides and detritus produced by anammox bacteria that handle ammonia-rich effluent. Many transcripts were detected in uncultivated *Chlorobi* and Planctomycetes phylum members, as well as previously identified but understudied transcripts from other phyla that might explain their role in these cells (Lawson et al., 2017). Transcriptome analysis of sludge microbiomes was used to figure out what function nitrifying microbes play in heavy crude oil breakdown (Sato et al., 2019; Chandran, Meena, and Sharma, 2020). The role of differentially expressed genes of *Pseudomonas aeruginosa* in crude oil degradation has been expressed using transcriptomics analysis (Ghosh et al., 2018; Yoneda et al., 2016).

6.5.4 METAPROTEOMICS

A proteome is the collection of proteins that develop within a cell, tissue, organ, or organism. Proteomics is a scientific discipline that analyses and investigates proteomes (Hart et al., 2018). The deciphering of molecular pathways, metabolic functions, post-translational alterations, and other cellular activities is aided by proteomics research (Meena et al., 2019). It has allowed the investigation and tracking of transcriptional activation in microorganisms living in contaminated areas as a result of human-caused activities (Chandran, Meena, and Sharma, 2020). The study of the complete proteome of a sample including a variety of microbial communities is known as metaproteomics. It includes protein extraction and purification, peptide extraction, digestion, and mass spectrometry and computational analysis (Hart et al., 2018). The protein complements are mapped to genomic databases as part of the computational analysis. These databases might be public (like NCBI), reference genomes, or the genomic equivalent of the sample in question. Metaproteomics helps researchers to identify active microbial pathways and link structures, relationships and functions in microbial communities (Mattarozzi et al., 2017). Metaproteomics is characterized as a functional genomics technique since it aids in the exploration of a single organism's protein expression pattern and the generation of a protein map of all expressed proteins by a single organism surviving in a certain environment (Wang et al., 2016). The viability of metaproteomics research is dependent on the effectiveness of protein extraction; the techniques employed for protein fractionation from a complex mixture; and the accurate prediction of partitioned peptides/proteins, databases search, and interpretation of data (Meena et al., 2019). External factors have an impact on an organism's metabolism. Changes in gene regulation are caused by external stimuli, and assessing these changes might be beneficial in remediation techniques (Mattarozzi et al., 2017; Fierer et al., 2013). Several studies have identified and quantified proteins expressed by microbes in a range of natural habitats, including soil, ocean, and freshwater settings, sediments, sludge, industrial effluent biofilms, human or animal microbiota, and plant-associated microbes (Wang et al., 2016; Karlsson et al., 2013; Plewniak et al., 2018).

The analysis of metaproteome data from such environments aids in the understanding microbiome, dynamics, and function. An anaerobic microbial population degrading toluene was allegedly studied using metaproteomics (Espinosa et al., 2020; Rodríguez et al., 2020). Siggins and his associates investigated the effect of temperature and trichloroethylene (TCE) exposure on the translational activation by a population of microorganisms in a laboratory-scale anaerobic environment (Pankhurst et al., 2012). Metaproteomic analysis, in combination with community genomics, is a dynamic approach for finding microbial ecology and differentiating closely related species in an acid mine discharge biofilm (Mishra et al., 2019). The first research to use metaproteome in a wastewater treatment facility found the protein components of housekeeping genes important in improved biological phosphorous removal, which had previously been identified using metagenomics. The phosphorous removal procedure used in this study was carried out in a sequencing batch reactor based on activated sludge (Meena et al., 2021). The optimization of this approach made it possible to identify critical microorganisms and biochemical pathways in multiple treatment processes (Salerno et al., 2016). Depending on the content of effluent and operating conditions of wastewater treatment facilities, the amount of produced proteins may differ from the number of genes that encode proteins, revealing critical treatment pathways. This strategy, when combined with other omics techniques, may also be used to detect functional pathways in uncultured microbes. Integrated omics research of a thermophilic anaerobic digester, metaproteomics was utilized to describe the digestive pathway of acetate and fatty acid intermediates possessed by uncultivated species of Planctomycetes and "candidate phylum *Atribacteria*". It also indicated syntrophic interactions among the bacteria in the digester and a significant abundance of methanogenic pathways (Hagen et al., 2017).

The omics approaches described above provide crucial data on how microbial populations in wastewater treatment facilities work (Aguiar-Pulido et al., 2016). In research, aromatic hydrocarbon catabolism routes in *Pseudomonas putida* KT 2440 were investigated using a multimodal proteomic approach that relies on 2DE/MS and cleavable isotope-coded epitope tag screening (Kubicki et al., 2020). Proteomics-based studies have been useful in determining changes in protein composition and concentration, as well as identifying important proteins involved in microbes' physiological responses to anthropogenic contaminants.

6.5.5 METABOLOMICS

The characterization and quantification of metabolites generated by microbes in a specific environment are known as metabolomics. A metabolome is an organism's total metabolites, and metabolomics is the study of a cell's metabolite profile under certain conditions (Beale et al., 2017). It can be used in conjugation with other omics approaches to determine microbial interactions with other organisms and the environment. Metabolomics is a thorough examination that identifies and quantifies all of a sample's metabolites. The metabolome is thought to be a valuable indication of an environment's health, and it may be used for pathway analysis, medication development, and pharmacogenomics (Aguiar-Pulido et al., 2016). When a cell is stressed, it generates a variety of primary and secondary metabolites, allowing us to better understand and evaluate the impact of environmental factors on an organism's metabolome (Malla et al., 2018a). Metabolomics also seeks to enhance our knowledge of the microbiome's function in the metabolism of nutrients and toxins and other abiotic factors that might disrupt the host environment's homeostasis (Narayanasamy et al., 2015b; Malik et al., 2021). These metabolites can be used as bioindicators to screen the biological consequences of hazardous waste for a deeper knowledge of the ecosystem. Analytical devices, statistical techniques, and bioinformatics tools have improved access to evaluating, extracting, and interpreting different metabolites, as well as elucidating their pathways (Madamwar et al., 2021). The metabolome can reveal how microbial interaction is mediated by signalling systems such as quorum sensing, which links transcriptomic response to variations in cell population density. Metabolomics data are quite different from metagenomics and meta-transcriptomics data, which are largely reliant on sequencing (Lu et al., 2019). Typically, metabolites are characterized and quantified using a perfect blend of chromatography (e.g., liquid chromatography and gas chromatography) and detection techniques (e.g., mass spectrometry and nuclear magnetic resonance), which produce spectra with patterns of peaks that enable both detection and quantification (Lu et al., 2019).

Metabolomics, in combination with bioinformatics tools and databases, has allowed researchers to get a more profound knowledge of the microbial population, catabolic pathways, and genes that encode catabolic enzymes. As a result, it's a great approach to find new metabolic pathways and characterize metabolic networks. Metabolic pathway repositories can be used to track, research, and disseminate information on metabolites and pathways. They provide a metabolic data library and help in converting complicated data into metabolic pathways. These datasets and archives may also be used to model metabolic pathways, which can be researched and simulated via mathematical modelling methods (Chandran, Meena, and Sharma, 2020; Jeevanandam and Osborne, 2021; Gutierrez et al., 2018).

6.5.6 Fluxomics

Fluxomics is a word that refers to a variety of methods for determining the rates of metabolic processes inside a biological organism. The fluxome, or collection of metabolic fluxes, provides a realistic representation of the phenotype because it preserves the metabolome in its functional linkages with the environment and the genome. Fluxomics offers numerous benefits over proteomics and functional genomics, including the fact that it is based on metabolite data, which is much less than the data from genes or proteins (Bharagava et al., 2019; Malla et al., 2018b). *Shewanella* sp., which is known to have co-metabolic mechanisms for remediation of toxic metals, radioisotopes, and heterocyclic aromatic compounds, was studied using fluxomics (Bharagava et al., 2019; Mishra et al., 2019). Fluxome refers to a cell's whole collection of metabolic fluxes, which gives information on a variety of cellular activities, which is a unique phenotypic feature of cells. Fluxome study analyses the metabolome's functional interactions with the genome and environment, revealing important phenotypic information (Bilal, Malik, and Hakeem, 2018). A successful fluxomics research requires the dependability of analytical metabolomics datasets, isotopes labelling observations, and the reconfiguration of metabolic networks establishing stoichiometry and regulation of metabolic processes (Chandran, Meena, and Sharma, 2020; Kitamura, Toya, and Shimizu, 2019). Flux balance analysis, also known as stoichiometric kinetics assessment, and tracer-based metabolomics, which tracks carbon distribution through various pathways using a stable isotope of carbon 13C, are preceded by identification and characterization using analytical techniques such as NMR (nuclear magnetic resonance) or MS (mass spectroscopy) (Kitamura, Toya, and Shimizu, 2019; Malla et al., 2018b).

6.6   BIOINFORMATICS TOOLS FOR OMICS DATA ANALYSIS

The omics techniques outlined so far have considerably improved our understanding of microorganisms' metabolic potential in wastewater treatment facilities. However, to predict how microbes will behave in their environment, a more comprehensive picture of metabolic activities is required to comprehend the effects of the hundreds of different processes occurring simultaneously in the cells of microbes (Kumar and Kumar, 2021; Malla et al., 2018a; Singh et al., 2021). To analyse or correlate molecular and omics data, a variety of *in silico* tools, pipelines, repositories, web resources, and algorithms are employed (Sharma and Shukla, 2020). Table 6.3 summarizes some of the most significant computational and bioinformatics applications used in meta-omics data processing. Bioinformatics is a branch of biology that deals with the use of computational and statistical tools to solve biological problems. It includes phylogenetic analysis, data mining, determining the closest molecular phylogeny, and systems biology, all of which help to simplify the bioremediation process (Chandran, Meena, and Sharma, 2020; Dhanjal and Sharma, 2018; Kumari and Kumar, 2021). *In silico* and computational methods may evaluate microbial interaction, degradation mechanisms, toxicity levels, pollutant chemical structure, and extracellular biological processes. PANTHER gateway, KEGG (Kyoto Encyclopedia of Genes and Genomes) PathPred (Yu and Zhang, 2012), , EAWAG-BBD pathway prediction system, remeDB, enviPath, (Madamwar et al., 2021), and other pathway databases and repositories store data of specific metabolic pathways, processes and enzymes, chemical structures, molar mass, the presence of metabolites in pathways, and so on (Kumari and Kumar, 2021). In bioremediation pathway prediction of environmental pollutants with different endpoints, a combination of *in silico* tools with computational techniques such as pathway prediction system, QSBR (quantitative structure-biodegradation relationship) models, ANN (artificial neural network), and GN (genetic algorithm) has been acquired (Singh et al., 2021). *In silico* techniques are extremely useful for determining newly introduced or persistent hazardous chemicals present in wastewater treatment plants such as pharmaceutical wastes based on various computer algorithms that anticipate the best probable pollutant biodegradation pathway. Aside from pathway prediction and QSBR modelling, there are many other *in silico* techniques for determining the binding mechanism and structural specificity of particular enzymes and their conformation and ligand binding, catalysis, and degradation behaviour (Rawat, Dhasmana, and Kumar, 2020).

TABLE 6.3

In Silico and Bioinformatics Tools for Omics Data Analysis

In Silico Tool/ Database	Description/Application	Web URL	References
remeDB	A new prediction method that uses metagenomic datasets to identify enzymes involved in bioremediation.	https://www.niot.res. in/Remedb/	Sankara Subramanian et al. (2020)
KEGG PathPred	It is a knowledge-based metabolic pathway prediction database that makes use of data from the Kyoto Encyclopaedia of Genes and Genomes.	https://www.genome. jp/tools/pathpred/	Yu and Zhang (2012)
enviPath	It is a database and prediction system for organic environmental pollutants' microbial biotransformation.	https://envipath.org/	Benfenati et al. (2019)
OxDBase	This database identifies the involvement of the oxygenase enzyme in bioremediation.	http://crdd.osdd. net/raghava/oxdbase/	Arora et al. (2009), Malla et al. (2022)
EAWAG-BBD pathway prediction system	This pathway prediction algorithm generates probable microbial degradation routes for organic molecules.	http://eawag-bbd.ethz. ch/predict/	Garg, Khan, and Dutt (2014)
MG-RAST	Metagenome annotation server	http://metagenomics.anl. gov/	Napp et al. (2018)
WebMGA	A web server that may be configured to do rapid metagenomic sequence analysis.	http://weizhongli-lab. org/metagenomic-analysis	Wu et al. (2011)
myPhyloDB	A local web server for the storage and analysis of metagenomic data.	http://www.ars.usda. gov/services/software/ download.htm? softwareid5472, http:// www.myphylodb.org	Manter et al. (2019)
FOAM	It was developed to analyse environmental metagenomic sequencing datasets and includes a new functional ontology that uses HMMs to categorize gene functions relevant to environmental microorganisms.	http://portal.nersc. gov/project/m1317/ FOAM/	Prestat et al. (2014)
UCHIME	For improved sensitivity and speedy chimera detection.	http://drive5.com/uchime	Edgar et al. (2011)
COGNIZER	A method for functionally annotating metagenomic datasets.	http://metagenomics.atc. tcs.com/cognizer, https://metagenomics. atc.tcs.com/function/ cognizer	Bose et al. (2015)

6.7   CONCLUSIONS

The enormous threat presented to the environment by anthropogenic activity has prompted researchers to consider innovative decontamination and clean-up techniques for wastewater bodies. Realizing and discovering the interaction between microbial populations in contaminated settings is challenging. The modern-age omics approaches such as metagenomics, metaproteomics, meta-transcriptomics, metabolomics, and fluxomics have broken down barriers to understanding the processes involved in diverse remediation pathways. Researchers may use these methods to look into the genetic and physiological diversity of an ecosystem, as well as the identity and potential physiological capabilities of its microbes. Such cutting-edge approaches are now essential for studying uncultivated wastewater microorganisms and the metabolism associated with wastewater treatment technology, which is critical for better bioreactor design, management, and interpretation. Additionally, utilizing a systems biology method to integrate meta-omics data, quantitative models for forecasting the reaction of natural or artificial systems to external upheaval may be built.

REFERENCES

Agrawal, N., V. Kumar, and S. K. Shahi. 2021. Biodegradation and detoxification of phenanthrene in *in-vitro* and *in-vivo* conditions by a newly isolated ligninolytic fungus *Coriolopsis byrsina* strain APC5 and characterization of their metabolites for environmental safety. *Environmental Science and Pollution Research.* Doi: 10.1007/s11356-021-15271-w.

Aguiar-Pulido, V., W. Huang, V. Suarez-Ulloa, T. Cickovski, K. Mathee, and G. Narasimhan. 2016. "Metagenomics, metatranscriptomics, and metabolomics approaches for microbiome analysis." *Evolutionary Bioinformatics* 12 (May): 5–16. Doi: 10.4137/EBO.S36436.

Akpor, O. B., D. A. Otohinoyi, D. T. Olaolu, and B. I. Aderiye, 2014. "Pollutants in wastewater effluents: Impacts and remediation processes." *International Journal of Environmental Research and Earth Science* Eprints.Lmu.Edu.Ng. http://eprints.lmu.edu.ng/1023/.

Araújo, W. J., J. S. Oliveira, S. C.S. Araújo, C. F. Minnicelli, R. C.B. Silva-Portela, M. M.B. da Fonseca, J. F. Freitas, et al., 2020. "Microbial culture in minimal medium with oil favors enrichment of biosurfactant producing genes." *Frontiers in Bioengineering and Biotechnology* 8. Doi: 10.3389/fbioe.2020.00962.

Arora, P. K., M. Kumar, A. Chauhan, G. P. Raghava, and R. K. Jain. 2009. "OxDBase: A database of oxygenases involved in biodegradation." *BMC Research Notes* 2. Doi: 10.1186/1756-0500-2-67.

Bashiardes, S., G. Zilberman-Schapira, and E. Elinav. 2016. "Use of metatranscriptomics in microbiome research." *Bioinformatics and Biology Insights* 10: BBI-S34610.

Baydoun, E., A. Wahab, N. Shoaib, and M. S. Ahmad. 2016. "Microbial transformation of contraceptive drug etonogestrel into new metabolites with *Cunninghamella blakesleeana* and *Cunninghamella echinulata.*" *Steroids* Elsevier. https://www.sciencedirect.com/science/article/pii/S0039128X1630099X.

Beale, D. J., A. V. Karpe, W. Ahmed, S. Cook, P. D. Morrison, C. Staley, M. J. Sadowsky, and E. A. Palombo. 2017. "A community multi-omics approach towards the assessment of surface water quality in an urban river system." *International Journal of Environmental Research and Public Health* 14 (3): 303. Doi: 10.3390/ijerph14030303.

Benfenati, E., A. Roncaglioni, A. Lombardo, and A. Manganaro. 2019. "Integrating QSAR, read-across, and screening tools: The VEGAHUB platform as an example." *Challenges and Advances in Computational Chemistry and Physics* 30: 365–81. Doi: 10.1007/978-3-030-16443-0_18.

Bharagava, R. N., G. Saxena, S. I. Mulla, and D. K. Patel. 2018. "Characterization and identification of recalcitrant organic pollutants (ROPs) in tannery wastewater and its phytotoxicity evaluation for environmental safety." *Archives of Environmental Contamination and Toxicology* 75 (2): 259–72. Doi: 10.1007/S00244-017-0490-X.

Bhatt, P., S. Gangola, G. Bhandari, W. Zhang, D. Maithani, S. Mishra, and S. Chen. 2021. "New insights into the degradation of synthetic pollutants in contaminated environments." *Chemosphere* Elsevier Ltd. Doi: 10.1016/j.chemosphere.2020.128827.

Bhatt, P., S. Gangola, P. Chaudhary, P. Khati, G. Kumar, A. Sharma, and A. Srivastava. 2019. "Pesticide induced up-regulation of esterase and aldehyde dehydrogenase in indigenous bacillus Spp." *Bioremediation Journal* 23 (1): 42–52. Doi: 10.1080/10889868.2019.1569586.

Bilal, T., B. Malik, and K. R. Hakeem. 2018. "Metagenomic analysis of uncultured microorganisms and their enzymatic attributes." *Journal of Microbiological Methods* 155 (December): 65–9. Doi: 10.1016/J.MIMET.2018.11.014.

Bose, T., M. M. Haque, R. Cvsk, and S. S. Mande. 2015. "COGNIZER: A framework for functional annotation of metagenomic datasets." *PLoS One* 10 (11). Doi: 10.1371/journal.pone.0142102.

Bursle, E., and J. Robson. 2016. "Non-culture methods for detecting infection." *Australian Prescriber* 39 (5): 171–75. Doi: 10.18773/AUSTPRESCR.2016.059.

Chandra, R., and V. Kumar. 2017a. Detection of *Bacillus* and *Stenotrophomonas* species growing in an organic acid and endocrine-disrupting chemicals rich environment of distillery spent wash and its phytotoxicity. *Environmental Monitoring and Assessment* 189: 26. Doi: 10.1007/s10661-016-5746-9.

Chandra, R. and V. Kumar. 2017b. Phytoextraction of heavy metals by potential native plants and their microscopic observation of root growing on stabilised distillery sludge as a prospective tool for In-situ phytoremediation of industrial waste. *Environmental Science and Pollution Research* 24: 2605–19 Doi: 10.1007/s11356-016-8022-1.

Chandran, H., M. Meena, and K. Sharma. 2020. "Microbial biodiversity and bioremediation assessment through omics approaches." *Frontiers in Environmental Chemistry* 1 (September). Doi: 10.3389/FENVC.2020.570326/FULL.

Chrzanowski, Ł., M. Owsianiak, A. Szulc, R. Marecik, A. Piotrowska-Cyplik, A. K. Olejnik-Schmidt, J. Staniewski, et al., 2011. "Interactions between rhamnolipid biosurfactants and toxic chlorinated phenols enhance biodegradation of a model hydrocarbon-rich effluent." *International Biodeterioration and Biodegradation* 65 (4): 605–11. Doi: 10.1016/j.ibiod.2010.10.015.

Chu, B. T.T., M. L. Petrovich, A. Chaudhary, D. Wright, B. Murphy, G. Wells, and R. Poretsky. 2018. "Metagenomics reveals the impact of wastewater treatment plants on the dispersal of microorganisms and genes in aquatic sediments." *Applied and Environmental Microbiology* 84 (5). Doi: 10.1128/AEM.02168-17.

Clark, M. M., M. D. Paxhia, J. M. Young, M. P. Manzella, and G. Reguera. 2021. "Adaptive synthesis of a rough lipopolysaccharide in *Geobacter sulfurreducens* for metal reduction and detoxification." *Applied and Environmental Microbiology* 87 (20): e00964–21.

Crini, G., and E. Lichtfouse. 2019. "Advantages and disadvantages of techniques used for wastewater treatment." *Environmental Chemistry Letters* 17 (1): 145–55. Doi: 10.1007/S10311-018-0785-9.

Datta, S., K. N. Rajnish, M. S. Samuel, A. Pugazlendhi, and E. Selvarajan. 2020. "Metagenomic applications in microbial diversity, bioremediation, pollution monitoring, enzyme and drug discovery. A review." *Environmental Chemistry Letters* Springer. Doi: 10.1007/s10311-020-01010-z.

de Graaff, D. R., M. C. van Loosdrecht, and M. Pronk. 2020. "Stable granulation of seawater-adapted aerobic granular sludge with filamentous thiothrix bacteria." *Water Research*, Elsevier. Accessed September 26, 2021. https://www.sciencedirect.com/science/article/pii/S0043135420302190.

Dhanjal, D. S., and D. Sharma. 2018, September. "Microbial metagenomics for industrial and environmental bioprospecting: The unknown envoy." *Microbial Bioprospecting for Sustainable Development*: 327–52. Doi: 10.1007/978-981-13-0053-0_18.

Edgar, R. C., B. J. Haas, J. C. Clemente, C. Quince, and R. Knight. 2011. "UCHIME improves sensitivity and speed of chimera detection." *Bioinformatics*, Academic.Oup.Com. https://academic.oup.com/bioinformatics/article-abstract/27/16/2194/255262.

Espinosa, M. J. C., A. C. Blanco, T. Schmidgall, A. K. Atanasoff-Kardjalieff, U. Kappelmeyer, D. Tischler, D. H. Pieper, H. J. Heipieper, and C. Eberlein. 2020. "Toward biorecycling: Isolation of a soil bacterium that grows on a polyurethane oligomer and monomer." *Frontiers in Microbiology* 11 (March). Doi: 10.3389/FMICB.2020.00404/FULL.

Fang, H., H. Zhang, L. Han, J. Mei, Q. Ge, Z. Long, and Y. Yu. 2018. "Exploring bacterial communities and biodegradation genes in activated sludge from pesticide wastewater treatment plants via metagenomic analysis." *Environmental Pollution*, Elsevier. https://www.sciencedirect.com/science/article/pii/S0269749118326253.

Fierer, N., J. Ladau, J. C. Clemente, J. W. Leff, S. M. Owens, K. S. Pollard, R. Knight, J. A. Gilbert, and R. L. McCulley. 2013. "Reconstructing the microbial diversity and function of pre-agricultural tallgrass prairie soils in the United States." *Science*, Sciencemag. Org. Accessed September 25, 2021. https://science.sciencemag.org/content/342/6158/621.abstract.

Fito, J., N. Tefera, and S. W. H. Van Hulle. 2019. "Sugarcane biorefineries wastewater: Bioremediation technologies for environmental sustainability." *Chemical and Biological Technologies in Agriculture* 6 (1). Doi: 10.1186/S40538-019-0144-5.

Gao, N., M. Xia, J. Dai, D. Yu, W. An, S. Li, S. Liu, P. He, L. Zhang, Z. Wu, and X. Bi. 2018. "Both widespread PEP-CTERM proteins and exopolysaccharides are required for floc formation of *Zoogloea resiniphila* and other activated sludge bacteria." *Environmental Microbiology* 20 (5): 1677–92, Wiley Online Library. Doi: 10.1111/1462–2920.14080.

Garg, V., S. Khan, and K. Dutt. 2014. "Systematic analysis of microbial degradation pathway of 1-naphthyl-N-methyl carbamate generated by EAWAG Biocatalysis/biodegradation database-pathway prediction system." *International Journal of Applied Science-Research and Review* 1 (2): 049–055.

Garrido-Sanz, D., M. Redondo-Nieto, M. Guirado, O. Pindado Jiménez, R. Millán, M. Martin, and R. Rivilla. 2019. "Metagenomic insights into the bacterial functions of a diesel-degrading consortium for the rhizoremediation of diesel-polluted soil." *Genes*: 456, Mdpi.Com. Doi: 10.3390/genes10060456.

Gaytán, I., A. Sánchez-Reyes, M. Burelo, M. Vargas-Suárez, I. Liachko, M. Press, S. Sullivan, M. J. Cruz-Gómez, and H. Loza-Tavera. 2020. "Degradation of recalcitrant polyurethane and xenobiotic additives by a selected landfill microbial community and its biodegradative potential revealed by proximity ligation-based metagenomic analysis." *Frontiers in Microbiology* 10 (January). Doi: 10.3389/FMICB.2019.02986/FULL.

Ghosh, S., and A. P. Das. 2018. "Metagenomic insights into the microbial diversity in manganese-contaminated mine tailings and their role in biogeochemical cycling of manganese." Nature.Com. Accessed August 1, 2021. https://www.nature.com/articles/s41598-018-26311-w.

Giri, B. S., S. Geed, K. Vikrant, S. S. Lee, K.-H. Kim, S. K. Kailasa, M. Vithanage, P. Chaturvedi, B. N. Rai, and R. S. Singh. 2021. "Progress in bioremediation of pesticide residues in the environment." *Environmental Engineering Research* 26 (6): 77–100.

Gupta, S. K., H. Shin, D. Han, H. G. Hur, and T. Unno. 2018. "Metagenomic analysis reveals the prevalence and persistence of antibiotic-and heavy metal-resistance genes in wastewater treatment plant." *Journal of Microbiology* 56 (6): 408–15. Doi: 10.1007/s12275-018-8195-z.

Gupta, V. K., I. Ali, T. A. Saleh, A. Nayak, and S. Agarwal. 2012. "Chemical treatment technologies for waste-water recycling—An overview." *RSC Advances* 2 (16): 6380–88.

Gutierrez, D. B., R. L. Gant-Branum, C. E. Romer, M. A. Farrow, J. L. Allen, N. Dahal, Y. W. Nei, et al., 2018. "An integrated, high-throughput strategy for multiomic systems level analysis." *Journal of Proteome Research* 17 (10): 3396–08. Doi: 10.1021/ACS. JPROTEOME.8B00302.

Gutleben, J., M. C. De Mares, J. D. Van Elsas, H. Smidt, J. Overmann, and D. Sipkema. 2018. "The multi-omics promise in context: From sequence to microbial isolate." *Critical Reviews in Microbiology* 44 (2): 212–229.

Ha, D. D., and O. T. Nguyen. 2020. "Degradation of P-chlorocresol by facultative *Thauera* sp. strain DO." *Biotech* 10 (2). Doi: 10.1007/S13205-019-2025-9.

Hagen, L. H., J. A. Frank, M. Zamanzadeh, V. G. H. Eijsink, P. B. Pope, S. J. Horn, and M. Arntzen. 2017. "Quantitative metaproteomics highlight the metabolic contributions of uncultured phylotypes in a thermophilic anaerobic digester." *Applied and Environmental Microbiology* 83 (2). Doi: 10.1128/AEM.01955-16.

Handelsman, J. 2004. "Metagenomics: Application of genomics to uncultured microorganisms." *Microbiology and Molecular Biology Reviews* 68 (4): 669–85. Doi: 10.1128/MMBR.68.4.669-685.2004.

Hart, E. H., C. J. Creevey, T. Hitch, and A. H. Kingston-Smith. 2018. "Of rumen microbiota indicates niche compartmentalisation and functional dominance in a limited number of metabolic pathways between abundant bacteria." *Scientific Reports* 8 (1): 1–11. https://www.nature.com/articles/s41598-018-28827-7.

Hemmat-Jou, M. H., A. A. Safari-Sinegani, A. Mirzaie-Asl, and A. Tahmourespour. 2018. "Analysis of microbial communities in heavy metals-contaminated soils using the metagenomic approach." *Ecotoxicology* 27 (9): 1281–91. Doi: 10.1007/S10646-018-1981-X.

Hesselsoe, M., S. Füreder, M. Schloter, L. Bodrossy, N. Iversen, P. Roslev, P. H. Nielsen, M. Wagner, and A. Loy. 2009. "Isotope array analysis of rhodocyclales uncovers functional redundancy and versatility in an activated sludge." *The ISME Journal* 3 (12): 1349–64. Doi: 10.1038/ismej.2009.78.

Hong, Y. H., C. C. Ye, Q. Z. Zhou, X. Y. Wu, J. P. Yuan, J. Peng, H. Deng, and J. H. Wang. 2017. "Genome Sequencing reveals the potential of *Achromobacter* sp. HZ01 for bioremediation." *Frontiers in Microbiology* 8 (Aug). Doi: 10.3389/FMICB.2017.01507/FULL.

Igiri, B. E., S. I. Okoduwa, G. O. Idoko, and E. P. Akabuogu. 2018, October. "Toxicity and bioremediation of heavy metals contaminated ecosystem from tannery wastewater: A review." *Journal of Toxicology*, https://www.hindawi.com/journals/jt/2018/2568038/.

Ismail, M. M., T. M. Essam, Y. M. Ragab, and F. E. Mourad. 2016. "Biodegradation of ketoprofen using a microalgal–bacterial consortium." *Biotechnology Letters* 38 (9): 1493–502. Doi: 10.1007/S10529-016-2145-9.

Jaiswal, S., D. K. Singh, and P. Shukla. 2019. "Gene editing and systems biology tools for pesticide bioremediation: A review." *Frontiers in Microbiology* 10 (Feb). Doi: 10.3389/FMICB.2019.00087/FULL.

Jamaly, S., A. Giwa, and S. W. Hasan. 2015. "Recent improvements in oily wastewater treatment: Progress, challenges, and future opportunities." *Journal of Environmental Sciences* 37 (Nov): 15–30. Doi: 10.1016/j.jes.2015.04.011.

Jeevanandam, V., and J. Osborne. 2021, August. "Understanding the fundamentals of microbial remediation with emphasize on metabolomics." *Preparative Biochemistry & Biotechnology*, 1–13. Doi: 10.1080/10826068.2021.1946694.

Jeffries, T. C., S. Rayu, U. N. Nielsen, K. Lai, A. Ijaz, L. Nazaries, and B. K. Singh. 2018. "Metagenomic functional potential predicts degradation rates of a model organophosphorus xenobiotic in pesticide contaminated soils." *Frontiers in Microbiology* 9 (Feb). Doi: 10.3389/FMICB.2018.00147/FULL.

Jiang, K., J. Sanseverino, A. Chauhan, S. Lucas, A. Copeland, A. Lapidus, T. G. Del Rio, et al., 2012. "Complete genome sequence of thauera aminoaromatica strain MZ1T." *Standards in Genomic Sciences* 6 (3): 325–35. Doi: 10.4056/SIGS.2696029.

Jin, H., X. Zhang, K. Li, Y. Niu, M. Guo, C. Hu, X. Wan, Y. Gong, and F. Huang. 2014. "Direct bio-utilization of untreated rapeseed meal for effective iturin a production by bacillus subtilis in submerged fermentation." *PLoS One* 9 (10): e111171. Doi: 10.1371/journal. pone.0111171.

Johnson, O. A., and A. C. Affam. 2019. "Petroleum sludge treatment and disposal: A review." *Environmental Engineering Research*, Koreascience.or.Kr. Doi: 10.4491/eer.2018.134.

Kang, A. J., A. K. Brown, C. S. Wong, Z. Huang, and Q. Yuan. 2018. "Variation in bacterial community structure of aerobic granular and suspended activated sludge in the presence of the antibiotic sulfamethoxazole." *Bioresource Technology*, 261:322–328. Doi: 10.1016/j.biortech.2018.04.054.

Karlsson, F. H., V. Tremaroli, I. Nookaew, and G. Bergström. 2013. "Gut metagenome in European women with normal, impaired and diabetic glucose control." *Nature*, 498, 99–103. Doi: 10.1038/nature12198.

Khan, S. T., Y. Horiba, M. Yamamoto, and A. Hiraishi. 2002. "Members of the family comamonadaceae as primary poly(3-Hydroxybutyrate-Co-3-Hydroxyvalerate)-degrading denitrifiers in activated sludge as revealed by polyphasic approach." *Applied and Environmental Microbiology* 68 (7): 3206–14. Doi: 10.1128/AEM.68.7.3206-3214.2002.

Kim, M., K. H. Lee, S. W. Yoon, B. S. Kim, J. Chun, and H. Yi. 2013. "Analytical tools and databases for metagenomics in the next-generation sequencing era." *Genomics & Informatics*, Ncbi.Nlm.Nih.Gov. Accessed September 25, 2021. https://www.ncbi.nlm. nih.gov/pmc/articles/PMC3794082/.

Kitamura, S., Y. Toya, and H. Shimizu. 2019. "13C-metabolic flux analysis reveals effect of phenol on central carbon metabolism in *Escherichia coli*." *Frontiers in Microbiology* 10 (May). Doi: 10.3389/FMICB.2019.01010.

Kubicki, S., I. Bator, S. Jankowski, K. Schipper, T. Tiso, M. Feldbrügge, L. M. Blank, S. Thies, and K. E. Jaeger. 2020. "A straightforward assay for screening and quantification of biosurfactants in microbial culture supernatants." *Frontiers in Bioengineering and Biotechnology* 8 (August): 1–11. Doi: 10.3389/fbioe.2020.00958.

Kumar, A., and A. Dubey. 2020. "Rhizosphere microbiome: Engineering bacterial competitiveness for enhancing crop production." *Journal of Advanced Research*, Cairo University. Doi: 10.1016/j.jare.2020.04.014.

Kumar, A., A. Dubey, M. A. Malla, and J. Dames. 2021. "Pyrosequencing and phenotypic microarray to decipher bacterial community variation in *Sorghum bicolor* (L.) moench rhizosphere." *Current Research in Microbial Sciences* Doi: 10.1016/j. crmicr.2021.100025.

Kumar A., M. B. Ravindran, S. Sarsaiya, H. Chen, S. Wainaina, E. Singh, T. Liu, et al., 2020. "Metagenomics for taxonomy profiling: Tools and approaches." 11 (1): 356–74, Taylor & Francis. Doi: 10.1080/21655979.2020.1736238.

Kumar, M., R. Mahajan, and H. S. Saini. 2020. "Evaluating metabolic potential of *Thauera* sp. M9 for the transformation of 4-chloroaniline (4-CA)." *Biocatalysis and Agricultural*, Elsevier. Accessed September 26, 2021. https://www.sciencedirect. com/science/article/pii/S1878818120308069.

Kumar, V. 2021. Phytoremediation of distillery effluent: Current progress, challenges, and future opportunities. In Saxena, G., V. Kumar, and M.P. Shah (Eds), *Bioremediation for Environmental Sustainability: Toxicity, Mechanisms of Contaminants Degradation, Detoxification and Challenges.* Elsevier. Doi: 10.1016/B978-0-12-820524-2.00014-6.

Kumar, V., I. S. Thakur, and M. P. Shah. 2020a. Bioremediation approaches for pulp and paper industry wastewater treatment: Recent advances and challenges. In: Shah, M.P. (Ed.), *Microbial Bioremediation & Biodegradation*. Springer, Singapore. Doi: 10.1007/978-981-15-1812-6_1.

Kumar, V., I. S. Thakur, and M. P. Shah. 2020c. Bioremediation approaches for pulp and paper industry wastewater treatment: Recent advances and challenges. In: Shah,

M.P. (Ed.), *Microbial Bioremediation & Biodegradation*. Springer, Singapore. Doi: 10.1007/978-981-15-1812-6_1.

Kumar, V., I. S. Thakur, A. K. Singh, and M. P. Shah. 2020b. Application of metagenomics in remediation of contaminated sites and environmental restoration. In: Shah M, S. Rodriguez-Couto, and S. S. Sengor (Eds.), *Emerging Technologies in Environmental Bioremediation*. Elsevier. Doi: 10.1016/B978-0-12-819860-5.00008-0.

Kumar, V., K. Singh, and M. P. Shah. 2021a. Advanced oxidation processes for complex wastewater treatment. In: Shah, M.P. (Eds.), *Advance Oxidation Process for Industrial Effluent Treatment*. Elsevier. Doi: 10.1016/B978-0-12-821011-6.00001-3.

Kumar, V., K. Singh, M. P. Shah, A. K Singh, A. Kumar, and Y. Kumar. 2021b. Application of omics technologies for microbial community structure and function analysis in contaminated environment. In Shah, M.P., A. Sarkar, and S. Mandal (Eds.), *Wastewater Treatment: Cutting Edge Molecular Tools, Techniques & Applied Aspects in Waste Water Treatment*. Elsevier. Doi: 10.1016/B978-0-12-821925-6.00013-7.

Kumar, V., and R. Chandra. 2018. "Characterisation of manganese peroxidase and laccase producing bacteria capable for degradation of sucrose glutamic acid-Maillard reaction products at different nutritional and environmental conditions." *World Journal of Microbiology and Biotechnology* 34: 32.

Kumar, V., and R. Chandra. 2020. "Metagenomics analysis of rhizospheric bacterial communities of *Saccharum arundinaceum* growing on organometallic sludge of sugarcane molasses-based distillery." *Biotech* 10(7): 316. Doi: 10.1007/s13205-020-02310-5.

Kumar, V., S. K. Shahi, and S. Singh. 2018. Bioremediation: An eco-sustainable approach for restoration of contaminated sites. In: Singh, J., D. Sharma, G. Kumar, and N. Sharma (Eds.), *Microbial Bioprospecting for Sustainable Development*. Springer, Singapore. Doi: 10.1007/978-981-13-0053-0_6.

Kumar, V., S. K. Shahi, L. F. R. Ferreira, M. Bilal, J. K. Biswas, and L. Bulgariu. 2021c. Detection and characterization of refractory organic and inorganic pollutants discharged in biomethanated distillery effluent and their phytotoxicity, cytotoxicity, and genotoxicity assessment using *Phaseolus aureus* L. and *Allium cepa* L. *Environmental Research* 201: 111551. Doi: 10.1016/j.envres.2021.111551.

Kumari, P., and Y. Kumar. 2021. "Bioinformatics and computational tools in bioremediation and biodegradation of environmental pollutants." *Bioremediation for Environmental Sustainability*, Elsevier. https://www.sciencedirect.com/science/article/pii/B9780128203187000198.

Lamilla, C., H. Schalchli, G. Briceño, B. Leiva, P. Donoso-Piñol, L. Barrientos, V. A. L. Rocha, D. M. G. Freire, and M. C. Diez. 2021. "A pesticide biopurification system: A source of biosurfactant-producing bacteria with environmental biotechnology applications." *Agronomy* 11 (4): 624. Doi: 10.3390/agronomy11040624.

Lawson, C. E., S. Wu, A. S. Bhattacharjee, J. J. Hamilton, K. D. McMahon, R. Goel, and D. R. Noguera. 2017. "Metabolic network analysis reveals microbial community interactions in anammox granules." *Nature Communications*, 8, 15416. Doi :10.1038/ncomms15416.

Lin, B., J. Lyu, X. Lyu, H. Yu, Z. Hu, and J. C. W. Lam. 2015. "Characterization of cefalexin degradation capabilities of two pseudomonas strains isolated from activated sludge." *Journal of Hazardous Materials*, Elsevier. https://www.sciencedirect.com/science/article/pii/S030438941400569X.

Lira, F., I. Vaz-Moreira, J. Tamames, and C. M. Manaia. 2020. "Metagenomic analysis of an urban resistome before and after wastewater treatment." *Scientific Reports, Nature. Com.* https://www.nature.com/articles/s41598-020-65031-y.

Lu, H., Y. Que, X. Wu, T. Guan, and H. Guo. 2019. "Metabolomics deciphered metabolic reprogramming required for biofilm formation." *Scientific Reports* 9 (1). Doi: 10.1038/S41598-019-49603-1.

Madamwar, D., M. Tsuda, I. Solyanikova, S. Chen, S. Mishra, Z. Lin, S. Pang, W. Zhang, and P. Bhatt. 2021. "Recent advanced technologies for the characterization of xenobiotic-degrading microorganisms and microbial communities." *Frontiers in Bioengineering and Biotechnology* 9: 632059. Doi: 10.3389/fbioe.2021.632059.

Malik, G., R. Arora, R. Chaturvedi, and M. S. Paul. 2021. "Implementation of genetic engineering and novel omics approaches to enhance bioremediation: A focused review." *Bulletin of Environmental Contamination and Toxicology*. Doi: 10.1007/S00128-021-03218-3.

Malla, M. A., A. Dubey, A. Kumar, S. Yadav, A. Hashem, and E. F. Abd-Allah. 2019. "Exploring the human microbiome: The potential future role of next-generation sequencing in disease diagnosis and treatment". *Frontiers in Immunology* 9, 2868. Doi: 10.3389/fimmu.2018.02868.

Malla, M. A., Dubey, A., Raj, A., Kumar, A., Upadhyay, N., & Yadav, S. 2022. "Emerging frontiers in microbe mediated pesticide remediation: Unveiling role of omics and in silico approaches in engineered environment". *Environmental Pollution* 299, 118851. Doi: 10.1016/j.envpol.2022.118851.

Malla, M. A., A. Dubey, S. Yadav, A. Kumar, A. Hashem, and E. F. Abd-Allah. 2018. "Understanding and designing the strategies for the microbe-mediated remediation of environmental contaminants using omics approaches." *Frontiers in Microbiology*. Doi: 10.3389/fmicb.2018.01132.

Manter, D. K., M. Korsa, C. Tebbe, and J. A. Delgado. 2016. "MyPhyloDB: A local web server for the storage and analysis of metagenomic data." *Database*, Academic.Oup.Com. Accessed September 25, 2021. https://academic.oup.com/database/article-abstract/doi/10.1093/database/baw037/2630297.

Maroli, A. S., T. A. Gaines, M. E. Foley, and S. O. Duke. 2018. "Omics in weed science: A perspective from genomics, transcriptomics, and metabolomics approaches." *Weed Science* 66 (6): 681–95, Cambridge.Org. Doi: 10.1017/wsc.2018.33.

Mattarozzi, M, M. Manfredi, B. Montanini, F. Gosetti, A. M. Sanangelantoni, E. Marengo, M. Careri, and G. Visioli. 2017. "A metaproteomic approach dissecting major bacterial functions in the rhizosphere of plants living in serpentine soil." *Analytical and Bioanalytical Chemistry* 409 (9): 2327–39, Springer. Doi: 10.1007/s00216-016-0175-8.

Meena, M., G. Yadav, P. Sonigra, and M.P. Shah. 2021, September. "A comprehensive review on application of bioreactor for industrial wastewater treatment." *Applied Microbiology*, Wiley Online Library. Doi: 10.1111/lam.13557.

Meena, M., K. Divyanshu, S. Kumar, P. Swapnil, A. Zehra, V. Shukla, M. Yadav, and R. S. Upadhyay. 2019. "Regulation of L-proline biosynthesis, signal transduction, transport, accumulation and its vital role in plants during variable environmental conditions." *Heliyon* 5 (12), Elsevier. https://www.sciencedirect.com/science/article/pii/S2405844019366113.

Miao, L., and Z. Liu. 2018. "Microbiome analysis and -omics studies of microbial denitrification processes in wastewater treatment: Recent advances." *Science China Life Sciences* 61 (7): 753–61. Doi: 10.1007/S11427-017-9228-2.

Mishra, A., K. Medhi, P. Malaviya, and I. S. Thakur. 2019. "Omics approaches for microalgal applications: Prospects and challenges." *Bioresource Technology*, Elsevier Ltd. Doi: 10.1016/j.biortech.2019.121890.

Misra, B. B., C. Langefeld, M. Olivier, and L. A. Cox. 2019. "Integrated omics: Tools, advances and future approaches." *Journal of Molecular Endocrinology* 62 (1): R21–45. Doi: 10.1530/JME-18-0055.

Mohan, H., S. S. Rajput, E. B. Jadhav, M. S. Sankhla, S. S. Sonone, S. Jadhav, and R. Kumar. 2021. "Ecotoxicity, occurrence, and removal of pharmaceuticals and illicit drugs from aquatic systems." *Biointerface Research in Applied Chemistry* 11 (5): 12530–46, Researchgate.Net. Doi: 10.33263/BRIAC115.1253012546.

Napp, A. P., J. E. S. Pereira, J. S. Oliveira, R. C. B. Silva-Portela, L. F. Agnez-Lima, M. C. R. Peralba, F. M. Bento, L. M. P. Passaglia, C. E. Thompson, and M. H. Vainstein. 2018. "Comparative metagenomics reveals different hydrocarbon degradative abilities from enriched oil-drilling waste." *Chemosphere* 209: 7–16. Doi: 10.1016/j.chemosphere. 2018.06.068.

Narayanasamy, S., E. E.L. Muller, A. R. Sheik, and P. Wilmes. 2015a. "Integrated omics for the identification of key functionalities in biological wastewater treatment microbial communities." *Microbial Biotechnology* 8 (3): 363–68. Doi: 10.1111/1751-7915.12255.

Narayanasamy, S., E. E. L. Muller, A. R Sheik, and P. Wilmes. 2015b. "Opinion integrated omics for the identification of key functionalities in biological wastewater treatment microbial communities." Doi: 10.1111/1751-7915.12255.

Ngara, T. R., and H. Zhang. 2018. "Recent advances in function-based metagenomic screening." *Genomics, Proteomics and Bioinformatics* 16 (6): 405–15. Doi: 10.1016/J. GPB.2018.01.002.

Otsuka, M., Y. Fukui, and Y. Ozaki. 2009. "Comparative evaluation of bioactivity of crystalline trypsin for drying by Fourier-transformed infrared spectroscopy." *Colloids and Surfaces B: Biointerfaces* 69 (2): 194–200. Doi: 10.1016/j.colsurfb.2008.11.016.

Pacwa-Płociniczak, M., J. Czapla, T. Płociniczak, and Z. Piotrowska-Seget. 2019. "The Effect of bioaugmentation of petroleum-contaminated soil with *Rhodococcus erythropolis* strains on removal of petroleum from soil." 169 (March): 615–22, Elsevier. https://www. sciencedirect.com/science/article/pii/S0147651318312260.

Padhi, S. K., S. Tripathy, R. Sen, A. S. Mahapatra, S. Mohanty, and N. K. Maiti. 2013. "Characterisation of heterotrophic nitrifying and aerobic denitrifying *Klebsiella pneumoniae* CF-S9 strain for bioremediation of wastewater." *International Biodeterioration & Biodegradation*, Elsevier. Accessed September 26, 2021. https:// www.sciencedirect.com/science/article/pii/S0964830513000036.

Pandey, A., P. H. Tripathi, A. H. Tripathi, S. C. Pandey, and S. Gangola. 2019. "Omics technology to study bioremediation and respective enzymes." In *Smart Bioremediation Technologies*, Elsevier, January, 23–43. https://www.sciencedirect.com/science/article/ pii/B9780128183076000020.

Pang, S., Z. Lin, W. Zhang, S. Mishra, P. Bhatt, and S. Chen. 2020. "Insights into the microbial degradation and biochemical mechanisms of neonicotinoids." *Frontiers in Microbiology* 11 (May). Doi: 10.3389/FMICB.2020.00868/FULL.

Pankhurst, L. J., C. Whitby, M. Pawlett, L. D. Larcombe, B. McKew, L. J. Deacon, S. L. Morgan, R. Villa, G. H. Drew, S. Tyrrel, and S. J. Pollard. 2012. "Temporal and spatial changes in the microbial bioaerosol communities in green-waste composting." *FEMS Microbiology Ecology* 79 (1): 229–39, Academic.Oup.Com. https://academic.oup.com/ femsec/article-abstract/79/1/229/671109.

Pant, G., D. Garlapati, U. Agrawal, R. G. Prasuna, T. Mathimani, and A. Pugazhendhi. 2020, November. "Biological approaches practised using genetically engineered microbes for a sustainable environment: A review." *Journal of Hazardous Materials*: 124631, Elsevier. https://www.sciencedirect.com/science/article/pii/S0304389420326212.

Parada, A. E., D. M. Needham, and J. A. Fuhrman. 2016. "Every base matters: Assessing small subunit RRNA primers for marine microbiomes with mock communities, time series and global field samples." *Environmental Microbiology* 18 (5): 1403–14. Doi: 10.1111/1462-2920.13023.

Plewniak, F., S. Crognale, S. Rossetti, and P. N. Bertin. 2018a. "A genomic outlook on bioremediation: The case of arsenic removal." *Frontiers in Microbiology* 9 (Apr). Doi: 10.3389/FMICB.2018.00820.

Popa Ungureanu, C., L. Favier, and G. Bahrim. 2016. "Screening of soil bacteria as potential agents for drugs biodegradation: A case study with clofibric acid." *Journal of Chemical Technology and Biotechnology* 91 (6): 1646–53. Doi: 10.1002/JCTB.4935.

Prestat, E., M. M. David, J. Hultman, N. Taş, R. Lamendella, J. Dvornik, R. Mackelprang, D. D. Myrold, A. Jumpponen, S. G. Tringe, and E. Holman. 2014. "FOAM (Functional ontology assignments for metagenomes): A hidden markov model (HMM) database with environmental focus." *Nucleic Acids Research*, Academic.Oup.Com. https://academic.oup.com/nar/article-abstract/42/19/e145/2902479.

Rahman, S. F., R. S. Kantor, R. Huddy, B. C. Thomas, A. W. van Zyl, S. T. L. Harrison, and J. F. Banfield. 2017. "Genome-resolved metagenomics of a bioremediation system for degradation of thiocyanate in mine water containing suspended solid tailings." *MicrobiologyOpen* 6 (3): 446. Doi: 10.1002/mbo3.446.

Raj, A., A. Kumar, and J. F. Dames. 2021. "Tapping the role of microbial biosurfactants in pesticide remediation: An eco-friendly approach for environmental sustainability." *Frontiers in Microbiology* 12: 791723–791723.

Rawat, G., A. Dhasmana, and V. Kumar. 2020. "Biosurfactants: The next generation biomolecules for diverse applications." *Environmental Sustainability* 3 (4): 353–69. Doi: 10.1007/s42398-020-00128-8.

Rawat, M., and S. Rangarajan. 2019. "Omics approaches for elucidating molecular mechanisms of microbial bioremediation." In *Smart Bioremediation Technologies*, Elsevier. Doi: 10.1016/B978-0-12-818307-6.00011-1.

Rodríguez, A., M. L. Castrejón-Godínez, E. Salazar-Bustamante, Y. Gama-Martínez, E. Sánchez-Salinas, P. Mussali-Galante, E. Tovar-Sánchez, and M. L. Ortiz-Hernández. 2020. "Omics approaches to pesticide biodegradation." *Current Microbiology*, Springer. Doi: 10.1007/s00284-020-01916-5.

Rodríguez, E., P. A. García-Encina, A. J. M. Stams, F. Maphosa, and D. Z. Sousa. 2015. "Meta-omics approaches to understand and improve wastewater treatment systems." *Reviews in Environmental Science and Biotechnology*, Kluwer Academic Publishers. Doi: 10.1007/s11157-015-9370-x.

Rosso, G. E., J. A. Muday, and J. F. Curran. 2018. "Tools for metagenomic analysis at wastewater treatment plants: Application to a foaming episode." *Water Environment Research* 90 (3): 258–68. Doi: 10.2175/106143017X15054988926352.

Rubio-Rincón, F. J., L. Welles, C. M. Lopez-Vazquez, M. Nierychlo, B. Abbas, M. Geleijnse, P. H. Nielsen, M. C. van Loosdrecht, and D. Brdjanovic. 2017. "Long-term effects of sulphide on the enhanced biological removal of phosphorus: The symbiotic role of thiothrix caldifontis." *Water Research*, Elsevier. Accessed September 26, 2021. https://www.sciencedirect.com/science/article/pii/S0043135417301835.

Salerno, C., D. Benndorf, S. Kluge, L. L. Palese, U. Reichl, and A. Pollice. 2016. "Metaproteomics applied to activated sludge for industrial wastewater treatment revealed a dominant methylotrophic metabolism of *Hyphomicrobium zavarzinii*." *Microbial Ecology* 72 (1): 9–13. Doi: 10.1007/S00248-016-0769-X.

Sankara Subramanian, S. H., K. R. S. Balachandran, V. R. Rangamaran, and D. Gopal. 2020. "RemeDB: Tool for rapid prediction of enzymes involved in bioremediation from high-throughput metagenome data sets." *Journal of Computational Biology* 27 (7): 1020–29. Doi: 10.1089/CMB.2019.0345.

Sato, Y., T. Hori, H. Koike, R. R. Navarro, A. Ogata, and H. Habe. 2019. "Transcriptome analysis of activated sludge microbiomes reveals an unexpected role of minority nitrifiers in carbon metabolism." *Communications Biology* 2 (1): 1–8, *Nature.Com*. https://www.nature.com/articles/s42003-019-0418-2.

Shah, A., and M. Shah. 2020, October. "Characterisation and bioremediation of wastewater: A review exploring bioremediation as a sustainable technique for pharmaceutical wastewater." *Groundwater for Sustainable Development*: 100383. https://www.sciencedirect.com/science/article/pii/S2352801X19302607.

Sharma, B., and P. Shukla. 2020. "Designing synthetic microbial communities for effectual bioremediation: A review." *Biocatalysis and Biotransformation* 38 (6): 405–14. Doi: 10.1080/10242422.2020.1813727.

Shi, L. D., Y. S. Chen, J. J. Du, Y. Q. Hu, J. P. Shapleigh, and H. P. Zhao. 2019. "Metagenomic evidence for a methylocystis species capable of bioremediation of diverse heavy metals." *Frontiers in Microbiology* 10 (Jan). Doi: 10.3389/FMICB.2018.03297/FULL.

Shi, X., K. Y. Leong, H. Y., and Ng. 2017. "Anaerobic treatment of pharmaceutical wastewater: A critical review." *Bioresource Technology.* https://www.sciencedirect.com/science/article/pii/S0960852417314645.

Silva, C. C., H. Hayden, T. Sawbridge, P. Mele, R. H. Kruger, M. V. N. Rodrigues, G. G. L. Costa, et al., 2012. "Phylogenetic and functional diversity of metagenomic libraries of phenol degrading sludge from petroleum refinery wastewater treatment system." *AMB Express* 2 (1): 1–13. Doi: 10.1186/2191-0855-2-18.

Silva, C. C., H. Hayden, T. Sawbridge, P. Mele, S. O. De Paula, L. C. F. Silva, P. M. P. Vidigal, et al., 2013. "Identification of genes and pathways related to phenol degradation in metagenomic libraries from petroleum refinery wastewater." *PLoS One* 8 (4). Doi: 10.1371/journal.pone.0061811.

Singh, A. K., M. Bilal, H. M. N. Iqbal, and A. Raj. 2021. "Trends in predictive biodegradation for sustainable mitigation of environmental pollutants: Recent progress and future outlook." *Science of The Total Environment*, Elsevier. Accessed June 30, 2021. https://www.sciencedirect.com/science/article/pii/S004896972038092X.

Sponza, D. T., and O. Gok. 2011. "Effects of sludge retention time and biosurfactant on the treatment of polyaromatic hydrocarbon (PAH) in a petrochemical industry wastewater." *Water Science and Technology* 64 (11): 2282–92. Doi: 10.2166/wst.2011.734.

Sraw, A., T. Kaur, Y. Pandey, A. Sobti, R. K. Wanchoo, and A. P. Toor. 2018. "Fixed bed recirculation type photocatalytic reactor with TiO$_2$ immobilized clay beads for the degradation of pesticide polluted water." *Journal of Environmental Chemical Engineering* 6 (6): 7035–43, Elsevier. https://www.sciencedirect.com/science/article/pii/S2213343718306717.

Staley, C., T. Unno, T. J. Gould, B. Jarvis, J. Phillips, J. B. Cotner, and M. J. Sadowsky. 2013. "Application of Illumina next-generation sequencing to characterize the bacterial community of the upper Mississippi river." *Journal of Applied Microbiology* 115 (5): 1147–58. Doi: 10.1111/JAM.12323.

Steliga, T., K. Wojtowicz, P. Kapusta, and J. Brzeszcz. 2020. "Assessment of biodegradation efficiency of polychlorinated biphenyls (PCBs) and petroleum hydrocarbons (TPH) in soil using three individual bacterial strains and their mixed culture." *Molecules* 25 (3): 709, Mdpi.Com. Doi: 10.3390/molecules25030709.

Suneja, G., and R. Sharma. 2020. "Insights into the microbial processes prevalent in wastewater treatment systems using omics approaches." In *Removal of Toxic Pollutants Through Microbiological and Tertiary Treatment*, 409–30, Elsevier. Doi: 10.1016/B978-0-12-821014-7.00016-2.

Techtmann, S. M., and T. C. Hazen. 2016. "Metagenomic applications in environmental monitoring and bioremediation." *Journal of Industrial Microbiology and Biotechnology*, Springer Verlag. Doi: 10.1007/s10295-016-1809-8.

Thies, S., S. C. Rausch, F. Kovacic, A. Schmidt-Thaler, S. Wilhelm, F. Rosenau, R. Daniel, W. Streit, J. Pietruszka, and K. Erich Jaeger. 2016. "Metagenomic discovery of novel enzymes and biosurfactants in a slaughterhouse biofilm microbial community." *Scientific Reports* 6 (1): 1–12. Doi: 10.1038/srep27035.

Tiwari, B., B. Sellamuthu, Y. Ouarda, P. Drogui, R. D. Tyagi, and G. Buelna. 2017. "Review on fate and mechanism of removal of pharmaceutical pollutants from wastewater using biological approach." *Bioresource Technology*, Elsevier. Doi: 10.1016/j.biortech.2016.11.042.

Uhlik, O., M. C. Leewis, M. Strejcek, L. Musilova, M. Mackova, M. B. Leigh, and T. Macek. 2013. "Stable isotope probing in the metagenomics era: A bridge towards improved bioremediation." *Biotechnology Advances*. Accessed September 25, 2021. https://www.sciencedirect.com/science/article/pii/S073497501200153X.

Wacławek, S., H. V. Lutze, K. Grübel, V. V. Padil, M. Černík, and D. D. Dionysiou. 2017. "Chemistry of persulfates in water and wastewater treatment: A review." 44–62, December. https://www.sciencedirect.com/science/article/pii/S1385894717312718.

Wang, D. Z., L. F. Kong, Y. Y. Li, and Z. X. Xie. 2016. "Environmental microbial community proteomics: Status, challenges and perspectives." *International Journal of Molecular* 17 (8), Mdpi.Com. Doi: 10.3390/ijms17081275.

Wang, S., and J. Wang. 2018. "Biodegradation and metabolic pathway of sulfamethoxazole by a novel strain *Acinetobacter* sp." *Applied Microbiology and Biotechnology* 102 (1): 425–32. Doi: 10.1007/S00253-017-8562-4.

Wang, Z., W. Li, H. Li, W. Zheng, and F. Guo. 2020. "Phylogenomics of rhodocyclales and its distribution in wastewater treatment systems." *Scientific Reports* 10 (1): 1–12, *Nature. Com.* https://www.nature.com/articles/s41598-020-60723-x.

Wang, Z., X. X. Zhang, X. Lu, B. Liu, Y. Li, C. Long, and A. Li. 2014. "Abundance and diversity of bacterial nitrifiers and denitrifiers and their functional genes in tannery wastewater treatment plants revealed by high-throughput sequencing." *PLoS One* 9 (11). Doi: 10.1371/journal.pone.0113603.

Wu, S., Z. Zhu, L. Fu, B. Niu, and W. Li. 2011. "WebMGA: A customizable web server for fast metagenomic sequence analysis." *BMC Genomics* 12 (September). Doi: 10.1186/1471-2164-12-444.

Xing, Z., T. Hu, Y. Xiang, P. Qi, and X. Huang. 2020. "Degradation mechanism of 4-chlorobiphenyl by consortium of *Pseudomonas* sp. Strain CB-3 and *Comamonas* sp. strain CD-2." *Current Microbiology* 77 (1): 15–23. Doi: 10.1007/S00284-019-01791-9.

Xu, X., W. Liu, S. Tian, W. Wang, Q. Qi, P. Jiang, X. Gao, F. Li, H. Li, and H. Yu. 2018. "Petroleum hydrocarbon-degrading bacteria for the remediation of oil pollution under aerobic conditions: A perspective analysis." *Frontiers in Microbiology* 9. Doi: 10.3389/FMICB.2018.02885/FULL.

Yadav, S., and A. Kapley. 2019. "Exploration of activated sludge resistome using metagenomics." *Science of the Total Environment* 692, 1155–1164.

Yang, Y., K. Yu, Y. Xia, F. T. K. Lau, D. T. W. Tang, W. C. Fung, H. H. P. Fang, and T. Zhang. 2014. "Metagenomic analysis of sludge from full-scale anaerobic digesters operated in municipal wastewater treatment plants." *Applied Microbiology and Biotechnology* 98 (12): 5709–18. Doi: 10.1007/S00253-014-5648-0.

Yoneda, A., W. R. Henson, N. K. Goldner, K. J. Park, K. J. Forsberg, S. J. Kim, M. W. Pesesky, M. Foston, G. Dantas, and T. S. Moon. 2016. "Comparative transcriptomics elucidates adaptive phenol tolerance and utilization in lipid-accumulating *Rhodococcus opacus* PD630." *Nucleic Acids Research* 44 (5): 2240–54.

Yu, K., and T. Zhang. 2012. "Metagenomic and metatranscriptomic analysis of microbial community structure and gene expression of activated sludge." *PLoS One* 7 (5). Doi: 10.1371/journal.pone.0038183.

7 Omics Insights into Quorum Sensing and Biofilm Formation

Edwin Hualpa-Cutipa
Universidad Nacional Mayor de San Marcos

Richard Andi Solórzano Acosta
Universidad César Vallejo

Daniela Landa-Acuña
National Agrarian University La Molina (UNALM)
Universidad Privada del Norte

María Elena Salazar Salvatierra and
Julio Reynaldo Ruiz Quiroz
Universidad Nacional Mayor de San Marcos

CONTENTS

7.1 INTRODUCTION

Biofilm is a bacterial community at large structural aggregation, can establish itself on both living and non-living hosts (Chen and Wen, 2011), is regulated by quorum sensing (QS), and is a mode of sending and receiving signal through inductor and regulator molecules (Brindhadevi et al., 2020). This particular phenomenon of bacterial ecology has become an emerging problem in everyday life because biofilm-forming microbes grow on living surfaces and leads to diseases of the respiratory tract, valvular endocarditis, eyes, teeth, wound infections, diabetic foot ulcers, periodontitis,

DOI: 10.1201/9781003247883-7

and urinary tract infections (Donlan, 2002); also, medical devices such as catheters, breast implants, skimming, synthetic valves, coronary stents, and neurosurgical, cochlear, and ocular devices support the growth of bacterial biofilms (Arciola et al., 2018). When bacteria form biofilms, the structural matrix of the bacteria changes, allowing them to develop resistance to a wide range of natural and synthetic molecules (Manner et al., 2017), which have a major impact on the management, treatment, and exploitation of this bacterial phenomenon; this is why the QS has received significant attention in the clinical field and has begun to be studied at the molecular level (Wu et al., 2020), where discovering the genetic and molecular foundations of the microbial community behavior related to QS and biofilm formation points to therapeutic targets that may provide a means for biofilm infection control (Sauer, 2003) that avoids the development of resistance, as with the use of antibiotics. Only limited information has been generated on issues related to the role of QS in food contamination and spoilage, biofilm establishment, or the growth and/or production of toxins in the food industry (Annous et al., 2009).

In another perspective, the use of QS signals in the fabrication of engineered biofilms with improved degradation kinetics in comparison with bioremediation technology is being investigated (Mangwani et al., 2016). While genetic and biochemical studies have identified some genetic regulators of biofilm formation, the quorum sensing and biofilm formation remain unexplored areas of research (Hegde, 2020), which have been addressed with omics technologies, with their attendant limitations and benefits. Advances in "omics" tools such as transcriptomics, proteomics, metabolomics, and multi-omics data integration represent the future of biofilm regulation research to gain better insights into biofilm phenotyping at the systems level, which may allow the development of effective strategies to control clinical, food, and environmental biofilms (Seneviratne et al., 2020). The study of QS and biofilms by conventional methods does not elucidate their deep mechanisms, so the use of omics analysis, such as transcriptomics, metabolomics, and proteomics, is important (Silva, Marques and Röder, 2021). This chapter highlights the application of omics sciences in the understanding of chemical communication through mechanisms such as bacterial QS and its influence on biofilm formation.

7.2   TRANSCRIPTOMIC INSIGHTS INTO QUORUM SENSING WITHIN BIOFILM

Transcriptomics is an "omics tool" that studies the set of RNAs produced when a cell, tissue, or organ is subjected to a specific condition (Fang et al., 2012). Transcriptomic analysis of an organism can give a first picture about gene regulation under specific conditions, for example, the transition from the planktonic to the biofilm state (Seneviratne et al., 2020). The traditional approaches to studying the transcriptome (EST, qPCR, high-throughput approaches, microarrays, RNA-Seq, SAGE, etc.) have evolved over time (Hrdlickova, Toloue, and Tian, 2017; Lowe et al., 2017). Some differential expression studies by RNA-Seq have been applied to *Pseudomonas aeruginosa* (Waite et al., 2006; Cao et al., 2017), *Vibrio tapetis* (Rodrigues et al., 2018), and *Porphyromonas gingivalis* (Sánchez et al., 2018).

QS is a key mechanism for communication in a microbial community associated with a biofilm, where virulence factors, biofilm maturation, and other factors are modulated, and the PqsE gene has been identified as a key regulator in *Pseudomonas aeruginosa's* environmental adaptation (Rampioni et al., 2010). Also, the regulatory function of the AbaM gene in the QS abaR/abaI system of *Acinetobacter baumannii* was described by this technique (López-Martín et al., 2021), and the elucidation of the regulation of QS in *Vibrio cholerae* by autoinducer-2 (AI-2), cholera autoinducer-1 (CAI-1), and 3,5-dimethylpyrazin-2-ol (DPO) was through the application of RNA-Seq (Herzog et al., 2019).

Multi-species consortia are common in nature and in infections, where concentration gradients of substrates and metabolic products influence the growth of microorganisms in biofilms (Joshi, Gunawan, and Mann, 2021). *Clostridioides difficile* and *Bacteroides fragilis* (Slater et al., 2019); *Candida albicans* and *Streptococcus mutans* (Sztajer et al., 2014); and a consortium of three bacterial species (*Variovorax* sp. WDL1, *Hyphomicrobium sulfonivorans* WDL6, and *Comamonas testosteroni* WDL7) are some examples of transcriptomics applications in consortia analysis. The search for new therapeutic alternatives to combat biofilm-related infections is one of today's biggest challenges, and QS disruption could be an intriguing strategy to control their progression. Transcriptomic analyses can assist in the verification of the mechanisms by which many substances exert their effects in biofilms, e.g. the antibacterial effect of *Pine honey* on biofilm formation and QS in In *P. aeruginosa* (Kafantaris et al., 2021). Also, S-phenyl-l-cysteine sulfoxide isolated from *Petiveria alliacea* functions as an analogue of kynurenine, the last one being a precursor for the synthesis of anthranilate critical for *P. aeruginosa* virulence (Kasper et al., 2016), A synthetic furanone (Furanone C- 30) acts on the QS of *P. aeruginosa* inhibiting its virulence factors. (Hentzer et al., 2003), coumarin which inhibits the formation of biofilms and blocks the QS (Zhang et al., 2018), and 6-gingerol (5-hydroxy-1-(4 -hydroxy-3-methoxyphenyl)-3-decanone) *in silico* and experimentally studies have been shown to reduce biofilm formation and additionally repress the expression of QS-dependent genes (specifically related to virulence factors). (Kim et al., 2015).

In methicillin-resistant *Staphylococcus aureus* (MRSA), the action of ursolic acid and resveratrol (Qin et al., 2014) and norlichexanthone (unreduced tricyclic polyketide) (Baldry et al., 2016) was investigated. Another example is the effect of three QS inhibitors (cinnamaldehyde, furanone C-30, and F-DPD) on uropathogenic *Escherichia coli* (Henly et al., 2021).

7.3 PROTEOMIC INSIGHTS INTO THE QS REGULATION FOR BIOFILMS FORMATION

Proteins are the main part of all organisms including eukaryotic and prokaryotic cells because they control many processes such as communication through signals among bacteria leading to changes in the collective behavior that results from the gene expression, for example from motile life to adhesion on surfaces and aggregate into clusters covered with a matrix of different molecules depending on the species. QS is a cell-to-cell communication that can detect high bacterial density and then release different kinds of diffusible molecules. The predominance of signal molecules differs among bacterial species. In gram-negatives, signal molecules used in QS are generally N-acyl homoserine lactones (AHLs); the AHL molecules can contain 4 and 14 carbons with an oxo or a hydroxyl substitution at the third position. Oligopeptides called autoinducers (AIs) are found in gram-positive bacteria, and the luxS-encoded AI-2 is present in gram-negative and gram-positive bacteria. The activated genes can induce competition response, bioluminescence, and virulence factors. The last one refers to pathogens that can modulate the formation of biofilms detecting the population density and assuring high number of cells inside it. Besides, during biofilm maturation, bacteria start to emit signals that result in the expression of genes for developing virulence factors such as biofilm formation. Bacteria within biofilms are more resistant to chemical treatments such as conventional antibiotics agents, and a lot of strategies are being evaluated to control it, and one of the most studied is the disruption of QS signals, which at the same time could improve the antibiotics action.

7.3.1   ENVIRONMENTAL ADAPTATION OF BIOFILMS

Gram-positive coagulase-negative staphylococci (CoNS) are a human microbiota colonizer, but can be associated with infections in immunosuppressed patients. The ability of CoNS to produce virulence factors such as biofilm can be regulated by a QS system. *S. aureus* is the only pathogen of this genus that produces fibronectin-binding protein, hemolysins, protein A, lipase, enterotoxins, as well as other virulence factors such as biofilm formation via the agr QS system (Ivanova et al., 2018), and gene expression activation needs to be accurately timed in sequence to be effective, increasing the virulence and complicating the treatment of infections caused by this bacterium. The inhibition of QS could provide a promising alternative because it would make this pathogen less virulent and sensitive to treatments. The agr QS or AgrBDCA operon has two promoters that are dissimilar, P2 and P3, associated with the RNA II and RNA III transcripts, respectively. The RNA II locus contains the four genes, *agrB, agrD, agrC,* and *agrA*. AgrB and agrD products have a coordinated function, and AgrB is a transmembrane endopeptidase that modifies the AgrD, an oligopeptide precursor that is secreted after a C-terminal cleavage and acts as an AI. Then a cascade of reactions begin, so AIs can be detected by a sensor histidine kinase, AgrC, that immediately transfers its phosphoryl group to the DNA-binding response regulator, AgrA, and the regulating RNA, RNA III, is expressed (Paulander et al., 2018). RNA III acts as a switch for the expression of different virulence factors such as surface proteins (Gomes-Fernandes et al., 2017) that could be a key to the bacterial establishment on a surface, but is also related to maturation and dispersal of biofilms (Gupta et al., 2016). So, at high density, the agr QS system decreased the production of factors associated with cell wall, causing the dispersion (Lee et al., 2020).

Vibrio spp. species have diverse QS systems that induce collective responses such as the release of luciferase and other molecules which act in intracellular regulation and other functions. Multiple QS systems were identified in Vibrio pathogen species, which realize similar functions. Studies in *Vibro harveyi.* a fish pathogen has identified luxN and lux R homologs that are involved in the control of virulence processes, e.g. biofilm production. Using the bio-orthogonal non-canonical amino acid tagging (BONCAT) method, different proteins patterns were identified in *V. harveyi*, which change during the transition of individual to group behavior (Bagert et al., 2016). The pathogen *Vibrio cholerae* also produces AIs as part of the QS system that controls biofilms; specifically, CAI-1 is restricted to *Vibrio* and is the product of the CqsA synthase that belongs to the PLP-dependent acyl-CoA transferase family, and AI-2 (4,5-dihydroxy-2,3-pentan edione), the product of the LuxS synthase, which is expressed in many bacteria; the receptors are CqsS (cholera QS sensor) and LuxPQ, respectively. When bacteria do not produce AI or at low density of cells, receptors work as kinases and phosphory-late LuxU, which shuttle the phosphate to LuxO. The phosphorylated LuxO then acti-vates the transcription of four genes encoding four small RNAs: Qrr-1 to Qrr-4, which induce AphA translation and repress HapR, a situation that promotes virulence fac-tors and surface biofilm formation (Jemielita et al., 2018). In high-density conditions, binding of CAI-1 and AI-2 to receptors takes place, the phosphorylation cascade is reverted, and the production of Qrr1 ceases; consequently, aphA translation is not acti-vated, inversely to hapR translation, which once activated represses genes required for virulence as biofilm formation (Papenfort et al., 2017). The 3,5-dimethylpyrazin-2-ol (DPO), a QS AI, and receptor VqmA have also been described in *V. cholera,* as well as synthase Tdh and VqmR regulator that take part in the control of population behav-iors such as biofilm formation. The synthesis of virulence factors and biofilm develop-ment by *Pseudomonas aeruginosa,* an opportunistic pathogen, could be QS-regulated, although QS-deficient strains also form biofilms, but with distinct structure. Four inter-connected QS systems are known in *P. aeruginosa*. The two major systems are *las* and *rhl*, which have HSL as AIs: N-3-oxododecanoyl-L-homoserine lactone (OdDHL) for *las* system and N-butyryl-L-homoserine lactone (BHL) for *rhl* system. The third QS system is a Pseudomonas quinolone signal (PQS)-based one, which depends on pqs-ABCDE operon that encodes the 2-alkyl-4-quinolones (AQs) synthesizing enzymes, and the last system is known as the integrated QS (Iqs) system that is related to an ambBCDE cluster, which synthesizes 2-(2-hydroxyphenyl)-thiazole-4-carbaldehyde, the QS signaling molecule. More than 10% of the population can be regulated by QS of *P. aeruginosa* genome, including virulence factors and biofilm formation genes. Each kind of QS system molecules binds to a specific receptor and triggers the activation of different genes for specific factors such as biofilm formation and other virulence fac-tors. The major systems, *las* and *rhl*, possibly are the principal mechanisms to control the virulence factors expression and biofilm development. In a study of the influence of these both systems on the protein pattern of *P. aeruginosa* PAO1 by comparing the wild type and the mutants for one or both systems and analyzing by two-dimensional gel electrophoresis (2DE) and MALDI/MS, it was concluded that the absence of the *rhl* or *las* gene interferes with virulence factors secretion, such as LasB (Sauvage and Hardouin, 2020).

Using sequential window acquisition of all theoretical fragment ion spectra mass spectrometry (SWATH-MS) to evaluate protein expression patterns of clinical isolates, it was shown that in biofilm growth, the proteome profiles were very different between strains, meanwhile in planktonic growth, they were very similar and both were not related to genetic background metabolic processes. Biofilm patterns were not linked to protein profiles under planktonic growth conditions, but an increase in certain proteins related to PQS and other virulence factors was found (Erdmann et al., 2019). Using the two-dimensional difference gel electrophoresis (2D-DiGE) method, the cell-associated and extracellular proteome in *lasI*, *rhlI*, and *pqsR* mutants was characterized. All mutants were unable to make AIs. In cell-associated and extracellular proteome, the high number of proteins corresponded to LasI mutant, confirming that it is the major system. The expression of extracellular protein decreased for all QS mutants. *P. aeruginosa* QS possibly is affected by some antibiotics. Sub-MIC azithromycin (AZM) affects a few proteins, but about half of them were related to QS (Swatton et al., 2016).

Understanding bacteria communication through QS signals is a relevant aspect because it is an important pathway that stimulates virulence factors such as biofilms formation and can provide the base for the development of strategies to control resistant pathogens.

7.4   METABOLOMIC INSIGHTS INTO THE QS WITHIN THE BIOFILM

Biofilms are complex microbial communities formed by bacteria or fungi. These microorganisms synthesize and secrete an extracellular matrix that protects them from external threats and allows firm adherence of the biofilm to the biotic or abiotic surface (Stoodley et al., 2002). Likewise, biofilms are continuously dynamic heterogeneous communities. They may consist of a single bacterial or fungal species, although polymicrobial communities are possible. At the most basic level, a biofilm can be described as a collection of bacteria embedded in a dense, viscous barrier of sugars and proteins (Dowd et al., 2008).

Different phases of biofilm formation are considered: The first phase is the reversible adhesion phase to the surface, which involves the response of microorganisms to form clusters under natural conditions with the production of extracellular components; at this level, adhesion is reversible (Stoodley et al., 2002). The next phase is the irreversible adhesion to the surface due to gene expression patterns that promote survival, as a result of bacterial self-induction known as QS (Horswill et al., 2007), and finally in the third phase, a protective matrix known as extracellular polymeric substance (EPS) is formed. This protective matrix is usually formed by polysaccharides, proteins, glycolipids, and bacterial DNA that confer protection advantages, although its composition varies according to the microorganisms that conform it (Rice et al., 2007).

So, how do microbial cells respond to environmental fluctuations? Bacterial cells may experience changes in their cell envelopes and the degree of fluidity of their membrane, and alterations in membrane lipids as an essential adaptive response to adverse environmental conditions (Baysse, 2005). Likewise, cells can recognize structural changes in their membranes in response to certain environmental changes that trigger adaptive cellular responses (Rumbaugh and Sauer, 2020). QS is a communication system that regulates the communal behavior of various microorganisms (Chen et al., 2017), through autoinducers that upon reaching a concentration that exceeds the threshold form a transcriptional complex that binds to the corresponding receptor and activates the expression of genes that trigger biofilm formation; among other types of responses (virulence factors, bioluminescence, and drug resistance), lactone derivatives, homoserine, and cyclic thiolactone have been identified as auto-inducers, in large negative and large positive bacteria (Dong and Zhang, 2005).

In this sense, QS has widely been investigated in substantial populations of negative bacteria such as *P. aeruginosa*. Some have three QS systems that are linked together. Two of them are based on N-acyl homoserine lactone (AHL), and these are the rhl (Camara et al., 2002). The third is the PQS system. These mediate the synthesis of 3-oxo-C12-HSL, C4-HSL, and PQS, respectively, where the rhl and pqs systems mediate them hierarchically. Among the N-Homoserine molecules generated by *P. aeruginosa* is that produced by the N-(3-oxododecanoyl)-L-homoserine lactone (3-oxo-C12-HSL), and the N-butanoyl-L-homoserine lactone (C4-HSL) synthesized by RhlI. Likewise, the molecule 2-heptyl-3-hydroxy-4-quinolone (PQS) is a third signal molecule that was originally described as a regulator of LasB protease production (D'Argenio et al., 2002). On the other hand, some rhl-dependent phenotypes, such as pyocyanin synthesis, depend on the presence of PQS at the onset of the stationary phase, whose production depends on the regulation by lasR such that PQS potentiates the expression of rhlI, rhlR, and the sigma factor rpoS (Diggle et al., 2003). Meanwhile, a proteomic study determined an increase in the level of RelA protein in *P. aeruginosa* under magnesium limitation, considering that membrane fluidity can be altered by the presence of ions such as Ca^{2+}, Mg^{2+}, and Sr^{2+} (Baysse, 2005).

Hordenine, a QS inhibitory agent, causes destruction of QS, by downregulating the expressions of genes involved in QS in *P. aeruginosa*, leading to QS malfunction with consequent alteration in the plasma membrane resulting in its permeability and disrupting the energy metabolism of these cells (Zhou et al., 2019). Another inhibitor, chlorolactone (CL), interferes with QS in *Chromobacterium violaceum* by inhibiting LasR/RhlR CviR homologues; likewise, other molecules such as chloro-thiolactone (CTL), meta-bromo-thiolactone (mBTL), and meta-chloro-thiolactone (mCTL) have been evaluated in strains of *P. aeruginosa*, because of activation by LasR:3OC12-HSL. The RhlR:C4-HSL complex activates the pyocyanin operon's expression, a virulence factor (Swem, 2009).

In contrast, in vivo studies have demonstrated that neither CL nor CTL inhibited pyocyanin production. Additionally, mCTL and mBTL inhibited pyocyanin without affecting *P. aeruginosa* growth (Dietrich et al., 2006). The mBTL peptide is an analog of native autoinducers of *P. aeruginosa* and inhibits the expression of genes involved in the pathogenicity pyocyanin, preventing the formation of biofilms, as both the quorum sensing receptors mBTL and mCTL inhibits LasR and RhlR to a certain extent (O'Loughlin et al., 2013).

7.5   DRAWBACKS AND FUTURE PERSPECTIVES OF THE OMICS TECHNOLOGIES FOR UNDERSTANDING QUORUM SENSING FOR BIOFILM FORMATION

a. Genomics and quorum sensing

Understanding the genomic diversity of bacterial strains and relevant bioactive products is important in today's science; genetic and genomic

studies have shed new light on the mechanisms governing biofilm development and the acquisition of unique biofilm phenotypes in vitro (Nobile and
Mitchell, 2006); moreover, genomic studies have identified a wide range of
genes that are differentially expressed throughout biofilm formation, outlin-
ing the complex nature of this growth stage (Sauer, 2003).

Gene regulation is known to involve cell-to-cell interaction of signaling molecules that modulate many physiological characteristics of
microorganisms, such as virulence expression, biofilm formation, bioluminescence, and conjugation, and these processes are generally composed
of three main QS systems, the autoinducer-1 (AI-1)/LuxR system, the
autoinducer-2 (AI-2)/LuxS system, and the autoinducer-3 (AI-3) system
(Zhou et al., 2016). A plethora of studies using QS systems have looked
into a variety of microbial behaviors (Papenfort and Bassler, 2016).
Understanding the genetic basis of QS systems could pave the way for
future research into microbial sociability in a variety of host and non-host
environments (Chen et al., 2020). At present, pan-genome and bacterial
genome analysis tools can also locate auxiliary genes and the core genome
SNPs associated with two different biofilm phenotypes, which play roles
in biofilm biosynthesis and QS (Redfern et al., 2021). The resolution of
the structure and function of exopolysaccharides (EPS) and demonstrating the significance of in vitro observations with recently developed in
vivo models are the main future challenges in genomics and QS research
(Nobile and Mitchell, 2006).

b. **Metagenomics and quorum sensing**

Although much research has been conducted to understand cell-to-cell
bacterial signaling systems, because most microorganisms in natural
environments are not culturable. There is still a large gap in our current
knowledge; however, metagenomics technology is capable of studying
microorganisms in the environment that are not yet culturable and represent
more than 99% of organisms in some environments; although much research
has been conducted to understand cell-to-cell bacterial signaling systems,
there is still a large gap in our current research. The goal of metagenomic
research has been to identify novel QS systems from uncultured bacteria
in environmental samples (Nasuno et al., 2012; Williamson et al., 2005).
Metagenomics can provide information on the functional dimensions of QS
systems in microbial communities and aid in understanding the functional
linkage between individual microbial species. This tool is still in its early
stages, with many limitations to overcome (Awasthi et al., 2020). A significant limitation of a metagenomic approach is the rate at which QS system
genes are discovered (the hit rate). Because it is statistically difficult to suc-
ceed with small insert libraries with clones smaller than 106, it would be a
technical challenge to find QS system genes in a complex metagenome. The
hit rate, on the other hand, is determined by a combination of vector selection, assay method, and heterologous gene expression efficiency in a surrogate host, and one of the most important of these factors is gene expression
efficiency (Kimura, 2014).

c. Transcriptomics and quorum sensing

The analysis of transcriptomes sheds light on QS modulation in response to anti-QS substances, QS regulation is activated (Ahmed et al., 2019; Asfahl and Schuster, 2018); as a result, studies have included: (1) total RNA extraction from bacterial samples, (2) reverse transcription of RNA into a complementary DNA (cDNA) library for transcriptome sequencing, and (3) comparative analysis of differentially expressed genes using various genetic techniques such as quantitative polymerase chain reaction (qPCR) or high-throughput approaches such as microarray (Lowe et al., 2017). Then, using bioinformatics approaches, differential expression analysis was used to uncover and finally characterize global gene expression patterns in response to anti-QS compounds (Lu et al., 2022). Transcriptomic techniques are relatively fast and inexpensive, and because they are widely available, several studies of microbial biofilms have used transcriptomic approaches to study processes such as biofilm development, biofilm-host interactions, and biofilm drug resistance mechanisms; however, only known sequences can be targeted, and only a few target genes can be investigated, and the results of these studies are not publicly available (Seneviratne et al., 2020).

d. Proteomics and quorum sensing

Genomics and transcriptomics have been the primary approaches used to investigate the various mechanisms of QS. Genome sequencing allowed comparative and functional proteomics techniques to emerge. Proteomics is an essential discipline for elucidating the mechanisms of regulation of a plethora of language signals that circulate through various microbial communities (Di Cagno et al., 2011) and is one of the most important omics approaches in the study of QS at the moment. Proteomics is the study of all proteins that are expressed in a given organism under various conditions, allowing the detection of proteins, functional entities of a cell, and post-translational protein modifications that mRNA expression analysis cannot predict (Sauer, 2003). Within this field, research has focused on: (1) how bacteria communicate (N-acyl-L-lactone homoserine, AHL; autoinducer peptide, AIP; and autoinducer-2, AI-2); (2) QS signals that induce various phenotypes (e.g., virulence and biofilm maturation); and (3) the crosstalk between bacteria within various food ecosystems, e.g., sourdough and fermented (Di Cagno et al., 2011).

Many of the gene products differentially expressed during biofilm formation have been identified through proteomic studies, revealing the sophistication of this growth stage process; however, only limited information on proteins of biofilm proteomics is currently available (Sauer, 2003). In this sense, a surprisingly high number of proteins are regulated by QS-inducing molecules such as N-acyl homoserine lactone (AHL) (Arevalo-Ferro, 2003).

Another novel approach has been the discovery of QSI, which used a two-step procedure involving initial quantitative 2D-DiGE and subsequent identification of proteins at each spot by mass spectrometry to measure the impact of various factors on a variety of QS-mediated proteomic and metabolomic changes in bacteria (Swatton et al., 2016). Despite the fact that comparative proteome analyses are being used, the number of studies using proprietary techniques of this technology, such as comparison of protein profiles of wild-type and mutant strains, 2D-DiGE, and the use of isobaric tags for relative and absolute quantification, is still limited (Eberl and Riedel, 2011).

e. Metabolomics and quorum sensing

In comparison with gene and protein, metabolites provide the most direct assessment of cellular phenotype, and metabolomics is the most appropriate technique for quantitative and dynamic tracking of variations in bacterial metabolism in response to specific environmental conditions (Shen et al., 2020). Metabolomics has emerged as a powerful tool for complementing other omics studies and investigating the massive metabolic diversity found in prokaryotes (Kiamco et al., 2018), such as understanding the efficiency of inhibitory functions or the alteration in some metabolic pathways under stress or antibiotic exposure (Li and Zhu, 2018).

Although many QS inhibitors have been discovered or developed through metabolomic studies from natural products, such as resveratrol and hordenine (Chen et al., 2017; Zhou et al., 2019), or through chemical synthesis, such as N-methyl anthranilate, few publications have reported a complete metabolomic analysis (Ma et al., 2021).

REFERENCES

Ahmed, S. A., Rudden, M., Smyth, T. J., Dooley, J. S., Marchant, R., & Banat, I. M. (2019). Natural quorum sensing inhibitors effectively downregulate gene expression of *Pseudomonas aeruginosa* virulence factors. *Applied Microbiology and Biotechnology* 103(8): 3521–35.

Annous, B. A., Fratamico, P. M., & Smith, J. L. (2009). Scientific status summary: Quorum sensing in biofilms: Why bacteria behave the way they do. *Journal of Food Science* 74(1): R24–37.

Arciola, C. R., Campoccia, D., & Montanaro, L. (2018). Implant infections: Adhesion, biofilm formation and immune evasion. *Nature Reviews Microbiology* 16(7): 397–409. Doi: 10.1038/s41579-018-0019-y.

Arevalo-Ferro, C., Hentzer, M., Reil, G., et al., (2003). Identification of quorum-sensing regulated proteins in the opportunistic pathogen *Pseudomonas aeruginosa* by proteomics. *Environmental Microbiology* 5(12): 1350–69.

Asfahl, K. L., & Schuster, M. (2018). Additive effects of quorum sensing anti-activators on *Pseudomonas aeruginosa* virulence traits and transcriptome. *Frontiers in Microbiology* 8: 2654.

Awasthi, M. K., Ravindran, B., Sarsaiya, S., Chen, H., Wainaina, S., Singh, E., Liu, T., Kumar, S., Pandey, A., Singh, L., & Zhang, Z. (2020). Metagenomics for taxonomy profiling: Tools and approaches. *Bioengineered* 11(1): 356–74. Doi: 10.1080/21655979.2020.1736238.

Bagert, J. D., Van Kessel, J. C., Sweredoski, M. J., et al., (2016). Time-resolved proteomic analysis of quorum sensing in *Vibrio harveyi*. *Chemical Science* 7: 1797–806. Doi: 10.1039/c5sc03340c.

Baldry, M., Nielsen, A., Bojer, M. S., et al., (2016). Norlichexanthone reduces virulence gene expression and biofilm formation in *Staphylococcus aureus*. *PloS One* 11(12): e0168305. Doi: 10.1371/journal.pone.0168305.

Baysse, C. (2005). Modulation of quorum sensing in *Pseudomonas aeruginosa* through alteration of membrane properties. *Microbiology* 151(8): 2529–42.

Brindhadevi, K., LewisOscar, F., Mylonakis, E., Shanmugam, S., Verma, T. N., & Pugazhendhi, A. (2020). Biofilm and quorum sensing mediated pathogenicity in *Pseudomonas aeruginosa*. *Process Biochemistry* 96: 49–57.

Camara, M., Williams, P. & Hardman, A. (2002). Controlling infection by tuning in and turning down the volume of bacterial small-talk. *The Lancet Infectious Diseases* 2: 667–676.

Cao, H., Lai, Y., Bougouffa, S., Xu, Z., & Yan, A. (2017). Comparative genome and transcriptome analysis reveals distinctive surface characteristics and unique physiological potentials of *Pseudomonas aeruginosa* ATCC 27853. *BMC Genomics* 18(1): 459. Doi: 10.1186/s12864-017-3842-z.

Chen, Y. J., He, G. C., Cheng, J. F., Lee, Y. T., Hung, Y. H., Chen, W. H., Huang, Y. T., & Liu, P. Y. (2020). Comparative genomics reveals insights into characterization and distribution of quorum sensing-related genes in *Shewanella* algae from marine environment and clinical sources. *Comparative Immunology, Microbiology and Infectious Diseases* 73: 101545. Doi: 10.1016/j.cimid.2020.101545.

Chen, T., Sheng, J., Fu, Y., Li, M., Wang, J., & Jia, A. Q. (2017). 1H NMR-based global metabolic studies of *Pseudomonas aeruginosa* upon exposure of the quorum sensing inhibitor resveratrol. *Journal of Proteome Research* 16(2): 824–30.

Chen, L., & Wen, Y. M. (2011). The role of bacterial biofilm in persistent infections and control strategies. *International Journal of Oral Science* 3(2): 66–73.

D'Argenio, D. A., Calfee, M. W., Rainey, P. B. & Pesci, E. C. (2002). Autolysis and auto-aggregation in *Pseudomonas aeruginosa* colony morphology mutants. *Journal of Bacteriology* 184: 6481–89.

Di Cagno, R., De Angelis, M., Calasso, M., & Gobbetti, M. (2011). Proteomics of the bacterial cross-talk by quorum sensing. *Journal of Proteomics* 74(1): 19–34.

Dietrich, L. E. P., Price-Whelan, A., Petersen, A., Whiteley, M., & Newman, D. K. (2006). The phenazine pyocyanin is a terminal signalling factor in the quorum sensing network of *Pseudomonas aeruginosa*. *Molecular Microbiology* 61(5): 1308–21.

Diggle, S. P., Winzer, K., Chhabra, S. R., Worrall, K. E., Camara, M., & Williams, P. (2003). The *Pseudomonas aeruginosa* quinolone signal molecule overcomes the cell density-dependency of the quorum sensing hierarchy, regulates rhl-dependent genes at the onset of stationary phase and can be produced in the absence of LasR. *Molecular Microbiology* 50: 29–43.

Dong, Y. H., & Zhang, L. H. (2005). Quorum sensing and quorum-quenching enzymes. *Journal of Microbiology* (Seoul, Korea) 43: 101–9.

Donlan, R. M. (2002). Biofilms: Microbial life on surfaces. *Emerging Infectious Diseases* 8(9): 881–90. Doi: 10.3201/eid0809.020063.

Dowd, S. E., Sun, Y., Secor, P. R., et al., (2008). Survey of bacterial diversity in chronic wounds using pyrosequencing, DGGE, and full ribosome shotgun sequencing. *BMC Microbiology* 8(1): 43.

Eberl, L., & Riedel, K. (2011). Mining quorum sensing regulated proteins–role of bacterial cell-to-cell communication in global gene regulation as assessed by proteomics. *Proteomics* 11(15): 3070–85.

Erdmann, J., Thoming, J. G., Pich, A., Lenz, C., & Haussler, S. (2019). The core proteome of biofilm-grown clinical *Pseudomonas aeruginosa* isolates. *Cells* 8: 1129. Doi: 10.3390/cells8101129.

Fang, Z., Martin, J., & Wang, Z. (2012). Statistical methods for identifying differentially expressed genes in RNA-seq experiments. *Cell & Bioscience* 2(1): 26. Doi: 10.1186/2045-3701-2-26.

Gomes-Fernandes, M., Laabel, M., Pagan, N., et al., (2017). Accessory gene regulator (Agr) functionality in *Staphylococcus aureus* derived from lower respiratory tract infections. *PLoS One* 12(4): e0175552. Doi: 10.1371/journal.pone.0175552.

Gupta, P., Sarkar, S., Das, B., Bhattacharjee, S., & Tribedi, P. (2016). Biofilm, pathogenesis and prevention--a journey to break the wall: A review. *Archives of Microbiology* 198(1): 1–15. Doi: 10.1007/s00203-015-1148-6.

Hegde, S. R. (2020). Computational identification of the proteins associated with quorum sensing and biofilm formation in *Mycobacterium tuberculosis. Frontiers in Microbiology* 10: 3011.

Henly, E. L., Norris, K. Rawson, N., et al., (2021). Impact of long-term quorum sensing inhibition on uropathogenic *Escherichia coli. The Journal of Antimicrobial Chemotherapy* 76(4): 909–19. Doi: 10.1093/jac/dkaa517.

Hentzer, M., Wu, H., Andersen, J. B., et al., (2003). Attenuation of *Pseudomonas aeruginosa* virulence by quorum sensing inhibitors. *The EMBO Journal* 22(15): 3803–15. Doi: 10.1093/emboj/cdg366.

Herzog, R., Peschek, N. Fröhlich, K. S., Schumacher, K., & Papenfort, K. (2019). Three autoinducer molecules act in concert to control virulence gene expression in *Vibrio cholerae. Nucleic Acids Research* 47(6): 3171–83. Doi: 10.1093/nar/gky1320.

Horswill, A. R., Stoodley. P., Stewart, P. S., & Parsek, M. R. (2007). The effect of the chemical, biological, and physical environment on quorum sensing in structured microbial communities. *Analytical and Bioanalytical Chemistry* 387(2): 371–80.

Hrdlickova, R., Toloue, M., & Tian, B. (2017). RNA-Seq methods for transcriptome analysis. *Wiley Interdisciplinary Reviews. RNA* 8(1). Doi: 10.1002/wrna.1364.

Ivanova, A., Ivanova, K., & Tzanov, T. (2018). "Inhibition of quorum-sensing: A new paradigm," in *Controlling Bacterial Virulence and Biofilm Formation, in Biotechnological Applications of Quorum Sensing Inhibitors*, ed. Kalia V. C., Berlin: Springer Nature, 5–10.

Jemielita, M., Wingreen, N. S., & Bassler, B. L. (2018). Quorum sensing controls *Vibrio cholera* multicellular aggregate formation. *Microbiology and Infectious Disease* 7: E42057. Doi: 10.7554/eLife.42057.

Joshi, R. V., Gunawan, C., & Mann, R. (2021). We are one: Multispecies metabolism of a biofilm consortium and their treatment strategies. *Frontiers in Microbiology* 12: 635432. Doi: 10.3389/fmicb.2021.635432.

Kafantaris, I., Tsadila, C., Nikolaidis, M., Tsavea, E., Dimitriou, T. G., Iliopoulos, I., Amoutzias, G. D., & Mossialos, D. (2021). Transcriptomic analysis of *Pseudomonas Aeruginosa* response to pine honey via RNA sequencing indicates multiple mechanisms of antibacterial activity. *Foods (Basel, Switzerland)* 10(5): 936. Doi: 10.3390/foods10050936.

Kasper, S. H., Bonocora, R. P., Wade, J. T., Musah, R. A. & Cady, N. C. (2016). Chemical inhibition of kynureninase reduces *Pseudomonas aeruginosa* quorum sensing and virulence factor expression. *ACS Chemical Biology* 11(4): 1106–17. Doi: 10.1021/acschembio.5b01082.

Kiamco, M. M., Mohamed, A., Reardon, P. N., et al. (2018). Respuestas estructurales y metabólicas de las biopelículas de *Staphylococcus aureus* al estrés hiperosmótico y antibiótico. *Biotechnology and Bioengineering* 115: 1594–603. Doi: 10.1002/bit.26572.

Kim, H.-S., S.-H. Lee, Y. Byun, & H.-D. Park. (2015). 6-Gingerol Reduces *Pseudomonas aeruginosa* biofilm formation and virulence via quorum sensing inhibition. *Scientific Reports* 5: 8656. Doi: 10.1038/srep08656.

Kimura, N. (2014). Metagenomic approaches to understanding phylogenetic diversity in quorum sensing. *Virulence* 5(3): 433–42.

Lu, L., Li, M., Yi, G., Liao, L., Cheng, Q., Zhu, J., Zhang, B., Wang, Y., Chen, Y. and Zeng, M. (2022). Screening strategies for quorum sensing inhibitors in combating bacterial infections. *Journal of Pharmaceutical Analysis* 12(1): 1–14. doi:10.1016/j.jpha.2021.03.009.

Lee, S. O., Lee, S., Lee, J. E., et al. (2020). Dysfunctional accessory gene regulator (agr) as a prognostic factor in invasive *Staphylococcus aureus* infection: A systematic review and meta-analysis. *Scientific Reports* 10: 20697. Doi: 10.1038/s41598-020-77729-0.

Li, H., & Zhu, J. (2018). Diferenciación de *Staphylococcus aureus* resistente a antibióticos mediante espectrometría de masas en tándem de ionización por electropulverización secundaria. *Analytical Chemistry* 90: 12108–15. Doi: 10.1021/acs.analchem.8b03029.

López-Martín, M., Dubern, J.-F. Alexander, M. R., & Williams, P. (2021). AbaM regulates quorum sensing, biofilm formation, and virulence in *Acinetobacter baumannii. Journal of Bacteriology* 203 (8): e00635–20. Doi: 10.1128/JB.00635-20.

Lowe, R., Shirley, N., Bleackley, M., Dolan, S., & Shafee, T. (2017). Transcriptomics technologies. *PLoS Computational Biology* 13 (5): e1005457. Doi: 10.1371/journal.pcbi.1005457.

Ma, Y., Shi, Q., He, Q., & Chen, G. (2021). Metabolomic insights into the inhibition mechanism of methyl N-methylanthranilate: A novel quorum sensing inhibitor and antibiofilm agent against *Pseudomonas aeruginosa. International Journal of Food Microbiology* 109402.

Mangwani, N., Kumari, S., & Das, S. (2016). Bacterial biofilms and quorum sensing: Fidelity in bioremediation technology. *Biotechnology and Genetic Engineering Reviews* 32 (1–2): 43–73.

Manner, S., Goeres, D. M., Skogman, M., Vuorela, P., & Fallarero, A. (2017). Prevention of staphylococcus aureus biofilm formation by antibiotics in 96-microtiter well plates and drip flow reactors: Critical factors influencing outcomes. *Scientific Reports* 7 Doi: 10.1038/srep43854.

Nasuno, E., Kimura, N., Fujita, M. J., Nakatsu, C. H., Kamagata, Y., & Hanada, S. (2012). Phylogenetically novel LuxI/LuxR-type quorum sensing systems isolated using a metagenomic approach. *Applied and Environmental Microbiology* 78 (22): 8067–74.

Nobile, C. J., & Mitchell, A. P. (2006). Genetics and genomics of *Candida albicans* biofilm formation. *Cellular Microbiology* 8 (9): 1382–91. Doi: 10.1111/J.1462-5822.2006.00761.X.

O'Loughlin, C. T., Miller, L. C., Siryaporn, A., Drescher, K., Semmelhack, M. F., & Bassler, B. L. (2013). A quorum-sensing inhibitor blocks *Pseudomonas aeruginosa* virulence and biofilm formation. *Proceedings of the National Academy of Sciences* 110(44): 17981–6.

Papenfort, K., & Bassler, B. L. (2016). Quorum sensing signal–response systems in Gram-negative bacteria. *Nature Reviews Microbiology* 14(9): 576–88.

Papenfort, K., Silpe, J. E., Schramma, K. R., Cong, J. P., Seyedsayamdost, M. R., & Bassler, B. L. (2017). A *Vibrio cholera* autoinducer-receptor pair that controls biofilm formation. 20. Doi:10.1038/NChemBio.2336.

Paulander, W., Varming, A. N., Bojer, M. S., Friberg, C., Bæk, K., & Ingmer, H. (2018). The agr quorum sensing system in *Staphylococcus aureus* cells mediates death of sub-population. *BMC Research Notes* 11(1): 503. Doi: 10.1186/s13104-018-3600-6.

Qin, N., X. Tan, Y. Jiao, L. Liu, W. Zhao, S. Yang, & A. Jia. (2014). RNA-Seq-based transcriptome analysis of methicillin-resistant *Staphylococcus aureus* biofilm inhibition by ursolic acid and resveratrol. *Scientific Reports* 4: 5467. Doi: 10.1038/srep05467.

Rampioni, G., Pustelny, C., Fletcher, M. P., Wright, V. J., Bruce, M., Rumbaugh, K. P., Heeb, S., Cámara, M., & Williams, P. (2010). Transcriptomic analysis reveals a global alkyl-quinolone-independent regulatory role for PqsE in facilitating the environmental adaptation of *Pseudomonas aeruginosa* to plant and animal hosts. *Environmental Microbiology* 12 (6): 1659–73. Doi: 10.1111/j.1462-2920.2010.02214.x.

Redfern, J., Wallace, J., van Belkum, A., Jaillard, M., Whittard, E., Ragupathy, R., Verran, J., Kelly, P., & Enright, M. C. (2021). Biofilm associated genotypes of multiple antibiotic resistant *Pseudomonas aeruginosa. BMC Genomics* 22(1) Doi: 10.1186/s12864-021-07818-5.

Rice, K. C., Mann, E. E., Endres, J. L., et al., (2007). The cidA murein hydrolase regulator contributes to DNA release and biofilm development in *Staphylococcus aureus*. *Proceedings of the National Academy of Sciences of the United States of America* 104 (19): 8113–18.

Rodrigues, S., Paillard, C., Van Dillen, S., Tahrioui, A., Berjeaud, J.-M. Dufour, A. & Bazire, A. (2018). Relation between biofilm and virulence in *Vibrio tapetis*: A transcriptomic study. *Pathogens (Basel, Switzerland)* 7 (4): E92. Doi: 10.3390/pathogens7040092.

Rumbaugh, K. P., & Sauer, K. (2020). Biofilm dispersion. *Nature Reviews Microbiology* 18 (10): 571–86. Doi: 10.1038/s41579-020-0385-0.

Sánchez, M. C., Romero-Lastra, P., Ribeiro-Vidal, H., Llama-Palacios, A., Figuero, E., Herrera, D., & Sanz, M. (2018). Comparative gene expression analysis of planktonic *Porphyromonas gingivalis* ATCC 33277 in the presence of a growing biofilm versus planktonic cells. *BMC Microbiology* 19 (1): 58. Doi: 10.1186/s12866-019-1423-9.

Sauer, K. (2003). The genomics and proteomics of biofilm formation. *Genome Biology* 4 (6): 1–5.

Sauvage, S., & Hardouin, J. (2020). Exoproteomics for better understanding *Pseudomonas aeruginosa* virulence. *Toxins* 12: 571. Doi: 10.3390/toxins12090571.

Seneviratne, C. J., Suriyanarayanan, T., Widyarman, A. S., Lee, L. S., Lau, M., Ching, J., Delaney, C., & Ramage, G. (2020). Multi-omics tools for studying microbial biofilms: Current perspectives and future directions. *Critical Reviews in Microbiology* 46 (6): 759–778. Doi: 10.1080/1040841X.2020.1828817.

Shen, F., Ge, C., & Yuan, P. (2020). Metabolomics study reveals inhibition and metabolic dysregulation in *Staphylococcus aureus* planktonic cells and biofilms induced by carnosol. *Frontiers in Microbiology* 11: 2315.

Silva, N. B. S., Marques, L. A., & Röder, D. D. B. (2021). Diagnosis of biofilm infections: Current methods used, challenges and perspectives for the future. *Journal of Applied Microbiology*, febrero. Doi: 10.1111/jam.15049.

Slater, R. T., Frost, L. R., Jossi, S. E., Millard, A. D., & Unnikrishnan, M. (2019). Clostridioides difficile luxS mediates inter-bacterial interactions within biofilms. *Scientific Reports* 9 (1): 9903. Doi: 10.1038/s41598-019-46143-6.

Stoodley, P., Sauer, K., Davies, D. G., & Costerton, J. W. (2002). Biofilms as complex differentiated communities. *Annual Review of Microbiology* 56: 187–209.

Swatton, J. E., Davenport, P. W., Maunders, E. A., Griffin, J. L., Lilley, K. S., & Welch, M. (2016). Impact of azithromycin on the quorum sensing-controlled proteome of *Pseudomonas aeruginosa*. *PloS One* 11(1): e0147698. Doi: 10.1371/journal.pone.0147698.

Swem, L. R., Swem, D. L., O'Loughlin, C. T., Gatmaitan, R., Zhao, B., Ulrich, S. M., & Bassler, B. L. (2009). A quorum-sensing antagonist targets both membrane-bound and cytoplasmic receptors and controls bacterial pathogenicity. *Molecular Cell* 35(2): 143–153.

Sztajer, H., Szafranski, S. P., Tomasch, J., Reck, M., Nimtz, M., Rohde, M., & Wagner-Döbler, I. (2014). Cross-feeding and interkingdom communication in dual-species biofilms of *Streptococcus mutans* and *Candida albicans*. *The ISME Journal* 8(11): 2256–71. Doi: 10.1038/ismej.2014.73.

Waite, R. D., Paccanaro, A., Papakonstantinopoulou, A., Hurst, J. M., Saqi, M., Littler, E., & Curtis, M. A. (2006). Clustering of *Pseudomonas aeruginosa* transcriptomes from planktonic cultures, developing and mature biofilms reveals distinct expression profiles. *BMC Genomics* 7: 162. Doi: 10.1186/1471-2164-7-162.

Williamson, L. L., Borlee, B. R., Schloss, P. D., Guan, C., Allen, H. K., & Handelsman, J. (2005). Intracellular screen to identify metagenomic clones that induce or inhibit a quorum-sensing biosensor. *Applied and Environmental Microbiology* 71 (10): 6335–6344.

Wu, S., Liu, J., Liu, C., Yang, A., & Qiao, J. (2020). Quorum sensing for population-level control of bacteria and potential therapeutic applications. *Cellular and Molecular Life Sciences* 77 (7): 1319–1343.

Zhang, Y., Sass, A., Van Acker, H., Wille, J., Verhasselt, B., Van Nieuwerburgh, F., Kaever, V., Crabbé, A., & Coenye, T. 2018. Coumarin reduces virulence and biofilm formation in *pseudomonas aeruginosa* by affecting quorum sensing, type III secretion and C-Di-GMP Levels. *Frontiers in Microbiology* 9: 1952. Doi: 10.3389/fmicb.2018.01952.

Zhou, J. W., Muhammad, J., Sun, B., Yang, R., Wadood, A., Wang, J. S., & Jia, A. Q. (2019). Metabolomic analysis of quorum sensing inhibitor hordenine on *Pseudomonas aeruginosa*. *Applied Microbiology and Biotechnology* 103 (15): 6271–6285.

8 Genome Editing Tools
Increasing Efficiency of Microbes for Remediating Contaminated Environment

Madhumita Barooah and Dibya Jyoti Hazarika
Assam Agricultural University

CONTENTS

8.1 INTRODUCTION

The rapid increase in world population and their corresponding food requirements has necessitated enhanced food production and supply (Cazalis et al., 2018; Drangert et al., 2018). One important way to achieve food security through increased food production is to use agrichemicals to prevent pest and pathogen attack (Schmidt-Jeffris and Nault, 2018). A wide range of pesticides, including herbicides, fungicides, insecticides, and rodenticides, are used globally to target particular pests and pathogens in agriculture, horticulture, aquaculture, and households (Alavanja, 2009; van de Merwe et al., 2018). Pesticides are categorized based on their target pests and display enormous effect in crop production and decrease the rate of damages from pests (Allmaras et al., 2018). The practice of using pesticides for controlling plant diseases offers a number of benefits, including effective food production and control of pathogens and vectors (such as fleas, ticks, and mosquitoes) (Cooper and Dobson, 2007; van de Merwe et al., 2018). However, the indiscriminate use of pesticides has resulted in the deterioration of soil health and biomagnification of pesticides such as DDT (d ichlorodiphenyltrichloroethane) in water bodies at each trophic level of the food web (Thomas et al., 2008; Plattner et al., 2018; Silva-Barni et al., 2018). Through bioaccumulation, the pesticidal compounds affect the metabolic functions of the biotic community, including human population. Pesticides such as azoxystrobin and atrazine show their effects on DNA, leading to the development of cancer and neurological diseases (Fatima et al., 2018; Singh et al., 2018, Vidart d'Egurbide Bagazgoïtia et al., 2018). Field-applied pesticides already made their way into the serving table and unknowingly became a part of our foodstuff (Bakirhan et al., 2018; Sarlio, 2018; Marete et al., 2020). The occurrence of pesticides in food supplements, milk, fruit juices, and other food items has widely been reported (Bedi et al., 2018; Farajzadeh et al., 2018; Hamedi et al., 2018; Kumar et al., 2018a).

In developing countries, particularly in the countries of South Asian tropical region, including India, Nepal, Bangladesh, Pakistan, and Sri Lanka, pesticides belonging to the organochlorine group have been testified persistently in higher concentrations in various urban environmental matrices (Ali et al., 2014; Nasir et al., 2014; Pokhrel et al., 2018). In a collaborative report of UNEP and World Health Organization (WHO), it was described that roughly 0.2 million people die worldwide and around three million people suffer from pesticide poisoning each year, among which a large number of reports are from developing countries (Pope et al., 1994; Peshin et al., 2014; Yadav et al., 2015). Serious health effects associated with pesticides such as dichlorodiphenyltrichloroethane (DDT) and isomers of hexachlorocyclohexane (HCH) have prompted many countries to either ban or restrict their application (Yadav et al., 2015; Leong et al., 2018). Despite this, many countries continue to use such chemicals. The classical example is the use of DDT whose use as a pesticide in agriculture was partially banned in 1985; however, it is still being used in many regions of some countries (including India and China) for controlling the malaria vector (Zhang et al., 2011; Yadav et al., 2015). In view of the health risk and environmental sustainability, the use of synthetic pesticides must be controlled through a paradigm shift toward organic farming (Tal, 2018).

Apart from synthetic pesticides, other environmental pollutants including industrial toxic wastes, petroleum-based hydrocarbons, and macro- and micro-plastics are becoming major threats to the biosphere (Kumar et al. 2021a, b, c; Chandra and Kumar 2017a, b). Majority of these pollutants are responsible for showing various toxic effects to the biotic communities, including human beings. These toxic effects include dermal toxicity, neurotoxicity, hepatotoxicity, carcinogenicity, nephrotoxicity, and respiratory toxicity (Ibanez et al., 2007). To overcome the health risks of these pollutants, eco-friendly detoxification strategies are highly essential in the present scenario. Microbe-derived biodegradation of toxic chemicals to their non-toxic forms has shown remarkable progress in the past decades. However, the efficiency of microbial degradation is limited to certain conditions and is greatly affected by external environmental factors. The recent advent of genome editing techniques has opened new avenues for increasing the efficiency of microbes to remediate contaminated sites. Genome editing allows the incorporation of novel metabolic capabilities into natural microbial isolates to biochemically and genetically modulate biodegradative pathways.

8.2 BIOREMEDIATION OF ENVIRONMENTAL POLLUTANTS

Bioremediation refers to the detoxification process that involves microbes to disseminate the toxic molecules from environment (Kumar et al., 2018b; Kumar and Chandra 2020; Jaiswal et al., 2019). It is an effective strategy for the degradation of toxic pollutants in the environment in a sustainable manner. The degradation and elimination of pesticide residues are carried out using conventional approaches of bioremediation (John et al., 2018; Moorman, 2018; Kumar et al. 2020a). Biotransformation of environmental contaminants into less toxic or non-toxic forms through *in situ* or *ex situ* microbiological processes is the primary goal of bioremediation. Microorganisms can assist the restoration of environment by binding, immobilizing, oxidizing, volatilizing, or transforming the pollutants. There is substantial attention toward such microbial-mediated bioremediation because of its simplicity of use, cost-effectiveness, and eco-friendliness compared to the more commonly used non-biological options, which involve the removal of contaminants by digging or pumping up and then transporting into a safer site (Lovely, 2003). Physicochemical methods for the remediation of pollutants, such as solidification, filtration, chemical precipitation, evaporation, oxidation and reduction, incineration, reverse osmosis, electrochemical treatment, and ion exchange approaches, have also been applied frequently for several decades (Porcelli and Judd, 2010; Erdem and Özverdi, 2011; Aksoy et al., 2014; Shi et al., 2014; Kanadasan and Razak, 2015). However, the use of these conventional approaches is limited by several factors, including high reagent requirements, high processing costs, large reaction volumes, and the formation of toxic secondary by-products. Limitations of these techniques are more apparent in the case of contaminated ground waters, mine tailings, effluents, and other wastewaters from industrial sources (Bhalara et al., 2014; Dasgupta et al., 2015).

Ideally, the strategies for bioremediation are designed based on the knowledge of the microbiome structure in the contaminated environments and their response to the changes in environmental conditions, rather than the metabolic capabilities of the microbes. Unfortunately, much of the necessary information is not readily available for the native microbial isolates in a particular contaminated environment. Recent advancements in metabolic engineering (ME) and systems biology (SB) tools have opened up new possibilities for increasing the efficiency of microbial strains for better biodegradative potential. The successful application of ME and SB approaches in various fields of biological sciences has influenced environmental scientists to utilize these approaches for bioremediation. In this chapter, we discuss the available techniques for genome editing of potential bacterial strains to metabolically engineer them for increased efficiency in removing environmental pollutants.

8.3   PRE-GENOMICS TECHNIQUES FOR BIOREMEDIATION

8.3.1   NON-MOLECULAR TECHNIQUES

In the present scenario, "treatability study" comes under the most applied microbiological investigations for bioremediation processes, where samples obtained from the polluted environment are treated in the laboratory and the rates of contaminant degradation or immobilization are evaluated (Rogers and McClure, 2003). These types of studies deliver an estimate of the metabolic potential of the native microbial community, but give minute information about the microbes that are actually responsible for the biodegradation of contaminants. Such studies do not provide direct evidence about the metabolic properties and mechanism of the bioremediation process to make further modifications to stimulate the bioactivity (Lovley, 2003).

Detailed investigation of bioremediation processes attempts to isolate the organism(s) responsible for degradation of the contaminants (Watanabe, 2001; Rogers and McClure, 2003). The isolation and characterization of organisms in pure cultures are crucial for the development and elucidation of molecular mechanisms in microbial ecology. Pure cultured microorganisms involved in the process of bioremediation play a significant role in understanding the underlying mechanism of bioremediation. Inspecting these isolates for their physiology and biochemistry offers the opportunity to understand their chemical reactions, along with other aspects that regulate their growth and bioactivity in contaminated environments. However, prior to the use of molecular techniques for bioremediation, the applicability of the isolated organisms for bioremediation *in situ* was uncertain. It was also not confirmed whether these microorganisms were "weeds" that show rapid growth during *in vitro* culture, but were not primarily responsible for the underlying biodegradation reaction.

8.3.2   THE 16S RRNA-BASED APPROACH

A significant revolution in microbial ecology started with the deduction of highly conserved gene sequences that are available in the genome of all microorganisms, particularly the 16S rRNA gene that contributed to the phylogenetic characterization of the microorganisms (Pace et al., 1986; Amann et al., 1995; Kumar et al., 2021c). By analyzing the 16S rRNA sequences, the abundance of bacterial isolates in contaminated environments along with their phylogenetic relationship could be investigated and linked to the bioremediation processes (Watanabe and Baker, 2000; Rogers and McClure, 2003). Advancements in the next-generation sequencing further simplified the study of bacterial community structure, which uses the 16S rRNA approach for environmental samples. From the 16S rRNA sequencing, sometimes it has been observed that the microbes that predominate during the process of bioremediation closely resemble microbes that can be cultured *in vitro* from the contaminated environments (Lovley, 2001). This contradicts the general predicament in environmental microbiology, which suggests that the most environmentally relevant organisms are difficult to recover in culture (Amann et al., 1995).

8.3.3 ANALYSIS OF FUNCTIONAL GENES IN BIOREMEDIATION

Genome sequencing projects yielded plenty of information regarding the genetic organization of the microorganisms along with the functional annotation of diverse metabolic pathways. Detection and expression analysis of the vital genes responsible for bioremediation provides more information on microbial processes as compared to the classical approach of 16S rRNA sequence analysis (Rogers and McClure, 2003). Metagenomic analyses described that there is a positive correlation between the relative abundance of bioremediation-related genes and the potential of microorganisms for the degradation of contaminants (Schneegurt et al., 1998; Rogers and McClure, 2003; Awasthi et al., 2020). However, the genes for bioremediation may not be expressed despite their presence in the genome. Therefore, there has been an emphasis on determining the levels of transcripts for key genes responsible for bioremediation. High mRNA concentrations can be correlated to higher degradation rates of the contaminants (Schneegurt et al., 1998).

8.4 GENOME EDITING TOOLS

Genome editing is a remarkable approach that allows the manipulation of DNA sequences within a genome with the help of engineered nucleases termed as molecular scissors. Molecular scissors have vast applications in a wide range of biology research related to animals, plants, and microbes (Butt et al., 2018). The techniques of editing a gene involves its recognition by a self-designed guide sequence complementary to the targeted gene sequence, followed by assisting a site-specific break and repairing it by homologous recombination that results in the modification (insertion or deletion) of the desired sequence fragment (Bier et al., 2018). The gene editing tools have tremendous potential to improve the efficiency of bioremediation processes, including the elimination of xenobiotics, biotransformation of toxic compounds to their less-toxic or non-toxic forms, and degradation of pesticides to simple components (Basu et al., 2018; Hussain et al., 2018). The main gene editing tools: ZFN (zinc-finger nuclease), TALEN (transcription activator-like effector nuclease), and CRISPR-Cas (clustered regularly interspaced short palindromic repeats) can probably accomplish the above expectations (Singh et al., 2018; Waryah et al., 2018; Wong, 2018). The combined activities of these gene editing tools introduce a double-stranded break (DSB) in the specific target site of the gene, which is then repaired using the DNA repair mechanisms such as HRD (homology-directed repair) and error-prone NHEJ (non-homologous end joining) pathway (Arazoe et al., 2018; Yadav et al., 2018). ZFNs and TALENs utilize artificial restriction enzymes to make DSBs in the specific target DNA sequence with the help of zinc-finger DNA-binding domain and the TAL effector DNA-binding domain, respectively (Urnov et al., 2010; Sun and Zhou, 2013; Banerjee et al., 2018; Shah et al., 2018). Gene editing tools have gained importance in the last decades for manipulating prokaryotic and eukaryotic genomes. These gene editing tools are employed to develop better microbial strains with a more complex genetic organization that leads to enhanced bioactivity of microorganisms with maximum potential (Basu et al., 2018; Dangi et al., 2019;

Arroyo-Olarte et al., 2021). It is the key alteration of the genetic makeup that differentiates the genome-edited microorganisms from their wild types in terms of obtaining the desirable functional properties (Dai et al., 2018; Stein et al., 2018).

8.4.1 Zinc-Finger Nucleases

Among the various gene editing tools, ZFNs are the most routinely used endonucleases for gene manipulation. These artificial restriction enzymes are capable of binding DNA at their target site and thereby creating DSBs in the DNA. ZFNs basically contain two parts: ZFPs (zinc-finger proteins) – the eukaryotic transcription factors that can act as DNA-binding domain – and a Fok1 (a nucleotide cleavage domain) first obtained from *Flavobacterium okeanokoites* (Hiroyuki et al., 1981; Bitinaite et al., 1998). Multiple ZFPs (usually 4–6) are associated with the cleavage domain to assist it to the target site. Each of the zinc fingers recognizes three base pairs to determine the target site in the genome (Carroll, 2011). These ZFPs have 18 bp target specificity that guides the cleavage domain to an accurate target site for site-specific gene editing. ZFPs are composed of 30 amino acids folded into a single alpha-helix and two antiparallel β-sheets around a central Zn ion (Chaible et al., 2017). This gene editing tool is widely employed with gene knockout (NHEJ) and knock-in (homology-directed repair: HDR) for effective gene editing in prokaryotic and eukaryotic organisms. Most of these studies have addressed clinical biology problems related to antibiotic resistance, bacterial virulence factor production, viral infections, and others (Jo et al., 2015; Dastjerdeh et al., 2016; Hosseini et al., 2020). There is immense potential for the use of this technology for targeting biodegradative pathways. Knockout of the genes encoding transcriptional repressors can be easily achieved through this technology for the enhanced production of bacterial biodegradative enzymes.

8.4.2 Transcription Activator-Like Effector Nucleases

TALEN (transcription activator-like effector nuclease) is an excellent tool for the manipulation of genes through site-directed mutagenesis (Sun and Zhao, 2013). TALENs employ TAL proteins for DNA binding. These proteins are derived originally from a plant pathogenic bacterial genus *Xanthomonas*. The DNA-binding effectiveness of TAL proteins are so high that they can recognize even very short sequences for binding, *i.e.*, 1–2 nucleotides. Furthermore, these nucleases consist of 34-amino-acid tandem repeats that ensure their binding to the target site (Juillerat et al., 2014; Jaiswal et al., 2019). Gene knock-in (HDR) and gene knockout (NHEJ) can be performed using TALENs. This system contains two protein domains, one of which recognizes and binds the very unique and specific target site and the second one is used for cleavage of a specific sequence at that unique site (Jaiswal et al., 2019). However, this technique is applied to manipulate the genomes of many eukaryotic targets such as mammalian cells, frog, zebrafish, mouse, rat, and chicken (Lei et al., 2012; Moore et al., 2012; Bloom et al., 2013; Qiu et al., 2013; Park et al., 2014; Chen et al., 2017).

8.4.3 CRISPR-Cas9 System

Among the various genome editing tools, clustered regularly interspaced short palindromic repeats (CRISPR)-Cas9 is the most effective and productive gene editing system. This is an RNA-guided DNA repair mechanism of the adaptive immune system, which can be utilized as a programmable gene and genome editing platform in living cells (Doudna and Charpentier, 2014; Hsu et al., 2014). The functional loci of CRISPR-Cas involve a CRISPR array of 30–40 bp identical repeats intercalated with invader DNA-targeting spacers that encode the crRNA components, along with an operon of *cas* genes that encodes the components of Cas protein. This system was first detected in *Escherichia coli* genome as a series of short direct repeats interspaced by short sequences (Ishino et al., 1987), and their presence was later detected in numerous bacterial and archaeal species (Mojica et al., 2000).

8.4.3.1 Genome Editing Using CRISPR-Cas System

After the establishment of functional role in DNA repair (Makarova et al., 2002, 2011), CRISPR-Cas9 system is gaining attention regarding targeted genome editing (Doudna and Charpentier, 2014). Because of high compatibility with archaeal and bacterial systems, the CRISPR-Cas platform is suitable for utilization in diverse fields, including industrial bioprocess technology and bioremediation. This platform concretes the way for the construction of more intricate, programmable, and efficient gene circuits (Selle and Barrangou, 2015; Sontheimer and Barrangou, 2015).

Many industrially important products have been produced by overexpressing heterogeneous pathways in different bacteria. The successful application of CRISPR-Cas technology for metabolite production includes n-butanol and 5-aminolevulinic acid production in *E. coli* (Ding et al., 2017; Heo et al., 2017), n-butanol production in *Clostridium saccharoperbutylacetonicum* (Jiménez-Bonilla et al., 2021), isopropanol-butanol-ethanol production in *Clostridium acetobutylicum* and *C. saccharoperbutylacetonicum* (Wasels et al., 2017; Wang et al., 2019), succinic acid production in *Synechococcus elongatus* (Li et al., 2016), and γ-aminobutyric acid (GABA) production in *Corynebacterium glutamicum* (Cho et al., 2017). In *Corynebacterium glutamicum*, Cas12a (formerly Cpf1)-based RNA-guided endonuclease was also used to enhance the natural proline production through site-directed mutagenesis to dismiss the product inhibition (Jiang et al., 2017). Success stories of CRISPR-Cas platform for bacterial metabolic engineering may expand its applicability in the remediation of natural pollutants under controlled environment.

8.4.3.2 Programmable Base Editing Using CRISPR-Cas

Programmable base editing in genomes is an indispensable approach to determine the functions of genes (Jinek et al., 2012; Gasiunas et al., 2012; Wang et al., 2016). Recently, a system termed as "Base Editor" (BE) has been reported, which employed the fusion of cytidine deaminase to a variant of CRISPR-associated Cas9, generating the ability to convert C (cytosine) to U (uracil) at its target site without DSBs, resulting in the conversion of C to T (thymine) or G (guanine) to A (adenine) substitution at the target sites in the dsDNA (Komor et al., 2016). This BE system directly replaces single nucleotides and thus eliminates the dependency of this system on homology-dependent DNA repair. Several derivations of the BE system have been developed, which include BE1, BE2, and BE3, and all of these derivations showed high efficiency of base editing in mammalian cells with negligible indel rates (Rees and Liu, 2018). These systems have shown promising results in human cell lines, mammalian cells, animals, and plants (Komor et al., 2016; Gaudelli et al., 2017; Kim et al., 2017; Kuscu et al., 2017; Zhang et al., 2017; Zong et al., 2017; Rees et al., 2018; Li et al., 2020). Other CRISPR-guided BE systems designed with the involvement of human activation-induced cytidine deaminase (AID) (Ma et al., 2016) or PmCDA1 from sea lamprey (an ortholog of human AID) (Nishida et al., 2016) have been tested in eukaryotic organisms (Hess et al., 2017; Shimatani et al., 2017). The BE3 can also be designed to enable premature coding termination for providing an alternative approach to the regular gene knockout systems (Billon et al., 2017; Kim et al., 2017; Kuscu et al., 2017).

CRISPR-Cas9-guided BE systems can be a promising alternative strategy to generate mutant bacterial strains with site-specific gene editing. Recently, a nickase Cas9-cytidine deaminase fusion protein has been utilized to induce the substitution of cytosine to thymine in prokaryotic cells. This CRISPR-Cas9-guided base editing system resulted in high frequencies of mutagenesis in *Brucella melitensis* and *E. coli* (Zheng et al., 2018). This type of specially designed gene editing system could help to generate enzymes with increased activity and efficiency toward the degradation of environmental pollutants. These systems can also be promising to enhance the enzyme stability toward selective external environmental parameters, thus increasing the enzymatic efficiency for pollutant remediation.

8.4.3.3   Transcriptional Modulation of Gene Expression through CRISPR-Cas

Transcriptional regulation administrates almost all cellular processes essential for life. In response to intracellular or extracellular signals, specific DNA sequences are recognized by transcription factors to facilitate gene activation or repression. Programmable activation and repression of genes at transcriptional level enable on-demand manipulation of specific biological processes and pathways without permanently altering the genomic sequences of a cell. Programmable activation of a gene enables precise regulation of endogenous and synthetic pathways for the regulation of cellular behavior. Recently, a bacterial CRISPR-Cas-mediated system has been developed for gene activation, where a novel CRISPR activator called dCas9-AsiA was designed to activate gene expression in bacteria. The system showed an efficiency of more than 200-fold overexpression of genome and plasmid genes (Ho et al., 2020).

Class 1 RNA-targeting CRISPR systems, including the Cmr complex (comprising the type III-B and type III-C) and Csm complex (comprising the type III-A and type III-D), have been extensively described, which contains multiple subunits (Hale et al., 2009; Staals et al., 2014; Tamulaitis et al., 2014). An RNA-targeting CRISPR-Cas (RCas) system has been designed using the oligonucleotide PAMmer encoding protospacer adjacent motif (PAM) that catalyzes a single cleavage event in the target RNA (O'Connell et al., 2014). Afterward, several Cas9 systems with PAM-independent RNA-targeting mechanisms have been described, which possess a naturally ambiguous DCas (DNA-targeting CRISPR-Cas) and RCas functionality (Strutt et al., 2018), which may also be utilized for RNA targeting. Unlike the regular Cas9, the type VI Cas13 has the function of targeting ssRNA exclusively (Makarova et al., 2018).

Additionally, whereas the DCas system exploits the endonuclease domains RuvC (named for a DNA repair protein of *Escherichia coli*) and a characteristic HNH (histidine and asparagine)-containing domain as the catalytic machineries to introduce a single break in the targeted dsDNA, the Cas13-crRNA complex (an RCas system) recognizes a target ssRNA containing a protospacer flanking sequence (PFS) (Smargon et al., 2020).

8.5 COMPUTATIONAL TOOLS USED FOR METABOLIC ENGINEERING

The accessibility to giant repositories associated with whole-genome sequencing as well as the understanding of previously decoded natural metabolic pathways allows redesigning of pathways through comparison with previously elucidated metabolic networks for remediation of toxic contaminants. Enzymes produced by different organisms are identified, and genes encoding those enzymes are assembled to construct novel metabolic pathways. Databases such as BRENDA (Placzek et al., 2017), KEGG (Kanehisa et al., 2017), MetaCyc (Caspi et al., 2016), and Rhea (Morgat et al., 2015) provide the information regarding the necessary enzymes for redesigning these pathways. Thus, already available pathways can be improved with additional enzymatic reactions for filling the gaps. These collective pathways are also called reference pathways and are very helpful for the comparison of various metabolic models of different organisms. BLAST (Altschul et al., 1990) – the sequence alignment program – enables the comparison of sequences by providing significant statistical similarities between the query sequences (nucleotide or protein) and the target database sequences. This approach identifies enzymes based on the fact that proteins with higher sequence homology are likely to perform similar functions.

A GSMM (genome-scale metabolic model) can be utilized for automated redesigning of metabolic networks. This tool has been found to be effective for the reconstruction of metabolic networks to remediate environmental pollutants. By using this model, microbial phenotypes can be predicted from genotypes (Henry et al., 2010). The GSMM utilizes the genome sequence information of microbial strains and constructs the metabolic network with the help of computational programs and data resources. The reconstruction of metabolic networks can also be performed using the ModelSEED (Henry et al., 2010), MG-RAST (Wilke et al., 2016), and MEGAN (Huson et al., 2007). MAPLE – another program – is employed for the analysis of large metagenomics data in diversity analyses (Takami et al., 2016). Likewise, COBRA (constraint-based reconstruction and analysis) is used to predict the improvement of metabolic rate and yield during the production of metabolites through the prediction of optimal genetic modification (King et al., 2015).

8.6 BACTERIAL GENOME EDITING FOR ENHANCED PHYTOREMEDIATION

In recent years, genome editing platforms such as CRISPR-Cas9 and CRISPR-Cpf1 have displayed their potential to improve agricultural traits and phytoremediation efficiency by improving plant-microbe interactions. Several microbial genes have been reported to contain phytoremediation attributes, *e.g.*, the genes that encode contaminant-biodegrading enzymes involved in 2,4-dinitrootoluene, arsenate, chlorobenzoic acid, or trichloroethylene degradation (Table 8.1). Such genes have displayed tremendous potential to diminish soil contaminants, and overexpression of those genes can be carried out using genome editing tools in plant growth-promoting microorganisms (PGPMs). Similarly, indole-3-acetic acid (IAA) production by rhizobacterial species enables metal stress resistance in their host plants and also improves nitrogen fixation. The application of editing tools for the customization of these rhizobacterial gene expressions results in improved IAA production in the rhizosphere. Previous works established that PGPMs such as *Bradyrhizobium elkanii, Rhizobium* spp., and *Sinorhizobium* sp. had abilities to ameliorate the effect of ethylene on root growth inhibition through the production of rhizobitoxine or ACC deaminase (Yuhashi et al., 2000; Ma et al., 2003; Di Gregorio et al., 2006; Ruzicka et al., 2007). Genome editing tools may be used to target such genes to enhance remediation of heavy metals and plant growth promotion simultaneously. Genome editing tools such as CRISPR-Cas suggest the subsequent opportunity for predicting long histories of the ancestry, adaptation, and molecular capabilities of bacterial cells. In short, the Cas9/sgRNA system and related gene editing tools can be effectively designed for the insertion or deletion of target genes to customize metabolic pathways in plants and PGPMs for improved phytoremediation (Mali et al., 2013).

8.7 RISKS ASSOCIATED WITH GENETICALLY MODIFIED ORGANISMS

Even though the application of genetically modified organisms (GMOs) for remediating environmental pollutants can be an efficient and environment-friendly approach, a few serious issues associated with the use of GMOs have restricted their applicability. The unpredictable effect of GMOs on the diverse life forms in the environment is the most important issue that needs to be addressed. Genetic contamination through the horizontal or vertical transfer of the genetic material to similar or other compatible species can have a negative impact on the whole ecosystem (Prakash et al., 2011). Ethical issues related to horizontal gene transfer have been raised on account of apparent threats to the integrity and fundamental value of the species and the ecosystems (Gautier, 2008). Moreover, there is a huge risk of competition among GMOs and other native species, which may allow the spreading of GMOs into new environments leading to ecological as well as economic damage. The release of GMOs into the environment can potentially affect the growth and sustainability of other available species, including plants, animals, and human beings. Some of such risks may be alleviated by employing genetic "kill switches" that could eliminate the GMOs from the environment readily after their task is complete or simply by introduction of a specific chemical (Chan et al., 2016). All of these issues need to be anticipated, and strict guidelines are to be employed to regulate the application of GMOs for bioremediation and any other use that involves their exposure to the environment. Researchers are working to address these issues so that some fruitful solutions can

TABLE 8.1

A List of Microbial Genes Used for the Bioremediation of Toxic Chemicals

Host Organism	Gene	Target Organism	Functional Properties	References
Arthrobacter globiformis, Burkholderia sp.	FcbB (chlorobenzoate dehalogenase)	Pseudomonas fluorescens	Degrades 2,4-dinitrotoluene	Peel and Wyndham (1999)
Burkholderia cepacia, Pseudomonas putida F1	Toluene-o-monooxygenase	P. fluorescens Deinococcus radiodurans	Removes trichloroethylene (TCE); effectively oxidizes toluene, chlorobenzene, 3,4-dichloro-1-butene, and TCE in highly irradiating environments	Barac et al. (2004), Mahendra and Alvarez-Cohen (2006)
Burkholderia sp.	bph operon (biphenyl)	Pseudomonas fluorescens	Improves the ability to degrade polychlorinated biphenyls (PCBs) and biphenyls	Villacieros et al. (2005)
Delftia acidovorans	pnp operon (transformation of p-nitrophenol into β-ketoadipate)	Pseudomonas putida	Degradation of paraoxon (an organophosphorus compound)	Kim et al. (2011)
Myceliophthora thermophila	mtL (laccase gene)	Saccharomyces cerevisiae	Degradation of lignin and polyaromatic hydrocarbons	Yang et al. (2017)
Pseudomonas aeruginosa	Ohb (ortho-dechlorination gene)	Comamonas testosteroni	Encodes enzymes to metabolize chlorobenzoic acids	Cho et al. (2014)
Pseudomonas putida	Polyphosphate kinase	Pseudomonas aeruginosa	Bioremediation of uranium	Renninger et al. (2004)
	Cytochrome P450$_{CAM}$	–	Oxidation of hexane and 3-methylpentane	Kelly and Kelly (2013)
	Muconate and chloromuconate cycloisomerases	Escherichia coli	Linearization of ring structure of aromatic compounds	Vollmer et al. (1998)
	Chlorobenzene dioxygenase (CDO) gene	Escherichia coli	Catalysis of cis-dihydroxylation of aromatic compounds (such as benzonitrile)	van der Meer et al. (1998)
	luc under Pu (P. putida) promoter and transcription activator	Escherichia coli	Estimation of toluene and toluene-like compounds in field water	Hernández-Sánchez et al. (2016)

(Continued)

TABLE 8.1 (*Continued*)
A List of Microbial Genes Used for the Bioremediation of Toxic Chemicals

Host Organism	Gene	Target Organism	Functional Properties	References
Ralstonia pickettii PKO1, *Pseudomonas mendocina* KR1	Toluene 4-monooxygenases, toluene 3-monooxygenase	*Burkholderia cepacia*	Degradation of non-aromatic N-nitrosodimethylamine (NDMA) – a water pollutant (carcinogenic for humans)	Fishman et al. (2004)
Rhodococcus erythropolis, *Chelatococcus sp.*	*dszA/B/C* encoding dibenzothiophene (DBT) monooxygenase	*Pseudomonas* sp.	Desulfurization of DBT, elimination of sulfur without hampering fuel content	Khairy et al. (2014)
Shigella flexneri	*MerR* (mercury resistance)	*Escherichia coli*	Confers resistance to Hg(II)	Qin et al. (2006)
Staphylococcus aureus	*ArsA/ArsB/ArsC* (genes for arsenic resistance)	*Escherichia coli*	Confers arsenate resistance by detoxification (reduction) of arsenate	Abbas et al. (2014)
Vitreoscilla sp.	*vgb* (bacterial hemoglobin gene)	*Escherichia coli*	Facilitates production of useful compounds and improves growth	Stark et al. (2015)

Source: Based on published reports.

be made and controllable use of GMOs under strict caution measures can be undertaken for the degradation of pesticides, agrochemicals, and other environmental pollutants. Thermo-sterilization-based bioreactors can be one of such solutions for the controlled biodegradation of toxic pollutants.

8.8   CONCLUSIONS

Bioremediation is an attractive alternative to conventional chemical and mechanical practices in terms of a sustainable environment. The success of a microbial bioremediation approach is greatly influenced by the versatile metabolic activity of the microbial strains and their ability to grow in a wide range of environmental conditions. External environmental factors such as temperature, pH, and inhibition and stability of the enzymes also have a large influence on the efficiency of the microbial strains. Recent advances in genetic engineering and genome editing have opened up new possibilities to overcome these problems. Efficient genome editing technologies such as CRISPR-Cas and ZFNs that employ site-directed mutagenesis to modify the target genes can be used to enhance the inherent pollutant-removing capabilities of microbes. In light of the issues presently involved with the use of GMO in the field due to perceived environmental risks, CRISPR-generated metabolically engineered efficient microbial strains have promising possibility in bioremediation.

REFERENCES

Abbas, S.Z., Riaz, M., Ramzan, N., Zahid, M.T., Shakoori, F.R. and Rafatullah, M., 2014. Isolation and characterization of arsenic resistant bacteria from wastewater. *Brazilian Journal of Microbiology, 45*, pp.1309–1315.

Aksoy, D.O., Aytar, P., Toptaş, Y., Çabuk, A., Koca, S. and Koca, H., 2014. Physical and physicochemical cleaning of lignite and the effect of cleaning on biodesulfurization. *Fuel, 132*, pp.158–164.

Alavanja, M.C., 2009. Introduction: Pesticides use and exposure, extensive worldwide. *Reviews on Environmental Health, 24*(4), pp.303–310.

Ali, U., Syed, J.H., Malik, R.N., Katsoyiannis, A., Li, J., Zhang, G. and Jones, K.C., 2014. Organochlorine pesticides (OCPs) in South Asian region: A review. *Science of the Total Environment, 476*, pp.705–717.

Allmaras, R.R., Wilkins, D.E., Burnside, O.C. and Mulla, D.J., 2018. Agricultural technology and adoption of conservation practices. In Pierce, F.J. and Frye, W.W. (Eds.), *Advances in Soil and Water Conservation* (pp. 99–158). Routledge, New York.

Altschul, S.F., Gish, W., Miller, W., Myers, E.W. and Lipman, D.J., 1990. Basic local alignment search tool. *Journal of Molecular Biology, 215*(3), pp.403–410.

Awasthi, M.K., Ravindran, B., Sarsaiya, S., Chen, H., Wainaina, S., Singh, E., Liu, T., Kumar, S., Pandey, A., Singh, L. and Zhang, Z., 2020. Metagenomics for taxonomy profiling: Tools and approaches. *Bioengineered, 11*(1), pp.356–374. Doi: 10.1080/21655979. 2020.1736238.

Amann, R.I., Ludwig, W. and Schleifer, K.H., 1995. Phylogenetic identification and in situ detection of individual microbial cells without cultivation. *Microbiological Reviews, 59*(1), pp.143–169.

Arazoe, T., Kondo, A. and Nishida, K., 2018. Targeted nucleotide editing technologies for microbial metabolic engineering. *Biotechnology Journal, 13*(9), p.1700596.

Arroyo-Olarte, R.D., Bravo Rodríguez, R. and Morales-Ríos, E., 2021. Genome editing in bacteria: CRISPR-cas and beyond. *Microorganisms*, *9*(4), p.844.

Bakirhan, N.K., Uslu, B. and Ozkan, S.A., 2018. The detection of pesticide in foods using electrochemical sensors. In Grumezescu, A.M., Holban, A.M. (Eds.), *Food Safety and Preservation* (pp. 91–141). Academic Press, Cambridge.

Banerjee, A., Banerjee, C., Negi, S., Chang, J.S. and Shukla, P., 2018. Improvements in algal lipid production: A systems biology and gene editing approach. *Critical Reviews in Biotechnology*, *38*(3), pp.369–385.

Barac, T., Taghavi, S., Borremans, B., Provoost, A., Oeyen, L., Colpaert, J.V., Vangronsveld, J. and van der Lelie, D., 2004. Engineered endophytic bacteria improve phytoremediation of water-soluble, volatile, organic pollutants. *Nature Biotechnology*, *22*(5), pp.583–588.

Basu, S., Rabara, R.C., Negi, S. and Shukla, P., 2018. Engineering PGPMOs through gene editing and systems biology: A solution for phytoremediation? *Trends in Biotechnology*, *36*(5), pp.499–510.

Bedi, J.S., Gill, J.P.S., Kaur, P. and Aulakh, R.S., 2018. Pesticide residues in milk and their relationship with pesticide contamination of feedstuffs supplied to dairy cattle in Punjab (India). *Journal of Animal and Feed Sciences*, *27*(1), pp.18–25.

Bhalara, P.D., Punetha, D. and Balasubramanian, K., 2014. A review of potential remediation techniques for uranium (VI) ion retrieval from contaminated aqueous environment. *Journal of Environmental Chemical Engineering*, *2*(3), pp.1621–1634.

Bier, E., Harrison, M.M., O'Connor-Giles, K.M. and Wildonger, J., 2018. Advances in engineering the fly genome with the CRISPR-Cas system. *Genetics*, *208*(1), pp.1–18.

Billon, P., Bryant, E.E., Joseph, S.A., Nambiar, T.S., Hayward, S.B., Rothstein, R. and Ciccia, A., 2017. CRISPR-mediated base editing enables efficient disruption of eukaryotic genes through induction of STOP codons. *Molecular Cell*, *67*(6), pp.1068–1079.

Bitinaite, J., Wah, D.A., Aggarwal, A.K. and Schildkraut, I., 1998. FokI dimerization is required for DNA cleavage. *Proceedings of the National Academy of Sciences*, *95*(18), pp.10570–10575.

Bloom, K., Ely, A., Mussolino, C., Cathomen, T. and Arbuthnot, P., 2013. Inactivation of hepatitis B virus replication in cultured cells and *in vivo* with engineered transcription activator-like effector nucleases. *Molecular Therapy*, *21*(10), pp.1889–1897.

Butt, H., Jamil, M., Wang, J.Y., Al-Babili, S. and Mahfouz, M., 2018. Engineering plant architecture via CRISPR/Cas9-mediated alteration of strigolactone biosynthesis. *BMC Plant Biology*, *18*(1), pp.1–9.

Carroll, D., 2011. Genome engineering with zinc-finger nucleases. *Genetics*, *188*(4), pp.773–782.

Caspi, R., Altman, T., Billington, R., Dreher, K., Foerster, H., Fulcher, C.A., Holland, T.A., Keseler, I.M., Kothari, A., Kubo, A. and Krummenacker, M., 2014. The MetaCyc database of metabolic pathways and enzymes and the BioCyc collection of pathway/genome databases. *Nucleic Acids Research*, *42*(D1), pp.D459–D471.

Cazalis, V., Loreau, M. and Henderson, K., 2018. Do we have to choose between feeding the human population and conserving nature? Modelling the global dependence of people on ecosystem services. *Science of the Total Environment*, *634*, pp.1463–1474.

Chaible, L.M., Kinoshita, D., Corat, M.A.F. and Dagli, M.L.Z., 2013. Genetically modified animal models. In Conn, P.M. (Ed.), *Animal Models for the Study of Human Disease* (pp. 811–831). Academic Press, Cambridge.

Chan, C.T., Lee, J.W., Cameron, D.E., Bashor, C.J. and Collins, J.J., 2016. 'Deadman' and 'Passcode' microbial kill switches for bacterial containment. *Nature Chemical Biology*, *12*(2), pp.82–86.

Chandra, R. and Kumar, V., 2017a. Detection of *Bacillus* and *Stenotrophomonas* species growing in an organic acid and endocrine-disrupting chemicals rich environment of distillery spent wash and its phytotoxicity. *Environmental Monitoring and Assessment*, *189*, p.26. Doi: 10.1007/s10661-016-5746-9.

Chandra, R. and Kumar, V., 2017b. Detection of androgenic-mutagenic compounds and potential autochthonous bacterial communities during in-situ bioremediation of post methanated distillery sludge. *Frontiers in Microbiology* 8, p.87. Doi: 10.3389/fmicb.2017.00887.

Chen, Y., Lu, W., Gao, N., Long, Y., Shao, Y., Liu, M., Chen, H., Ye, S., Ma, X., Liu, M. and Li, D., 2017. Generation of obese rat model by transcription activator-like effector nucleases targeting the leptin receptor gene. *Science China Life Sciences*, *60*(2), pp.152–157.

Cho, J.S., Choi, K.R., Prabowo, C.P.S., Shin, J.H., Yang, D., Jang, J. and Lee, S.Y., 2017. CRISPR/Cas9-coupled recombineering for metabolic engineering of *Corynebacterium glutamicum*. *Metabolic Engineering*, *42*, pp.157–167.

Cho, S.W., Kim, S., Kim, Y., Kweon, J., Kim, H.S., Bae, S. and Kim, J.S., 2014. Analysis of off-target effects of CRISPR/Cas-derived RNA-guided endonucleases and nickases. *Genome Research*, *24*(1), pp.132–141.

Cooper, J. and Dobson, H., 2007. The benefits of pesticides to mankind and the environment. *Crop Protection*, *26*(9), pp.1337–1348.

Dai, Z., Zhang, S., Yang, Q., Zhang, W., Qian, X., Dong, W., Jiang, M. and Xin, F., 2018. Genetic tool development and systemic regulation in biosynthetic technology. *Biotechnology for Biofuels*, *11*(1), pp.1–12.

Dangi, A.K., Sharma, B., Hill, R.T. and Shukla, P., 2019. Bioremediation through microbes: Systems biology and metabolic engineering approach. *Critical Reviews in Biotechnology*, *39*(1), pp.79–98.

Dasgupta, J., Sikder, J., Chakraborty, S., Curcio, S. and Drioli, E., 2015. Remediation of textile effluents by membrane based treatment techniques: A state of the art review. *Journal of Environmental Management*, *147*, pp.55–72.

Dastjerdeh, M.S., Kouhpayeh, S., Sabzehei, F., Khanahmad, H., Salehi, M., Mohammadi, Z., Shariati, L., Hejazi, Z., Rabiei, P. and Manian, M., 2016. Zinc finger nuclease: A new approach to overcome beta-lactam antibiotic resistance. *Jundishapur Journal of Microbiology*, *9*(1), p.e29384.

Di Gregorio, S., Barbafieri, M., Lampis, S., Sanangelantoni, A.M., Tassi, E. and Vallini, G., 2006. Combined application of Triton X-100 and *Sinorhizobium* sp. Pb002 inoculum for the improvement of lead phytoextraction by *Brassica juncea* in EDTA amended soil. *Chemosphere*, *63*(2), pp.293–299.

Ding, W., Weng, H., Du, G., Chen, J. and Kang, Z., 2017. 5-Aminolevulinic acid production from inexpensive glucose by engineering the C4 pathway in *Escherichia coli*. *Journal of Industrial Microbiology and Biotechnology*, *44*(8), pp.1127–1135.

Doudna, J.A. and Charpentier, E., 2014. The new frontier of genome engineering with CRISPR-Cas9. *Science*, *346*(6213), p.1258096.

Drangert, J.O., Tonderski, K. and McConville, J., 2018. Extending the European union waste hierarchy to guide nutrient-effective urban sanitation toward global food security—Opportunities for phosphorus recovery. *Frontiers in Sustainable Food Systems*, 2, p.3.

Erdem, M. and Özverdi, A., 2011. Environmental risk assessment and stabilization/solidification of zinc extraction residue: II. Stabilization/solidification. *Hydrometallurgy*, *105*(3–4), pp.270–276.

Farajzadeh, M.A., Mohebbi, A., Mogaddam, M.R.A., Davaran, M. and Norouzi, M., 2018. Development of salt-induced homogenous liquid–liquid microextraction based on isopropanol/sodium sulfate system for extraction of some pesticides in fruit juices. *Food Analytical Methods*, *11*(9), pp.2497–2507.

Fatima, S.A., Hamid, A., Yaqub, G., Javed, A. and Akram, H., 2018. Detection of volatile organic compounds in blood of farmers and their general health and safety profile. *Nature Environment and Pollution Technology*, *17*(2), pp.657–660.

Fishman, A., Tao, Y. and Wood, T.K., 2004. Toluene 3-monooxygenase of *Ralstonia pickettii* PKO1 is a para-hydroxylating enzyme. *Journal of Bacteriology*, *186*(10), pp.3117–3123.

Gasiunas, G., Barrangou, R., Horvath, P. and Siksnys, V., 2012. Cas9–crRNA ribonucleo-protein complex mediates specific DNA cleavage for adaptive immunity in bacteria. *Proceedings of the National Academy of Sciences, 109*(39), pp.E2579–E2586.

Gaudelli, N.M., Komor, A.C., Rees, H.A., Packer, M.S., Badran, A.H., Bryson, D.I. and Liu, D.R., 2017. Programmable base editing of A T to G• C in genomic DNA without DNA cleavage. *Nature, 551*(7681), pp.464–471.

Gautier, M., 2008. Ethical issues raised by genetically modified microorganisms. *Laboratory of Microbiology and Food Hygiene, UMR STLO, 1253.*

Hale, C.R., Zhao, P., Olson, S., Duff, M.O., Graveley, B.R., Wells, L., Terns, R.M. and Terns, M.P., 2009. RNA-guided RNA cleavage by a CRISPR RNA-Cas protein complex. *Cell, 139*(5), pp.945–956.

Hamedi, R., Aghaie, A.B.G. and Hadjmohammadi, M.R., 2018. Magnetic core micelles as a nanosorbent for the efficient removal and recovery of three organophosphorus pesti-cides from fruit juice and environmental water samples. *Journal of Separation Science, 41*(9), pp.2037–2045.

Henry, C.S., DeJongh, M., Best, A.A., Frybarger, P.M., Linsay, B. and Stevens, R.L., 2010. High-throughput generation, optimization and analysis of genome-scale metabolic models. *Nature Biotechnology, 28*(9), pp.977–982.

Heo, M.J., Jung, H.M., Um, J., Lee, S.W. and Oh, M.K., 2017. Controlling citrate synthase expression by CRISPR/Cas9 genome editing for n-butanol production in *Escherichia coli. ACS Synthetic Biology, 6*(2), pp.182–189.

Hernández-Sánchez, V., Molina, L., Ramos, J.L. and Segura, A., 2016. New family of biosensors for monitoring BTX in aquatic and edaphic environments. *Microbial Biotechnology, 9*(6), pp.858–867.

Hess, G.T., Tycko, J., Yao, D. and Bassik, M.C., 2017. Methods and applications of CRISPR-mediated base editing in eukaryotic genomes. *Molecular Cell, 68*(1), pp.26–43.

Hiroyuki, S. and Susumu, K., 1981. New restriction endonucleases from *Flavobacterium okeanokoites* (FokI) and *Micrococcus luteus* (MluI). *Gene, 16*(1–3), pp.73–78.

Ho, H.I., Fang, J.R., Cheung, J. and Wang, H.H., 2020. Programmable CRISPR-Cas transcrip-tional activation in bacteria. *Molecular Systems Biology, 16*(7), p.e9427.

Hosseini, N., Khanahmad, H., Esfahani, B.N., Bandehpour, M., Shariati, L., Zahedi, N. and Kazemi, B., 2020. Targeting of cholera toxin A (ctxA) gene by zinc finger nuclease: Pitfalls of using gene editing tools in prokaryotes. *Research in Pharmaceutical Sciences, 15*(2), p.182.

Hsu, P.D., Lander, E.S. and Zhang, F., 2014. Development and applications of CRISPR-Cas9 for genome engineering. *Cell, 157*(6), pp.1262–1278.

Huson, D.H., Auch, A.F., Qi, J. and Schuster, S.C., 2007. MEGAN analysis of metagenomic data. *Genome Research, 17*(3), pp.377–386.

Hussain, I., Aleti, G., Naidu, R., Puschenreiter, M., Mahmood, Q., Rahman, M.M., Wang, F., Shaheen, S., Syed, J.H. and Reichenauer, T.G., 2018. Microbe and plant assisted-remediation of organic xenobiotics and its enhancement by genetically modified organ-isms and recombinant technology: A review. *Science of the Total Environment, 628*, pp.1582–1599.

Ibanez, J.G., Hernandez-Esparza, M., Doria-Serrano, C., Fregoso-Infante, A. and Singh, M.M., 2007. Effects of pollutants on the biosphere: Biodegradability, toxicity, and risks. *Environmental Chemistry: Fundamentals*, pp.198–236.

Ishino, Y., Shinagawa, H., Makino, K., Amemura, M. and Nakata, A., 1987. Nucleotide sequence of the iap gene, responsible for alkaline phosphatase isozyme conversion in *Escherichia coli*, and identification of the gene product. *Journal of Bacteriology, 169*(12), pp.5429–5433.

Jaiswal, S., Singh, D.K. and Shukla, P., 2019. Gene editing and systems biology tools for pes-ticide bioremediation: A review. *Frontiers in Microbiology, 10*, p.87.

Jiang, Y., Qian, F., Yang, J., Liu, Y., Dong, F., Xu, C., Sun, B., Chen, B., Xu, X., Li, Y. and Wang, R., 2017. CRISPR-Cpf1 assisted genome editing of *Corynebacterium glutamicum*. *Nature Communications*, 8(1), pp.1–11.

Jiménez-Bonilla, P., Feng, J., Wang, S., Zhang, J., Wang, Y., Blersch, D., de-Bashan, L.E., Gaillard, P., Guo, L. and Wang, Y., 2021. Identification and investigation of autolysin genes in *Clostridium saccharoperbutylacetonicum* strain N1–4 for enhanced biobutanol production. *Applied and Environmental Microbiology*, 87(7), pp.e02442–20.

Jinek, M., Chylinski, K., Fonfara, I., Hauer, M., Doudna, J.A. and Charpentier, E., 2012. A programmable dual-RNA–guided DNA endonuclease in adaptive bacterial immunity. *Science*, 337(6096), pp.816–821.

Jo, Y.I., Kim, H. and Ramakrishna, S., 2015. Recent developments and clinical studies utilizing engineered zinc finger nuclease technology. *Cellular and Molecular Life Sciences*, 72(20), pp.3819–3830.

John, E.M., Varghese, E.M., Krishnasree, N. and Jisha, M.S., 2018. In situ bioremediation of chlorpyrifos by *Klebsiella* sp. isolated from pesticide contaminated agricultural soil. *International Journal of Current Microbiology and Applied Sciences*, 7(3), pp.1418–1429.

Juillerat, A., Dubois, G., Valton, J., Thomas, S., Stella, S., Maréchal, A., Langevin, S., Benomari, N., Bertonati, C., Silva, G.H. and Daboussi, F., 2014. Comprehensive analysis of the specificity of transcription activator-like effector nucleases. *Nucleic Acids Research*, 42(8), pp.5390–5402.

Kanadasan, J. and Razak, H.A., 2015. Engineering and sustainability performance of self-compacting palm oil mill incinerated waste concrete. *Journal of Cleaner Production*, 89, pp.78–86.

Kanehisa, M., Furumichi, M., Tanabe, M., Sato, Y. and Morishima, K., 2017. KEGG: New perspectives on genomes, pathways, diseases and drugs. *Nucleic Acids Research*, 45(-D1), pp.D353–D361.

Kelly, S.L. and Kelly, D.E., 2013. Microbial cytochromes P450: Biodiversity and biotechnology. Where do cytochromes P450 come from, what do they do and what can they do for us? *Philosophical Transactions of the Royal Society B: Biological Sciences*, 368(1612), p.20120476.

Khairy, H., Wübbeler, J.H. and Steinbüchel, A., 2015. Biodegradation of the organic disulfide 4, 4′-dithiodibutyric acid by *Rhodococcus* spp. *Applied and Environmental Microbiology*, 81(24), pp.8294–8306.

Kim, K., Ryu, S.M., Kim, S.T., Baek, G., Kim, D., Lim, K., Chung, E., Kim, S. and Kim, J.S., 2017. Highly efficient RNA-guided base editing in mouse embryos. *Nature Biotechnology*, 35(5), pp.435–437.

Kim, K., Tsay, O.G., Atwood, D.A. and Churchill, D.G., 2011. Destruction and detection of chemical warfare agents. *Chemical Reviews*, 111(9), pp.5345–5403.

King, Z.A., Lloyd, C.J., Feist, A.M. and Palsson, B.O., 2015. Next-generation genome-scale models for metabolic engineering. *Current Opinion in Biotechnology*, 35, pp.23–29.

Komor, A.C., Kim, Y.B., Packer, M.S., Zuris, J.A. and Liu, D.R., 2016. Programmable editing of a target base in genomic DNA without double-stranded DNA cleavage. *Nature*, 533(7603), pp.420–424.

Kumar, D., Rai, D., Porwal, P. and Kumar, S., 2018a. Compositional quality of milk and its contaminants on physical and chemical concern: A review. *International Journal of Current Microbiology and Applied Sciences*, 7(5), pp.1125–1132.

Kumar, V. and Chandra, R., 2020. Bioremediation of melanoidins containing distillery waste for environmental safety. In: Bharagava, R., Saxena, G. (Eds.), *Bioremediation of Industrial Waste for Environmental Safety*. Springer, Singapore. Doi: 10.1007/978-981-13-3426-9_20.

Kumar, V., Ferreira, L.F.R., Sonkar, M. and Singh, J. 2021b. Phytoextraction of heavy metals and ultrastructural changes of *Ricinus communis* L. grown on complex organometallic sludge discharged from alcohol distillery. *Environmental Technology & Innovation 22*, p.101382. Doi: 10.1016/j.eti.2021.101382.

Kumar, V., Shahi, S.K., Ferreira, L.F.R., Bilal, M., Biswas, J.K. and Bulgariu, L., 2021a. Detection and characterization of refractory organic and inorganic pollutants discharged in biomethanated distillery effluent and their phytotoxicity, cytotoxicity, and genotoxicity assessment using *Phaseolus aureus* L. and *Allium cepa* L. *Environmental Research 201*, p.111551. Doi: 10.1016/j.envres.2021.111551.

Kumar, V., Shahi, S.K. and Singh, S., 2018b. Bioremediation: An eco-sustainable approach for restoration of contaminated sites. In: Singh, J., Sharma, D., Kumar, G., and Sharma, N. (Eds.), *Microbial Bioprospecting for Sustainable Development*. Springer, Singapore. https://doi.org/10.1007/978-981-13-0053-0_6

Kumar, V., Singh, K., Shah, M.P., Singh, A.K, Kumar, A. and Kumar, Y., 2021c. Application of omics technologies for microbial community structure and function analysis in contaminated environment. In Shah, M.P., Sarkar, A. and Mandal, S., (Ed.), *Wastewater Treatment: Cutting Edge Molecular Tools, Techniques & Applied Aspects in Waste Water Treatment*. Elsevier. Doi: 10.1016/B978-0-12-821925-6.00013-7.

Kumar, V., Thakur, I.S. and Shah, M.P., 2020a. Bioremediation approaches for pulp and paper industry wastewater treatment: Recent advances and challenges. In: Shah, M.P. (Ed.), *Microbial Bioremediation & Biodegradation*. Springer, Singapore. Doi: 10.1007/978-981-15-1812-6_1.

Kumar, V., Thakur, I.S., Singh, A.K. and Shah, M.P., 2020b. Application of metagenomics in remediation of contaminated sites and environmental restoration. In: Shah, M., Rodriguez-Couto, S., and Sengor, S.S. (Eds), *Emerging Technologies in Environmental Bioremediation*. Elsevier. Doi: 10.1016/B978-0-12-819860-5.00008-0.

Kuscu, C., Parlak, M., Tufan, T., Yang, J., Szlachta, K., Wei, X., Mammadov, R. and Adli, M., 2017. CRISPR-STOP: Gene silencing through base-editing-induced nonsense mutations. *Nature Methods, 14*(7), pp.710–712.

Lei, Y., Guo, X., Liu, Y., Cao, Y., Deng, Y., Chen, X., Cheng, C.H., Dawid, I.B., Chen, Y. and Zhao, H., 2012. Efficient targeted gene disruption in Xenopus embryos using engineered transcription activator-like effector nucleases (TALENs). *Proceedings of the National Academy of Sciences, 109*(43), pp.17484–17489.

Leong, Y.H., Ariff, A.M., Khan, H.R.M., Rani, N.A.A. and Majid, M.I.A., 2018. Paraquat poisoning calls to the Malaysia National Poison Centre following its ban and subsequent restriction of the herbicide from 2004 to 2015. *Journal of Forensic and Legal Medicine, 56*, pp.16–20.

Li, H., Shen, C.R., Huang, C.H., Sung, L.Y., Wu, M.Y. and Hu, Y.C., 2016. CRISPR-Cas9 for the genome engineering of cyanobacteria and succinate production. *Metabolic Engineering, 38*, pp.293–302.

Li, X., Qian, X., Wang, B., Xia, Y., Zheng, Y., Du, L., Xu, D., Xing, D., DePinho, R.A. and Lu, Z., 2020. Programmable base editing of mutated TERT promoter inhibits brain tumour growth. *Nature Cell Biology, 22*(3), pp.282–288.

Lovley, D.R., 2001. Anaerobes to the rescue. *Science, 293*(5534), pp.1444–1446.

Lovley, D.R., 2003. Cleaning up with genomics: Applying molecular biology to bioremediation. *Nature Reviews Microbiology, 1*(1), pp.35–44.

Ma, W., Sebestianova, S.B., Sebestian, J., Burd, G.I., Guinel, F.C. and Glick, B.R., 2003. Prevalence of 1-aminocyclopropane-1-carboxylate deaminase in *Rhizobium* spp. *Antonie Van Leeuwenhoek, 83*(3), pp.285–291.

Ma, Y., Zhang, J., Yin, W., Zhang, Z., Song, Y. and Chang, X., 2016. Targeted AID-mediated mutagenesis (TAM) enables efficient genomic diversification in mammalian cells. *Nature Methods, 13*(12), pp.1029–1035.

Mahendra, S. and Alvarez-Cohen, L., 2006. Kinetics of 1, 4-dioxane biodegradation by monooxygenase-expressing bacteria. *Environmental Science & Technology, 40*(17), pp.5435–5442.

Makarova, K.S., Aravind, L., Grishin, N.V., Rogozin, I.B. and Koonin, E.V., 2002. A DNA repair system specific for thermophilic archaea and bacteria predicted by genomic context analysis. *Nucleic Acids Research*, *30*(2), pp.482–496.

Makarova, K.S., Haft, D.H., Barrangou, R., Brouns, S.J., Charpentier, E., Horvath, P., Moineau, S., Mojica, F.J., Wolf, Y.I., Yakunin, A.F. and Van Der Oost, J., 2011. Evolution and classification of the CRISPR–Cas systems. *Nature Reviews Microbiology*, *9*(6), pp.467–477.

Makarova, K.S., Wolf, Y.I. and Koonin, E.V., 2018. Classification and nomenclature of CRISPR-Cas systems: Where from here? *The CRISPR Journal*, *1*(5), pp.325–336.

Mali, P., Esvelt, K.M. and Church, G.M., 2013. Cas9 as a versatile tool for engineering biology. *Nature methods*, *10*(10), pp.957–963.

Marete, G.M., Shikuku, V.O., Lalah, J.O., Mputhia, J. and Wekesa, V.W., 2020. Occurrence of pesticides residues in French beans, tomatoes, and kale in Kenya, and their human health risk indicators. *Environmental Monitoring and Assessment*, *192*(11), pp.1–13.

Mojica, F.J., Díez-Villaseñor, C., Soria, E. and Juez, G., 2000. Biological significance of a family of regularly spaced repeats in the genomes of archaea, bacteria and mitochondria. *Molecular Microbiology*, *36*(1), pp.244–246.

Moore, F.E., Reyon, D., Sander, J.D., Martinez, S.A., Blackburn, J.S., Khayter, C., Ramirez, C.L., Joung, J.K. and Langenau, D.M., 2012. Improved somatic mutagenesis in zebrafish using transcription activator-like effector nucleases (TALENs). *PloS One*, *7*(5), p.e37877.

Moorman, T.B., 2018. Pesticide degradation by soil microorganisms: Environmental. Ecological and management effects. In Hatfield, J.L., Stewart, B.A., (Eds), *Soil biology: Effects on Soil Quality* (pp. 121–153). CRC Press, Boca Raton.

Morgat, A., Axelsen, K.B., Lombardot, T., Alcántara, R., Aimo, L., Zerara, M., Niknejad, A., Belda, E., Hyka-Nouspikel, N., Coudert, E. and Redaschi, N., 2015. Updates in Rhea—a manually curated resource of biochemical reactions. *Nucleic Acids Research*, *43*(D1), pp.D459–D464.

Nasir, J., Wang, X., Xu, B., Wang, C., Joswiak, D.R., Rehman, S., Lodhi, A., Shafiq, S. and Jilani, R., 2014. Selected organochlorine pesticides and polychlorinated biphenyls in urban atmosphere of Pakistan: Concentration, spatial variation and sources. *Environmental Science & Technology*, *48*(5), pp.2610–2618.

Nishida, K., Arazoe, T., Yachie, N., Banno, S., Kakimoto, M., Tabata, M., Mochizuki, M., Miyabe, A., Araki, M., Hara, K.Y. and Shimatani, Z., 2016. Targeted nucleotide editing using hybrid prokaryotic and vertebrate adaptive immune systems. *Science*, *353*(6305), p.aaf8729.

O'Connell, M.R., Oakes, B.L., Sternberg, S.H., East-Seletsky, A., Kaplan, M. and Doudna, J.A., 2014. Programmable RNA recognition and cleavage by CRISPR/Cas9. *Nature*, *516*(7530), pp.263–266.

Pace, N.R., Stahl, D.A., Lane, D.J. and Olsen, G.J., 1986. The analysis of natural microbial populations by ribosomal RNA sequences. In Marshall, K.C. (Ed), *Advances in Microbial Ecology*, pp.1–55. Springer, Boston, MA.

Park, T.S., Lee, H.J., Kim, K.H., Kim, J.S. and Han, J.Y., 2014. Targeted gene knockout in chickens mediated by TALENs. *Proceedings of the National Academy of Sciences*, *111*(35), pp.12716–12721.

Peel, M.C. and Wyndham, R.C., 1999. Selection of CLC, CBA, and FCB chlorobenzoate-catabolic genotypes from groundwater and surface waters adjacent to the Hyde Park, Niagara Falls, chemical landfill. *Applied and Environmental Microbiology*, *65*(4), pp.1627–1635.

Peshin, S.S., Srivastava, A., Halder, N. and Gupta, Y.K., 2014. Pesticide poisoning trend analysis of 13 years: A retrospective study based on telephone calls at the National Poisons Information Centre, All India Institute of Medical Sciences, New Delhi. *Journal of Forensic and Legal Medicine*, *22*, pp.57–61.

Placzek, S., Schomburg, I., Chang, A., Jeske, L., Ulbrich, M., Tillack, J. and Schomburg, D., 2016. BRENDA in 2017: New perspectives and new tools in BRENDA. *Nucleic Acids Research*, p.gkw952.

Plattner, J., Kazner, C., Naidu, G., Wintgens, T. and Vigneswaran, S., 2018. Pesticide and microbial contaminants of groundwater and their removal methods: A mini review. *Journal of Jaffna Science Association*, *1*(1), pp.12–18.

Pokhrel, B., Gong, P., Wang, X., Khanal, S.N., Ren, J., Wang, C., Gao, S. and Yao, T., 2018. Atmospheric organochlorine pesticides and polychlorinated biphenyls in urban areas of Nepal: Spatial variation, sources, temporal trends, and long-range transport potential. *Atmospheric Chemistry and Physics*, *18*(2), pp.1325–1336.

Pope, J.V., Skurky-Thomas, M. and Rosen, C.L., 1994. Toxicity: Organochlorine pesticides (pp.259–278). Medscape Education, New York.

Porcelli, N. and Judd, S., 2010. Chemical cleaning of potable water membranes: A review. *Separation and Purification Technology*, *71*(2), pp.137–143.

Prakash, D., Verma, S., Bhatia, R. and Tiwary, B.N., 2011. Risks and precautions of genetically modified organisms. *International Scholarly Research Notices*, *2011*, p.369573.

Qin, J., Song, L., Brim, H., Daly, M.J. and Summers, A.O., 2006. Hg (II) sequestration and protection by the MerR metal-binding domain (MBD). *Microbiology*, *152*(3), pp.709–719.

Qiu, Z., Liu, M., Chen, Z., Shao, Y., Pan, H., Wei, G., Yu, C., Zhang, L., Li, X., Wang, P. and Fan, H.Y., 2013. High-efficiency and heritable gene targeting in mouse by transcription activator-like effector nucleases. *Nucleic Acids Research*, *41*(11), pp.e120–e120.

Rees, H.A. and Liu, D.R., 2018. Base editing: Precision chemistry on the genome and transcriptome of living cells. *Nature Reviews Genetics*, *19*(12), pp.770–788.

Renninger, N., Knopp, R., Nitsche, H., Clark, D.S. and Keasling, J.D., 2004. Uranyl precipitation by *Pseudomonas aeruginosa* via controlled polyphosphate metabolism. *Applied and Environmental Microbiology*, *70*(12), pp.7404–7412.

Rogers, S.L. and McClure, N., 2003. The role of microbiological studies in bioremediation process optimization. In: Head, I.M., Singleton, I., and Milner, M. (Eds), *Bioremediation: A Critical Review* (pp.27–59). Horizon Scientific Press, Wymondham.

Ruzicka, K., Ljung, K., Vanneste, S., Podhorská, R., Beeckman, T., Friml, J. and Benková, E., 2007. Ethylene regulates root growth through effects on auxin biosynthesis and transport-dependent auxin distribution. *The Plant Cell*, *19*(7), pp.2197–2212.

Sarlio, S., 2018. "When Enough Is Not Enough": Our Food Systems Are Badly Out of Balance. In: *Towards Healthy and Sustainable Diets*. SpringerBriefs in Public Health. Springer, Cham. Doi: 10.1007/978-3-319-74204-5_2.

Schmidt-Jeffris, R.A. and Nault, B.A., 2018. Crop spatiotemporal dominance is a better predictor of pest and predator abundance than traditional partial approaches. *Agriculture, Ecosystems & Environment*, *265*, pp.331–339.

Schneegurt, N.A. and Kulpa Jr, C.F., 1998. The application of molecular techniques in environmental biotechnology for monitoring. *Biotechnology and Applied Biochemistry*, *27*, pp.73–79.

Selle, K. and Barrangou, R., 2015. Harnessing CRISPR–Cas systems for bacterial genome editing. *Trends in Microbiology*, *23*(4), pp.225–232.

Shah, T., Andleeb, T., Lateef, S. and Noor, M.A., 2018. Genome editing in plants: Advancing crop transformation and overview of tools. *Plant Physiology and Biochemistry*, *131*, pp.12–21.

Shi, X., Tal, G., Hankins, N.P. and Gitis, V., 2014. Fouling and cleaning of ultrafiltration membranes: A review. *Journal of Water Process Engineering*, *1*, pp.121–138.

Shimatani, Z., Kashojiya, S., Takayama, M., Terada, R., Arazoe, T., Ishii, H., Teramura, H., Yamamoto, T., Komatsu, H., Miura, K. and Ezura, H., 2017. Targeted base editing in rice and tomato using a CRISPR-Cas9 cytidine deaminase fusion. *Nature Biotechnology*, *35*(5), pp.441–443.

Silva-Barni, M.F., Gonzalez, M., Wania, F., Lei, Y.D. and Miglioranza, K.S.B., 2018. Spatial and temporal distribution of pesticides and PCBs in the atmosphere using

XAD-resin based passive samplers: A case study in the Quequén Grande River watershed, Argentina. *Atmospheric Pollution Research*, 9(2), pp.238–245.

Singh, N.S., Sharma, R., Parween, T. and Patanjali, P.K., 2018. Pesticide contamination and human health risk factor. In: Oves, M., Khan, M.Z., and Ismail, I.M.I. (Eds), *Modern Age Environmental Problems and Their Remediation* (pp. 49–68). Springer, Cham.

Smargon, A.A., Shi, Y.J. and Yeo, G.W., 2020. RNA-targeting CRISPR systems from metagenomic discovery to transcriptomic engineering. *Nature Cell Biology*, 22(2), pp.143–150.

Sontheimer, E.J. and Barrangou, R., 2015. The bacterial origins of the CRISPR genome-editing revolution. *Human Gene Therapy*, 26(7), pp.413–424.

Staals, R.H., Zhu, Y., Taylor, D.W., Kornfeld, J.E., Sharma, K., Barendregt, A., Koehorst, J.J., Vlot, M., Neupane, N., Varossieau, K. and Sakamoto, K., 2014. RNA targeting by the type III-A CRISPR-Cas Csm complex of *Thermus thermophilus*. *Molecular Cell*, 56(4), pp.518–530.

Stark, B.C., Pagilla, K.R. and Dikshit, K.L., 2015. Recent applications of Vitreoscilla hemoglobin technology in bioproduct synthesis and bioremediation. *Applied Microbiology and Biotechnology*, 99(4), pp.1627–1636.

Stein, H.P., Navajas-Pérez, R. and Aranda, E., 2018. Potential for CRISPR genetic engineering to increase xenobiotic degradation capacities in model fungi. In *Approaches in Bioremediation* (pp. 61–78). Springer, Cham.

Strutt, S.C., Torrez, R.M., Kaya, E., Negrete, O.A. and Doudna, J.A., 2018. RNA-dependent RNA targeting by CRISPR-Cas9. *elife*, 7, p.e32724.

Sun, N. and Zhao, H., 2013. Transcription activator-like effector nucleases (TALENs): A highly efficient and versatile tool for genome editing. *Biotechnology and Bioengineering*, 110(7), pp.1811–1821.

Takami, H., Taniguchi, T., Arai, W., Takemoto, K., Moriya, Y. and Goto, S., 2016. An automated system for evaluation of the potential functionome: MAPLE version 2.1. 0. *Dna Research*, 23(5), pp.467–475.

Tal, A., 2018. Making conventional agriculture environmentally friendly: Moving beyond the glorification of organic agriculture and the demonization of conventional agriculture. *Sustainability*, 10(4), p.1078.

Tamulaitis, G., Kazlauskiene, M., Manakova, E., Venclovas, Č., Nwokeoji, A.O., Dickman, M.J., Horvath, P. and Siksnys, V., 2014. Programmable RNA shredding by the type III-A CRISPR-Cas system of *Streptococcus thermophilus*. *Molecular Cell*, 56(4), pp.506–517.

Thomas, J.E., Ou, L.T. and Al-Agely, A., 2008. DDE remediation and degradation. *Reviews of Environmental Contamination and Toxicology*, 194, pp.55–69.

Urnov, F.D., Rebar, E.J., Holmes, M.C., Zhang, H.S. and Gregory, P.D., 2010. Genome editing with engineered zinc finger nucleases. *Nature Reviews Genetics*, 11(9), pp.636–646.

van de Merwe, J.P., Neale, P.A., Melvin, S.D. and Leusch, F.D., 2018. *In vitro* bioassays reveal that additives are significant contributors to the toxicity of commercial household pesticides. *Aquatic Toxicology*, 199, pp.263–268.

van der Meer, J.R., Werlen, C., Nishino, S.F. and Spain, J.C., 1998. Evolution of a pathway for chlorobenzene metabolism leads to natural attenuation in contaminated groundwater. *Applied and Environmental Microbiology*, 64(11), pp.4185–4193.

Vidart d'Egurbide Bagazgoitia, N., Bailey, H.D., Orsi, L., Lacour, B., Guerrini-Rousseau, L., Bertozzi, A.I., Leblond, P., Faure-Conter, C., Pellier, I., Freycon, C. and Doz, F., 2018. Maternal residential pesticide use during pregnancy and risk of malignant childhood brain tumors: A pooled analysis of the ESCALE and ESTELLE studies (SFCE). *International Journal of Cancer*, 142(3), pp.489–497.

Villacieros, M., Whelan, C., Mackova, M., Molgaard, J., Sánchez-Contreras, M., Lloret, J., Aguirre de Carcer, D., Oruezábal, R.I., Bolanos, L., Macek, T. and Karlson, U., 2005. Polychlorinated biphenyl rhizoremediation by *Pseudomonas fluorescens* F113 derivatives, using a *Sinorhizobium meliloti* nod system to drive *bph* gene expression. *Applied and Environmental Microbiology*, 71(5), pp.2687–2694.

Vollmer, M.D., Hoier, H., Hecht, H.J., Schell, U., Gröning, J., Goldman, A. and Schlömann, M., 1998. Substrate specificity of and product formation by *Muconate cycloisomerases*: An analysis of wild-type enzymes and engineered variants. *Applied and Environmental Microbiology*, 64(9), pp.3290–3299.

Wang, H., La Russa, M. and Qi, L.S., 2016. CRISPR/Cas9 in genome editing and beyond. *Annual Review of Biochemistry*, 85, pp.227–264.

Wang, P., Feng, J., Guo, L., Fasina, O. and Wang, Y., 2019. Engineering *Clostridium saccharoperbutylacetonicum* for high level Isopropanol-Butanol-Ethanol (IBE) production from acetic acid pretreated switchgrass using the CRISPR-Cas9 system. *ACS Sustainable Chemistry & Engineering*, 7(21), pp.18153–18164.

Waryah, C.B., Moses, C., Arooj, M. and Blancafort, P., 2018. Zinc fingers, TALEs, and CRISPR systems: A comparison of tools for epigenome editing. In: Jeltsch, A., and Rots, M.G. (Eds), Epigenome Editing (pp.19–63). Humana (Springer), New York.

Wasels, F., Jean-Marie, J., Collas, F., López-Contreras, A.M. and Ferreira, N.L., 2017. A two-plasmid inducible CRISPR/Cas9 genome editing tool for *Clostridium acetobutylicum*. *Journal of Microbiological Methods*, 140, pp.5–11.

Watanabe, K., 2001. Microorganisms relevant to bioremediation. *Current opinion in Biotechnology*, 12(3), pp.237–241.

Watanabe, K. and Baker, P.W., 2000. Environmentally relevant microorganisms. *Journal of Bioscience and Bioengineering*, 89(1), pp.1–11.

Wilke, A., Bischof, J., Gerlach, W., Glass, E., Harrison, T., Keegan, K.P., Paczian, T., Trimble, W.L., Bagchi, S., Grama, A. and Chaterji, S., 2016. The MG-RAST metagenomics database and portal in 2015. *Nucleic Acids Research*, 44(D1), pp.D590–D594.

Wong, D.W., 2018. Gene targeting and genome editing. In *The ABCs of Gene Cloning* (pp. 187–197). Springer, Cham.

Yadav, I.C., Devi, N.L., Syed, J.H., Cheng, Z., Li, J., Zhang, G. and Jones, K.C., 2015. Current status of persistent organic pesticides residues in air, water, and soil, and their possible effect on neighboring countries: A comprehensive review of India. *Science of the Total Environment*, 511, pp.123–137.

Yadav, R., Kumar, V., Baweja, M. and Shukla, P., 2018. Gene editing and genetic engineering approaches for advanced probiotics: A review. *Critical Reviews in Food Science And Nutrition*, 58(10), pp.1735–1746.

Yang, J., Li, W., Ng, T.B., Deng, X., Lin, J. and Ye, X., 2017. Laccases: Production, expression regulation, and applications in pharmaceutical biodegradation. *Frontiers in Microbiology*, 8, p.832.

Yuhashi, K.I., Ichikawa, N., Ezura, H., Akao, S., Minakawa, Y., Nukui, N., Yasuta, T. and Minamisawa, K., 2000. Rhizobitoxine production by *Bradyrhizobium elkanii* enhances nodulation and competitiveness on *Macroptilium atropurpureum*. *Applied and Environmental Microbiology*, 66(6), pp.2658–2663.

Zhang, W., Jiang, F. and Ou, J., 2011. Global pesticide consumption and pollution: With China as a focus. *Proceedings of the International Academy of Ecology and Environmental Sciences*, 1(2), p.125.

Zhang, Y., Qin, W., Lu, X., Xu, J., Huang, H., Bai, H., Li, S. and Lin, S., 2017. Programmable base editing of zebrafish genome using a modified CRISPR-Cas9 system. *Nature Communications*, 8(1), pp.1–5.

Zheng, K., Wang, Y., Li, N., Jiang, F.F., Wu, C.X., Liu, F., Chen, H.C. and Liu, Z.F., 2018. Highly efficient base editing in bacteria using a Cas9-cytidine deaminase fusion. *Communications Biology*, 1(1), pp.1–6.

Zong, Y., Wang, Y., Li, C., Zhang, R., Chen, K., Ran, Y., Qiu, J.L., Wang, D. and Gao, C., 2017. Precise base editing in rice, wheat and maize with a Cas9-cytidine deaminase fusion. *Nature Biotechnology*, 35(5), pp.438–440.

9 Recent Advancements in Microbial Degradation of Xenobiotics by Using Proteomics Approaches

Neha Sharma, Smriti Shukla,
Kartikeya Shukla, and Ajit Varma
Amity University

Vineet Kumar
G D Goenka University

Menaka Devi Salam and Arti Mishra
Amity University

CONTENTS

DOI: 10.1201/9781003247883-9

9.1 INTRODUCTION

Xenobiotics are foreign chemical compounds present in the surrounding. These compounds are man-made (for example, pesticides) and are added to the soil to remove harmful pests. Various examples of xenobiotics are fuels, pesticides, alkanes, solvents, polycyclic aromatic hydrocarbons (PAHs), antibiotics, pollutants, synthetic azo dyes, and chlorinated PAHs (Sinha et al. 2009; Chandra and Kumar 2015). However, most of the mixtures are very harmful to humans due to their persistent nature, resulting in various deleterious effects. The use of microbes in eliminating pollutants from the soil or sediments by their complete conversion into harmless substances and release of by-products such as carbon dioxide and water is the key concept of the bioremediation process (Singh et al. 2016). These xenobiotic compounds enter the food chain through aquatic bodies and interfere in various metabolic reactions. Both natural and human activities are responsible for their persistent nature (Gienfrada and Rao 2008). Organophosphorus (OP) pesticides are xenobiotics that are used as pesticides and petroleum essences. Long-term exposure to these xenobiotics can cause various diseases such as depression, loss of appetite, anxiety, and other diseases, so their early removal is important. Conventional remediation approaches are more expensive, less effective, and time-consuming (Azad et al. 2014). Bioremediation is an important and natural method that helps in the removal of xenobiotic compounds from the environment by using microorganisms (Kumar et al. 2018; Bhat et al. 2018). It is a more beneficial method. Various natural attenuation (NA) methods help in decontamination and have great potential for spot remediation (Verma et al. 2019). Microbes use these compounds for their survival and convert them into least toxic forms (Singh et al. 2008). The microbial species that are found to bio-fix a wide range of xenobiotic chemicals include both aerobic, e.g. *Pseudomonas, Rhodococcus, Pandoraea, Sphingobium, Escherichia, Gordonia, Bacillus, Moraxella,* and *Micrococcus,* and anaerobic types, e.g. *Pelatomaculum, Desulfotomaculum, Syntrophobacter, Syntrophus, Desulfovibrio Methanospirillum,* and *Methanosaeta* (Chowdhury et al. 2008). Several characteristics of the compounds (including chemical and physical) affect the rate of biodegradation. They include environmental factors such as structure, solubility, concentration, soil type, pH, temperature, salinity, and soil microbial biomass (Negi et al. 2014). Xenobiotic remediation by using microbial-based technologies is also one of the effective methods (Stolz et al. 2001; Robinson et al. 2001; Bürger and Stolz 2010; Oturkar et al. 2013). In bioremediation, a polluted site can be treated by using bacteria, fungi, and yeasts. Bacterial reductive, hydrolytic, dehydro- and thiolyticdehalogenase and some mono- and dioxygenases, methyltransferases enzymes that take part in the degradation of organic xenobiotic compounds. Algae also help in pesticides degradation apart from bacteria and fungi. Various green and blue-green algae, isolated from soil or water, help in the degradation of organophosphorus insecticides chlorpyriphos, monocrotophos, and quinalphos (Mukherjee et al. 2004). White rot fungi, which play a crucial role in lignin degradation, also show the ability to degrade various chemicals having structural similarity to lignin. *Phanerochaete chrysosporium,* a fungal species, has also shown the capability to degrade various persistent organic pollutants such as Dichlorodiphenyltrichloroethane (DDT), lindane, benzopyrene, azo-dyes, and dioxin. However, the metabolic pathway for bioremediation differs from one organism to other. The process of bioremediation has been studied by many researchers (Negi et al. 2014). It was investigated in a previous research that microbes synthesize some common intermediates through their metabolism that downstream activate various cellular metabolic pathways involved in the remediation process. Different groups of researchers have revealed the occurrence of xenobiotic-degrading enzymes, genes, and major metabolites formed because of this process. The metabolization and degradation of pollutants mainly depend on the bioavailability of these pollutants to the microorganisms (Antizar-Ladislao 2010; Singh et al. 2018). However, it has been observed that microbes have an important role in bioremediation, but our knowledge about the changes that occur in microbial communities remains incomplete and the microbial community is still treated as a "black box" (Iwamoto and Nasu 2001; Dua et al. 2002). It will be important to understand the biochemical and physiological aspects of bioremediation. It will be helpful in imparting the knowledge and tools to improve these processes and make it more reliable. Recent advancements in molecular techniques, along with genomic information, are significantly helping microbiologists in the identification of such microorganisms and elucidating genes involved in the process. Novel strategies are being explored in designing new biocatalysts necessary for bioremediation. Microorganisms may be modified for heavy metal bioremediation by expressing metal-binding proteins or enzymes that convert these metals into least toxic forms. In this chapter, we focus on the microbial degradation of xenobiotics and also discuss the molecular mechanism involved in bioremediation by using proteomics approaches.

9.2 XENOBIOTIC COMPOUNDS

9.2.1 Different Classes of Xenobiotic Compounds

Xenobiotic compounds are classified into the following six classes.

9.2.1.1 Halocarbons

Halocarbons have different numbers of halogen in place of hydrogen atoms (e.g. Br, F, and I) Halocarbons include chloroform or trichloromethane ($CHCl_3$), Ferons, trichlorofluoromethane (CCl_3F), dichlorodifluoromethane (CCl_2F_2), chlorotrifluoromethane ($CClF_3$), carbon tetrafluoride (CF_4), and insecticides such as DDT, benzene hexachloride (BHC), and lindane (Jha et al. 2015).

9.2.1.2 Polychlorinated Biphenyls (PCBs)

This includes a group of man-made chemicals. It contains various compounds with two covalently linked benzene ring along with halogen substituting for a hydrogen atom. In industries, they are widely used as insulator coolants in transformers, plasticizers, heat exchange fluids, or hydraulic fluids because of their low combustion rate and high insulating properties (Pieper and Seeger 2008; Xing et al. 2010; Jha et al. 2015). They are inert (both chemically and biologically) to certain extent. Their inert nature increases with the increasing number of chlorine atoms (Cl) in the molecule. Worldwide, they are among the 12 most important persistent organic pollutants (POPs) (Stockholm Convention on Persistent Organic Pollutants). They have capacity to bioaccumulate and biomagnify because of their persistent, high lipophilic, and non-biodegradable nature

(Loganathan and Lam 2012; Bedard et al. 2007). They can cause various problems such as a high risk to foetuses and various neurological problems.

9.2.1.3 Synthetic Polymers

Plastics are artificially synthesized long-chain polymeric molecules. It consists of a variety of semi-synthetic, synthetic, organic, and inorganic compounds (Saminathan et al. 2014). Synthetic polymers are derived from petroleum oil. Various examples of synthetic polymers are polyethylene, nylons, polyvinyl chloride, and polystyrene. These are used in wrapping materials, garments, etc. These compounds are mostly accumulative in nature because of their insoluble nature (Jha et al. 2015). They pollute the aquatic bodies because of their toxic nature. These non-biodegradable plastics impose a serious threat to the environment by accumulating in large amounts because of improper waste management and uncontrolled disposal. Thus, they become a serious threat to our planet (USEPA 2005; Sharma and Dhingra 2016; Krueger et al. 2015) even though burying in landfill, incineration, and recycling are some of the plastic waste disposal methods.

9.2.1.4 Alkylbenzene Sulphonates

These are a class of anionic surfactants. They have sulphonate hydrophilic head at one end and hydrophobic alkylbenzene tail group at the other end. They are one of the oldest and most commonly used synthetic detergents (Jha et al. 2015). They are used in numerous personal care products such as soaps and shampoo. They are resistant to biodegradation, because of which they are called hard detergents. Alkylbenzene sulphonate is responsible for various environmental problems by causing persistent foam in sewage treatment plants. They are basically of two types depending on their chain structure: branched and linear chain. The linear chain structure is more preferred in detergent industry than the branched chain structure because it is more biodegradable. Linear alkylbenzene sulphonate (LAS) is made by sulphonation of linear alkylbenzene. It is used in place of branched dodecylbenzene sulphonates. Branched chain ABS is non-biodegradable or degrade very slowly, because of which they form stable foams. Detergents in biological systems disturb the bio-membrane, state, and quality of the protein and alter the biological properties (Jensen 1999; Ivanković and Hrenović 2010). They have detrimental effects on reproduction, growth, and various physiological activities of aquatic organisms. Their toxic effects on aquatic life have been reported. These anionic surfactants enter fish bodies via gills. They also accumulate in water bodies and thus harm other aquatic life forms (Zhou et al. 2018).

9.2.1.5 Oil Mixtures

This class include a number of hydrocarbons derived from crude oil, including polyaromatic compounds, heterocyclic aromatics, n-alkanes, and some aliphatic compounds (Agrawal et al. 2021). These compounds are released during drilling of wells and transportation and are highly hazardous. PAHs are also a class of organic hydrocarbons. They are also hazardous because of their low degradability rate and highly persistent nature (Mishra et al. 2019). These compounds are generally derived from natural and man-made activities. All crude oil derivatives contain saturated hydrocarbons, aromatic hydrocarbons, and a portion of trace metals. Petroleum hydrocarbons are categorized into two groups, which include aliphatic and aromatic hydrocarbons. Biodegradation of hydrocarbon can occur aerobically as well as anaerobically. Aerobic degradation can occur by a variety of bacteria with ring fission, while anaerobic degradation occurs via syntrophy in which the activity of one microbe is dependent on other microbes. In the environment, these compounds are released by spilling of petroleum products, release of industrial effluents, and various anthropogenic activities. Oil enters the soil and adsorbs to its particles during land contamination. It results in the decrease in soil quality. Hydrocarbons can cause contamination by entering groundwater (Xu et al. 2018). They have harmful effects on aquatic life. Oil breaks down into small drops because of airflow and forms emulsion in the water body. Its presence in the aquatic body stops the transfer or uptake of oxygen by aquatic life. In case of a large amount of oil spillage, various problems occur. It is necessary to degrade or remove these oil spills. Microbes can easily degrade oil spills if it is in a small amount. But their large amount can become recalcitrant and highly toxic. Because they are water-insoluble and have toxic properties, it becomes difficult to degrade them. Oil spills can cause immediate and widespread toxic effects on the environment. The widespread contamination occurs by oil spills and can have social, community, and environmental impacts (Murphy et al. 2016).

9.2.1.6   Other Xenobiotic Compounds

This class includes various pesticides, which are compounds used to control pests. Plenty of pesticides are of aliphatic, cyclic structure, which contains various groups such as nitro, sulphonate, methyl, and amino groups (Jha et al. 2015). These xenobiotic compounds also contain halogens. In modern agriculture, they are preferentially used to control pests, to avoid low yield and low-quality farm products. Most of these groups are responsible for the recalcitrant nature because of their non-biodegradable nature. Although they have beneficial effects, their extensive application is highly toxic to the environment. Recently, some new synthetic chemicals have been introduced in order to control the spread of disease.

9.2.2   Harmful Impacts of Xenobiotic Compounds

Xenobiotic compounds are very hazardous. They affect lower as well as higher eukaryotes. Various diseases such as skin diseases and cancer are caused by prolonged exposure to them. Their persistent nature and increasing concentration in the environment have raised concerns for their harmful effects (Crinnion et al. 2010; Kim et al. 2013; Embrandiri et al. 2016; Tsaboula et al. 2016; Dhakal et al. 2017). Xenobiotics are non-degradable, and they remain persistent in the environment because of their bioaccumulation nature. Bioaccumulation is the process by which any non-degradable substances get accumulated and may enter food chain, leading to disturbance of metabolic activities (Bharadwaj et al. 2018). For example, an increase in minute levels of DDT can disturb the metabolism of fish, birds, and mammals. Xenobiotics can exert toxic as well as late effects. Late effects include congenital defects, cancer, and allergies. In animals, the increasing level of DDT leads to premature birth and, reduction in semen quality and duration of lactation in human beings. DDT is a human carcinogen. It has the potential to cause pancreatic cancer.

It has been revealed by experimental studies on rats that it causes hepatotoxicity, cell necrosis, hyperplasia, hypertrophy, and increased activity of serum liver enzymes (Kostka et al. 1996). Endosulfan is another organochlorine insecticide that kills a wide range of arthropod pests and insects. A prolonged contact with this pesticide can cause congenital birth defects, immunosuppression, neurological disorders, mental retardation, chromosomal abnormalities, and amnesia. Endosulfan is made up of two isomers: α and β. The latter is more harmful to insects as well as mammals. It is highly toxic. In animal studies, the use of endosulfan has shown reduced count of spermatozoa and testosterone inhibition.

Organophosphate compounds are a cluster of pesticides that contain several toxic substances used in agriculture (Sidhu et al. 2019). Laboratory as well as field studies have revealed that a small quantity of these pesticides might lead to endocrine and reproductive disturbance in invertebrates and vertebrates such as fishes, birds, reptiles, and mammals. These pesticides are progressive and neurotoxic. They prevent the enzyme acetyl cholinesterase, which is involved in the hydrolysis of the neurotransmitter acetylcholine that occurs in both the peripheral and the central nervous system. These pesticides are termed as strong alkylating agents because they cause genotoxic effects. It is observed that their occupational exposure considerably leads to chromosomal damage. Profenofos is one of the organophosphate pesticides. It is exceptionally toxic for aquatic organisms as well as for animals (Soares et al. 2021). It is also used in agriculture for controlling lepidopteron pests of cotton and tobacco.

Carbamate pesticides are also toxic similar to organophosphate insecticides. They kill insects, but their harmful impact is for a shorter duration. It is because of the reversibility of the inhibition of nervous tissue. Some of the carbamate pesticides are toxic and carcinogenic (Yadav and Devi 2017). Even small exposure to them can result in reversible neurological disorders because they inhibit acetyl cholinesterase.

Heavy metals involve a group of metals having atomic density five times more than that of water. Heavy metals are basically of two types, which include trace essential and toxic heavy metals. The former are needed in very small quantities. They include copper (Cu), zinc (Zn), chromium (Cr), iron (Fe), and cobalt (Co). They are important for the proper activities of enzymes, haemoglobin formation, and vitamin synthesis in humans (Nagajyoti et al. 2010). While the latter are not required by the body, only a minute of exposure to them can be deleterious. Various toxic heavy metals are lead (Pb), cadmium (Cd), and mercury (Hg). They have a harmful impact on health as they enter food chain causing accumulation. They are responsible for interfering in various metabolic reactions and cause oxidative stress by the release of reactive oxygen and nitrogen species. Various diseases such as osteoporosis, birth defects, and autoimmune disorders (Pratush et al. 2018) are caused by their accumulation.

9.3   STRATEGIES FOR REMOVING XENOBIOTIC COMPOUNDS

9.3.1   DIFFERENT METHODS

Xenobiotics are harmful to human health. They disturb multiple cellular communication pathways, which are involved in growth, normal physiological function, and development. These compounds are hazardous and can have adverse effects on lower

TABLE 9.1

Different Types of Bioremediation Techniques (Kensa et al. 2011)

Technology	Examples	Benefits	Limitations
In situ	In situ bioremediation	Most cost-effective	• Environmental constraints
	Bioventing	Less harmful	• Monitoring difficulties
	Bioaugmentation	Relatively passive	• Increased time consumption
	Biosparging	Natural attenuation process	
Ex situ	Landfarming	Can be done on site	• Space requirement
	Biopiles	Low cost	• Bioavailability limitation
	Composting	Cost-effective	• Extended treatment time
			• Need to control abiotic loss
Bioreactors	Slurry reactors	Fast degradation	• Soil requires excavation
	Aqueous reactors	Enhanced mass transfer	• Relatively high operating cost
		Effective use of inoculants and surfactants	• Relatively high capital cost

as well as higher eukaryotes. There are biological and non-biological remediation methods for the degradation of xenobiotic (Table 9.1). Natural attenuation, phytoremediation, biosparging, composting, bioventing, bioreactors, biopiling, landfarming, and bioslurping are biological remediation methods whereas non-biological degradation methods consist of thermal treatment (*in situ* and *ex situ*), which is suitable for xenobiotic degradation (Tegene et al. 2020). Landfilling and composting are the most suitable methods as they use both the wild type and genetically modified bacterial strains. Natural attenuation method is a process that is used to control the flow of contaminants and also minimize their concentration at contaminated sites. Phytoremediation is carried out by using plants along with microorganisms for the remediation of the polluted area, and bioreactors use selected bacteria to biodegrade the pollutants (Kumar et al. 2018). Thermal treatment (*in situ*) includes five technologies, i.e. steam injection, electrical resistance heating and extraction, conductive heating, radiofrequency heating, and vitrification, while thermal treatment (*ex situ*) includes removal of pollutants through exposure to high temperature in treatment cells or combustion chambers. There are various advantages and disadvantages of using these techniques.

In microbial-based degradation, pollutants are removed from the contaminated environment in a most effective way (Kumar et al. 2020). It is an environment-friendly way. It is also better than other physicochemical methods. There are some potential microorganisms that are capable of mineralizing various toxic compounds present in nature. Various microbes help in the degradation of toxic xenobiotics (Table 9.2). Nowadays, new compounds are being synthesized and released by chemical industries that are hard to be degraded by natural microbial mechanisms (Huang et al. 2018). Also, microbes don't have the ability to degrade every compound. Therefore, it is required to understand the process of xenobiotic degradation for elucidating the microbial metabolic diversity.

TABLE 9.2

List of Xenobiotic Compounds and Degrading Bacterial Genera

Target Compounds	Bacteria Degrading the Compounds	References
	Pesticides	
Endosulfan compounds	*Mycobacterium* sp.	Sutherland et al. (2002)
Endosulphate compounds	*Arthrobacter* sp.	Weir et al. (2006)
HCH	*Pseudomonas* sp.	Banezet et al. (1973)
2,4-D	*Alcaligenes eutrophus*	Don and Pemberton (1981)
DDT	*Dehalospirillum multivorans*	
	Phthalate	
Phthalate	*Burkholderia cepacia* DB01	Chang and Zylstra (1999)
	PAH Compounds	
Naphthalene	*Pseudomonas putida*	Habe and Omori (2003)
3-CBA	*Arthrobacter* sp.	Pignatello et al. (1983)
1,4-DCB	*Alcaligenes* sp.	Don and Pemberton (1981)
2,3,4-Chloroaniline	*Pseudomonas* sp.	Spain and Nishino (1987)
2,3,5-T	*Pseudomonas* sp.	Latorre et al. (1984)
Pyrene	*Mycobacterium* PYR-1	Kanaly and Harayama (2000)
	Sphingomonas paucimobilis	Habe and Omori (2003)
	Halogenated Organic Compounds	
Vinyl chloride	*Dehalococcoides* sp.	He et al. (2003)
PCE	*Dehalococcoides ethenogenes* 195	Magnuson et al. (2000)
Atrazine	*Pseudomonas* sp.	Bruhn et al. (1988)
	Other Compounds	
PCB	*Rhodococcus* RHA1	Kimbara et al. (2005)
Dioxins	*Dehalococcoides* sp.	Bunge et al. (2003)
Benzene	*Dechloromonas* sp.	Coates et al. (2001)
Petroleum products	*Achromobacter* sp.	Austin et al. (1977)
	Acinetobacter sp.	
	Micrococcus sp.	
	Flavobacterium sp.	
Azo dyes	*Bacillus* sp.	Dykes at al. (1994)
	Pseudomonas sp.	Stolz et al. (2001)
	Sphingomonas sp.	Reife et al. (2000)
	Xanthomonas sp.	

9.3.2 ANTIOXIDANT MECHANISM FOR REMEDIATION OF XENOBIOTIC COMPOUNDS

Xenobiotics cause oxidative stress by affecting various protective proteins and their expressions and are also responsible for the production of reactive nitrogen and reactive oxygen species. Overproduction of these species results in DNA, lipid, and protein damage. Reactive oxygen and nitrogen (ROS and RON) species cause various diseases such as cancer, coronary heart disease (CHD), and osteoporosis. Cellular reactions to xenobiotic-induced stress can signal proliferation, homeostasis, apoptosis, or necrosis. The cell tackles free radical generation including antioxidants and antioxidant enzymes by various means. The defence mechanisms differ from species to species (Haider et al. 2020). Oxidative stress can be counterbalanced by various antioxidants. Most xenobiotics cross cell membranes via simple diffusion. Small water-soluble molecules (up to a molecular weight of about 600) move by aqueous pores, while water-insoluble molecules move by the lipid layer of membranes. Plants also have such a detoxification system that helps counterbalance the phytotoxicity effects caused by a wide range of natural and synthetic xenobiotics present in the environment (Zhang and Yang 2021). Chemical modification of xenobiotics is one of the important detoxification mechanisms.

9.4 MAJOR METABOLIC PATHWAYS

Metabolic reactions are chemical reactions that occur inside the body. These metabolic reactions can be catabolic and anabolic. Catabolic reactions are involved in the breakdown of molecules, while anabolic reactions are involved in the biosynthesis of molecules. These reactions are required for the sustainability of life. The metabolization of xenobiotics occurs through three phases. These phases are modification, conjugation, and excretion (Croom 2012). All these steps occur in order to expel the xenobiotic compounds from the cells. The modification phase is also known as phase I (Figure 9.1). In this phase, various reactive and polar groups are added to the substrate by the help of enzymes. The first phase is also known as biotransformation, and it involves major reactions, including hydroxylation and oxidation. The structure of a xenobiotic is being modified from the absorbed (lipophilic) to the one that can be readily removed by the body (hydrophilic or water-soluble). This is mostly done by enzymes that catalyse monooxygenase and hydrogen atom, adding one atom of oxygen to a xenobiotic and generating a molecule of water in the form of by-product. The most prominent group involved in this phase is the cytochrome P450 family (Zhou

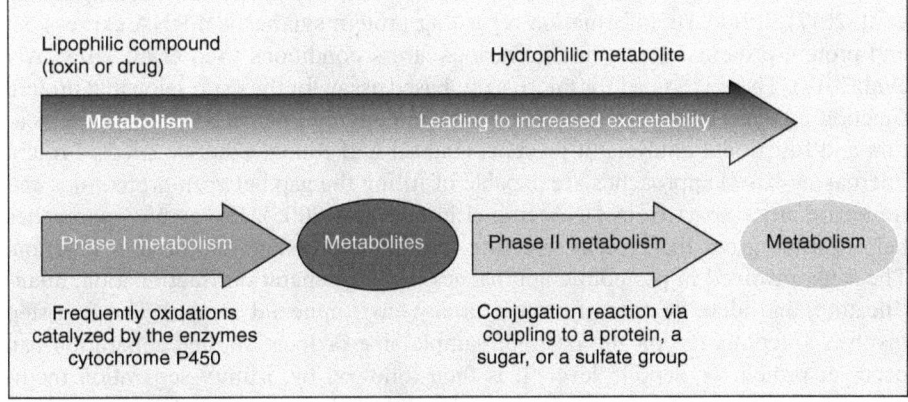

FIGURE 9.1 Mechanism of xenobiotic metabolism in body.

et al. 2006). In medicine, most of the drugs are xenobiotics and their absorption is essential. Some drugs are chemically altered by the patient's metabolism and become active. This is called bioactivation or biotransformation (Lakshmanan 2019). The products of biotransformation are called metabolites. This biotransformation mainly occurs in liver. The enzymes of liver facilitate various reactions, including oxidation, reduction, and hydrolysis. Various other tissues for xenobiotic metabolization are the small and large intestines, tissue of the nasal mucosa, and lung, which play vital functions in the metabolism of drugs administered through an aerosol spray.

In phase II, enzymes attach substances to the xenobiotic-producing reactions called conjugations (David Josephy et al. 2005). The modified xenobiotics are termed as xenobiotic conjugates. The chemicals present in this phase contain glycine, glutathione, and glucuronic acid. These chemicals are mostly anions carrying negative charges. The resulting conjugate may undergo further reactions for detoxification. Finally, the conjugates are expelled out from the cell. This phase is called phase III. In this, negatively charged groups allow conjugates to bind with protein carrier molecules and transport them across the cellular membrane. The resultant compound may further be removed from the body in the form of sweat, urine, or faeces. The phase is also known as excretion.

9.5 PROTEOMICS

Proteomics is a most important tool in omics approach, especially in microbiology field. It allows the investigation of the complete profile of protein obtained from the microbial population on a contaminated site (Williams et al. 2013; Arsene-Ploetze et al. 2015; Gong et al. 2016). The practice of proteomics in the process of bioremediation gives a detailed view of the protein structures of the microbial cells and also addresses the molecular mechanism (Kim et al. 2004). The expression of proteins also differs with changing environmental conditions. These proteins can be identified by using 2D polyacrylamide gel electrophoresis (2DE). Proteomics can be used to identify the microbial communities in different environments such as soil, activated sludge, marine, ground water sediment, and wastewater treatment plants (Williams et al. 2013; Colatriano et al. 2015; Grob et al. 2015; Bastida et al. 2016; Jagadeesh et al. 2017). It imparts information regarding protein synthesis, mRNA expression, and protein-protein interactions in various stress conditions (Seo et al. 2013; Wei et al. 2017). There is a need for microarray-based assay for the expression and protein function analysis. Proteome microarray technology can be used for the identification and functional analysis of proteins (Labaer and Ramachandran 2005). Protein microarray-based approaches are capable of filling the gap between proteomics and transcriptomics apart from DNA chip (Liu and Zhu 2005). Proteomic approaches help in finding out the protein structure, protein expression, and protein function. The steps involved in proteomic approaches include separation/fractionation, quantification, and identification of protein in any environmental sample. The first step involves fractionation on the basis of sample size or location, and separation can occur at protein or peptide level. It is then followed by affinity separation methods such as one- or two-dimensional polyacrylamide gel electrophoresis (1DE or 2DE). 2DE is a more popular and relatively low-cost equipment with low sample

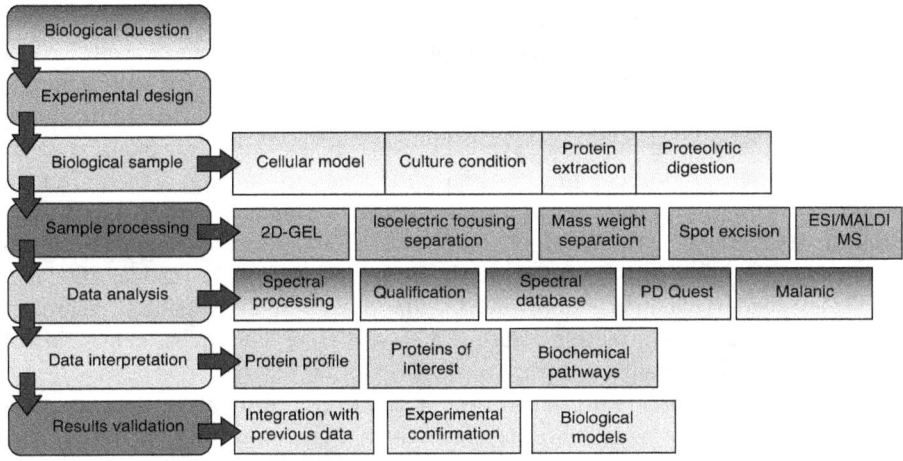

FIGURE 9.2 Steps of proteomics research.

requirements. Various applications of proteomics are identification of enzymes and proteins involved in the metabolism of xenobiotic compounds and community structure analysis during *in situ* remediation. The proteome composition of *P. putida* KT 2440 revealed about 80 unique proteins, including different dioxygenases, thiolases, hydrolases, benzoate, p-hydroxybenzoate, and vanillin (Kim et al. 2006). The important techniques that are used in proteomics analysis are 2DE and mass spectrometry (MS) (Figure 9.2). They give important information regarding the changes that occur in microbes in the presence of pollutants (Kim et al. 2004). It has been observed that if the site is polluted with heavy metals, the aromatic hydrocarbon degradation activity of microbe is inhibited.

In recent years, the proteomic study of *Pseudomonas* has also been published. Various proteins involved in biofilm formation of *P. aeruginosa*, particularly those localized in extracellular matrix, were identified (Toyofuku et al. 2012). In another study, proteomic changes of *P. putida* in the presence of heavy metals such as cadmium were studied (Manara et al. 2012).

In addition, many proteomic studies have been described in relation to hydrocarbon degradation by the genus *Pseudomonas*. One of them is described by Hemamalini and Khare (2014) about the effect of alkanes such as cyclohexane, octadecane, and dodecane on the outer membrane proteome of *P. aeruginosa* PseA. They observed differential regulation of different porins (as an example, OprD and OprF were downregulated, while OprE and OprH were upregulated in the presence of alkanes). PAH- and alkane-utilizing bacteria show a rise in cell surface hydrophobicity while growing on various hydrophobic compounds (Al-Tahhan et al. 2000). For example, increasing cell surface hydrophobicity of *P. aeruginosa* growing on n-hexadecane results in the complete reduction of the expressed lipopolysaccharide (Al-Tahhan et al. 2000). Similarly, in the presence of low temperature, *Rhodococcus* sp. strain Q15 shows an increase in cell surface hydrophobicity during their growth on diesel fuel as revealed by electron microscopic examination (Whyte et al. 1999). Membrane proteins are of great interest as they are involved in the bioremediation of aromatic pollutants such as NACs and organophosphates and altered during the process of bioremediation (Blackstock and Weir 1999; Wang and Yuan 2005). Caprolactam is an industrially important compound, which is basically used in the production of Nylon 6. Nylon 6 is an important synthetic polymer applied in mechanical parts, fabrics, and utensils. Various microbes have the tendency to degrade caprolactam, including strains of *Alcaligenes*, *Acinetobacter*, and *Pseudomonas* (Kulkarni and Kanekar 1998; Rajoo et al. 2013; Baxi and Shah 2002). The caprolactam biodegradation steps in the bacterium *P. jessenii* strain GO3 were studied. It was identified that an ATP-dependent lactamase is involved in the process of conversion of caprolactam to 6-ACA, and ω-aminotransferase is responsible for subsequent conversion of 6-ACA to 6-oxohexanoate.

Generally, xenobiotic biotransformation is carried out through a series of enzymatic reactions having more substrate specificities (Parkinson and Ogilvie 2001). A group of enzymes involved in the methylation, oxidoreduction, transportation across membrane, and stress resistance were significantly overexpressed during TPhP treatment as indicated by proteomic profile. To identify the key role of proteins in TPhP biodegradation, a total of 23 upregulated proteins were selected for further analysis.

9.6   OTHER OMICS APPROACHES

Multiomics approaches such as genomics, metagenomics, transcriptomics, and metabolomics deliver an insight into the molecular mechanisms occurring inside microbes and investigation of expression of various genes, proteins, and mRNA at different environmental conditions (Kumar et al. 2021a, b). The latest progress in molecular technologies has helped environmental microbiologists in recognizing those microbial communities, which can be used to test the polluted sites. It occurs by detecting changes taking place in microbial populations in the presence of pollutants (Desai et al. 2009). Genomics approaches help in the identification of different genes, promoters, and the various pathways involved. This information can be used in the construction of such highly efficient microbial strains for the purpose of controlling the pollution (Figure 9.3).

Genomics is an important and powerful tool that gives information about the full genetic make-up of microbial cells. Data of genomics also tell about the details of metabolic ability of microbes in a given environment and about the entire DNA sequence and fine-scale genetic mapping of microorganisms (Awasthi et al. 2020; Aziz et al. 2021). Using this tool will helpful in understanding the metabolism and interaction of microbes with pollutant and in microbial communities. The whole genome sequencing and next-generation sequencing (NGS) are the most important tools of genomics to get insights into the bioremediation process. By using this approach, a large number of microorganisms can be identified based on DNA sequencing such as 16s and 18s rDNA in a single run. It will be important to exploit the bacterial genomic information for at least two levels (1) To explore efficient strains having capability with a wide catabolic diversity (2) To explain the roles of genes with unknown enzymes (Vilchez-Vargas et al. 2010). Therefore, this approach is an important tool for identifying and exploring novel biocatalysts. Genomic approach is an important tool in

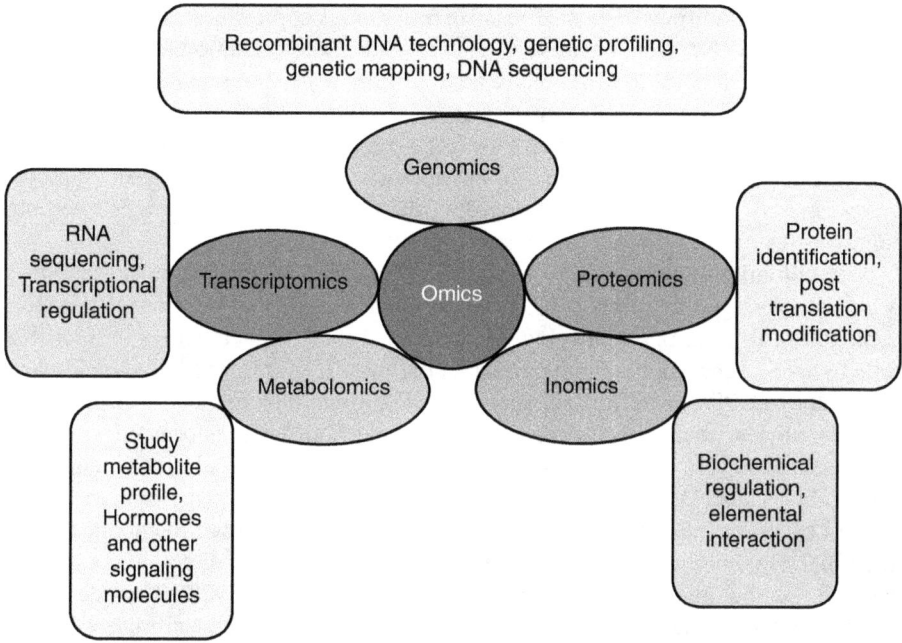

FIGURE 9.3 Classification of omics approaches.

providing the complete genomic information of *Deinococcus*, *Dehalococcoides*, *Pseudomonas*, and other microbes involved in bioremediation. The identification of various genes, promoters, and pathways of bioremediation can be used in the production of more competent pollutant-degrading microbes to tackle the problem.

Metagenomics approach is called environmental genomics. It involves genomic DNA isolation directly from the environmental sample followed by genome analysis and elucidation by sequencing of target gene for the purpose of identification of genes or pathways involved (i.e. sequence metagenomics) (Awasthi et al. 2020). It involves the overall analysis of genome of whole community of microorganisms within a definitive environment (Kumar et al. 2020; Kumar and Chandra 2020). Recent advancements in metagenomics lead to de novo sequencing of microbial metagenomes, which delivers efficient and low-cost methods to generate microbial community profiles consecutively from different samples of the environment (Hamady et al. 2008). On a wide scale, metagenomics approaches are showing increasing trend. These approaches are being used to find out the taxonomic and functional composition of microbes from clinical and agriculture settings.

The transcriptome is the set of all RNA transcripts that include the coding and noncoding RNA. Microarray is one of the important tools. It gives information about the mRNA expression of every gene of an organism. DNA microarray has been used for determining bacterial species and their quantitative analysis of genes, which are upregulated or downregulated in stressed environments (Muffler et al. 2002; Singh and Nagaraj 2006). This tool can be used for the screening of bacterial expression in various stress conditions. Advancements in multiomics approaches enabled researchers to extract mRNA from bacterial, archaeal, and eukaryotic microbial cells. It also enabled researchers to obtain the complete profile of gene expression of microbial communities. It is known as "meta-transcriptome". cDNA microarray offers an important tool for monitoring the transcriptional activities of the microbial communities (Urich et al. 2008). Meta-transcriptomic approaches have been employed by researchers to get information regarding the function and structure of microbial communities of rhizospheric soil (Urich et al. 2008).

Metabolomics includes complete analysis of metabolites of a biological sample. It is a newly emerged field that gives information about unknown genes (Villas-Bôas et al. 2005). It helps in quantifying the functional roles of metabolites in microbial cells by using separation and analytical techniques. High-performance liquid chromatography (HPLC), gas chromatography (GC), direct injection mass spectrometry (DIMS), nuclear magnetic resonance (NMR), and Fourier transform infrared (FT-IR) spectroscopy are some of the important techniques for the analysis of metabolites (Chakraborty and Das 2017). It has become easier to visualize the bioremediation process by using these approaches. Several studies have reported the application of microbial metabolome analysis for studying the bioremediation of pollutants. A comparative metabolome study was carried out by Keum et al. (2008). They described the metabolome analysis of *Sinorhizobium* sp. C4 during phenanthrene degradation and compared the results with those from natural carbon sources. The profiles of metabolites containing fatty acids and polar metabolites were studied by using untargeted metabolome analysis. It was observed that about 60% of the 207 peaks obtained in the GC-MS analysis were of polar metabolite fraction of *Sinorhizobium* sp. C4. Various intermediates such as branched chain amino acids and intermediates of glycolysis accumulated in *Sinorhizobium* sp. C4 as indicated by metabolic profiles in a study than other 70% of the identified metabolites reduced during the degradation of phenanthrene.

9.7   CONCLUSIONS AND FUTURE PERSPECTIVES

Xenobiotics can have hazardous effects on the environment. It is necessary to remove these compounds from the environment. The best method is microbial-based remediation. Microbes uptake these xenobiotics for carbon and nitrogen sources. Some microbes even have the potential to survive in harsh conditions and have great potential to biodegrade pollutants. It will be helpful to understand microbial physiology and to find out the major metabolites involved in the bioremediation pathways. In the present document, applications of proteomics approach in the field of xenobiotic degradation were reviewed. Proteomics approach helps in providing an integrative understanding of the proteins involved in the biodegradation process and also helps in the identification of the key proteins in microbes during a particular physiological state (like in stresses conditions). Microbial communities help in the breakdown of xenobiotics by metabolizing organic material and recycling of nutrients. The mechanism underlying the bioremediation process can be tracked by utilizing omics tools. Proteomics methods can be used for in-depth inspection of alterations in the structure of proteins and their identification and characterization in the presence of pollutants during the bioremediation process. Proteomics approach can also be used in the detection of biomarkers for identifying environmental pollutants such as PAHs (Aardema and McGregor 2002). During the biodegradation of PAHs, alterations occur on the microbial surface and can be easily traced by using proteomics approaches.

REFERENCES

Aardema, M.J. and MacGregor, J.T. 2003. Toxicology and genetic toxicology in the new era of "toxicogenomics": impact of "-omics" technologies. *Mutation Research* 499(1):13–25.

Agrawal, N., Kumar, V. and Shahi, S.K. 2021. Biodegradation and detoxification of phenanthrene in *in-vitro* and *in-vivo* conditions by a newly isolated ligninolytic fungus *Coriolopsis byrsina* strain APC5 and characterization of their metabolites for environmental safety. *Environ Sci Pollut Res* Doi: 10.1007/s11356-021-15271-w.

Al-Tahhan, R.A, Sandrin, T.R., Bodour, A.A. and Maier, R.M. 2000. Rhamnolipid-induced removal of lipopolysaccharide from *Pseudomonas aeruginosa*: Effect on cell surface properties and interaction with hydrophobic substrates. *Appl Environ Microbiol* 668: 3262–3268.

Antizar-Ladislao, B. (2010). Bioremediation: Working with bacteria. *Elements* 6: 389–394.

Arsene-Ploetze, F., Bertin, P.N. and Carapito, C. 2015. Proteomic tools to decipher microbial community structure and functioning. *Environ Sci Pollut Res* 22: 13599–13612.

Austin, B, Calomiris, J.J., Walker, J.D. and Colwell, R.R. 1977. Numerical taxonomy and ecology of petroleum degrading bacteria. *Appl Environ Microbiol* 34: 60–68.

Awasthi, M.K., Ravindran, B., Sarsaiya, S., Chen, H., Wainaina, S., Singh, E., Liu, T., Kumar, S., Pandey, A., Singh, L. and Zhang, Z. 2020. Metagenomics for taxonomy profiling: Tools and approaches. *Bioengineered*, 11(1): 356–374. Doi: 10.1080/21655979.2020.1736238.

Azad, M.A.K, Amin, L. and Sidik, N.M. 2014. Genetically engineered organisms for bioremediation of pollutants in contaminated sites. *Chin Sci Bull* 59: 703–714.

Aziz, A., Yasmeen, T., Tariq, M., Arif, M.S., Shahzad, S.M., Riaz, M., Ali, S. and Rizwan, M. 2021. Genomics in understanding bioremediation of inorganic pollutants. In Hasanuzzaman, M. and Prasad, M.N.V. (Eds.) *Handbook of Bioremediation: Physiological, Molecular and Biotechnological Interventions* 397–410. Academic Press.

Bastida, F., Jehmlich, N., Lima, K., Moris, B.E., Richnow, H.H., Hernandez, T. et al. 2016. The ecological and physiological responses of the microbial community from a semi-arid soil to hydrocarbon contamination and its bioremediation using compost amendment. *J Proteomic* 135: 162–169.

Baxi, N. and Shah, A. 2002. E-Caprolactam-degradation by *Alcaligenes faecalis* for bioremediation of wastewater of a nylon-6 production plant. *Biotechnol Lett* 24: 1177–1180.

Bedard, D.L., Ritalahti, K.M. and Loffler, F.E. 2007. The *Dehalococcoides* population in sediment-free mixed cultures metabolically dechlorinates the commercial polychlorinated biphenyl mixture Aroclor 1260. *Appl Environ Microbiol* 738: 2513–2521.

Bharadwaj, A. 2018. Bioremediation of xenobiotic: An eco-friendly cleanup approach. In: Parmar, V., Malhotra, P., Mathur, D. (eds.) *Green Chemistry in Environmental Sustainability and Chemical Education*. Springer, Singapore. Doi: 10.1007/978-981-10-8390-7_1.

Bhat, S.A., Singh, S., Singh, J., Kumar, S. and Vig, A.P. 2018. Bioremediation and detoxification of industrial wastes by earthworms: Vermicompost as powerful crop nutrient in sustainable agriculture. *Bioresource Technol*, 252: 172–179. Doi: 10.1016/j.biortech. 2018.01.003.

Blackstock, W.P. and Weir, M.P. 1999. Proteomics: Quantitative and physical mapping of cellular proteins. *Trends Biotechnol* 173: 121–127.

Bruhn, C., Batley, R.C. and Knockmues, H.J. 1988. The *in-vivo* construction of 4-chloro-2-nitro phenol assimilatory bacteria. *Arch Microbiol* 150: 171–177.

Bunge, M., Adrian, L., Kraus, A., Lorenz, W.G., Andreesen, J.R., Gorisch, H. and Lechner, U. 2003. Reductive dehalogenation of chlorinated dioxins by the anaerobic bacterium *Dehalococcoides ethenogenes* genes sp. strain CBDBI. *Nature* 421: 357–360.

Bürger, S. and Stolz, A 2010. Characterization of the favin-free oxygen-tolerant dyes. *Appl Microbiol Biotechnol* 56: 69–80.

Chakraborty, J. and Das, S. 2017. Application of spectroscopic techniques for monitoring microbial diversity and bioremediation. *Appl Spectrosc Rev* 521: 1–38.

Chandra, R. and Kumar, V. 2015. Biotransformation and biodegradation of organophosphates and organohalides. In: Chandra, R. (Ed.) *Environmental Waste Management*. Boca Raton, FL: CRC Press. Doi: 10.1201/b19243-17.

Chang H.K. and Zylstra G.J. 1999. Characterization of the phthalate permease ophD from *Burkholderia cepacia* DBO1. *J Bacteriol* 181: 6197–6199.

Chowdhury, A., Pradhan, S., Saha, M. and Sanyal, N. 2008. Impact of pesticides on soil microbiological parameters and possible bioremediation strategies. *J Ind Microbiol* 48: 114–127.

Coates, J.D., Chakraborty, R., Lack, J.G., O'Connor, S.M., Cole, K.A., Bender, K.S. and Achenbach, L.A. 2001. Anaerobic benzene oxidation coupled to nitrate reduction in pure culture by two strains of *Dechloromonas*. *Nature* 411: 1039–1043.

Colatriano, D., Ramachandran, A., Yergeau, E., Maranger, R., Gelinas, Y. and Walsh, D.A. 2015. Metaproteomics of aquatic microbial communities in a deep and stratified estuary. *Proteomics* 15: 3566–3579.

Crinnion, W.J. 2010. Toxic effects of the easily avoidable phthalates and parabens. *Alternative Med Rev* 15: 190–196.

Croom, E. 2012. Metabolism of xenobiotic of human environments. *Prog Mol Bio Trans Sci* 112: 31–88.

David Josephy, P., Peter Guengerich, F. and Miners, J.O. 2005. "Phase I and Phase II" drug metabolism: Terminology that we should phase out? *Drug Metab Rev* 374: 575–580.

Desai, C., Parikh, R.Y., Vaishnav, T., Shouche, Y.S. and Madamwar, D. 2009. Tracking the influence of long-term chromium pollution on soil bacterial community structures by comparative analyses of 16S rRNA gene phylotypes. *Res Microbiol* 160: 1–9.

Dhakal, K., Gadupadi, G.S. and Robertson, L.W. 2017. Sources and toxicities of phenolic polychlorinated biphenyls (OH-PCBs). *PCBs Risk Eval Environ Protect* 25: 16277–16290. Doi: 10.1007/s11356-017-9694-x.

Don, R.H. and Pemberton, J.M. 1981. Properties of six pesticide degradation plasmids isolated from *Alcaligenes paradoxus* and *Alcaligenes eutrophus*. *J Bacteriol* 145: 681–686.

Dua, M., Singh, A., Sethunathan, N. and Johri, A.K. 2002. Biotechnology and bioremediation: Successes and limitations. *Appl Microbiol Biotechnol* 59: 143–152.

Dykes, G.A., Timm, R.G. and Von, H.A. 1994. Azoreductase activity in bacteria associated with the greening of instant chocolate puddings. *Appl Environ Microbiol* 60: 3027–3029.

Embrandiri, A, Kiyasudeen, S.K., Rupani, P.F. and Ibrahim, M. H. 2016. "Environmental xenobiotics and its effect on natural ecosystem", In *Plant Response to Xenobiotic* Vol. 1: 18. Springer, Singapore.

Gienfrada, L. and Rao, M.A. 2008. Interactions between xenobiotics and microbial and enzymatic soil activity. *Crit Rev Environ Sci Technol* 38: 269–310.

Gong, H, Bao, M., Pi, G., Li, Y., Wang, A. and Wang, Z. 2016. Dodecanol-modified petroleum hydrocarbon degrading bacteria for oil spill remediation: Double effect on dispersion and degradation. *ACS Sustain Chem Eng* 41: 169–176.

Grob, C., Taubert, M., Howat, A.M., Burns, O.J., Dixon, J.L., Richnow, H.H. et al. 2015. Combining metagenomics with metaproteomics and stable isotope probing reveals metabolic pathways used by a naturally occurring marine methylotroph. *Environ Microbiol* 17: 4007–4018.

Habe, H. and Omori, T. 2003. Genetics of polycyclic aromatic hydrocarbon degradation by diverse aerobic bacteria. *Biosci Biotechnol Biochem* 67: 225–243.

Haider, K., Haider, M.R., Neha, K. and Yar, M.S. 2020. Free radical scavengers: An overview on heterocyclic advances and medicinal prospects. *Eur J Med Chem* 204: 112–607.

Hamady, M., Walker, J., Harris, K.J., Gold, N.J. and Knight, R. 2008. Error-correcting barcoded primers for pyrosequencing hundreds of samples in multiplex. *Nat Methods* 5: 235–237.

He, J., Ritalahti, K.M., Yang, K.L., Koenigsberg, S.S. and Löffler, F.E. 2003. Detoxification of vinyl chloride to ethene coupled to an anaerobic bacterium. *Nature* 424: 62–65.

Hemamalini, R. and Khare, S. 2014. A proteomic approach to understand the role of the outer membrane porins in the organic solvent-tolerance of *Pseudomonas aeruginosa* PseA. *PLoS One*. 9: e103788.

Huang, Y., Xiao, L., Li, F., Xiao, M., Lin, D., Long, X. and Wu, Z. 2018. Microbial degradation of pesticide residues and an emphasis on the degradation of cypermethrin and 3-phenoxy benzoic acid: A review. *Molecules* 239: 2313.

Ivanković, T. and Hrenović, J. 2010. Surfactants in the environment, *Arh Hig Rada Toksikol.* 61: 95–110.

Iwamoto, T. and Nasu, M. 2001. Current bioremediation practice and perspective. *J Biosci Bioeng* 92: 1–8.

Jagadeesh, D.S., Kannegundla, U. and Reddy R.K. 2017. Application of proteomic tools in food quality and safety. *Adv Anim Vet Sci* 5: 213–225.

Jensen, J. 1999. Fate and effects of linear alkyl benzene sulfonates (LAS) in the terrestrial environment. *Sci Total Environ* 226: 93–111.

Jha, S.K, Jain, P. and Sharma, H.P. 2015. Xenobiotic degradation by bacterial enzymes. *Int J Curr Microbiol App Sci* 46: 48–62.

Kanaly, R.A. and Harayama, S. 2000. Biodegradation of high-molecular weight polycyclic aromatic hydrocarbons by bacteria. *J Bacterial* 182: 2059–2067.

Kensa, V.M. 2011. Bioremediation-An overview. *J Ind Pollut* 272: 161–168.

Keum, Y-S., Seo, J-S., Li, Q.X. and Kim, J-H. 2008. Comparative metabolomic analysis of *Sinorhizobium* sp. C4 during the degradation of phenanthrene. *Applied Microbiology and Biotechnology* 80:863–872.

Kim, K.H., Jahan, S.A., Kabir, E. and Brown, R.C.B. 2013. A review of airborne polycyclic aromatic hydrocarbons (PAHs) and their human health effect. *Environ Int* 60: 71–80. Doi: 10.1016/j.envint.2013.07.019.

Kim, S.J., Jones, R.C. and Cha, C.J. 2004. Identification of proteins induced by polycyclic aromatic hydrocarbon in *Mycobacterium vanbaalenii* PYR-1 using two-dimensional polyacrylamide gel electrophoresis and de novo sequencing methods. *Proteomics* 4: 3899–3908.

Kim, Y.H., Cho, K., Yun, S.H., Kim, J.Y., Kwon, K.H., Yoo, J.S. and Kim, S.I. 2006. Analysis of aromatic catabolic pathways in *Pseudomonas putida* KT 2440 using a combined proteomic approach: 2DE/MS and cleavable isotope-coded affinity tag analysis. *Proteomics* 6: 1301–1318.

Kimbara, K. 2005. Recent developments in the study of microbial aerobic degradation of polychlorinated biphenyls. *Microbes Environ* 20: 127–134.

Kostka, G, Kopeć-Szlęzak, J. and Palut, D. 1996. Early hepatic changes induced in rats by two hepatocarcinogenic organohalogen pesticides: Bromopropylate and DDT. *Carcinogenesis* 173: 407–412.

Krueger, M.C., Hofmann, U., Moeder, M. and Schlosser, D. 2015. Potential of wood-rotting fungi to attack polystyrene sulfonate and its depolymerisation by *Gloeophyllum trabeum* via hydroquinone-driven Fenton chemistry. *PLoS One* 107: 0131773.

Kulkarni, R.S. and Kanekar, P.P. 1998. Bioremediation of ε-caprolactam from Nylon-6 waste water by use of *Pseudomonas aeruginosa* MCMB-407. *Curr Microbiol* 37: 191–194.

Kumar, V. and Chandra, R. 2020. Metagenomics analysis of rhizospheric bacterial communities of *Saccharum arundinaceum* growing on organometallic sludge of sugarcane molasses-based distillery. *3 Biotech* 10(7): 316. Doi: 10.1007/s13205-020-02310-5.

Kumar, V., Shahi, S.K. and Singh, S. 2018. Bioremediation: An eco-sustainable approach for restoration of contaminated sites. In *Microbial Bio Prospecting for Sustainable Development* 115–136. Springer, Singapore. Doi: 10.1007/978-981-13-0053-0_6.

Kumar, V., Singh, K., Shah, M.P., Singh, A.K, Kumar, A. and Kumar, Y. 2021a. Application of omics technologies for microbial community structure and function analysis in contaminated environment. In Shah, M.P., Sarkar, A. and Mandal, S., (Eds.), *Wastewater Treatment: Cutting Edge Molecular Tools, Techniques & Applied Aspects in Waste Water Treatment*. Elsevier. Doi: 10.1016/B978-0-12-821925-6.00013-7.

Kumar, V., Thakur, I.S. and Shah, M.P. 2020a. Bioremediation approaches for pulp and paper industry wastewater treatment: Recent advances and challenges. In Shah, M.P. (Ed.) *Microbial Bioremediation & Biodegradation*. Springer, Singapore. Doi: 10.1007/978-981-15-1812-6_1.

Kumar, V., Thakur, I.S., Singh, A.K. and Shah, M.P. 2020b. Application of metagenomics in remediation of contaminated sites and environmental restoration. In Shah, M., Rodriguez-Couto, S. and Sengor, S.S. (Eds.), *Emerging Technologies in Environmental Bioremediation*. Elsevier. Doi: 10.1016/B978-0-12-819860-5.00008-0.

Labaer, J. and Ramachandran, N. 2005. Protein microarrays as tools for functional proteomics. *Curr Opin Chem Biol* 9: 14–9.

Latorre, J., Reineke, W. and Knackmuss, H.J. 1984. Microbial metabolism of *chloroanilines*: Enhanced evolution by natural genetic exchange. *Arch Microbiol* 140: 159–165.

Lakshmanan, M. 2019. Drug metabolism. *Basic Clin Pharmacol Toxicol* 99–116. Springer, Singapore.

Liu, W.T. and Zhu L. 2005. Environmental microbiology-on-a-chip and its future impacts. *Trends Biotechnol.* 234: 174–179.

Loganathan, B.G. and Lam, P.K.S. eds. 2011. *Global Contamination Trends of Persistent Organic Chemicals*, Vol. 23, 174–179, CRC Press: Boca Raton, Florida.

Magnuson, J.K., Romine M.F., Burris, D.R. and Kingsley, M.T. 2000. Trichloroethene reductive dehalogenase from *Dehalococcoides ethenogenes*: Sequence of tceA and substrate range characterization. *Appl Environ Microbiol* 66: 5141–5147.

Manara, A., DalCorso, G., Baliardini, C., Farinati, S., Cecconi, D. and Furini, A. 2012. *Pseudomonas putida* response to cadmium: Changes in membrane and cytosolic proteomes. *J Proteome Res* 11: 4169–4179.

Mishra, A., Rathour, R., Singh, R., Kumari, T. and Thakur, I.S. 2020. Degradation and detoxification of phenanthrene by actinobacterium *Zhihengliuella* sp. ISTPL4. *Environ Sci Pollut Res*. 2722: 27256–27267.

Muffler, A., Bettermann, S., Haushalter, M., Hörlein, A., Neveling, U., et al. 2002. Genome-wide transcription profiling of *Corynebacterium glutamicum* after heat shock and during growth on acetate and glucose. *J Biotechnol* 982: 255–268.

Mukherjee, I., Gopal, M. and Dhar, D.W. 2004. Disappearance of chlorpyrifos from cultures of *Chlorella vulgaris*. *Bull Environ Contam Toxicol* 73: 358–363.

Murphy, D, Gemmell B, Vaccari L, Li C, Bacosa H, Evans M, Gemmell C, Harvey T, Jalali M. and Niepa T.H. 2016. An in-depth survey of the oil spill literature since 1968: Long term trends and changes since deepwater horizon. *Mar Pollut* 1131: 371–379.

Nagajyoti, P.C, Lee, K.D. and Sreekanth, T.V.M. 2010. Heavy metals, occurrence and toxicity for plants: A review. *Environ Chem Lett* 83: 199–216.

Negi, G., Pankaj, S.A. and Sharma, A. 2014. *In situ* biodegradation of endosulfan, imidacloprid, and carbendazim using indigenous bacterial cultures of agriculture fields of Uttarakhand, India. *Int J Biol Food Vat Agric Eng*, 89: 935–943.

Oturkar, C.C., Othman, M.A., Kulkarni, M., Madamwar, D. and Gawai, K.R. 2013. Synergistic action of flavin containing NADH dependent azoreductase and cytochrome P450 monooxygenase in azoaromatic mineralization. *RSC Adv* 3: 3062–3070.

Parkinson, A. and Ogilvie, B.W. 2008. Biotransformation of xenobiotics. *Casarett Doull's Toxicol: Basic Sci Poisons* 7: 161–304.

Pieper, D.H. and Seeger, M. 2008. Bacterial metabolism of polychlorinated biphenyls. *J Mol Microbiol Biotechnol* 152–3: 121–138.

Pignatello, J.J., Martinson, M.M., Stelert, J.G., Carison, R.E. and Crawford, R.L. 1983. Biodegradation and photolysis of pentachlorophenol in artificial fresh water streams. *Appl Environ Microbiol* 46: 1024–1031.

Pratush, A, Kumar, A. and Hu, Z. 2018. Adverse effect of heavy metals (As, Pb, Hg, and Cr) on health and their bioremediation strategies: A review. *Int Microbiol* 213: 97–106.

Rajoo, S., Ahn, J.O., Lee, H.W. and Jung, J.K. 2013. Isolation and characterization of a novel ε-caprolactam-degrading microbe, *Acinetobacter calcoaceticus*, from industrial wastewater by chemostat-enrichment. *Biotechnol Lett* 35: 2069–2072.

Reife, A. and Freeman H.S. 2000. Pollution prevention in the production of dyes and pigments. *Text Chem Color Am Dyes Rep* 32: 56–60.

Robinson, T., McMullan, G., Marchant, R. and Nigam, P. 2001. Remediation of dyes in textile effluent: A critical review on current treatment technologies with a proposed alternative. *Bioresour Technol* 77: 247–255.

Saminathan, P., Sripriya, A., Nalini, K., Sivakumar, T. and Thangapandian, V. 2014. Biodegradation of plastics by *Pseudomonas putida* isolated from garden soil samples. *J Adv Bot Zool* 13: 34–38.

Seo, J.S, Keum, Y.S. and Li, Q.X. 2013. Metabolomic and proteomic insights into carbaryl catabolism by *Burkholderia* sp. C3 and degradation of ten N-methylcarbamates. *Biodegradation* 24: 795–811.

Sharma, M. and Dhingra, H.K. 2016. *Br Microbiol Res J* 143: 1–11. Doi: 10.9734/BMRJ/2016/25430.

Sidhu, G.K., Singh, S., Kumar, V., Dhanjal, D.S., Datta, S. and Singh, J. 2019. Toxicity, monitoring and biodegradation of organophosphate pesticides: A review. *Crit Rev Environ Sci Technol* 4913: 1135–1187.

Singh, A., Chaudhary, S., Dubey, B. and Prasad, V. 2016. Microbial-mediated management of organic xenobiotic pollutants in agricultural lands. In: Singh, A., Prasad, S., Singh, R. (eds.) *Plant Responses to Xenobiotics*, 211–230. Springer, Singapore. Doi: 10.1007/978-981-10-2860-1_9.

Singh, D.K. 2008. Biodegradation and bioremediation of pesticide in soil: Concept, method and recent developments. *Ind J Microbiol* 48: 35–40.

Singh, O.V. and Nagaraj, N.S. 2006. Transcriptomics, proteomics and interactomics: Unique approaches to track the insights of bioremediation. *Brief Funct Genomics* 44: 355–362.

Singh, V.K., Singh, A.L. and Singh, R. 2018. Iron oxidizing bacteria: Insights on diversity, mechanism of iron oxidation and role in management of metal pollution. *J Environ Sustain* 1: 221–231.

Sinha, S., Chattopadhyay, P., Pan, I., Chatterjee, S., Chanda, P., Bandyopadhyay, D., Das, K. and Sen, S.K. 2009. Microbial transformation of xenobiotic for environmental bioremediation. *Afr J Biotechnol* 8: 22.

Soares, P.R.S., Birolli, W.G., Ferreira, I.M. and Porto, A.L.M. 2021. Biodegradation pathway of the organophosphate pesticides chlorpyrifos, methyl parathion and profenofos by the marine-derived fungus *Aspergillus sydowii* CBMAI 935 and its potential for methylation reactions of phenolic compounds. *Marine Pollut Bullet* 166: 112185.

Spain, J. and Nishino, S.F. 1987. Degradation of 1,4 dichlorobenzene by a *Pseudomonas* sp. *Appl Environ Microbiol* 53: 1010–1019.

Stolz, A. 2001. Basic and applied aspects in the microbial degradation of azo azoreductase from *Xenophilus azovorans* KF46F in comparison to favin containing azoreductases. *Appl Microbiol Biotechnol* 87: 2067–2076.

Sutherland, T.D., Horne, I., Russell, R.J. and Oakeshott, J.G. 2002. Isolation and characterization of a *Myobacterium* strain that metabolizes the insecticide endosulfan. *J Appl Microbiol* 93: 380–389.

Tack, F.M. and Bardos, P. 2020. Overview of soil and groundwater remediation. In *Soil and Groundwater Remediation Technologies*. Yong Sik Ok, Jörg Rinklebe, Deyi Hou, Daniel C.W. Tsang, Filip M.G. Tack (Eds.). Vol. 1, 11. CRC Press.

Tegene, B.G. and Tenkegna T.A. 2020. Mode of action, mechanism and role of microbes in bioremediation service for environmental pollution management. *J Biotechnol Bioinform Research*: 39–50 SRC/JBBR-112. Doi: 10.47363/JBBR/2020 2116.

Toyofuku, M., Roschitzki, B., Riedel, K. and Eberl, L. 2012. Identification of proteins associated with the *Pseudomonas aeruginosa* biofilm extracellular matrix. *J Proteome Res* 1110: 4906–4915.

Tsaboula, A., Papadakis, E.N., Vrvzas Z. and Kotopoulou A. 2016. Environmental and human risk hierarchy of pesticide: A monitoring hazard assessment and environmental fate. *Environ Int* 78–93. Doi: 10.1016/j.envint.2016.02.008.

United States Environmental Protection Agency (USEPA), Municipal solid waste. 2005. Available at: http://www.epa.gov/epaoswer/nonhw/muncpl/facts.htm.

Urich, T., Lanzen, A., Qi, J., Huson, D.H., Schleper, C. and Schuster. F.S.C. 2008. Simultaneous assessment of soil microbial community structure and function through analysis of the meta-transcriptome. *PLoS One* 3(e2527): 1–13.

Verma, S. and Kuila, A. 2019. Bioremediation of heavy metals by microbial process. *Environ Technol Innov* 14: 100369.

Vilchez-Vargas, R., Junca, H. and Pieper, D.H. 2010. Metabolic networks, microbial ecology and 'omics' technologies: Towards understanding *in situ* biodegradation processes. *Environ Microbiol* 12: 3089–3104.

Villas-Bôas, S.G., Rasmussen, S. and Lane, G.A. 2005. Metabolomics or metabolite profiles? *Trends Biotechnol* 238: 385–386.

Wang D.Z., Kong, L.F., Li, Y.Y. and Xie, Z.X. 2016. Environmental microbial community proteomics: Status, challenges and perspectives. *Int J Mol Sci* 17: 1275.

Wei, K., Yin, H., Peng, H., Liu, Z., Lu, G. and Dang, Z. 2017. Characteristics and proteomic analysis of pyrene degradation by *Brevibacillus brevis* in liquid medium. *Chemosphere* 178: 80–87.

Weir, K.M., Sutherland, T.D., Horne, I., Russell, R.J. and Oakeshott, J.G. 2006. A single moonoxygenases is involved in the metabolism of the organochlorides endosulfan and endosulphate in an *Arthrobacter* sp. *Appl Environ Microbiol* 72: 3524–3530.

Whyte, L.G., Slagman, S.J., Pietrantonio, F., Bourbonniere, L., Koval, S.F., Lawrence, J.R., Innis, W.E. and Greer C.W. 1999. Physiological adaptations involved in alkane assimilation at a low temperature by *Rhodococcus* sp. strain Q15. *Appl Environ Microbiol* 657: 2961–2968.

Williams, T.J., Wilkins, D., Long, E., Evans, F., DeMaere, M.Z., Raftery, M.J. et al. 2013. The role of planktonic *Flavobacteria* in processing algal organic matter in coastal East Antarctica revealed using metagenomics and metaproteomics. *Environ Microbiol* 15: 1302–1317.

Xing, G.H., Wu, S.C. and Wong, M.H. 2010. Dietary exposure to PCBs based on food consumption survey and food basket analysis at Taizhou, China–the world's major site for recycling transformers. *Chemosphere* 8110: 1239–1244.

Xu, X., Liu, W., Tian, S., Wang, W., Qi, Q., Jiang, P. and Yu, H. 2018. Petroleum hydrocarbon-degrading bacteria for the remediation of oil pollution under aerobic conditions: A perspective analysis. *Front Microbiol* 9: 2885. Doi: 10.3389/fmicb.2018.02885.

Yadav, I.C. and Devi, N.L. 2017. Pesticides classification and its impact on human and environment. *Front Environ Sci Eng* 6: 140–158.

Zhang, J.J. and Yang, H. 2021. Metabolism and detoxification of pesticides in plants. *Sci Total Environ* 790: 148034.

Zheng, Y., Yanful, E.K. and Bassi, A.S. 2005. A review of plastic waste biodegradation. *Crit Rev Biotechnol* 254: 243–250.

Zhou, C., Tabb, M.M., Nelson, E.L., Grün, F., Verma, S., Sadatrafiei, A., Lin, M., Mallick, S., Frman, B.M., Thummel, K.E. and Blumberg, B. 2006. Mutual repression between steroid and xenobiotic receptor and NF-κB signaling pathways links xenobiotic metabolism and inflammation. *J Clin Invest* 1168: 2280–2289.

Zhou, J., Wu, Z., Yu, D., Pang, Y., Cai, H. and Liu, Y. 2018. Toxicity of linear alkylbenzene sulfonate to aquatic plant *Potamogeton perfoliatus* L. *Environ Sci Pollut Res* 25: 32303–32311.

10 Importance of Genetically Engineered Microbes (GEMs) in Bioremediation of Environmental Pollutants
Recent Advances and Challenges

Wilgince Apollon, Héctor Flores-Breceda, and Gerardo Méndez-Zamora
Universidad Autónoma de Nuevo León

Juan Florencio Gómez-Leyva
TecNM-Instituto Tecnológico de Tlajomulco (ITTJ)

Alejandro Isabel Luna-Maldonado
Universidad Autónoma de Nuevo León

Sathish-Kumar Kamaraj
TecNM-Instituto Tecnológico El Llano Aguascalientes (ITEL)

CONTENTS

DOI: 10.1201/9781003247883-10

10.1   INTRODUCTION

For decades, human activity has resulted in the accelerated pollution of the environment, bringing about the accumulation of substances highly dangerous to health. On various occasions, different components of the chemical structures of most of these contaminants have exceeded the bioremediation capacity of the microbes used in clean-up (Benjamin et al., 2019). Hence, in order to eliminate pollutants, various strategies have been adopted by different sectors, i.e. public or private, as well as researchers and companies from different countries, dismayed by this precarious situation. Due to the considerable efforts made by researchers during the last decades, a large portion of these pollutants and materials containing heavy metals have been eliminated from the environment. However, completely reducing the toxic waste that causes ecological problems has not been an easy task.

Pollutants are known to cause a serious imbalance between human concerns and nature itself, and the conventional processes for the eradication of said contaminators are very expensive, due to the high consumption of energy and numerous reagents. For efficient and inexpensive pollutant removal, traditional methods are used, which has resulted in a series of bioremediation techniques. The bioremediation process includes the following levels: (1) without any human intervention, native microbes reduce pollutants; (2) oxygen and nutrients are supplied to the systems, and biostimulation is used; and (3) microbes are added to the systems during the process of bioaugmentation (Joutey et al., 2013). According to Diez (2010), the added organisms must have more capacity than the native flora to comply with the degradation process of the polluting area; in addition, they use this process as an energy source simultaneously.

Furthermore, among the microorganisms used in the bioremediation process, there are genetically engineered microbes (GEMs). These are the result of great advances in genetic engineering, which have given researchers in the areas of molecular biology and biochemistry a great deal of insight into the manipulation of these modified microbes. The use of GEMs to remove metals from contaminated areas has attracted the attention of many ecologists and environmentalists, not only because of its efficiency, but also because it is a very promising technology for today's environmental concerns. GEMs are closely monitored when they are released into the environment (Benjamin et al., 2019).

There are two main aspects to consider during the pollutant removal process using GEMs: (1) the rapid adaptation capacity of the microorganisms and (2) the time taken for the efficient removal of pollutants of interest (Seo et al., 2009). Also, many other factors should be taken into account, such as the pH, temperature, and nitrogen (N) and phosphorus (P) available during the process. Finally, the genetic potential and the degree of degradation of microorganisms are additional important factors (Fritsche and Hofrichter, 2008). GEMs, therefore, have a high degradation capacity, and their applications in bioremediation of various pollutants have been tested under well-defined conditions (Menn et al., 2008). With that in mind, the main purposes of this chapter are first to present recent advances in genetic engineering to create the GEMs; second, to discuss the prominence of environmental pollutants rescue when using GEMs; thirdly, to elaborate on the field application of GEMs; and finally, to present an overview of the stability and risk of GEMs in the field of application.

10.2 RECENT GENETIC ENGINEERING PROGRESS IN THE USE OF GEMS FOR THE BIOREMEDIATION OF POLLUTANTS

10.2.1 MODIFICATION OF GENES FOR THE BIODEGRADATIVE ENZYMES

Genetic engineering such as molecular biology has proven to be a mainstay in eradicating environmental pollution caused by human negligence. Several strategies have been adopted to find sustainable solutions to this problem that affects world health so greatly. According to environmental biotechnology, filamentous fungi, yeasts, and microorganisms such as bacteria have the ability to eliminate toxic elements in aqueous solutions. This is because the metabolic potential of microorganisms is a highly beneficial and safe technology for the eradication of contaminants found in contaminated sites. For this reason, microorganisms have been modified in the laboratory through the use of recombinant DNA technology, for their subsequent release in affected or contaminated areas. The creation of these modified genes aims to efficiently degrade pollutants. This recombinant DNA technology takes a very important step towards being able to design several degradative pathways, which can carry out a partial or complete degradation of any site contaminated by toxic elements (Azad et al., 2014).

Table 10.1 shows previous studies related to recombinant DNA technology (gene modification) using laboratory techniques. Here, it can be seen that GEMs have been applied in biodegradation processes for some time. Figure 10.1 illustrates the schematic representation of the mechanism for GEM creation in a laboratory. Genes modified by DNA techniques were used to evolve both monooxygenases and dioxygenases in all bioremediation processes (Leungsakul et al., 2005). Furthermore, RNA technology has also been developed for the elimination of toxic elements in the environment (Poretsky et al., 2005). RNA technology allows not only the acquisition of nitrogen (N) and assimilation of Cl compounds, but also the oxidation of sulphur (S) as an abundant element in the Earth's crust. With this technology, 16S rRNA genes are used that have a high capacity to allow the characterization of microorganisms, which can then be exploited for the bioremediation of polluted areas (Stewart et al., 2010).

TABLE 10.1

Previous Developments in Recombinant DNA Technologies for Bioremediation of Pollutants

Modification of Gene	Bioremediation Process	Reference
Coal tar waste container detection by FISH	Messenger ribonucleic acid (mRNA) transcripts connected to naphthalene dioxygenase and tyramide signal amplification	Bakersmans and Madsen (2002)
Destructive or suicidal genetically engineered microbes	Killer–anti-killer genes vulnerable to programmed cell death	Pandey et al. (2005)
Carbamate pesticide- and organophosphate-degrading genetically engineered microbes	Methyl parathion hydrolase encoding gene and cognate regulator	Liu et al. (2006)

FISH, fluorescence *in situ* hybridization; Hg, mercury.

However, there is no doubt that the efficient degradation of contaminated sites is highly reliant on the affinity and particularity of the enzyme. Oxygenase enzymes participate in the degradation process, catalysing the reduction of oxygen (O_2) with the incorporation of monooxygenases or dioxygenases or all of this, within the substrate that undergoes the oxidation process. Protein oxygenase engineering is one of the molecular techniques used to successfully improve the oxidative degradation of contaminated sites (Urgun-Dermitas et al., 2006). Normally, oxygenase catalysts are efficiently implicated in the degradation of many aromatic compounds in relation not only to substrate relaxation, but also to high rates of enhancement of the degradation of pollutants (Furukawa et al., 2004).

10.2.2 Gene Transfer and/or Pathways to the Heterologous Host

Gene transfer comprises two main aspects: (1) a catabolic pathway with the use of engineering and (2) microbes developed for the improvement of the bioremediation process of both mixed waste and metals. The cause of water and soil contamination is due to aromatic compounds, which allow ecologists to partially or completely resolve this problem through bioremediation with the use of microbe engineering. Monti et al. (2005) designed 17,400 genes in *Pseudomonas fluorescens* ATCC with specificity to encode the degradation of the 2,4-dinitrotoluene (DNT) pathway of *Burkholderia* sp. strain DNT. According to these researchers, the recombinant strain is the main source of N that can accept (receive) 2,4-DNT, completely degrading the compound, resulting in carbon (C) being utilized metabolically by the cell. The transgenic strain was found to be more efficient in degrading DNT when compared to *Burkholderia* at cold temperatures. Wu et al. (2006) used the bacterium *E. coli* as a host for the expression and replication of *Comamonas* sp. CNB-1 strains have the ability to encode a new partial reducing pathway for both nitrobenzene and 4-chloronitrobenzene, with the same conditions as the aforementioned case.

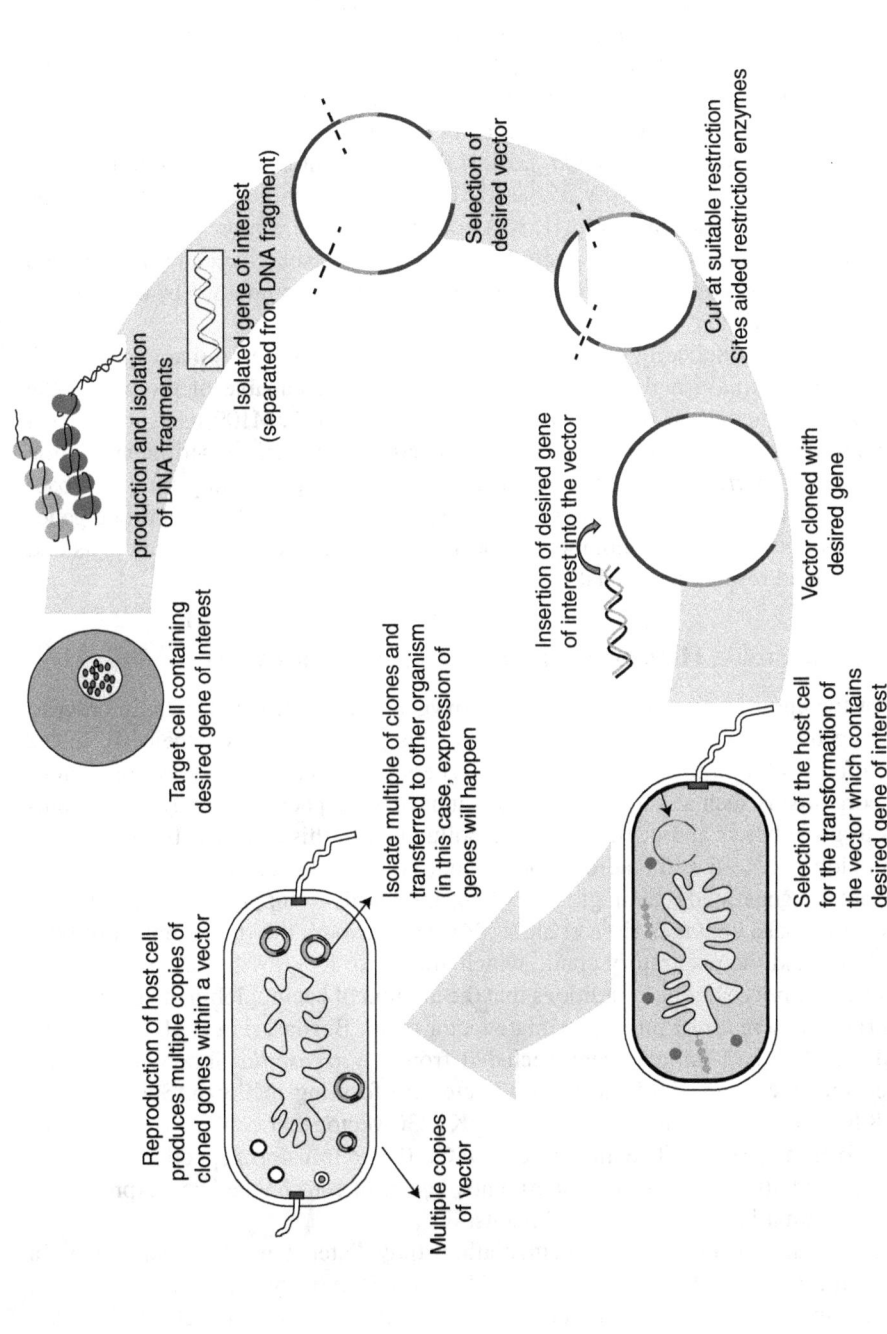

FIGURE 10.1 Schematic diagram of the mechanism of production of genetically engineered microorganisms (GEMs) for the bioremediation process of pollutants.

The bacterium *E. coli* has also been employed as a host in heavy metal bioreme-diation processes since genetic engineering has put forward this type of bacteria for its high bioremediation capacity. According to Crameri et al. (1997), the recombinant *E. coli* bacterium was shown to have developed a certain resistance against arsenate with the ability to detoxify this element by the reduction reaction. Subsequently, Qin et al. (2006) used genetically modified *E. coli* for mercury detoxification. The said bacterium contained an encoded sequence of the binding domain (MBD) of Hg(II) from the MerR protein of *Shigella flexneri*. Recombinant cells were found to have indicated an enormous quantity of the MBD binding domain on the outside of the cell, to which 6.1 times more Hg(II) was bound. These were more resistant to Hg(-II) compared to parental cells without MBD. However, researchers have ascertained that recombinant cells also by preference bind to Hg(II) accompanied by other heavy metals, e.g. Cd(II) and Zn(II).

In summary, genetically modified bacteria (transgenic bacteria) that express both polyphosphate kinase and metallothionein enzymes are capable of promoting the effective bioremediation of Hg (Ruiz et al., 2011). *E. coli* JM109 has shown great potential when eliminating Hg in contaminated soil, water, or sediments, while *Pseudomonas putida* and *Deinococcus radiodurans* are genetically modified bac-teria capable of degrading organics in polluted areas. The use of the modified genes in the bioremediation of pollutants has been a great achievement of ecologists and environmental researchers to date.

10.2.3 BACTERIAL HAEMOGLOBIN TECHNOLOGY FOR POLLUTANTS

One of the engineering solutions is to add oxygen to contaminated sites, in order to speed up the degradation process. However, this process is very expensive. In the search for a solution to this problem, researchers have opted for the development of microorganisms such as aerobic bacteria that have the capacity under hypoxic condi-tions to grow better and degrade organic pollutants. In this case, the bacterium vgb haemoglobin gene, i.e. the aerobic bacterium, is the result of genetic engineering. The bacteria generated [haemoglobin (VHb)] are mainly to supply oxygen in addition to the oxygenase enzymes (Lee et al., 2004). This favours the improvement of both its activity and the respiratory chain, which allows for its growth. Researchers have shown under microaerobic conditions that the amount of haemoglobin in vgb-bearing bacteria can increase by an approximate factor of 10. Bagdasarian et al. (1981) indi-cated that *Vgb* had initially been secluded from a *Vitreoscilla* library and finally cloned into the "vector" of the *E. coli* bacterium forming pUC8:16. Subsequently, pUC8:16 was enclosed into the so-called pKT230 vector that contained a high host range, forming pSC160. The main role of pSC160 is to transform specific bacteria in the bioremediation process of contaminants. *Vgb* has been changed and expressed in plants, mammalian cells, fungi, and plants.

Additionally, in a pollutant bioremediation study, Patel et al. (2000) succeeded in transforming the Burkholderia strain DNT with pSC160 (forming the YV1 strain) and also monitored the degradation of 2,4-DNT at the same time. The YV1 strain is capable of carrying approximately 15 copies of vgb/cell, but can express levels of VHb around three times lower than that of *Vitreoscilla* and around 15–25 times smaller than the carrier of recombinant *E. coli* located in a vector of many copies. These genes and enzymes are mainly responsible for the total mineralization of the 2,4-DNT strain through the oxidative pathway; its characterization has been very successful. The presence of O_2 is essential to the first three steps of this process. With the use of VHb, the degradation of the 2,4-DNT strain was improved by between 50% and 60% approximately. Both the improvement of the three steps mentioned above and the improvement of the degradation of 2,4-DNT depend on oxygen in the degradation of the DNT strain pathway and the increase in the intracellular concentration of VHb, respectively.

Based purely on the points discussed above, it can be argued that haemoglobin bacteria technology has contributed greatly to the bioremediation of contaminated regions. However, the expression of VHb in recombinant hosts has proved to be a great challenge for ecologists as it affects both the respiration rates and the metabolism (in some cases) of the microbes. That being said, increases of up to 20% in respiration rates (using VHb) have been observed relative to cells without the presence of *vgb*.

10.2.4 Tracking of GEMs with the Aid of Molecular Tools and Techniques

After the production of GEMs in the laboratory, these must be released in contaminated places. However, in complex samples, it is necessary to develop detection and enumeration methods for these organisms, in order to determine how their release affects the environment. Four parameters must be taken into account during this process: (1) dispersion of released GEMs, (2) number, (3) activity, and (4) survival. There are traditional GEM tracking methods that depend on how the colonies are grown in Petri dishes (Widada et al., 2002). However, there is a certain disadvantage to using these simple methods because they not only are cumbersome, but are also of limited sensitivity. To find a resolution to these limitations, it is necessary to apply molecular techniques that incorporate nucleic acid expansion via the polymerase chain reaction (PCR). After extraction and purification of the DNA of a strain or a species, the specific fragment obtained is amplified and sequenced. These techniques include the amplification of particular DNA sequences to a specific strain and reverse transcription PCR (RT-PCR) that amplifies exclusively the RNA. These sequences are capable of detecting and quantifying metabolically active cells in a sample (Widada et al., 2002).

Other types of PCR that have been used for GEM tracking are competitive PCR (cPCR) and the most-probable-number PCR (MPN-PCR). These allow the estimation of the number of cells that make up a soil sample, as demonstrated by Chandler (1998). Therefore, when using cPCR this normally involves the co-amplification of a DNA sample called "target" with known amounts of other competing DNA whose specificity is to share most of the nucleotide sequence with the "target"; in this case, any parameter (whether predictable or unpredictable) that tends to affect amplification using PCRs has a similar effect on the two molecular species. In contrast, the use of MPN-PCR implies the combination between conventional MPN culture and the PCR technique directed to the so-called uidA gene (Picozzi et al., 2004).

This indicates that molecular methods are promising because they allow the cultivation of microbes with a high bioremediation capacity. The development of GEMs through genetic engineering has been a great achievement because these organisms are more efficient compared to native microbes. Brockman (1995) demonstrated that a combination of traditional methods with novel molecular methods results in a more robust method of efficient quantification of GEMs, which will have positive effects on the natural microbial population. GEM production and release methods must be (1) very efficient and (2) highly sensitive where there is a low number of GEMs present in the sample.

10.3 THE PROMINENCE OF ENVIRONMENTAL POLLUTANTS RESCUE USING GEMS

10.3.1 HEAVY METALS

The symbiotic techniques developed for the bioremediation of contaminants caused by heavy metals have received much attention as promising techniques for future use. By nature, heavy metals are impossible to destroy biologically (i.e. no degradation); therefore, they can only go into a state of oxidation (Jaysankar et al., 2008). The main heavy metals that constitute a danger to the environment and the ecosystem, in general, are mercury (Hg), cobalt (Co), cadmium (Cd), silver (Ag), nickel (Ni), strontium (Sr), vanadium (V), copper (Cu), tin (Sn), caesium (Cs), chromium (Cr), thallium (Tl), arsenic (As), and zinc (Zn). These heavy metals are capable of damaging all the vital organs of any living organism, including the vital organs of human beings (Pant et al., 2021). Because of the high costs of traditional techniques and also the transfer of pollutants by the same methods, that is, generating waste through incineration, ecologists have opted for less expensive methods. For example, the use of genetically modified microbes in real-time heavy metal bioremediation is an inexpensive engineering mechanism (Figure 10.2). The idea behind the use of GEMs is to comply with an environmentally friendly bioremediation process and to be efficient at the same time.

10.3.2 DYES

Dyes are considered the main sources of environmental pollution in the world, and the main cause is the huge discharge of colourants as effluents as demonstrated by Wang et al. (2017). In addition, these contaminants are very harmful to both human health and ecosystems. As dyes play a crucial function in various industries such as textiles and pulp and paper, they are widely used, leading to the presence of colourants preventing not only the growth of aquatic flora and fauna, but also the solubility of the gas found in the aquatic ecosystem (Luo et al., 2018). Currently, there are different methods of bioremediation of sites polluted by these types of contaminants. These include biological, physical (sedimentation, flocculation, and coagulation), and chemical (redox, neutralization, etc.) interventions. Another method used to detoxify areas contaminated by the dye is the application of genetically modified microbes

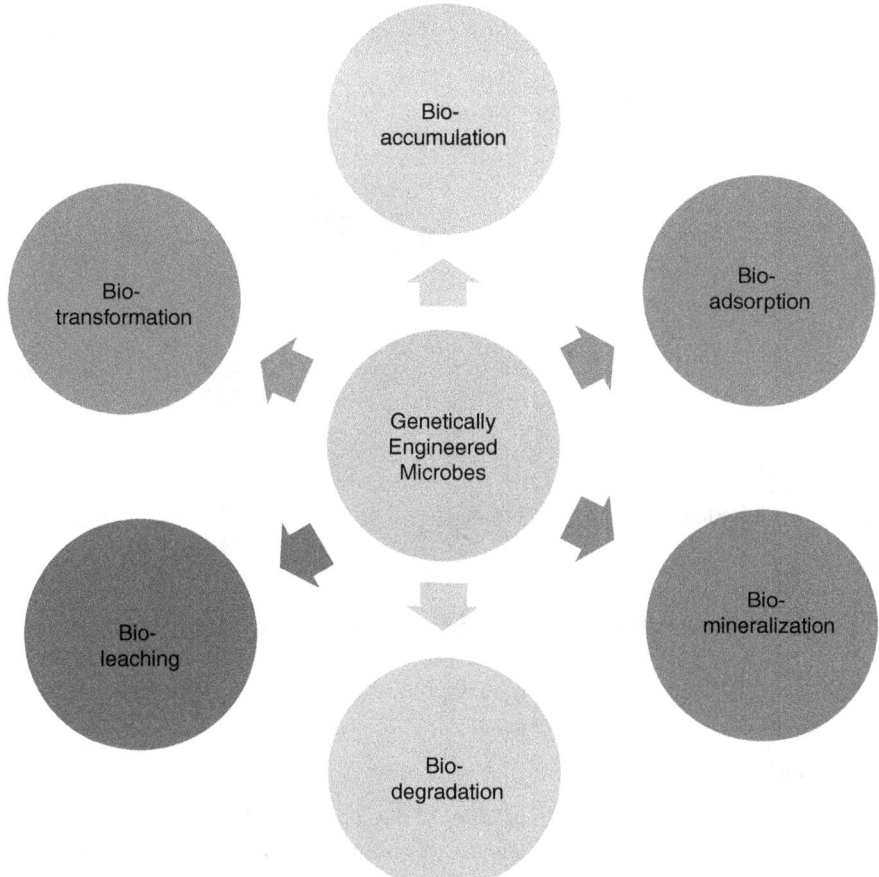

FIGURE 10.2 Mechanism of bioremediation of heavy metals using genetically engineered microbes (GEMs).

(Figure 10.3). This method is of great importance due to its easy handling, lower expense, and promise compared to conventional methods.

10.3.3 XENOBIOTICS

Xenobiotics are synthetic chemicals and highly persistent in the ecosystem for the simple reason that they have a very complex organic structure. Xenobiotics are pollutants with a high degree of danger (direct risk) that have had a comparatively brief time interval of exhibition in the ecosystem. In addition, they are considered the main sources of pollutants associated with induced toxicity (e.g. chemical substances) by human activity (Peeters et al., 2019). The treatment of xenobiotics has been the subject of great challenge for environmental researchers or ecologists. To meet these challenges, the technique of genetic engineering has been used, by which microbes

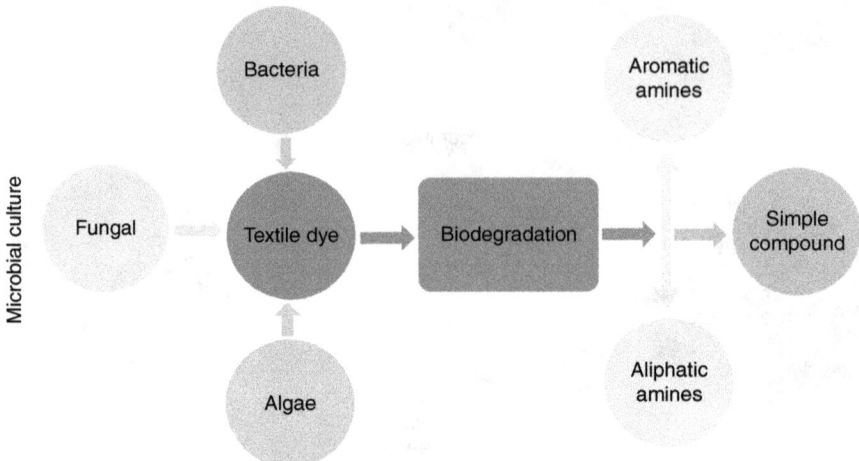

FIGURE 10.3 Biodegradation process of dyes through genetically engineered microbes (GEMs).

with targeted or enhanced degradation specificity are created. These microbes have a high degradation capacity for xenobiotic pollutants. GEMs are a novel detoxification technique for sites contaminated by such pollutants.

10.3.4 PESTICIDES

Due to large populations of different types of pests that invade environmental spaces, homes, production fields, etc., it is necessary to find appropriate solutions in order to remove them. But sometimes mistakes have been made by using substances that are not environmentally friendly in controlling these "plagues" of insects. Global companies that manufacture high-risk chemicals such as pesticides (e.g. atrazine, hexachlorocyclohexane, and dichlorodiphenyltrichloroethane known as DDT) to combat the invasion of pests contribute largely to air pollution. The utilization of these categories of products to eradicate the issues caused by pests of all kinds, for instance, diseases of both crops and man caused by them, contribute to a large extent to the accelerated pollution today. Pesticides have indiscriminately harmful effects on ecosystems because they are very stable in the soil and also soluble in water (Zhang et al., 2016). Currently, farmers and researchers in the area of agricultural engineering are looking for alternatives to decrease the utilization of pesticides through the application of genetically modified organisms (e.g. fungi and beneficial insects) and plant extracts (such as repellents) as shown in Figure 10.4. However, genetic engineering technologies have been successful in the degradation processes of pesticides.

10.3.5 ORGANIC COMPOUNDS

Organic compounds, as well as other pollutants, are also one of the main sources of ambient pollution. These pollutants originate from heavy metal contamination. Over

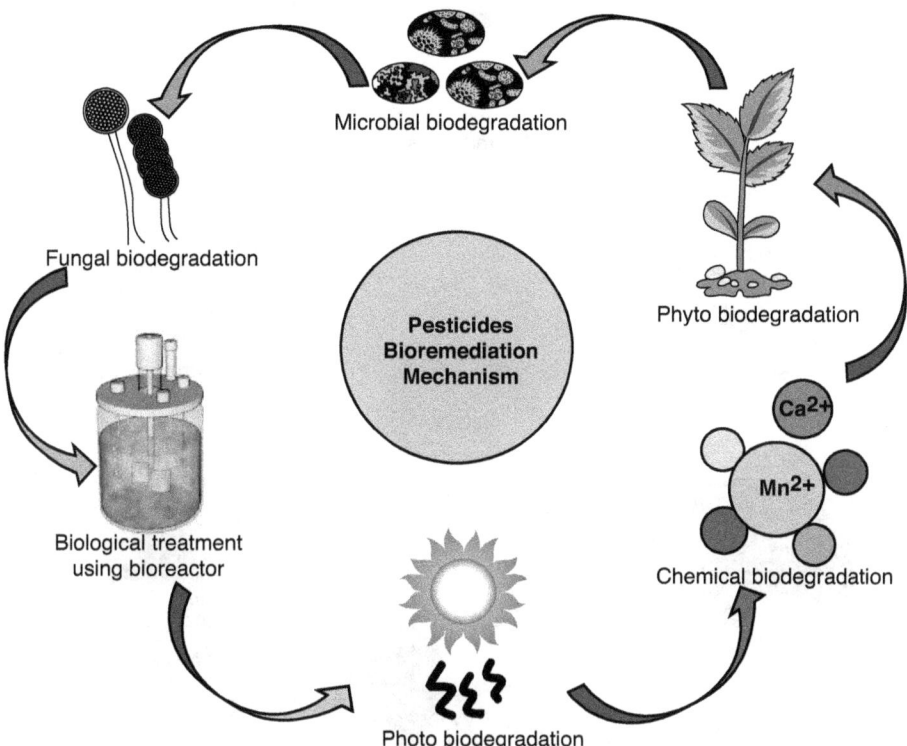

FIGURE 10.4 Bioremediation of pesticides in the contaminated environment by using diverse practices. (Modified from Pant et al., 2021.)

several decades, diverse conventional methods have been developed to remedy toxic wastes (from organic compounds). However, these methods instead of completely eradicating the contamination tend to produce toxic intermediates (Chen et al., 2016); they are also not very successful as a degradation mechanism. The creation of micro-organisms such as *Mycobacteria* sp. remediating contaminated areas has, however, been very successful in the scientific community. These bacteria have the potential to detoxify sites contaminated by polycyclic aromatic hydrocarbons (PAHs).

10.3.6 Oil Components

For decades, petrochemical industries in the world have represented a great danger to the atmosphere due to the release of petroleum hydrocarbons into both rivers and the sea. In addition, other industries that use sewer pipes leading to the sea or rivers also contribute to this great harm that acts as "gangrene" to the environment and marine ecosystems. These oil spills in the ocean have dire consequences on the life in the ocean (Pant et al., 2021). The bioremediation mechanism of these contaminants by GEMs has been implemented with great success due to the high remediation and even degradation capacities of these microbes (see Figure 10.5). With the application

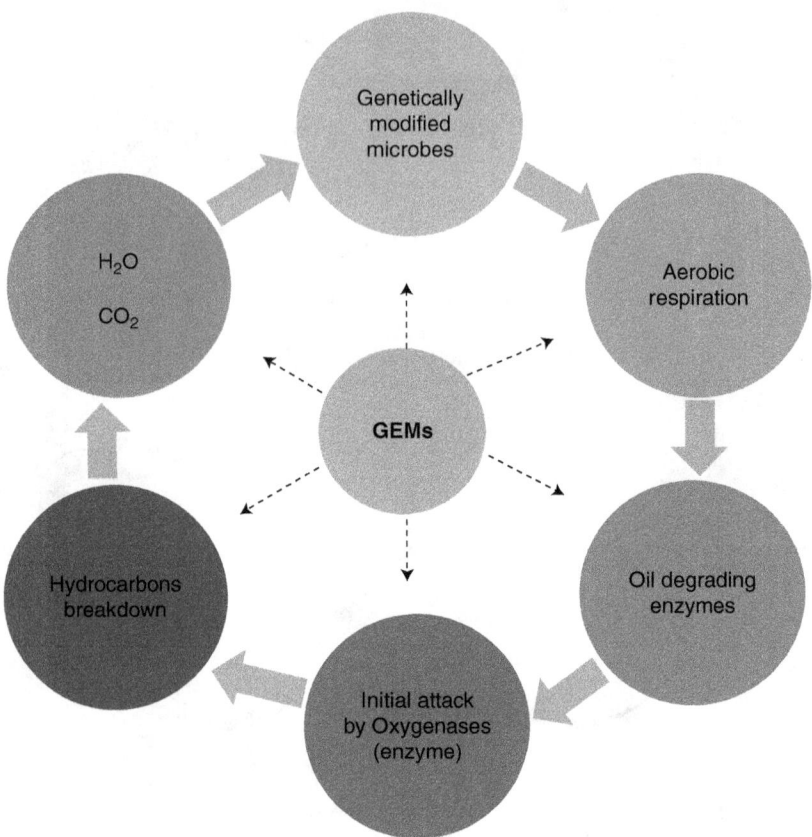

FIGURE 10.5   Oil degradation process using genetically engineered microbes (GEMs).

of GEMs, all toxic aggregates found in oil can be removed after being extracted from the ground (crude oil).

10.3.7   RADIOACTIVE COMPOUNDS

The accumulation of radioactive waste is a major source of environmental contamination. This type of contamination originates, for example, from various types of institutions that are dedicated to research (research organizations), health centres (hospitals), etc. (Pant et al., 2021). The use of genetically modified organisms is the best alternative for the elimination of radioactive compounds because natural (wild) microbes are not efficient in carrying out this process successfully. GEMs not only have high survival power, but also are more resistant to consuming and digesting heavy metals (highly radioactive) such as toluene (aromatic class hydrocarbon) and ionic Hg (Gogada et al., 2015). Using genetic engineering to degrade radioactive waste is safer and more environmentally friendly compared to conventional methods.

TABLE 10.2

Genetically Engineered Microbes (GEMs) Used in the Bioremediation of Pollutants

Genetically Engineered Microbes	Gene Modified	Compound Eliminated	Reference
Pseudomonas fluorescens HK4	*lux CDABE*	Naphthalene	Sayler and Ripp (2000)
Pseudomonas fluorescens F113rifpcbrrnBP1::gfp-mut3	operon *bph*, *gfp*	Chlorinated biphenyls	Boldt et al. (2004)
Pseudomonas putida KT2442	pNF142 plasmid, gfp	Naphthalene	Filonov et al. (2005)
Pseudomonas putida PaW85	pWW0 plasmid	Petroleum	Jussila et al. (2007)
Pseudomonas putida PaW340 (pDH5)	pDH5 plasmid	4-Chlorobenzoic acid	Massa et al. (2009)

Source: Table modified from Kumar et al. (2018).

10.4 FIELD APPLICATIONS OF GEMS

As mentioned in the previous sections, GEMs have significant potential for the bioremediation of polluted sites. The use of these microbes has been a leap forward for genetic engineering. However, their use in the field has had great limitations because they simultaneously cause risks, that is, the horizontal transfer of genetic material (Urgun-Demirtas et al., 2006). The first genetically modified microorganism that has been created for soil bioremediation in the field is known as *Pseudomonas fluorescens* HK44 specific to degrade PAHs (Ripp et al., 2000). This strain was authorized by the US Environmental Protection Agency. By extensively studying the HK44 strain for PAH degradation, it was found to have high survivability in the field of over 2 years. Table 10.2 indicates the frequent use *of Pseudomonas* sp. in the bioremediation of pollutants. GEMs have also been efficiently applied in the domain of both rhizoremediation and phytoremediation for pollutant removal from ecological areas (Kumar et al., 2018).

10.5 STABILITY AND RISK ASSESSMENT OF GEMS IN THE FIELD

The stability of GEMs has raised big questions among researchers around the world as to the transfer of unwanted genes and their possible decline during their application in the field. To avoid the transfer of all unwanted genes, the researchers opted for the use of vectors, in this case mini-transposons. By combining mini-transposons with the selection of non-antibiotic resistance systems, the introduction of harmful genes (for both animals and humans) into the environment is eliminated. Several strategies were adopted for the safe release of GEMs at contaminated sites. For example, when there are gene transfers at one site (where microbes are released), which also affect native microbes present at the site, GEMs transfer is not significant (Urgun-Demirtas et al., 2006). In addition, assessing the environmental impacts of GEMs after release is another aspect that genetic engineering must consider, but these assessments are difficult to perform. Recombinant DNA GEMs behave differently in the environment compared to natural microbes. Therefore, the use of GEMs can be justified when they are the only option for the degradation of hazardous pollutants. In summary, field experiments are very important in assessing or determining the risks posed by GEMs when released into the environment.

10.6   CONCLUSIONS

The modified microorganisms by recombinant DNA degrade pollutants partially, or completely, at any site contaminated by toxic elements in aqueous solutions. They evolve both monooxygenases and dioxygenases in all bioremediation processes. Bacteria have been designed with the specificity to encode the degradation of 2,4-dinitrotoluene, as well as to encode a new partial reducing pathway for both nitrobenzene and 4-chloronitrobenzene. Furthermore, it was observed that the recombinant *bacteria* detoxify arsenate. Transgenic bacteria are also capable of promoting effective bioremediation of Hg and degrading organics in contaminated sites. On the other hand, RNA technology eliminates toxic elements in the environment. Certain bacteria can supply oxygen in addition to the oxygenase enzymes.

In complex samples, it is necessary to develop detection and enumeration methods. Molecular methodologies are promising because they allow the cultivation of microbes with a high bioremediation capacity. GEMs are shown to be more efficient organisms when compared to native microbes. The combination of traditional methods with novel molecular methods results in a more robust method of efficient quantification of GEMs. The symbiotic methods developed for the bioremediation of contaminants caused by heavy metals have, justifiably, received much attention as promising methods. The use of genetically modified microbes in real-time heavy metal bioremediation is an inexpensive engineering mechanism. It has been demonstrated that dyes, xenobiotics, pesticides, organic compounds, oil components, and radioactive compounds can be degraded by GEMs. Despite the successful findings, some GEMs have high survivability in the field of over 2 years and risk instability. For the above, more research is required.

ACKNOWLEDGEMENTS

WA extends his thanks for the PhD Scholarship from the National Council for Science and Technology (CONACYT, for its acronym in Spanish), as well as the Universidad Autónoma de Nuevo León (UANL) through the Subdirección de Estudios de Posgrado of Faculty of Agronomy (FA-UANL) for his acceptance in the PhD programme. Particularly, WA also thanks his Doctoral Thesis Committee. In addition, AILM and WA thank PAICYT (Programa de Apoyo a la Investigación Científica y Tecnológica-CT1519–21) from Universidad Autónoma de Nuevo León for their support. Finally, KSK would like to acknowledge the Proyectos de Desarrollo Tecnológico e Innovación-2021-TecNM – Instituto Tecnológico El Llano Aguascalientes (IT19F245).

REFERENCES

Azad MAK, Amin L, Sidik NM. 2014. Genetically engineered organisms for bioremediation of pollutants in contaminated sites. *Chinese Sci Bull* 59(8):703–714. Doi: 10.1007/s11434-013-0058-8.

Bagdasarian M, Lurz R, Ruckert B. et al. 1981. Broad host range, high copy number, RSFS1010-derived vectors, and a host-vector system for gene cloning in *Pseudomonas*. *Gene* 16:237–247. Doi: 10.1016/0378-1119(81)90080-9.

Bakersmans C, Madsen EL. 2002. Detection in coal tar based contaminated ground water of mRNA transcripts related to naphthalene dioxygenase by FISH with tyramide signal amplification. *J Microbiol Methods* 2002(50):75–84. Doi: 10.1016/s0167-7012(02)00015-5.

Benjamin SR, de Lima F, Rathoure AK. 2019. Genetically engineered microorganisms for bioremediation processes: GEMs for bioremediaton. In Management Association (Ed.), *Biotechnology: Concepts, Methodologies, Tools, and Applications*. IGI Global, 1607–1634. Doi: 10.4018/978-1-5225-8903-7.ch067.

Boldt TS, Sørensen J, Karlson U, Molin S, Ramos C. 2004. Combined use of different Gfp reporters for monitoring single-cell activities of a genetically modified PCB degrader in the rhizosphere of alfalfa. *FEMS Microbiol Ecol* 48(2):139–148. Doi: 10.1016/j.femsec.2004.01.002.

Brockman FJ. 1995. Nucleic-acid based methods for monitoring the performance of *in situ* bioremediation. *Mol Ecol* 4:567–578. Doi: 10.1111/j.1365-294X.1995.tb00257.x.

Crameri A, Dawes G, Rodriquez E, Silver S, Stemmer WPC. 1997. Molecular evolution of an arsenate detoxification pathway by DNA shuffling. *Nat Biotechnol* 15:436–438. Doi: 10.1038/nbt0597-436.

Cheng Y, He H, Yang C, Zeng G, Li X, Chen H, Yu G. 2016. Challenges and solutions for biofiltration of hydrophobic volatile organic compounds. *Biotechnol Adv* 34(6):1091–1102. Doi: 10.1016/j.biotechadv.2016.06.007

Diez MC. 2010. Biological aspects involved in the degradation of organic pollutants. *J Soil Sci Plant Nutr.* 10(3):244–267. Doi: 10.4067/S0718-95162010000100004.

Filonov AE, Akhmetov LI, Puntus IF, Esikova TZ, Gafarov AB, Izmalkova TY, Sokolov SL, Kosheleva IA, Boronin AM. 2005. The construction and monitoring of genetically tagged, plasmid-containing, naphthalene-degrading strains in soil. *Microbiology* 74(4):453–458. Doi: 10.1007/s11021-005-0088-6

Fritsche W, Hofrichter M. 2008. Aerobic degradation by microorganisms in biotechnology set, 2nd Ed. H-J Rehm and G Reed (Eds.), Wiley-VCH Verlag GmbH, Weinheim, Germany, 144–167. Doi: 10.1002/9783527620999.ch6m.

Furukawa K, Suenaga H, Goto M. 2004. Biphenyl dioxygenases: Functional versatilities and directed evolution. *J Bacteriol* 186:5189–5196. Doi: 10.1128/JB.186.16.5189-5196.2004.

Gogada R, Singh SS, Lunavat SK. et al. 2015. Engineered *Deinococcus radiodurans* R1 with NiCoT genes for bioremoval of trace cobalt from spent decontamination solutions of nuclear power reactors. *Appl Microbiol Biotechnol* 99:9203–9213. Doi: 10.1007/s00253-015-6761-4.

Jaysankar D, Ramaiah N, Vardanyan L. 2008. Detoxification of toxic heavy metals by marine bacteria highly resistant to mercury. *Mar Biotechnol* 10:471–477. https://goo.gl/JBfhJp.

Joutey NT, Bahafid W, Sayel H, El Ghachtouli N. 2013. Biodegradation: Involved microorganisms and genetically engineered microorganisms. *Biodegrad Life Sci* Doi: 10.5772/56194.

Jussila MM, Zhao J, Suominen L, Lindström K. 2007. TOL plasmid transfer during bacterial conjugation *in vitro* and rhizoremediation of oil compounds *in vivo*. *Environ Pollut* 146:510–524. Doi: 10.1016/j.envpol.2006.07.012.

Kumar NM, Muthukumaran C, Sharmila G, Gurunathan B. 2018. Genetically modified organisms and its impact on the enhancement of bioremediation. In: Varjani S., Agarwal A., Gnansounou E., Gurunathan B. (Eds.), *Bioremediation: Applications for Environmental Protection and Management. Energy, Environment, and Sustainability.* Springer, Singapore, 53–76. Doi: 10.1007/978-981-10-7485-1_4.

Lee SY, Stark BC, Webster DA. 2004. Structure-function studies of the *Vitreoscilla* hemoglobin D-region. *Biochem Biophys Res Commun* 316(4):1101–1106. Doi: 10.1016/j.bbrc.2004.02.154.

Leungsakul T, Keenan BG, Smets BF, Wood TK. 2005. TNT and nitroaromatic compounds are chemoattractants for *Burkholderia cepacia* R34 and *Burkholderia* sp. Strain DNT. *Appl Microbiol Biotechnol* 69:321–325. Doi: 10.1007/s00253-005-1983-5.

Liu Z, Jainhong Q, Xu J-H, Jun W, Li S-P. 2006. Construction of a genetically engineered organism for degrading organo phosphate and carbamate pesticides. *Int Biodeterioration Biodegrad* 58:65–69. Doi: 10.1016%2Fj.ibiod.2006.07.009.

Luo Q, Chen Y, Xia J. et al. 2018. Functional expression enhancement of *Bacillus pumilus* CotA-laccase mutant WLF through site-directed mutagenesis. *Enzyme Microb Technol* 109:11–19. Doi: 10.1016/j.enzmictec.2017.07.013.

Massa V, Infantino A, Radice F. et al. 2009. Efficiency of natural and engineered bacterial strains in the degradation of 4-chlorobenzoic acid in soil slurry. *Int Biodeteriorat Biodegrad* 63(1):112–115. Doi: 10.1016/j.ibiod.2008.07.006.

Menn F-M, Easter JP, Sayler GS. 2008. Genetically engineered microorganisms and bioremediation. *Biotechnology* 441–463. Doi: 10.1002/9783527620951.ch21.

Monti MR, Smania AM., Fabro G, Alvarez ME, Argarana CE. 2005. Engineering *Pseudomonas fluorescens* for biodegradation of 2,4-dinitrotoluene. *Appl Environ Microbiol* 71: 8864–8872. Doi: 10.1128/aem.71.12.8864-8872.2005.

Pandey G, Paul D, Jain RK. 2005. Suicidal genetically engineered microorganisms for bioremediation-need and perspectives. *Bio Essays* 25:563–573. Doi: 10.1002/bies.20220.

Pant G, Garlapati D, Agrawal U, Prasuna RG, Mathimani T, Pugazhendhi A. 2021. Biological approaches practised using genetically engineered microbes for a sustainable environment: A review. *J Hazard Mater* 405:124631. Doi: 10.1016/j.jhazmat.2020.124631.

Patel SM, Stark BC, Hwang, K-W, Dikshit, KL, Webster, DA. 2000. Cloning and expression of *Vitreoscilla* hemoglobin gene in *Burkholderia* sp. Strain DNT for enhancement of 2,4-dinitrotoluene degradation. *Biotechnol Prog* 16(1):26–30. Doi: 10.1021/bp9901421.

Picozzi C, Foschino R, Pilato P. 2004. Exploiting an MPN-PCR technique to quantify *Escherichia coli* in minced meat. *Ann Microbiol* 54(2):343–349. https://citeseerx.ist.psu.edu/viewdoc/download?doi=10.1.1.502.276&rep=rep1&type=pdf.

Poretsky RS, Bano N, Buchan A. et al. 2005. Analysis of microbial gene transcripts in environmental samples. *Appl Environ Microbiol* 71:4121–4126. Doi: 10.1128/aem.71.7.4121-4126.2005.

Qin J, Song L, Brim H, Daly MJ, Summers AO. 2006. Hg (II) sequestration and protection by the MerR metal-binding domain (MBD). *Microbiology* 152:709–719. Doi: 10.1099/mic.0.28474-0.

Ripp S, Nivens DE, Ahn Y. et al. 2000. Controlled field release of a bioluminescent genetically engineered microorganism for bioremediation process monitoring and control. *Environ Sci Technol* 34:846–853. Doi: 10.1021/es9908319.

Ruiz ON, Alvarez D, Gonzalez-Ruiz G, Torres C. 2011. Characterization of mercury bioremediation by transgenic bacteria expressing metallothionein and polyphosphate kinase. *BMC Biotechnol* 11:1–8. Doi: 10.1186/1472-6750-11-82.

Sayler GS, Ripp S. 2000. Field applications of genetically engineered microorganisms for bioremediation processes. *Curr Opin Biotechnol* 11:286–289. Doi: 10.1016/s0958-1669(00)00097-5.

Seo J-S, Keum Y-S, Li Q. 2009. Bacterial degradation of aromatic compounds. *Int J Env Res Pub He* 6(1):278–309. Doi: 10.3390/ijerph6010278.

Stewart F, Ottesen E, DeLong E. 2010. Development and quantitative analyses of a universal rRNA-subtraction protocol for microbial metatranscriptomics. *ISME J* 4(7):896–907. Doi: 10.1038/ismej.2010.18.

Urgun-Demirtas M, Stark B, Pagilla K. 2006. Use of genetically engineered microorganisms (GEMs) for the bioremediation of contaminants. *Crit Rev Biotechnol* 26(3):145–164. Doi: 10.1080/07388550600842794.

Wang J, Lu L, Feng F. 2017. Improving the Indigo carmine decolorization ability of a *Bacillus amyloliquefaciens* laccase by site-directed mutagenesis. *Catalysts* 7(275):1–10. Doi: 10.3390/catal7090275.

Widada J, Nojiri H, Omori T. 2002. Recent developments in molecular techniques for identification and monitoring of xenobiotic degrading bacteria and their catabolic genes in bioremediation. *Appl. Microbiol Biotechnol* 60:45–59. Doi: 10.1007/s00253-002-1072-y.

Wu J-F, Jiang C, Wang B, Liu Z, Liu S. 2006. Novel partial reductive pathway for 4-chloronitrobenzene and nitrobenzene degradation in *Comamonas* sp. strain CNB-1. *Appl Environ Microbiol* 72:1759–1765. Doi: 10.1128/aem.72.3.1759-1765.2006.

Zhang R, Xu X, Chen W, Huang Q. 2016. Genetically engineered *Pseudomonas putida* X3 strain and its potential ability to bioremediate soil microcosms contaminated with methyl parathion and cadmium. *Appl Microbiol Biotechnol* 100:1987–1997. Doi: 10.1007/s00253-015-7099-7.

11 Role of Indigenous Microbial Community in Bioremediation
Recent Advances, Challenges, and Future Outlook

Bhupendra Pushkar and Pooja Sevak
University of Mumbai

CONTENTS

11.1 INTRODUCTION

Environmental pollution by various hazardous pollutants is a serious international issue (Zhang et al., 2019; Wainaina et al., 2020). Rapid economic development without much focus on environmental sustainability has resulted in environmental damage (Zhao et al., 2019; Kumar et al., 2019; Bartholameuz et al., 2020; Mishra et al.,

2021). The recalcitrant and hazardous pollutants are a threat to the environment and need immediate attention (Wang et al., 2018; Singh et al., 2021). Pollutants such as polycyclic aromatic hydrocarbons (PAHs), aromatic hydrocarbons (AHs), trichloro-ethylene (TCE), tiabendazole (TBZ), polychlorinated biphenyls (PCBs), and heavy metals discharged from various industries and domestic activities have long-term hazardous effects on living beings and the quality of the ecosystem (Khan et al., 2010; Dutta et al., 2021). Increasing levels of pollution in the current time have reduced bio-diversity and destroyed natural habitats of various regions along with the activities of the associated ecosystem (Gouveia et al., 2018; Bartholameuz et al., 2020). Thus, there is an urgent need to address the issue of environmental pollution to avoid its long-term negative impact on the ecosystem (Zhang et al., 2019; Singh et al., 2021).

Physicochemical treatment methods are not efficient in solving the problem of environmental pollution and may further worsen the environmental condition by generating secondary pollutants (Yu et al., 2019). Hence, there is an urgent need for an effective and eco-friendly solution for the remediation of pollutants (Roy et al., 2018). Bioremediation is known for its non-intrusive, cost-effective, and eco-friendly approach to the conventional environment clean-up methods (Lee and Ulrich, 2021). Bioremediation involves the mineralization or transformation of toxic pollutants into less toxic forms by various groups of microbes (Jaime et al., 2016). Microbial consor-tia with a variety of enzymatic activities can effectively serve the purpose (Gouveia et al., 2018). The treatment of environmental pollution using biological means has long been an interest for environmental studies as bioremediation besides pollution control can also restore the polluted site ecologically (Zhang et al., 2021).

Indigenous microbes are the driving force for the nutrient cycle and geochemical dynamics of elements (Chen et al., 2020). Microbes exposed to high concentrations of pollutants adapt to such toxic conditions and can perform in situ bioremediation of respective pollutants (Yu et al., 2016). Therefore, naturally occurring microbial communities play a crucial role in the biodegradation process (Ribeiro et al., 2018). Bioremediation using indigenous microbes has advantages of their natural adaptabil-ity to the particular site, less competition for survival, and no disturbance to the eco-system (Gouveia et al., 2018). Naturally occurring indigenous microbial communities can change their composition with changing environmental conditions, and microbial communities with different compositions accomplish corresponding functions. Thus, indigenous microbial communities have high plasticity and ideal tool for bioremedia-tion (Wang et al., 2018). Thus, bioremediation using indigenous microbial communi-ties is recognized as the foremost strategy to remediate hazardous compounds from the environment (Ribeiro et al., 2018). In situ biostimulation is an important bioreme-diation strategy that can occur naturally or can be stimulated externally by biostimu-lation or bioaugmentation methods. Biostimulation involves stimulation of catabolic activities of indigenous microbes by the addition of nutrient minerals, supplying oxy-gen or other electron acceptors, and maintaining suitable conditions of temperature, pH, and moisture (Jaime et al., 2016). However, bioaugmentation involves the addition of microbial population or consortia capable of bioremediation (Jaime et al., 2016).

This chapter aims to discuss the indigenous microbial communities and their application in bioremediation. The level of pollutants and nutritional and physico-chemical conditions prevailing on site define the specific composition of the microbial community, and their bioremediation potential are addressed (Sarkar et al., 2016). The knowledge of the functional activities of the microbial community can help in designing and developing the biodegradation process using microbial communities at various polluted sites (Tikariha and Purohit, 2020). Also, the insights into the metabolic interconnection among the microbial community can help in better understanding the dynamics of the bioremediation and will be useful in developing an efficient bioremediation strategy (Sarkar et al., 2016). The richness and abundance of microbial functional genes are associated with various environmental factors and levels of pollutants. Therefore, the correlations between microbial community structure and functions in ecosystems are discussed to provide insights into microbial in situ bioremediation process (Xu et al., 2010).

11.2 EFFECTS OF POLLUTION ON INDIGENOUS MICROBIAL COMMUNITY

Environmental pollution affects plants, humans, and animals, as well as microbial communities that are at the tertiary level of the ecosystem and responsible for the decomposition of waste and help in recycling of nutrients (Yu et al., 2016; Paliya et al., 2021). Naturally occurring microorganisms are sensitive towards any ecosystem distress such as exogenous toxic as well as non-toxic substances that can alter the properties of their habitat (Chen et al., 2020; Li et al., 2018). Microbial communities residing at different polluted sites are directly exposed to a plethora of toxic pollutants such as polycyclic aromatic hydrocarbons (PAHs), aromatic hydrocarbons (AHs), trichloroethylene (TCE), heavy metals, and dyes. These pollutants severely impact the composition and structure of the indigenous microbial communities (Gouveia et al., 2018; Li et al., 2018).

Ecological sites without any pollutants exhibit high taxonomical richness and diversity of microbial communities as compared to the polluted sites (Zhang et al., 2021). Various studies have reported that the presence of pollutants significantly affects the diversity of the living being. For example, the presence of nitrogen-containing organic pollutants such as indole is reported to reduce the richness and diversity of the microbial communities (Zhang et al., 2021). Aromatic pollutants such PAHs and TBZ are known to shape the microbial community composition as well as their activity and can significantly reduce the α-diversity of soil bacteria (Papadopoulou et al., 2018). Few sites such as mine tailing hardly have any microflora due to the presence of toxic heavy metals and low pH conditions (Zhang et al., 2017). Long-term and short-term exposure to heavy metals can reduce the diversity of microbial communities significantly and also decreases their number and activities. Soil bacteria and actinomycetes reduce notably in the presence of heavy metals, while fungal cells are somewhat less affected. Microbe shows sensitivity towards heavy metals in the following order: fungi < actinomycetes < bacteria (Khan et al., 2010). Vanadium at low concentration acts as a nutrient and support the growth of bacteria; however, it is detrimental at higher concentrations. Vanadium shifts the microbial communities, especially aquatic, by stimulating the growth and proliferation of specific microbes, with an overall decrease in richness and diversity (Zhang et al., 2019).

The change in the composition of indigenous microbial community also alters the metabolic and functional activity of the polluted site. Inflection in the carbon metabolic process of soil tends to shift the microbial communities and metabolites composition of the soil (Li et al., 2020). The functional profile of the microbial population is influenced by hydrocarbon degradation, which proceeds according to redox potential value and the availability of terminal electron acceptor (Lee and Ulrich, 2021). A study by Chen et al. (2020) showed that PCBs (polychlorinated biphenyls) are highly toxic and dramatically affect the number of microorganisms. PCB toxicity decreases the relative abundance of *Methanosarcina* and *Viridibacillus*, which may be due to the competition for electron donors among dechlorinating and fermenting microorganisms. However, the co-metabolism among microbial communities helps in overcoming the toxicity of the PCBs (Chen et al., 2020). Pollutants such as hydrocarbons alter the molecular composition of polluted sites by decreasing the abundance of hydrocarbon metabolic genes and proteins, which further hamper the survival of indigenous microbes (Liu et al., 2019). Heavy metals markedly affect the acid phosphatase, urease, and microbial biomass carbon (MBC) activities of the microbial community (Khan et al., 2010). Vanadium pollution reduces the rate of soil basal respiration by accumulating pentavalent vanadium V^{5+} reducer at contaminated sites (Zhang et al., 2019).

Along with toxic pollutants, different environmental factors also contribute to the alterations in the structure and function of microbial communities (Yu et al., 2016). Generally, the composition of microbial communities of any ecosystem is decided by the typical properties such as pH, temperature, salinity, type of sediment, and clay content (Gouveia et al., 2018; Liu et al., 2019; Ribeiro et al., 2018). pH and salinity are important factors that determine the indigenous microbial community and their distribution. Salinity majorly determines the structure of microbial community in original saline-alkali soil. Petroleum hydrocarbon (PHC) along with other factors such as soil salinity, clay content, and anaerobic conditions majorly affects the microbial community composition present at varying soil depths (0–30, 30–60, and 60–80 cm) (Liu et al., 2019). Further, the availability of nutrients, organic matter, and total nitrogen and sulphur also significantly affects and shapes the microbial communities in the presence of pollutants (Khan et al., 2010; Zhang et al., 2019). The effects of pollution on microbial community are schematically presented in Figure 11.1.

11.3   ANALYSIS OF MICROBIAL COMMUNITY COMPOSITION AND STRUCTURE

Microbial community structure changes to acclimatize and engage themselves in the degradation of pollutants (Chen et al., 2020). This kind of response is the adaptation mechanism of the microbes, which involves physiological and genetic evolution (Miao et al., 2021). Indigenous microbial communities respond dynamically to different stress conditions. Microbial communities change from their original composition after undergoing stress, which to a certain extent is part of adaptation and can be quantified (Tikariha and Purohit, 2020). The analysis of microbial community structure provides a link between microbial community structure and changes in their activities in response to exposure to various pollutants (Khan et al., 2010). Hence, it is very crucial to analyse

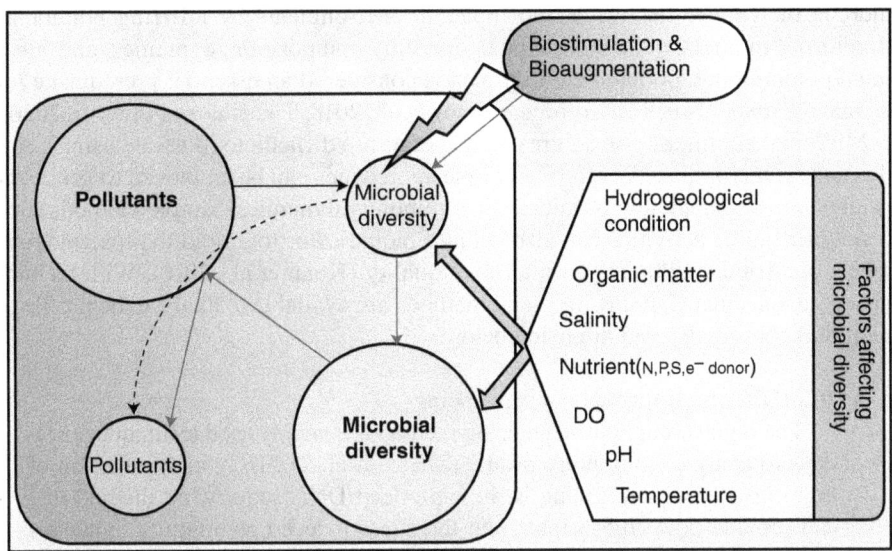

FIGURE 11.1   Effect of environmental pollution on the diversity of indigenous microbial communities and response of microbes against pollutants. Various environmental factors also influence the diversity of indigenous microbial communities. The indigenous microbial communities can be stimulated for bioremediation using biostimulation and bioaugmentation strategies.

the dynamics of native microbial communities in response to different organic and inorganic pollutants in order to understand the degradation process of pollutants and the ecological and environmental role of native microbial communities (Chen et al., 2020).

Analysing the impact of pollutants and other abiotic stress on microbial communities may provide significant insights into the biogeochemical processes (Miao et al., 2021). Biodegradation efficiency can be optimized with the knowledge of the microbial community, where the dominating microbes in the community possess specific growth rates that can resist the stress and at the same time also consume the target pollutant (Tikariha and Purohit, 2020). Microbial analysis can give information about appropriate strategies to be utilized as a tool for attaining a mechanistic understanding of mutual effects among microbiomes, pollutants, and remediation methods. Knowledge of the dynamics and function of the underlying microbial community can be useful to achieve rapid pollution control via the bioremediation process in the least invasive manner (Miao et al., 2021).

Analyses of diversity, metabolic potential, and response of microbial community towards biostimulatory agents are crucial to develop a suitable bioremediation technology (Sarkar et al., 2016). Microbial analysis of such sites will also provide in-depth information on microbial community diversity and their dynamics during biostimulation and bioaugmentation processes (Papadopoulou et al., 2018).

Environment factors and nutrient availability also alter the microbial communities and can shift community structure and composition, which thereby affects the bioremediation process majorly by influencing the potential microbes, where certain microbe thrive and survive in incompatible environments by utilizing pollutants. Monitoring the indigenous microbial community composition, dynamics, and functional potential in a polluted environment is considered an essential prerequisite for the management of such environments (Roy et al., 2018; Tikariha and Purohit, 2020).

Microbial community structure and activities are difficult to illustrate using a single analysis technique. Therefore, multiple approaches can be employed to get better insights into and depiction of microbial community dynamics. Simple methods such as soil enzymatic activities can also act as a bioindicator to unfold the toxicological effects of various pollutants on microbial quality (Khan et al., 2010). With technological advancements, many high-end methods are available to analyse the profile of microbial community and are listed below.

a. 16S rDNA high-throughput sequencing

The high-throughput sequencing method is a widely used technique nowadays to analyse community profile (Zhang et al., 2021). A higher version of high-throughput sequencing is the ultra-deep DNA sequencing method that can provide powerful insights into the stress-induced community dynamics of sample microbiome during the pollutant removal process. It can also reveal the existence and importance of even low abundant rare microbial groups for preserving ecosystem functions in varied habitats (Roy et al., 2018). This technique relies on high-throughput sequencing of universal 16S rRNA of the V4-V5 variable region of bacteria and internal transcribed spacer (ITS) region of the fungal gene (Li et al., 2020). The genomic DNA is directly extracted from the sample to be analysed using cetyltrimethylammonium bromide (CTAB) or sodium dodecyl sulphate (SDS) methods. The V3-V4 region of 16S rRNA conserved DNA can be amplified using fusion primers (Chen et al., 2020; Li et al., 2020). The DNA libraries are sequenced on the Illumina HiSeq 2500 or IonS5 TM XL platform, and the phylogenetic diversity of the microbial community can be analysed using available bioinformatics tools (Li et al., 2020; Zhang et al., 2021).

b. Denaturing gradient gel electrophoresis (DGGE)

DGGE is a commonly employed molecular technique for rapid analysis of composition, diversity, and dynamics of microbial communities. It is based on the separation of double-stranded polymerase chain reaction (PCR) products of the same length, but varying sequence composition. DGGE utilizes the physicochemical fundamental of DNA base pairing. The genomic DNA obtained from environmental samples can be used for PCR amplification using appropriate primers. In DGGE, the double-stranded DNA obtained after PCR shows distinct mobility upon partial denaturation in polyacrylamide matrix. The PCR amplicons after separation on partial denaturing polyacrylamide matrix produce a complex banding pattern representing the predominant in situ microbial populations. The primary advantages of DGGE method are rapidity, economy, and analysis of multiple samples at a time (Neufeld et al., 2012).

Generally, short DNA fragments of length 150–300 bp are ideal for DGGE analysis as longer DNA fragments are more resistant to denaturation, need higher gradients, and most importantly, produce long amplicons, which may result in the formation of chimeras and undesired artefacts. Although, DGGE, an electrophoretic technique, can discriminate between highly similar DNA sequences, indicating differences as small as a single nucleotide and few highly similar rDNA sequences is difficult to analyse using DGGE (Gonzalez and Sainz-Jimenez, 2004).

c. Metagenomics sequencing

Metagenomics is used for studying the genetic material obtained directly from a sample such as soil, sludge, and water (Zhang et al., 2021). Metagenomics combined with high-throughput sequencing can give the functional profile of the microbial communities (Yu et al., 2016). The organization and relation of taxonomical and functional composition and associated physiological intelligence of the microbial communities can be determined using metagenomics in assistance with different sets of bio-informatics tools. Besides the diversity and functional aspects of specific habitats, metagenomics also provides metabolic activities and can anno-tate the degradation capacity of individual microbes for specific pollut-ants (Tikariha and Purohit, 2020). Functional gene array is used to study the microbial communities, and it contains genes encoding key enzymes involved in biogeochemical cycling processes. GeoChip 2.0 covers more than 10,000 genes involved in critical biogeochemical processes mediated by microbes (Xu et al., 2010).

d. Automated ribosomal intergenic spacer analysis (ARISA)

The ARISA method relies on variability in length of intergenic spacer (IGS) between small (16S) and large (23S) subunit rRNA genes in the rrn operon. IGS region shows heterogeneity in length and nucleotide sequence depending on the microbial species. ARISA contributes towards advanced microbial ecology, and it is well suited for a general overview of variation in the abundance of microbial types. In the ARISA technique, the 16S-23S intergenic spacer region present in the rRNA operon can be amplified using the forward primer (ITSF 5′-GTCGTAACAAGGTAGCCGTA-3′) and reverse primer (ITSReub 5′-GCCAAGGCATCCACC-3′) set that ampli-fies the ITS1 (internal transcribed spacer 1) region in rRNA. Microbial diversity and richness index can be determined using the ARISA profile to analyse the ecological details of microbial communities of the sample. Electropherogram obtained from ARISA is used to calculate the total number of unique operational taxonomic units (OTUs), which further pro-vide information on the richness of microbes. ARISA technique is useful in microbial community analysis, especially under controlled conditions (Gouveia et al., 2018)

e. Ester-linked fatty acid methyl esters (EL-FAMEs) analysis

The EL-FAMEs analysis is also called phospholipid fatty acid (PLFA) analysis. It is a fast, cost-effective, sensitive, and reproducible technique for the analysis of the microbial composition of a sample. PLFAs present in the cell wall are specific to different microbial groups and act as a biomarker. These biomarkers are used to track the relative differences in microbial communities. For analysing bacteria, i15:0, a15:0, 15:0, i16:0, 16:1ω7c, 17:0, i17:0, cy17:0, 18:1ω7c, and cy19:0 fatty acids are used, while for analysing fungi, 18:2ω6 9c fatty acids are used. PLFAs for bacteria are further categorized into gram-positive (i-C15:0, a-C15:0, i-C16:0, and i-C17:0) and gram-negative bacterial (cy17:0, cy19:0, 16:1ω7c, and 18:1ω7c). The physiological or nutritional stress in the microbial communities is indicated by the ratio of saturated to monounsaturated phospholipid fatty acids, where the ratio of (cy17:0 + cy19:0)/(16:1ω7c+18:1ω7c) is used as an indicator of stress in bacteria. The differentiation in the microbial communities at different stages of pollutant degradation can be analysed using EL-FAMEs analysis (Li et al., 2018).

11.4 MICROBIAL DIVERSITY AND COMPOSITION IN RESPONSE TO POLLUTION

Each microbial community has a unique structure, composition, and function specific to their environment, which majorly depends on the available nutrients, electron donors/acceptors, and physicochemical properties (Yu et al., 2016). Besides these factors, the level and type of pollutants have a major effect on microbial community composition and structure (Gouveia et al., 2018). The succession of a particular group of microbe in the ecosystem is dependent on the nutrient available, type of pollutants, and also the stage of degradation (Tikariha and Purohit, 2020; Zhang et al., 2021). The shift in the microbial communities is majorly attributed to the degradation of the respective pollutant. Rare microbial taxa may replace the dominant or metabolically similar taxa in a short period of time due to the presence of stress exerted by the pollutants. However, co-metabolic processes support the growth of diverse microbes independent of the target pollutants. The presence of pollutant indicates that diverse groups of microbial communities are thriving and responsible for the production of various intermediates. Generally, a balance of target pollutant-metabolizing and target pollutant-co-metabolizing microbes is preferred for efficient bioremediation (Miao et al., 2021). The microbes remediating the same pollutant may vary for different environmental conditions such as aerobic and halophile, but certain groups are highly prominent against various pollutants (Tikariha and Purohit, 2020; Zhang et al., 2019).

The Proteobacteria are one of the widely distributed and dominant bacteria in different ecosystems, as shown in Table 11.1 (Tikariha and Purohit, 2020; Zhang et al., 2019). Geobacter (Proteobacteria) is frequently found at polluted sites and can biodegrade numerous organic pollutants such as benzene, benzoate, phenol, toluene, and naphthalene, to CO_2 (Zhao et al., 2019). Firmicutes have high bioremediation efficacy and also dominate among the microbes degrading the pollutants (Yu et al., 2019). The bacteria, such as *Pseudomonas*, unclassified genera of the Bacteroidetes phylum, and Candidatus Cloacimonas class, and methanogenesis archaea use many substrates for the growth and are present widely (Wang et al., 2018). The microbial communities exposed to mixed pollutants are more diverse and exhibit more variations (Feng and Chen, 2018).

Indigenous microbial communities have innate plasticity character, and the composition of microbe changes with respect to environmental conditions and also

TABLE 11.1

Details of Microbial Communities Active for Bioremediation of Different Pollutants

Name of Pollutant	Dominant Microbes	Dominant Phyla/Class	Ecosystem	Biostimulation/ Bioaugmentation	References
Crude oil	*Pseudomonas*.g, *Marinobacter*.g, *Lactococcus*.g, *Fusarium*.g, and *Aspergillus*.g	*Proteobacteria*.p, *Firmicutes*.p, *Bacteroidetes*.p, *Chloroflexi*.p, *Actinobacteria*.p, and *Acidobacteriota*.p	Oil field soil, saline-alkaline soil	Crude oil	Liu et al. (2019)
Oil	*Pseudomonas*.g, *Rhodocyclales*.o, *Desulfurococcales*.o, *Desulfurococcaceae*.f, *Methanothrix*.g, *Methanomethylovorans*.g, *Methanoculleus*.g, *Methanolinea*.g, and *Methanospirillum*.g	*Firmicutes*.p, *Betaproteobacteria*.c, *Cloacimonetes*.p, *Euryarchaeota*.p	Heavy oil-polluted water treatment plant	-	Wang et al. (2018)
Total petroleum hydrocarbon	*Azovibrio*.g, *Pseudoxanthomonas*.g, and *Comamonadaceae*.f	Gamma- and Deltaproteobacteria.c, and Euryarchaeota.p	Petroleum refinery waste sludge	Nitrate	Sarkar et al. (2016)
Total petroleum hydrocarbon	*Dietzia*.g, *Halomonas*.g, and *Salinimicrobium*.g	Actinobacteria.p, Proteobacteria.p, Bacteroidetes.p Firmicutes.p, and *Deinococcus-Thermus*.p	Soil	*Phragmites*	Yu et al. (2019)

(Continued)

TABLE 11.1 (*Continued*)
Details of Microbial Communities Active for Bioremediation of Different Pollutants

Name of Pollutant	Dominant Microbes	Dominant Phyla/Class	Ecosystem	Biostimulation/ Bioaugmentation	References
Total petroleum hydrocarbon and PAHs	*Bacillus*.g, *Coprothermobacter*.g, *Rhodobacter*.g, *Pseudomonas*.g, *Achromobacter*.g, *Desulfitobacter*.s, *Desulfosporosinus*.g, *Methanobacterium*.g, *Methanosaeta*.s, *Alcaligenaceae*.f, and *Comamonadaceae*.f	Gammaproteobacteria.c, Betaproteobacteria.c, Firmicutes.p, Bacteroidetes.p, Actinobacteria.p, Euryarchaeota.p, and Chloroflexi.p	Oily sludge sample	Phosphate, nitrate	Roy et al. (2018)
Petroleum hydrocarbon (C19) and polycyclic aromatic hydrocarbons (pyrene)	*Methanomethylovorans*.g, *Desulfobulbus*.g, *Desulfobacca*.g, *Sulfuricurvum*.g, *Geobacter*.g, and *Dechloromonas*.g	Euryarchaeota.p and Proteobacteria.p	Urban river sediment	Methanol	Zhao et al. (2019)
PAHs – phenylalanine, pyrene, and benzo(a)pyrene	*Micrococcales*.o, *Verrucomicrobiae*.f, *Bacillus*.g, and *Pseudarthrobacter*.g, *Lasiosphaeriaceae*.f, and *Zopfiella*	Actinobacteria.p, Bacteroidetes.p, Firmicutes.p, Patescibacteria.p, Proteobacteria.p, Verrucomicrobia.p, and Deltaproteobacteria.c	Agricultural land soil	Biochar and rhizosphere soils	Li et al. (2020)
Benzene	*Geobacter*.g (iron-reducing (genus) Peptococcaceae.f (nitrate-reducing) (family)	Proteobacteria.p Firmicutes.p	Sediment	SO_4^{2-}, Fe^{3+}, and NO_3^- SO_4^{2-}, Fe^{3+}, and NO_3^-	Lee and Ulrich (2021)
Trichloroethylene (TCE) and nitrate	*Paludibacter*.g and *Chitinophaga*.g	Bacteroidetes.p	Soil from contaminated sites	Methanol	Feng and Chen (2018)

(Continued)

TABLE 11.1 (*Continued*)
Details of Microbial Communities Active for Bioremediation of Different Pollutants

Name of Pollutant	Dominant Microbes	Dominant Phyla/Class	Ecosystem	Biostimulation/ Bioaugmentation	References
Thiabendazole	*Sphingobacteriales.o, Rhizobiales.s, Rubrobacterales.o, Micrococcales.o, Planctomycetales.o, Cytophagales.o, Bacillales.o, Acidobacteriales.o, and Xanthomonadales.o*	*Proteobacteria.p, Bacteroidetes.p, and Actinobacteria.p*	Wastewater soil	Microbial consortia from wastewater disposal site	Papadopoulou et al. (2018)
1,4-Dioxane	*Mycobacterium.g, Methyloversatilis.g, and Clostridia.c*	Actinobacteria.p, β-Bacteroidetes.p, and Firmicutes.p	–	*Propane, Pseudonocardia dioxanivorans* CB1190, and Rhodococcus ruber ENV425	Miao et al. (2021)
Quinolone, pyridine, and indole	*Dokdonella.g, Comamonas.g, and Pseudoxanthomonas.g*	Proteobacteria.p	Sludge	*Comamonas* sp. Z1 and *Acinetobacter* sp. JW	Zhang et al. (2021)
Pentachlorophenol (PCP)	*Pandoraea.f, Clostridium.g, Desulfitobacterium.g, unclassified Veillonellaceae.f, and Cupriavidus.g*	Proteobacteria.p and Firmicutes.p	–	Lactate and anthraquinone-2,6-disulphonate (AQDS)	Chen et al. (2020)

(Continued)

TABLE 11.1 (*Continued*)
Details of Microbial Communities Active for Bioremediation of Different Pollutants

Name of Pollutant	Dominant Microbes	Dominant Phyla/Class	Ecosystem	Biostimulation/ Bioaugmentation	References
Mine tailing bioremediation (Fe)	*Desulfosporosinus.g*, *Desulfotomaculum.g*, *Sulfobacillus.g*, *Acidithiobacillus.g*, and *Acidiphilium.g*	Firmicutes.p and Proteobacteria.p	Mine tailings	Yeast extract, tryptone	Zhang et al. (2017)
Vanadium	*Bacillus.g* and *Thauera.g*	Proteobacteria.p and Firmicutes.p	Surface water and sediment sample	Mineral salt solution with glucose	Zhang et al. (2019)
Chromium (VI)	*Geobacter.g*, *Clostridium.g*, and *Desulfosporosinus.g*	Chloroflexi.p, Firmicutes.p, and Proteobacteria.p	Sediment samples from industrial discharge	-	Yu et al. (2016)
Uranium	*Desulfovibrio.g*, *Geobacter.g*, *Anaeromyxobacter.g*, *Shewanella.g*, *Rhodopseudomonas.g*, and *Pseudomonas.g*	Alphaproteobacteria.c, Gammaproteobacteria.c, and Deltaproteobacteria.c	Groundwater	Ethanol	Xu et al. (2010)

Note: Indicate the information not available; c, class; f, family; g, genus; o, order; p, phylum; s, species.

during various stages of the bioremediation process to adapt well to their corresponding environment (Yu et al., 2019). Certain microbes modulate similar activities and can also alter the community composition over a period of time, while the biogeochemical process remains unchanged (Miao et al., 2021). Gram-negative bacteria with specific PLFAs are correlated with hydrocarbon pollution degradation. Gram-negative bacteria communities increase drastically in diesel oil contamination and biostimulation treatment process; however, gram-positive bacteria and fungi do not show much variation in the phospholipid fatty acid during the pollution treatment and do not change actively during PHC stress or degradation (Li et al., 2018). In contrast, a study by Jaime et al. showed that gram-negative bacteria reduce, while gram-positive bacteria, Actinobacteria, and fungi thrive due to their resistance against osmotic stress. Also, the gasoline pollution and toxic light hydrocarbons may cause the extinction of the few members of the gram-negative bacterial population at polluted sites. An increase in certain fungi acts as a biomarker for the biodegradation of hydrocarbons. The relative increase in fungal markers shows their hydrocarbonoclastic and adaptive abilities (Jaime et al., 2016). In the case of heavy metals, bacteria are much sensitive followed by the actinomycetes and fungi (Yu et al., 2016).

Gram-negative bacteria are most abundant among all phyla and are actively involved in the bioremediation process. Proteobacteria, Firmicutes, and Actinobacteria are most influential in the total petroleum hydrocarbon (TPH) bioremediation process. Nutrients amendment positively impacts the Firmicutes and Deinococcus-Thermus. Firmicutes can efficiently degrade TPH under an optimum ratio of C/N/P (Yu et al., 2019).

Betaproteobacteria class of bacteria is highly abundant in organic waste treatment. Another class present in abundance is Rhodocyclales, mainly of genera *Azonexus*, *Azoarcus*, *Zoogloea*, and an unclassified genus that belongs to the family Rhodocyclaceae. The phylum Cloacimonetes is also prominent in the anaerobic organic treatment process. Archaeal methanogenic genera, namely *Methanothrix*, *Methanomethylovorans*, *Methanoculleus*, *Methanolinea*, and *Methanospirillum*, are helpful in biological oxygen demand (BOD) reduction (Wang et al., 2018). The nitrogen waste is reported to majorly consist of Proteobacteria, Firmicutes, Bacteroidetes, and Gemmatimonadetes. Alphaproteobacteria and Gammaproteobacteria are predominate groups of Proteobacteria in the nitrogen-containing waste. Other native microbes, namely *Dokdonella*, *Comamonas*, and *Pseudoxanthomonas* of activated sludge, are very much efficient in removing pollutants (Zhang et al., 2021). Proteobacteria and Euryarchaeota together covered 80% and are among the major bacterial and archaebacterial phyla in oil-contaminated sites (Sarkar et al., 2016). Deltaproteobacteria are anaerobes, sulphate reducers, and hydrocarbon degraders, mostly found in the native sample and highly present in hydrocarbon-polluted sites. The sludge community is dominated by the Proteobacteria, but with the addition of hydrocarbon, and other groups such as Alcaligenaceae and Comamonadaceae of Betaproteobacteria are also found in increased number. Chloroflexi increases drastically in the presence of hydrocarbon. Chloroflexi and Deltaproteobacteria degrade alkanes and other hydrocarbons by fermentative-anaerobic metabolism under anaerobic sulphate-reducing conditions and are connected to methanogenesis via reverse electron transport (Roy et al., 2018).

Bacilli, Clostridia, Alphaproteobacteria, Betaproteobacteria, Deltaproteobacteria, Gammaproteobacteria, and Methanobacteria are the dominant classes found during pentachlorophenol (PCP) treatment. Clostridia, Betaproteobacteria, and Deltaproteobacteria are the mostly dominant among the anaerobes that use chlorophenol as an electron acceptor during PCP remediation. Therefore, a change in these three classes is associated with the PCP dechlorination, thus indicating the adaptation of these classes to the PCP stress by active biotransformation of PCP or its product. Microbes such as *Clostridium, Desulfitobacterium,* and *Methanobacterium* are known as dehalorespiring microorganisms and metabolize PCP as an electron acceptor under anaerobic conditions. Dechlorinating and iron-reducing microbial groups such Clostridium, Desulfitobacterium, Pandoraea, and unclassified Veillonellaceae get stimulated and are involved in PCP dechlorination, while some dominant genera present in the uncontaminated sample inhibit the PCP dechlorination (Chen et al., 2020).

Certain amendments are helpful in bioremediation, but also alter the composition of microbial communities. The combination of biochar and rhizosphere causes an increase in bacterial responders. Bacteroidetes, Alphaproteobacteria, and Sphingomonas are involved in PAH degradation in the presence of biochar, whereas in rhizosphere, *Actinobacteria, Chloroflexi, Firmicutes, Patescibacteria, Verrucomicrobia, Bacilli, Deltaproteobacteria, Micrococcales, Verrucomicrobiae, Bacillus,* and *Pseudarthrobacter* contribute majorly to PAH degradation. However, in the rhizosphere with biochar, *Actinobacteria, Bacteroidetes, Firmicutes, Patescibacteria, Proteobacteria, Verrucomicrobia, Deltaproteobacteria, Micrococcales, Verrucomicrobiae, Bacillus,* and *Pseudarthrobacter* are the main players in PAH degradation. Fungus, *Lasiosphaeriaceae,* and *Zopfiella* also play an important role in PAH degradation in the rhizosphere with or without biochar amendment (Li et al., 2020).

The sulphur-reducing bacteria (SRB) and sulphur-oxidizing bacteria (SOB) are actively involved in the sulphur cycle and also participate in petroleum hydrocarbons (PHs) and PAH bioremediation. The chemolithoautotrophic SOB such as Sulfuricurvum are enriched when total organic carbon depletes and are positively correlated with PHs and PAHs degradation efficacies. Sulfuricurvum is a SOB that grows anaerobically and microaerobically by reduced sulphur species, such as sulphide, elemental sulphur, and thiosulphate. Being an anaerobe, Smithella generally exists in sediment and is directly involved in anaerobic hydrocarbon degradation in the methanogenic ecosystem. However, another anaerobe, *Dechloromonas* can degrade benzene, toluene, and ethylbenzene. *Dechloromonas* and *Geobacter* have great bioremediation potential and are responsible for alkane degradation (Zhao et al., 2019).

The microbial community structure changes during the TCE bioremediation process, where the number of dechlorinating bacteria reduces and denitrifying bacteria proliferate. TCE and nitrate affect the abundance of denitrifying *Acinetobacter* (changes from >19% to >1.7%), while other denitrifying microbes such as *Paludibacter* and *Chitinophaga* proliferate in the presence of nitrate (Feng and Chen, 2018).

The microbial diversity and composition also vary with the concentration of pollutants. Soil with lower TBZ contamination has β-diversity, while a higher concentration of TBZ (12,000 mg/kg) in soil exhibits a very low diversity of microbes and is majorly dominated by Beta- and Gammaproteobacteria. The naturally contaminated sites have majorly Proteobacteria and Actinobacteria among β-diversity. Sites contaminated by TBZ for long term show reduced microbial diversity and dominance of Beta- and Gammaproteobacteria (Papadopoulou et al., 2018).

Microbes such as Alphaproteobacteria (*Rhodopseudomonas*), Gammaproteobacteria (*Pseudomonas* and *Shewanella*), and Deltaproteobacteria (*Anaeromyxobacter, Desulfovibrio*, and *Geobacter*) have metal-resistant properties and appear in response to metal pollution (Xu et al., 2010). The presence of chromium shows the abundance of Proteobacteria groups such as *Geobacter* followed by *Clostridium, Desulfotomaculum*, and *Desulfosporosinus* due to their Cr^{6+} reduction potential (Yu et al., 2016). On the other hand, vanadium shows the dominance of Bacteroidetes, Proteobacteria, and Firmicutes having V^{5+} reduction properties. Bacillus and Thauera are enriched in the presence of vanadium due to their V^{5+} to V^{4+} biotransformation property. The metal-resistant microbial communities differ due to varying geological conditions (Zhang et al., 2019). For a better understanding of microbial dynamics, including composition, structure, and function as well as their linkages with environmental factors, in situ analysis and laboratory incubation can be combined (Yu et al., 2016). The microbial diversity can be studied using the below-given metrics.

a. Shannon index (microbial richness and uniformity)
b. Simpson's diversity index
c. Chao1 index (microbial richness)
d. Abundance-based coverage estimator (ACE)
e. Evenness index.

11.5 POTENTIAL OF INDIGENOUS MICROBIAL COMMUNITY FOR BIOREMEDIATION

It is well accepted that in situ bioremediation process has a relatively low impact on the environment and is economically friendly as compared to other chemical and physical treatment methods (Miao et al., 2021). Pollutants are highly diverse, and thus, it becomes extremely difficult to devise a universal technology for different types of pollutants. Bioremediation using indigenous microbial communities is thoughtful as the indigenous microbial communities are selectively enriched according to the environmental conditions to degrade the pollutant (Wang et al., 2018). Indigenous microbial communities respond differently to various pollutants, which is correlated to their degradation capacity (Gouveia et al., 2018). Microbes sensitive to toxic pollutants diminish upon exposure to high concentrations of pollutants, while the tolerant ones survive and increase in abundance. The physiological adaptation and genetic modifications are involved in the enhancement of abundance of tolerant microbes by replacing the sensitive microbes (Khan et al., 2010). The microbial community has a much specialized hierarchical structure where a decentralized metabolic network works with each other. Such organization enables great plasticity with high resilience power, which undergoes continuous changes with changing environmental conditions (Tikariha and Purohit, 2020).

There is a direct relationship between the concentration of pollutants and the population density of heterotrophic bacteria in the polluted ecosystem. The microbial community adapted to stress can efficiently utilize pollutants. The prior encounter of microbes with pollutant induces physiological selection and acclimatization of microbial community capable of effective catabolism. The advantage of in situ bioremediation is that certain important uncultivable microbes also remain active in the degradation process, which may be lost if grown in laboratory (Jaime et al., 2016). Further, the rate of bioremediation can be accelerated by nutrients supplement, which induces the indigenous microbial community and its metabolic interplay. Sometimes, the biostimulation process is hampered due to the scarcity of efficient bioremediating microbes. However, this can be overcome by adding bioremediating microbes. However, sometimes bioaugmentation also fails due to strong competition and pillage of autochthonous microorganisms on the exogenously introduced strains. Hence, the use of indigenous microbes is preferred due to their easy adaptability and acclimatization to the polluted site (Roy et al., 2018).

A co-occurrence network of bacteria and fungi shows the majority of functions related to each other. The bacterial function involves carbohydrate, amino acid, lipid, and energy metabolism, and the fungal function is strongly correlated to bacterial metabolism. Generally, functions associated with bacterial metabolism dominate in the community as bacteria are predominant in the microbial community. Microbial community is responsible for the distribution of metabolites in the soil, which support the growth of other microbes and improve bioremediation (Li et al., 2020). However, certain indigenous microbes are highly efficient in bioremediation, and sometimes, they outcompete the inoculated culture used for treatment (Zhang et al., 2021).

Microbes consume PHCs as a sole source of carbon and energy and can be obtained from polluted sites (Ribeiro et al., 2018). Using indigenous microbial communities for bioremediation of PHCs and their derivatives from polluted sites is an effective alternative method. Bioaugmentation of autochthonous microbes can make the oil removal faster, achieving 80% of hydrocarbons in 60 days as per the laboratory experiment (Gouveia et al., 2018). Amendment with specific nutrient selectively enriches the hydrocarbon-degrading microbes. Even the addition of N and NS amendments decreases the diversity, but has a positive impact on the TPH degradation (Sarkar et al., 2016). Autotrophic bacteria and heterotrophic SRB have different nutritional utilization characteristics, which changes the microbial community structure to perform bioremediation more efficiently (Zhang et al., 2017). The anaerobic bacterial communities degrade long-chain alkanes to methane. Carbon sulphur metabolic processes of microbe metabolism can be used for remediation of urban river sediment (Zhao et al., 2019). Indigenous microbes can efficiently treat polluted sites containing PAHs, TPH, 1,4-dioxane, nitrogen-containing organic waste, etc. (Miao et al., 2021; Zhang et al., 2021).

The recalcitrant chemical oxygen demand (COD) is converted to BOD in the anaerobic stage, which plays a major role in enhancing the biodegradation of heavy oil-produced water. The BOD is then mineralized aerobically. The microbial communities shift during different treatment stages and have the combined action of bacterial and archaeal microbial communities in anaerobic conditions to attain the corresponding function. The microbial community shows cooperation among methanogenesis in the anaerobic treatment process and *Halomonas*, *Bacilli* class, *Oceanobacillus*, and *Planococcus* in the aerobic stage. Rhodocyclales are highly abundant during all processes and can be described as diazotrophic, denitrifying, or nitrifying strains that are crucial in the nitrogen cycle (Wang et al., 2018). Certain functional soil microbes actively accelerate the PCP dechlorination in the presence of natural interference. Indigenous microbes have versatile metabolism for utilizing electron acceptors and donors. They also have high adaptation capacity and are more capable of degrading a wide variety of chlorinated compounds. Hence, such microbes are enriched fast in PCP-polluted sites and further help in the bioremediation of PCP (Chen et al., 2020).

Particular microbial community is strongly correlated to metabolites such as PAHs or degraded PAH products present in their surroundings. Rhizosphere in combination with biochar has high PAH removal efficacy and has prominent PAH degraders, namely *Sphingomonas*, *Pseudarthrobacter*, *Bacillales*, *Bacilli*, *Firmicutes*, *Gemmatimonadetes*, and *Actinobacteria*. The microbes consume soil metabolites and help in carbon flow during PAH degradation. Microbes produce trehalose, which usually increases the solubility of hydrocarbon compounds such as PAHs, thus promoting biodegradation (Li et al., 2020). The microbial community structure transforms during the bioremediation process of TCE adsorbed on poly-aluminium chloride (PAC), and the population of denitrification bacteria increases while the dechlorinating bacteria decline. Native microbial communities play a key role in the dechlorination process when methanol and TCE act as electron donors and acceptors. A direct relation between ammonium accumulation and nitrate reduction indicates the sensitivity of microbes against pH and temperature associated with competition between denitrification and dissimilatory nitrate reduction to ammonium (DNRA) (Feng and Chen, 2018).

Certain indigenous microbial communities are tolerant of heavy metals such as Cr^{6+} and can be employed as a tool for bioremediation (Yu et al., 2016). Long-term studies have also confirmed the possibility of metal bioremediation using native microbes, within acid-soluble fractions. Microbes with V^{5+}-detoxifying functional activity increase in abundance in high concentration of V^{5+} to replace the sensitive microbes. The functional species with reduction ability increase in abundance and also dominate the microbial communities in the V^{5+}-polluted site, thus indicating the relevance of indigenous microbes for the bioremediation of heavy metals (Zhang et al., 2019). A demonstration study showed that groundwater contaminated with high levels of uranium can be treated using in situ bioremediation to keep uranium levels below the threshold values. Uranium bioremediation occurs in the environment through U^{6+} reduction. Microbial populations stimulated in the presence of heavy metals such as uranium are involved directly or indirectly in the reduction of U^{6+} and maintaining a low concentration of uranium in the ecosystem (Xu et al., 2010). Bioremediation of different pollutants by diverse microorganisms is schematically presented in Figure 11.2.

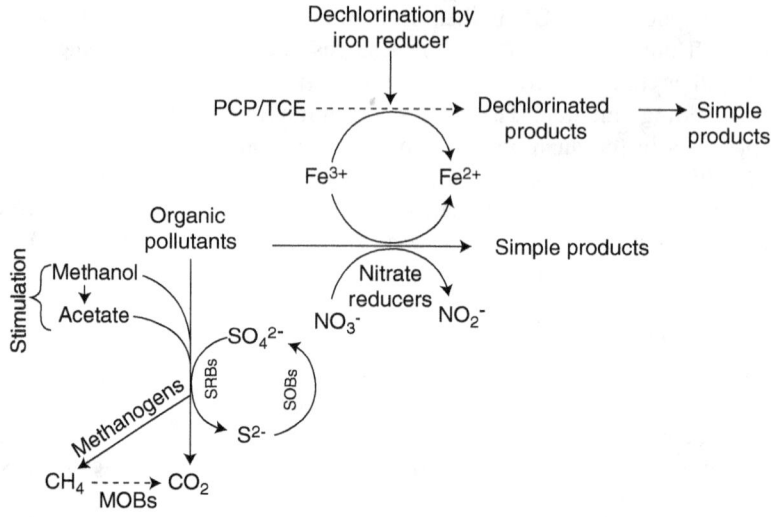

FIGURE 11.2   Biodegradation reaction of organic pollutants by different microorganisms.

11.6   BIOREMEDIATION STRATEGIES OF INDIGENOUS MICROBIAL COMMUNITY

Bioremediation of pollutants by indigenous microbial communities is the result of collective metabolic activities and interaction among individual microbial assemblages within diverse microbial groups. Several microbes evolve themselves to utilize or fortuitously degrade a variety of pollutants and their by-products. Thus, it is very crucial to understand the metabolic interaction between and within microbial functional guilds and further establish the relationship between functions involved in bioremediation. Various stress response mechanisms in the microbes may benefit or maintain the fundamental functions first and then activate the mechanisms that enable the biodegradation of target pollutants (Miao et al., 2021). Using these mechanisms, microbes either completely degrade or convert it to a less toxic form (Wang et al., 2018).

Various biotic and abiotic stresses cause a myriad of effects on the microbial community, which leads to cross-communication between the members of the community to change the pattern for sharing the sources. The presence of toxic pollutants above a certain limit generally causes rearrangement of the shared metabolic pathways and activates mechanisms responsible for the alleviation of toxic pollutants. With time, the microbial population shift to more tolerant of inhibitory intermediates and these new members acquire the necessary metabolic capabilities through various interspecies genetic events. With the increase in adaptation among microbes, the rate of bioremediation also increases (Tikariha and Purohit, 2020). Microorganisms utilize the following mechanism to deal with the stress induced by the pollutants.

a. Biosorption of pollutants

The biosorption mechanism of bioremediation is very popular and widely reported. Microbes biosorb pollutants on their surface molecules. Exopolysaccharide produced by microbes also helps in biosorption. Biosorption is non-specific and involves electrostatic force, ionic bond hydrophobic interaction, etc., for adsorption. Biosorption using microbial communities is simple, efficient, and very much flexible and can be used for in situ bioremediation of pollutants such as heavy metals and organic compounds (Rizvi et al., 2020; Zhang et al., 2019). Biosorption can be metabolism dependent or independent and can be performed using live or dead biomass of microorganisms. It is an innate and entirely non-enzymatic mechanism adopted by microbes that basically use binding or chelation. The surface molecules with functional groups such as hydroxyl, amino, carboxylate, and phosphoryl are involved in interaction with pollutants. Biosorption is a passive process and may be reversible. However, the microbes may transfer the biosorbed pollutants into the cell for internal metabolism. The biosorption capacity of microbes decreases with time due to saturation of the binding site. Therefore, biosorption occurs maximum at lower concentrations of pollutants as there is a complete collision between pollutants and binding sites (Rizvi et al., 2020).

b. Membrane transport of pollutants

The cellular membrane of the microbes separates the outside environment from the cytoplasm. The transport of various pollutants across the microbial cell membrane influences the rate of bioremediation. Cellular membrane plays a key role in the bioavailability of nutrients as well as pollutants for metabolism. Degradation of pollutants mostly takes place in the cytoplasm due to the localization of degrading enzymes in the cytoplasm. Therefore, the target pollutant should be transported inside for active bioremediation (Hua and Wang, 2014). The pollutants are transported inside the cell via different pathways. For example, phenols at higher concentrations enter the cell of bioremediating microbes by passive transport. However, at lower concentrations of phenol, microbes take up phenol using an active transport system (Hua and Wang, 2014). The active or passive uptake rate of pollutants has a major effect on bioremediation. Energy-dependent transport of the pollutants across the cell membrane in microbes determines the biotransformation rate (Chen et al., 2019). The microbial membrane also performs efflux mechanisms, in which the toxic compounds are moved out of the cell into the surrounding or in periplasm after its detoxification inside the cytoplasm. Efflux is widely used for heavy metal bioremediation in microbes. Besides bioremediation, efflux is employed for different functions such as cell homeostasis and resistance to antibiotics, heavy metals, and salts, which enable microbes to survive under extreme conditions (Pushkar et al., 2021).

c. Biosurfactant production by microbes

Microbes produce oil-dispersive amphiphilic compounds such as biosurfactants, which can accelerate the biodegradation of toxic pollutants

(Li et al., 2018). Biosurfactants produced by microbes reduce the surface and interface tension between the liquid and solid substance and thus form emulsions in liquids, which facilitate the intracellular uptake and further degradation of hydrocarbons by microbes (Roy et al., 2018). Biosurfactants aid in hydrocarbon bioremediation by either improving the nutrient availability for microbial growth, or increasing the interaction of cell surface with hydrophobic pollutants by increasing the surface hydrophobicity, thus allowing the binding of hydrophobic substrates more easily to microbial cells (Li et al., 2018). Bacillus strains, which are biosurfactant-producing and hydrocarbon-utilizing indigenous microbes, upon bioaugmentation along with nutrients can remove 57%–75% of TPH (Roy et al., 2018).

d. Redox reaction in microbes

Oxidation and reduction reactions play a very important role in the biodegradation as well as detoxification of the pollutants. The delivery of the electron donor and the resulting reduction or oxidation during the bioremediation process is related to subsurface hydrology. The reduction mechanism of microbes where highly soluble toxic metal is reduced to insoluble form is a promising technique for effective remediation of uranium-contaminated water ecosystem (Xu et al., 2010).

The Fe^{3+}-reducing cultures have high pollutant removal potential maybe due to the high redox energy potential of Fe^{3+} ($\Delta E0 = +36\,V$) as a terminal electron acceptor and also due to the low toxicity of Fe^{2+}. Therefore, *Geobacter* (iron-reducing) and Peptococcaceae (nitrate-reducing) microbes are present in the clay or sand and known as primary benzene degrader taxa, which show the highest degradation in the Fe^{3+}-reducing and NO_3^--reducing microcosms (Lee and Ulrich, 2021). *Geobacter, Clostridium, Desulfosporosinus,* and *Desulfosporosinus* detoxify the Cr^{6+} by reduction. Chloroflexi phylum-related population plays a role in Fe^{3+} reduction, which increases in the presence of Cr^{6+} (Yu et al., 2016). U^{6+} reduction takes place under controlled hydro-geochemical conditions, especially by manipulating the electron donor for the growth of U^{6+} reducers and maintaining the hydrological conditions. *Bacillus* and *Thauera* are enriched in the presence of vanadium (V^{5+}) and bioreduce it to less toxic V^{4+}. Microbial V^{5+} reduction can occur via two pathways. First is the respiration of V^{5+} through electron transfer and microbial-based reduction for detoxification, where vanadium binds to reductases of other electron acceptors. V^{4+} produced after reduction from V^{5+} is less mobile and toxic, thus alleviating the toxicity in water with the help of indigenous microbes (Zhang et al., 2019).

Electron acceptors such as ethanol, sulphate, and nitrate also help the metabolism in the microbial community. Electron donors such as ethanol support the increase in *Desulfovibrio* sp. at uranium-contaminated sites. Ethanol also stimulates the U^{6+}-reducing microbial communities and the main functional genes of respective microbes. Slow-release hydrogen compounds (HRCs) provide electron donors and are the source of carbon to the growth of indigenous microbes and further help in the reduction of the available electron acceptor such as Cr^{6+} or U^{6+} for a longer period (Xu et al., 2010; Yu et al., 2016).

Anthraquinone-2,6-disulphonate (AQDS) supplement has an important role in stimulating the special indigenous microbes for indirect dechlorination of the chlorinated pollutants in the paddy soil, where AQDS chiefly works as a redox shuttle to accelerate the degradation rate of PCP by their redox property (Chen et al., 2020). The degradation of organics pollutants such as PH and PAHs can be coupled with sulphate reduction (Zhao et al., 2019).

e. Biofilm formation in microbes

The microbial communities under extreme stress conditions resort to protective mechanisms such as biofilm formation. In the biofilm, microorganisms live in cooperation and benefit each other through different ecological niches. The properties of biofilm are governed to some extent by structure, diffusion of nutrients, and physiological activity of the cells. EPSs are rich in water, nucleic acids, proteins, carbohydrates, and uronic acids, which play a considerable role in biofilm formation. Biofilm acts as a physical barrier for cells, which allows the cell to perform its basic metabolism for growth and active remediation (Shukla et al., 2014). Biofilm formation supports the survival of microbes and simultaneously promotes pollutant degradation (Tikariha and Purohit, 2020). The biofilm formation in microbes is beneficial for the remediation process and positively affects nitrate adsorption and denitrification (Feng and Chen, 2018).

Microbes in biofilm express specific genes in coordination to combat pollution. Biofilm formation enables higher resistance against lethal pollutants (Shukla et al., 2014). Microbes in biofilms communicate and coordinate through various signalling molecules that form a system called quorum sensing. Quorum sensing regulates the changes in the biofilm as per the need of the surrounding environment. Biofilm also promotes gene transfer among microbes, which increases the rate of DNA and plasmid transfer. The rapid exchange of resistant genes among microbes in biofilm promotes resistance in microbes and also increases the rate of bioremediation (Singh et al., 2006). Moreover, bioremediation using biofilms offers an alternative strategy (Shukla et al., 2014).

11.7   FUNCTIONAL GENE ANALYSIS OF MICROBIAL COMMUNITY

In a natural ecosystem, there is variation in the distribution of different metabolic genes in each member of the community or the presence of a similar set of metabolic genes with varying arrangements in different microbes. The presence of functional genes is majorly responsible for the selection of dominant members in the community and thus broadly impacts community functioning. The other microbes in the community have a chance to acquire or complement the related genes to become a part of a functional community network. Bacteria generally tend to acquire and incorporate genes in their genome according to stress conditions, which in turn provide them with a competitive edge over other members and increase their survival in various stress conditions (Tikariha and Purohit, 2020). Analyses of microbial community and functional profile can help to uncover the pitfalls in the bioremediation process. They can provide the information required to better evaluate and strategize the bioremediation process to attain maximum remediation and accurate adjustment upon encountering unexpected issues in engineered remediation work (Miao et al., 2021).

The functional genes are important for xenobiotic biodegradation and metabolism increase in abundance after exposure of microbes to the respective pollutants (Zhang et al., 2021). Moreover, mobile genetic elements can be bioaugmented, which involves the horizontal transfer of genes responsible for bioremediation from exogenous strains to indigenous microbes (Zhang et al., 2021). The functional profile at a broader range is more redundant and thus mostly buffered against the changes in the taxonomy of microbial community caused by biotic or abiotic factors. The engineered systems also show the decoupling between microbial function and community compositions, where dominant microbes shift, but the overall functionality is unchanged. For example, nitrogen shifts the wood-inhabiting fungal communities; however, the functional redundancy is preserved (Miao et al., 2021).

The taxonomical abundance can be transformed to metabolic capabilities, where the classified taxa can be used to envision the metabolic pathways using Kyoto Encyclopedia of Genes and Genomes (KEGG) by the "tax4fun2" package in R (v3.5.0). The package depends on the OTU assignment to reference sequences in the NCBI RefSeq database. This can give information about the relative abundance of enzyme-coding genes in KEGG pathways and orthologues (Miao et al., 2021).

The level of expression of functional genes corresponds to pollutants degradation capacity. Glutathione metabolism is generally linked to the stress tolerance system in rhizosphere soil. However, the carbohydrate metabolism responsible for energy supply and compound transformation for microbes is usually inhibited by pollution stress. The labile carbon resources are beneficial for growth, shift the microbial community, improve pollutants bioavailability, and enrich the functional microbes and microbial functions that work in association and help in removing the organic pollutants from the ecosystem (Li et al., 2020).

The organic pollutant-degrading and metal-resistant gene groups are very important. The genes involved in aromatic and chlorinated compound degradation are increased in the presence of respective pollutants. The benzoyl-CoA reductase gene from *Rhodopseudomonas palustris* is abundant in organic polluted sites. The predicted microbial functional analysis indicated a shift and interrelationships between the genes involved in anaerobic and aerobic degradation of PAHs and the dissimilatory nitrate-reducing pathways (denitrification and dissimilatory nitrate reduction to ammonium – DNRA) (Ribeiro et al., 2018). The upregulation of functional genes such as dioxygenase and dehydrogenase is associated with the degradation of PAHs. Amino acid metabolism also plays an important role in the microbial detoxification process. The expression of PAH degradation genes improves upon modulation of carbohydrate and amino acid metabolisms (Li et al., 2020). The high abundance of nitrogen-containing organic pollutants-degrading functional genes is majorly contributed by microbial communities such as *Dokdonella, Comamonas, Pseudoxanthomonas, Achromobacter,* and *Thermomonas,* which also confirms the role of these taxa in the removal of nitrogen-containing organic pollutants (Zhang et al., 2021).

Functional degrading genes encode metabolic enzymes such as alkane hydroxylase genes, ring-hydroxylating dioxygenase α-subunit (RHDα) genes, alkylsuccinate synthase gene *(assA)*, and benzylsuccinate synthase gene *(bssA)*, which are important in the aerobic and anaerobic degradation of PHCs. A well-studied alkane mono-oxygenase is a non-haem integral membrane protein encoded by alkB gene that activates the initial step of aerobic aliphatic hydrocarbon metabolism. The phenol monooxygenase systems and naphthalene dioxygenase encoded by phe and nah genes are involved in the metabolism of PAHs by catalysing the single or two atoms of molecular oxygen to aromatic rings. Oil-polluted sites exhibit higher abundance of genes and transcripts associated with aromatic compound degradation, mainly meta- and ortho-pathways as compared to the pristine sites. Catabolic genes alkB, nah, and phe increase in relative abundance in oil pollution in the following order: alkB > nah > phe. The intrinsic characters of monooxygenase and the requirement of oxygen and cofactors cause the alkane-degrading microbes having alkB, nah, and phe genes to accumulate majorly at surface soil than in the deeper layer where conditions are more anaerobic and anoxic (Liu et al., 2019).

Hydrocarbon-degrading communities at the refinery show genes for aerobic and anaerobic alkane metabolism *(alkB* and *bssA)*, methanogenesis *(mcrA)*, denitrification *(nirS* and *narG)*, and N_2 fixation *(nifH)*. The functional efficiency of the microbial population can be validated by the abundance of these genes in the metagenome. The gene encoding a fragment of nitrogenase enzyme responsible for N_2 fixation *(nifH)* is present in the community, which is similar to the gene from archaeal origin, namely *Methanosarcina*, *Methanocella*, and *Methanolinea*. The aerobic degradation of n-alkane is catalysed by hydroxylation (at terminal or sub-terminal methyl group) through hydroxylase [methane monooxygenases (MMOs), cytochrome P450 monooxygenases (pMMOs), and alkane and alkane-like hydroxylase (alkB and alkB-like)]. *mcrA, bssA,* and *alkB* genes commonly present in hydrocarbon biodegradation system especially in nitrogen supplied microcosm that confirm the existence of abundance indigenous microbial population with these genes abundant (Sarkar et al., 2016).

The functional genes responsible for carbon cycling, namely *aceB, acet, amyA, xylA, pcc,* and pmoA, and nitrogen cycling, such as *ureC, narG, nirK, nir-S,* and *nifH,* and the resistance against metals are increased significantly upon heavy metal exposure. Gene *aceB* encodes the malate synthase A is involved in the glyoxylate cycle and *pmoA* is involved in methane consumption. Some genes involved in nitrogen fixation (e.g., nifH), N mineralization (e.g., ureC), and denitrification (e.g., *narG, nirS,* and *nirK*) are associated with carbon and nitrogen metabolism, increase due to chromium, and fulfil the energy requirement of the microbes during metal resistance and reduction. Chromium increases the key functional genes related to metal resistance and unique OTUs/population related to metal and sulphate reduction. The microbial population involved in Cr^{6+} reduction are important for chromium bioremediation. Genes such as chrA are majorly responsible for Cr^{6+} reduction. *czcA* and *chrA* genes are popularly known to transport chromate out of bacteria and are involved in chromium bioremediation. An increase in the abundance of two genes and a decrease in Cr^{6+} indicate the reduction of Cr^{6+} by microbes. Further, copA is responsible for resistance to Cu through translocation and/or maintenance of Cu homeostasis. Also, certain heavy metal transporters express in the presence of Cr^{6+} (Yu et al., 2016).

The abundance of functional genes is directly correlated with the degradation of respective pollutants. Hence, the quantification of catabolic genes gives insights into microbial responses against PHC pollution and also acts as a biomarker for analysing the oil degradation potential of the indigenous microbes. The functional metabolic genes such as *alkB*, *nah*, *phe*, *MMOs*, *bssA*, *mcrA*, *nirS*, *chrA*, and *narG* have conserved topologies and oligonucleotide primers. DNA probes targeting these genes can be designed to characterize their abundance and diversity by molecular biological techniques such as real-time quantitative PCR (RT-qPCR), hybridization, and GeoChip. The connection between functional metabolic genes and the rRNA phylotypes can reasonably predict the microbial response to PHC pollution (Liu et al., 2019).

11.8   BIOSTIMULATION OF INDIGENOUS MICROBIAL COMMUNITY

The catabolic activity of indigenous microorganisms in pollutant-rich environments is severely hampered in nutrient-deficient conditions, which further limits the rate of intrinsic bioremediation (Sarkar et al., 2016). The lack of suitable and readily available nutrients at polluted sites demands the implementation of engineered bioremediation strategies. Biostimulation is considered a constructive approach for the remediation of polluted areas (Roy et al., 2018). Biostimulation is the process of artificially adding nutrients to polluted sites to enhance the biodegradation of pollutants. It helps in overcoming the stress and supports the microbes with the required metabolic and competitive capacity to flourish by expressing the corresponding genes involved in metabolic pathways of the target pollutant (Miao et al., 2021). Nutrient amendment is of great importance to microbial assimilation and dissimilation processes and helps considerably in bioremediation. Improved bioremediation by nutrient addition can timely control the adverse effects of pollutants on the environment and allude to further move or leach the pollutants to other ecosystems (Li et al., 2018). It is widely accepted that biostimulation is more efficient to achieve sustainable bioremediation as compared to bioaugmentation (Yu et al., 2019). However, biostimulation and bioaugmentation approaches can be combined for effective bioremediation (Roy et al., 2018).

Nutrient supplement enhances the functional strains for improved bioremediation. Soil microbes can be stimulated with nutrient supplements, which increases the richness and diversity of the indigenous microbes despite the high concentration of pollutants such as vanadium (Zhang et al., 2019). The paucity of basic nutrients such as nitrogen, phosphorous, and terminal electron acceptor or donor has a critical impact on microbial metabolism and causes natural attenuation. The lack of available electron acceptors and/or inadequate nitrogen availability majorly slows down the rate of intrinsic biodegradation of hydrocarbons. Biostimulation by nutrient amendment causes activation of metabolizing enzymes as well as proteins critical to bioremediation (Sarkar et al., 2016). Some non-biological materials such as nutrients (nitrogen and phosphate), biochar coke, and phenol can also stimulate the activated sludge for efficient treatment (Zhang et al., 2021). Biochar possesses abundant accessible nutrients that help in bioremediation by shifting the structure of the microbial community

(Li et al., 2020). Addition of water also stimulates bioremediation, as water increases the availability of metabolites and nutrients (Roy et al., 2018). Agricultural waste such as phragmites (weed) can be used for biostimulation as it enhances the biodegradation efficiency and considerably regulates the microbial community structure and functional activities during the remediation process (Yu et al., 2019).

Attaining the appropriate ratio of the nutrients is one of the most common methods of biostimulation (Yu et al., 2019). However, excessive nutrients can also hamper bioremediation activity and have negative effects on biodegradation. Nutrients in excessive amounts cause microbial communities to decline rapidly and shorten the stable phase. An excess nutrient supplement can stimulate the activities of microbes not involved in pollutant bioremediation. Thus, it indicates that nutrient at high concentration is unfavourable for microbial stability and bioremediation (Li et al., 2018; Zhang et al., 2017).

Besides nutrients, certain environmental factors such as temperature, moisture content, nutrient status, water, soil texture, pH, salinity, alkalinity, and oxygen content majorly dictate the microbial response to pollutants and in situ biodegradation efficiency (Liu et al., 2019). Hence, indigenous microbes can be stimulated for effective bioremediation by optimizing these environmental factors (Li et al., 2018).

a. Organic nutrients

   Organic nutrients from different sources such as rice straw, vegetable waste, and food waste are rich in nitrogen and phosphorous and stimulate the growth and activity of specific microbial communities (Li et al., 2018; Tikariha and Purohit, 2020). Humic substances constitute a major part of soil organic matter and are responsible for inducing anaerobic reactions by accelerating the electron transfer in soil due to their redox-active function. Organic acids and humic substances are added to promote the growth of the indigenous microbes that have PCP degradation capacity and accelerate biodegradation of PCP (Chen et al., 2020). Addition of yeast extract and tryptone stimulates microbial groups, which can improve the mine tailing environment (Zhang et al., 2017). Nitrogen addition is beneficial in the area where N is scarce and organic compounds rich in carbons are high (Sarkar et al., 2016). The bio-friendly and biocompatible rhamnolipids support higher metabolic activities of native microbes to assimilate hydrocarbon (Li et al., 2018; Chen et al., 2020). Indigenous microbes consume methanol as a source of carbon for their growth and at a high dose efficiently degrade long-chain hydrocarbons. Methanol also stimulates hydrolytic acidifying bacteria and thereby increases the hydrolysis acidification of organic matter and the methanogenesis process. Further, methanol stimulates methanogenesis and sulphate reduction. Methanol can be directly consumed by SRB and methanogens, or indirectly metabolized by other microbes, such as homoacetogens, and converted to acetate, which can later be used by SRB or methanogens. Methanol also increases the relative abundance of Geobacter and Dechloromonas, which are potential degraders of PHs and PAHs (Zhao et al., 2019). Indigenous microbial populations can degrade toxic 1,4-dioxane efficiently on the addition of propane and nutrient, which also increase the abundance of *Mycobacterium* and Methyloversatilis taxa and expressions of propane monooxygenase gene, prmA (Miao et al., 2021). A comparative analysis of dissolved organic carbon sources indicated that acetate is highly beneficial in V^{5+} bioreduction (Zhang et al., 2019).

Protein sources such as yeast extract and tryptone (more than 1.6 g/L) stimulate SRB and inhibit acidophilic bacteria, whereas glucose supports acidophilic bacteria more as compared to SRB (Zhang et al., 2017). Yeast extract is reported to stimulate the growth of Firmicutes phylum, including Desulfosporosinus and Desulfotomaculum, while glucose promotes the growth of the Proteobacteria phylum, especially the key acidophilic bacteria, *Acidithiobacillus*, present in mine tailings. The increasing concentration of yeast extract decreases the autotrophic acidophilic bacteria and increases the heterotrophic bacteria significantly (Zhang et al., 2017).

b. Inorganic nutrients

Microbial growth and bioremediation require inorganic nutrients. The amendment of inorganic nutrients is a good means of biostimulation to achieve high bioremediation. Addition of inorganic nutrient sources can enhance the indigenous microbiome (Miao et al., 2021). The amendment of inorganic nutrients shifts the native microbial community to fermentative, sulphate-reducing, hydrocarbon-degrading, syntrophic, and methanogenic populations that can actively perform bioremediation (Roy et al., 2018). Active Fe^{2+} produced by the indigenous microorganisms using the iron reduction process is involved in the PCP dechlorination, and Fe amendment helps in the biodegradation process (Chen et al., 2020). Geobacter and Peptococcaceae are globally reported as primary benzene degraders and can be utilized as a biomarker for contaminated sites. The presence of these microbes indicates the natural attenuation, and these microbes are biostimulated by Fe^{3+} and NO_3^- (Lee and Ulrich, 2021).

Nitrogen amendment in the form of nitrate increases TPH reduction by increasing the number of hydrocarbon-degrading nitrate reducers, whereas a combination of nitrogen and phosphate has the highest degradation potential, indicating the synergistic effects of nutrient addition on indigenous biodegrading microorganisms. Nitrate is used as an electron acceptor in microbial metabolism, and phosphate is a favourable nutrient that supports energy (ATP) generation and cell mass synthesis. Nitrate in microaerophilic and anaerobic conditions acts not only as a terminal electron acceptor, but also as an activator of alkanes assisting in their biodegradation by indigenous microbial communities in an oxygen-deficient niche. Bioavailable phosphate increases the catabolic potential of autochthonous microbes (Roy et al., 2018). Also, inorganic phosphate along with sulphate and propane supplement can significantly remove 1,4-dioxane. Further, the addition of ammonium as an inorganic source of nitrogen can enhance the indigenous microbiome (Miao et al., 2021).

c. Effects of pH

pH has a strong impact on biochemical reactions, and with changing pH, the rate of reaction also changes. The pH influences the solubility of pollutants and their bioavailability to microbes for bioremediation (Zhang et al., 2017). The pH also has a huge impact on the growth and metabolism of microorganisms. The optimum pH can help in effective bioremediation by indigenous microbes. Pollution usually alters the pH of the contaminated site, which affects the microbe at that site (Shukla et al., 2014). Therefore, pH modulation is another chief strategy to achieve effective in situ bioremediation. Heavy metals such as uranium can be bioremediated using biostimulated indigenous microbial communities where the structure and performance of microbes are manipulated by adjusting geochemical and hydrological conditions such as pH. Biosorption of heavy metals on microbes is highly dependent on the pH, and maintaining optimum pH at the site can promote the rate of bioremediation. The pH of any site can be adjusted using soda lime or calcium carbonate (Xu et al., 2010).

d. Temperature

The properties of hydrocarbon pollutants, as well as microbial community composition, are influenced by temperature, which in turn affects the hydrocarbon degradation level. The microbial denitrification activity is sensitive to varying temperatures. The temperature 20°C–25°C is optimum for microbial community bioremediation, but may also vary with the type of microbe (Feng and Chen, 2018)

e. Electron donor/acceptor and shuttle

The amendment of electron donors and shuttles has considerable effects on microbial communities residing at contaminated sites. Electron donors strongly influence the diversity of the microbial community, while electron shuttle has a lesser impact. Electron donors and shuttles stimulate the functional groups in the microbial community and change the redox potential directly or indirectly in order to enhance the bioremediation processes (Chen et al., 2020). Different compounds such as methanol, hydrogen, formate, acetate, lactate, and glucose can be used as electron donors. Nitrate, sulphate, and heavy metals act as electron acceptors. The selection of electron donor/shuttle should be done wisely to accelerate the bioremediation of pollutants (Feng and Chen, 2018).

Lactate acts as a carbon source as well as an electron donor and promotes the dechlorination of TCE and PCP by microbial communities (Chen et al., 2020; Feng and Chen, 2018). AQDS (anthraquinone-2,6-disulphonate) acts as an electron shuttle and changes the redox potential of the system by generating reduced ADQS. AQDS increases the transfer of electrons between iron minerals and microbes to promote iron reduction. Lactate or AQDS can stimulate the dechlorination pathway of PCP by ortho-dehalogenase of bacteria and also further improve the release of dechlorination enzymes (Chen et al., 2020). The microbial population can use ethanol as an electron donor and grow on ethanol or its derived product. Ethanol and its product (acetate) stimulate microbial communities differently and form a unique composition that is actively involved in bioremediation (Xu et al., 2010). Toxic heavy metals (such as Ur^{6+} and Cr^{6+}) can be reduced to less toxic forms through in situ biostimulation of indigenous microbes by providing adequate electron donors. The optimal electron donor can be mixed thoroughly with solid matrices or injected directly into groundwater for on-site bioreduction of heavy metals in the geological environment by indigenous microbes, whereas the electron acceptor can decrease the heavy metal reduction as they are more efficient electron acceptor (Zhang et al., 2019).

f. Dissolved oxygen

Oxygen is essential for the active metabolism of aerobic microbes. It helps the indigenous microbial communities in the oxidation of pollutant in the presence of a primary carbon source with the use of monooxygenase. Microbial diversity can change with a sufficient supply of oxygen to a previously anoxic site. With the change in condition from anaerobic to aerobic, the minor microbial group increases in number. Oxygen also increases the biodiversity of microbes with or without the addition of nutrients despite microbial competition, lag responses to dissolved oxygen availability, and different oxygen demands between organisms (Miao et al., 2021). Microbial community diversity and structure respond differently to varying levels of dissolved oxygen (Wang et al., 2018). 1,4-Dioxane can be efficiently remediated by introducing oxygen. A decrease in dissolved oxygen during the bioremediation process indicates that microbes are performing biodegradation (Miao et al., 2021).

11.9   BIOAUGMENTATION

Bioaugmentation through inoculation of efficient strains has gained considerable attention in recent years (Zhang et al., 2021). The process of addition of indigenous or exogenous microbial culture to any system for bioremediation purposes is known as bioaugmentation. It is one of the strategies used to achieve effective remediation of pollutants using microorganisms (Miao et al., 2021). Bioaugmentation increases the resistance of microbes by reducing the shock of pollutants. It also boosts the metabolic functions of indigenous microbial communities and aid the indigenous communities to elevate their capacities for xenobiotics removal (Zhang et al., 2021). Bioaugmentation is an effective solution for recovering the sites polluted with persistent chemicals such as TBZ, PAH, and PH (Papadopoulou et al., 2018).

Bioaugmentation enhances the bioremediation performance of indigenous microbial communities by shifting the metabolizing and co-metabolizing microbial community structure that forms an effective microbial community (Miao et al., 2021; X. Zhang et al., 2021). However, it does not majorly alter the microbial community structure. For example, bioaugmentation with TBZ-degrading consortium does not induce any significant alteration in α- and β-diversity of the bacterial community. Thus, the potential of inoculated microbial consortia to attain the highest removal of the target pollutants without inducing major and persistent effects on the native microbial community of the treatment site is considered an important character of bioaugmentation (Papadopoulou et al., 2018). The changes in the native microbial community due to bioaugmentation can be analysed using molecular fingerprinting. However, the molecular fingerprinting approach may sometimes give conflicting results and fail to identify significant and persistent shifts in the structure of microbial communities (Papadopoulou et al., 2018).

The metabolism of carbohydrates, amino acids, lipids, xenobiotics, and polyketides increases after the addition of *Pseudonocardia dioxanivorans* CB1190 and *Rhodococcus ruber* ENV425. These strains can effectively remove 1,4-dioxane and reduce the level of 1,4-dioxane below the detectable limit (< 5 µg/L) within 7 days. *R. ruber* ENV425 can degrade 1,4-dioxane in the absence of any primary substrate, which indicates that native microbes can support the degradation process under starvation conditions (Miao et al., 2021). Bioaugmented SRB enrich the functional strains that are beneficial for the treatment of wastewater containing pollutants such as indole, pyridine, and quinoline. Flavobacterium and Alcaligenes are versatile aromatic-degrading microbes that can be used for the purpose of bioaugmentation.

The count of augmented microbe may decrease due to the lack of growth substrate and competition with native microbes, but the bioaugmented microbe may continue to perform bioremediation, as functional redundancy helps in the development of a stable microbiome, restores the structure and function of microbes to the original level, and induces the decoupling between basic metabolic functions and taxonomy (Miao et al., 2021).

The bioaugmented bioremediation experiment shows a higher microbial count as compared to the biostimulation approach due to the addition of microbes in the bioaugmentation process. Also, the augmented strategy shows high bioremediation potential (Roy et al., 2018). It is considered a promising approach for environmental clean-up, especially for the pollutants that tend to resist biodegradation by the native microbial communities. Besides its advantages, there are certain constraints associated with the bioaugmentation strategy. Bioremediation using the augmentation method depends on the bioavailability of the target pollutants and also on the capacity of inoculated microbes to survive in extremely polluted sites. Hence, the concentration of pollutants is also a key factor limiting the success of bioaugmentation (Papadopoulou et al., 2018). The application of exogenous microbes may work in optimum environmental conditions and may not acclimatize to other conditions. The commercially available or genetically engineered microbes can also be used for large-scale bioremediation purposes, but it will increase the handling cost and also endangering environmental safety (Wang et al., 2018).

11.10 CHALLENGES ASSOCIATED WITH THE BIOREMEDIATION BY INDIGENOUS MICROBIAL COMMUNITY

Bioremediation using indigenous microbial communities has certain challenges that should be identified and addressed for its implementation for environmental clean-up (Miao et al., 2021). Based on a thorough analysis of the available studies, challenges associated with bioremediation by the indigenous microbial community are identified and presented here for the future outlook.

Physicochemical properties of the polluted site and the concentration of pollutants majorly affect native microbe bioremediation activity and reduce the success of their application as a biotechnological tool (Gouveia et al., 2018). Biodegrading indigenous microorganisms require suitable habitat for effective performance. Physicochemical properties such as imbalanced nutrients and/or unfavourable environmental factors such as pH, salinity, temperature, total organic carbon (TOC), particle size, clay content, and aerobic and anaerobic conditions prevailing within the polluted site majorly hamper on-site bioremediation using indigenous microbes (Liu et al., 2019). The scarcity of nutrients severely impairs the catabolic activities of the indigenous microbes at polluted sites and decreases the rate of in situ bioremediation (Sarkar et al., 2016). In particular, the lack of electron donor and/or acceptor often suppresses the bioremediation potential of native microbes when the concentration of pollutants is high (Roy et al., 2018). Even bioaugmented microbes do not work in shortage of nutrients, and the number of inoculated microbes decreases due to microbial competition (Miao et al., 2021). Sometimes, the primary substrate may act as a competitive inhibitor in the biodegradation of pollutants targeted for co-metabolic degradation. For example, 1,4-dioxane degradation can be more effective when the primary substrate propane is scarce (Miao et al., 2021). Therefore, the balance of nutrients at the treatment site is very crucial for optimum remediation.

Additionally, the bioavailability of the pollutant to microbes for degradation is directly associated with the biodegradation efficacy (Li et al., 2018). The organic matter at polluted sites hampers the bioavailability of hydrocarbons and prevents biodegradation regardless of the presence of active degrading microbes. A mixture of pollutants also affects the structure of the microbial community; for example, the presence of metal influences the microbial community and has deleterious effects on hydrocarbon degradation processes (Gouveia et al., 2018). Thus, it is very crucial to attain optimum environmental conditions, nutrient availability, geological structure, and operating conditions while undertaking on-site bioremediation (Zhang et al., 2019).

Generally, the lack of effective native degradative microbes at polluted sites has low remediation outcomes as a low population of biodegrading microbes have low or zero activity to sustain the removal (Lee and Ulrich, 2021). The storage of bioaugmentation culture reduces the density of heterotrophic microbes and thus affects the overall bioremediation rate (Jaime et al., 2016). Therefore, an appropriate number of active bioremediating microbes should be maintained to achieve effective bioremediation. The potential biodegrading microbes may require repeated inoculation to overcome the issue of competition, thus increasing the cost of the process. Another alternative is to use immobilized microbes on different carriers to achieve the purpose of bioremediation without losing the cell number (Zhang et al., 2021).

Till date, bioremediation of pollutants using exogenous or autochthonous microbes has mostly been studied in an artificially contaminated microcosm, which does not reflect the conditions and challenges associated with in situ bioremediation of naturally polluted sites where pollutants availability becomes limited over time due to ageing (Papadopoulou et al., 2018).

11.11 FUTURE PROSPECTS OF UTILIZING INDIGENOUS MICROBIAL COMMUNITY FOR BIOREMEDIATION PURPOSE

The benefits associated with bioremediation have inveigled many to analyse and further apply it to attain environmental clean-up. Indigenous microbes have great potential for bioremediation and can be a tool of choice to deal with environmental pollution. On the basis of available studies, certain future prospects are highlighted to further enhance the utility of indigenous microbes for bioremediation.

i. The information collected on the response of indigenous microbial communities against pollutants can be analysed in detail and used to develop advanced strategies for bioremediation. The more effective assessment and monitoring of the bioremediation of polluted areas can help to generate a culture capable of robust biodegradation for bioaugmentation efforts. Culture having high bioremediation potential can be used for bioaugmentation according to the suitability of the physicochemical parameters (Lee and Ulrich, 2021).

ii. The high potential performing community may have the potential to regain the active metabolism rapidly and smoothly after any type of disturbance in the bioremediation process. The correlation studies between community efficacy and various operational environmental factors across a time scale need to be conducted to select a community with high resilience power and member with high kinetic efficiency. Future studies should be dedicated to designing and engineering microbial communities based on the knowledge of community intelligence at a particular scenario with optimization of bioreactor operational conditions (Tikariha and Purohit, 2020).

iii. Additionally, with the binning and functional algorithm, metagenomics can support the identification of novel species with potency to act as biodegrading keystone species (Tikariha and Purohit, 2020). Many studies are available on bioremediation, but further research needs to be conducted to confirm the identity of primary degraders in different environmental conditions (Lee and Ulrich, 2021).

iv. The microbial communities capable of degrading the specific pollutants should be preserved in a georeferenced microbial consortia bank, which can be reused in future at the native geographic region from which the microbial consortia were obtained for bioremediation (Gouveia et al., 2018).

v. New biostimulating compounds such as PAC need to be searched, which will protect the suspended biomass from degradation-related toxicity without affecting the microbial biodegradation activity (Feng and Chen, 2018).

vi. Studies should be directed to develop cheap substrates such as municipal solid waste compost (MSWC) for the cultivation of microbial inoculants for the purpose of bioremediation (Jaime et al., 2016).

ABBREVIATIONS

ACE	Abundance-based coverage estimator
AH	Aromatic hydrocarbon
AQDS	Anthraquinone-2,6-disulphonate
ARISA	Automated ribosomal intergenic spacer analysis
BOD	Biological oxygen demand
COD	Chemical oxygen demand

CTAB	Cetyltrimethylammonium bromide
DGGE	Denaturing gradient gel electrophoresis
DNRA	Dissimilatory nitrate reduction to ammonium
EL-FAMEs	Ester-linked fatty acid methyl esters
HRC	Slow-release hydrogen compound
IGS	Intergenic spacer
ITS1	Internal transcribed spacer 1
KEGG	Kyoto Encyclopedia of Genes and Genomes
MBC	Microbial biomass carbon
MMOs	Methane monooxygenases
MSWC	Municipal solid waste compost
OTUs	Operational taxonomic units
PAC	Poly-aluminium chloride
PAH	Polycyclic aromatic hydrocarbon
PCB	Polychlorinated biphenyl
PCP	Pentachlorophenol
PCR	Polymerase chain reaction
PHC	Petroleum hydrocarbon
PLFA	Phospholipid fatty acid
pMMOs	Cytochrome P450 monooxygenases
RHDα	Ring-hydroxylating dioxygenase α-subunit
RT-PCR	Real-time quantitative PCR
SDS	Sodium dodecyl sulphate
SOB	Sulphur-oxidizing bacteria
SRB	Sulphur-reducing bacteria
TBZ	Tiabendazole
TCE	Trichloroethylene
TOC	Total organic carbon
TPH	Total petroleum hydrocarbon

REFERENCES

Bartholameuz, E.M., Hettiaratchi, J.P.A., Steele, M., & Kumar, S., 2020. Reaction kinetic analysis of manganese peroxidase augmented aerobic waste degradation. *Journal of Hazardous, Toxic, and Radioactive Waste*, 24(4), 04020043.

Chen, M., Tong, H., Qiao, J., Lv, Y., Jiang, Q., Gao, Y., & Liu, C. (2020). Microbial community response to the toxic effect of pentachlorophenol in paddy soil amended with an electron donor and shuttle. *Ecotoxicology and Environmental Safety*, 205, 111328. Doi: 10.1016/j.ecoenv.2020.111328.

Chen, S., Zhang, K., Jha, R. K., Chen, C., Yu, H., Liu, Y., & Ma, L. (2019). Isotope fractionation in atrazine degradation reveals rate-limiting, energy-dependent transport across the cell membrane of gram-negative *rhizobium* sp. CX-Z. *Environmental Pollution*, 248, 857–864. Doi: 10.1016/j.envpol.2019.02.078.

Dutta, D., Arya, S., & Kumar, S., 2021. Industrial wastewater treatment: Current trends, bottlenecks, and best practices. *Chemosphere*, 285, 131245. Doi: 10.1016/j.chemosphere.2021.131245.

Feng, C., & Chen, N. (2018). Anaerobic bioremediation performance and indigenous microbial communities in treatment of trichloroethylene/nitrate-contaminated groundwater. *Environmental Engineering Science*, 35(4), 311–322. Doi: 10.1089/ees.2017.0124.

Gonzalez, J. M., & Sainz-Jimenez, C. (2004). Microbial diversity in biodeteriorated monuments as studied by denaturing gradient gel electrophoresis. *Journal of Separation Science*, 27(3), 174–180. Doi: 10.1002/JSSC.200301609.

Gouveia, V., Almeida, C. M. R., Almeida, T., Teixeira, C., & Mucha, A. P. (2018). Indigenous microbial communities along the NW Portuguese coast : Potential for hydrocarbons degradation and relation with sediment contamination. *Marine Pollution Bulletin*, 131, 620–632. Doi: 10.1016/j.marpolbul.2018.04.063

Hua, F., & Wang, H. Q. (2014). Uptake and trans-membrane transport of petroleum hydrocarbons by microorganisms. *Biotechnology & Biotechnological Equipment*, 28(2), 165–175. Doi: 10.1080/13102818.2014.906136.

Jaime, A., Montes, E., Lopes, P., Daniela, A., Júlio, L., Cássia, R., De, Fernandes, R., Chaer, A., & Rogério, M. (2016). Changes in the microbial community during bioremediation of gasoline-contaminated soil. *Brazilian Journal of Microbiology*, 48(2), 342–351. Doi: 10.1016/j.bjm.2016.10.018.

Khan, S., Hesham, A. E. L., Qiao, M., Rehman, S., & He, J. Z. (2010). Effects of Cd and Pb on soil microbial community structure and activities. *Environmental Science and Pollution Research*, 17(2), 288–296. Doi: 10.1007/s11356-009-0134-4.

Kumar, S., Dhar, H., Nair, V.V., Govani, J., Arya, S., Bhattacharya, J.K., Vaidya, A.N., & Akolkar, A.B., 2019. Environmental quality monitoring and impact assessment of solid waste dumpsites in high altitude sub-tropical regions. *Journal of Environmental Management*, 252, 109681. Doi: 10.1016/j.jenvman.2019.109681.

Lee, K., & Ulrich, A. (2021). Indigenous microbial communities in Albertan sediments are capable of anaerobic benzene biodegradation under methanogenic, sulfate-reducing, nitrate-reducing, and iron-reducing redox conditions. *Water Environment Research*, 93(4), 524–534. Doi: 10.1002/wer.1454.

Li, X., Fan, F., Zhang, B., Zhang, K., & Chen, B. (2018). Biosurfactant enhanced soil bioremediation of petroleum hydrocarbons: Design of experiments (DOE) based system optimization and phospholipid fatty acid (PLFA) based microbial community analysis. *International Biodeterioration and Biodegradation*, 132, 216–225. Doi: 10.1016/j.ibiod.2018.04.009.

Li, X., Song, Y., Bian, Y., Gu, C., Yang, X., Wang, F., & Jiang, X. (2020). Insights into the mechanisms underlying efficient Rhizodegradation of PAHs in biochar-amended soil: From microbial communities to soil metabolomics. *Environment International*, 144, 105995. Doi: 10.1016/j.envint.2020.105995.

Liu, Q., Tang, J., Liu, X., Song, B., Zhen, M., & Ashbolt, N. J. (2019). Vertical response of microbial community and degrading genes to petroleum hydrocarbon contamination in saline alkaline soil. *Journal of Environmental Sciences*, 81, 80–92. Doi: 10.1016/j.jes.2019.02.001.

Miao, Y., Heintz, M. B., Bell, C. H., Johnson, N. W., Polasko, A. L. P., Favero, D., & Mahendra, S. (2021). Profiling microbial community structures and functions in bioremediation strategies for treating 1,4-dioxane-contaminated groundwater. *Journal of Hazardous Materials*, 408, 124457. Doi: 10.1016/j.jhazmat.2020.124457.

Mishra, D., Kumari, S., Jaiswal, A., Arya, S., Yadav, B.R., Thul, S.T., Kumar, S., Pandey, R., Kumar, R., & Pandey, A., 2021. Evaluation of distillery sludge as a soil amendment for improving soil quality and sugarcane (CO-265) yield. *Environmental Technology & Innovation*, 23, 101624. Doi: 10.1016/j.eti.2021.101624.

Neufeld, J.D., Leigh, M.B., & Green, S.J. (2012). Denaturing gradient gel electrophoresis (DGGE) for microbial community analysis. In *Hydrocarbon and Lipid Microbiology Protocols* (pp. 77–99). http://www.springerreference.com/index/chapterdbid/168321.

Paliya, S., Mandpe, A., Kumar, M.S., & Kumar, S., 2021. Aerobic degradation of decabrominated diphenyl ether through a novel bacterium isolated from municipal waste dumping site: Identification, degradation and metabolic pathway. *Bioresource Technology*, 333, 125208. Doi: 10.1016/j.biortech.2021.125208.

Papadopoulou, E.S., Genitsaris, S., Omirou, M., Perruchon, C., Stamatopoulou, A., Ioannides, I., & Karpouzas, D.G. (2018). Bioaugmentation of thiabendazole-contaminated soils from a wastewater disposal site: Factors driving the efficacy of this strategy and the diversity of the indigenous soil bacterial community. *Environmental Pollution*, *233*, 16–25. Doi: 10.1016/j.envpol.2017.10.021.

Pushkar, B., Sevak, P., Parab, S., & Nilkanth, N. (2021). Chromium pollution and its bioremediation mechanisms in bacteria: A review. *Journal of Environmental Management*, *287*(November 2020), 112279. Doi: 10.1016/j.jenvman.2021.112279.

Ribeiro, H., de Sousa, T., Santos, J.P., Sousa, A.G.G., Teixeira, C., Monteiro, M.R., Salgado, P., Mucha, A.P., Almeida, C.M.R., Torgo, L., & Magalhães, C. (2018). Potential of dissimilatory nitrate reduction pathways in polycyclic aromatic hydrocarbon degradation. *Chemosphere*, *199*, 54–67. Doi: 10.1016/j.chemosphere.2018.01.171.

Rizvi, A., Ahmed, B., Zaidi, A., & Khan, M.S. (2020). Biosorption of heavy metals by dry biomass of metal tolerant bacterial biosorbents: An efficient metal clean-up strategy. *Environmental Monitoring and Assessment*, *192*(12), 1–21. Doi: 10.1007/s10661-020-08758-5.

Roy, A., Dutta, A., Pal, S., Gupta, A., Sarkar, J., Chatterjee, A., Saha, A., Sarkar, P., Sar, P., & Kazy, S.K. (2018). Biostimulation and bioaugmentation of native microbial community accelerated bioremediation of oil refinery sludge. *Bioresource Technology*, *253*(November 2017), 22–32. Doi: 10.1016/j.biortech.2018.01.004.

Sarkar, J., Kazy, S.K., Gupta, A., & Dutta, A. (2016). Biostimulation of Indigenous Microbial Community for Bioremediation of Petroleum Refinery Sludge. *Frontiers in Microbiology*, *7*, 1407. Doi: 10.3389/fmicb.2016.01407.

Shukla, S.K., Mangwani, N., Rao, T.S., & Das, S. (2014). Biofilm-mediated bioremediation of polycyclic aromatic hydrocarbons. In *Microbial Biodegradation and Bioremediation* (pp. 201–230). Doi: 10.1016/B978-0-12-800021-2.00008-X.

Singh, E., Kumar, A., Mishra, R., You, S., Singh, L., Kumar, S., & Kumar, R., 2021b. Pyrolysis of waste biomass and plastics for production of biochar and its use for removal of heavy metals from aqueous solution. *Bioresource Technology*, *320*, 124278. Doi: 10.1016/j.biortech.2020.124278.

Singh, R., Paul, D., & Jain, R.K. (2006). Biofilms: Implications in bioremediation. *Trends in Microbiology*, *14*(9), 389–397. Doi: 10.1016/j.tim.2006.07.001.

Tikariha, H., & Purohit, H.J. (2020). Unfolding microbial community intelligence in aerobic and anaerobic biodegradation processes using metagenomics. *Archives of Microbiology*, *202*(6), 1269–1274. Doi: 10.1007/s00203-020-01839-6.

Wainaina, S., Awasthi, M.K., Sarsaiya, S., Chen, H., Singh, E., Kumar, A., Ravindran, B., Awasthi, S.K., Liu, T., Duan, Y., & Kumar, S., 2020. Resource recovery and circular economy from organic solid waste using aerobic and anaerobic digestion technologies. *Bioresource Technology*, *301*, 122778. Doi: 10.1016/j.biortech.2020.122778.

Wang, X., Jiang, L., Gai, Z., Tao, F., Tang, H., & Xu, P. (2018). The plasticity of indigenous microbial community in a full-scale heavy oil-produced water treatment plant. *Journal of Hazardous Materials*, *358*, 155–164. Doi: 10.1016/j.jhazmat.2018.06.049.

Xu, M., Wu, W., Wu, L., He, Z., Van Nostrand, J.D., Deng, Y., Luo, J., Carley, J., Ginder-vogel, M., Gentry, T.J., Gu, B., Watson, D., Jardine, P.M., Marsh, T.L., Tiedje, J.M., Hazen, T., Criddle, C.S., & Zhou, J. (2010). Responses of microbial community functional structures to pilot-scale uranium in situ bioremediation. *The ISME Journal*, *4*(8), 1060–1070. Doi: 10.1038/ismej.2010.31.

Yu, W., Song, J., Chen, C., Song, Q., Zhao, C., Li, X., & Chen, H. (2019). *Phragmites* stimulation and natural attenuation of indigenous microbial community accelerated bioremediation of oil contaminated soil. *Petroleum Science and Technology*, *37*(8), 882–888. Doi: 10.1080/10916466.2019.1570251.

Yu, Z., He, Z., Tao, X., Zhou, J., Yang, Y., Zhao, M., Zhang, X., Zheng, Z., Yuan, T., Liu, P., Chen, Y., Nolan, V., & Li, X. (2016). The shifts of sediment microbial community phylogenetic and functional structures during chromium (VI) reduction. *Ecotoxicology, 25*(10), 1759–1770. Doi: 10.1007/s10646-016-1719-6.

Zhang, B., Wang, S., Diao, M., Fu, J., Xie, M., Shi, J., & Liu, Z. (2019). Microbial community responses to vanadium distributions in mining geological environments and bioremediation assessment. *Journal of Geophysical Research: Biogeosciences, 124*(2), 601–615. Doi: 10.1029/2018JG004670.

Zhang, M., Liu, X., Li, Y., Wang, G., Wang, Z., & Wen, J. (2017). Microbial community and metabolic pathway succession driven by changed nutrient inputs in tailings: Effects of different nutrients on tailing remediation. *Scientific Reports, 7*(1), 1–10. Doi: 10.1038/s41598-017-00580-3.

Zhang, X., Song, Z., Tang, Q., Wu, M., Zhou, H., Liu, L., & Qu, Y. (2021). Performance and microbial community analysis of bioaugmented activated sludge for nitrogen-containing organic pollutants removal. *Journal of Environmental Sciences, 101*, 373–381. Doi: 10.1016/j.jes.2020.09.002.

Zhao, Y., Bai, Y., Guo, Q., Li, Z., Qi, M., Ma, X., Wang, H., Kong, D., Wang, A., & Liang, B. (2019). Bioremediation of contaminated urban river sediment with methanol stimulation : Metabolic processes accompanied with microbial community changes. *Science of the Total Environment, 653*, 649–657. Doi: 10.1016/j.scitotenv.2018.10.396.

12 Functional Metagenomics in Environmental Bioremediation
Recent Advances, Challenges and Future Outlook

*Menaka Devi Salam, Shalini Porwal,
Arti Mishra, and Ajit Varma*
Amity University

CONTENTS

12.1 FUNCTIONAL METAGENOMICS AND ITS IMPORTANCE IN ENVIRONMENTAL ENGINEERING

Functional metagenomics is the sequence-based study focusing on functional genes isolated through a metagenomic approach and screening for functions of interest by the expressed genes. It not only helps in the identification of microbial species present in the environment, but also helps in elucidating the metabolic activities in the given environment. The functional activity of unculturable microbes can be expressed in a routine host, and the desired function can be enhanced for its use in bioremediation. This approach has greatly helped in the discovery of novel enzymes with potential applications in bioremediation and in environmental monitoring. It has also expanded the understanding of the effect of pollutants on microbial communities in various environmental sites, e.g., heavy metal-contaminated groundwater and soils, hydrocarbon-contaminated sites, and electronic waste-contaminated sites (Chandra and Kumar, 2017a, b). There are two important strategies for functional metagenomics: (1) sequence-based and (2) phenotype-based strategies. Sequence-based metagenomics relies on the sequence data generated and on the known sequences in databases, whereas phenotype-based metagenomics relies on the screening of important enzymes leading to the discovery of novel genes. The methods adopted for screening are agar-based screening method, microtiter plate screening, and high-throughput methods such as FACS method, cell-based or in vivo method using reporter gene strategy. With the advancements of next-generation sequencing platforms and computational biology, sequence analysis has become much easier and economical in recent times, but since the sequence-based metagenomics approach depends on the sequence similarity with already known and characterized genes in public databases such as GenBank, a significant number of genes discovered are tagged as hypothetical genes. The alternative approach, i.e., the functional or phenotype-based approach, becomes quite indispensable in spite of having challenges in library preparation, expression, and screening.

Functional metagenomics can help us identify gene clusters involved in the degradation of toxic pollutants, remediation of heavy metals, etc. In wastewater treatment plants, the metagenomics approach helps us to understand the microbial community involved in the treatment system. Identification of genes involved in the microbial degradation pathway through functional metagenomics will help us have a better insight into how the degradation takes place and the enzymes involved in the process.

12.2 METAGENOMIC LIBRARIES CONSTRUCTED FROM DIFFERENT ENVIRONMENTAL SAMPLES (RECENT ADVANCES)

It has been observed as a general phenomenon that soil and groundwater samples contaminated with heavy metals influence the microbial community and diversity, and so far, a number of evidences have been found, which show this influence (Awasthi et al., 2020). For example, sediments contaminated with heavy metal(loid)s such as lead, arsenic, cadmium, and mercury showed that *Proteobacteria*, *Bacteroidetes*, and *Firmicutes* were the dominant bacterial phyla present in such environments as there are many bacterial types in these phyla that can resist the presence of heavy metals (Chen et al., 2018). It was also found that the above-mentioned phyla possessed heavy metal resistance genes at the maximum. Similar results were also observed by Hemme et al. (2010), in which the microbial community and diversity were greatly altered in a groundwater sample having long-term exposure to heavy metals, nitric acid, and organic solvents. Sequence-based functional metagenomic analysis revealed the abundance of many resistance genes to the contaminants such as mercury, nitrate, and chlorinated hydrocarbons. Abundance of copper resistance genes involved in pumping out the heavy metals from the cell wall, and therefore, elucidation of the mechanism of heavy metal resistance has also been reported using Illumina high-throughput metagenomic sequencing method to examine the phylogenetic and functional profile of copper-contaminated soils (Forouzan et al., 2020). Many novel industrially important enzymes, such as lipase, glucosidase, and esterases, have also been identified through the agar plate screening method of metagenomic clones prepared using soil samples from polluted environments (Ngara et al., 2018). This opens up a new avenue to discover novel biocatalysts that can be used in the industries. Metagenomic library clones prepared from aromatic compounds polluted soil when screened on indole-containing agar plates for indigo-forming activity identified genes coding for monooxygenase, hydroxylase, and reductase, which are involved in the degradation of aromatic compounds (Nagayam et al., 2015). Similar studies using contaminated samples showed metagenomic clones having the ability to degrade aromatic compounds, and a number of novel genes encoding enzymes involved in the degradative pathways have been identified (Ngara et al., 2018). However, the agar plate screening method has low sensitivity, therefore, improved methods of screening having high sensitivity with high throughput is required. Chromogenic substrates such as para-nitrophenyl substrates and 4-methylumbelliferyl substrates have higher sensitivity and give better results. Biochemical assays using highly sensitive indicators can be applied in microtiter plates, and thus, the screening methods can be improved, for example the use of tetrazolium dye, WST-1 {4-[3-(4-iodophenyl)-2-(4-nitrophenyl)-2H-5-tetrazolio]-1,3-benzene disulfonate}, as respiration indicator for the utilization of aromatic compounds by the metagenomic library clones prepared from bioreactor sludge of a petroleum refinery wastewater treatment plant, which has led to the identification of a number of enzymes involved in the degradation of aromatic compounds such as phenol and benzoate (Silva et al., 2013).

Hydrocarbon pollution is very common in industrial sites, mainly the petrochemical industries, and environmental samples collected from such polluted sites provide a source for the discovery of novel enzymes involved in the degradation of hydrocarbons. Most of the microorganisms thriving in such environment cannot be cultured in the laboratory, and the metagenomics approach is a possible alternative for the identification of novel genes and subsequently for bioprospecting novel enzymes for hydrocarbon degradation that can be applied in bioremediation (Kumar et al., 2020, 2021). Such investigations have led to the discovery of novel genes involved in *n*-alkane degradation, for example *alk*B, *alm*A, and *lad*A genes (Gacesa et al., 2018).

The knowledge about marine microbial ecosystems has greatly been enhanced through metagenomic data, and much more information can still be explored using metagenomics. So far, sequencing projects on large scale to unravel the vast complexity and diversity of marine microbial ecosystem have been carried out through major projects such as Global Ocean Survey (Venter et al., 2004), Tara Oceans (Pesant et al., 2015), and GEOTRACES (Anderson et al., 2014; Biller et al., 2018), and considerable progress has been made in understanding the diversity and complexity of the microbial populations in the marine ecosystem. Plastic waste is a major pollutant in the marine environment. Metagenomic analyses of plastic wastes from coastal regions of the marine environment have shown the effect of major and long-term plastic waste contamination on the benthic microbial community. Metagenomic library and sequence analysis of the biofilms on plastic waste surfaces revealed the dominance of sulfate-reducing bacteria with an enriched gene pool of enzymes involved in plastic degradation, such as esterases and depolymerases (Pinnell and Turner, 2019).

12.3 CHALLENGES IN FUNCTIONAL METAGENOMICS AND LIBRARY CONSTRUCTION

Functional metagenomic assessments can be carried out on metagenomic libraries using DNA that is isolated, purified, amplified, and then cloned in a suitable vector. Heterologous expression of the vector comprising of the gene of interest in a suitable surrogate host, generally *E. coli*, and evaluation of transformants are the important steps in this approach. To identify the expression of a particular phenotype presented in the host by the inserted gene, screening of metagenomic libraries may be performed on various clones simultaneously on a fixed matrix in which the whole group is assayed with a suitable indicator to disclose the existence of a phenotypically relevant clone. This type of strategy requires the functional protein to be secreted by the host cell, and in many cases, this poses a challenge in function-based screenings. Metagenomic clones can also be grown on particular indicator media, which permit visual identification of active clone, e.g., lipolytic activity (Henne et al., 2000), hemolytic activity on blood agar (Rondon et al., 2000), and dye-degrading activity (Qu et al., 2018). In a different direction, the existence of a zone of inhibition in soft agar cover assays using indicator microbes that expose inhibitory or antimicrobial agents formed by an active clone can also be utilized (Tannieres et al., 2013; Iqbal et al., 2014). Selection can also be based on the ability to metabolize a substrate as the sole carbon source (Entcheva et al., 2001), the ability to grow in an antimicrobial agent (Donato et al., 2010), or the ability to resist any toxic heavy metal (Staley et al., 2015).

An alternate approach for the detection of novel genes was established by Uchiyama et al. (2005), which was referred to as substrate-induced gene expression (SIGEX) screening. This screening depends on the principle that catabolic gene expression is triggered by a specific substrate or metabolite of catabolic enzymes and is measured by regulatory elements situated in the periphery of those genes. In this method, the community DNA insert is attached with a reporter gene encrypting green fluorescent protein (GFP) on operon-trap vector and induced by a target substrate. The system is merged with fluorescence activated cell sorting (FACS) for the high-throughput selection of GFP-expressing clones. This method removes the incorporation of clones comprising self-ligated plasmids and those which are constitutively expressing GFP.

Notwithstanding the possible effectiveness of such structures, phenotype-based functional metagenomic approach confronts a few obstacles to which potential determinations are currently being developed. For identifying valuable genes and protein candidates, a series of sequential steps in the screening and cloning must arise efficiently. Whole gene transcription, translation of mRNA, proper protein folding, and secretion of the active protein from the surrogate host should be attained before initiation of functional screening. Appropriate and cost-effective screening approaches should be used for distinguishing the occurrence of any gene of interest within the metagenomic library. High-throughput screening procedures may advance the chances of getting active clones by enabling a higher number of clones to be screened at the same time.

There are two different approaches used in metagenomics for the construction of libraries. The preferable approach for the construction of libraries is to prepare large insert libraries using cosmid, fosmid, or bacterial artificial chromosomes so that the maximum amount of the accessible genetic resources in the sample can be captured. The other approach is to construct small insert expression libraries, especially those made in lambda phage vectors, for activity screening of the sample. There are many difficult steps in library construction. First, the isolated DNA must be of adequate length and of high quality for effective packaging into lambda phage heads (Parks and Graham, 1997). The extraction typically utilizes gentle lysis to avoid shearing of DNA samples (Zhou et al., 1996), but even then, it could be difficult to achieve large fragments of DNA (Kakirde et al., 2010). Earlier studies showed that starting with crude DNA extracts that contain at least ~75 kb fragments led to high-quality libraries, and it is extremely important to check the fragment size range by pulsed-field elec-trophoresis before continuing. An especially useful and inexpensive molecular ladder for pulsed-field gels is self-ligated lambda DNA, which could be easily prepared and which results in bands at approximately 50–150 kb. A freeze-grinding step before the extraction can significantly improve cell lysis (Lee and Hallam, 2009). This step could fragment DNA, but it does not interfere with library construction (Brady, 2007).

Isolated metagenomic DNA is frequently contaminated with composites such as humic acid and other organic acids that generally co-purify with DNA and requires supplementary purification steps that could lead to DNA damage. The most frequent toxin in soil sample DNA extract is humic acid that interferes with enzymatic reac-tions. Non-linear electrophoresis is efficient for contaminant elimination and produces purified and concentrated DNA suitable for amplification using polymerase chain reaction for further metagenomic analysis, but it requires specialized equipment (Pel et al., 2009; Engel et al., 2012). Humic acids can be avoided by allowing them to run off the gel during the electrophoresis of crude extract as they migrate faster than large DNA fragments. Another possibility, in order to prevent contamination of DNA, is by the circulating buffer, electrophoresis can be halted after humic acids have formed a front, the part of the gel containing the humic acids contamination can be excised, and then that region can be supplanted with fresh gel (Cheng et al., 2014). Scientists have also discovered that the contaminating nucleases are inhibited by treating extracted DNA in an agarose plug with sodium chloride and formamide (Liles et al., 2008). In recent years, high-quality soil metagenomic DNA isolation kits are available that give a purified form of DNA and are ideal for downstream processing.

Once the extracted DNA is purified and size-selected, it has to be end-repaired and ligated to a de-phosphorylated, blunt-end vector and this process is a challenging step. To guarantee the appropriate size range prior to the ligation, DNA is checked for co-migration with the largest band of a lambda-HindIII ladder on agarose gel (Brady, 2007) or run on a pulsed-field gel for precise size for the estimation. There is a requirement of a small amount of the ligation to transform E. coli prior to the expensive packaging step; subsequent transformants indicate the presence of circular DNA molecules evolving from the ligation of successfully blunt-ended fragments. Even though ligation conditions might not favor the creation of circular molecules, this is the very best substitute for effective end repair.

Additional challenges involve susceptibility of packaging extracts and purified digested and de-phosphorylated vector DNA for ligation step. Even though bril-liant commercial products are available, still in-house vector preparation may be necessary when the specific expression hosts are intended to be used in functional screening outside the host range of available commercial vectors. The crowning step in library construction constitutes the transduction of *E. coli*, and while it is possible to thousands of clones with the first effort, troubleshooting is required to improve the size of the library. When transduction results in an unfortunate small number of transductants, it becomes hard to determine the cause for the same.

In fact, metagenomic library construction in several ways is an art that demands time and training to master. Provided the substantial challenges and costs associated with library construction, as well as potential difficulties in obtaining rare ecological samples, a clear result is that we must find ways to maximize these valuable resources for shared benefits.

12.4 APPLICATIONS OF FUNCTIONAL METAGENOMICS IN ENVIRONMENTAL BIOREMEDIATION

Metagenomics approach has greatly been used in understanding the effect of toxic pollutants on soil microbial composition. It has significantly increased our knowledge about the microorganisms that can resist and utilize the toxic substances present in contaminated sites (Kumar et al., 2020; Kumar and Chandra, 2020). Thus, remediation strategies can be designed with either *in situ* or *ex situ* methods. Bioremediation strategies are usually based on the indigenous microbial consortia present in contaminated sites, and *in situ* method is desirable. Through functional metagenomics, genes and enzymes involved in the degradation or detoxification process of the hazardous component have been identified, further expanding the scope of bioremediation. For example, novel microbial enzymes that confer resistance to acrylate were discovered through functional metagenomics screening (Curson et al., 2014).

Functional profiling of samples collected from an area in northwest Iberian Peninsula coast, which is susceptible to oil spills, revealed the presence of genera having hydrocarbon-degrading genes. These genera were *Alcanivorax*, *Thalassospira*, and *Pseudomonas* spp., and they had the potential to degrade a number of aromatic compounds such as benzoate, chlorobenzene, naphthalene, and xylene (Bôto et al., 2021). Metabolic potential of microbes for bioremediation was clearly observed in whole metagenome shotgun analysis of polluted river samples through the presence of sequence homologs involved in polyethylene terephthalate and plastics biodegradation along with genes involved in heavy metal tolerance, thus opening new avenues in microbial bioremediation (Kumar et al., 2018; Breton-Deval et al., 2020). Through functional metagenomics, Cu resistance clones were identified on a solid medium amended with copper (Cu), and the genes responsible for responding to Cu stress were identified after sequencing and bioinformatics analysis (Xing et al., 2020). Such evidence proves that a vast majority of novel genes and enzymes further need to be explored and that functional metagenomics help in developing novel bioremediation strategies. Environmental monitoring can also be mediated through functional metagenomics, for example, the discovery of novel cellulase genes that can degrade biofilms, improved pathogen detection through identification of functional characteristics possessed by the pathogens, and comparison of environmental samples on a temporal and spatial basis (Hong et al., 2020).

12.5 STRATEGIES FOR IMPROVEMENT IN FUNCTIONAL METAGENOMICS APPROACH AND FUTURE PROSPECTS

The discovery of new genes through functional metagenomics depends on factors such as the abundance of the extracted DNA, size of the gene, host-vector system, expression of the gene in a surrogate host, and the screening method used. In order to overcome host-biased challenges, alternative hosts can be tried besides *E. coli*, through which expression and screening of the phenotype can be successfully achieved. This can be made possible by using broad host range vectors that help in replication, expression, and screening, for example improved cosmid vectors pJC8 and pJC24 which can replicate in diverse genera of *Proteobacteria* and also can be used for functional screening in multiple hosts (Cheng et al., 2014). Often, metagenomic libraries are made using *E. coli* hosts because of its high transforming efficiency as compared to other bacterial hosts. In this case, the broad-host-range vectors or shuttle vectors can be used and the recombinant DNA can be transferred to a suitable host to carry out functional screening. For example, broad-host-range vector (pKS13S) was used successfully in various hosts such as *Pseudomonas putida*, *E. coli*, and *Burkholderia multivorans* for the study of oxygenase involved in dye degradation from a metagenomic library constructed using polluted soil samples (Nagayama et al., 2015). Smarter molecular systems can be created using synthetic biology that can also improve the screening methods. High-throughput screening methods help in the identification of functional clones much easier, and so far, a number of such methods have been developed. Examples include nanostructure-initiator mass spectrometry (NIMS), probing enzymes with click-assisted NIMS (PECAN), self-assembled monolayers for matrix-assisted desorption/ionization mass spectrometry (SAMDI MS), and biosensor-based screening (Robinson et al., 2021). In order to understand the control and the regulatory units of functional genes, the functional metagenomics search can be expanded for identifying the regulatory and promoter regions, too. This will help in understanding the expression of the novel functional genes identified through metagenomics. Further, in order to understand the complete genetic makeup of degradation pathways, operon systems, and gene clusters, high-capacity vectors are preferable in order to construct large insert DNA libraries.

Although a number of improved vectors have come up, which can meet the challenges faced in functional metagenomics, more innovations are still required through which manipulations can be facilitated and the discovery of new enzymes for bioremediation can be made (Kumar and Chandra, 2018). To engineer vectors that can prove efficient in functional metagenomics approach, the main points that have to be considered are low copy number to improve performance, compatibility of vectors in case of using multiple vectors in a single host, broad or narrow host range (according to the requirement), and stability of the vector. In recent years, newer vector systems have been developed using synthetic biology and vectors have been designed using synthetic regulatory units through which they can be used for expression in both gram-positive and gram-negative bacteria, and also in yeast (Yang et al., 2018).

An encouraging approach using synthetic biology to meet the challenges faced in metagenomic library for the discovery of novel biocatalysts is the use of biological systems as those comparable to electronic devices. It involves designing a new gene circuit based on the requirement, modelling, and selection of the molecular components using computational simulations, application of the model by assembling the molecular components, and finally testing of the circuit in vivo (Guazzaroni et al., 2015). Some strategies have been adopted to improve the level of gene expression in the host cell, for example, by introducing transcriptional machinery (i.e., co-expression of sigma factor of RNA polymerase) that can recognize broad promoter range (Rhodius et al., 2013), and by using high-efficiency expression system (Terrón-González et al., 2013). The gene expression can also be manipulated at the level of mRNA translation by co-expressing suitable proteins that help in mRNA recognition and finally increase the capability of foreign gene expression by the host.

12.6 CONCLUSIONS

Functional metagenomics has its own advantages, especially in the discovery of new enzymes; therefore, it is a popularly used method of metagenomics. In spite of many biases, it is an important approach in the field of bioremediation as it makes easy-to-screen novel enzymes and degradative pathways exhibited by microorganisms. Therefore, alongside sequence-based metagenomics, activity-based functional metagenomics is an equally important approach in bioremediation. The challenges of functional metagenomics can be overcome by designing innovative vectors to increase heterologous expression. An important approach is to apply synthetic biology in functional metagenomics to increase expression and therefore obtain detectable enzyme activity, which is required for activity screening. The development of new activity screening methods should be aimed for high throughput and high generalizability. Overall, new tools still need to be developed in order to meet the challenges of functional metagenomics and to adapt for gene expression and activity screening from diverse environments.

REFERENCES

Anderson, R., Mawji, E., Cutter, G., Measures, C., Jeandel, C. 2014. GEOTRACES: Changing the way we explore ocean chemistry. *Oceanography* 27: 50–61.
Awasthi, M.K., Ravindran, B., Sarsaiya, S., Chen, H., Wainaina, S., Singh, E., Liu, T., Kumar, S., Pandey, A., Singh, L., Zhang, Z. 2020. Metagenomics for taxonomy profiling: Tools and approaches. *Bioengineered* 11(1): 356–374. Doi: 10.1080/21655979.2020.1736238.
Biller, S., Berube, P., Dooley, K., et al. 2018. Marine microbial metagenomes sampled across space and time. *Scientific Data* 5: 180176.
Bôto, M.L., Magalhães, C., Perdigão, R., et al. 2021. Harnessing the potential of native microbial communities for bioremediation of oil spills in the Iberian Peninsula NW coast. *Frontiers of Microbiology* 12:633659. Doi: 10.3389/fmicb.2021.633659.
Brady, S.F. 2007. Construction of soil environmental DNA cosmid libraries and screening for clones that produce biologically active small molecules. *Nature Protocols* 2:1297–1305.
Breton-Deval, L., Sanchez-Reyes, A., Sanchez-Flores, A., Juárez, K., Salinas-Peralta, I., Mussali-Galante, P. 2020. Functional analysis of a polluted river microbiome reveals a metabolic potential for bioremediation. *Microorganisms* 8: 554.
Chandra, R., Kumar, V. 2017b. Detection of androgenic-mutagenic compounds and potential autochthonous bacterial communities during *in-situ* bioremediation of post methanated distillery sludge. *Frontiers in Microbiology* 8: 87. Doi: 10.3389/fmicb.2017.00887.

Chandra, R., Kumar, V. 2017a. Detection of *Bacillus* and *Stenotrophomonas* species growing in an organic acid and endocrine-disrupting chemicals rich environment of distillery spent wash and its phytotoxicity. *Environmental Monitoring and Assessment* 189: 26. Doi: 10.1007/s10661-016-5746-9.

Chen, Y., Jiang, Y., Huang, H., Mou, L., Ru, J., Zhao, J., Xiao, S. 2018. Long-term and high-concentration heavy-metal contamination strongly influences the microbiome and functional genes in yellow river sediments. *Science of Total Environment* 637–638: 1400–1412.

Cheng, J., Pinnell, L., Engel, K., Neufeld, J.D., Charles, T.C. 2014. Versatile broad-host-range cosmids for construction of high quality metagenomic libraries. *Journal of Microbiology Methods* 99: 27–34.

Curson, A.R.J., Burns, O.J., Voget, S., et al. 2014. Screening of metagenomic and genomic libraries reveals three classes of bacterial enzymes that overcome the toxicity of acrylate. *PLoS ONE* 9(5): e97660.

Donato, J.J., Moe, L.A., Converse, B.J., et al. 2010. Metagenomic analysis of apple orchard soil reveals antibiotic resistance genes encoding predicted bifunctional proteins. *Applied Environmental Microbiology* 76: 4396–4401.

Engel, K., Pinnell, L., Cheng, J., Charles, T.C., Neufeld, J.D. 2012. Nonlinear electrophoresis for purification of soil DNA for metagenomics. *Journal of Microbiological Methods* 88: 35–40.

Entcheva, P., Liebl, W., Johann, A., Hartsch, T., Streit, W.R. 2001. Direct cloning from enrichment cultures, a reliable strategy for isolation of complete operons and genes from microbial consortia. *Applied Environmental Microbiology* 67: 89–99.

Forouzan, E., Karkhane, A.A., Yakhchali, B. 2020. Exploring metal resistance genes and mechanisms in copper enriched metal ore metagenome. *bioRxiv* 184564. Doi: 10.1101/2020. 07.02.184564.

Gacesa, R., Baranasic, D., Starcevic, A., et al. 2018. Bioprospecting for genes encoding hydrocarbon-degrading enzymes from metagenomic samples isolated from northern Adriatic sea sediments. *Food Technology and Biotechnology* 56(2): 270–277.

Guazzaroni, M.E., Silva-Rocha, R., Ward, R.J. 2015. Synthetic biology approaches to improve biocatalyst identification in metagenomic library screening. *Microbial Biotechnology* 8(1): 52–64.

Hemme, C.L., Deng, Y., Gentry, T.J., et al. 2010. Metagenomic insights into evolution of a heavy metal-contaminated groundwater microbial community. *ISME Journal* 4(5): 660–672.

Henne, A., Schmitz, R.A., Bomeke, M., Gottschalk, G., Daniel, R. 2000. Screening of environmental DNA libraries for the presence of genes conferring lipolytic activity on *Escherichia coli*. *Applied Environmental Microbiology* 66: 3113–3116.

Hong, P.Y., Mantilla-Calderon, D., Wang, C. 2020. Metagenomics as a tool to monitor reclaimed-water quality. *Applied Environmental Microbiology* 86(16): e00724–e00720.

Iqbal, H.A., Craig, J.W., Brady, S.F. 2014. Antibacterial enzymes from the functional screening of metagenomic libraries hosted in *Ralstonia metallidurans*. *FEMS Microbiology Letters* 354: 19–26.

Kakirde, K.S., Parsley, L.C., Liles, M.R. 2010. Size does matter: Application-driven approaches for soil metagenomics. *Soil Biology and Biochemistry* 42: 1911–1923.

Kumar, V., Chandra, R. 2018. Characterisation of manganese peroxidase and laccase producing bacteria capable for degradation of sucrose glutamic acid-Maillard reaction products at different nutritional and environmental conditions. *World Journal of Microbiology and Biotechnology* 34: 32.

Kumar, V., Chandra, R. 2020. Metagenomics analysis of rhizospheric bacterial communities of *Saccharum arundinaceum* growing on organometallic sludge of sugarcane molasses-based distillery. *3 Biotech* 10(7): 316. Doi: 10.1007/s13205-020-02310-5.

Kumar, V., Shahi, S.K., Singh, S. 2018. Bioremediation: An eco-sustainable approach for restoration of contaminated sites. In: Singh, J., Sharma, D., Kumar, G., Sharma, N., (Eds.), *Microbial Bioprospecting for Sustainable Development.* Springer, Singapore. Doi: 10.1007/978-981-13-0053-0_6.

Kumar, V., Singh, K., Shah, M.P., Singh, A.K, Kumar, A., Kumar, Y. 2021. Application of omics technologies for microbial community structure and function analysis in contaminated environment. In Shah, M.P., Sarkar, A., Mandal, S., (Eds.), *Wastewater Treatment: Cutting Edge Molecular Tools, Techniques & Applied Aspects in Waste Water Treatment.* Elsevier. Doi: 10.1016/B978-0-12-821925-6.00013-7.

Kumar, V., Thakur, I.S., Singh, A.K., Shah, M.P. 2020. Application of metagenomics in remediation of contaminated sites and environmental restoration. In: Shah, M., Rodriguez-Couto, S., Sengor, S.S., (Eds.), *Emerging Technologies in Environmental Bioremediation.* Elsevier. Doi: 10.1016/B978-0-12-819860-5.00008-0.

Lee, S., Hallam, S.J. 2009. Extraction of high molecular weight genomic DNA from soils and sediments. *Journal of Visualized Experiment* 33: 1569. Doi: 10.3791/1569

Liles, M.R., Williamson, L.L., Rodbumrer, J., Torsvik, V., Goodman, R.M., Handelsman, J. 2008. Recovery, purification, and cloning of high-molecular-weight DNA from soil microorganisms. *Applied Environmental Microbiology* 74: 3302–3305.

Nagayama, H., Sugawara, T., Endo, R., et al. 2015. Isolation of oxygenase genes for indigo-forming activity from an artificially polluted soil metagenome by functional screening using *Pseudomonas putida* strains as hosts. *Applied Microbiology and Biotechnology* 99: 4453–4470.

Ngara, T.R., Zhang, H. 2018. Recent advances in function-based metagenomic screening. *Genomics, Proteomics & Bioinformatics* 16(6): 405–415.

Parks, R.J., Graham, F.L. 1997. A helper-dependent system for adenovirus vector production helps define a lower limit for efficient DNA packaging. *Journal of Virology* 71: 3293–3298.

Pel, J., Broemeling, D., Mai, L., et al. 2009. Nonlinear electrophoretic response yields a unique parameter for separation of biomolecules. *Proceedings of the National Academy of Sciences* 106: 14796–14801.

Pesant, S., Not, F., Picheral, M., et al. 2015. Tara oceans consortium coordinators. Open science resources for the discovery and analysis of Tara Oceans data. *Scientific Data* 2: 150023–150016.

Pinnell, L.J., Turner, J.W. 2019. Shotgun metagenomics reveals the benthic microbial community response to plastic and bioplastic in a coastal marine environment. *Frontiers in Microbiology* 10: 1252.

Qu, W., Liu, T., Wang, D., Hong, G., Zhao, J. 2018. Metagenomics-based discovery of malachite green-degradation gene families and enzymes from mangrove sediment. *Frontiers in Microbiology* 9: 2187.

Rhodius, V.A., Segall-Shapiro, T.H., Sharon, B.D., et al. 2013. Design of orthogonal genetic switches based on a crosstalk map of σs, anti-σs, and promoters. *Molecular Systems Biology* 9: 702.

Robinson, S.L., Piel, J., Sunagawa, S. 2021. A roadmap for metagenomic enzyme discovery. *Natural Product Reports* 38(11): 1994–2023. Doi: 10.1039/D1NP00006C.

Rondon, M.R., August, P.R., Bettermann, A.D., et al. 2000. Cloning the soil metagenome: a strategy for accessing the genetic and functional diversity of uncultured microorganisms. *Applied Environmental Microbiology* 66(6): 2541–2547.

Silva, C.C., Hayden, H., Sawbridge, T., et al. 2013. Identification of genes and pathways related to phenol degradation in metagenomic libraries from petroleum refinery wastewater. *PLoS One* 8(4): e61811.

Staley, C., Johnson, D., Gould, T.J., et al. 2015. Frequencies of heavy metal resistance are associated with land cover type in the Upper Mississippi River. *Science of the Total Environment* 511: 461–468.

Tannieres, M., Beury-Cirou, A., Vigouroux, A., et al. 2013. A metagenomic study highlights phylogenetic proximity of quorum-quenching and xenobiotic-degrading amidases of the AS-family. *PLoS ONE* 8(6): e65473.

Terrón-González, L., Medina, C., Limón-Mortés, M., et al. 2013. Heterologous viral expression systems in fosmid vectors increase the functional analysis potential of metagenomic libraries. *Scientific Reports* 3: 1107.

Uchiyama, T., Abe, T., Ikemura, T., Watanabe, K. 2005. Substrate-induced gene-expression screening of environmental metagenome libraries for isolation of catabolic genes. *Nature Biotechnology* 23: 88–93.

Venter, J.C., Remington, K., Heidelberg, J.F., et al. 2004. Environmental genome shotgun sequencing of the Sargasso Sea. *Science* 304(5667): 66–74.

Xing, C., Chen, J., Zheng, X., Chen, L., Chen, M., Wang, L., Li, X. 2020. Functional metagenomic exploration identifies novel prokaryotic copper resistance genes from the soil microbiome. *Metallomics* 12(3): 387–395.

Yang, S., Liu, Q., Zhang, Y., Du, G., Chen, J., Kang, Z. (2018) Construction and characterization of broad-spectrum promoters for synthetic biology. *ACS Synthetic Biology* 7: 287–291.

Zhou, J., Bruns, M.A., Tiedje, J.M. 1996. DNA recovery from soils of diverse composition. *Applied Environmental Microbiology* 62: 316–322.

13 The Use of Microalgae and Cyanobacteria for Wastewater Treatment and the Sustainable Production of Biomass

Celestino García-Gómez,
Julia Mariana Márquez-Reyes,
Juan Antonio Vidales-Contreras,
Juan Nápoles-Armenta, and
Alejandro Isabel Luna-Maldonado
Universidad Autónoma de Nuevo León

CONTENTS

DOI: 10.1201/9781003247883-13

13.1 INTRODUCTION

Today, water scarcity is one of the main challenges to be solved for the growing world population. Considerable amounts of wastewater and agricultural and industrial waste are discharged untreated into water resources causing contamination, which puts the supply of usable water at risk (Dutta et al., 2021). A centralized system called a wastewater treatment plant is used to treat wastewater; this system is composed of primary treatments such as tanks, secondary treatments such as biofilters or activated sludge, and a tertiary disinfection treatment; however, they are designed to remove mainly organic matter and fail to remove nutrients. In addition, the number of facilities in many countries is not enough to control the growing pollution problem. These untreated or partially treated discharges contain various organic and inorganic nutrients, which are precursors of eutrophication processes in water bodies (Li et al., 2021).

The use of microalgae or cyanobacteria for the removal of organic and inorganic compounds is known as phytoremediation (Brar et al., 2020). This process involves the capture of nitrogen, phosphorus, and carbon among other compounds contained in wastewater and that generates the growth of biomass that can serve as a raw material in different industrial applications with high economic value, including food, feed, fertilizers/stimulants, pharmaceuticals, cosmetics, and biofuels (Moshood et al., 2021). The elimination of nutrients by microalgae and cyanobacteria is economical, sustainable, simple, and environmentally friendly; these microorganisms exhibit a greater efficiency in the uptake of nutrients compared to others. Microalgae and cyanobacteria can be found in a wide variety of biosystems, including rivers, lakes, and lagoons, and these microorganisms possess a great diversity in their morphology and metabolism, which makes them a repertoire of different biocomposites. Significant progress around microalgae and cyanobacteria cultures along with wastewater treatment has resulted in an improvement in biomass production, which has made it a topic of current interest to the scientific community (Leng et al., 2021).

This chapter describes the usefulness of microalgae and cyanobacteria in wastewater treatment and biomass production for various industrial applications.

13.2 MICROALGAE AND CYANOBACTERIA IN WASTEWATERS

Microalgae are classified as prokaryotes and eukaryotes. Prokaryotic microalgae lack a membrane-bound nucleus, making them single-celled organisms. Under this category are photosynthetic cyanobacteria found in saltwater or freshwater, which also have chlorophyll and phycobilins. Among them, we find species such as *Spirulina (Arthrospira) maxima* or *platensis*. On the other hand, eukaryotic microalgae have a nucleus surrounded by a membrane and intracellular organelles, thus being uni- or multicellular freshwater organisms. The main classes are green algae (Chlorophyta), red algae (Rhodophyta), and diatoms (Bacillariophyta), all of which contain pigments such as chlorophylls, carotenoids, and phycobiliproteins.

Nowadays, new sources of biomolecules are sought for different applications and the use of wastewater is emerging as a new potential alternative for the sustainable generation of biomass from microalgae and cyanobacteria. With these microorganisms,

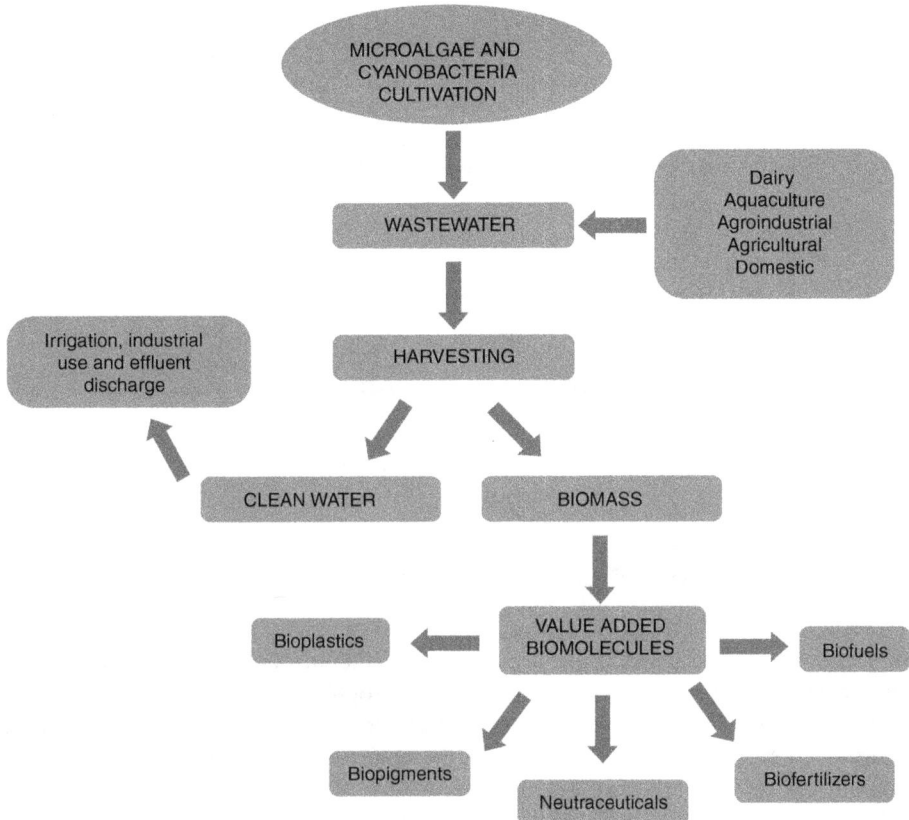

FIGURE 13.1 Synergistic approaches toward wastewater treatment by microalgae and cyanobacteria.

wastewater can be treated and generate biomass that can contain high levels of biochemical composition such as proteins, carbohydrates, and lipids, which can be used in various industrial applications (Figure 13.1).

The biochemical composition of microalgae and cyanobacteria is composed of 10%–30% lipids, around 50% proteins, 10%–50% carbohydrates, and other compounds accounting for approximately 5%. The composition of some microalgae and cyanobacteria species is listed in Table 13.1.

However, different factors, such as the quantity and quality of nutrients, light intensity and its photoperiod, carbon dioxide concentration, temperature, pH, agitation/turbulence, and salinity, should be considered for an effective microalgae culture (Herrera et al., 2021).

Municipal wastewater and agro-industrial effluents are characterized by their high content of carbon and nutrients such as nitrogen (ammonium $\left(NH_4^+\right)$) with low concentrations of nitrite and nitrate) and phosphorus. In addition, heavy metals such as Cu, Pb, Cd, Cr, or Zn are also present. Microalgae and cyanobacteria are capable of growing in

TABLE 13.1

Composition of Various Microalgae and Cyanobacteria Species

Microalgae and Cyanobacteria Species	Composition (% Dry Matter)		
	Proteins	Carbohydrates	Lipids
Chlorella vulgaris	51–58	12–17	14–22
Dunaliella salina	57	32	6
Haematococcus pluvialis	48	27	15
Porphyridium cruentum	28–39	40–57	9–14
Scenedesmus obliquus	50–56	10–17	12–14
Spirulina maxima	60–71	13–16	6–7

different types of wastewater that usually have a pH of 7–9, which coincides with the optimal range for the growth of these microorganisms (Chai et al., 2021).

13.2.1 DEVELOPMENT OF A CULTURE OF MICROALGAE AND CYANOBACTERIA

The conditions of a culture define the growth and production yields of biomolecules in microalgae and cyanobacteria. The biomass development of these microorganisms can be carried out in a photoautotrophic, heterotrophic, mixotrophic, and photoheterotrophic way. In a phototrophic culture, light and inorganic carbon are used as a source of energy and carbon, respectively. By using autotrophic cultivation, the problem of contamination that happens in other forms of development is avoided. When microalgae use organic carbon as a source of energy and carbon, they experience a heterotrophic development. Several studies have shown a higher lipid and cellular content when using this culture development with various sources of carbon, such as glycerol, glucose, fructose, sucrose, lactose, galactose, and mannose. In a mixotrophic culture, microalgae consume both organic and inorganic carbon as a source of carbon for growth, thus justifying the use of wastewater by combining an autotrophic and heterotrophic system. Finally, when microalgae use light with an organic carbon source, it is known as photoheterotrophic culture (Manhaeghe et al., 2020).

Microalgae and cyanobacteria can use organic carbon in addition to inorganic N and P, so wastewater, being rich in these components, makes it an ideal substrate for the growth of these microorganisms. Several studies have used wastewater using oxidation ponds with mechanical mixture called raceway pond, and it has been observed that they have high efficiencies of use of wastewater constituents, thus reducing eutrophication hazards caused by excess nutrients in wastewater discharges. Also, microalgae and cyanobacteria have proven to be efficient in the recovery of metals, becoming the main critical point; however, cost savings in wastewater treatment have become its main advantage (Danouche et al., 2021).

Studies have been reported on the use of mixotrophic systems, reaching high biomass conversions by microalgae and cyanobacteria compared to a system in autotrophic or heterotrophic mode. In a recent study, it was reported that the growth of microalgae biomass *Asterarcys* sp. SCS-1881 was 0.75 g/L for heterotrophic growth using glucose to 3.71 g/L for mixotrophic conditions, with increased biosynthesis of triacylglycerol, carbohydrates, and pigments in biomass compared to when mixotrophic mode was applied (Li et al., 2021).

Bhatnagar et al. (2010) showed that the use of glucose in the mixotrophic growth of *Chlamydomonas globosa*, *Chlorella minutissima*, and *Scenedesmus bijuga* increased biomass production by 3–10 times more compared to autotrophic conditions. Also in the work of Abreu et al. (2012), it was shown that *Chlorella vulgaris* under mixotrophic conditions exhibited a higher final concentration of biomass (3.58 g/L) than that grown under autotrophic conditions (1.22 g/L).

13.2.2 Use of Microalgae and Cyanobacteria for Wastewater Treatment

In a conventional activated sludge process, which is commonly used in wastewater treatment systems, complete nitrification is not achieved in most cases, so an additional process for the complete removal of nitrogen and phosphorus is required before the discharge of effluents into water-receiving bodies. As mentioned above, the use of microalgae and cyanobacteria is an efficient and sustainable alternative for the removal of nitrogen and phosphorus, using these nutrients for the growth of microorganisms. One of the main difficulties occurs in effluents with high COD or DBO load since it inhibits the growth of a culture of microalgae and cyanobacteria, making it less efficient as a removal system. An alternative is to implement the use of microalgae and cyanobacteria as a tertiary treatment using effluents with the reduction of organic compounds after having undergone a conventional treatment of activated sludge or diluting properly so as not to inhibit the culture because of high levels of organic load. The efficiency of the use of microalgae and cyanobacteria depends on the species to be used and the factors that benefit the development of the culture.

Several species of microalgae and cyanobacteria have been reported, which have shown that the application of these microorganisms is efficient for the uptake of nitrogen and phosphorus. Examples of nutrient removal can be cited, such as the study with *C. vulgaris* (75% N and 90% P) and *Chlorella salina* (73% N and 82% P) (El-Sheekh et al., 2016), that with the consortium of *Chlorella* and *Phormidium* sp. (94% N and 90% P) (Choudhary et al., 2017), and that with *Spirulina platensis* (99% N and 99% P) and *Mucidosphaerium pulchellum* (66% N and 33% P) (Almomani et al., 2019). High-rate algal ponds (HRAPs), photobioreactors (PBs), or hybrid systems are commonly employed for biomass growth. An HRAP is an open system that provides the right conditions for the low-cost production of microalgae biomass and cyanobacteria. Photosynthesis, nutrient uptake, and degradation of organic compounds are the mechanisms that are carried out for the growth of biomass in an HRAP system. Additionally, the incorporation of CO_2 allows a cost-effective process; in addition, the bacteria decompose the waste and then assimilate into a new biomass of microalgae and cyanobacteria, which is removed by sedimentation. PBs irradiated with LED lights are considered closed systems and can eliminate bacterial contamination problems. PBs facilitate the capture of CO_2 to produce high value-added biomass in various applications. More recently, the culture of microalgae and hybrid cyanobacteria has been developed and high rates have been shown in terms of microalgae production and rich biochemical composition. In these crops, efficient biomass growth and by-product development phases are achieved, while contamination is also avoided (Bani et al., 2021).

13.3   CULTIVATION AND HARVESTING

The production of biomass from microalgae and cyanobacteria requires an efficient separation process for the cell growth generated. The recovery of biomass and subsequent obtaining of bioproducts is based on a cost-effective separation method, since the collection cost reaches around 30% of the total cost, so choosing the right method is of utmost importance. Sedimentation, centrifugation, filtration, and flocculation/coagulation processes, whether organic, inorganic, biological, or electrochemical, are part of the current strategies that can be applied to recover the biomass of a culture of microalgae and cyanobacteria.

Sedimentation is based on the separation of particles by gravity. Microalgae and cyanobacteria have various sedimentation behaviors. More than 3% recovery in TSS in sedimentation tanks has been reported (Chatsungnoen & Chisti, 2016).

Centrifugation has become the most used practice for collecting, mainly because of its high efficiency, with concentrations of 96% reported for *Nannochloropsis* sp. (Dassey & Theegala, 2013). Although it is effective and fast, this system has the disadvantage of high cost when it comes to its large-scale application.

Biomass filtration involves the use of filter membranes for harvesting, which separates microalgae and cyanobacteria in the form of paste; it is an expensive method, but efficient. Şirin et al. (2012) reported the harvest of *C. vulgaris* of 98% using ultrafiltration membranes.

Flocculation of microalgae and cyanobacteria is based on the addition of a flocculant to form flocs and the subsequent aggregation of them. It can be physical, chemical, or biological in nature. The process is based on the dispersion of charges, with these microorganisms being of negative charge. They are neutralized, and the application of substances with positive charge makes them floccule and sediment, later having the advantage of a lower energy consumption. Vandamme et al. (2012) reported an attractive alternative of low cost, low energy consumption, non-toxic to microalgae, and that does not require the use of flocculants, which allows a reuse of the medium, using autoflocculation of *C. vulgaris* with a pH adjustment with NaOH where a recovery of 98% was achieved. Bioflocculation is related to flocculation caused by secreted biopolymers, and it has been reported that *Chlorella sorokiniana* has successfully been removed by 99% by chitosan (Xu et al., 2013). On the other hand, the harvest of *C. vulgaris* has been shown to be 97% efficient by the fungus *Aspergillus oryzae* (Zhou et al., 2013). However, co-cultivation of microalgae with fungi or bacteria results in microbiological contamination that interferes with the purity of the biomass for future applications. Another option is the electric approach to biomass collection; although these methods are not widespread, they are versatile, non-selective, and environmentally friendly since they do not require the addition of chemicals. In the work reported by Lee et al. (2013), the harvest of 91% of *Tetraselmis* sp. was achieved by a method of electroflocculation combined with sedimentation.

Chemical coagulation/flocculation is the main application toward the economic optimization of biomass collection processes. A wide variety of salts have been tested as coagulants for collecting microalgae and cyanobacteria. Multivalent metal salts, such as $FeCl_3$, $Al_2(SO_4)_3$, and $Fe_2(SO_4)_3$, have effectively been tested. Şirin et al. (2012) reported harvesting *Phaeodactylum tricornutum* with 83% recovery.

13.3.1 HARVEST BY FLOCCULATION OF *CHLORELLA VULGARIS*

Microalgae are potentially a sustainable source of different applications for society; however, the commercial production of different algae-based products is currently expensive. The efficiency of the harvested medium must be studied, to reduce costs and facilitate the profitable production of products using algae as a raw material. To address this issue, the sustainable harvesting of the microalgae *C. vulgaris* was analyzed. The microalgae were grown in bold culture medium, the pH was evaluated for five inorganic flocculation processes and one organic through a chitosan biopolymer. The harvesting efficiency for algae culture and the corresponding composition of pigments, total sugars, and lipids were studied, before and after the flocculation process.

The autoflocculation efficiency of *C. vulgaris* was analyzed in 100 mL of algae culture in a beaker at different time intervals for 30 minutes. The stationary phase microalgae were used a biomass concentration of 0.5 g/L (dry weight) for all experiments. Inorganic flocculants (i.e., $Ca(OH)_2$, $MgCl_2$, alum, $ZnSO_4$, $Al_2(SO_4)_3$, and chitosan) were added to a concentration of 150 mg/L.

The effect of pH-induced flocculation on *C. vulgaris* was tested without adding the flocculants. The pH of the culture was adjusted from 4 to 11 by adding the appropriate amount of HCl 0.1 M and NaOH 1 M.

The effect of pH on the flocculation process was verified by keeping the flocculant concentration constant (150 mg/L) at pH of 4–11 to the different compounds evaluated. The appropriate range for pH was selected based on the single-factor flocculation experiments used for the flocculation experiment (Figure 13.2). During the addition of the flocculant and the pH adjustment, the cultures were kept in stirring mode and, after adjustment, a vigorous stirring at 250 rpm for 3 minutes at room temperature and another slow stirring at 40 rpm for 20 minutes to form the flocs. The sedimentation time was standardized at 10 minutes, at which point self-flocculation was negligible. Subsequently, flocculation efficiency was evaluated by measuring optical density (OD 680).

13.3.1.1 pH-Induced Autoflocculation of *Chlorella vulgaris*
Before carrying out the chemical flocculation experiments, a simple and low-cost separation process of *C. vulgaris* was evaluated for 10–30 minutes by pH-induced autoflocculation. The autoflocculation was <15% from 10 to 30 minutes at a pH less than 9 and was considered negligible. An increase of up to 35% was observed at pH of 10, decreasing again when a pH of 11 was used. Chemical flocculation tests were carried out over 20 minutes to compensate for the effect of autoflocculation. The autoflocculation of algae cells is variable and depends largely on the morphology, cell size, and cell wall structure of the microorganism (Augustine et al., 2017). Greater self-flocculation is directly related to cell size and a subsequent sedimentation stage that increases with larger cell size and due to the small cell size (2–10 µm) *C. vulgaris* is resistant to autoflocculation (Vergini et al., 2016).

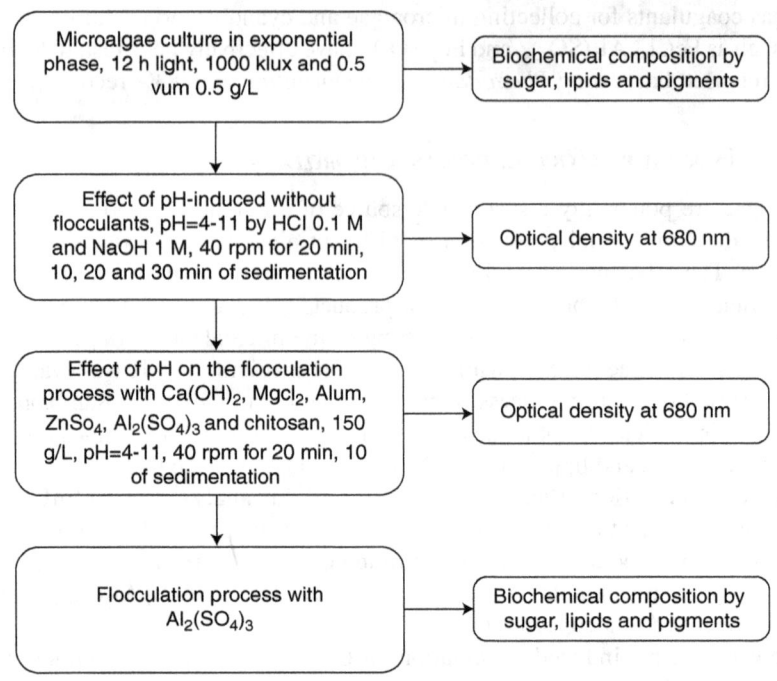

FIGURE 13.2 Flowchart of the flocculation of *Chlorella vulgaris* in this study.

The pH induced the flocculation of *C. vulgaris* (Figure 13.3) performed in a range of pH values of 4–11, and the effect is similar with sedimentation time, since no differences between 10 and 30 minutes were observed. By simply changing the pH value of the medium, the algae cells achieved a flocculation efficiency of 0%–14.29% at a change pH of 4–9 during the minutes evaluated and increased up to 32.54% at a pH of 10 over a period of 10 minutes of sedimentation time. There was little difference in flocculation efficiency over a period of 20 minutes of sedimentation time under the same condition.

Sukenik and Shelef (1984) indicated that pH-induced microalgae autoflocculation increased by raising the pH to more than 8.5 or reducing the pH to less than 3.5. The pH-induced autoflocculation study in *Nannochloropsis oculata* and *Isochrysis* sp. showed 30% flocculation efficiency by pH adjustment (Harith et al., 2009). But on the other hand, in diatom species such as *Chaetoceros calcitrans*, a greater flocculation efficiency was achieved, >90% increasing to a pH of 10.2 (Cheng et al., 2011).

Cheng et al. (2011) reported that autoflocculation in *Chlorella* sp. at low pH is due to the lower cell surface load of pH 5.5. This is explained given that, at low pH, the microalgae cell wall is composed of extracellular polysaccharides, proteins, and lipids with an amine group, dissociated and carboxylic group protects dissociation, then a negative surface charge is weakened, and the algae cell wall promotes flocculation by charge neutralization. With an increase in pH, microalgae cells become more negative avoiding this pH-induced flocculation (Vandamme et al., 2012). It has been

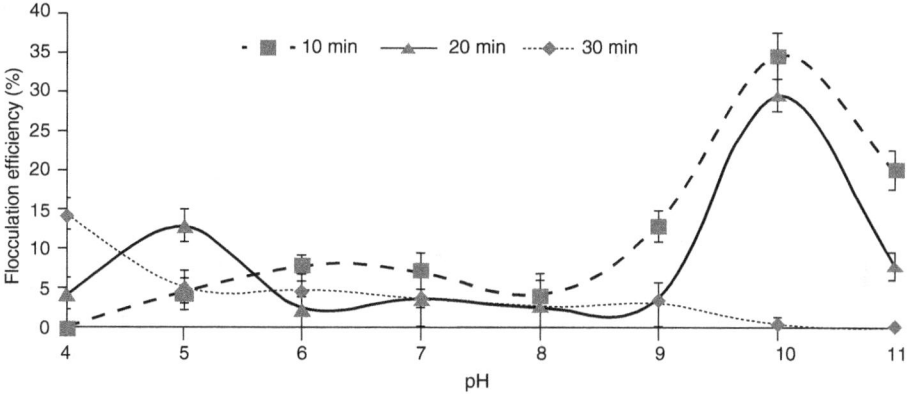

FIGURE 13.3   The effect of different values of pH on the flocculation efficiency of *Chlorella vulgaris*.

reported that with a high pH, a chemical precipitation of the calcium and magnesium salts present in the medium is generated instead of the neutralization of the charge, and then this precipitate has a positive surface charge which can produce flocculation (Knuckey et al., 2006).

Also at higher pH values, cells can release extracellular polysaccharides that reduce the load on the cell surface by promoting the flocculation process.

13.3.2   EFFECT OF pH ON CHEMICAL FLOCCULATION EFFICIENCY OF *CHLORELLA VULGARIS*

Specific standardization of microalgae or cyanobacteria is required in the chemical flocculation process for efficient and rapid collection since the optimal pH value varies depending on the type of flocculant and the species of biomass. In this study, a pH optimization of a factor was used keeping the concentration of flocculant and microalgae biomass constant on changing pH values (Figure 13.4). The results showed a significant effect in terms of pH, where it is observed that at a pH <7, there is a high flocculation efficiency (> 95%) for aluminum sulfate. Calcium hydroxide ions generated a flocculation efficiency > 80% at pH between 7 and 9. The organic flocculant chitosan presented a removal > 60% when the pH was 4. The other flocculants such as alum, $MgCl_2$, and $ZnSO_4$ showed > 20% flocculation efficiency at a pH > 9. The flocculation efficiency of metal ions improved with increasing pH values which were low to neutral or slightly acidic pH values. Vandamme et al. (2012) reported that, at a higher pH, metal ions in the medium hydrolyzed to form positive precipitates and, where negatively charged, microalgae cells are neutralized.

The chitosan-based flocculation efficiency of *C. vulgaris* varied with pH, which may be due to structural changes in chitosan. At an acidic pH, chitosan produces smaller flocs. On the other hand, at neutral pH, microalgae and cyanobacteria cells have a negative charge, which decreases the viscosity of chitosan and helps improve flocculation efficiency as a function of electrostatic forces (Sukenik & Shelef, 1984). Also, in alkaline

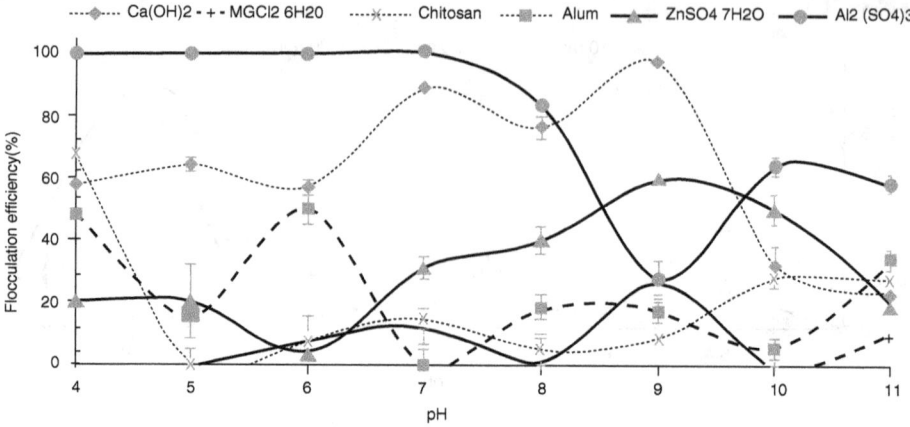

FIGURE 13.4   The effect of pH on the chemical flocculation processes of *Chlorella vulgaris* as a function of acidity or alkalinity of the medium.

conditions, chitosan loses its positive charge generating large and dense flocs. In the present study, flocculation efficiency for *C. vulgaris* increased under slightly alkaline conditions of pH 10 and 11. However, the highest flocculation efficiency of chitosan for *Chlorella sorokiniana* was at acidic pH values as it resulted in our study using a pH of 4; in other species such as *Spirulina, Oscillatoria,* and *Chlorella,* it has been reported that at neutral pH values, the highest recovery efficiencies were obtained (Xu et al., 2013). This optimal pH behavior for flocculation with chitosan is mainly due to the characteristic of the flocculation medium, the deacetylation of chitosan, the molecular weight, and the cell surface characteristic of microorganisms.

On the other hand, in Figure 13.5, it is evident how aluminum sulfate exhibited the highest removal efficiency by presenting the highest value. On the other hand, calcium hydroxide showed removals close to, but less than those obtained by aluminum sulfate.

The biochemical profiles of pigments, total sugars, and lipids were evaluated before and after a chemical flocculation, to obtain information on how the composition could be affected in the microalgae. The microalgae were obtained in stationary phase growth and under optimal growth conditions in previous works (12 hours light, 1000 klux, and 0.5 vvm), and this was considered for the experimentation and initial analysis of biochemical composition. After this analysis, the composition was as shown in Table 13.2, where there was no evidence of significant difference before or after the harvest evaluated with $Al_2(SO_4)_3$ in this study.

13.4   BIOPRODUCTS OF HIGH COMMERCIAL VALUE

Microalgae and cyanobacteria have the potential to reproduce in wastewater because it is rich in carbon, nitrogen, and phosphorus, which makes it an appropriate medium for the sustainable production of biomass from these microorganisms and subsequent extraction of bioproducts for industry. Some species of microalgae and cyanobacteria

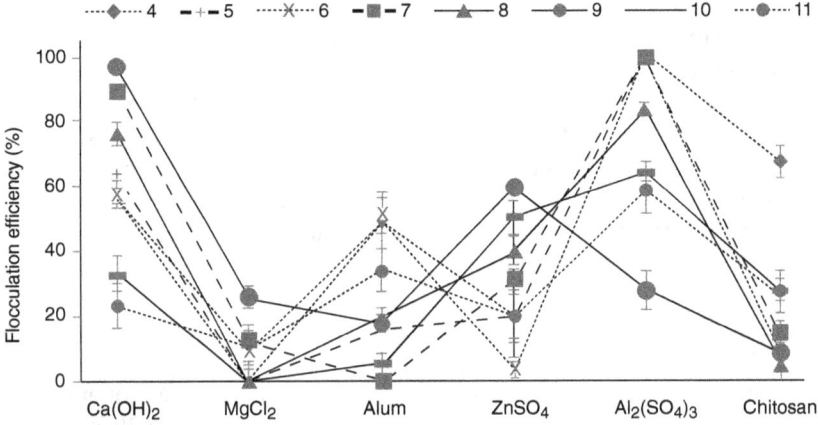

FIGURE 13.5 The effect of pH on the chemical flocculation processes of *Chlorella vulgaris* as a function of the flocculant.

TABLE 13.2
Biochemical Profile of the Microalgae *Chlorella vulgaris*

Biochemical Profile	No Flocculation	With Flocculation
Total sugars (%)	22.13 ± 1.03	21.02 ± 0.96
Total lipids (%)	13.28 ± 0.86	11.88 ± 0.65
Chlorophyll a (mg/L)	1.23 ± 0.09	1.12 ± 0.04
Chlorophyll b (mg/L)	0.89 ± 0.05	0.82 ± 0.05
Carotenoids (mg/L)	0.19 ± 0.01	0.16 ± 0.01

with optimal operational and nutritive parameters produce lipids at a high concentration of about 80% of dry weight. Microalgae and cyanobacteria have good expectations in the future for the generation of biofuels due to their CO_2 capture capacity and a high percentage of lipid content. Piligaev et al. (2015) reported that *C. vulgaris A1123 and S. abundans* A1175 showed a saturated fatty acid (SFA) and monounsaturated fatty acid (MFA) content of 67% and 72%, respectively, showing the potential for biofuel processing.

Microalgae and cyanobacteria can be applied to increase the nutritional value of animal or human feed; this is due to their high level of enzymes, pigments, lipids, carbohydrates, vitamins, and sterols. The cyanobacterium *Arthrospira platensis* contains about 42%–70% dry weight protein in addition to amino acids that the human body cannot generate and must be consumed (Danouche et al., 2021).

Various bioactive biomolecules of microalgae and cyanobacteria have been studied in the pharmaceutical industry for the generation of drugs such as antimicrobials, antivirals, therapeutic proteins, drugs, antioxidants, and antifungals that can be derived from algae (Mehariya et al., 2021). In this way, the biomass of microalgae and cyanobacteria contains several biocompounds with a high demand in the market of the nutraceutical, cosmetic, and pharmaceutical sector for various commercial purposes (Abu-Ghosh et al., 2021).

A current trend in agribusiness is the evaluation of biofertilizers, which can increase crop yields at the same time as soil regeneration, thus exercising modern agriculture. Inoculation of microalgae and cyanobacteria biofertilizers can save more than 40% of chemical fertilizer. Swarnalakshmi et al. (2013) reported that inoculation of *Anabaena* sp. in wheat cultivation increased the N content of the soil to 57% compared to chemical fertilizers as a control.

13.5 FUTURE TRENDS

Various species of microalgae and cyanobacteria can make use of nutrient-rich wastewater to develop and produce biomass. This biomass can be collected and used as a source of profitable raw material in various industries as they are friendly to the environment with the potential to replace other non-renewable and harmful products.

The treatment of industrial or domestic wastewater could use species of microalgae and cyanobacteria and propose a viable and conventional method in the future. Research has currently shown the remediation of water from various sources of generation and where biomolecules can replace traditional products at a lower cost of production if nutrient recovery is integrated, so in the future, it could be a remediation tool and producer of high-value biomolecules. On the other hand, it is necessary to isolate, cultivate, and develop new species of microalgae and cyanobacteria that can be used in wastewater with toxic molecules; also, it is necessary to identify that wastewater can potentiate the production of specific biomolecules. Also, the production of biomolecules must be optimized under the correct ratio of nutrients and operational parameters. In this prospect, phytoremediation and biorefinery of microalgae and cyanobacteria become an emerging biotechnology with great environmental and commercial potential for the industry.

13.6 CONCLUSIONS

Microalgae and cyanobacteria are a class of microorganisms that can use wastewater as a source of nutrients to generate sustainable biomass, considered as a raw material rich in biomolecules for the commercialization of high-value products in the market. This integration is an alternative for the integral management of water and an economic option by reducing nutrient costs for culture development, being at the same time an environmentally friendly choice to the phytoremediation of nutrients that remain in water discharges that can generate eutrophication. Constant efforts and new generation of knowledge will make the use of microalgae and cyanobacteria a potential way for the generation and commercialization of green products in the market.

REFERENCES

Abreu, A. P., Fernandes, B., Vicente, A. A., Teixeira, J., & Dragone, G. (2012). Mixotrophic cultivation of *Chlorella vulgaris* using industrial dairy waste as organic carbon source. *Bioresource Technology, 118*, 61–66. Doi: 10.1016/j.biortech.2012.05.055.

Abu-Ghosh, S., Dubinsky, Z., Verdelho, V., & Iluz, D. (2021). Unconventional high-value products from microalgae: A review. *Bioresource Technology, 329*(January). Doi: 10.1016/j.biortech.2021.124895.

Almomani, F., Judd, S., Bhosale, R. R., Shurair, M., Aljaml, K., & Khraisheh, M. (2019). Intergraded wastewater treatment and carbon bio-fixation from flue gases using *Spirulina platensis* and mixed algal culture. *Process Safety and Environmental Protection, 124,* 240–250. Doi: 10.1016/j.psep.2019.02.009.

Augustine, A., Kumaran, J., Puthumana, J., Sabu, S., Bright Singh, I. S., & Joseph, V. (2017). Multifactorial interactions and optimization in biomass harvesting of marine picoalga *Picochlorum maculatum* MACC3 with different flocculants. *Aquaculture, 474*(2016), 18–25. Doi: 10.1016/j.aquaculture.2017.03.020.

Bani, A., Fernandez, F. G. A., D'Imporzano, G., Parati, K., & Adani, F. (2021). Influence of photobioreactor set-up on the survival of microalgae inoculum. *Bioresource Technology, 320*(November 2020). Doi: 10.1016/j.biortech.2020.124408.

Bhatnagar, A., Bhatnagar, M., Chinnasamy, S., & Das, K. C. (2010). *Chlorella minutissima* - A promising fuel alga for cultivation in municipal wastewaters. *Applied Biochemistry and Biotechnology, 161*(1–8), 523–536. Doi: 10.1007/s12010-009-8771-0.

Brar, A., Kumar, M., Singh, R. P., Vivekanand, V., & Pareek, N. (2020). Phycoremediation coupled biomethane production employing sewage wastewater: Energy balance and feasibility analysis. *Bioresource Technology, 308*(April), 123292. Doi: 10.1016/j.biortech.2020.123292.

Chai, W. S., Tan, W. G., Halimatul Munawaroh, H. S., Gupta, V. K., Ho, S. H., & Show, P. L. (2021). Multifaceted roles of microalgae in the application of wastewater biotreatment: A review. *Environmental Pollution, 269,* 116236. Doi: 10.1016/j.envpol.2020.116236.

Chatsungnoen, T., & Chisti, Y. (2016). Harvesting microalgae by flocculation-sedimentation. *Algal Research, 13,* 271–283. Doi: 10.1016/j.algal.2015.12.009.

Cheng, Y. S., Zheng, Y., Labavitch, J. M., & Vandergheynst, J. S. (2011). The impact of cell wall carbohydrate composition on the chitosan flocculation of chlorella. *Process Biochemistry, 46*(10), 1927–1933. Doi: 10.1016/j.procbio.2011.06.021.

Choudhary, P., Prajapati, S. K., Kumar, P., Malik, A., & Pant, K. K. (2017). Development and performance evaluation of an algal biofilm reactor for treatment of multiple wastewaters and characterization of biomass for diverse applications. *Bioresource Technology, 224,* 276–284. Doi: 10.1016/j.biortech.2016.10.078.

Danouche, M., El Ghachtouli, N., & El Aroussi, H. (2021). Phycoremediation mechanisms of heavy metals using living green microalgae: Physicochemical and molecular approaches for enhancing selectivity and removal capacity. *Heliyon, 7*(7), e07609. Doi: 10.1016/j.heliyon.2021.e07609.

Dassey, A. J., & Theegala, C. S. (2013). Harvesting economics and strategies using centrifugation for cost effective separation of microalgae cells for biodiesel applications. *Bioresource Technology, 128,* 241–245. Doi: 10.1016/j.biortech.2012.10.061.

Dutta, D., Arya, S., & Kumar, S. (2021). Industrial wastewater treatment: Current trends, bottlenecks, and best practices. *Chemosphere 285,* 131245. Doi: 10.1016/j.chemosphere.2021.131245.

El-Sheekh, M. M., Farghl, A. A., Galal, H. R., & Bayoumi, H. S. (2016). Bioremediation of different types of polluted water using microalgae. *Rendiconti Lincei, 27*(2), 401–410. Doi: 10.1007/s12210-015-0495-1.

Harith, Z. T., Yusoff, F. M., Mohamed, M. S., Mohamed Din, M. S., & Ariff, A. B. (2009). Effect of different flocculants on the flocculation performance of microalgae, *Chaetoceros calcitrans,* cells. *African Journal of Biotechnology, 8*(21), 5971–5978. Doi: 10.5897/ajb09.569.

Herrera, A., D'Imporzano, G., Acién Fernandez, F. G., & Adani, F. (2021). Sustainable production of microalgae in raceways: Nutrients and water management as key factors influencing environmental impacts. *Journal of Cleaner Production, 287:* 125005. Doi: 10.1016/j.jclepro.2020.125005.

Knuckey, R. M., Brown, M. R., Robert, R., & Frampton, D. M. F. (2006). Production of microalgal concentrates by flocculation and their assessment as aquaculture feeds. *Aquacultural Engineering, 35*(3), 300–313. Doi: 10.1016/j.aquaeng.2006.04.001.

Lee, A. K., Lewis, D. M., & Ashman, P. J. (2013). Harvesting of marine microalgae by electro-flocculation: The energetics, plant design, and economics. *Applied Energy, 108*, 45–53. Doi: 10.1016/j.apenergy.2013.03.003.

Leng, L., Li, W., Chen, J., Leng, S., Chen, J., Wei, L., ... Huang, H. (2021). Co-culture of fungi-microalgae consortium for wastewater treatment: A review. *Bioresource Technology, 330*, 125008. Doi: 10.1016/j.biortech.2021.125008.

Li, Y., Shang, J., Zhang, C., Zhang, W., Niu, L., Wang, L., & Zhang, H. (2021). The role of freshwater eutrophication in greenhouse gas emissions: A review. *Science of the Total Environment, 768.* Doi: 10.1016/j.scitotenv.2020.144582.

Manhaeghe, D., Blomme, T., Van Hulle, S. W. H., & Rousseau, D. P. L. (2020). Experimental assessment and mathematical modelling of the growth of *Chlorella vulgaris* under photoautotrophic, heterotrophic and mixotrophic conditions. *Water Research, 184*, 116152. Doi: 10.1016/j.watres.2020.116152.

Mehariya, S., Goswami, R. K., Karthikeysan, O. P., & Verma, P. (2021). Microalgae for high-value products: A way towards green nutraceutical and pharmaceutical compounds. *Chemosphere, 280*, 130553. Doi: 10.1016/j.chemosphere.2021.130553.

Moshood, T. D., Nawanir, G., & Mahmud, F. (2021). Microalgae biofuels production: A systematic review on socioeconomic prospects of microalgae biofuels and policy implications. *Environmental Challenges, 5*(July), 100207. Doi: 10.1016/j.envc.2021.100207.

Piligaev, A. V., Sorokina, K. N., Bryanskaya, A. V., Peltek, S. E., Kolchanov, N. A., & Parmon, V. N. (2015). Isolation of prospective microalgal strains with high saturated fatty acid content for biofuel production. *Algal Research, 12*, 368–376.

Şirin, S., Trobajo, R., Ibanez, C., & Salvadó, J. (2012). Harvesting the microalgae *Phaeodactylum tricornutum* with polyaluminum chloride, aluminium sulphate, chitosan and alkalinity-induced flocculation. *Journal of Applied Phycology, 24*(5), 1067–1080. Doi: 10.1007/s10811-011-9736-6.

Sukenik, A., & Shelef, G. (1984). Algal autoflocculation—verification and proposed mechanism. *Biotechnology and Bioengineering, 26*(2), 142–147. Doi: 10.1002/bit.260260206.

Swarnalakshmi, K., Prasanna, R., Kumar, A., Pattnaik, S., Chakravarty, K., Shivay, Y. S., ... Saxena, A. K. (2013). Evaluating the influence of novel cyanobacterial biofilmed biofertilizers on soil fertility and plant nutrition in wheat. *European Journal of Soil Biology, 55*, 107–116. Doi: 10.1016/j.ejsobi.2012.12.008.

Vandamme, D., Foubert, I., Fraeye, I., Meesschaert, B., & Muylaert, K. (2012). Flocculation of *Chlorella vulgaris* induced by high pH: Role of magnesium and calcium and practical implications. *Bioresource Technology, 105*, 114–119. Doi: 10.1016/j.biortech.2011.11.105.

Vergini, S., Aravantinou, A. F., & Manariotis, I. D. (2016). Harvesting of freshwater and marine microalgae by common flocculants and magnetic microparticles. *Journal of Applied Phycology, 28*(2), 1041–1049. Doi: 10.1007/s10811-015-0662-x.

Xu, Y., Purton, S., & Baganz, F. (2013). Chitosan flocculation to aid the harvesting of the microalga *Chlorella sorokiniana. Bioresource Technology, 129*, 296–301. Doi: 10.1016/j.biortech.2012.11.068.

Zhou, W., Min, M., Hu, B., Ma, X., Liu, Y., Wang, Q., ... Ruan, R. (2013). Filamentous fungi assisted bio-flocculation: A novel alternative technique for harvesting heterotrophic and autotrophic microalgal cells. *Separation and Purification Technology, 107*, 158–165. Doi: 10.1016/j.seppur.2013.01.030.

14 Bioprospecting of Microbial Diversity for Sustainable Agriculture and Environment

Hiren K. Patel, Nensi K. Thumar,
Priyank D. Patel, and Azaruddin V. Gohil
P P Savani University

CONTENTS

14.1 INTRODUCTION

Soil is the uppermost thin and fine layer on earth also known as pedosphere having various different biotic and abiotic components. Biotic components include various organisms of macro-, meso-, microfauna, and various floras. All the microbes that cohabitate and interact with plants collectively are known as plant microbiota

DOI: 10.1201/9781003247883-14

(Bulgarelli et al. 2013). Abiotic components include water, gases, different minerals, and organic compounds. Soil formation takes place through the interaction of various earths' spheres – atmosphere; hydrosphere, a watery part; lithosphere, a rocky portion, and biosphere, which is a combination of all spheres (Nortcliff et al. 2000). Soil has various sized particles, and based on their size, they are divided in to sand, silt, and clay. Particle sizes of sand, silt, and clay include 63–2000, 2–63, and less than 2 µm, respectively. Fine-sized pores, which are having diameter less than 0.2 µm, can hold water molecules so strongly that they are not freely available for plants to absorb. While medium-sized pores of soil having a diameter ranging between 0.2 and 50 µm can hold water with such efficiency that it is available to roots of plants and other microbes (Nortcliff et al. 2000). Various factors such as soil texture, pH, temperature, moisture content, nutritional availability, and decomposition will determine the microbial community structure of particular soil. Microbes present in soil greatly affect the fertility of the soil and determine which type of plant or crop yield can be obtained from soil. Soil has various functions; for example, it provides protection, habitat, various regulatory functions, waste removal system, anchoring to plants, and raw materials. Soil also provides food for all living organisms. To fulfil the need for food for increasing population day by day, humans have developed food production in agriculture system through various methods and techniques. Modern agricultural system includes applications of chemical fertilizers and pesticides.

Excess applications of chemical fertilizers and pesticides are the main causes of soil degradation. Although pesticides kill pests in agricultural sector, they have adverse effects on soil quality and also on its normal microbial flora. Pesticides used in agricultural sector can either promote or retard growth of microbes. Sometimes, pesticides can decrease biomass of plant growth-enhancing microbes, due to non-tolerance capacity of pesticides by microbes (Johnsen et al. 2001). The applied target-specific pesticides may result in an imbalance in a number of other microbial groups; that is, fungicides hinder the growth of fungi, resulting in an increased number of bacteria (Chen et al. 2001). Fungicidal pesticides (mefenoxam and metalaxyl) were reported to inhibit nitrogen-fixing bacteria (Monkiedje and Spiteller 2002). Pesticides can reduce total biological nitrogen fixation in soil by adversely affecting the nodulation efficiency of microbes, or they directly affect growth of plant negatively. DDT, 2,4,5-T, and 2,4-D were reported to reduce nitrogen fixation through inhibition of signalling mechanism between *Rhizobium* and plant and through inhibition of nod expression (Fox et al. 2001; McLachlan 2001).

Chemical fertilizers change soil pH, due to which the microbial community of soil gets affected. Acidification of soil also results in decreased solubilization of phosphate and less availability to plants (Kumar and Prakash 2019). Chemical fertilizers having heavy metals sometimes have irreversible toxic effects on soil, which is further transferred and magnified at the level of the food chain through absorption of these heavy metals by crops and plants (Sonmez et al. 2007). According to Worldometer, the total population till March 2020 was 7.8 billion. In ancient times, people used to produce their own food, but due to modernization and non-availability of land for agriculture, nowadays, self-farming is decreased. So, due to the daily food requirement, agricultural land reclamation and soil quality improvement are an urgent need for maximum crop yield in a different environment.

Microbes interact with plants by colonizing into different areas or regions. Generally, microbes colonize into regions where nutrition, protection, and other suitable optimum environmental conditions are available. Sometimes, colonization also depends on host cell or host tissue specificity. If microbes colonize into internal tissues of the plant, then they are known as endophytic microbes. When the colonization area is root or nearby root, they are known as rhizospheric microbes. When microbes interact with plants through colonizing on surface portion, they are known as epiphytic microbes. Microbes interacting with plants have the ability to produce phytohormones required for plant growth, development, fruiting, and other reactions necessary to complete the life cycle (Yadav 2017).

The interaction between microbes and agricultural crops can be beneficial, antagonistic, or neutral. The symbiotic relation between plants and soil microbe provides macro- and microelements to plants, protection from pathogens, and other nutrition. Microbes can colonize in different ways with the plant (Kumar and Chandra 2020). Microbes having the capacity of plant growth induction by accumulation and solubilization of various elements provide protection from pathogens that damage agricultural crops. So, to nullify and reduce the adverse effects of chemical pesticides and synthetic fertilizers, one can replace them with biofertilizers and biocontrol agents (Glick 2012). This chapter deals with the exploration of microbial diversity as the microbial population in soil considerably affects agricultural crop growth, development, and protection and also controls various balancing systems of water and nutritional elements. Microbes present in soil recirculate and reinforce nutrition through various cycles of nitrogen, phosphorous, sulphur, and carbon. Microbes have various effects via association with plant root, which can be either beneficial or harmful to crops. These associations may be symbiotic, parasitic, or mutualistic. Plants also have an impact on the growth of microbes as plants and microbes can co-evolve *via* association. By screening effects of microbes on agricultural crops, one can utilize this knowledge to gain more yield of crop, can prevent crop damage, and can utilize microbes as biofertilizers to prevent environmental pollution.

14.2 MICROBIAL INTERACTIONS AND THEIR POTENTIAL ROLES IN PLANT GROWTH

As microbes are omnipotent, they are present everywhere – in the water, soil, atmosphere, and also the host body. Roots of plants extend in various directions and to different depths. The area that surrounds the root is known as the rhizosphere. Microbes associate and colonize with plant, and plant root may have different roles in plant growth through various conducting mechanisms such as nutritional element provision, pollutants and xenobiotics removal, remediation and degradation, phytohormones production, confrontation, and protection from phytopathogens and stress conditions. Microbes facilitate furnishing of nutrients to plant through mineralizing organic compounds, solubilization of phosphate, fixation of nitrogen, siderophores production for ion availability, indole-3-acetic acid (IAA) production, which is a source of auxin, and production of 1-amino-cyclopropane-1-carboxylase (ACC) deaminase enzyme. Growth enhancement by making nutrients available for plants

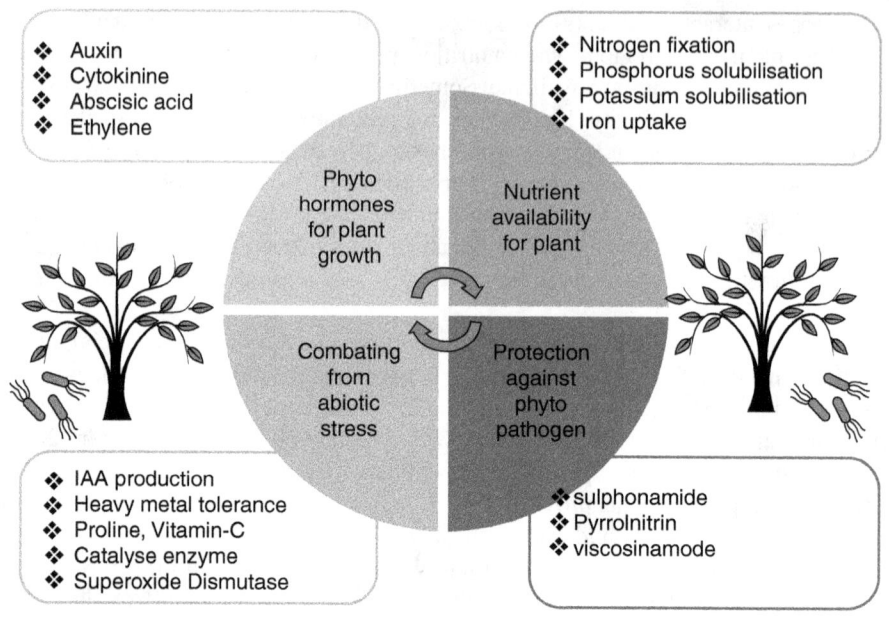

FIGURE 14.1 Role of microbial interaction for plant growth.

is the direct effect of microbial interaction, while stress management and protection against phytopathogens are indirect effects of microbes interacting with agricultural crops (Gouda et al. 2018) (Figure 14.1). As microbes colonizing the rhizosphere work through different mechanisms and help in plant growth through various ways, these microbes are known as plant growth-promoting rhizobacteria (PGPR) (Figure 14.2).

14.2.1 Nutrients Availability for Plant

Rhizobacteria may be symbiotic or non-symbiotic. PGPR may be intracellular, mainly inhabiting plants through nodule formation, and extracellular, inhabiting the rhizosphere or the inner space of cortex and cell (Martinez-Viveros et al. 2010). Intracellular or endogenous PGPR commonly include *Allorhizobium*, *Bradyrhizobium*, *Frankia*, *Mesorhizobium*, and *Rhizobium*. Extracellular microbial PGPR include the genera of *Agrobacterium*, *Arthrobacter*, *Bacillus*, *Burkholderia*, *Caulobacter*, *Flavobacterium*, *Micrococcus*, *Pseudomonas*, and *Serratia* (Bhattacharyya and Jha 2012).

14.2.1.1 Nitrogen Fixation

Biological nitrogen is an essential component for agricultural crops and is required for the biosynthesis of proteins, enzymes, growth factors, nucleic acids, hormones, cell components, and different pigments. Nitrogen is also a compulsory component of fertilizers. From the environment, nitrogen is lost due to the leaching mechanism. Atmospheric nitrogen (N_2) is made available to plants in reduced form (NH_3) by

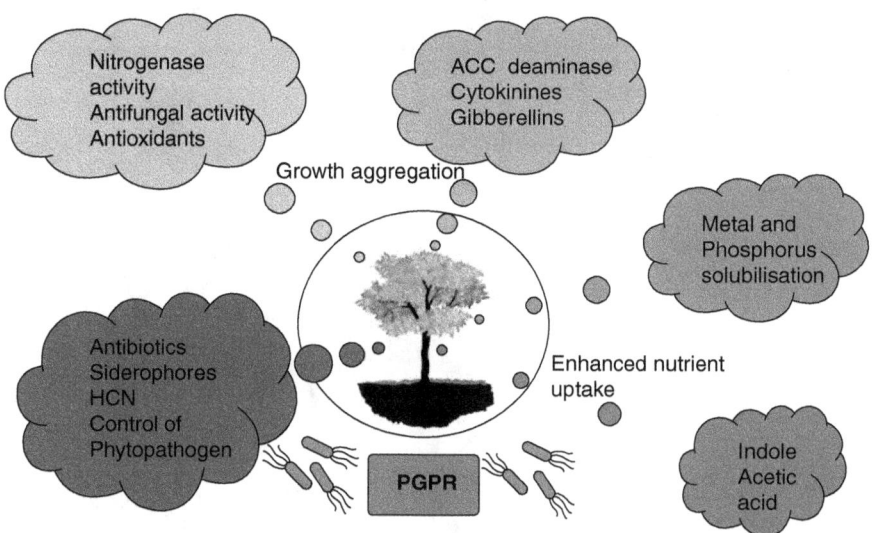

Nitrogenase
activity
Antifungal activity
Antioxidants

ACC deaminase
Cytokinines
Gibberellins

Growth aggregation

Metal and
Phosphorus
solubilisation

Antibiotics
Siderophores
HCN
Control of
Phytopathogen

Enhanced nutrient
uptake

PGPR

Indole
Acetic
acid

FIGURE 14.2   Mechanism of plant growth promotion by plant growth-promoting rhizobacteria.

microbial actions, as plants cannot utilize molecular nitrogen directly. Microbes carrying out nitrogen fixation are known as diazotrophs. Nitrogen fixation is a part of the nitrogen cycle that involves the steps of fixation, ammonification, nitrification, denitrification, and assimilation. Nodule formation surrounding root is characteristic of symbiotic microbes for nitrogen fixation in legumes (Zahran 1999) (Figure 14.3). Microbes fix atmospheric nitrogen into soil through enzyme nitrogenase, which requires molybdenum and iron as cofactors. Nitrogenases are very sensitive towards the presence of oxygen. Due to this reason, nitrogen fixation is mostly carried out in anaerobic environmental conditions. Leghemoglobin is a protein that helps in the maintenance of anaerobic conditions (Robson and Postgate 1980). Nitrogenase enzyme complex is encoded by a complex of genes known as *nif* (Gaby and Buckley 2011). Overall, the microbial nitrogen fixation can be represented as follows (Postgate 1998).

$$N_2 + 16ATP + 8e^- + 8H^+ 2NH_3 + H_2 + 16ADP + 16P_i$$

Generally, at the phyla scale, Archaea, Cyanobacteria, Firmicutes, and Proteobacteria are associated with nitrogen fixation in soil. *Azospirillum*, which is microaerophilic, and *Herbaspirillum* are found to be specific to rhizosphere and help in nitrogen fixation (Rashid et al 2016). Symbiotic nitrogen-fixing rhizobacteria include *Azoarcus* sp., *Beijerinckia* sp., *Klebsiella pneumoniae*, and *Rhizobium* sp. (Ahemad and Kibret 2014). Other bacterial species involved in nitrogen fixation are *Azotobacter chroococcum*, *Bacillus megaterium*, *B. mucilaginosus*, *Bradyrhizobium japonicum* UCM B-6018, *Burkholderia* sp., *Pantoea agglomerans*, *Pseudomonas aeruginosa* BS8, *P. alcaligenes* PsA15, *P. fluorescens* C7, and *Rhizobium leguminosarum* (Rashid et al 2016). *Rhizobium leguminosarum* is a symbiotic mutualistic bacterium that fixes

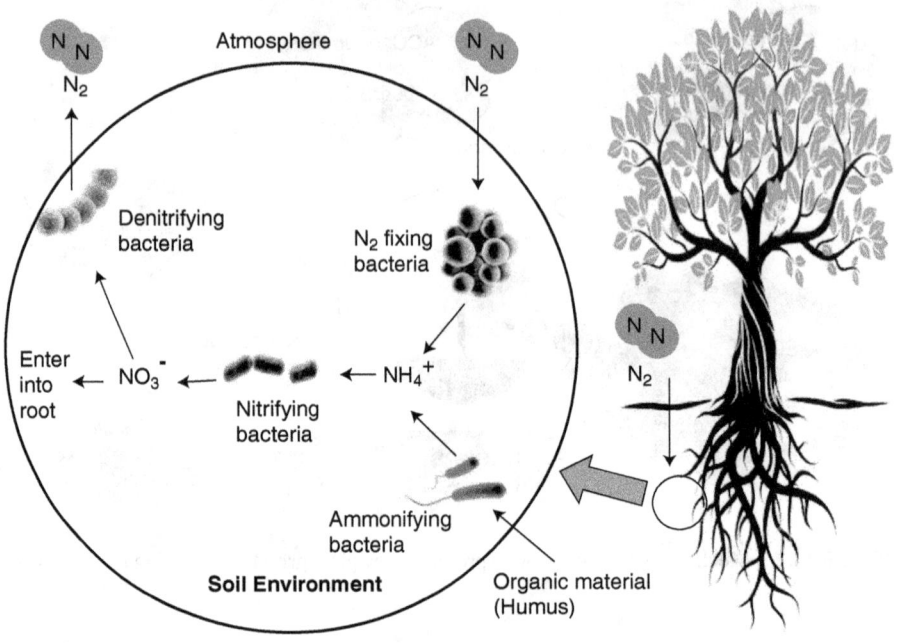

FIGURE 14.3   Mechanism of nitrogen fixation by bacteria.

nitrogen with nodule formation. Strains of *Rhizobium leguminosarum* have many plasmids that are host specific, and only one plasmid from each strain is involved in symbiosis through expression of nod, sym, and fix genes. Plasmid transfer can improve symbiotic nitrogen fixation (Kalloo 1993). *Azotobacter chroococcum* produce melanin during metabolism, which helps in the protection of nitrogenase enzyme from oxygen (Shivprasad and Page 1989). The fungi that are reported to be indirectly involved in nitrogen fixation are *Claroideoglomus claroideum, C. etunicatum, Glomus mosseae, G. viscosum,* and *Rhizophagus intraradices*. *Glomus viscosum* is arbuscular mycorrhizal symbiotic fungi that take part in the improvement of nutrient uptake and provide protection against drought stress to crops (Bidartondo et al. 2002). Fungi in the soil directly do not fix environmental nitrogen, but they provide nutrients to bacteria that are involved in nitrogen fixation. Mycelium of fungi also provides protection of bacterial enzymes against oxygen for successful nitrogen fixation. In symbiotic association of fungi and bacteria, fungi provide carbon and phosphorous to bacteria for growth. Arbuscular fungi also enhance the growth of non-symbiotic nitrogen fixer bacteria by supplementation of carbon (Jones and Oburger 2011). Both symbiotic and non-symbiotic microbes can be used for nitrogen replenishment in nutrient-deprived agricultural soil for plant growth.

14.2.1.2   Phosphorus Solubilization and Mobilization

Phosphorus is very essential for most of the reactions of cell, e.g. nucleic acid synthesis, enzyme activation or repression, energy source, respiration, signal transduction

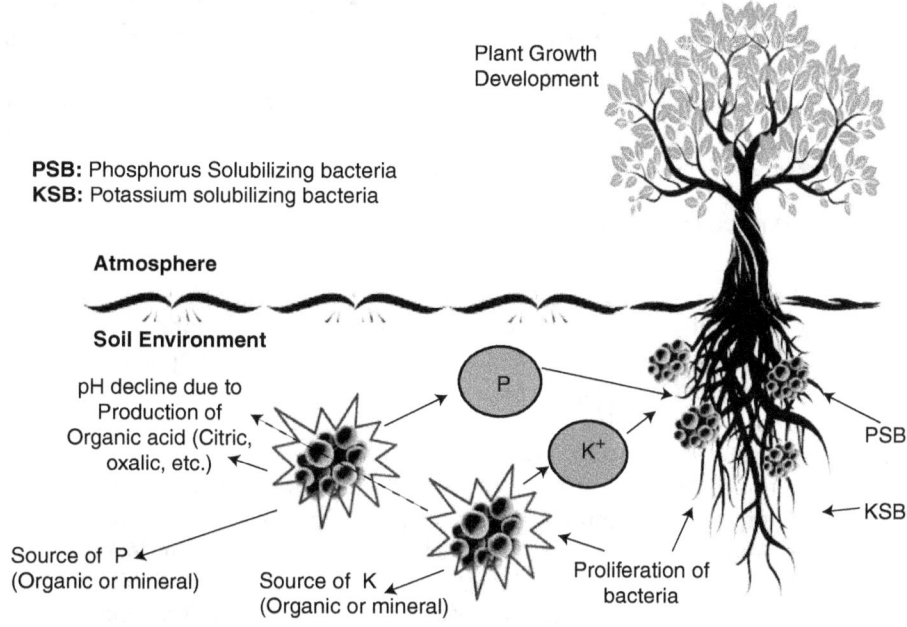

FIGURE 14.4   Mechanism of phosphorus and potassium solubilization.

process, and photosynthesis, and has an important role in other metabolic process regulations (Anand et al. 2016). There are 95%–99% of unavailable phosphorous for agricultural crops due to its precipitated and insoluble form in nature. Phosphorous is highly reactive with calcium and aluminium, which form precipitates. Monobasic phosphate and dibasic phosphate are available and absorbable forms for plants. PGPR solubilize and mobilize the unavailable precipitated phosphate through the production and release of organic acids having low molecular weight and also by enzymatic reactions (Sharma et al. 2003; Owen et al. 2015). There are various mechanisms by which microbes make phosphorous available to plants (Figure 14.4). The direct process of ligand exchange involves low pH for production by organic anions, which are exchanged for phosphorous and production of phytase or phosphatase enzymes for the hydrolysis of the source of organic phosphates. Indirect mechanisms involve the production of carbonic acids by the release of carbon dioxide during the respiration process, which lowers the pH, enhances proton release, and regulates and maintains phosphate equilibrium through assimilation and release of it (Rashid et al. 2016). The genera that are reported to solubilize phosphate with PGPR activity are *Arthrobacter*; free-living *Beijerinckia*; facultative anaerobe *Enterobacter* and *Serratia*; opportunistic pathogens *Pseudomonas* and *Flavobacterium*; nodule-forming bacteria *Bacillus*, *Enterobacter,* and *Rhizobium*; and rod-shaped *Erwinia* (Otieno et al. 2015). Some of the other reported phosphate solubilizer and mobilizer PGPR are *Achromobacter* sp., *Azospirillum brasilense, Azospirillum* sp., *Bacillus megaterium, Bacillus mucilaginosus, Micrococcus* sp., *Pseudomonas aeruginosa* BS8, *Pseudomonas alcaligenes,*

Pseudomonas natatu, Rhizobium leguminosarum, and *Streptomyces* sp. (Rashid et al. 2016). According to Gouda et al. (2018), the other microbes that help in phosphorous bioavailability through solubilization are *Azotobacter chroococcum, Bradyrhizobium japonicum, Bacillus circulans,* and *Enterobacter agglomerans.* The reported phosphate-solubilizing archaea that are halophilic are *Haloferax volcanii, Haloarcula marismortui,* and *H. vallismortis* (Yadav et al. 2015a).

14.2.1.3 Potassium Solubilization

Besides nitrogen and phosphorous, potassium is the third vital macroelement required for agricultural plants and other crops. In soil, the potassium range is between 0.04% and 3.0%. Out of this, 90%–98% is unavailable to plants due to its insoluble form in rock as phyllosilicate (Shelobolina et al. 2014). Potassium is an important cofactor of metabolic and regulatory processes in plants, such as glycolysis, stomatal opening, and closing. The deficiency of potassium shows some abrasions on leave and poor development of seeds and roots. The use of chemical fertilizers can replenish potassium only for plants, but the use of PGPR can fulfil potassium to soil also. Similar to phosphorous solubilization, the availability of potassium to plant is possible with the production of organic acids from PGPR microbes (Figure 14.4). Among potassium-solubilizing bacteria (KSB), rhizospheric microbes can solubilize potassium more efficiently than non-rhizospheric one. In the presence or absence of oxygen, rhizospheric PGPR microbes can solubilize phosphate. Some of the reported KSB are *Acidithiobacillus* sp., *Arthrobacter* sp., *Bacillus edaphicus, B. mucilaginosus, Burkholderia* sp., *Enterobacter hormaechei, Ferroxidans* sp., *Paenibacillus glucanolyticus, P. mucilaginosus,* and *Pseudomonas* sp. (Meena et al. 2016; Liu et al. 2012). The moulds *Fusarium, Aspergillus,* and *Penicillium* can solubilize immobilized potassium from rocks. *Aspergillus niger* is the most efficient organic acid producer fungi, and this organic acid production is dependent on pH. Inoculation of two arbuscular mycorrhiza *Glomus intraradices* and *G. mosseae* enhance potassium uptake by releasing protons (Verma et al. 2017).

Acidophilic extremophiles such as *Bacillus aerophilus, B. atrophaeus, Lysinibacillus fusiformis,* and *Planomicrobium* sp. can solubilize potassium at pH 3 (Yadav et al. 2015a). At alkaline pH conditions, some alkalophilic solubilizers were also isolated from the soil of wheat-, sugarcane-, and tobacco-cultivated land and they were *Agrobacterium, Duganella, Exiguobacterium, Lysinibacillus, Psychrobacter, Stenotrophomonas,* and *Variovorax* (Verma et al. 2017). Saline condition in soil exerts two times more osmotic pressure on agricultural crops. Psychrophiles, which could be used as biofertilizers having the ability to provide potassium to agricultural crops, are *Aspergillus awamori, Achromobacter piechaudii, Bacillus horikoshii, Phanerochaete chrysosporium,* and *Trichoderma viride* (Verma et al. 2017). Water stress-tolerant, xerophilic, potassium-solubilizing extremophiles *Duganella violaceusniger, Planococcus salinarum, Pseudomonas thivervalensis, Psychrobacter fozii,* and *Sporosarcina* sp., which belong to phylum Gammaproteobacteria, were also reported (Verma et al. 2017). Some thermophiles having the ability of PGPR for potassium solubility were also detected, such as *Bacillus altitudinis,* which belongs to phylum Firmicutes, and *Methylobacterium mesophilicum, Salmonella bongori,* and *Delfia acidovorans,* which belongs to phylum Betaproteobacteria. Inoculation of potassium-solubilizing bacteria and fungi showed growth enhancement in different agricultural crops of tomato, maize, banana, potato, cotton, peanut, and chilli (Verma et al. 2017).

14.2.1.4 Iron Uptake

Iron is also one of the necessary nutrient elements for plants, which conducts many cellular processes such as electron transport, photosynthesis, and nitrogen fixation. Iron deficiency in plants causes chlorosis and makes them pathogen sensitive. Economic loss arises due to phytopathogenic disease in plants, in the state of deficiency of iron. Anaemia in humans also occurs due to the consumption of food that is iron deficient. The growth phase of plant requires 5–10 kg/ha of iron generally (Sayyed et al. 2013). In soil, iron remains in oxidized form, which is insoluble for uptake. To solubilize iron, microbes in soil produce and secrete siderophores, which have the ability to chelate iron and are of low molecular weight compounds (Rashid et al. 2016). Iron-regulated outer membrane proteins are present on the cell surface of bacteria and able to produce and secrete siderophores such as FpvA, which has receptor ferripyoverdine; PupA, which has receptor protein Pseudobactin 38; and FhuE, which has receptor ferrioxamine E. Hydroxamate, peptide, catecholates, mycobactin, and citrate-hydroxamate are classes of bacterial siderophores, while coprogens, ferrichromes, rhizoferrin, rhodotoluric acid, and fusarinines are fungal siderophores (Sayyed et al. 2013). Agrobactin is produced by *Agrobacterium tumefaciens*, ferribactin and pyroverdin are produced by *Pseudomonas fluorescens*, and alcaligen E is produced by *Alcaligenes eutrophus*. Some fungi were also reported to produce siderophores. *Aspergillus ochraceus, Curvularia lunata, Penicillium parvum, Rhizopus microsporus*, and *Rhodotorula piliminae* produce fungal siderophores ferrichrome; asperchromes A, B, and C; rhizoferrin; neocoprogens I and II; and rhodotoluric acid, respectively (Sayyed et al. 2013). The other reported siderophore-producing PGPR are *Pseudomonas aeruginosa* BS8, *P. strain* GRP3, *P. fluorescens* C7, and *Rhizobium leguminosarum* (Rashid et al. 2016). Ferreira et al. (2019) carried out a comparative study of siderophore production using five different bacteria at alkaline pH. Excepting for *Azotobacter vinelandii*, the other four bacteria *Bacillus subtilis*, *Bacillus megaterium, Pantoea allii*, and *Rhizobium radiobacter* were able to produce siderophore between 24 and 48 hours, almost at stationary phase of growth. Catechol-type and hydroxamate-type iron chelators were produced by *Bacillus subtilis* and *Bacillus megaterium*, and *Rhizobium radiobacter* and *Pantoea allii*, respectively, at pH 9. Both types of siderophores were produced by *Azotobacter vinelandii*. From the five isolates, *Bacillus megaterium* was with the highest ability for iron chelator siderophores production.

14.2.2 Phytohormones for Plant Growth

Phytohormones are organic compounds produced in one region of the plant and affect another region by passing through transportation. Phytohormones mainly regulate physiological processes that are definite and specific, such as growth, flowering, and development. A specific amount of phytohormones is an a priori necessity of plants for their growth and completion of life cycle too. Rhizospheric microbes and other agricultural crops-associated microbes have a dual role in the enhancement of phytohormone production and inhibition of phytohormone inhibitory hormones. Phytohormones synthesis also gives plants the ability to combat biotic stress and oxidative stress and the ability to increase nutrition absorbance (Kudoyarova et al. 2019).

Auxin mainly affects root development in plants. It also promotes elongation, differentiation, and division process of cell. Tryptophan is a precursor compound of auxin (IAA). Stressed conditions such as the presence of heavy metals and salt negatively affect the auxin-synthesizing capacity as this enhances IAA oxidase enzyme of agricultural crops. Biotic and abiotic stress tolerant and resistant microbes can synthesize auxin and provide growth enhancement conditions for plants (Egamberdieva et al. 2017). From tryptophan precursor, auxin or IAA is produced by different intermediary metabolite pathways. Intermediate indole-3-acetamide-synthesizing pathways were reported in *Agrobacterium*, *Rhizobium* and *Pseudomonas* (Theunis et al. 2004). The pathway involving indole-3-pyruvate intermediate was detected in PGPR of *Rhizobium* and *Bradyrhizobium*. In *Bacillus* sp., the intermediate metabolite tryptamine gets directly converted to IAA through the activity of enzyme amine oxidase (Maheshwari et al. 2015). The concentration of auxin also affects root growth in inhibitory way. Applications of auxin-producing Rhizobacteria decreased root elongation in *Beta vulgaris*. Opposite to this, when seedlings of canola were treated with auxin deficient *Pseudomonas putida* GR12-2 mutant, the elongation of root was successfully enhanced. Another finding suggests that dicotyledonous crops are more sensitive towards auxin than monocotyledonous crops. This auxin concentration tolerance difference may be due to the presence of endogenous auxin level (Kudoyarova et al. 2019). One study also indicates that increased auxin level also enhances the level of growth-retardant phytohormone ethylene. The increased level of ethylene can be controlled or decreased through ACC (1-aminocyclopropane-1-carboxylic) deaminase enzyme (Chen et al. 2013). According to Li et al. (2000), *Enterobacter cloacae* UW4 can produce both auxin and ACC and enhance the elongation of root in canola, while a mutant of ACC deaminase could not exert root elongation effect, which indicates that PGPR microbes are required for root growth. Different genes *aldA*, *hisC1*, *ipdC*, *iaaM*, *nit*, *y4wE*, *oxdRG*, *nthAB*, *yhcX*, and Nha1 were reported from various microbes. *Azospirillum brasilense* Yu62, *Bacillus amyloliquefaciens* FZB42, *Corynebacterium* sp. C5, *E. cloacae* FERM BP-1529, *Ralstonia solanacearum*, and *Rhodococcus globerulus* A-4 were reported to synthesize auxin to promote root development (Maheshwari et al. 2015).

Cytokinin-synthesizing ability of PGPR was also studied. Cytokinin is required for germination of seed, expansion of leaves, and delay in senescence. It also enhances cell division. Isopentenyladenosine monophosphate is the precursor of cytokinin, which is derived from dimethylallyl phosphate and adenosine monophosphate. The active form of cytokinin is zeatin (Maheshwari et al. 2015). From *Coleus*, two *Pseudomonas*, i.e. *P. stutzeri* and *P. putida*, and one *Stenotrophomonas*, i.e. *Stenotrophomonas maltophilia*, were isolated, which have the potential to synthesize cytokinin (Patel and Saraf 2017). Root cells can absorb more cytokinin in their free form bases, and they can be more accumulated in shoot than root. This shows more rapid export is possible with root-derived ribosides of cytokinin in shoot (Kudoyarova et al. 2019). *Pseudomonas* sp., *Agrobacterium* sp., and the genera such as *Escherichia*, *Klebsiella*, *Proteus*, and *Xanthomonas* were reported to synthesize cytokinin (Maheshwari et al. 2015).

Abscisic acid is a natural phytohormone that affects the dormancy of seed, adaptation towards stress, and abscission of leaf. It contains sesquiterpenoids and is synthesized *via* a direct or indirect pathway. In the direct pathway, isopentenyl pyrophosphate was synthesized through mevalonate pathway and it is mostly operated in plant disease-causing fungi. In indirect pathway, isopentenyl pyrophosphate is synthesized from precursor compound methylerythritol. *Azospirillum brasilense* and *Bradyrhizobium japonicum* were reported for the production of abscisic acid in soil (Maheshwari et al. 2015). Drought resistance was adapted in *Saccharum officinarum* with inoculation of *Gluconacetobacter diazotrophicus* through activation of abscisic acid-dependent signalling genes. The study also indicates that in *in vitro* conditions, *Bacillus pumilus* can synthesize five times higher abscisic acid than *Pseudomonas* sp. Stomatal closing process is enhanced by abscisic acid to prevent water loss by transpiration. Thus, in drought conditions, abscisic acid prevents extra water loss from agricultural crops. Dried soil conditions can induce abscisic acid synthesis. By metabolizing abscisic acid, bacteria can decrease its concentration. As abscisic acid and ethylene have antagonistic relation, decreased concentration can lead to induction of ethylene, which diminishes microbial ethylene production need (Kudoyarova et al. 2019). *Variovorax paradoxus*, which is Proteobacteria 5C-2, can *in vitro* decrease abscisic acid 40%–60% in pea plant (Jiang et al. 2012).

Ethylene is the gaseous form phytohormone required for the ripening of fruit. Besides this, it is also responsible for gravitropism of root and permeability of membrane. S-adenosylmethionine is the precursor of ethylene. *Pseudomonas syringae* was reported for ethylene bacterial production (Maheshwari et al. 2015). *Azospirillum* sp. can produce ethylene with concentration of 0.17 μmol/g of dry weight in artificial media with the addition of L-methionine (Cassan et al. 2014). Another phytohormone gibberellic acid is necessary for elongation of stem, seed dormancy inhibition, and flowering at an early period of plants. Gibberellic acids may be free, conjugated, or bound in the state. Mevalonic acid is the precursor component of gibberellic acid. *Acetobacter diazotrophicus*, *Bacillus* sp., *Herbaspirillum seropedicae*, and *Rhizobium meliloti* were reported for gibberellic acid production. During saline stressed conditions, *Pseudomonas putida* could enhance the growth of soybean. Endophytic *Bacillus sphingomonas* can increase ethylene production in tomatoes for growth (Maheshwari et al. 2015).

Patel and Sharaf (2017) studied the isolation of plant growth-promoting phytohormone-producing rhizobacteria. From 70 different strains, *P. putida* MTP 50, *P. stutzeri* MTP 40, and *Stenotrophomonas maltophilia* MTP 42 possessed phytohormones production ability. From these, three isolates of *S. maltophilia* had maximum auxin production potential with a concentration of 240 μg/mL. *P. stutzeri* had an efficiency of maximum production of gibberellic acid and cytokinin among all the three isolates with a concentration of 34 and 13 μg/mL, respectively.

14.2.3 COMBATING ABIOTIC STRESS

Stress is a state of condition that negatively affects living things surrounding it. Some organisms are resistance or have tolerance capacity towards stress, while some are sensitive. In the association between plant and microbes, microbes have a

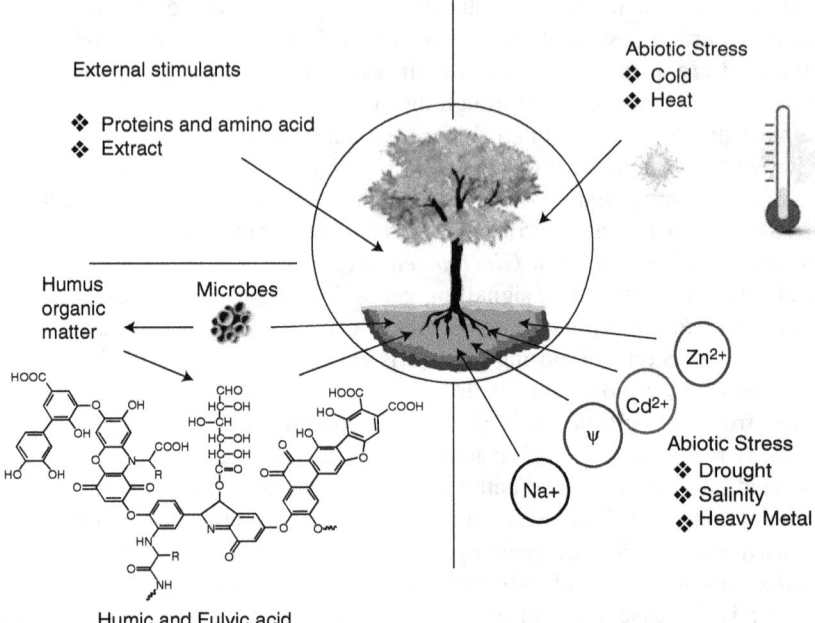

External stimulants

❖ Proteins and amino acid
❖ Extract

Abiotic Stress
❖ Cold
❖ Heat

Humus organic matter

Microbes

Zn²⁺

Cd²⁺

ψ

Na+

Abiotic Stress
❖ Drought
❖ Salinity
❖ Heavy Metal

Humic and Fulvic acid

FIGURE 14.5　Plant-microbes interactions and their role in the mitigation of abiotic stress.

great adaptation capacity towards stresses and changing environment. This versatile ability of microbes can be explored for plant growth even in harsh conditions. The production of reactive oxygen species (ROS) is increased during stress conditions, a reason to harm the lipid and protein of organisms (Ramegowda and Senthil-Kumar 2015). Microbes associated with plants helps to mitigate various stressed conditions such as water stress, temperature variation, heavy metal accumulation above the acceptable level, pesticides, and other xenobiotics. The ability to cope up with and tolerance capacity towards stress can be quantified using the detection of various compounds which concentration get varies highly like proline protein, catalyse enzyme, vitamin C, ascorbic acid, and superoxide dismutase (Figure 14.5). The concentration of these compounds is higher in stressed plants as it is a multigenic trait (Agami et al. 2016).

Bacteria having PGPR activity also show resistance towards heavy metals such as Cd, Zn, Cu, and Pb. These bacteria are *Agrobacterium tumefaciens, Bradyrhizobium, Enterobacter cloacae, Mesorhizobium, Rhizobium leguminosarum, R. sullae, Rhizobium* sp. CCNWSX0481, *Pseudomonas* sp., *Pseudomonas fluorescens*, and *Sinorhizobium*. They exhibit PGPR activity and can tolerate 4.1 mM of cadmium, *Enterobacter cloacae* can tolerate 4.05 mM of Pb, *Mesorhizobium* can tolerate 2.2 mM of copper, and *Bradyrhizobium* can tolerate 5.1 mM of zinc (Jebara et al. 2019). The auxin production for root cell elongation and division was also reported in *Brassica juncea, Zea mays*, and *Vinca rosea* through PGPR *Achromobacter xylosoxidans, Leifsonia* sp., and *Bacillus megaterium* for copper, cadmium, and nickel stressed conditions, respectively (Egamberdieva et al. 2017). In case of *Medicago sativa* and *Pseudomonas fluorescens* help to sequester cadmium and also degrade trichloroethylene (Ramadan et al. 2016).

Various species of genus *Arthrobacter* have the ability to reduce 4-chlorophenol from soil (Westerberg et al. 2000). *Acinetobacter faecalis*, *Arthrobacter* sp., *Bacillus licheniformis*, *Ensifer garamanticus*, *Marinobacterium* sp., *Rhizobium* sp., *Sinorhizobium* sp., *Serratia plymuthica*, *Streptomyces geysiriensis*, and *Pseudomonas extremorientalis* were reported for auxin synthesis even in abiotic salt stressed conditions for root development in *Sulla carnosa*, *Triticum aestivum*, *Cucumis sativus*, *Zea mays*, *Glycine max*, and *Trifolium repens*. *Azospirillum* sp. and *Arthrobacter* sp. are able to produce shoot-developing cytokinin at a high concentration of salt for *Glycine* max. Fungi *Trichoderma asperellum* can efficiently synthesize auxin, gibberellic acid, and abscisic acid in saline soil for *Cucumis sativus* (Egamberdieva et al. 2017).

Pseudomonas putida along with *B. megaterium* was reported positive for IAA production in water stressed conditions in *T. repens* crop. Abiotic conditions of drought can decrease cytokinin anabolism and retard growth of leaves. *Micrococcus luteus* is also capable of cytokinin synthesis in *Zea mays* in water stressed soil (Egamberdieva et al. 2017). *Bacillus aryabhattai* can synthesize phytohormones gibberellic acid, auxin, and abscisic acid at high temperature for the development of *Glycine max* (Park et al. 2017).

14.2.4 Protection against Phytopathogen

Besides symbiotic and mutualistic behaviour, microbes show antagonistic behaviour under some conditions such as stress or nutrition deprivation. As soil has a very less amount of soluble and available nutrients, some microbes display antagonistic behaviour to obtain nutrients and inhabit the place. This kind of antagonistic nature can replace the need for synthetic pesticides. Many PGPR *Pseudomonas* sp. can produce and secrete antibiotic extracellular. Some antifungal antibiotics produced by *Pseudomonas* sp. are sulphonamide, pyrrolnitrin, and viscosinamide, produced antiviral antibiotic karalicine (Ramadan et al. 2016). Ribosomal antibiotics (subtilosin A) and non-ribosomal antibiotics (rhizocticins and mycobacillin) targeting fungi can be produced by *Bacillus* sp. and *Paenibacillus polymyxa*, which help to prevent fungal disease in *Sesamum indicum*. *Sclerotinum rolfsii* causing disease in *Phaseolus vulgaris* can be fought through *Pseudomonas cepacia*. Prevention of *Fusarium culmorum* and damping of cotton in *Triticum aestivum* and *Gossypium hirsutum* are possible with the inoculation of *Pseudomonas fluorescens*. *Serratia marcescens* and *Pseudomonas putida* help *Cucumis sativus* to cease cucumber anthracnose (Gouda et al. 2018).

14.3 EXTREMOPHILES FOR SUSTAINABLE AGRICULTURE

As environmental conditions are changing due to various natural and man-made processes, conditions in soil, which is ideal for microbial growth and development, also changed. Some microbes have the ability to tolerate altered and harsh environmental conditions, which are known as extremophiles. Extremophiles are unicellular, different from other microbes in terms of adaptation towards various extreme conditions due to some special cellular mechanisms and some extraordinary metabolic pathways and genes (Kulkarni et al. 2019). The composition of lipid is also different in extremophiles from other normal condition-surviving microbes. Hydrophobic proteins and osmolytes are the main compounds that attribute extremophiles' adaption capacity towards thriving environmental conditions. Enzymes that are produced by extremophiles are known as extremozymes (Salgaonkar and Bragança 2019). These extreme abiotic soil conditions are pH, pressure, temperature, and concentration of salt or osmolarity (Kulkarni et al. 2019). At prokaryotic level, extremophiles include eukarya, bacteria, and archaea. Of these three prokaryotes, archaea can tolerate extreme conditions better than the other two. Cyanobacteria form bacterial group, and fungi from eukarya are more stable and adapted towards harsh environmental conditions (Rampelotto 2013).

Salinization due to salt accumulation in soil is one of the major threats and obstacles in agricultural process. According to World Atlas of Desertification, soil which was naturally salinated (primary salinization) is around 1 billion hectors and 77 million hectors was salinated due to man-made activity (secondary salinization). Excess salt can be accumulated at root zone of agricultural crops due to the evaporation of irrigated water. Excess salt deposition around root hinders the water absorption of agricultural plants. Excess amount of sodium ion lowers down the potassium level in agricultural crops due to competitive effect. This will create an imbalance of sodium-to-potassium ratio and affect various cellular processes such as protein synthesis, enzyme activities, and stomata opening and closing (Sharma et al. 2016). These hurdles negatively affect total crop yield quantitatively as well as qualitatively. As agricultural crop is not sufficiently produced, it will also decrease the economic gain. Hypersaline regions have mostly salt concentration greater than 15%. Halophilic microbes can tolerate a high range of sodium chloride, and high salt concentration is a prerequisite of halophiles for their growth (Yadav and Saxena 2018). Halophiles are divided into three categories. True halophilic microbes can tolerate salt concentration of sodium chloride ranging from 15% to 30%, moderate halophilic microbes can tolerate salt between 3% and 15%, and slight halophilic microbes can tolerate salt ranging between 1% and 2% (Yadav et al. 2015b).

Management of such salinated agricultural field using low-cost input is necessary for better crop yield to fulfil food need as free lands are not available any more or capable of agricultural crop production. So, halophilic microbes, which are part of natural ecosystem and nutrient cycling process, can be used to overcome such stressed conditions, as they require salt for their survival and growth. Sort out halophiles belong to phyla such as Actinobacteria, Bacteroidetes, Euryarchaeota, Firmicutes, Proteobacteria, and Spirochetes (Yadav and Saxena 2018). Archaea belongs to *Halobacteriaceae* family and has plant growth-promoting ability due to their various processes such as production of siderophores, ammonia, deamination capacity, fixation of nitrogen, and production and secretion of various phytohormones, e.g. auxin and cytokinin. They have another immense characteristic of balancing of external environment salinity through accumulation of salts as osmolytes, and this process is known as salt-in (Sleator and Hill 2002). *Pseudomonas fluorescens* MSP 393 can carry out salt-in process with the activity of plant growth promoter production of various growth-enhancing amino acids (Paul and Nair 2008). Other halophiles provide antioxidant enzymes and protection of agricultural crops against various pathogens. Polluted soil reclamation can also be possible using remediating halophiles. Isolated halophile *Bacillus polymyxa* BcP26 is able to fix nitrogen and acts as a biofertilizer as well as a biocontrol agent, *Pseudomonas alcaligenes* PsA15 can remediate and degrade pesticides and PAHs, and *Mycobacterium phlei* MbP18 has the benefit of adaptation to high salt and high temperature conditions. So, they can be used for hypersaline and arid agricultural soil quality improvement (Egamberdiyeva 2007). *Pseudomonas simiae* AU has a salt tolerance capacity of 100 mM of sodium chloride with positive effect on shoot root growth and seedlings of soybean (Vaishnav et al. 2015). Retarded growth of *Arachis hypogaea* could be recovered using diazotrophic *Brachybacterium saurastrense* (Shukla et al. 2012).

Ethylene is a volatile growth hormone of crops and plants at its low concentration, but stressed-induced high concentration retards root growth. Some halophile and halotolerant microbes produce an enzyme ACC deaminase that lowers the level of precursor of ethylene by hydrolysing aminocyclopropane-1-carboxylate into alpha-ketoglutarate and ammonia, a source of nitrogen. This enzyme was detected in various halophilic microbial genera of *Arthrobacter, Bacillus, Brevibacterium, Corynebacterium, Exiguobacterium,* and *Zhihengliuella. Burkholderia* sp. And *Herbaspirillum seropedicae* mobilizes phosphate for crop availability, and this positively affects plant growth with an increase in weight of 1.5%–20%. The amount of ROS increases in plant cells in case of high stress conditions. This active compound damages cellular mechanisms and metabolism, which leads to the death of plant after a specific time. Some halophiles produce various antioxidant enzymes in high salt stress conditions, which have scavenging properties of ROS; that is, *R. leguminosarum* and *S. proteamaculans* are able to produce catalase and superoxide dismutase enzymes (Sharma et al. 2016). Some novel halophilic microbes able to produce various high salt-tolerant enzymes are *Aquisalimonas asiatica, Halomonas marisflavae, Haloprofundus marisrubri, Prauserella isguenensis,* and *Natrinema versiforme.* Besides bacteria and archaea, some fungi able to survive in and tolerate high salt concentration were also found, e.g. *Aspergillus, Cladosporium, Myrothecium, Piriformospora,* and *Stemphylium. Brevibacterium iodinum, Bacillus licheniformis,* and *Zhihengliuella alba* RS111 help agricultural crops in the uptake of nutrients with the reduction in saline stressed condition. *Mesorhizobium* sp. MBD26 helps in increasing the biomass with enhanced nodulation. *Paenibacillus xylanexedens* and *Enterobacter cloacae* induce root growth and nutrient uptake (Yadav and Saxena 2018).

Another abiotic extreme condition is high temperature in arid and semi-arid area. At very high temperatures, water from soil evaporates very rapidly. Without water availability, most of the metabolic activities of plants ceases, resulting in the death of the plant. This creates a huge loss in an agricultural area. At high temperature, many changes in crops lead to their damage and death. Thermophiles are microbes having capacity to withstand high temperatures and provide various nutrients required by agricultural crops. Saharan and Verma (2014) isolated *Bacillus licheniformis* having the ability to perform PGPR activities such as IAA production, solubilization of phosphate, and heavy metal remediation. As it deposits nutrients, it also acts as a biofertilizer. Along with *Bacillus licheniformis, B. pumilus* was reported for the production of phytohormone gibberellins. *Bacillus circulans, B. thuringiensis*, and *B. subtilis* can degrade chitin. So, they can be used as a biocontrol agent for pests in agriculture field. *Bacillus subtilis* BHUJP-H1, *Bacillus* sp. BHUJP-H2, and *B. licheniformis* BHU-H3 can grow up to 60°C with positive catalase and cellulase activity. All of these three strains also have phosphate solubilization activity and IAA production ability with 24.85%, 22.27%, and 34.76% for 300 µg/mL of supplied tryptophan after 72 hours, respectively. Besides other two isolated strains, *Bacillus licheniformis* BHU-H3 was negative for siderophore and hydrogen cyanide production (Verma et al. 2018). This shows the importance of thermophiles in PGPR activity to enhance agricultural crops growth and to help in the combat of high temperature.

14.4   ARBUSCULAR MYCORRHIZAL FUNGI (AMF)

Arbuscular mycorrhizal fungi are symbiotic fungi associated with plants through mutualistic relation. AMF use plant as a host to complete their life cycle as they are compulsory biotrophs (Ferrol et al. 2004). AMF complete their life cycle using the plant as a host; in turn, they give many benefits to the plant, such as the regulation of photosynthesis, controlled water loss and absorption, ability to combat stressed conditions, osmoregulation, and protection from phytopathogens (Ahanger et al. 2014). Most AMF belong to Mucoromycota phylum. There are four orders of AMF: (1) Archaeosporales, (2) Glomerales, (3) Diversisporales, and (4) Paraglomerales. These fungi can produce various structures such as hyphae, spores, vesicles, and arbuscules. AMF can be used as biofertilizers as they affect carbon dioxide fixation for the growth of photosynthetic microbes and plants, decompose organic matter through hyphae, and thus increase soil nutrients level by sink mechanism (Begum et al. 2019).

Mixtures of naturally synthesized compounds used to improve soil quality and fertility for crop production are known as biofertilizers. Applications of synthetic artificial fertilizers and pesticides affect soil quality and crop yield negatively after a long time which is hard or sometimes not possible for reclamation of soil. Studies indicate that AMF help to uptake, mobilize, and stimulate nutrients in the soil for plant growth. AMF also help in the availability of micronutrients Cu and Zn. AMF also assist in the supply and exchange of phosphorous to agricultural crop (Begum et al. 2019). The quantity of nitrogen and phosphorous was improved in *Chrysanthemum morifolium* after having an inoculation of AMF (Wang et al. 2017). The ratio of calcium to sodium also gets balanced through AMF colonization. It also enhances K+ transporter efficiency in *Lotus japonicas* roots. The amount of potassium, iron, and nitrogen also increased in drought environmental conditions through AMF in *Pelargonium graveolens* L. (Begum et al. 2019).

AMF also help plants and agricultural crops to fight against abiotic stresses such as high salinity, temperature, the presence of toxic heavy metals, and drought. Drought increases oxidative stress, decreases transpiration, and negatively affects the uptake of nutrition. Symbiotic association of AMF with agricultural plants helps to increase plant biomass by improving leaf area index and root size. *Funneliformis mosseae* increases proline level and photosynthesis in *Glycine max* L. during drought conditions. *Glomus mosseae, G. fasciculatum*, and *Gigaspora decipiens* increase total chlorophyll content, produce antioxidant enzymes, and balance osmotic potential during drought in *Triticum aestivum* L., and *Rhizophagus intraradices* BGCBJ09 enhances the uptake of nutrients such as nitrogen, potassium, phosphate, and magnesium in *Zea mays*. In lettuce and tomato, the production of abscisic acid and strigolactones was found to be higher during drought in the presence of *Rhizophagus irregularis*. The content of sugar, mainly soluble and enzyme acid phosphatase, was found to increase with the reduction in proline in *Vigna subterranea* during abiotic drought stressed condition by *Glomus intraradices*, *Gigaspora gregaria*, and *Scutellospora gregaria* (Begum et al. 2019).

Salinity is one of the major threats and problems of abiotic stress, which reduces total crop yield in agricultural sector due to negative effects on plant vegetation and the production of ROS. Inorganic nutrient and the content of salicylic acid and jasmonic acid can be increased with AMF associated with crops. *Claroideoglomus etunicatum* increases the PSII activity of photosynthesis and conductance of stomata in *Oryza sativa* L., enhances the production of free amino acids, and increases sodium and potassium uptake, and dry biomass of root and shoot. *Glomus fasciculate* increases the quantity of zinc, copper, and phosphorous in *Acacia nilotica* (Begum et al. 2019). Ubiquitous polyamine tools were also altered for the growth of plants, which is AMF-mediated during stressed condition of excess salt (Kapoor et al. 2013). Area of leaf and biomass were also enhanced with AMF-inoculated *Allium sativum* in saline soil.

Heavy metals in soil are also culprit for plant growth and development. Hyphae of AMF stop the spreading of heavy metals by immobilization in cell wall region and collecting into vacuole. Thus, heavy metals are unavailable and cannot cause toxic effects and mortality of agricultural crops. *Acaulospora laevis*, *Gigaspora nigra*, *Glomus clarum*, and *G. monosporum* increase malondialdehyde and antioxidants in the presence of cadmium in *Trigonella foenum-graecum* L. AMF also regulate inorganic nutrient uptake. AMF enhance silicon uptake in *Glycine max* (Begum et al. 2019). AMF improve cadmium tolerance in *Medicago sativa* L (Wang et al. 2012).

Temperature above or below tolerable scale causes various damage to agricultural crops, such as burning or wilting of leaves, fruits decolouration, and induced oxidative stressed condition abscission of leaves, buds, and fruits. AMF help crops to fight and survive against the stress of hot or cold temperature. In cold condition, AMF induce chlorophyll content of plants for the improvement of photosynthesis. Alteration of protein content in tomato crops was also reported (Latef and Chaoxing 2011). *Funneliformis geosporum* and *F. mosseae* enhance chlorophyll content and thus increase the overall photosynthesis in *Zea mays*. Aquaporin and phosphorylation were positively enhanced with the hydraulic conductivity of root in *Solanum lycopersicum* in the presence of AMF *Rhizophagus irregularis* (Begum et al. 2019).

Phytopathogens are infectious agents that are specific to plants and cause damage or even death of them. Phytopathogens reduce agricultural crop yield and negatively affect the economic status of countries. For example, *Phytophthora fragariae* causes infection of the root of strawberry (Norman et al. 1996). Endoparasitic nematodes *Radopholus* sp. and *Pratylenchus* sp. feed on root cortex area and cause damage to plant. AMF provide protection against various phytopathogens. As AMF are also a type of microbe, plants can distinguish between pathogenic microbe and AMF through microbe-associated molecular patterns (MAMPs). MAMPs were detected through their recognition receptors, and MAMP-triggered immunity (MTI) response of plant gets activated as the first line of immune defence (Jones and Dangl 2006). Initially, plants perceive AMF as biotropic putative phytopathogens and lead to MTI response in plant, inducing transcriptional and phytohormone synthesis changes at symbiosis establishment. For example, the pattern of transcription of two fungal species *F. mosseae* and *R. irregularis* was different for colonization in the root of tomato plant. Thirty-five per cent similarity was found for transcription profile screening. Compared to *R. irregularis*, weak colonization was observed in *F. mosseae* due to the induction of jasmonate-conjugated isoleucine and 9-LOX pathway (Lopez-Raez et al. 2010).

AMF contrive protection from phytopathogenic fungi to agricultural crops. Symptoms of *Fusarium oxysporum* such as external damage and necrosis were reduced in cultivated banana treated with *Glomus intraradices* and other *Glomus* sp. *Glomus mosseae* was also reported for the protection of soybean plants from the infection of *Pseudomonas syringae*. Infection of *Pseudomonas lachrymans* could be reduced with the treatment of *Glomus macrocarpum* in cucumber plant and eggplant. Phytoalexins, which are produced in plants as a result of infection, act as a toxin for phytopathogen and give protection to host plant. Increased levels of phytoalexins such as total phenol, beta-glucosidase, and phenylalanine were reported in tomato when treated with *Glomus mosseae* to fight against *Fusarium oxysporum*.

AMF provide protection to agricultural crops against viral infection. For example, *Funneliformis mosseae* provides protection from infection of Tomato yellow leaf curl *Sardinia virus* and Cucumber mosaic virus to *Cucumis sativus* L.cv. and *Solanum lycopersicum* L., respectively. In contrast to this, *Rhizophagus intraradices* increased titre to *Solanum tuberosum* L. and *Nicotiana tabacum* cv. from the infection of *Potato virus Y* and *tobacco mosaic virus*, respectively (Hao et al. 2019). In some cases, it has been observed that AMF promote virus infection to plant. For example, *Funneliformis macrocarpa* increased virus titre of *tomato aucuba mosaic virus* and *Potato virus X* in tomato. *Arabic mosaic virus* titre was induced positively in strawberry in the presence of *Funneliformis macrocarpa* (Miozzi et al. 2019).

Rocha et al. (2019) experimented with chickpea production using AMF *R. irregularis*. Experiment was carried out using seed coating with single *R. irregularis* BEG 140 and consortium of five BEG 141, BEG 236, DAOM 197198, KW, and AS AFM *R. irregularis*. Consortium-inoculated seeds showed increased growth compared to control. Pods and grains numbers were induced by 160% and 148%, respectively. The colonization rate of *R. irregularis* was higher towards root for a consortium of five fungi compared to a single one.

14.5   ROLE OF MICROBIAL CELL SIGNALLING AND COMMUNICATION

14.5.1   QUORUM SENSING

Bacterial cell-to-cell communication for the regulation of expression of genes mainly based on microbial population density is known as quorum sensing. Auto-inducers are synthesized and released by bacteria as a chemical signalling molecule. These signalling molecules affect the expression of specific genes mainly by a positive feedback system. Quorum sensing helps bacteria in the formation of biofilm, in sporulation, for motility, and in the expression of virulence factor (Pan and Ren 2009). Microbial quorum sensing also benefits agricultural plants and other crops *via* fixation of nitrogen, the production of siderophores, secretion of hydrolytic enzymes which are extracellular, and antibiotic production. In rhizospheric region of plants, microbes are positively affected by exudates of root; in return, these affected microbes enhance development and growth of plants (Raghavendra 2017). Thus, through quorum sensing, whole microbial population modifies physiological behaviour as a single unit.

N-acyl-L-homoserine lactones (AHLs) are molecules involved in quorum sensing widely produced by gram-negative bacteria. Plants also produce and secrete compounds similar to AHLs, which affect bacterial quorum sensing. For example, plant-secreted natural N-acylethanolamines (NAEs) are similar to bacterial AHLs. In response to elicitors, N-acylethanolamines production is enhanced in leaves and gets accumulated in seeds that are desiccated. The application of alkamide, which is another bacterial AHL similar to signalling molecules and NAEs, to seedlings of *Arabidopsis* affects morphogenesis in a concentration-dependent way by altering the growth of root and shoot. Mutant *Pseudomonas fluorescens* 2P24 having the deletion of AHLs-encoding gene pcoI failed to colonize the rhizosphere of wheat, while pcoI complementation could successfully restore the wild-type activity. Systemic defence against *Alternaria alternata* fungi in tomato could be achieved through the enhanced synthesis of ethylene and salicylic acid by AHLs-secreting bacteria in soil (Ortiz-Castro et al. 2009). Two different but conserved homoserine-based quorum sensing systems were detected in PGPR *Paraburkholderia* sp., *i.e.* BpI.1/R.1 and BpI.2/R.2. BpI.1/R.1 has an important role in colonization for *Arabidopsis thaliana*. The present study indicates that BpI.1/R.1 quorum sensing system has a vital role in biofilm production control via regulation of expression of ecf26.1 gene, which is responsible for the encoding of sigma factor, having an extra cytoplasmic function (Zuniga et al. 2017).

14.5.2 Volatile Organic Compounds

Organic compounds that can vaporize in the atmosphere at specific pressure are known as volatile organic compounds (VOCs). VOCs include compounds such as aldehyde (CHO), ketone (C=O), alcohol (C-OH), and hydrocarbon (H-C) having a molecular weight less than 300 g/mol. Both plants and associated microbes can synthesize and evaporate VOCs, but microbial VOCs are more complex than those synthesized by plants. Volatile organic compounds synthesized and secreted by exudates of root were recognized by PGPR. In turn, microbes also synthesized VOCs that have a role in plant growth through the stimulation of the production of phytohormones, bioprotection, biotolerance of biotic and abiotic stress, and biocontrol. Thus, VOCs can be synthesized and volatilized by both plants and PGPR microbes, and also the effect is in a bidirectional way to both. VOCs such as ammonia, alcohol, and hydrogen cyanide produced by Rhizobacteria and other bacteria have antifungal properties via negative effects on mycelium growth and inhibition of spore formation. Moreover, the response of every VOC is specific in terms of host and concentration level (Ortiz-Castro et al. 2009).

Systematic resistance pathway can be activated in *Arabidopsis thaliana* by VOCs produced and secreted from *Bacillus amyloliquefaciens* IN937a and *Bacillus subtilis* GB03 from protection against *Erwinia carotovora*. Most of the induced systematic resistance pathways are dependent on ethylene and jasmonate and are independent of salicylic acid. Differential gene expression was reported in *Bacillus subtilis* GB03-treated Arabidopsis for metabolism, modifications of cell wall, and response towards stress condition as it affects morphogenesis through homeostasis of auxin (Zhang et al. 2007). Moreover, *Bacillus subtilis* GB03 can secrete various VOCs such as ketones, volatile compounds having sulphur, alcohol, and aldehyde, induce phytohormones production in *Arabidopsis thaliana,* and also enhance photosynthesis. VOCs induce systemic tolerance to abiotic stress condition. For example, VOCs of *Bacillus subtilis* GB03 induce tolerance towards sodium chloride by downregulation of high-affinity K^+ transporter 1 (HKT1) responsible for the lower accumulation of salt in plants (Ortiz-Castro et al. 2009).

14.6　TOOLS AND TECHNIQUES FOR SCREENING AND TRAIT ALTERATION

A wide range of techniques can be used to study plant-microbe interactions, microbe populations in the rhizosphere, and microbe effects on plants. The impact of microbial communities on plants and crops, whether endogenously or exogenously, revealed variations in microbiota due to external factors, development of resistance, and alteration of microbial genetic makeup.

Metagenomics also known as genomics of community is the study of genetic material derived directly from the sample. It is a next-generation sequencing-based omics approach mainly used to understand abundance and interaction of microbes with agricultural traits at the community scale. Through metagenomics study, non-culturable microbes can also be studied. Mainly two types of screening are possible with metagenomic data. One is taxonomical screening based on conserved sequence analysis, and the other is functional screening and can be done through cloning and expression (Schloss and Handelsman 2003). Tools of metagenomics can be used for the analysis and study of agricultural soil microbial populations. Orellana et al. (2018) studied the functional and taxonomical dynamic nature of microbial community in the aspect of seasonal variation for the year of agricultural soil and the effect of fertilizer on ammonia oxidizer microbes of agricultural soil. Samples were taken from a different depth of two locations having sandy type soil of Havana and silt textured soil of Urbana. Functional and taxonomic diversity was positively screened out for two different locations. But, for different depths at the same site, no major changes were observed for a year timeline. The activity of ammonia-oxidizing genes was increased after the application of synthetic nitrogen fertilizer in Havana, which was observed six times higher in archaea than in bacteria. Naturally present nitrogen had no major positive impact on the nitrification process of soil microbes. Two novel nitrifiers *Nitrospirae* and *Thaumarchaeota* were also reported through this metagenomic screening. Thus, from this study, it can be concluded that seasonal variation has no major role in changing agricultural microbial community or in magnification of nitrogen fixers, although provision of nutrient source in the form of fertilizers can enhance adaptation and growth of plant beneficial microbes which are responsible for micro- and macroelement deposition in soil.

Holobiome analysis of grapevine rhizosphere indicates that the microbial community of plant and soil surrounding it was affected by factors such as pH, nutritional source, and carbon-to-nitrogen ratio. Moreover, cultivation practice and methods also influence the abundance of soil microbiota (Zarraonaindia et al. 2015). Not only environmental factors and climatic conditions, but genotype of plants also has a great impact on the microbiome associated with plants (Walters et al. 2018). Mendes et al. (2011) studied the microbiome of the sugar beet plant and concluded that crucial rhizospheric bacterial communities such as Firmicutes, Actinobacteria, and Proteobacteria suppress the activity of pathogenic fungi. Peptide synthesis relying on non-ribosomal sources is the main mechanism of fungal-attributed disease suppression from Gammaproteobacteria.

Genes are hereditary unit of any life form on earth. By modification of genes associated with specific traits, one can enhance or hinder genes expression. In agriculture practices, genetic engineering is used for the alteration or modification of plant or plant-associated microbial traits. Genetic alteration includes gene deletion, edition, recombination, and modification. Genetic engineering and genetic modification employ mainly restriction endo and exonucleases act as molecular scissors and ligases act as the glue in genetic tools (Gaj et al. 2013). Site-specific or targeted genetic engineering is mainly carried out through nucleases that are specific to site (SSN), effecter nucleases that are activators of transcription process (TALEN), clustered regularly interspaced short palindromic repeats (CRISPR) Cas9, and nucleases that work with zinc finger motif (ZFN). CRISPR Cas system is more beneficial in terms of higher specificity and work efficiency (Brandt and Barrangou 2019). CRISPR Cas system mainly targets UTR, ORF, and non-coding RNAs. Based on the activation of either homologous repair or non-homologous repair system, random nucleotide base substitution, addition, deletion (error-prone mechanisms), modifications (error-free mechanism), or knock-in functions can be expressed in target. SpCas9-VQR, SpCas9-VRER, dCas9, and nCas9 were derived from *Streptococcus pyogenes* and were utilized for crop species of rice, tomato, maize, wheat, and *Arabidopsis* (Wang et al. 2019). Single-stranded RNA of rice and tobacco was targeted by Cas13a (C2c2), which has a source of *Leptotrichia shahii* (Aman et al. 2018; Abudayyeh et al. 2017). AMLT9 gene of tomato was targeted using CRISPR Cas system for decreased production of malate. Similarly, *SSADH, GAD2, GAD3, CAT 9, GABA-TP1, GABA-TP2*, and *GABA-TP 3* genes were targeted for increased quantity of GABA production (Nonaka et al. 2017; Li et al. 2018). The digestion ability of maize was improved using the increased production of amylopectin. The inhibition of expression of granule-bound starch synthase enzyme in endosperm of maize decreases amylase synthesis, and this led to a higher amount of amylopectin (Zhang et al. 2018). Moreover, besides genetic modifications, CRISPR is also used for the analysis of fundamental interaction process between microorganisms and plants. For example, Evolve R is used for diversity generation through new allele creation at target site. It can also be applied for identification and determination of microbial genes involved in the interaction with agricultural crops (Shelake et al. 2019).

ZFN recognizes triplet nucleotide sequence, restriction enzyme cleaves DNA sequence, and then fingers region of ZFN attach to a particular specific sequence (Urnov et al. 2010). ZmIPK1 gene was edited through the addition of cassette of gene of Phosphinothricin N-acetyltransferase for tolerance of herbicide in maize (Shukla et al. 2009). OsQQR gene of rice was edited using ZFN for arrangement of traits. The main limitation of using ZFN for crop improvement is its low efficiency.

Transcription activator-like effector nucleases (TALENs) are a combination of repeats of transcriptional activator and restriction enzyme, mainly FokI. TALEN was first applied for the production of bacterial blight resistance rice. The efficiency of saccharification of sugarcane was enhanced using TALEN-driven mutation (Zhang et al. 2018). Targeting FAD2-1A, FAD2-1B, and FAD3A genes through TALEN linoleic acid content can be lower with increasing amount of oleic acid in soybean by non-homologous end joining (Haun et al. 2014; Demorest et al. 2016). Haploidy was induced in maize using *ZmMTL* gene modification (Kelliher et al. 2017).

Conventional genetic modification techniques require insertion of DNA into host cell, mostly in coding portion of the gene. Integration at random site besides specific site, degradation of DNA, and non-efficient expression poses many problems, including regulatory concerns, undesirable changes, sometimes polar effects, and non-target expression. To overcome these limitations in genome edition, agricultural crop improvement was advanced for DNA-free techniques. Particle bombardment and conversion of nucleotide, for example C-to-T base conversion, helps in facilitating genome editing, which is DNA-free for plant breeding. Identification of gene function is the most recent challenge in agricultural crop plants after successful technological implementation of genome sequencing of many crops. The Knockout of functional genes in agriculturally significant crops with mutant library preparation gives enormous advantage of traits function identification and improvement (Zhang et al. 2018).

Gene regulation in crops can also be achieved at the transcription stage of the cell process. Regulation at transcription is obtained through activators and repressors having two domains. One is for DNA binding, and the other is for regulation process. Exogenous DNA insertion is not required in the case of transcriptional regulation as this mechanism occurs at gene level internally to the cell (Qi et al. 2013). Gupta et al. (2012) studied the effect of designed activator of transcription through fusion of DNA attaching domain of ZFN and scaffold domain of VP16. This construct was transferred to *Brassica napus* using *Agrobacterium* for the regulation of fatty acid synthesis using transcription regulation. Successful transgenic crops show elevated transcription levels of KASII as well as less production of palmitic acid. Variability and adaptations of agricultural crops are important for better quality yield. According to crop type, product need, and system efficiency, one can choose either DNA-based or DNA-free genome editing tool. Metagenomic techniques are very useful for understanding the effects of microbes on the plant, change in microbes due to plant interactions as well as due to environment effect, and their possible interaction outcome.

14.7 MICROBIOME SCREENING OF IMPORTANT AGRICULTURAL CROPS

Microbiome analysis of agriculturally important crops allows getting insight into the importance of them, which is not possible through culture-dependent processes. As microbiome is a culture-independent study, it allows screening of very rare and less abundant, but important microbes, which contribute to immense activity in agricultural and economically pearled crops. Table 14.1 indicates the selected microbiome of some important crops used worldwide for food.

Verma and Suman (2018) studied the microbiome of *Triticum aestivum* L. contributes the total one-third of grain food worldwide. Forty-nine per cent Proteobacteria were dominant at the phyla level, followed by Firmicutes, Actinobacteria, and Bacteroidetes with 32%, 12%, and 7% abundance, respectively. The relative distribution pattern of the wheat microbiome suggests most of them are classified as epiphytic, endophytic, and rhizospheric. Some bacteria can occupy only a specific sphere, and some can occupy more than one sphere. *Brevundimonas diminuta*, *Methylobacterium phyllosphaerae*, and *Pseudomonas fuscovaginae* were found completely occupying epiphytic space on wheat, while *Delftia lacustris*, *Ochrobactrum intermedium*, and *Pseudomonas monteilii* were completely endophytic. *Azotobacter tropicalis*, *Bacillus siamensis*, *B. thuringiensis*, and *Rhodobacter sphaeroides* were found as rhizospheric bacteria.

Eyre et al. (2019) studied microbiome seeds of six different *Oryza sativa* with each of four parts of grain, outer region of grain, husk, and outer region of husk. Core microbiome analysis of all samples indicates an abundance of Alphaproteobacteria. At the genus level, *Sphingomonas*, *Methylobacterium*, and *Aureimonas* were found as key microorganisms. The presence of *Enterobacter* indicates PGPR activity with nitrogen and phosphate supply to rice. *Curtobacterium* and *Methylobacterium* were common for the outer region of the husk and outer region of grain. *Rhizobium* was core for all three regions of rice seed excluding the grain region. Fungus *Ascomycota* was found to associate with all four regions of seed. *Epicoccum* was dominant in the outer region of husk. Other fungi such as *Hannaella*, *Papiliotrema*, and *Tremellales* were also found for all regions of rice seed. *Alternaria* was found to be common for grain, outer region of grain, husk, and outer region of husk. Raj et al. (2019) screened endogenous microbiome of rice seed from hotspot area of Indo-Burma. Endogenous microbiome analysis indicates the presence of vitamin B12-synthesizing pathway and abundance of animal gut microflora *Bifidobacterium*, *Faecalibacterium*, and *Lactobacillus* in rice seed. These bacteria were associated with PGP activities such as phosphate solubilization, reduction of nitrate for nitrogen availability, and production of siroheme for iron chelation.

Beirinckx et al. (2020) studied whether root microbiome can enhance the growth of *Zea mays* L. crop at low temperature or not. A slight but impactful difference was

TABLE 14.1

Microbiome Applications in Agricultural Crops Worldwide

Agricultural Crop	Phylum	Microorganisms	Response Specifications	References
Triticum aestivum L.	Actinobacteria, Bacteroidetes, Firmicutes, Proteobacteria	*Arthrobacter nicotianae, Bacillus sphaericus, Flavobacterium psychrophilum, Pseudomonas azotoformans, Pseudomonas rhodesiae, Rhodobacter capsulatus*	Nitrogen fixation, siderophore production, antibiotics and phytohormone synthesis	Verma and Suman (2018)
Oryza sativa	Actinobacteria, Firmicutes, Proteobacteria	*Exiguobacterium indicum, Leucobacter chromiiresistens, Methylobacterium radiotolerans, Microbacterium testaceum, Pantoea ananatis, Pseudomonas parafulva, Pseudomonas psychrotolerans, Staphylococcus sciuri*	Different PGPR activities	Midha et al. (2016)
Zea mays L. (teosinte)	Actinobacteria, Bacteroidetes, Proteobacteria	*Flavobacterium* sp., *Sandaracinobacter* sp., *Skermanella* sp.	Nutrient recruitment	Brisson et al. (2019)
Carica papaya L.	Actinobacteria, Bacteroidetes, Cyanobacteria, Firmicutes, Planctomycetes, Proteobacteria	*Acinetobacter* sp., *Candidatus tremblaya, Chryseobacterium* sp., *Pseudomonas* sp., *Sphingomonas* sp.	Endosymbiosis for nutritional benefit	Megaladevi et al. (2020)
Solanum melongena L.	Actinobacteria, Bacteroidetes, Cyanobacteria, Firmicutes, Proteobacteria	*Acinetobacter* sp., *Candidatus tremblaya, Chryseobacterium* sp., *Mesorhizobium* sp., *Ochrobactrum* sp., *Pseudomonas* sp., *Sphingomonas* sp.	Endosymbiosis for nutritional benefit	Megaladevi et al. (2020)
Solanum lycopersicum cv. MG	Acidobacteria, Bacteroidetes, Proteobacteria, Verrucomicrobia	*Arthrobacter* sp., *Bradyrhizobium* sp., *Hyphomicrobium* sp., *Limnobacter* sp., *Massilia* sp., *Sphingomonas* sp.	Rhizospheric microbiome contributes to growth and productivity	Cheng et al. (2020)
Brassica napus	Actinobacteria, Bacteroidetes, Chloroflexi, Proteobacteria, Verrucomicrobia	*Balneimonas* sp., *Candidatus nitrososphaera, Cryptococcus* sp., *Fusarium* sp., *Kaistobacter* sp., *Luteimonas* sp., *Rhodoplanes* sp.	Assimilate carbon in rhizosphere	Gkarmiri et al. (2017)

observed between microbiome of maize crop cultivated at room temperature and under low temperature. *Acinetobacteria* at phylum scale were decreased to a significant level in the endosphere of root of maize at chilling condition. *Enterobacteriaceae, Flavobacteriaceae, Methylophilaceae, Sphingobacteriaceae*, and *Pseudomonadaceae* were found enhanced under chilling conditions. *Caulobacteraceae, Comamonadaceae, Flavobacteriaceae*, and *Streptomycetaceae* were found common for both conditions of room temperature and chilling temperature. *Oxalobacteraceae* and *Bradyrhizobiaceae* both were PGPR, but not responsive for lower temperature.

Ndour et al. (2017) resolved the effect of root exudation on rhizospheric microbiome, which is based on the genetic type of *Pennisetum glaucum* (pearl millet). Exudation of root affects soil aggregation and allocation of carbon. Rhizospheric aggregation was analysed using the ratio of RAS/RT, which is the ratio of dry mass of adhering soil of root to dry mass tissue of root. From screening, it can be concluded that there was variation for aggregation of rhizospheric soil, which is significant between nine lines of pearl millet. Members of *Bacillales* and *Rhizobiales* were isolated having exopolysaccharide production ability. But no correlation was found between RAS/RT and exopolysaccharide-producing microbes. Besides this, the study also indicated that inbreeding of pearl millet can modulate abundance of rhizospheric bacterial taxa such as Actinobacteria, Acidobacteria, Bacteroidetes, Firmicutes, and Proteobacteria, and other phyla were common, but have different relative abundance in contacts of distribution pattern in 9 inbred lines.

Xu et al. (2018) screened the effect of drought on microbiome of *Sorghum bicolour.* The study indicated that Actinobacteria were dominant in drought stressed plant, while Proteobacteria were dominant in normal conditioned control plant. Drought condition also has a significant impact on Cyanobacteria as they did not majorly appear after the application of water stressed condition. Minamisawa et al. (2019) carried out metagenome screening of *Bradyrhizobium*, which is correlated with fixation of nitrogen. The nitrogen-fixing capacity was screened out for root, stem, and leaves of sorghum for four lines. The nitrogen-fixing ability was found highest for KM1 and KM2 lines in the root. Genome and proteome analysis suggested that NifHDK from *Bradyrhizobium* were consistently present in all samples. The study also revealed that Bradyrhizobia, which is unique in roots of sorghum, was similar to *Bacillus oligotrophicum* S58^T having photosynthetic activity and with *Bradyrhizobium* sp. S23321, which is non-nodule-forming bacteria.

Mareque et al. (2018) analysed the effect of chemical-based nitrogen fertilizer on endophytic microbiota of sorghum. The study revealed that at phyla level, the presence of nitrogen-containing chemical fertilizer increases the relative abundance of Proteobacteria and decreased the relative abundance of Firmicutes, while at class scale, the presence of chemical fertilizer decreases the relative abundance of Gammaproteobacteria, Alphaproteobacteria, and Bacilli. The relative abundance of Betaproteobacteria was increased in the presence of chemical fertilizer having nitrogen, while Actinobacteria had no major significant difference and chemical nitrogen fertilizer had effects on bacterial community not on diversity.

14.8 CONCLUSIONS AND FUTURE PROSPECTS

The increasing population of the world demands more food with better quality. As pollution increases day by day and due to excess modernization, soil pollution also increases, which negatively affects the agricultural crop production. Microbial communities present in the soil help to maintain various mechanisms such as nutrient recycling, pollutant removal, and growth enhancement of plants. But they are also affected by exudates of plants secreted mainly from root and by climate change. Sometimes, plants also affect microbes present within the rhizosphere. This effect may be negative, positive, or neutral. Pathogenesis controlling PGPR and AMF can be used as a biofertilizer and as biocontrol agents. This is not the same for all time because some AMF also induce growth and biomass of phytopathogens. So, before applying any microbe in the field, one should go through all screening parameters of selected microbial agents for induction and betterment of agricultural yields and preventing negative side effects.

Microbial signalling also plays an important role in the establishment of colonization in rhizosphere. As a novel approach, one can utilize farming-friendly extremophiles having the ability to amplify plant growth also in extreme conditions of soil environment. By using molecular tools, one can optimize demanding agricultural product through gene amplification or modification. Some important limitations associated with molecular-level techniques are the need of high economical support and less chance of success compared to others. Elsewhere in the aspect of time, it is a very fast approach. Metagenomic tools provide insight into microbial community associated and interacting with plant and soil. So, one can explore data analysis knowledge of metagenome directly without culturing microbes. As metagenome is a culture-independent area, it gives a fast and rapid screening facility with accuracy. So, one can use this study to strengthen the knowledge about sustainability of agricultural soil for the upregulation of various plant-interacting microbes and for enhancing yield.

REFERENCES

Abudayyeh OO, Gootenberg JS, Essletzbichler P, Han S, Joung J, Belanto JJ et al. (2017) RNA targeting with CRISPR–Cas13. *Nature* 550:280–284.

Agami R, Medani R, Abd El-Mola I, Taha R (2016) Exogenous application with plant growth promoting rhizobacteria (PGPR) or proline induces stress tolerance in basil plants (*Ocimum basilicum* L.) exposed to water stress. *Int J Environ Agri Res* 2:78–92.

Ahanger MA, Hashem A, Abd-Allah EF, Ahmad P (2014) Arbuscular mycorrhiza in crop improvement under environmental stress. In: Ahmad P, Rasool S (Eds.), *Emerging Technologies and Management of Crop Stress Tolerance.* (pp 69–95) Academic Press, San Diego, CA. Doi: 10.1016/B978-0-12-800875-1.00003-X.

Ahemad M, Kibret M (2014) Mechanisms and applications of plant growth promoting rhizobacteria: Current perspective. *J King Saud Univ Sci* 26:1–20.

Aman R, Ali Z, Butt H, Mahas A, Aljedaani F, Khan MZ et al. (2018) RNA virus interference via CRISPR/Cas13a system in plants. *Genome Biol* 19:1. Doi: 10.1186/s13059-017-1381-1.

Anand K, Kumari B, Mallick MA (2016) Phosphate solubilizing microbes: An effective and alternative approach as biofertilizers. *J Pharm Pharm Sci* 8:37–40.

Begum N, Qin C, Ahanger MA, Raza S, Khan MI, Ahmed N, Zhang L (2019). Role of arbuscular mycorrhizal fungi in plant growth regulation: Implications in abiotic stress tolerance. *Front Plant Sci* 10:1068.

Beirinckx S, Viaene T, Haegeman A, Debode J, Amery F, Vandenabeele S, De Tender C (2020). Tapping into the maize root microbiome to identify bacteria that promote growth under chilling conditions. *Microbiome* 8:1–13.

Bhattacharyya PN, Jha DK (2012). Plant growth-promoting rhizobacteria (PGPR): Emergence in agriculture. *World J Microbiol Biotechnol* 28:1327–1350.

Bidartondo MI, Redecker D, Hijri I, Wiemken A, Bruns TD, Domínguez L, Read DJ (2002). Epiparasitic plants specialized on arbuscular mycorrhizal fungi. *Nature* 419:389–392.

Brandt K, Barrangou R (2019). Applications of CRISPR technologies across the food supply chain. *Annu Rev Food Sci Technol* 10:133–150.

Brisson VL, Schmidt JE, Northen, TR, Vogel, JP, Gaudin AC (2019). Impacts of maize domestication and breeding on rhizosphere microbial community recruitment from a nutrient depleted agricultural soil. *Sci Rep* 9:1–14.

Bulgarelli D, Schlaeppi K, Spaepen S, Van Themaat EVL, Schulze-Lefert P (2013). Structure and functions of the bacterial microbiota of plants. *Annu Rev Plant Biol* 64:8 07–838.

Cassan F, Vanderleyden J, Spaepen S (2014). Physiological and agronomical aspects of phytohormone production by model plant-growth-promoting rhizobacteria (PGPR) belonging to the genus *Azospirillum*. *J Plant Growth Regul* 33:440–459.

Chen L, Dodd IC, Theobald JC, Belimov AA, Davies WJ (2013). The rhizobacterium *Variovorax paradoxus* 5C-2, containing ACC deaminase, promotes growth and development of *Arabidopsis thaliana* via an ethylene-dependent pathway. *J Exp Bot* 64:1565–1573.

Chen SK, Edwards CA, Subler S (2001). Effects of the fungicides benomyl, captan and chlorothalonil on soil microbial activity and nitrogen dynamics in laboratory incubations. *Soil Biol Biochem* 33:1971–1980.

Cheng Z, Lei S, Li Y, Huang W, Ma R, Xiong J, Tian B (2020). Revealing the variation and stability of bacterial communities in tomato rhizosphere microbiota. *Microorganisms* 8:170. Doi: 10.3390/microorganisms8020170.

Demorest ZL, Coffman, A, Baltes NJ, Stoddard TJ, Clasen BM, Luo S, Mathis L (2016). Direct stacking of sequence-specific nuclease-induced mutations to produce high oleic and low linolenic soybean oil. *BMC Plant Biol* 16:225.

Egamberdiyeva D (2007). The effect of plant growth promoting bacteria on growth and nutrient uptake of maize in two different soils. *Appl Soil Ecol* 36:184–189.

Egamberdieva D, Wirth SJ, Alqarawi AA, Abd_Allah EF, Hashem A (2017). Phytohormones and beneficial microbes: Essential components for plants to balance stress and fitness. *Front Microbiol* 8:2104.

Eyre AW, Wang M, Oh Y, Dean RA (2019). Identification and characterization of the core rice seed microbiome. *Phytobiomes J* 3:148–157.

Ferreira CM, Vilas-Boas A, Sousa CA, Soares HM, & Soares EV (2019). Comparison of five bacterial strains producing siderophores with ability to chelate iron under alkaline conditions. *AMB Express* 9:78.

Ferrol N, Azcon-Aguilar C, Bago B, Franken P, Gollotte A, Gonzalez-Guerrero M, & Gianinazzi-Pearson V (2004). Genomics of arbuscular mycorrhizal fungi. Applied mycology and biotechnology. *Fungal Genom* 4:379–403. Doi: 10.1016/S1874-5334(04)80019-4.

Fox JE, Starcevic M, Kow KY, Burow ME, McLachlan JA (2001). Nitrogen fixation: Endocrine disrupters and flavonoid signalling. *Nature* 413:128–130.

Gaby JC & Buckley DH (2011). A global census of nitrogenase diversity. *Environ Microbiol* 13:1790–1799.

Gaj T, Gersbach CA, Barbas III CF (2013). ZFN, TALEN, and CRISPR/Cas-based methods for genome engineering. *Trends Biotechnol* 31:397–405.

Gkarmiri K, Mahmood S, Ekblad A, Alström S, Hogberg N, Finlay R (2017). Identifying the active microbiome associated with roots and rhizosphere soil of oilseed rape. *Appl Environ Microbiol* 83:e01938–17.

Glick BR (2012). Plant growth-promoting bacteria: Mechanisms and applications. *Scientifica*. Doi: 10.6064/2012/963401.

Gouda S, Kerry RG, Das G, Paramithiotis S, Shin HS, Patra JK (2018). Revitalization of plant growth promoting rhizobacteria for sustainable development in agriculture. *Microbiol Res* 206:131–140.

Gupta M, DeKelver RC, Palta A, Clifford C, Gopalan S, Miller JC, Flook J (2012). Transcriptional activation of *Brassica napus* β-ketoacyl-ACP synthase II with an engineered zinc finger protein transcription factor. *Plant Biotechnol J* 10:783–791.

Hao Z, Xie W, Chen B (2019). Arbuscular mycorrhizal symbiosis affects plant immunity to viral infection and accumulation. *Viruses* 11:534. Doi: 10.3390/v11060534.

Haun W, Coffman A, Clasen BM, Demorest ZL, Lowy A, Ray E, Mathis L (2014). Improved soybean oil quality by targeted mutagenesis of the fatty acid desaturase 2 gene family. *Plant Biotechnol J* 12:934–940.

Jebara SH, Ayed SA, Chiboub M, Fatnassi IC, Saadani O, Abid G, Jebara M (2019). Phytoremediation of cadmium-contaminated soils by using legumes inoculated by efficient and cadmium-resistant plant growth-promoting bacteria. In *Cadmium Toxicity and Tolerance in Plants* (pp. 479–493). Academic Press. Doi: 10.1016/B978-0-12-814 864-8.00019-X.

Jiang F, Chen L, Belimov AA, Shaposhnikov AI, Gong F, Meng X, Dodd IC (2012). Multiple impacts of the plant growth-promoting rhizobacterium *Variovorax paradoxus* 5C-2 on nutrient and ABA relations of *Pisum sativum. J Exp Bot* 63:6421–6430.

Johnsen K, Jacobsen CS, Torsvik V, Sorensen J (2001). Pesticide effects on bacterial diversity in agricultural soils–a review. *Biol Fertil Soils* 33:443–453.

Jones DL, Oburger E (2011). Solubilization of phosphorus by soil microorganisms. In *Phosphorus in Action* (pp. 169–198). Springer, Berlin, Heidelberg. Doi: 10.1007/978-3-642-15271-9_7.

Jones JD, Dangl JL (2006). The plant immune system. *Nature* 444:323–329.

Kalloo G (1993). Pea: *Pisum sativum* L. In *Genetic Improvement of Vegetable Crops* (pp. 409–425). Pergamon. Doi: 10.1016/B978-0-08-040826-2.50033-3.

Kapoor R, Evelin H, Mathur P, Giri B (2013). Arbuscular mycorrhiza: Approaches for abiotic stress tolerance in crop plants for sustainable agriculture. In *Plant Acclimation to Environmental Stress* (pp. 359–401). Springer, New York. Doi: 10.1007/978-1-4614-5001-6_14.

Kelliher T, Starr D, Richbourg L, Chintamanani S, Delzer B, Nuccio ML, Liebler T (2017). MATRILINEAL, a sperm-specific phospholipase, triggers maize haploid induction. *Nature.* 542:105–109.

Kudoyarova G, Arkhipova TN, Korshunova T, Bakaeva M, Loginov O, Dodd IC (2019). Phytohormone mediation of interactions between plants and non-symbiotic growth promoting bacteria under edaphic stresses. *Front Plant Sci* 10:1368.

Kulkarni S, Dhakar K, Joshi A (2019). Alkaliphiles: Diversity and bioprospection. In *Microbial Diversity in the Genomic Era* (pp. 239–263). Academic Press. Doi: 10.1016/B978-0-12-814849-5.00015-0.

Kumar R and Prakash O (2019). The impact of chemical fertilizers on our environment and ecosystem. In: Poonam Sharma (ed.), *Research Trends in Environmental Sciences* (pp. 69–86). New Delhi, India: AkiNik Publications.

Kumar, V, Chandra, R 2020. Metagenomics analysis of rhizospheric bacterial communities of *Saccharum arundinaceum* growing on organometallic sludge of sugarcane molasses-based distillery. *3 Biotech* 10(7):316. Doi: 10.1007/s13205-020-02310-5.

Latef AAHA, Chaoxing H (2011). Arbuscular mycorrhizal influence on growth, photosynthetic pigments, osmotic adjustment and oxidative stress in tomato plants subjected to low temperature stress. *Acta Physiol Plant* 33:1217–1225.

Li J, Ovakim DH, Charles TC, Glick BR (2000). An ACC deaminase minus mutant of *Enterobacter cloacae* UW4No longer promotes root elongation. *Curr Microbiol* 41:101–105.

Li R, Li R, Li X, Fu D, Zhu B, Tain H, Luo Y (2018). Multiplexed CRISPR/Cas9-mediated metabolic engineering of γaminobutyric acid levels in *Solanum lycopersicum*. *Plant Biotechnol. J* 16:415–427.

Liu D, Lian B, Dong H (2012). Isolation of *Paenibacillus* sp. and assessment of its potential for enhancing mineral weathering. *Geomicrobiol J* 29:413–421.

Lopez-Raez JA, Verhage A, Fernandez I, Garcia JM, Azcon-Aguilar C, Flors V, Pozo MJ (2010). Hormonal and transcriptional profiles highlight common and differential host responses to arbuscular mycorrhizal fungi and the regulation of the oxylipin pathway. *J Exp Bot* 61:2589–2601.

Maheshwari DK, Dheeman S, Agarwal M (2015). Phytohormone-producing PGPR for sustainable agriculture. In *Bacterial Metabolites in Sustainable Agroecosystem* (pp. 159–182). Springer, Cham. Doi: 10.1007/978-3-319-24654-3_7.

Mareque C, da Silva TF, Vollú RE, Beracochea M, Seldin L, Battistoni, F (2018). The endophytic bacterial microbiota associated with sweet sorghum (*Sorghum bicolor*) is modulated by the application of chemical N fertilizer to the field. *Int J Genomics* Doi: 10.1155/2018/7403670.

Martinez-Viveros O, Jorquera MA, Crowley DE, Gajardo GMLM, Mora ML (2010). Mechanisms and practical considerations involved in plant growth promotion by rhizobacteria. *J Soil Sci Plant Nutr* 10:293–319.

McLachlan JA (2001). Environmental signaling: What embryos and evolution teach us about endocrine disrupting chemicals. *Endocr Rev* 22(3):319–341.

Meena VS, Maurya BR, Verma, J. P., Meena RS (Eds.) (2016). *Potassium Solubilizing Microorganisms for Sustainable Agriculture*. Springer, New Delhi. Doi: 10.1007/978-81-322-2776-2.

Megaladevi P, Kennedy JS, Jeyarani S, Nakkeeran S, Balachandar D (2020). Metagenomic exploration of the bacterial endosymbiotic microbiome diversity of papaya mealybug *Paracoccus marginatus* from different host plants. *J Entomol Zool Stud* 8:429–439.

Mendes R, Kruijt M, De Bruijn I, Dekkers E, van der Voort M, Schneider JH, Raaijmakers JM (2011). Deciphering the rhizosphere microbiome for disease-suppressive bacteria. *Science* 332:1097–1100.

Midha S, Bansal K, Sharma S, Kumar N, Patil PP, Chaudhry V, Patil PB (2016). Genomic resource of rice seed associated bacteria. *Front Microbiol* 6:1551.

Minamisawa K, Hara S, Morikawa T, Wasai S, Kasahara Y, Koshiba T, Tokunaga T (2019). Identification of nitrogen-fixing *Bradyrhizobium* associated with roots of field-grown sorghum by metagenome and proteome analyses. *Front Microbiol* 10:407.

Miozzi L, Vaira AM, Catoni M, Fiorilli V, Accotto GP, Lanfranco L (2019). Arbuscular mycorrhizal symbiosis: Plant friend or foe in the fight against viruses? *Front Microbiol* 10:1238.

Monkiedje A, Spiteller M (2002). Effects of the phenylamide fungicides, mefenoxam and metalaxyl, on the microbiological properties of a sandy loam and a sandy clay soil. *Biol Fertil Soils* 35:393–398.

Ndour P, Gueye M, Barakat M, Ortet P, Bertrand-Huleux M, Pablo AL, Vigouroux Y (2017). Pearl millet genetic traits shape rhizobacterial diversity and modulate rhizosphere aggregation. *Front Plant Sci* 8:1288.

Nonaka S, Arai C, Takayama M, Matsukura C, Ezura H (2017). Efficient increase of γ-aminobutyric acid (GABA) content in tomato fruits by targeted mutagenesis. *Sci Rep* 7:1–14.

Norman JR, Atkinson D, Hooker JE (1996). Arbuscular mycorrhizal fungal-induced alteration to root architecture in strawberry and induced resistance to the root pathogen *Phytophthora fragariae*. *Plant Soil* 185:191–198.

Nortcliff S, Hulpke H, Bannick CG, Terytze K, Knoop, Bredemeier M, Schulte-Bisping H (2000). Soil, 1. Definition, function, and utilization of soil. In *Ullmann's Encyclopedia of Industrial Chemistry*. Doi: 10.1002/14356007.b07_613.pub3.

Orellana LH, Chee-Sanford JC, Sanford RA, Loffler FE, Konstantinidis KT (2018). Year-round shotgun metagenomes reveal stable microbial communities in agricultural soils and novel ammonia oxidizers responding to fertilization. *Appl Environ Microbiol* 84:e01646–17.

Ortiz-Castro R, Contreras-Cornejo HA, Macias-Rodriguez L, Lopez-Bucio J (2009). The role of microbial signals in plant growth and development. *Plant Signal Behav* 4:701–712.

Otieno N, Lally RD, Kiwanuka S, Lloyd A, Ryan D, Germaine KJ, Dowling DN (2015). Plant growth promotion induced by phosphate solubilizing endophytic *Pseudomonas* isolates. *Front Microbiol* 6:745.

Owen D, Williams AP, Griffith GW, Withers PJ (2015). Use of commercial bio-inoculants to increase agricultural production through improved phosphrous acquisition. *Appl Soil Ecol* 86:41–54.

Pan J, Ren D (2009). Quorum sensing inhibitors: A patent overview. *Expert Opin Therap Patents* 19:1581–1601. Doi: 10.1517/13543770903222293.

Park YG, Mun BG, Kang SM, Hussain A, Shahzad R, Seo CW, Lee IJ (2017). *Bacillus aryabhattai* SRB02 tolerates oxidative and nitrosative stress and promotes the growth of soybean by modulating the production of phytohormones. *PLoS One* 12:e0173203. Doi: 10.1371/journal.pone.0173203.

Patel T, Saraf M (2017). Biosynthesis of phytohormones from novel rhizobacterial isolates and their in vitro plant growth-promoting efficacy. *J Plant Interact* 12:480–487.

Paul D, Nair S (2008). Stress adaptations in a plant growth promoting rhizobacterium (PGPR) with increasing salinity in the coastal agricultural soils. *J Basic Microbiol* 48:378–384.

Postgate J (1998). The origins of the unit of nitrogen fixation at the University of Sussex. *Notes Records Royal Soc London* 52:355–362. Doi: 10.1098/rsnr.1998.0055.

Qi LS, Larson MH, Gilbert LA, Doudna JA, Weissman, JS, Arkin AP, Lim WA (2013). Repurposing CRISPR as an RNA-guided platform for sequence-specific control of gene expression. *Cell* 152:1173–1183.

Raghavendra MP (2017). Quorum sensing in plant microbe interaction. In *Agriculturally Important Microbes for Sustainable Agriculture* (pp. 87–110). Springer, Singapore. Doi: 10.1007/978-981-10-5589-8_5.

Raj G, Shadab M, Deka S, Das M, Baruah J, Bharali R, Talukdar NC (2019). Seed interior microbiome of rice genotypes indigenous to three agroecosystems of Indo-Burma biodiversity hotspot. *BMC Genom* 20:924.

Ramadan EM, AbdelHafez AA, Hassan EA, Saber FM (2016). Plant growth promoting rhizobacteria and their potential for biocontrol of phytopathogens. *Afr J Microbiol Res* 10:486–504.

Ramegowda V, Senthil-Kumar M (2015). The interactive effects of simultaneous biotic and abiotic stresses on plants: Mechanistic understanding from drought and pathogen combination. *J Plant Physiol* 176:47–54.

Rampelotto PH (2013). Extremophiles and extreme environments. *Life* 3:482–485. Doi: 10.3390/life3030482.

Rashid MI, Mujawar LH, Shahzad T, Almeelbi T, Ismail IM, Oves M (2016). Bacteria and fungi can contribute to nutrients bioavailability and aggregate formation in degraded soils. *Microb Res* 183:26–41.

Robson RL, Postgate JR (1980). Oxygen and hydrogen in biological nitrogen fixation. *Annu Rev Microbiol* 34:183–207.

Rocha I, Duarte I, Ma Y, Souza-Alonso P, Latr A, Vosatka M, Oliveira RS (2019). Seed coating with arbuscular mycorrhizal fungi for improved field production of chickpea. *Agronomy* 9:471.

Saharan BS, Verma S (2014). Potential plant growth promoting activity of *Bacillus lichenifor-mis* UHI (II) 7. *Int J Microbial Res Technol* 2:22–27.

Salgaonkar BB, Bragança JM (2019). Production of polyhydroxyalkanoates by extremophilic microorganisms through valorization of waste materials. In *Advances in Biological Science Research* (pp. 419–443). Academic Press. Doi: 10.1016/B978-0-12-817497-5.00026-4.

Sayyed RZ, Chincholkar SB, Reddy MS, Gangurde NS, Patel PR (2013). Siderophore producing PGPR for crop nutrition and phytopathogen suppression. In *Bacteria in Agrobiology: Disease Management* (pp. 449–471). Springer, Berlin, Heidelberg. Doi: 10.1007/978-3-642-33639-3_17.

Schloss PD, Handelsman J (2003). Biotechnological prospects from metagenomics. *Curr Opin Biotechnol* 14:303–310.

Sharma A, Johri BN, Sharma AK, Glick BR (2003). Plant growth-promoting bacterium *Pseudomonas* sp. strain GRP3 influences iron acquisition in mung bean (*Vigna radiata* L. Wilzeck). *Soil Biol Biochem* 35:887–894.

Sharma A, Vaishnav A, Jamali H, Srivastava AK, Saxena AK, Srivastava AK (2016). Halophilic bacteria: Potential bioinoculants for sustainable agriculture and environment manage-ment under salt stress. In *Plant-Microbe Interaction: An Approach to Sustainable Agriculture* (pp. 297–325). Springer, Singapore. Doi: 10.1007/978-981-10-2854-0_14.

Shelake RM, Pramanik D, Kim JY (2019). Exploration of plant-microbe interac-tions for sustainable agriculture in CRISPR era. *Microorganisms* 7:269. Doi: 10.3390/microorganisms7080269.

Shelobolina E, Roden E, Benzine J, Xiong MY (2014). *Using Phyllosilicate-Fe(ii)-Oxidizing Soil Bacteria to Improve Fe and K Plant Nutrition*. U.S. Patent Application No. US20160046535A1.14/209, 509. https://patents.google.com/patent/US20160046535A1/en.

Shivprasad S, Page WJ (1989). Catechol formation and melanization by Na$^+$-dependent *Azotobacter chroococcum*: A protective mechanism for aeroadaptation? *Appl. Environ Microbiol* 55:1811–1817.

Shukla PS, Agarwal PK, Jha B (2012). Improved salinity tolerance of *Arachishypogaea* (L.) by the interaction of halotolerant plant-growth-promoting rhizobacteria. *J Plant Growth Regul* 31:195–206.

Shukla VK, Doyon Y, Miller JC, DeKelver RC, Moehle EA, Worden SE, Choi VM (2009). Precise genome modification in the crop species Zea mays using zinc-finger nucleases. *Nature* 459:437–441.

Sleator RD, Hill C (2002). Bacterial osmoadaptation: The role of osmolytes in bacterial stress and virulence. *FEMS Microbiol Rev* 26:49–71.

Sonmez I, Kaplan M, Sonmez S (2007). Investigation of seasonal changes in nitrate contents of soils and irrigation waters in greenhouses located in Antalya-Demn region. *Asian J Chem* 19:5639.

Theunis M, Kobayashi H, Broughton WJ, Prinsen E (2004). Flavonoids, NodD1, NodD2, and nod-box NB15 modulate expression of the y4wEFG locus that is required for indole-3-acetic acid synthesis in *Rhizobium* sp. strain NGR234. *Mol Plant Microbe Interact* 17:1153–1161.

Urnov FD, Rebar EJ, Holmes MC, Zhang HS, Gregory PD (2010). Genome editing with engi-neered zinc finger nucleases. *Nat Rev Genet* 11:636–646.

Vaishnav A, Kumari S, Jain S, Varma A, Choudhary DK (2015). Putative bacterial vola-tile-mediated growth in soybean (Glycine max L. Merrill) and expression of induced proteins under salt stress. *J Appl Microbiol* 119:539–551.

Verma JP, Jaiswal DK, Krishna R, Prakash S, Yadav J, Singh V (2018). Characterization and screening of thermophilic *Bacillus* strains for developing plant growth promoting consortium from hot spring of Leh and Ladakh region of India. *Front Microbiol* 9:1293.

Verma P, Suman A (2018). Wheat microbiomes: Ecological significances, molecular diversity and potential biore-sources for sustainable agriculture. *EC Microbiol* 14:641–665.

Verma P, Yadav AN, Khannam KS, Saxena AK, Suman A (2017). Potassium-solubilizing microbes: Diversity, distribution, and role in plant growth promotion. In *Microorganisms for Green Revolution* (pp. 125–149). Springer, Singapore. Doi: 10.1007/978-981-10-6241-4_7.

Walters WA, Jin Z, Youngblut N, Wallace JG, Sutter J, Zhang W, Knight R (2018). Large-scale replicated field study of maize rhizosphere identifies heritable microbes. *Proc Nat Acad Sci* 115:7368–7373. Doi: 10.1073/pnas.1800918115.

Wang T, Zhang H, Zhu H (2019). CRISPR technology is revolutionizing the improvement of tomato and other fruit crops. *Hortic Res* 6:1–13.

Wang Y, Huang J, Gao Y (2012). Arbuscular mycorrhizal colonization alters subcellular distribution and chemical forms of cadmium in *Medicago sativa* L. and resists cadmium toxicity. *PLoS One* 7:e48669. Doi: 10.1371/journal.pone.0048669.

Wang Y, Wang M, Li Y, Wu A, Huang J (2017). Enhancements of arbuscular mycorrhizal fungi on growth and nitrogen acquisition of *Chrysanthemum morifolium* under Salt Stress. Doi: 10.20944/preprints201710.0183.v1.

Westerberg K, Elväng AM, Stackebrandt E, Jansson JK (2000). *Arthrobacter chlorophenolicus* sp. nov., a new species capable of degrading high concentrations of 4-chlorophenol. *Int J Syst Evol Microbiol* 50:2083–2092.

Xu L, Naylor D, Dong Z, Simmons T, Pierroz G, Hixson KK, Gao C (2018). Drought delays development of the sorghum root microbiome and enriches for monoderm bacteria. *Proc Nat Acad Sci* 115:E4284–E4293. Doi: 10.1073/pnas.1717308115.

Yadav AN (2017). Beneficial role of extremophilic microbes for plant health and soil fertility. *J Agric Sci Bot* 1:30–33.

Yadav AN, Saxena AK (2018). Biodiversity and biotechnological applications of halophilic microbes for sustainable agriculture. *J Appl Biol Biotechnol* 6:48–55.

Yadav AN, Sharma D, Gulati S, Singh S, Dey R, Pal KK, Saxena AK (2015a). Haloarchaea endowed with phosphorus solubilization attribute implicated in phosphorus cycle. *Sci Rep* 5:1–10.

Yadav AN, Verma P, Kumar M, Pal KK, Dey R, Gupta A, Prasanna R (2015b). Diversity and phylogenetic profiling of niche-specific *Bacilli* from extreme environments of India. *Ann Microbiol* 65:611–629.

Zahran HH (1999). Rhizobium-legume symbiosis and nitrogen fixation under severe conditions and in an arid climate. *Microbiol Mol Biol Rev* 63:968–989.

Zarraonaindia I, Owens SM, Weisenhorn P, West K, Hampton-Marcell J, Lax S, van der Lelie D (2015). The soil microbiome influences grapevine-associated microbiota. *MBio* 6:e02527-14. Doi: 10.1128/mBio.02527-14.

Zhang H, Kim MS, Krishnamachari V, Payton P, Sun Y, Grimson M, Pare PW (2007). Rhizobacterial volatile emissions regulate auxin homeostasis and cell expansion in arabidopsis. *Planta* 226:839. Doi: 10.1007/s00425-007-0530-2.

Zhang Y, Massel K, Godwin ID, Gao C (2018). Applications and potential of genome editing in crop improvement. *Genome Biol* 19:210.

Zuniga A, Donoso RA, Ruiz D, Ruz GA, Gonzalez B (2017). Quorum-sensing systems in the plant growth-promoting bacterium *Paraburkholderia phytofirmans* PsJN exhibit cross-regulation and are involved in biofilm formation. *Mol Plant Microbe Interact* 30:557–565.

15 A Consortium of Sulfate-Reducing Bacteria Used for Lead, Copper, and Cadmium Bioremediation

Julia Mariana Márquez-Reyes, Julián Gamboa-Delgado, Fatima Estefanía Soto-Zamora, Juan Antonio Vidales-Contreras, Humberto Rodríguez-Fuentes, Alejandro Isabel Luna-Maldonado and Celestino García-Gómez
Universidad Autónoma de Nuevo León

CONTENTS

15.1 INTRODUCTION

The proposals for remediation of contaminated waters and soils impacted by mining and metallurgical activities are established to reduce the mobility of potentially toxic elements (PTEs) such as heavy metals toward surface water bodies and groundwater tables. However, recent proposals have both economic and environmental drawbacks. The latter is associated with the removal and depletion of nutrients in the soil, which prevents the natural restoration of the impacted site in a short time. Therefore,

DOI: 10.1201/9781003247883-15

it is convenient to consider technological alternatives that allow control of effluent heavy metal stabilization operations using sulfate-reducing bacteria (SRB). These microorganisms have been studied in different systems such as upflow anaerobic reactors, stirred tank reactors, fluidized bed reactors, and packed reactors (Kaksonen et al., 2004a, b, Costa et al., 2007). The above-mentioned systems use low molecular weight and easily degradable organic compounds as a source of carbon and energy for the SRB, such as acetate, lactate, or glucose, produced by dissimilatory sulfur reduction, with biogenic sulfide (Cao et al., 2012; Kaksonen et al., 2004a, b). However, other studies use more complex and high molecular weight carbon sources that can generate hydrogen sulfide in anaerobic systems. The anaerobic degradation of complex carbon sources occurs by the microbial action of consortia with a wide metabolic diversity, which can promote the hydrolysis of the compound and generate simpler forms of carbon (from C_2 to C_4) as a carbon source for SRB or methanogenic bacteria. Contaminated water or soils are treated with the generation of low-solubility precipitates, such as metal sulfides, because metals cannot be destroyed. Although this stabilization reaction is purely chemical, the generation of sulfides from sulfates by microbial means is of great relevance in removing metals from solutions. It is common for metals-contaminated effluents to contain metallic sulfates and organic molecules and allow the availability of suitable substrates for the reduction reaction. These factors can be coupled to carry out effluent remediation with both types of pollutants, becoming a competitive alternative for other conventional chemical treatments (Yang et al., 2021). This chapter describes the kinetic behavior of SRB in the presence of heavy metals for future applications in the treatment of polluted water.

15.2   HEAVY METAL CONTAMINATION

Heavy metals are metallic elements with a relatively high density and considered toxic even in low concentrations, causing damage to the environment (Tun-Canto et al., 2017; Chandra and Kumar, 2018). These highly toxic pollutants, capable of generating adverse effects on human health, are generally referred to as potentially toxic elements (PTEs), a reason why heavy metal contamination is one of the biggest challenges for the environment. Most heavy metals tend to bioaccumulate and biomagnify in trophic networks, so they can persist within ecosystems. Among PTEs, arsenic, cadmium, copper, chromium, iron, manganese, mercury, nickel, lead, and zinc stand out. PTEs can be absorbed and assimilated by plants, in addition to reducing the soil quality and productivity due to their low mobility. The mobility and bioavailability of PTEs will depend on certain properties of the location site and its physicochemical properties such as pH, solubility, vapor pressure, density, cation exchange capacity, and competition between soluble ions (Luo et al., 2022). In addition, they are not degradable, which generates a serious pollution problem.

15.3   SOURCES OF PTES GENERATION IN WATER

Mining and smelting metallurgical processes are important sources of metals and metalloids (As, Pb, Cd, Hg, Se, Sb, Cu, Zn, among others) that contribute significantly to air, water, and soil pollution (Li et al., 2018a). One of the main causes of soil contamination in mining areas is the deficiency and inefficiency in waste handling containing PTEs in high concentrations. The atmospheric dust and smelting emissions represent the largest source of contamination from metallurgical smelting processes and, when dispersed, can seriously affect the ecosystem. However, soil contamination stands out as the main receptor and regulator on the transfer of PTEs to the atmosphere, hydrosphere, and biota.

The environmental pollution problems in historical mining sites show that although there is natural attenuation at these sites, the concentrations of PTEs can remain as sources of dispersion for up to hundreds of years, generating health problems, destruction, and alteration of habitats near mining sites. The mining activity sites are developed for the exploitation and benefit of polymetallic mineralization, which can experience arsenic and heavy metals contamination due to: (1) the dispersion of mineral particles from the waste deposits (tailings), promoted by the wind and/or water action (rain, rivers, and streams), (2) the mobility of dissolved metals during weathering or alteration of minerals. In the latter, the acid mine drainage (AMD) stands out (Dold, 2014).

AMD is generated from the oxidation of sulfide minerals and the leaching of associated metals of sulfurous rocks when exposed to air and water. The development of AMD is a time-dependent process and involves chemical and biological oxidation processes and physicochemical phenomena, including precipitation and encapsulation. The classical development of AMD is metal-rich and presents low pH values; however, the chemistry of water gradually becomes more acidic increasing metals concentrations, and it can turn over time from slightly alkaline to almost neutral, and finally acidic. These waters are always associated with a yellowish-ocher coloration of the affected river and lake beds. They contain a large number of suspended solids, total dissolved solids, radioactive nuclides, nutrients, acidity, and a high content of sulfate and dissolved metals (Fe, Al, Mn, Zn, Cu, Pb, etc.).

It should be considered that not all sulfurous minerals are reactive, nor does acidity occur in the same proportion, and not all are potentially generators of AMD. The oxidation of pyrite is mainly responsible for the AMD generation, which is favored in mining areas due to the air easy approach to sulfides (through access mining tasks and existing pores in the sterile and waste piles), as well as the increase in the contact surface of the particles. The volume, concentration, grain size, and spatial distribution of pyrite are identified as factors that mostly affect the acid generation (Kwong, 1993). In the first stages of AMD generation, neutralization is immediate. While the oxidation of sulfurous minerals occurs, there is enough alkalinity available ($CaCO_3$) to neutralize the acidity and precipitate iron in the form of hydroxide and increase in calcium, magnesium, or other metals, depending on the rocks spent for acid neutralization. Considerations about the kinetics of iron, redox, and precipitation reactions are important for the water quality problems such as AMD, where pyrite (FeS_2), chalcopyrite ($CuFeS_4$), enargite (Cu_3AsS_4), galena (PbS), sphalerite (ZnS), and arsenopyrite ($FeAsS$), among others, are rapidly oxidized, generating acidity and soluble Fe_3+ at pH values of 2 or 3 (Dold, 2014). The classic stoichiometric representation of AMD can be expressed by the following reactions.

$$4FeS_{2(s)} + 14O_2 + 4H_2O \leftrightarrow 4Fe^{2+} + 8H^+ + 8SO_4^{2-} \qquad (15.1)$$

$$4Fe^{2+} + 4H^+ + O_2 \leftrightarrow 4Fe^{3+} + 2H_2O \qquad (15.2)$$

$$4Fe^{3+} + 12H_2O \leftrightarrow 4Fe(OH)_{3(S)} + 12H^+ \qquad (15.3)$$

The net result of these reactions is that 4 moles of pyrite is oxidized to produce 4 moles of Fe $(OH)_{3(s)}$, causing the water to turn into yellowish-brown. Twelve moles of H^+ are produced, reacting with the calcareous minerals in the soil, which increases the hardness of the water and leads to a high content of total solids. Another important reaction of AMD is the oxidation of $FeS_2(s)$ with ferric ion (Fe^{3+}). The stoichiometry of this reaction is as follows (Ritchie et al., 1994).

$$FeS_{2(s)} + 14Fe^{3+} + 8H_2O \leftrightarrow 15Fe^{2+} + 2SO_4^{2-} + 16H^+ \qquad (15.4)$$

The chemical oxidation of pyrite (equation 15.4) is a relatively rapid reaction with pH values less than 4.5, but slower with pH levels greater than this value. The rate of oxygen supply is the element that primarily controls the rate of chemical oxidation. At pH levels below 4.5, ferric iron oxidation becomes the dominant oxidation process (Dold, 2010).

However, several microorganisms such as *Thiobacillus thiooxidans*, *Thiobacillus ferrooxidans*, and *Ferrobacillus ferrooxidans* have been able to catalyze the oxidation of the ferrous ion (Bryner, 1967; Johnson, 2003). The action of these microorganisms increases reaction rates from 10 to a million times more than those generated by chemical oxidation. However, this increase depends on other physicochemical parameters such as pH, oxygen concentration, carbon sources, and available nutrients, as well as the surface area of the exposed sulfide (Benner et al., 2002).

Thiobacillus ferrooxidans catalyzes a half-redox reaction of Fe^{2+} (equation 15.5).

$$Fe^{2+} + 0.25O_{2(aq)} + 2.5H_2O \leftrightarrow Fe(OH)_3 + 2H^+ \qquad (15.5)$$

In addition, the oxidation reactions of sulfides can be due to different oxidizing agents such as O_2 or Fe^{3+}, or catalyzed by different microorganisms. Reactions can occur completely (equation 15.6) or incompletely (equation 15.7), in which different oxidizing agents promote the release of metal ions into the aqueous phase for the sulfate formation (Alloway, 1995; Costello, 2003).

$$MeS + 8Fe^{3+} + 4H_2O \rightarrow Me^{2+} + SO_4^{2-} + 8Fe^{2+} + 8H^+ \qquad (15.6)$$

$$MeS_{(s)} + 0.5O_{2(ac)} + 2H^+ \rightarrow Me^{2+} + S^0 + H_2O \qquad (15.7)$$

Here, Me represents any metal ion, or arsenic, for example As^{3+}, Pb^{2+}, Cd^{2+}, Cu^{2+}, Ni^{2+}, Zn^{2+}, among others. In the future (decades or centuries) after the start of the acids generation, the rate will decrease with the complete oxidation of the more reactive sulfides, the pH will increase until the rock becomes only slightly reactive, and the pH of the drain water will not be affected. Rather than waiting for the ecosystem to try to cushion the damage, it is more convenient to promote mitigation measures to control the ecosystem in less time, thorough the knowledge about the AMD production and its consequences.

15.4 LEAD, CADMIUM, AND COPPER TOXICITY

Heavy metals can be divided into two types: The first are considered trace elements and correspond to essential metals that help maintain an adequate metabolism in living beings, particularly in human body, such as copper, selenium, or zinc. However, when these are found in relatively high concentrations, they can cause intoxication and specific symptoms, depending on the element (Figure 15.1). The second type corresponds to non-essential metals such as cadmium and lead, as well as metalloids (i.e., arsenic and mercury), all considered highly toxic to humans, wildlife, and plants (Tun-Canto et al., 2017).

Lead can too mimic the action of calcium, generating an interruption of signal transduction that can affect synaptic transmission, calcium channels, calmodulin-dependent enzyme system, neuronal differentiation, permeability of brain capillaries, neuroendocrine function, protein phosphorylation, and catecholamine synthesis, among others. In addition, lead binds to proteins, particularly to those of sulfhydryl groups, less frequently to phosphate and carboxyl groups; therefore, its structure and function can be altered, or it may compete with other metals on the linking sites. Also, there have been reports of enzyme inhibition in the heme group synthesis, which can result in anemia, decreased availability of cytochromes for the respiratory chain, and also, accumulation of toxic metabolites, such as the δ-aminolevulinic acid (Matte, 2003). On the other hand, lead is also an oxidizing substance, so it can affect different organs producing oxidative stress. In the reproductive tract, lead has a negative effect on spermatogenesis (Rubio et al., 2006).

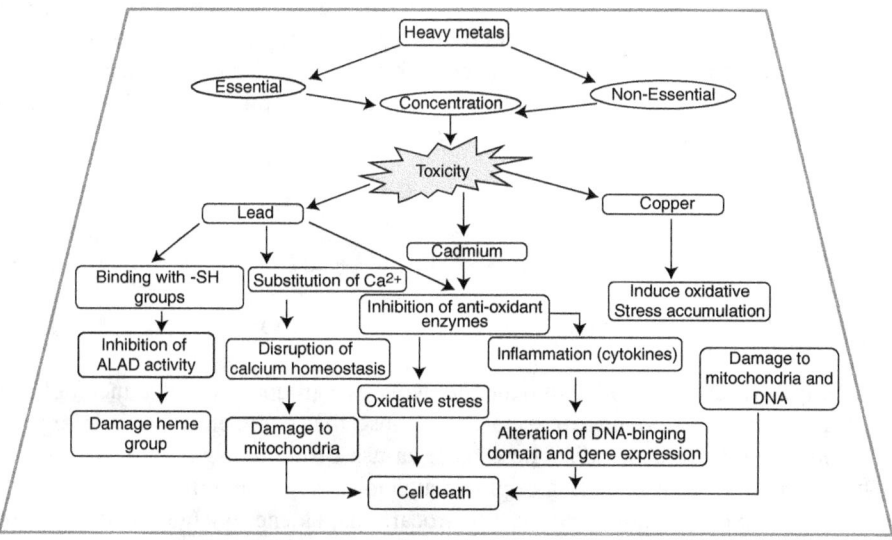

FIGURE 15.1 Mechanisms of lead, copper, and cadmium toxicity.

The main toxic effects of cadmium (Cd) are the generation of chemical pneumonitis, renal dysfunction with proteinuria and microproteinuria, and emphysema. Exposure to cadmium has been associated with inflammation involving the production and release of cytokines such as interleukin-1ß (IL-1ß), interleukin-6 (IL-6), interleukin-8 (IL-8), and tumor necrosis factor alpha (TNF-a) (Martínez-Flores et al., 2013). Cd is nephrotoxic; since kidney damage manifests with proteinuria, there is an increase in low molecular weight proteins, such as transferrin, albumin, N-acetyl-ß-D-glucosaminidase (NAG), and ß2-microglobulin (Nogué et al., 2004). Exposure to Cd has been associated with high mortality due to cancer, and it has recently been identified that Cd regulates the expression of the AEG-1 gene, whose encoded protein could be important for tumor development, progression, and metastasis in breast cancer cells (Luparello et al., 2012).

Although copper is essential for protein formation, in its free state, it can become toxic; due to this reason, the apoproteins regulate its accumulation in the body. When copper is absorbed in a higher concentration than metabolically required, the bile is responsible for excreting it. Long-term exposure to copper can cause irritation in the nose, mouth, and eyes; headache; dizziness; nausea; and diarrhea (Feoktistov et al., 2018).

15.5 SULFATE-REDUCING BACTERIA

The sulfate-reducing bacteria (SRB) are anaerobic, so they use sulfate as the final electron acceptor and decompose organic matter into biogas. SRB transforms the sulfate anion into hydrogen sulfide through a dissimilatory pathway (equation 15.8). Then, the hydrogen sulfide reacts with the dissolved metal ions resulting in the precipitation of the metal (equation 15.9), thus reducing its toxicity and adjusting the pH of the medium where these are found due to the formation of bicarbonates produced by microbial metabolism, which in turn can react with protons to form carbon dioxide and water (equation 15.10). SRB grows in reducing environments and in the absence of oxygen. They grow best under slightly alkaline conditions, where the pH range is from 7.0 to 7.8. Most SRB are mesophiles, with optimal growth in the temperature range from 25°C to 40°C.

$$2CH_2O + SO_4^{2-} = 2HCO_{3-} + H_2S \tag{15.8}$$

$$H_2S + M^{2+} \rightarrow MS \left(\downarrow\right) + 2H^+ \tag{15.9}$$

$$HCO_{3-} + H^+ \rightarrow CO_2 \left(g\right) + H_2O \tag{15.10}$$

Currently, SRB can be divided into the following two groups: those that incompletely degrade organic compounds to acetate and those that completely degrade organic compounds to carbon dioxide. These bacteria use different organic compounds of low molecular weight, including aliphatic alcohols, mono- and di-carboxylic acids, polar aromatic compounds, and even hydrocarbons, as energy sources. Among the organic compounds most used as substrates are lactate, acetate, pyruvate, ethanol, propanol, and glucose (Suzuki et al., 2007). Lactate is known to be the most used carbon source by a majority of SRB species; however, most bacterial species partially oxidize it to acetate and CO_2, thus requiring a large amount of substrate to reduce sulfate. This requirement generates effluents with a high chemical oxygen demand (COD). The minimum ratio required to theoretically achieve total sulfate removal is 0.67. This means that the conversion of 1 g of sulfate requires 0.67 g of COD of the organic compound found in the water to be treated (Oude Elferink et al., 1994).

The sulfate reduction process is a valuable biotechnological tool for metal removal in mine leachate and industrial effluents. It is potentially superior to other biological processes due to its ability to produce alkalinity and neutralize the pH of acidic waters, besides having the power to simultaneously remove organic matter, sulfates, heavy metals, metalloids, radioactive isotopes, and cyanides (Li et al., 2018a). The enormous advantage of SRB bioprecipitation is the ability to selectively separate the formed metal sulfides. These sulfides are highly insoluble at neutral pH, while some of them, such as copper sulfide, are insoluble when the pH is 2. It has been shown that each metal precipitates at a unique sulfide concentration or potential sulfide, which is directly related to the solubility of the metal sulfide product. By controlling the pH value, pure metal sulfides can be selectively precipitated and reused.

The evolution of the microbial culture over time follows a typical curve called Monod or batch growth (Monod, 1949). As shown in Figure 15.2, an initial phase is observed where there is practically no cell division, but there is an increase in the individual mass of microorganisms ("lag" phase or lag phase). Then, a stage appears where growth occurs at a specific maximum rate (q) and constant, q_{max} (exponential phase). At the end of this phase, the maximum microbial concentration is reached. Followed by a rapid deceleration period where $q \rightarrow 0$ and the stationary phase is entered, this is caused by the depletion of nutrients (the limiting substrate) or by accumulation of inhibitors. During this phase, the microbial (or biomass) concentration remains constant.

Finally, the biomass concentration decreases due to autolysis or as a consequence of endogenous metabolism (decay phase). The lag phase depends on the growth phase

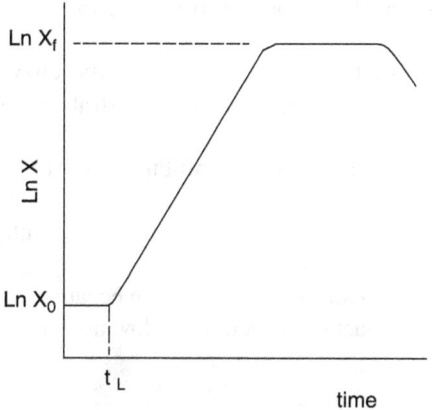

FIGURE 15.2 Bacterial growth curve: ln X vs time.

in which the cells are, at the time of being seeded, as well as on the composition of the culture medium in which they were grown. The mathematical description of the exponential phase and stationary growth (r_x) is as follows.

$$r_x = q \cdot X, \tag{15.11}$$

where X = biomass concentration. We have seen that q (microbial growth rate) varies during cultivation, being a constant and maximum value in the exponential phase (q_{max}) and null in the stationary phase. Monod proposes a very simple relationship between the value of q and the substrate concentration S (equation 15.12).

$$q = q_{max} \frac{S}{K_s + S}, \tag{15.12}$$

where K_s is the constant affinity to substrate and gives an idea of the affinity that the microorganism has for the substrate used as a carbon source; therefore, the lower it is, the higher the affinity. Normally, it has very small values ($10^{-2} - 10^{-3}$ g/L), so relatively small concentrations of S are sufficient to have the following:

$$q = q_{max} \tag{15.13}$$

Replacing equation (15.12) in (15.13) gives:

$$r_x = q_{max} \frac{S}{K_s + S} \cdot X. \tag{15.14}$$

Another important parameter to study is the inhibition of microorganisms (Muloiwa et al., 2020) by some environmental or metabolic factors (generated by the microorganisms themselves). The types of inhibitions that exist are the same as those demonstrated in enzyme kinetics (Walsh, 2018).

Competitive: The inhibitor is structurally related to the substrate; these substrate analogs compete with the actual substrate for the active site of the enzyme; the degree of inhibition depends on the inhibitor-to-substrate ratio rather than on the concentration of the inhibitor.

Uncompetitive: The inhibitor reacts only with the enzyme-substrate complex. This type of inhibition occurs frequently in bi-substrate reactions; it is very rare in single-substrate reactions.

Non-competitive: The binding of the inhibitor with the enzyme takes place at a different site than its active center; the inhibitor can react with the free enzyme or with the enzyme-substrate complex. Non-competitive inhibition often occurs in enzymes containing -SH groups.

This work focuses on non-competitive inhibition because the affected variable is the specific speed of sulfur production, which will allow us to evaluate the said behavior.

$$q = -\frac{q_{max} \cdot S}{(K_s + S)\left(1 + \dfrac{I}{K_I}\right)}, \tag{15.15}$$

where

 q = Sulfide production rate (mmol H_2S/mg SSV h)
 q_{max} = Maximum sulfide production rate (mmol H_2S/mg SSV h)
 S = Substrate concentration q_{max} (g DQO/L)
 K_s = Substrate affinity constant (g/DQO L)
 I = Inhibitor concentration (mg/L).

15.6 HYDROGEN SULFIDE PRODUCTION KINETICS IN THE PRESENCE OF PEAT MOSS

A SRB consortium isolated from a site that was used for mining activities from the late 19th century to 1993 was used. In batch reactors, the SRB were conditioned in a modified Postgate nutrient medium (1 g/L NH_4Cl; 0.5 g/L KH_2PO_4; 4.5 g/L Na_2SO_4; 0.04 g/L $CaCl_2 \cdot 2H_2O$; 0.06 g/L $MgSO_4 \cdot 7H_2O$; 0.004 g/L $FeSO_4 \cdot 7H_2O$; 0.6 g/L $Na_3C_6H_5O_7$; 2.3 mL/L $C_3H_5NaO_3$; 2.5 g/L $C_2H_3NaO_2$) at a temperature of 35°C and a pH of 6.6. The production of hydrogen sulfide was quantified using UV-Vis spectrophotometry, while biomass concentration was determined by the mass of suspended volatile solids (SSVs). During the conditioning phase, a specific concentration of an average production of 40 mM H_2S/mg SSV was quantified during 20 days of daily sampling.

To lower the operating costs of a sulfate-reducing system, the use of different concentrations of peat moss (0.5, 1, 1.5, 2, and 2.5 g/L) was proposed as the main source of carbon, thus eliminating all common sources in the culture medium (Postgate). Peat moss is a slowly degrading carbon source that, although composed of complex organic matter (lignin, cellulose, and humic substances), can be used by SRB communities as an alternative carbon source (Márquez-Reyes et al., 2013). The specific concentration of hydrogen sulfide produced in the presence of peat moss increased as the concentration of peat moss in the medium also increased (28.78 mM H_2S/mg SSV and 30.34 mM H_2S/mg SSV with 2 and 2.5 g/L, respectively), while at 0.5 and 1.5 g/L, concentrations of 17.29 mM H_2S/mg SSV and 24.38 mM H_2S/mg SSV were reached, respectively.

To explain microbial growth, the Monod equation (equation 15.12) was used, which relates to the specific growth rate and is mathematically analogous to the Michaelis-Menten expression for enzyme kinetics (Moo-Young and Butler, 2011). This mathematical expression can be applied to determine the μ_{max} and K_s within a batch reactor. It is essential to consider that the nutrients are in excess and that in the exponential growth phase, the estimation of the specific growth rate needs to be carried out, which is equal to the maximum specific growth rate (Doran, 2013). The μ_{max} production speed was 0.39 mmol H_2S/g SSV^{-h}, while the affinity constant to the substrate K_s was 453 mg DQO/L. The K_s constant represents the biomass affinity for a certain substrate. The term affinity is given by the relationship between the specific growth rate and the constant of affinity to the substrate; that is, high values of K_s correspond to a low affinity for the substratum. Previous studies that have used peat moss as a carbon source obtained K_s values of 740 mg DQO/L and μ_{max} values of 1.13 mmol H_2S/g SSV^{-h} (Márquez-Reyes et al., 2013). In the present study, a DQO / SO_4^{2-} ratio equal to 0.413 was estimated. However, in similar tests, the DQO / SO_4^{2-} ratios varied in the range from 0.85 to 1.25, favoring the reducing sulfate process and improving its efficiency, in addition to preventing the coexistence of methanogenic communities that compete for the substrate (Sahinkaya and Gungor, 2010, Gu et al., 2021). Now, biological diversity, source and concentration of carbon, sulfate concentration, the presence of trace metals, and environmental factors such as pH and temperature can be decisive in the kinetics of hydrogen sulfide production, which directly influences the formation of metallic precipitates (Patidar and Tare, 2005).

15.7   PRODUCTION OF HYDROGEN SULFIDE IN THE PRESENCE OF CADMIUM, COPPER, AND LEAD

The SRB consortium isolated from the contaminated site was established in batch reactors that contained salts (1 g/L NH_4Cl; 0.5 g/L KH_2PO_4; 4.5 g/L Na_2SO_4; 0.04 g/L $CaCl_2\cdot2H_2O$; 0.06 g/L $MgSO_4\cdot7H_2O$; 0.004 g/L $FeSO_4\cdot7H_2O$), 1 g of peat moss, and different concentrations of Cd, Cu, and Pb (5, 10, 15, 20, and 50 mg/L). In the presence of cadmium, the highest production of hydrogen sulfide occurred at 5 mg/L (0.55 mM H_2S/mg SSV); however, a decrease in the specific production of hydrogen sulfide was observed as the cadmium concentration increased (Figure 15.3). When hydrogen sulfide reacts with free cadmium, a yellow precipitate characteristic of cadmium sulfide is formed (Espinosa-Rosas, 2010). It has been previously reported that

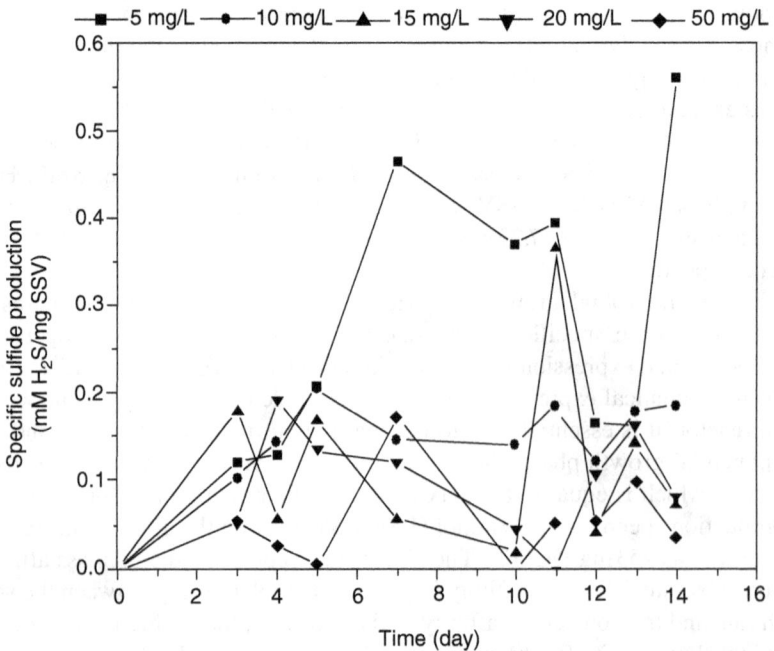

FIGURE 15.3   The specific production of H_2S by sulfate-reducing bacteria (SRB) exposed to different concentrations of cadmium, using peat moss as a carbon source.

the sulfate reduction process and the growth of SRB can be inhibited with cadmium concentrations higher than 4 and 20 mg/L, respectively (López-Pérez et al., 2015).

When hydrogen sulfide reacts with copper, a bluish copper sulfide precipitate forms. Depending on the tonality, it can be related to the ion concentration in the solution. For example, a blue/black precipitate is generated at intermediate concentrations ($\geq$ 10 mM), while a dark blue precipitate with a greenish tint occurs at the highest concentrations (50 mM) of copper in solution (Lewis, 2010). In the presence of copper, the production of H_2S (Figure 15.4) showed a slight increase when the metal concentration reached 5 mg/L (1.35 mM H_2S/mg SSV). During the last experimental days, it was observed that the higher the concentration of copper, the lower the production of hydrogen sulfide. Willis-Poratti (2016) concluded that copper can inhibit the growth of a consortium with sulfate-reducing capacity at a concentration of 100 mg/L. In similar studies in which a pure culture of D. vulgaris exposed to 0.9 mg Cu/L was used, an inhibitory effect was observed, directly related to the decrease in microbial growth (Cabrera et al., 2006).

The reaction of soluble lead with hydrogen sulfide produces lead sulfide precipitates, which have an ocher tonality. The H_2S production increased in the presence of lead, regardless of the concentration (Figure 15.5); however, from day 20 until the end of the experimental period, the hydrogen sulfide production decreased for all

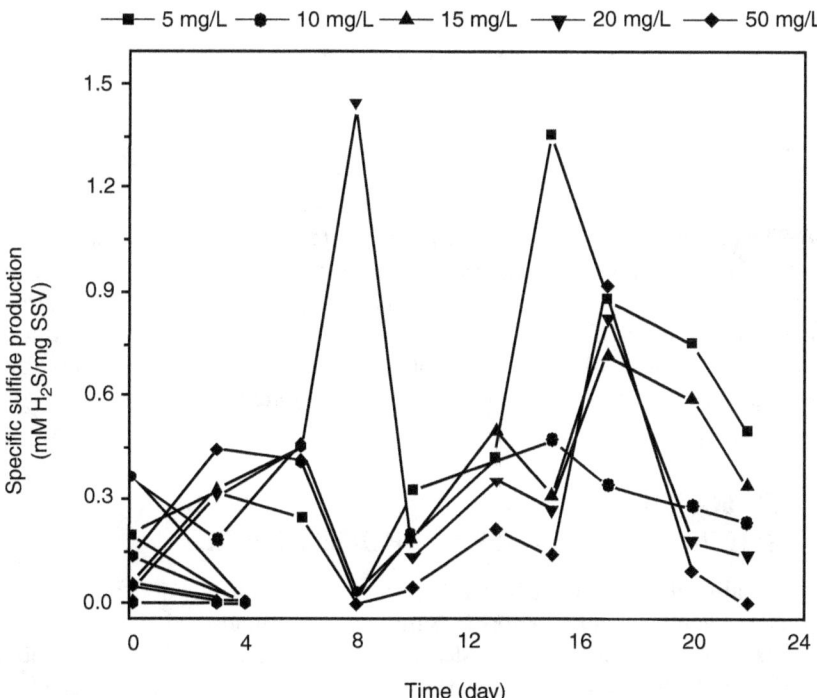

FIGURE 15.4 The specific production of H_2S by sulfate-reducing bacteria (SRB) exposed to different concentrations of copper, using peat moss as a carbon source.

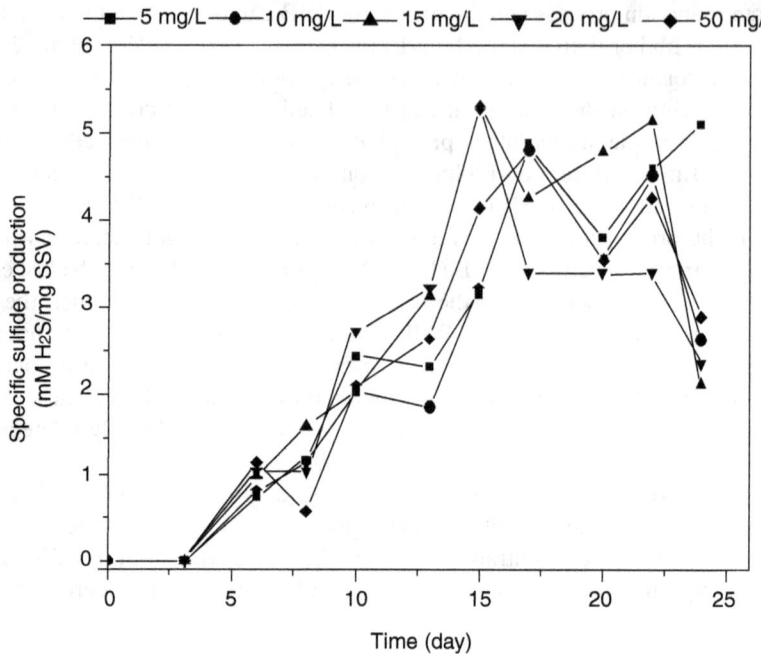

FIGURE 15.5   The specific production of H₂S by sulfate-reducing bacteria (SRB) exposed to different concentrations of lead, using peat moss as a carbon source.

treatments, except for the concentration of 5 mg/L where it continued to increase. This behavior is different from that previously reported, where regardless of the lead concentration, the production of hydrogen sulfide does not decrease (Martins et al., 2009). The decreasing effect must be directly related to the microbial diversity present in the consortium, since it showed a higher sensitivity at longer exposure times; similar results were obtained by Hwang and Jho (2018). On the other hand, the production of hydrogen sulfide continued to increase with the lower concentration of lead, indicating that this is a tolerable concentration for the SRB of the consortium since the production of hydrogen sulfide was not inhibited.

15.8   EFFECT OF THE CU, CD, AND PB MIXTURE IN THE PRODUCTION OF HYDROGEN SULFIDE

The same saline medium described above was used as control, but 0.5 g of peat moss was added, where a total of 14 mM H₂S/mg SSV was quantified. Then, 10 mg/L of each metal (Cu, Cd, and Pb) was added to the rest of the experiments, quantifying 4 mM H₂S/mg SSV (Figure 15.6). Hydrogen sulfide production was reduced by 71% in the medium compared to the control. Insoluble metal ions and metal sulfides have been reported to be inhibitory or toxic agents for SRB. This is because metals can deactivate enzymes and denature proteins, causing negative impacts on growth and

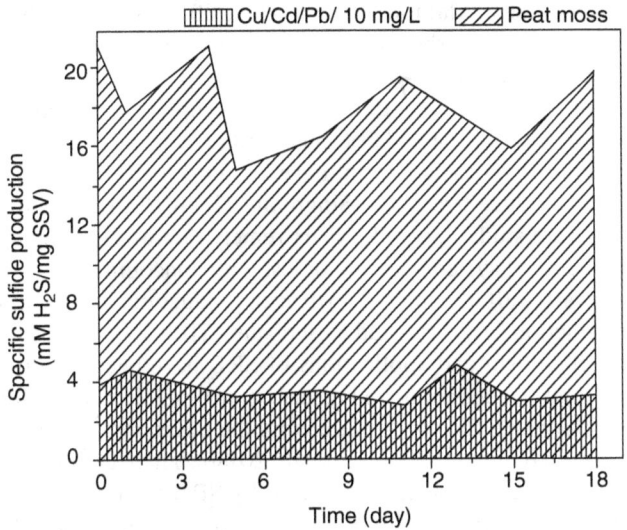

FIGURE 15.6 The specific production of H₂S by sulfate-reducing bacteria (SRB) exposed to 10 mg/L concentration of mixed copper, cadmium, and lead, using peat moss (0.5 mg/L).

bacterial activity. Other studies have shown that the addition of metals in a mixture generates a toxicity-reducing effect in a synergistic manner between metals, since SRB is inhibited at higher concentrations of heavy metals (Costa et al., 2021).

Sufficient hydrogen sulfide must be present to react with dissolved metal ions and allow the formation of insoluble metal complexes, avoiding toxicity from excess metal ions in the sulfate reduction process. When analyzing the production of hydrogen sulfide in the presence of each metal and a mixture, it is clear that the mixture has a synergistic effect between the metals that reduces the degree of toxicity. Higher production of hydrogen sulfide was quantified in the presence of the metal mixture than individually for copper and cadmium. Peat moss acts as a natural adsorbent for metal ions in solution, since it has a slow degradation capacity and generates a slow release of the adsorbed metal, thus controlling toxicity in the culture medium.

15.9 FUTURE TRENDS

Sulfate-reducing bacteria have proven to be an efficient and sustainable alternative in the bioremediation of different types of contaminants. The knowledge on the different enzymatic and genetic mechanisms that SRB use to stabilize or immobilize pollutants in a natural or synthetic environment has been deepened. New designs and modifications to bioreactors have favored the use of SRB, improving the maintenance conditions of microbial cultures and increasing the removal efficiency of all types of contaminants. There remain several challenges in the implementation of this remediation technique (seasonal temperature variations, maintenance of optimum sulfate concentrations, regulating the concentration and types of iron, and precipitation of trace metals) that need to be overcome for successful treatment technologies in actual field conditions (Alam and McPhedran, 2019). Different types of organic waste have been tested as a carbon source for bacteria, but there is still a wide variety of the agro-industrial waste that has the opportunity to be used and would reduce the operating costs of bioreactors. Among the widely employed microbial technologies, certain bacteria species are capable of generating electrons from their cell surface while decomposing organic matter and have been explored to generate bioelectricity or modified to produce hydrogen (Singh and Singh, 2021). The sulfate-reducing microorganisms were studied, which are extremely interesting biocatalysts for multiple bioelectrochemical systems technologies, with concomitant CO_2 fixation. They can accomplish water sulfate removal, hydrogen production, and in some cases, even biochemical production (Agostino and Rosenbaum, 2018; González-Paz et al., 2020).

Studies on metabolism and applications of SRB have been conducted, they still remain incompletely understood and even controversial, and understanding the metabolism of SRB paves the way for allowing the microorganisms to provide more beneficial services in bioremediation (Li et al., 2018b). There are many gaps in theory about the precise biological mechanisms used by SRB to counteract the toxicity of pollutants, the main chemical and biological inhibitors, and the intermediate products generated in the different reaction processes, which could cause worse problems to the environment and population health.

Considering these facts, the use of SRB will continue to be an attractive alternative to be used in the bioremediation of contaminated soils and waters and multiple bioelectrochemical systems technologies.

15.10 CONCLUSION

Sulfate-reducing bacteria can produce hydrogen sulfide, which reacts with dissolved metal ions in water to form stable metal precipitates, reducing their mobility and toxicity. The microbial diversity of the SRB consortium is important as some microorganisms can be inhibited by metal toxicity, but others can still produce hydrogen sulfide. The use of slow degradation carbon sources as adsorbents helps to control the toxicity of the metal since they are slowly released it into the environment, favoring the continuous production of hydrogen sulfide. The use of SRB represents an efficient pollution control strategy, which can be implemented in the treatment of contaminated effluents by heavy metals.

REFERENCES

Agostino, V., and Rosenbaum, M.A. 2018. Sulfate-reducing electro autotrophs and their applications in bioelectrochemical systems. *Frontiers in Energy Research*, 6, 55.

Alam, R., and McPhedran, K. 2019. Applications of biological sulfate reduction for remediation of arsenic – A review. *Chemosphere*, 222, 932–944.

Alloway, B. J. 1995. *Heavy Metals in Soils: Blackie Academic and Professional*. Chapman and Hall: London.

Benner, S.G., Blowes, D.W., Ptacek, C.J., and Mayer, K.U. 2002. Rates of sulfate reduction and metal sulfide precipitation in a permeable reactive barrier. *Applied Geochemistry*, 17, 301–320.

Cabrera, G., Pérez, R., Gómez, J.M., Ábalos, A., and Cantero, D. 2006. Toxic effects of dissolved heavy metals on *Desulfovibrio vulgaris* and *Desulfovibrio* sp. Strains. *Journal of Hazardous Materials*, 135, 40–46.

Cao, J., Zhang, G., Mao, Z.S., Li, Y., Fang, Z., and Yang, C. 2012. Influence of electron donors on the growth and activity of sulfate-reducing bacteria. *International Journal of Mineral Processing*, 106, 58–64.

Candra, R. and Kumar, V., 2018. Phytoremediation: agreen sustainable technology for industrial waste management. In: Chandra, R., Dubey, N.K., Dumar, V. (Eds.), *Phytoremediation of Environmental Pollutants*. CRC Press, Boca Raton, 1–42.

Costa, M.C., Martins, M., Jesus, C., Duarte, J. C. 2007. Treatment of acid mine drainage by sulphate-reducing bacteria using low cost matrices. *Water, Air, and Soil Pollution*, 189 (1–4), 149–162.

Costa, R.B., Gouvea, L.A., Maluf, A.F., Palladino, T., Bevilaqua, D. 2021. Sulfate removal rate and metal recovery as settling precipitates in bioreactors: Influence of electron donors. *Journal of Hazardous Materials*, 403(5), 123622.

Costello, C. 2003. *Acid Mine Drainage: Innovative Treatment Technologies*, US, EPA, Office of Solid Waste and Emergency Response. https://clu-in.org/download/studentpapers/costello_amd.pdf.

Dold, B. 2010. Basic concepts in environmental geochemistry of sulfide mine-waste management. In *Waste Management*, Kumar, S., (Ed.), InTech: Rijeka, Croatia, 173–198.

Dold, B. 2014. Evolution of acid mine drainage formation in sulphidic mine tailings. *Minerals*, 4(3), 621–641.

Doran, P.M. 2013. *Homogeneous Reactions in Bioprocess Engineering Principles* (2nd Ed.), Elsevier Science & Technology Books: London.

Feoktistov, L., Yulia, V., and Feoktistov, C. 2018. Metabolism of copper. Its consequences for human health. *Medisur*, 16(4), 579–587.

González-Paz, J.R., Ordaz, A., Jan-Roblero, J., Fernández-Linaresand, L.C., and Guerrero-Baraja, C. 2020. Sulfate reduction in a sludge gradually acclimated to acetate as the sole electron donor and its potential application as inoculum in a microbial fuel cell. *Revista Mexicana de Ingeniería Química*, 19(3), 1053–1069.

Gu, W., Cui, M., Tian, C., Wei, C., Zhang, L., Zheng, D., and Li, D. 2021. Carboxylic acid reduction and sulfate-reducing bacteria stabilization combined remediation of Cr (VI)-contaminated soil. *Ecotoxicology and Environmental Safety*, 218, 112263.

Hwang, S., and Jho, E.H. 2018. Heavy metal and sulfate removal from sulfate-rich synthetic mine drainages using. *Science of the Total Environment*, 635, 1308–1316.

Johnson, D.B., and Hallberg, K.B. 2003. The microbiology of acidic mine waters. *Research in Microbiology*, 154, 466–473.

Kaksonen, A.H., Franzman, P.D., and Puhakka, J.A. 2004b. Effects of hydraulic retention time and toxicity on etanol and acetate oxidation in sulphate-reducing metal-precipitating fluidized-bed reactor. *Biotechnology Bioengineering*, 86, 333–342.

Kaksonen, A.H., Plumb J.J., Franzmann, P.D., and Puhakka J.A. 2004a. Simple organic electron donors support diverse sulfate-reducing communities in fluidized-bed reactors treating acidic metal- and sulfate-containing wastewater. *FEMS Microbiology Ecology*, 47, 279–289.

Kwong, Y.T.J. 1993. *Prediction and Prevention of Acid Rock Drainage from a Geological and Mineralogical Perspective; MEND Project 1.32.1*. Mine Environment Neutral Drainage (MEND): Ottawa, Canada. http://mend-nedem.org/wp-content/uploads/1.32.1.pdf.

Lewis, A. 2010. Review of metal sulphide precipitation. *Hydrometallurgy*, 104, 222–234.

Li, X., Lan, S., Zhu, Z., Zhang, C., Zeng, G., Liu, Y., Cao, W., Song, B., Yang, H., Wang, S., and Wu, S. 2018a. The bioenergetics mechanisms and applications of sulfate-reducing bacteria in remediation of pollutants in drainage: A review. *Ecotoxicology and Environmental Safety*, 158, 162–17.

López Pérez, P.A., Aguilar López, R., and Neria González, M.I. 2015. Cadmium removal at high concentration in aqueous medium: Mediated by *Desulfovibrio alaskensis*. *International Journal of Environmental Science and Technology*, 12, 1975–1986.

Luo, Y., Zheng, Z., Wu, P., and Wu, Y. 2022. Effect of different direct revegetation strategies on the mobility of heavy metals in artificial zinc smelting waste slag: Implications for phytoremediation. *Chemosphere*, 286, 131678.

Luparello, C., Longo, A., and Vetrano, M. 2012. Exposure to cadmium chloride infuences astrocyte-elevated gene-1 (AEG-1) expression in MDA-MB231 human breast cáncer cells. *Biochimie*, 94, 207–213.

Márquez-Reyes, J.M., López-Chuken, U.J., Valdez-González, A., and Luna-Olvera, H.A. 2013a. Removal of chromium and lead by a sulfate-reducing consortium using peat moss as carbon source. *Bioresource Technology*, 144, 128–134.

Martínez-Flores, K., Souza-Arroyo, V., Bucio-Ortiz, L., Gómez-Quiroz, L.E., and Gutiérrez-Ruiz M.C. 2013. Cadmio: efectos sobre la salud. Respuesta celular y molecular. *Acta toxicológica argentina*, 21, 33–49.

Martins, M., Faleiro, M.L., Barros, R.J., Veríssimo, A.R., Barreiros, M.A., and Costa, M.C. 2009. Characterization and activity studies of highly heavy metal resistant sulphate-reducing bacteria to be used in acid mine drainage decontamination. *Journal of Hazardous Materials*, 166, 706–713.

Matte, T.D. 2003. Efectos del plomo en la salud de la niñez. *Salud Pública de México*, 45, 220–224.

Monod, J. 1949. The growth of bacterial cultures. *Annual Reviews in Microbiology*, 3(1), 371–394.

Moo-Young, M., and Butler, M. 2011. *Comprehensive Biotechnology: Engineering Fundamentals of Biotechnology*. Elservier: Amsterdam.

Muloiwa, M., Nyende-Byakika, S., and Dinka, M. 2020. Comparison of unstructured kinetic bacterial growth models. *South African Journal of Chemical Engineering*, 33, 141–150.

Nogué, S., Sanz-Gallén, P., Torras, A., and Boluda, F. 2004. Chronic overexposure to cadmium fumes as-sociated with IgA mesangial glomerulonephritis. *Occupational Medicine*, 54, 265–267.

Oude Elferink, S.J.W.H., Visser, A., Hulshoff Pol, L.W., and Stams, A.J.M. 1994. Sulfate reduction in methanogenic bioreactors. *FEMS Microbiology Reviews*, 15(2–3), 199–136.

Patidar, S.K., and Tare, V. (2005). Effect of molybdate on methanogenic and sulfidogenic activity of biomass. *Bioresource Technology*, 96(11), 1215–22.

Ritchie, A.I.M. 1994. Sulfide oxidation mechanisms: Controls and rates of oxygen transport. In *Short Course Handbook on Environmental Geochemistry of Sulfide Mine-Waste*, Jambor, J.L. and Blowes, DW (Eds.), Environmental geochemistry of sulfide mine-wates, Vol 22. Mineralogical Association of Canada: Nepean, 201–245.

Rubio, J., Riqueros, M.I., Gasco, M., Yucra, S., Miranda, S., and Gonzales, G.F. 2006. Lepidium meyenii (Maca) reversed the lead acetate induced damage on reproductive function in male rats. *Food and Chemical Toxicology*, 44, 1114–1122.

Sahinkaya, E., and Gungor, M. 2010. Comparison of sulfidogenic up-flow and down-flow fluidized-bed reactors for the biotreatment of acidic metal-containing wastewater. *Bioresource Technology*, 101, 9508–9514.

Singh, N. K., and Singh, R. 2021. A sequential approach to uncapping of theoretical hydrogen production in a sulfate-reducing bacteria-based bio-electrochemical system. *International Journal of Hydrogen Energy*, 46(39), 20397–20412.

Suzuki, D., Ueki, A., Amaishi, A., and Ueki, K. 2007. *Desulfobulbus japonicus* sp. nov., a novel Gram-negative propionate-oxidizing, sulfate-reducing bacterium isolated from an estuarine sediment in Japan. *International Journal of Systems Science*, 57(4), 849–855.

Tun-Canto, G.E., Álvarez-Legorreta, T., Zapata-Buenfil, G., and Sosa-Cordero, E. 2017. Metales pesados en suelos y sedimentos de la zona cañera del sur de Quintana Roo, México. *Revista Mexicana de Ciencias Geológicas*, 3, 157–169.

Walsh, R. 2018. Comparing enzyme activity modifier equations through the development of global data fitting templates in Excel. *Peer Journal*, 6, e6082.

Willis-Poratti. 2016. *Biorremediación de metales pesados por sulfidogénesis utilizando comunidades y microorganismos sulfato-reductores*. PhD Thesis. Universidad Nacional de La Plata, Facultad de Ciencias Exactas: La Plata, 273.

Yang, Z., Liu, Z., Dabrowska, M., Debiec-Andrzejewska, K., Stasiuk, R., Yin, H., and Drewniak, L. 2021. Biostimulation of sulfate-reducing bacteria used for treatment of hydrometallurgical waste by secondary metabolites of urea decomposition by *Ochrobactrum* sp. POC9: From genome to microbiome analysis. *Chemosphere*, 282, 131064.

16 Omics Reflection on the Bacterial Escape from the Toxic Trap of Metal(loid)s
Cracking the Code of Contaminants Stress, Resistance Repertoire, and Remediation

Jayanta Kumar Biswas and Monojit Mondal
University of Kalyani

Vineet Kumar
G D Goenka University

Meththika Vithanage
University of Sri Jayewardenepura

Rangabhashiyam Selvasembian
SASTRA Deemed University

Balram Ambade
National Institute of Technology

Manish Kumar
Guru Ghasidas Vishwavidyalaya

CONTENTS

DOI: 10.1201/9781003247883-16

16.1 METAL(LOID)S CONTAMINATION

Environmental contamination arises when an extraneous substance (element, organic and inorganic compound, or ion) enters the environment and causes some adverse impacts on its physicochemical and biological attributes (Kumar et al. 2021a). Contamination by toxic metal(loid)s (TMs) is a ubiquitous environmental problem of global concern. The TM contamination of water and soil can cause toxicity in living organisms of an ecosystem and degrade the water/soil quality and adversely affect ecosystem health and services, which has significant negative connotations for soil fertility, agricultural productivity, and human health (Bolan et al. 2022). The Agency for Toxic Substance and Disease Registry (ATSDR 2019) lists 275 compounds as priority substances, with As, Pb, and Hg being the three toppers. The widespread contamination by TMs has been caused as a combined result of rapid urbanization, industrialization, intensive agriculture, mining, industrial production, municipal waste(water) discharge, irrigation, and excessive use of fertilizers, fungicides, herbicides, and pesticides (Bolan et al. 2021; Kumar et al. 2020, 2021b; Prabha et al. 2021). TM contamination may have one or a combination of sources falling under the following natural and anthropogenic categories.

16.1.1 NATURAL

The occurrence of TMs in the natural environment depends on the local geology, hydrogeology, and physicochemical regimes (Wang and Mulligan 2006). The major source is weathering of parent materials, including igneous and sedimentary rocks and coal. The original source of TM contamination of the soil and water body is the weathering of sedimentary rocks such as limestone, shale, dolomite, and sandstone.

Some major elements may be introduced in the environment as a result of the interaction between water and igneous rocks (e.g. basalt, granite, gabbro, andesite, and ultramafic). The dissolution of specific minerals or ores that contribute to increased levels of TMs are magnetite, goethite, hematite, siderite, calcite, malachite, cuprite, azurite, chromite, kaolinite, montmorillonite, orpiment, arsenic trioxide, arsenopyrite, smithsonite, pyrolusite, calamine, and rhodochrosite (Borrok et al. 2008). In sulphide-bearing mineral deposits, gold mineralization, and hydrous iron oxides ores, TMs such as As are also found (Saha and Paul 2016). TMs such as Cd, Co, and Mn occur in the earth crust along with other minerals (Järup 2003). Several TMs such as Hg, Ni, and Pb get deposited into the aquatic system and soil as a dry or wet fallout of atmospheric aerosols arising from windborne dust, volcanic emissions, forest fires, and vegetational sources (Saha and Paul 2016).

16.1.2 ANTHROPOGENIC

Primarily, anthropogenic activities associated with industrial processes, manufacturing, and discharge of domestic and industrial wastes are the major source of metal(loid)s contamination in the environment (Bolan et al. 2022; Nie et al. 2020). The non-ferrous mining, mineral extraction, combustion of fossil fuels, feed additives, pesticides, wood preservatives, roasting of arsenious gold ores, and municipal and industrial wastes are the major contributors to As in the environmental pool (Wang and Mulligan 2006; Saha and Paul 2016). The phosphate compounds contain an array of metal(loid)s (Wuana and Okieimen 2011). Cd is emitted during smelting and mining and from alloy-involving industries, such as plastics and batteries (Rashid et al. 2013). Considerable amount of Cd is emitted from tobacco smoking (Mohapatra et al. 2014). Cd in agricultural soil primarily originates from phosphatic fertilizers. In agriculture, horticulture, and animal husbandry, large quantities of Cu are used in many fungicides as a trace element, and as growth promoters in piggery and poultry units (Wightwick et al. 2013). In agricultural soil, the accumulation of Cu results from the use of Cu fungicides and application of biosolids (Wightwick et al. 2010). In the environment, Ni comes from corroded metal pipes, containers, paints, petrol additives, and precipitation of aerosols formed during the high-temperature industrial processes such as coal combustion, smelting, and cement production (Saha and Paul 2016).

16.2 METAL(LOID)S STRESS ON BACTERIA

Heavy metals such as Zn, Cu, Fe, Mn, Ni, Mo, and Co in low concentration are necessary for several important biomolecules, which play important roles in different metabolic processes of organisms (Nagajyoti et al. 2010). Contrarily, some heavy metal(loid)s such as Cd, Hg, Pb, and As are considered as non-essential elements since they are not a part of any important biomolecule nor required for any biochemical process (Etesami 2018). The long-term exposure of microorganisms to such essential metals at elevated concentration and non-essential metal(loid)s even at low concentration may cause biological toxicity (Khan et al. 2015; Etesami 2018) and disrupt the functions and diversity of microbial communities, which may bring about substantial changes in ecological dynamics (Ahemad and Khan 2012). The TMs contamination in soils has adverse effects on the microbial diversity, community composition, and functions in the rhizosphere (Fuke et al. 2021; Li et al. 2006). The TMs pollution can reduce 99.9% diversity of bacterial populations, mostly affecting the functionally significant rarer taxa (Gans et al. 2005).

To explain the impact of TMs on the diversity of microbial community, two theories have been put forward. The first theory suggested that the TMs stressed systems are more stable compared to stress-free systems, due to their acquired physiological adaptations that empower them to encounter additional stresses (Bamborough and Cummings 2009). The second theory suggested that non-stressed systems have higher stability because they have more resources at their disposal, which can be employed to maintain ecosystem services in the face of stress (Tobor-Kapłon et al. 2005). As microorganisms present in wastewater contaminated with TMs are subjected to continuous exposure to the metal(loid)s stress condition, they can develop adaptation and resistance to those TMs over time (Frey et al. 2006).

Microbes interact with the TMs in two different ways: The first is biosorption, which is a fast metabolism-independent binding of metal ions to bacterial cell walls, and the second is bioaccumulation; however, it is a slow metabolism-dependent uptake of TMs inside the cells (Li et al. 2004). In the contaminated environment, bacteria respond to the TMs involving different processes such as biosorption to the cell walls, transportation through the plasma membrane, entrapment in extracellular capsules, complexation, precipitation, and redox transformations (Kumar et al. 2019; Singh et al. 2010). In the process, they develop diverse acclimatization, adaptation, and resistance mechanisms at hierarchical levels – cellular (molecular/biochemical/genetic), population (growth/development/functional activity), and community levels – to withstand TM-induced stress and toxicity (Jarosławiecka and Piotrowska-Seget 2014). The toxic TMs interfere with the uptake and distribution of nutrients (Rajkumar et al. 2012; Etesami 2018). They can displace essential metals from their original binding sites of the enzymes or other essential biomolecules; for example, As and Cd compete with P and Zn, respectively (Sharma and Archana 2016). In addition to enzymatic inhibition, they can also disrupt nucleic acid structure and gene expression (Nagajyoti et al. 2010; Sharma and Archana 2016).

16.3 METAL(LOID)S RESISTANCE MECHANISMS IN BACTERIA

Some water and soil bacteria possess tolerance/resistance to ambient TMs (Bruins et al. 2000). Microorganisms developed diverse types of chromosomal-, transposon-, and plasmid-mediated resistance mechanisms to overcome TMs stress. The resistance genes are either plasmid or chromosome encoded and are highly specific. Seven different types of TMs resistance mechanisms (Figure 16.1) are known in microorganisms, which include (1) exclusion of TMs by permeability barrier; (2) extracellular sequestration of TMs by protein/chelator binding; (3) intracellular sequestration of TMs by protein/chelator binding; (4) enzymatic detoxification of TMs to the less toxic form; (5) active transport of TMs; (6) passive tolerance; and (7) reduction in TMs sensitivity of cellular targets to the TMs ions (Bruins et al. 2000; Wheaton et al. 2015; Etesami 2018). Bacteria can possess one or a combination of these resistance mechanisms.

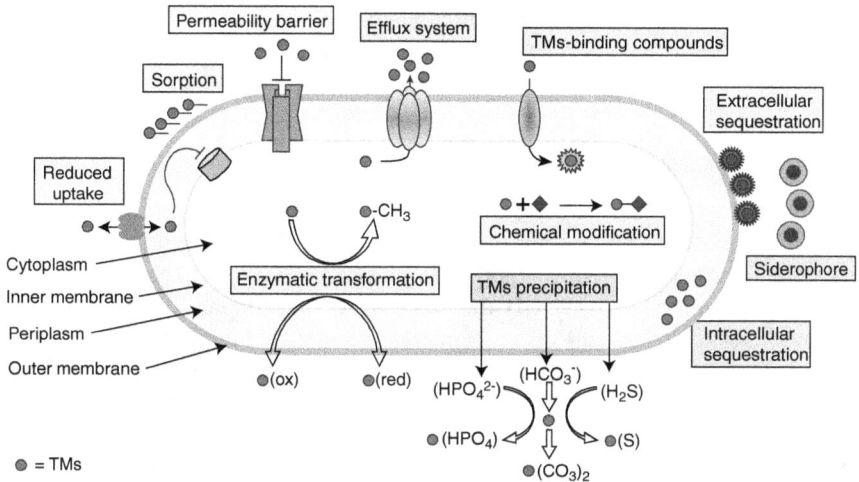

FIGURE 16.1 Different TMs resistance mechanisms adopted by bacteria. (Modified after Lemire et al. 2013; Ahemad 2019; Etesami 2018.)

16.3.1 Exclusion of TMs by Permeability Barrier

Bacteria protect their essential cellular component from TMs through metal exclusion by alteration in their envelope, cell wall, membrane, or surface layer (S-layer). Dissolved TMs are passively adsorbed by bacterial cell wall or envelope through charge-mediated attraction (Mohamed 2001). The outer membrane, S-layer, envelope, lipopolysaccharides (LPS), extracellular polymeric substances (EPS), proteins, carbohydrates, and humic substances have non-specific TMs binding potential and can prevent the entry of those TMs into the bacterial cell (Bruins et al. 2000; Wheaton et al. 2015). The anionic functional groups such as sulphhydryl, carboxyl, hydroxyl, sulphonate, amine and amide present in the membrane or wall, and extracellular matrix adsorb the TMs, leading to immobilization and extracellular detoxification (Bruins et al. 2000; Rajkumar et al. 2010) as reported in *Pseudomonas putida*, *Klebsiella aerogenes*, *Arthrobacter viscosus*, and *Bacillus sphaericus* (Bruins et al. 2000).

16.3.2 Extracellular Sequestration of TMs by Protein/Chelator Binding

Extracellular sequestration and precipitation of TMs are mediated by metabolites which efficiently bind and detoxify TMs through complex formation, thus saving the cell from toxic assault (Rajkumar et al. 2010; Biswas et al. 2018a, 2019). The bacterial siderophores and iron plaque can bind and chalet metals such as Al, Cd, Cu, Ga, In, Ni, Pb, and Zn (Tank and Saraf 2009; Tripathi et al. 2014; Etesami 2018; Biswas et al. 2020); therefore, those metals become immobilized. Many bacteria have the potential to produce low molecular weight organic acids such as gluconic acid, citric acid, oxalic acid, and succinic acid, which have a strong affinity for TMs and form metal-oxalate crystal (Gao et al. 2010; Archana et al. 2012; Biswas et al.

2018b). The bacteria can also achieve extracellular sequestration TMs through the production of biosurfactants, surface-active substances, EPS, etc. (Pacheco et al. 2010; Pacwa-Płociniczak et al. 2011; Etesami 2018; Banerjee et al. 2019).

16.3.3   Intracellular Sequestration of TMs by Protein/Chelator Binding

Intracellular sequestration is the accumulation of metals within the cytoplasm to protect important cellular components from exposure. To defend themselves against metals, many bacteria have developed a cytoplasmic sequestration mechanism. Some TMs resistant bacteria can detoxify the metals by compartmentalizing or converting them into less toxic forms, allowing them to accumulate at high intracellular concentrations in their cells (Haferburg and Kothe 2007). The TMs-resistant bacteria may bind toxic metals using metallothioneins (MTs), low-molecular-mass cysteine-rich proteins, and metallochaperones to render them non-bioavailable (Etesami 2018). These intracellular proteins can lower free ion concentrations (reduced availability) in the cytoplasm and sequester TMs (e.g. Cd, Cu, Ag, and Hg) as a result of being associated with heavy metals, resulting in metal detoxification (Silver and Phung 2005). The TMs-resistant bacteria such as *Pseudomonas aeruginosa*, *P. putida*, and *Anabaena* sp. have been shown to be capable of producing MTs, which function as cytoplasmic metal cation-binding proteins (Blindauer et al. 2002). Under TMs stress condition, these proteins are mainly produced (Etesami 2018). Mainly two genes such as *smtA* and *smtB* are responsible for TMs resistance in *Synechococcus* sp.; the *smtA* encodes a MT that binds to Cd(II) and Zn(II) (Bruins et al. 2000). The high level of Cd(II)-, Zn(II)-, and Cu(II)-induced *smtA* repressed the product of *smtB* gene that acts as a transacting transcriptional repressor turning off *smtA* expression and metallothionein production (Silver et al. 1989; Bruins et al. 2000). The Cu(II) is also sequestered intracellularly in *Mycobacterium scrofulaceum* through black copper sulphate precipitation (Etesami 2018).

16.3.4   Enzymatic Detoxification of TMs to the Less Toxic Form

Bacteria adopt enzymatic detoxification of TMs as an important defence strategy accomplished through oxidation, reduction, methylation, and demethylation (Etesami 2018). The TMs-reducing bacteria decrease the mobility and/or toxicity of TMs, such as Cr^{6+} to Cr^{3+}, U^{6+} to U^{4+}, and Se^{4+} to Se^0 (McLean and Beveridge 2001; Finneran et al. 2002; Di Gregorio et al. 2005; Chatterjee et al. 2009). Anaerobic sulphate-reducing bacteria minimize toxicity by generation of less soluble metal sulphides and phosphates (CdS and PbS) (Sharma et al. 2000). Bacteria can transform some TMs such as Se, Sn, Te, and Pb into the gaseous state via the addition of methyl groups (methylation), which promote volatilization of methylated TMs out of the bacterial cell (Meyer et al. 2007).

Moreover, the Hg resistance in bacteria also involves enzymatic detoxification by two enzymes, mercuric ion reductase (*merA*) and organomercurial lyase (*merB*), which are encoded by a set of genes that form *mer* operon (Etesami 2018). This operon can transport Hg^{2+} and self-regulate the detoxification process. The same set of genes also encode membrane-associated transport protein (genes of *merT*, *merP*, and *merC*) as well as periplasmic-binding protein (Bruins et al. 2000). The Hg^{2+} is collected in the surrounding environment by periplasmic-binding proteins, and it is transported to the cytoplasm for detoxification by transport proteins. Then the mercuric ion reductase reduces Hg^{2+} to Hg^0, which then diffuses in the surrounding environment through the cell membrane (Nies 1999; Bruins et al. 2000).

Similarly, the As resistance is a plasmid-mediated enzymatic detoxification system, which is reported in many bacteria such as *Bacillus subtilis, Staphylococcus aureus, P. aeruginosa,* and *Escherichia coli* (Etesami 2018). These bacteria can reduce As(V) to As(III) inside the cell using arsenate reductase enzyme and exclude As(III) by As (III) efflux pump (Bruins et al. 2000). Bacterial oxidation of As(III) to As(V) is also observed and is mediated by As(III) oxidase (Silver and Phung 2005). These enzymes are encoded by a cluster of genes known as *ars* operon (Etesami 2018).

16.3.5 ACTIVE TRANSPORT OF TMS

The toxic TMs are mainly eliminated from the bacterial cell through active transport or efflux systems, which are either chromosomal or plasmid encoded (Etesami 2018). The non-essential metals and metalloids enter the cell through regular nutrition transport systems, but are rapidly expelled through active transport pathways. These efflux systems are very specific to the cations and anions they export, and they might be ATPase-linked or non-ATPase (Silver et al. 1989; Nies and Silver 1995).

Seven major types of efflux pumps are known: two ATPases, (1) P-type ATPases (generally single polypeptide determinants, with a covalently phosphorylated – from ATP – intermediate) and (2) ABC ATPases (ATP-binding cassette without phosphorylated intermediates); a cytoplasmic membrane-associated ATPase subunit; one or two membrane-embedded pump channel subunits; a periplasmic space ion-binding protein; and an outer membrane porin protein. The ABC ATPase family includes both uptake pumps and efflux pumps.

The three additional efflux pump classes are chemiosmotic ion/proton exchangers: (3) the single membrane polypeptides of the major facilitator superfamily (MFS; generally a single polypeptide) that transverses the membrane 12 or 14 times primarily in an alpha helical structure (Saier et al. 1999); (4) the cation diffusion facilitator (CDF) family (including CzcD for cadmium, zinc, and cobalt) (Haney et al. 2005); and (5) the CBA family of three polypeptide chemiosmotic antiporters such as CzcCBA with an inner membrane (A) protein of over 1000 amino acids in length and an outer membrane (C) protein and a coupling (B) protein connecting the two in the periplasmic space.

However, the name resistance-nodulation-cell division (RND) superfamily is widely used for these efflux transporters as different examples are involved in resistance to such toxic compounds as cations and small organic compounds, nodulation in *Rhizobium*, or cell division such as in *E. coli* (Tseng et al. 1999; Valencia et al. 2013). Two additional families of chemiosmotic transport systems that are narrower in substrate range are the (6) ChrA (CHR) chromate efflux system and (7) ArsB, the arsenite efflux system that is currently unique in that it can function alone as a chemiosmotic efflux system or with a second ArsA subunit functions with energy from ATP (Table 16.1). Examples of some of these transporters are summarized in Table 16.1 and Figure 16.2.

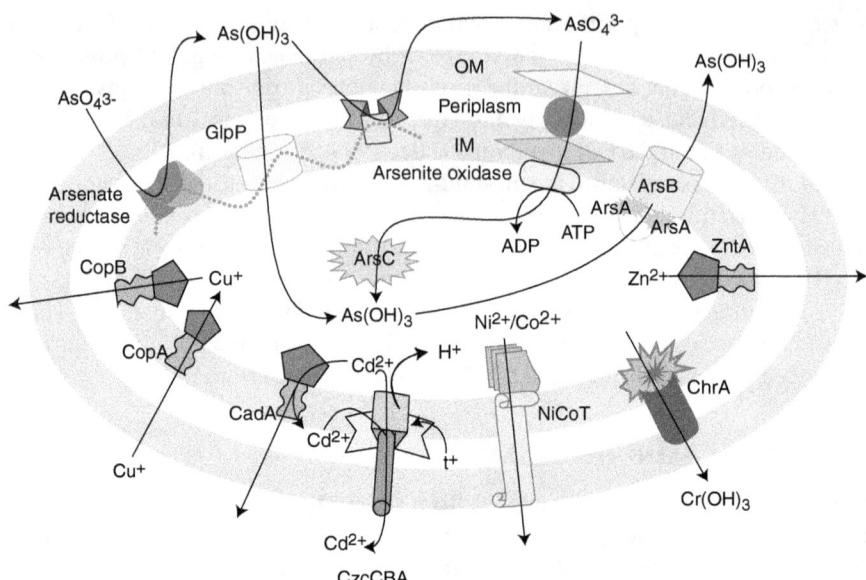

FIGURE 16.2   Overview of membrane-associated uptake, efflux, reduction, and oxidation of metal(loid) ions in bacteria. (Modified after Silver and Phung 2005.)

Resistance to Zn(II), Cd(II), and Pb(II) in bacteria is mostly reliant on active metal ion efflux to prevent harmful effects on the cell, which are mediated by chromosomes, plasmids, or transposons (Bruins et al. 2000). An ATPase efflux pump is encoded by the *cop* operon (genes of *copA*, *copB*, *copZ*, and *copY*) (Bruins et al. 2000) (Table 16.1). The *copA* gene encodes for a Cu(II) uptake ATPase, while *copB* gene encodes for a P-type efflux ATPase (Etesami 2018). Plasmid-encoded As (V) resistance and antimonite resistance mediated by the *ars* operon are two more notable instances of this resistance system (Etesami 2018). The *ars* operon (genes of *arsA*, *arsB*, *arsC*, *arsD*, and *arsR*) codes for an ATPase efflux pump as well as the arsenate reductase-detoxifying enzyme (Bruins et al. 2000; Etesami 2018).

There are three patterns found for Cd efflux, and these are (with the names of examples) CzcD, a single membrane polypeptide chemiosmotic efflux pump; CzcCBA, a three-polypeptide chemiosmotic complex with CzcA (a large inner membrane protein), CzcC (a smaller outer membrane protein), and CzcB (a periplasmic coupling protein that connects CzcA and CzcB and forms a continuous channel from the cytoplasm to the outside of the cell) (Figures 16.2 and 16.3b) (Nies 1995, 2003); and CadA, a large single polypeptide P-type ATPase (Figure 16.3a) (Table 16.1). All three of these proteins are membrane-embedded Cd^{2+} efflux pumps. The three-polypeptide chemiosmotic CzcCBA complex that functions as an ion/proton exchanger to efflux Cd^{2+}, Zn^{2+}, and Co^{2+} is a member of the metal resistance family within the larger superfamily of chemiosmotic pumps called RND (for resistance, nodulation, and cell division, since some members are involved in bacterial nodulation formation by *Rhizobium* and others are involved in cell division, for example in *E. coli*). They have also been

TABLE 16.1

Metals and Metalloids Resistance Systems and Their Mechanisms in Bacteria (Silver and Phung 2005)

Toxic Ions	Gene Mnemonic	Protein Function
Ag^+	*sil*	Silver resistance and binding
AsO_4^{3-} and $As(OH)_3$	*ars*	Arsenate reductase and transport
$As(OH)_3$	*aso*	Arsenite oxidase and transport
AsO_4^{3-}	*arr*	Respiratory arsenate reductase
Cd^{2+}	*cad*	P-type efflux ATPase
$Cd^{2+}, Zn^{2+}, Co^{2+}$	*czc*	CBA efflux permease
Co^{2+}, Ni^{2+}	*cnr*	CBA efflux permease
CrO_4^{2-}	*chr*	Chromate efflux permease
Cu^{2+}, Cu^+	*cop and pco*	Copper resistance and transport
Hg^{2+} and organomercurials	*mer*	Mercuric reductase and transport
$Ni^{2+}, Co^{2+}, Cd^{2+}$	*ncc*	CBA efflux permease
Ni^{2+}	*nre*	CBA efflux permease
Pb^{2+}	*pbr*	Lead resistance and efflux

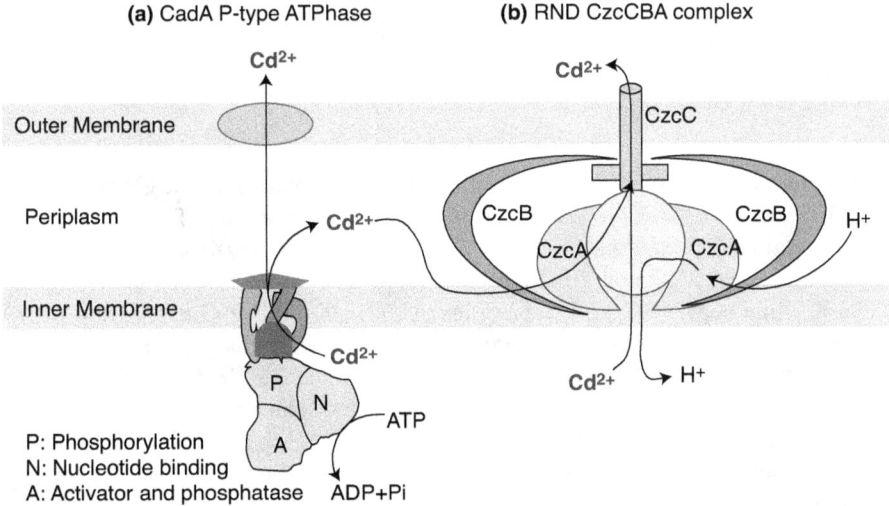

(a) CadA P-type ATPhase　　　**(b) RND CzcCBA complex**

FIGURE 16.3 Molecular models of efflux (a) P-type ATPase and (b) RND chemiosmotic antiporter. (Modified after Yu et al. 2003; Eswaran et al. 2004; Silver and Phung 2005.)

called CBA transporters to distinguish them from the ABC (ATPase-binding cassette) family of ATPases. Genes for related RND divalent cation efflux systems have been isolated in *Ralstonia metallidurans* strains CH34 and 31A, *czr* (Cd^{2+} and Zn^{2+} resistance), *cnr* (Co^{2+} and Ni^{2+} resistance), and *ncc* (Ni^{2+} and Co^{2+} resistance) (Hassan et al. 1999; Legatzki et al. 2003; Mergeay et al. 2003; Nies 2003).

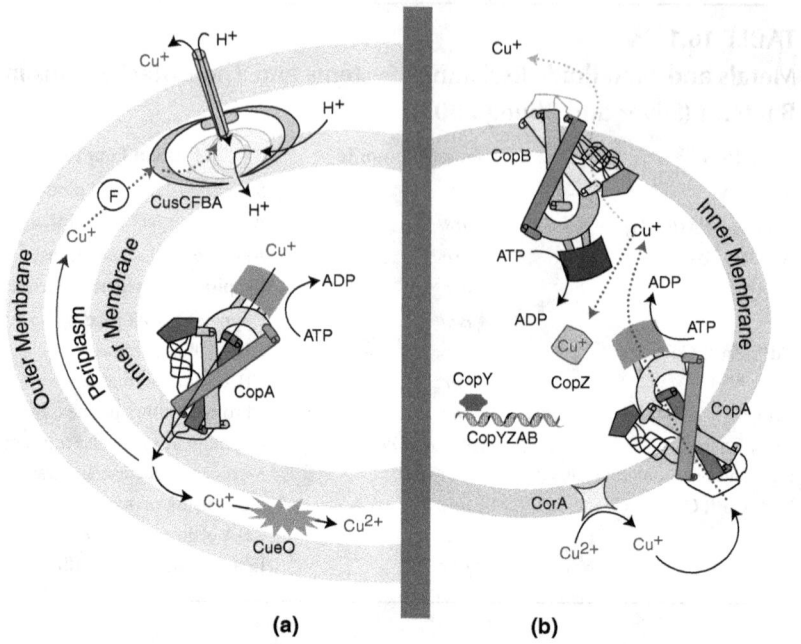

FIGURE 16.4   Current view of copper transport and resistance in (a) gram-negative bacteria and (b) gram-positive bacteria. (Modified after Rensing and Grass 2003; Solioz and Stoyanov 2003; Silver and Phung 2005.)

Another type of bacterial ATPase efflux mechanism exists for the essential metal ion Cu(II). In bacteria, Cu resistance is mediated by *cop* operon (Table 16.1), which encodes mainly four genes such as *copA*, *copB*, *copZ*, and *copY* (Rensing and Grass 2003; Solioz and Stoyanov 2003) (Figure 16.4). The *copA* encodes Cu(II) uptake ATPase, while copB encodes P-type efflux ATPase (González-Guerrero et al. 2010) (Figure 16.4b). The cop operon is regulated by the gene products *copY* and *copZ*. In gram-negative bacteria, CueO multi-copper oxidase and CusCFBA multicomponent efflux transport system present in the extracellular periplasmic space are protect the cells from damage (Singh et al. 2004; Djoko et al. 2010) (Figure 16.4a). An additional copper resistance determinant (*pco*) has been identified in some *E. coli* (Rensing and Grass 2003).

Metals such as Co, Ni, and Zn are required as micronutrients, but are toxic in excess, without being subject to bacterial redox chemistry. In bacterial cells, the Ni^{2+} uptake and efflux pumps are frequently shared with regard to substrate and referred to as Ni^{2+} Co^{2+} transporters (Mergeay et al. 2003; Hebbeln and Eitinger 2004) (Table 16.1). The Ni^{2+} uptake can occur either by a single polypeptide chemiosmotic carrier, or by a five-component ABC ATPase (Mulrooney and Hausinger 2003). The five gene products of ABC Ni^{2+} uptake transporter are NikA (a soluble periplasmic Ni^{2+}-binding protein), NikB and NikC (membrane proteins that form the transmembrane channel), and NikD and NikE (which form the intracellular ATPase part of the complex) (Mulrooney and Hausinger 2003). Another Ni^{2+} uptake system also presents in high nickel requiring bacteria known as NiCoT, which can uptake Ni^{2+} together with Co^{2+} (Hebbeln and Eitinger 2004). The uptake of Zn^{2+} is mainly regulated by different types of membrane uptake pumps such as ZnuABC, MntH, ZupT, CorA, and the elimination was by other membrane efflux pumps such as ZiaA, CadA, CzcCBA, ZntCBA, CzcD, and ZitB (Blencowe et al. 2003; Nies 2003). Some of these uptake and efflux systems are relatively specific to Zn^{2+} as substrate, but others are less specific and also can pump Cd^{2+} and Co^{2+}.

The bacteria *Pseudomonas putida*, *Bacillus licheniformis*, *Bacillus cereus*, *Brevibacillus laterosporus*, *Trachelophylum* sp. *Peranema*, *Adispica* sp., etc., have the resistance to chromium, which is related to the presence of chromosomal- or plasmid-encoded genes such as *chrA*, *chrB*, *chrC*, *chrD*, *chrE*, and *chrF* (Ramirez-Diaz et al. 2008; Kamika and Momba 2013) (Table 16.1). The chromate efflux system is encoded by the *chrA* gene. The chrA protein belongs to the CHR superfamily of transporters (Diaz-Perez et al. 2007). The *chrB* gene encodes a membrane-bound protein necessary for the regulation of chromate resistance (Branco and Morais 2013). The *chrC* gene encodes a protein almost similar to iron-containing superoxide dismutase, while the *chrE* gene encodes a gene product that is a rhodanese-type enzyme (Branco and Morais 2013); *chrF* encodes a repressor for chromate-dependent induction (Díaz-Pérez et al. 2007).

16.3.6 Passive Tolerance

Most of the TMs are more soluble at an acidic pH; bacteria that are particularly acidophilic must be tolerant of high metal concentrations. Acidophilic bacteria also thrive in situations with high concentration of sulphate ions, which can complex metal cations at acidic pH levels; therefore, entry into the bacterial cell is restricted (Dopson et al. 2014). Passive metal tolerance mechanism is also based on the internal positive membrane potential, which prevents protons and metals from entering the cell by establishing a chemiosmotic gradient (Baker-Austin and Dopson 2007; Slonczewski et al. 2009). Another passive metal tolerance mechanism is a result of competition between protons and TM cations for metal-binding sites on the surface of bacteria (Mangold et al. 2012). Bacteria can also form biofilm that usually includes EPS that can sorb TMs, thereby reducing the direct contact of microbes with TMs (Dopson et al. 2014).

16.3.7 Reduction in TMs Sensitivity of Cellular Targets

Some bacteria adapt to the presence of TMs by modifying the sensitivity of critical cellular components, providing some natural protection, which is achieved through mutation (Bruins et al. 2000). To remain ahead of metal inactivation, mutations that reduce sensitivity without affecting fundamental function or boosting synthesis of a specific cellular component are used. By the DNA repair systems, genomic and plasmid DNA are also protected to some extent. In an attempt to avoid sensitive components, the bacteria may also defend themselves by creating TM-resistant components or alternative pathways (Etesami 2018).

16.4 APPLICATIONS

16.4.1 BIOREMEDIATION

Once TMs are introduced and contaminate the environment, they may persist for long period of time depending on the type of metal(loid) and the local environment, because they are non-degradable in nature. The bioremediation methods used for the clean-up of TMs-contaminated sites may *be in situ* (on-site) or *ex situ* (off-site). Most of the TM resistant/tolerant bacterial species have potentials to remediate TMs. Bioremediation of TMs can be achieved through different methods such as bioaugmentation, biostimulation, and bioattenuation.

16.4.1.1 Bioaugmentation

Bioaugmentation is a technique for bioremoval of TM contaminants from polluted areas that involve the introduction of specialized bacteria or genetically modified bacteria that are capable of combating the TM contaminations. This is a very efficient and sustainable *in situ* technique that has a high degree of substrate specificity (Mrozik and Piotrowska-Seget 2010). On-site intervention of bioaugmented bacteria is frequently affected by abiotic and biotic stressors (Rahman et al. 2020). High concentration of TMs at polluted areas can prevent rapid growth and activity of allochthonous bacteria (Nanda et al. 2019). Another limiting factor for the growth of bioaugmented bacteria is nutrient shortage (Rahman et al. 2020). In this account, Mondal et al. (2019) showed that a wastewater bacterial strain *Bacillus* sp. KUJM2 has greatly immobilized multiple metal(loid)s (As, Cd, Cu, and Ni) in the soil. Another study used the bioaugmentation approach to remove Hg through volatilization from polluted soils using *Sphingobium* SA2 and a nutrient supplement (Mahbub et al. 2016).

16.4.1.2 Biostimulation

Biostimulation is a method of enhancing bioremediation by providing native bacteria with settings that stimulate their TMs-resistant potentials (Atagana 2008). This entails adding nutrients (phosphorus, nitrogen, oxygen, and carbon) to promote the growth of native bacterial species for biodegradation (Bundy et al. 2012). This method is applicable for polluted locations with a large population of native microorganisms that can interact with TMs. Many researchers have looked into biostimulation for bioremediation of various TMs such as, Fe, Cu, Cd, Fe, and Cr (Fulekar et al. 2012; Kanmani et al. 2012). Cr(VI) is removed from a contaminated soil in a microcosm experiment using acetate as an electron donor to promote the microbial reduction process (Lara et al. 2017). Carlos et al. (2016) also showed that nutritionally stimulated microbes have a beneficial effect in the remediation of Cr(VI)-contaminated sites.

The biostimulation process is very slow and less effective for highly contaminated sites. To overcome these limitations, recent studies have integrated the biostimulation with bioaugmentation to expand their possibility for effective treatment (Rahman et al. 2020). In this context, Wang et al. (2014) used the synergistic effects of UV-mutant *Bacillus subtilis* 38 and NovoGro (a biofertilizer) to immobilize multiple metals (Cd, Cr, Hg, and Pb) in the soil.

16.4.1.3 Bioattenuation

Bioattenuation, also known as natural attenuation, is the spontaneous elimination of heavy metal pollutants by the local bacterial community and does not involve any intervention from humans (Mulligana and Yong 2004; Ying 2018). The metabolic diversity of intrinsic microorganisms existing at the polluted site play main roles in this process (Abatenh et al. 2017). These indigenous bacteria can degrade, detoxify, neutralize, or transform the TMs based on their metabolic activities. This process includes aerobic and anaerobic biodegradation, sorption, volatilization, chemical or biological stabilization, and transformation of TMs (Mulligana and Yong 2004). The period of time it takes for pollutants to be naturally attenuated varies from site to site, depending on the circumstances, type of contaminants, and the degrading microbial flora present at the site. It is a slow but low-cost process and used if there is a low concentration of contaminants and no other bioremediation techniques will work (Azubuike et al. 2016). Bioattenuation could be sped up in combination with biostimulation or bioaugmentation (Li et al. 2010).

16.4.2 Bioleaching

The process of solubilization of metals and metalloids with the help of acidophiles, including chemolithoautotrophic and heterotrophic microbes, is referred to as bioleaching. It is a simple procedure for extracting and removing metal(loid)s from ores, contaminated soil, sediments, and sewage sludge (Fu et al. 2008; Lee et al. 2015). Bioleaching is a cost-effective and eco-friendly technique. Some research also reported that biological leaching of metals and metalloids is more effective than chemical leaching (Liu et al. 2003; Deng et al. 2013). Bioleaching is useful for *ex situ* treatment since increased mobility of metal(loid)s at natural locations may spread pollution further (Rahman et al. 2020). Furthermore, excessive levels of TMs, organic matter, and solid content in contaminated sources might prevent the development and/or activity of microbes (Cho et al. 2002; Rahman et al. 2020).

Mainly, sulphate-reducing bacteria and iron-reducing bacteria, such as *Acidithiobacillus ferrooxidans*, are implicated in the oxidation of insoluble sulphide and iron minerals of TMs to their soluble forms (Lee et al. 2015; Park et al. 2014). These bacteria can also oxidize S^0 to SO^{4-} and Fe^{2+} to Fe^{3+}, resulting in acid generation, which causes bioleaching (Zhang et al. 2007; Velgosová et al. 2013). Some heterotrophic bacteria belonging to the genera of *Acidophilum*, *Arthrobacter*, *Acetobacter*, *Pseudomonas*, etc., have the potential to produce organic acids and chelating agents, which are used for the recovery of metals and metalloids (Pathak et al. 2009).

16.4.3 Phytoremediation Assisted by TMs-Resistant Bacteria

Phytoremediation is usually limited due to the low TMs availability, uptake, translocation, resistance potential, and biomass. Microbe-assisted phytoremediation of TMs represents a promising method for the remediation of contaminated soil (Rajkumar et al. 2012). Even under TMs stress conditions, soil microbial association induces plant growth. The plant growth-promoting TMs-resistant microbes can protect plants from the adverse effects of TMs and also increase TMs uptake by hyperaccumulator plants (Weyens et al. 2009). Microbe-assisted phytoremediation involves several mechanisms such as biosorption, intracellular accumulation, biomineralization, enzymatic transformation, bioleaching, and redox reactions (Lloyd 2002). Plant growth and TMs accumulation are stimulated by TMs-resistant rhizobacteria through the production of indole acetic acid (IAA), mono-aminocyclopropane-1-carboxylate (ACC) deaminase, siderophores, etc. (Rajkumar et al. 2012). Moreover, the increased TMs resistance in plants may be attributed to the TMs-resistant rhizobacteria through induction of thiol compounds, metallothionein, and superoxide dismutase (Courbot et al. 2004; Ramesh et al. 2009; Vallino et al. 2009). The inoculation of *Burkholderia* sp. Z-90, *Pseudomonas* sp., and *Bacillus* sp. E1S2 has been reported to enhance TMs uptake and removal efficiency in many plants (Ma et al. 2015; Yang et al. 2016). The microbial association in plants also results in several other beneficial effects, including enhanced uptake of minerals, stomatal regulation, and osmotic adjustment (Compant et al. 2005). The mobility and bioavailability of TMs are increased by soil bacteria due to lowering of soil pH, altering soil redox conditions, and producing TMs-chelating and plant growth-promoting compounds, e.g. organic acids, siderophores, and biosurfactants (Ullah et al. 2015; Khalid et al. 2017; Ahemad 2019).

16.4.4 PLANT GROWTH PROMOTION BY TMS-RESISTANT BACTERIA

Plant-microbe interactions play an important role in the survival and adaptation of both the partners in the environment, even in soil contaminated with TMs (Rajkumar et al. 2012). Environmental stress can be alleviated in plants by its associated microbes (Gopalakrishnan et al. 2015). The metal(loid)s-resistant plant growth-promoting bacteria belonging to the genera of *Pseudomonas, Azotobacter, Azospirillum, Achromobacter, Bacillus, Enterobacter, Klebsiella, Aeromonas, Variovorax*, etc., are capable of inducing plant growth under environmental stress conditions mediated by the synthesis of IAA, 1-aminocyclopropane-1-carboxylic acid (ACC) deaminase, siderophores, organic acids, biosurfactants, exopolymers, biogenic compounds, and phosphate solubilization (Glick 2010; Rajkumar et al. 2012; Ma et al. 2011, 2015; Meena et al. 2017).

16.5 CONCLUSIONS AND FUTURE PROSPECTS

This chapter featured the TMs contamination sources, different mechanisms utilized by bacteria to resist/tolerate them, and the application of TMs-resistant bacteria. It uncovered the TMs stress on bacteria and the detailed defence mechanism adopted by the bacteria to resist TMs. Bacteria have an intrinsic biological system that allows them to survive and remove TMs from the environment. Microorganisms developed diverse types of chromosomal-, transposon-, and plasmid-mediated resistance mechanisms to overcome TMs stress. The resistance genes are either plasmid or chromosome encoded and are highly specific. Microbes can resist the toxicity of TMs through exclusion by permeability barrier, extracellular or intracellular sequestration, enzymatic detoxification, active or passive tolerance, and alteration of cellular targets. The resistant bacteria have the potential to be utilized in the management of TMs, such as bioremediation, bioleaching, phytoremediation, and plant growth promotion.

To study the community structure of bacteria present at the contaminated site and to research TMs-resistant genes for cleaning up various TMs by enhancing the degrading bacterial strains, the 'omic' techniques must be taken into consideration. A global analysis of available studies indicates that careful attention should be paid to 'omic' data analysis because genes that are transcriptionally upregulated are not always identified as differentially represented at the protein level. This incoherence between the genomic, transcriptomic, and proteomic data might be explained by the incompleteness of proteomic datasets. The existing 'omic' studies underline that TMs exposure appears to be associated with both general stress and specific TMs responses. Nevertheless, the use of mutants, which permits the validation at the cellular level of the results obtained with 'omic' approaches, will result in a better understanding of the complexity of cell responses to TMs toxicity.

REFERENCES

Abatenh, E., Gizaw, B., Tsegaye, Z., Wassie, M., 2017 Application of microorganisms in bioremediation-review. *J. Environ. Microbiol.* 1, 02–09.

Ahemad, M., Khan, M.S., 2012. Effect of fungicides on plant growth promoting activities of phosphate solubilizing *Pseudomonas putida* isolated from mustard (*Brassica compestris*) rhizosphere. *Chemosphere* 86, 945–950.

Ahemad, M., 2019. Remediation of metalliferous soils through the heavy metal resistant plant growth promoting bacteria: Paradigms and prospects. *Arab. J. Chem.* 12(7), 1365–1377.

Archana, G., Buch, A., Kumar, G.N., 2012. Pivotal role of organic acid secretion by rhizobacteria in plant growth promotion. In: Satyanarayana, T., Johri, B., Prakash, A. (Eds.), *Microorganisms in Sustainable Agriculture and Biotechnology*. Springer, Dordrecht, pp. 35–53.

Atagana, H.I., 2008. Compost bioremediation of hydrocarbon-contaminated soil inoculated with organic manure. *Afr. J. Biotechnol.* 7 (10), 1516–1525.

ATSDR (Agency for Toxic Substances and Disease Registry) 2019. US department of health and human services. Substance Priority List. https://www.atsdr.cdc.gov/spl/#2019spl (accessed 13.09.21).

Azubuike, C.C., Chikere, C.B., Okpokwasili, G.C., 2016. Bioremediation techniques-classification based on site of application: Principles, advantages, limitations and prospects. *World J. Microbiol. Biotechnol.* 32(11), 1–18.

Baker-Austin, C., Dopson, M., 2007. Life in acid: pH homeostasis in acidophiles. *Trends Microbiol.* 15, 165–171.

Bamborough. L., Cummings, S.P., 2009. The impact of increasing heavy metal stress on the diversity and structure of the bacterial and actinobacterial communities of metallophytic grassland soil. *Biol. Fertil. Soils.* 45(3), 273–280.

Banerjee, A., Biswas, J.K., Pant, D., Sarkar, B., Chaudhuri, P., Rai, M. and Meers, E., 2019. Enteric bacteria from the earthworm (*Metaphire posthuma*) promote plant growth and remediate toxic trace elements. *J. Environ. Manage.* 250, 109530.

Biswas, J.K., Banerjee, A., Rai, M., Naidu, R., Biswas, B., Vithanage, M., Dash, M.C., Sarkar, S.K., Meers, E., 2018a. Potential application of selected metal resistant phosphate solubilizing bacteria isolated from the gut of earthworm (*Metaphire posthuma*) in plant growth promotion. *Geoderma* 330, 117–124.

Biswas, J.K., Banerjee, A., Rai, M.K., Rinklebe, J., Shaheen, S.M., Sarkar, S.K., Dash, M.C., Kaviraj, A., Langer, U., Song, H., Vithanage, M., 2018b. Exploring potential applications of a novel extracellular polymeric substance synthesizing bacterium (*Bacillus licheniformis*) isolated from gut contents of earthworm (*Metaphire posthuma*) in environmental remediation. *Biodegradation* 29(4), 323–337.

Biswas, J.K., Banerjee, A., Majumder, S., Bolan, N., Seshadri, B., Dash, M.C., 2019, June. New extracellular polymeric substance producing enteric bacterium from earthworm, *Metaphire posthuma*: Modulation through culture conditions. *Proc. Zoolog. Soc.* 72(2), 160–170.

Biswas, J.K., Banerjee, A., Sarkar, B., Sarkar, D., Sarkar, S.K., Rai, M., Vithanage, M., 2020. Exploration of an extracellular polymeric substance from earthworm gut bacterium (*Bacillus licheniformis*) for bioflocculation and heavy metal removal potential. *Appl. Sci.* 10(1), 349.

Blencowe, D.K., Morby, A.P., 2003. Zn(II) metabolism in prokaryotes. *FEMS Microbiol. Rev.* 27, 291–311.

Blindauer, C.A., Harrison, M.D., Robinson, A.K., Parkinson, J.A., Bowness, P.W., Sadler, P.J., Robinson, N.J., 2002. Multiple bacteria encode metallothioneins and SmtA-like zinc fingers. *Mol. Microbiol.* 45, 1421–1432.

Bolan, N., Hoang, S.A., Beiyuan, Jingzi, J., Gupta, S., Hou, D., Karakoti, A., Joseph, S., Jung, S., Kim, K.-H., Kirkham, M.B., Kua, H.W., Kumar, M., Kwon, E.E., Ok, Y.S., Perera, V., Rinklebe, J., Shaheen, S.M., Sarkar, B., Sarmah, A.K., Singh, B.P., Singh, G., Tsang, D.C.W., Vikrant, K., Vithanage, M., Vinu, A., Wang, H., Wijesekara, H., Yan, Y., Younis, S.A., Zwieten, L.V., 2021. Multifunctional applications of biochar beyond carbon storage. *Int. Mater. Rev.* 1–51. Doi: 10.1080/09506608.2021.1922047.

Bolan, N., Kumar, M., Singh, E., Kumar, A., Singh, L., Kumar, S., Keerthanan, S., Hoang, S.A., El-Naggar, A., Vithanage, M., Sarkar, B., 2022. Antimony contamination and its risk management in complex environmental settings: A review. *Environ. Int.* 158, 106908.

Borrok, D.M., Nimick, D.A., Wanty, R.B., Ridley, W.I., 2008. Isotopic variations of dissolved copper and zinc in stream waters affected by historical mining. *Geochim. Cosmochim. Acta* 72(2), 329–344.

Branco, R., Morais, P., 2013. Identification and characterization of the transcriptional regulator ChrB in the chromate resistance determinant of *Ochrobactrum tritici* 5bvl1. *PLoS One* 8, e77987.

Bruins, M.R., Kapil, S., Oehme, F.W., 2000. Microbial resistance to metals in the environment. *Ecotoxicol. Environ. Saf.* 45(3), 198–207.

Bundy, J.G., Paton, G., Campbell, C.D., 2012. Microbial communities in different soils types do not converge after diesel contamination. *J. Appl. Microbiol.* 92, 276–288.

Carlos, F.S., Giovanella, P., Bavaresco, J., de Souza, B.C., de Oliveira, C.F.A., 2016. A comparison of microbial bioaugmentation and biostimulation for hexavalent chromium removal from wastewater. *Water Air Soil Pollut.* 227(6), 175.

Chatterjee, S., Sau, G.B., Mukherjee, S.K., 2009. Plant growth promotion by a hexavalent chromium reducing bacterial strain, *Cellulosimicrobium cellulans* KUCr3. *World J. Microbiol. Biotechnol.* 25, 1829–1836.

Cho, K.S., Ryu, H.W., Lee, I.S., Choi, H.M., 2002. Effect of solids concentration on bacterial leaching of heavy metals from sewage sludge. *J Air Waste Manage. Assoc.* 52(2), 237–243.

Compant, S., Reiter, B., Sessitsch, A., Nowak, J., Clement, C., Ait Barka, E., 2005. Endophytic colonization of *Vitis vinifera* L. by plant growth-promoting bacterium *Burkholderia* sp. Strain PsJN. *Appl. Environ. Microbiol.* 71(4), 1685–1693.

Courbot, M., Diez, L., Ruotolo, R., Chalot, M., Leroy, P., 2004. Cadmium-responsive thiols in the ectomycorrhizal fungus *Paxillus involutus*. *Appl. Environ. Microbiol.* 70(12), 7413–7417.

Deng, X., Chai, L., Yang, Z., Tang, C., Wang, Y., Shi, Y., 2013. Bioleaching mechanism of heavy metals in the mixture of contaminated soil and slag by using indigenous *Penicillium chrysogenum* strain F1. *J. Hazard. Mater.* 248, 107–114.

Di Gregorio, S., Lampis, S., Vallini, G., 2005. Selenite precipitation by a rhizospheric strain of *Stenotrophomonas* sp. isolated from the root system of *Astragalus bisulcatus*: A biotechnological perspective. *Environ. Inter.* 31, 233–241.

Díaz-Pérez, C., Cervantes, C., Campos-García, J., Julián-Sánchez, A., Riveros-Rosas, H., 2007. Phylogenetic analysis of the chromate ion transporter (CHR) superfamily. *FEBS J.* 274(23), 6215–6227.

Djoko, K., Chong, L., Wedd, A., Xiao, Z., 2010. Reaction mechanisms of the multicopper oxidase CueO from *Escherichia coli* support its functional role as a cuprous oxidase. *J. Am. Chem. Soc.* 132, 2005–2015.

Dopson, M., Ossandon, F.J., Lövgren, L., Holmes, D.S., 2014. Metal resistance or tolerance? Acidophiles confront high metal loads via both abiotic and biotic mechanisms. *Front. Microbiol.* 5, 157.

Eswaran, J., Koronakis, E., Higgins, M.K., Hughes, C., Koronakis, V., 2004. Three's company: Component structures bring a closer view of tripartite drug efflux pumps. *Curr. Opin. Struct. Biol.* 14, 741–747.

Etesami, H., 2018. Bacterial mediated alleviation of heavy metal stress and decreased accumulation of metals in plant tissues: Mechanisms and future prospects. *Ecotoxicol. Environ. Saf.* 147, 175–191.

Finneran, K.T., Housewright, M.E., Lovley, D.R., 2002. Multiple influences of nitrate on uranium solubility during bioremediation of uranium-contaminated subsurface sediments. *Environ. Microbiol.* 4, 510–516.

Frey, B., Stemmer, M., Widmer, F., Luster, J., Sperisen, C., 2006. Microbial activity and community structure of a soil after heavy metal contamination in a model forest ecosystem. *Soil Biol. Biochem.* 38(7), 1745–1756.

Fu, B., Zhou, H., Zhang, R., Qiu, G., 2008. Bioleaching of chalcopyrite by pure and mixed cultures of *Acidithiobacillus* spp. and *Leptospirillum ferriphilum*. *Int. Biodeterior. Biodegrad.* 62(2), 109–115.

Fuke, P., Kumar, M., Sawarkar, A.D., Pandey, A., Singh, L., 2021. Role of microbial diversity to influence the growth and environmental remediation capacity of bamboo: A review. *Indus. Crops Prod.*, 167, 113567.

Fulekar, M.H., Sharma, J., Tendulkar, A., 2012. Bioremediation of heavy metals using biostimulation in laboratory bioreactor. *Environ. Monit. Assess.* 184 (12), 7299–7307.

Gans, J., 2005. Computational improvements reveal great bacterial diversity and high metal toxicity in soil. *Science* 309(5739), 1387–1390.

Gao, Y., Miao, C., Mao, L., Zhou, P., Jin, Z., Shi, W., 2010. Improvement of phytoextraction and antioxidative defense in *Solanum nigrum* L. under cadmium stress by application of cadmium-resistant strain and citric acid. *J. Hazard. Mater.* 181, 771–777.

Glick, B.R., 2010. Using soil bacteria to facilitate phytoremediation. *Biotechnol. Adv.* 28(3), 367–374.

González-Guerrero, M., Raimunda, D., Cheng, X., Argüello, J.M., 2010. Distinct functional roles of homologous Cu+ efflux ATPases in *Pseudomonas aeruginosa*. *Mol. Microbiol.* 78, 1246–1258.

Gopalakrishnan, S., Sathya, A., Vijayabharathi, R., Varshney, R.K., Gowda, C.L.L., Krishnamurthy, L., 2015. Plant growth promoting rhizobia: Challenges and opportunities. *3 Biotech.* 5(4), 355–377.

Haferburg, G., Kothe, E., 2007. Microbes and metals: Interactions in the environment. *J. Basic Microbiol.* 47(6), 453–467.

Hassan, M.T., van der Lelie, D., Springael, D., Romling, U., Ahmed, N., Mergeay, M., 1999. Identification of a gene cluster, czr, involved in cadmium and zinc resistance in *Pseudomonas aeruginosa*. *Gene* 238, 417–425.

Hebbeln, P., Eitinger, T., 2004. Heterologous production and characterization of bacterial nickel/cobalt permeases. *FEMS Microbiol. Lett.* 230, 129–135.

Jarosławiecka, A., Piotrowska-Seget, Z., 2014. Lead resistance in micro-organisms. *Microbiology* 160(1), 12–25.

Järup, L., 2003. Hazards of heavy metal contamination. *Br. Med. Bull.* 68(1), 167–182.

Kamika, I., Momba, M., 2013. Assessing the resistance and bioremediation ability of selected bacterial and protozoan species to heavy metals in metal-rich industrial wastewater. *BMC Microbiol.* 13(1), 1–4.

Kanmani, P., Aravind, J., Preston, D., 2012. Remediation of chromium contaminants using bacteria. *Int. J. Environ. Sci. Technol. (Tehran)* 9, 183–193.

Khalid, S., Shahid, M., Niazi, N.K., Murtaza, B., Bibi, I., Dumat, C., 2017. A comparison of technologies for remediation of heavy metal contaminated soils. *J. Geochem. Explor.* 182, 247–268.

Khan, A., Khan, S., Khan, M.A., Qamar, Z., Waqas, M., 2015. The uptake and bioaccumulation of heavy metals by food plants, their effects on plants nutrients, and associated health risk: A review. *Environ. Sci. Pollut. Res.* 22(18), 13772–13799.

Kumar, M., Kumar, M., Pandey, A., Thakur, I.S., 2019. Genomic analysis of carbon dioxide sequestering bacterium for exopolysaccharides production. *Sci. Rep.* 9(1), 1–12.

Kumar, M., Xiong, X., Sun, Y., Yu, I.K., Tsang, D.C., Hou, D., Gupta, J., Bhaskar, T., Pandey, A., 2020. Critical review on biochar-supported catalysts for pollutant degradation and sustainable biorefinery. *Adv. Sustain. Syst.* 4(10), 1900149.

Kumar, M., Bolan, N.S., Hoang, S.A., Sawarkar, A.D., Jasemizad, T., Gao, B., Keerthanan, S., Padhye, L.P., Singh, L., Kumar, S. Vithanage, M., 2021a. Remediation of soils and sediments polluted with polycyclic aromatic hydrocarbons: To immobilize, mobilize, or degrade? *J. Hazard. Mater.* 420, 126534.

Kumar, V., Singh, K., Shah, M.P., Kumar, M., 2021b. Phytocapping: An eco-sustainable green technology for environmental pollution control. In *Bioremediation for Environmental Sustainability* (pp. 481–491). Elsevier. Doi: 10.1016/B978-0-12-820318-7.00022-8.

Lara, P., Morett, E., Juárez, K., 2017. Acetate biostimulation as an effective treatment for cleaning up alkaline soil highly contaminated with Cr(VI). *Environ. Sci. Pollut. Res.* 24(33), 25513–25521.

Lee, E., Han, Y., Park, J., Hong, J., Silva, R.A., Kim, S., Kim, H., 2015. Bioleaching of arsenic from highly contaminated mine tailings using *Acidithiobacillus thiooxidans*. *J. Environ. Manag.* 147, 124–131.

Legatzki, A., Grass, G., Anton, A., Rensing, C., Nies, D.H., 2003. Interplay of the Czc system and two P-type ATPases in conferring metal resistance to *Ralstonia metallidurans*. *J. Bacteriol.* 185, 4354–4361.

Lemire, J.A., Harrison, J.J., Turner, R.J., 2013. Antimicrobial activity of metals: mechanisms, molecular targets and applications. *Nat. Rev. Microbiol.* 11(6), 371–384.

Li, Q., Wu, S., Liu, G., Liao, X., Deng, X., Sun, D., Hu, Y., Huang, Y., 2004. Simultaneous biosorption of cadmium (II) and lead (II) ions by pretreated biomass of *Phanerochaete chrysosporium*. *Sep. Purif. Technol.* 34(1–3), 135–142.

Li, Z., Xu, J., Tang, C., Wu, J., Muhammad, A., Wang, H., 2006. Application of 16S rDNA-PCR amplification and DGGE fingerprinting for detection of shift in microbial community diversity in Cu-, Zn-, and Cd-contaminated paddy soils. *Chemosphere* 62(8), 1374–1380.

Li, C.H., Wong, Y.S., Tam, N.F., 2010. Anaerobic biodegradation of polycyclic aromatic hydrocarbons with amendment of iron (III) in mangrove sediment slurry. *Bioresour. Technol.* 101, 8083–8092.

Liu, H.L., Chiu, C.W., Cheng, Y.C., 2003. The effects of metabolites from the indigenous *Acidithiobacillus thiooxidans* and temperature on the bioleaching of cadmium from soil. *Biotechnol. Bioeng.* 83(6), 638–645.

Lloyd, J.R., 2002. Bioremediation of metals; The application of micro-organisms that make and break minerals. *Microbiol. Today* 29, 67–69.

Ma, Y., Prasad, M.N.V., Rajkumar, M., Freitas, H., 2011. Plant growth promoting rhizobacteria and endophytes accelerate phytoremediation of metalliferous soils. *Biotechnol. Adv.* 29(2), 248–258.

Ma, Y., Oliveira, R.S., Nai, F., Rajkumar, M., Luo, Y., Rocha, I., Freitas, H., 2015. The hyperaccumulator *Sedum plumbizincicola* harbors metal-resistant endophytic bacteria that improve its phytoextraction capacity in multi-metal contaminated soil. *J. Environ. Manag.* 156, 62–69.

Mahbub, K.R., Krishnan, K., Megharaj, M., Naidu, R., 2016. Bioremediation potential of a highly mercury resistant bacterial strain *Sphingobium* SA2 isolated from contaminated soil. *Chemosphere* 144, 330–337.

Mangold, S., Potrykus, J., Björn, E., Lövgren, L., Dopson, M., 2012. Extreme zinc tolerance in acidophilic microorganisms from the bacterial and archaeal domains. *Extremophiles* 17, 75–85.

McLean, J., Beveridge, T.J., 2001. Chromate reduction by a pseudomonad isolated from a site contaminated with chromated copper arsenate. *Appl. Environ. Microbiol.* 67, 1076–1084.

Meena, K.K., Sorty, A.M., Bitla, U.M., Choudhary, K., Gupta, P., Pareek, A., Sahu, P.K., Gupta, V.K., Singh, H.B., 2017. Abiotic stress responses and microbe-mediated mitigation in plants: The omics strategies. *Front. Plant Sci.* 8, 172.

Mergeay, M., Monchy, S., Vallaeys, T., Auquier, V., Benotmane, A., Bertin, P., Taghavi, S., Dunn, J., Van Der Lelie, D., Wattiez, R., 2003. *Ralstonia metallidurans*, a bacterium specifically adapted to toxic metals: Towards a catalogue of metal-responsive genes. *FEMS Microbiol. Rev.* 27, 385–410.

Meyer, J., Schmidt, A., Michalke, K., Hensel, R., 2007. Volatilisation of metals and metalloids by the microbial population of an alluvial soil. *Syst. Appl. Microbiol.* 30(3), 229–238.

Mohamed, Z.A., 2001. Removal of cadmium and manganese by a non-toxic strain of the freshwater cyanobacterium *Gloeothece magna*. Water Res. 35(18), 4405–4409.

Mohapatra, P., Preet, R., Das, D., Satapathy, S.R., Siddharth, S., Choudhuri, T., Wyatt, M.D., Kundu, C.N., 2014. The contribution of heavy metals in cigarette smoke condensate to malignant transformation of breast epithelial cells and *in vivo* initiation of neoplasia through induction of a PI3K–AKT–NFκB cascade. *Toxicol. Appl. Pharmacol.* 274(1), 168–179.

Mondal, M., Biswas, J.K., Tsang, Y.F., Sarkar, B., Sarkar, D., Rai, M., Sarkar, S.K., Hooda, P.S., 2019. A wastewater bacterium *Bacillus* sp. KUJM2 acts as an agent for remediation of potentially toxic elements and promoter of plant (Lens culinaris) growth. *Chemosphere* 232, 439–52.

Mrozik, A., Piotrowska-Seget, Z., 2010. Bioaugmentation as a strategy for cleaning up of soils contaminated with aromatic compounds. *Microbiol. Res.* 165, 363–375.

Mulligana, C.N., Yong, R.N., 2004. Natural attenuation of contaminated soils. *Environ. Int.* 30, 587–601.

Mulrooney, S.B., Hausinger, R.P., 2003. Nickel uptake and utilization by microorganisms. *FEMS Microbiol. Rev.* 27, 239–261.

Nagajyoti, P.C., Lee, K.D., Sreekanth, T.V.M., 2010. Heavy metals, occurrence and toxicity for plants: A review. *Environ. Chem. Lett.* 8(3), 199–216.

Nanda, M., Kumar, V., Sharma, D.K., 2019. Multimetal tolerance mechanisms in bacteria: The resistance strategies acquired by bacteria that can be exploited to 'clean-up' heavy metal contaminants from water. *Aquat. Toxicol.* 212, 1–0.

Nie, J., Sun, Y., Zhou, Y., Kumar, M., Usman, M., Li, J., Shao, J., Wang, L., Tsang, D.C., 2020. Bioremediation of water containing pesticides by microalgae: Mechanisms, methods, and prospects for future research. *Sci. Total Environ.* 707, 136080.

Nies, D.H., 1995. The cobalt, zinc, and cadmium efflux system CzcABC from *Alcaligenes eutrophus* functions as a cationproton antiporter in *Escherichia coli*. *J. Bacteriol.* 177, 2707–2712.

Nies, D. H., Silver, S., 1995. Ion efflux systems involved in bacterial metal resistances. *J. Ind. Microbiol.* 14, 189–199.

Nies, D.H., 1999. Microbial heavy-metal resistance. *Appl. Microbiol. Biotechnol.* 51(6), 730–750.

Nies, D.H., 2003. Efflux-mediated heavy metal resistance in prokaryotes. *FEMS Microbiol. Rev.* 27, 313–339.

Pacheco, G.J., Ciapina, E.M.P., Gomes, Ed.B., Pereira Junior, N., 2010. Biosurfactant production by *Rhodococcus erythropolis* and its application to oil removal. *Braz. J. Microbiol.* 41, 685–693.

Pacwa-Płociniczak, M., Płaza, G.A., Piotrowska-Seget, Z., Cameotra, S.S., 2011. Environmental applications of biosurfactants: Recent advances. *Inter. J. Mol. R. Sci.* 12, 633–654.

Park, J., Han, Y., Lee, E., Choi, U., Yoo, K., Song, Y., Kim, H., 2014. Bioleaching of highly concentrated arsenic mine tailings by *Acidithiobacillus ferrooxidans*. *Sep. Purif. Technol.* 133, 291–296.

Pathak, A., Dastidar, M.G., Sreekrishnan, T.R., 2009. Bioleaching of heavy metals from sewage sludge: A review. *J. Environ. Manag.* 90(8), 2343–2353.

Prabha, J., Kumar, M., Tripathi, R., 2021. Opportunities and challenges of utilizing energy crops in phytoremediation of environmental pollutants: A review. *Bioremed. Environ. Sustain.* 383–396. Doi: 10.1016/B978-0-12-820318-7.00017-4.

Rahman, Z., Singh, V.P., 2020. Bioremediation of toxic heavy metals (THMs) contaminated sites: Concepts, applications and challenges. *Environ. Sci. Pollut. Res.* 27(22), 27563–27581.

Rajkumar, M., Ae, N., Prasad, M.N.V., Freitas, H., 2010. Potential of siderophore-producing bacteria for improving heavy metal phytoextraction. *Trends Biotechnol.* 28(3), 142–149.

Rajkumar, M., Sandhya, S., Prasad, M.N.V., Freitas, H., 2012. Perspectives of plant-associated microbes in heavy metal phytoremediation. *Biotechnol. Adv.* 30(6), 1562–1574.

Ramesh, G., Podila, G.K., Gay, G., Marmeisse, R., Reddy, M.S., 2009. Different patterns of regulation for the copper and cadmium metallothioneins of the ectomycorrhizal fungus *Hebeloma cylindrosporum*. *Appl. Environ. Microbiol.* 75(8), 2266–2274.

Ramirez-Diaz, M., Diaz-Perez, C., Vargas, E., Riveros-Rosas, H., Campos-Garcia, J., Cervantes, C., 2008. Mechanisms of bacterial resistance to chromium compounds. *Biometals* 21, 321–332.

Rashid, K., Sinha, K., Sil, P.C., 2013. An update on oxidative stress-mediated organ pathophysiology. *Food Chem. Toxicol.* 62, 584–600.

Rensing, C., Grass, G., 2003. *Escherichia coli* mechanisms of copper homeostasis in a changing environment. *FEMS Microbiol. Rev.* 27, 197–213.

Saha, P., Paul, B., 2016. Assessment of heavy metal pollution in water resources and their impacts: A review. *J. Basic Appl. Eng. Res.* 3(8), 671–675.

Saier Jr, M.H., Beatty, J.T., Goffeau, A., Harley, K.T., Heijne, W.H., Huang, S.C., Jack, D.L., Jahn, P.S., Lew, K., Liu, J., Pao, S.S., 1999. The major facilitator superfamily. *J. Mol. Microbiol. Biotechnol.* 1, 257–279.

Sharma, P.K., Balkwill, D.L., Frenkel, A., Vairavamurthy, M.A., 2000. A new *Klebsiella planticola* strain (Cd-1) grows anaerobically at high cadmium concentrations and precipitates cadmium sulfide. *Appl. Environ. Microbiol.* 66(7), 3083–3087.

Sharma, R.K., Archana, G., 2016. Cadmium minimization in food crops by cadmium resistant plant growth promoting rhizobacteria. *Appl. Soil Ecol.* 107, 66–78.

Silver, S., Nucifors, G., Chu, L., Misra, T.K., 1989. Bacterial resistance ATPases: Primary pumps for exporting toxic cations and anions. *Trends Biochem. Sci.* 14, 76–80.

Silver, S., Phung, L.T., 2005. A bacterial view of the periodic table: Genes and proteins for toxic inorganic ions. *J. Ind. Microbiol. Biotechnol.* 32(11–12), 587–605.

Singh, S.K., Grass, G., Rensing, C., Montfort, W.R., 2004. Cuprous oxidase activity of CueO from *Escherichia coli*. *J. Bacteriol.* 186, 7815–7817.

Singh, V., Chauhan, P.K., Kanta, R., Dhewa, T., Kumar, V., 2010. Isolation and characterization of *Pseudomonas* resistant to heavy metals contaminants. *Int. J. Pharm. Sci. Rev. Res.* 3(2), 164–167.

Slonczewski, J.L., Fujisawa, M., Dopson, M., Krulwich, T.A., 2009. Cytoplasmic pH measurement and homeostasis in bacteria and archaea. *Adv. Microb. Physiol.* 55, 1–79.

Solioz, M., Stoyanov, J., 2003. Copper homeostasis in *Enterococcus hirae*. *FEMS Microbiol. Rev.* 27, 183–195.

Tank, N., Saraf, M., 2009. Enhancement of plant growth and decontamination of nickel-spiked soil using PGPR. *J. Basic Microbiol.* 49, 195–204.

Tobor-Kapłon, M.A., Bloem, J., Römkens, P.F., de Ruiter, P.D., 2005. Functional stability of microbial communities in contaminated soils. *Oikos* 111(1), 119–129.

Tripathi, R.D., Tripathi, P., Dwivedi, S., Kumar, A., Mishra, A., Chauhan, P.S., Norton, G.J., Nautiyal, C.S., 2014. Roles for root iron plaque in sequestration and uptake of heavy metals and metalloids in aquatic and wetland plants. *Metallomics* 6, 1789–1800.

Tseng, T.T., Gratwick, K.S., Kollman, J., Park, D., Nies, D.H., Goffeau, A., Saier Jr, M.H., 1999. The RND permease superfamily: An ancient, ubiquitous and diverse family that includes human disease and development proteins. *J. Mol. Microbiol. Biotechnol.* 1, 107–125.

Ullah, A., Heng, S., Munis, M.F.H., Fahad, S., Yang, X., 2015. Phytoremediation of heavy metals assisted by plant growth promoting (PGP) bacteria: A review. *Environ. Exp. Bot.* 117, 28–40.

Valencia, Y.E., Braz, S.V., Guzzo, C., Marques, V.M., 2013. Two RND proteins involved in heavy metal efflux in *Caulobacter crescentus* belong to separate clusters within proteobacteria. *BMC Microbiol.* 13, 79.

Vallino, M., Greppi, D., Novero, M., Bonfante, P., Lupotto, E., 2009. Rice root colonisation by mycorrhizal and endophytic fungi in aerobic soil. *Ann. Appl. Biol.* 154(2), 195–204.

Velgosová, O., Kaduková, J., Marcinčáková, R., Palfy, P., Trpčevská, J., 2013. Influence of H_2SO_4 and ferric iron on Cd bioleaching from spent Ni–Cd batteries. *Waste Manag.* 33(2), 456–461.

Wang, S., Mulligan, C.N., 2006. Occurrence of arsenic contamination in Canada: Sources, behavior and distribution. *Sci. Tot. Environ.* 366(2–3), 701–721.

Wang, T., Sun, H., Mao, H., Zhang, Y., Wang, C., Zhang, Z., Wang, B., Sun, L., 2014. The immobilization of heavy metals in soil by bioaugmentation of a UV mutant *Bacillus subtilis* 38 assisted by NovoGro biostimulation and changes of soil microbial community. *J. Hazard. Mater.* 278, 483–490.

Weyens, N., van der Lelie, D., Taghavi, S., Newman, L., Vangronsveld, J., 2009. Exploiting plant–microbe partnerships to improve biomass production and remediation. *Trends Biotechnol.* 27(10), 591–598.

Wheaton, G., Counts, J., Mukherjee, A., Kruh, J., Kelly, R., 2015. The confluence of heavy metal biooxidation and heavy metal resistance: Implications for bioleaching by extreme thermoacidophiles. *Minerals* 5(3), 397–451.

Wightwick, A.M., Salzman, S.A., Reichman, S.M., Allinson, G., Menzies, N.W., 2010. Inter-regional variability in environmental availability of fungicide derived copper in vineyard soils: An Australian case study. *J. Agric. Food Chem.* 58(1), 449–457.

Wightwick, A.M., Salzman, S.A., Reichman, S.M., Allinson, G., Menzies, N.W., 2013. Effects of copper fungicide residues on the microbial function of vineyard soils. *Environ. Sci. Pollut. Res.* 20(3), 1574–1585.

Wuana, R.A., Okieimen, F.E., 2011. Heavy metals in contaminated soils: A review of sources, chemistry, risks and best available strategies for remediation. *ISRN Ecol.* 2011, 402647.

Yang, Z., Zhang, Z., Chai, L., Wang, Y., Liu, Y., Xiao, R., 2016. Bioleaching remediation of heavy metal-contaminated soils using *Burkholderia* sp. Z-90. *J. Hazard. Mater.* 301, 145–152.

Ying, G.G., 2018. Remediation and mitigation strategies. In: Maestroni, B., Cannavan, A. (Eds.), *Integrated Analytical Approaches for Pesticide Management*. Academic, Cambridge, pp 207–217.

Yu, E.W., Aires, J.R., Nikaido, H., 2003. AcrB multidrug efflux pump of *Escherichia coli*: Composite substrate-binding cavity of exceptional flexibility generates its extremely wide substrate specificity. *J Bacteriol*. 185, 5657–5664.

Zhang, J., Zhang, X., Ni, Y., Yang, X., Li, H., 2007. Bioleaching of arsenic from medicinal realgar by pure and mixed cultures. *Process Biochem*. 42(9), 1265–1271.

17 Omics Approaches for Microalgal Remediation of Wastewater

Recent Advances, Challenges, and Future Outlook

Edwin Hualpa-Cutipa
Universidad Nacional Mayor de San Marcos

Richard Andi Solórzano Acosta
Universidad César Vallejo

Xiomara Gisela Mendoza Beingolea
Universidad Nacional Mayor de San Marcos

Ingrid Maldonado Jimenez and Evelin Yana-Neira
Universidad Nacional del Altiplano de Puno

Isabel Navarro Zabarburú
Universidad Nacional Mayor de San Marcos

CONTENTS

DOI: 10.1201/9781003247883-17

17.1   INTRODUCTION

The water demand of the world has increased, and it is expected that by 2030, the water deficit will be around 40%, posing a challenge for all countries. This increase is linked to increased pollution of water resources by urban, agricultural, and industrial activities (Li et al., 2019). Strategies for the reuse of wastewater for human benefit have been implemented by applying treatment methods and techniques (Hussain et al., 2021; Chandra and Kumar, 2017; Kumar et al., 2021a). Among the most common strategies are bioremediation, phytoremediation, and restoration, removing contaminating molecules through a combination of physical, chemical, and biological processes from effluents (Olguín et al., 2007; Kumar et al., 2018). In this regard, microalgae have sparked significant attention due to their versatility and competence to colonize a wide range of photoautotrophic habitats (Barsanti and Gualtieri, 2018). In recent years, progress in the investigation of molecular mechanisms has been achieved using different advanced molecular techniques and strategies (omics) to study biological processes in microalgae and other living organisms.

Omics tools allow the understanding and analysis of basic knowledge of biological processes, as well as optimizing microalgae growth, development, and yield (Perera et al., 2019; Kumar et al., 2021b). Microalgae grow in aquatic ecosystems, but can also be found in soils and rocks because of their ability to adapt to a wide range of environmental conditions such as temperature, salt content, pH, and nutrient levels. Properties such as CO_2 reduction, biofuel production, application as biofertilizers because they favor nitrogen fixation, wastewater treatment, and bioremediation, associated with their ability to remove phosphates, nitrates, ammonium, and heavy metals, and hence improving the quality of contaminated effluents, have been reported to be linked to their biotechnological potential (Vega, 2014; Kumar and Thakur, 2020a, b). The performance of these biological systems is determined by their ability to clean domestic wastewater, wastewater from aquaculture, wastewater from the chemical industry, and other polluted effluents (Katam and Bhattacharyya, 2019). A microalgae-based system has also been introduced to remove contaminants from pharmaceutical drugs and personal care products, demonstrating their higher efficacy to isolate pollutants (Leng et al., 2020). Technological breakthroughs in omics technologies are allowing for the exploration and understanding of cellular functions at the molecular level. The transcriptome, metabolome, and proteome studies have made possible the understanding of microalgae cellular responses to multiple environmental stressor components. The advantages of omics analysis allow applications at the laboratory scale (Wan et al., 2020). The purpose of this chapter is to highlight the application of omics strategies in the study of microalgae with potential in the treatment and recycling of domestic, industrial, and storm wastewater, as well as the integration of these omics to generate global application platforms for the understanding of molecular behavior in the biological processes of microalgae.

17.2 OMICS TECHNOLOGIES APPLIED IN WASTEWATER TREATMENT

In recent years, wastewater generation and discharge have been a source of concern across the world (Kumar et al., 2021c). Among the potential solutions, the use of microalgal species for bioremediation has received much interest, and a thorough understanding of their structure, behavior, and interaction is required (Perera et al., 2019). The amount of pollutants in wastewater, the increased cost of microalgae, and difficulties in large-scale production are the most significant barriers to effective microalgae-water coupling (El-Sheekh et al., 2021). However, microalgae are a promising species for wastewater treatment, as those who recurrently produce biomass by utilizing the nutrients dissolved in the wastewater, so wastewater recovery using microalgae has a clear advantage over their treatment and utilization (Jiang et al., 2021). Biotechnological interventions, particularly those involving omics sciences, can boost nutrient removal and vegetative growth by microalgae and photosynthetic bacteria (Mishra et al., 2019), It also has the potential to provide a clear picture of protein, gene, and DNA changes (Schnurr et al., 2013), and there has been background on the study of municipal, domestic, and wastewater from the pharmaceutical industry to date (El-Sheekh et al., 2021; Kumar et al., 2020; Kumar and Chandra, 2020).

17.2.1 GENOMICS OF MICROALGAE FOR WASTEWATER TREATMENT

At the current time, the use of genomics to detect single-cell genetic sequences and the expression of functional genes is a very effective molecular tool (Mishra et al., 2019). The genomic technique has the advantage of being fast and providing species composition in consortia, in addition to identifying genetic variations in consortia and then allowing comparison between species; however, it requires specific expertise, there is a lack of genome libraries, and it is difficult to isolate low abundance species in consortia (Kermarrec et al., 2013). A few of the genetic analyses of microalgal species in wastewater treatment are shown in Table 17.1.

Microalgae's attributes, such as their ability to assimilate nutrients and tolerate stress conditions caused by pollutants in the environment, make them ideal candidates for removing nutrients from wastewater (Mishra et al., 2019; Hodges et al., 2017). Furthermore, since its energy source is carbon and the load of nutrients contained in nature, as well as also in wastewater, its cultivation is inexpensive, resulting in biofuel and other benefits. As a result, researchers are interested in the use of microalgae in the treatment of contaminated effluents because they have the ability to reduce nitrogen and phosphorus as well as remove other toxic elements such as heavy metals (Hallmann, 2007). The genome is defined as a living organism's or cell's entire set of genes or nucleotide sequences (Giani et al., 2019); this nuclear structure consists of coding segments (genes) and non-coding

TABLE 17.1

Genomic Analysis of Microalgae for Wastewater Treatment

Genomic Technique	Analysis	Source	Reference
PCR-DGGE combined with 16S rDNA sequencing	Abundance of bacteria in a microalgae-bacteria consortium.	Textile wastewater	Su et al. (2009)
Whole-genome shotgun on HiSeq platform	To determine the underlying molecular mechanisms involved in the variation of growth rates in three species of *Chlorella*.	Wastewater from the swine farm	Wu et al. (2019)
16S rDNA phylogeny	Bacterial diversity in a microalgae-bacterial consortium.	Anaerobic synthetic wastewater	Vasseur et al. (2012)

TABLE 17.2

Microalgae Genomic Information Databases

Item	Name	Link
1	*Cyanidioschyzon merolae* Genome Project	http://merolae.biol.s.u-tokyo.ac.jp/
2	EST National Center for Biotechnology Information (NCBI)	https://www.ncbi.nlm.nih.gov/genbankdbest/
3	NCBI Organelle Database	https://www.ncbi.nlm.nih.gov/genome/organelle/
4	The Organelle Genome Megasequencing Program	https://megasun.bch.umontreal.ca/ogp/welcome2.html
5	Database of *Nannochloropsis*	http://nandesyn.single-cell.cn/

segments (regulatory sequences that activate or deactivate genes associated with growth, nutrient assimilation, as well as adaptation to different environmental conditions within the cytoplasm) (Teng et al., 2020). Genomic information helps us to understand the structure of metabolic pathways, such as inorganic carbon absorption, and to identify species with biotechnological potential (bioremediation and bio-factory, among others). All this genomic information is located and stored in various databases (Table 17.2).

Genomic studies include DNA isolation, amplification, sequencing, assembly, quality assessment (see Figure 17.1), and most importantly, structural and functional annotation of the genome using various bioinformatics programs (Mehta et al., 2019).

Research on the use of microalgae with specific genomic characteristics to remove toxic compounds or nutrients from the environment, through metagenomics microalgae from the genus *Chlorella* was applied to wastewater treatment (in a CO_2 bubbling system) leading to an increase of the biomass autotrophic biomass, as well as the number of mutualistic bacteria, resulting in a decrease in phosphorus and nitrogen concentrations in the effluent (Ye et al., 2016).

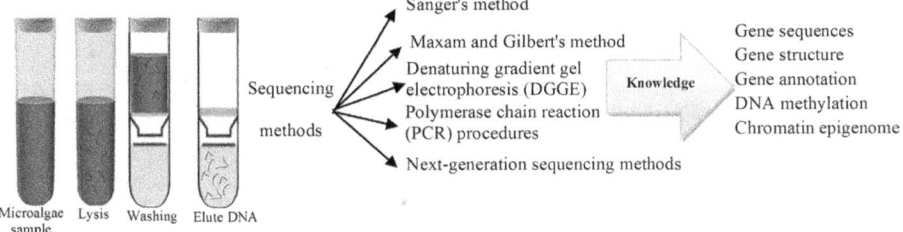

Sanger's method

Maxam and Gilbert's method

Denaturing gradient gel
electrophoresis (DGGE)

Polymerase chain reaction
(PCR) procedures

Next-generation sequencing methods

Sequencing methods

Knowledge

Gene sequences
Gene structure
Gene annotation
DNA methylation
Chromatin epigenome

Microalgae sample   Lysis   Washing   Elute DNA

FIGURE 17.1   Procedures, methods, and general knowledge obtained from genomic sequencing

17.2.2 Microalgae Metagenomics in Wastewater Treatment

Metagenomic studies, according to Piampiano et al. (2019), have started to characterize microalgae-associated microbial communities; however, there is little information available about their continuity and adaptability (KrohnMolt et al., 2017; Sambles et al., 2017). Also, Ebenezer et al. (2012) analyzed the phycospheric bacterial communities associated with different cultures of *Tetraselmis suecica* F&M-M33, a green marine microalga with several industrial applications, using a metagenomic approach, and then they identified a "core" bacterial community, which comprises at least 13 families and accounts for 70% of the total bacterial community.

17.2.3 Transcriptomics of Microalgae for Wastewater Treatment

Transcriptomics, as compared to genomics, provides insights into gene expression patterns generated by interactions between bacterial consortia and microalgae (Cooper and Smith, 2015). The transcriptome is defined as the total amount of mRNA from the expression of genes under a specified condition, which may be analyzed by microarray hybridization, RNA sequencing (RNA-Seq), and qRT-PCR (Gonzalez-Ballester et al., 2011). Transcriptomic phenomena such as photosynthesis, energy metabolism, and starch metabolism of diploid microalgae have been studied in the case of microalgae (Kwak et al., 2017). Also, based on the comparison of consortia and axenic cultures, transcriptomic studies allow differentiation of the functional characteristics of genes and their regulation in expression. RNA-Seq captures everything, both individual and consortia sequences; however, RNA-Seq and qRT-PCR can detect the quantity and abundance of transcripts in very low quantities, but they are time-consuming. Microarrays, on the other hand, can only quantify transcripts from organisms that have a sequenced genome (Durham et al., 2015). Transcriptomic studies on microalgae are abundant; however, in wastewater, they are limited. Some of them are shown in Table 17.3.

Hence, transcriptomics is a widely used tool for functional studies of microbial genes expressed under environmental conditions, for subsequent selection of autotrophic microorganisms with the capacity to bioremediate wastewater (Figure 17.2).

17.2.4 Proteomics of Microalgae for Wastewater Treatment

Proteomics has the advantage of high-throughput proteome analysis, illustrating key genes relationships with biological mechanisms of organisms in consortia and

TABLE 17.3

Transcriptomic Analysis of Microalgae for Wastewater Treatment

Species	Situation	Approach	Results	Reference
Chlorella pyrenoidosa	Nanodiamonds with capacity of induced phytotoxic effect and molecular mechanism.	Integration of regular and transcriptomic analyses.	In response to oxidative damage, genes involved in amino acid metabolism were positively regulated.	Zhang et al. (2021a)
Various wildlife species	Autotrophic microorganisms with high lipid content helped to fund advanced wastewater treatment.	Lipidomics	Prolonged abiotic stress promoted the growth of microalgae and increased the lipid content.	Arif et al. (2020)
Chlorella vulgaris	Phosphorus starvation and phosphorus uptake in green microalgae.	Lipidomics	Phosphorus deficiency reduces photosynthesis and causes phospholipids to be replaced by sulfo- and betaine lipids.	Solovchenko et al. (2019)

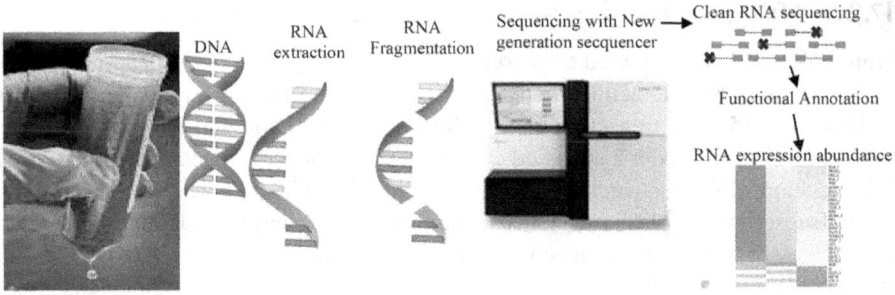

FIGURE 17.2   Gene expression of microalgae through a transcriptomic approach

processes in connection, as well as their linkage in enzymatic studies; however, it can present troubles and complexity in data analysis (VerBerkmoes et al., 2009); despite this, efforts are underway. Patel et al. (2015) introduced the first non-ecological proteomic analysis of microalgae on wastewater, characterizing changes in the proteome of *Chlamydomonas reinhardtii* after culture in a synthetic wastewater medium using MS-based proteomics. Burch et al. (2021), by comparing *Phaeodactylum tricornutum* culture in dilute DMW versus synthetic media, discovered modifications in protein regulations associated with protein metabolism, signal transduction, transcription, protein trafficking, and oxidative stress management pathways, providing insight into how *P. tricornutum* reconfigures its own proteome in dilute DMW versus artificial media. In the study by Elleuch et al. (2021), on the other hand, the alga

Dunaliella sp. was subjected to a comparative proteomic analysis using nano-HPLC coupled to LC-MS/MS on cells from pre- and post-zinc treatments for zinc removal in wastewater, and it was demonstrated that the target proteins are involved in various metabolic processes; photosynthesis and antioxidant defensive systems are the most important.

The importance of proteomic analysis lies in the understanding of the mechanisms used by microalgae for pollutant removal in wastewater, as demonstrated by Liu et al. (2015), who discovered that after municipal wastewater treatment, the expression of glutamine synthetase, an enzyme responsible for ammonium assimilation, was upregulated 6.4-fold at the proteome level and 18.0-fold at the transcript level. Similarly, Patel et al. (2015) identified 2358 proteins in *Chlamydomonas reinhardtii* cells in response to nitrogen source and nitrogen availability, as well as their differential expression in wastewater and synthetic solutions, using label-free shotgun proteomic analysis.

17.2.5 Metabolomics of Microalgae for Wastewater Treatment

The analysis of metabolic interactions within microbial consortia as well as the impacts of environmental changes on these consortia should be studied (Mishra et al., 2019). Metabolites are currently an important part of biological research because they can be influenced by environmental changes, biological states of organisms, and genetic changes (Chen et al., 2017). The qualitative and quantitative measurement of metabolites is highly reproducible, and metabolome study techniques have many advantages: easy and simple sample preparation applying high-throughput detection of primary and secondary metabolites by nuclear magnetic resonance (NMR), and rapid detection with high resolution in ultrahigh-performance supercritical fluid chromatography. There are few examples, but Zhang et al. (2021a) detected the effects of four drugs on *Chlorella pyrenoidosa* and demonstrated that metabolomic analysis supplemented with diclofenac could be an effective approach to understanding the mechanism of molecular evolution in *C. pyrenoidosa* for microalgal biomass and wastewater bioenergy in biological research. By working with JSC strain 4 in mariculture wastewater, Zhang et al. (2021b) demonstrated that sea salt is a better trigger to enhance cell growth and lipid accumulation.

17.2.6 Interactomics of Microalgae for Wastewater Treatment

Interaction relationships between bacteria and microalgae have attracted attention in several studies, although there are certain limitations in the algae growth in wastewater, because the presence of bacteria that influence and interact in the final efficiency of the process through parasitism or competition, generating products that reduce the performance of the algae. Mutualism relationships between algae and bacteria improve microalgae productivity, bio-flocculation, and wastewater treatment efficiency (Jiang et al., 2021). The algae-bacterial symbiosis method for wastewater treatment dates back to 1950, and over the last ten years, there has been a consistent increase (Mu et al., 2021), toward interactions for the formation of relationships between communities (Luo et al., 2019), which can be used to validate

TABLE 17.4

Interatomic Analysis of Microalgae for Wastewater Treatment

Bacterial Species	Algae Species	Interaction	Reference
Sulfitobacter sp.	*Pseudo-nitzschia multiseries*	The yield of organic sulfur molecules excreted by diatoms and ammonia excreted by bacteria was increased by nutrient exchange.	Amin et al. (2015)
Mesorhizobium loti	*Lobomonas rostrata*	The bacteria supplied vitamin B12, while the algae supplied photosynthates.	Grant et al. (2014)
Flavobacterium, Hyphomonas, Rhizobium, Sphingomonas	*Chlorella vulgaris*	The bacterial consortium enhanced algal biomass and lipid productivity significantly.	Cho et al. (2015)
Botryococcus braunii	*Nostoc muscorum*	Increased biomass, lipids, and production of a phytohormone.	Gautam et al. (2019)
Chlamydomonas reinhardtii	*Synechococcus* sp.	The bacteria provide acetate to the algae for lipid production.	Therien et al. (2014)
Bacillus licheniformis	*Chlorella* sp.	Increased phosphorus and nitrogen removal rate.	Liang et al. (2013)

the mechanism of effective wastewater treatment by algae-bacterial consortiums; the main interactions described to date are discussed in Table 17.4.

17.2.7 EPIGENOMICS OF MICROALGAE FOR WASTEWATER TREATMENT

To understand the complexity of biological systems, one must rely on epigenetic mechanisms based on gene regulation by methylation processes, small RNA and chromatin modifications are primary basic mechanisms of the epigenetic regulation (Singh et al., 2019). In the field of microalgae applied to wastewater treatment, this is still a less studied line of research; however, it can be mentioned that there is unexpected complexity of the RNAs population in diatoms and features that were not appreciated previously, generating new data on the diversity of ncRNA-based processes in eukaryotes (Rogato et al., 2014). Similarly, Fournet and Roussakis (2018) reported on the diatom species *Phaeodactylum tricornutum* and *Thalassiosira pseudonana* that silicon work as epigenomic element that seems to influence the expression of fifteen target genes in the polymerase group. Changes in chromatins and histones are also mentioned, which are found to be crucial in plant development, but whose implications in algae are unknown. In the industrial oleaginous microalga *Nannochloropsis* spp., a detailed ChIP protocol representing a reliable approach for histone modification analysis was developed, which generated a genome-wide histone modification event profile (Wei and Xu, 2018).

17.3 DRAWBACKS AND FUTURE PERSPECTIVES

The use of omics technology in microalgae is very promising because it allows us to learn about their potential as multipurpose raw materials with applications in a wide range of biotechnological, industrial, biofuel, and biomedical fields (Guarnieri, 2014). In addition, the set of omics technologies can provide ample opportunities to build new methods of algal metabolism; these omics strategies can be used to describe biological molecules such as DNA, RNA, and proteins, allowing for a better insight into proteomics-based approaches to microalgae (Anand et al., 2017). However, omics presents issues that must be addressed, such as bias, statistics, complex methodology, and method misuse (Lay, 2006). Also, it is a high-cost technique that is rarely used (Muñoz et al., 2018). The extraction of relevant compounds from microalgae, on the other hand, is hampered by their resistant and thick cell wall; consequently, alternative methods that are optimal while not degrading the extracted compounds are preferred. If heat is used for extraction, thermolabile compounds such as pigments or proteins will be less suitable. Shotgun proteomic analysis was chosen for one study because of its selectivity to chemical treatment caused by reactive oxygen species (Zocher et al., 2019). To have a better management of microalgae in wastewater treatment, different methodological options should be addressed; having knowledge of the genome is one of the options that gives us a metabolic approach to be able to perform some modification in the genomic DNA. As a result, more research is required to overcome these production-related challenges, including addressing large-scale issues (Sevda et al., 2019). The fast application of omics strategies will further guide in the application of metabolic changes of microalgae for sustainable, cost-effective, and renewable feedstock for biofuels, pharmaceuticals, wastewater, and metal phytoremediation (Perera et al., 2019).

17.4 RECENT ADVANCES, CHALLENGES, AND FUTURE OUTLOOK

Currently, the use of omics approaches, tools, and computer databases necessitates ongoing efforts to produce economically viable microalgae that can be applied in a variety of fields. To that end, improved microalgae strains are being sought, with improved characteristics such as higher biomass production, lipid accumulation, CO_2 utilization, and pollution remediation capacity, as in wastewater treatment. This is where metagenomics and genomics come in, providing molecular insight into metabolic fluxes. "Omics" applications in the study of microalgae help us understand the regulation and integration of metabolic initiators, intermediates, and end products, allowing us to identify the processes that regulate changes in metabolic pathways (Mishra et al., 2019). Furthermore, systems biology approaches such as proteomics, transcriptomics, and metabolomics have become critical to understanding the metabolism of microalgae to better understand the post-genomic techniques of this process. Moreover, details on the impact of environmental factors on genes, transcriptomes, metabolomes, protein expression, and regulation in bacterial and microalgal consortia have been discovered using omics approaches, providing greater insights into the functionality and adaptations of such consortia for industrial and environmental benefits.

REFERENCES

Amin, S. A., Hmelo, L. R., Van Tol, H. M., Durham, B. P., Carlson, L. T., Heal, K. R., ... Armbrust, E. V. (2015). Interaction and signalling between a cosmopolitan phytoplankton and associated bacteria. *Nature*, 522(7554), 98–101. Doi: 10.1038/nature14488.

Anand, V., Singh, P. K., Banerjee, C., & Shukla, P. (2017). Proteomic approaches in microalgae: Perspectives and applications. *3 Biotech*, 7(3), 197. Doi: 10.1007/s13205-017-0831-5.

Arif, M., Bai, Y., Usman, M., Jalalah, M., Harraz, F. A., Al-Assiri, M. S., ... Zhang, C. (2020). Highest accumulated microalgal lipids (polar and non-polar) for biodiesel production with advanced wastewater treatment: Role of lipidomics. *Bioresource Technology*, 298 Doi: 10.1016/j.biortech.2019.122299.

Barsanti, L., & Gualtieri, P. (2018). Is exploitation of microalgae economically and energetically sustainable? *Algal Research*, 31, 107–115. Doi: 10.1016/j.algal.2018.02.001.

Burch, A. R., Yothers, C. W., Salemi, M. R., Phinney, B. S., Pandey, P., & Franz, A. K. (2021). Quantitative label-free proteomics and biochemical analysis of *phaeodactylum tricornutum* cultivation on dairy manure wastewater. *Journal of Applied Phycology*, 33(4), 2105–2121. Doi: 10.1007/s10811-021-02483-3.

Chandra, R., & Kumar, V. (2017). Detection of androgenic-mutagenic compounds and potential autochthonous bacterial communities during in-situ bioremediation of post methanated distillery sludge. *Frontiers in Microbiology* 8, 87. Doi: 10.3389/fmicb.2017.00887.

Chandra, R., & Kumar, V. (2018). Phytoremediation: A green sustainable technology for industrial waste management. In: Chandra, R., Dubey, N., & Kumar, V., (Eds.), *Phytoremediation of Environmental Pollutants*. CRC Press, Boca Raton, FL. Doi: 10.1201/9781315161549–1

Chen, T., Zhao, Q., Wang, L., Xu, Y., & Wei, W. (2017). Comparative metabolomic analysis of the green microalga *Chlorella sorokiniana* cultivated in the single culture and a consortium with bacteria for wastewater remediation. *Applied Biochemistry & Biotechnology*, 183(3), 1–14.

Cho, D. H., Ramanan, R., Heo, J., Lee, J., Kim, B. H., Oh, H. M., & Kim, H. S. (2015). Enhancing microalgal biomass productivity by engineering a microalgal-bacterial community. *Bioresource Technology*, 175, 578–585. Doi: 10.1016/j.biortech.2014.10.159.

Cooper, M. B., & Smith, A. G. (2015). Exploring mutualistic interactions between microalgae and bacteria in the omics age. *Current Opinion in Plant Biology*, 26, 147–153.

Durham, B. P., Sharma, S., Luo, H., Smith, C. B., Amin, S. A., Bender, S. J., ... Moran, M. A. (2015). Cryptic carbon and sulfur cycling between surface ocean plankton. *Proceedings of the National Academy of Sciences*, 112(2), 453–457.

Ebenezer, V., Medlin, L. K., & Ki, J. (2012). Molecular detection, quantification, and diversity evaluation of microalgae. *Marine Biotechnology*, 14(2), 129–142. Doi: 10.1007/s10126-011-9427-y.

El-Sheekh, M., El-Dalatony, M. M., Thakur, N., Zheng, Y., & Salama, E. (2021). Role of microalgae and cyanobacteria in wastewater treatment: Genetic engineering and omics approaches. *International Journal of Environmental Science and Technology*, Doi: 10.1007/s13762-021-03270-w.

Elleuch, J., Ben Amor, F., Chaaben, Z., Frikha, F., Michaud, P., Fendri, I., & Abdelkafi, S. (2021). Zinc biosorption by *dunaliella* sp. AL-1: Mechanism and effects on cell metabolism. *Science of the Total Environment*, 773 Doi: 10.1016/j.scitotenv.2021.145024.

Fournet, J., & Roussakis, C. (2018). Silicon coordinates DNA replication with transcription of the replisome factors in diatom algae. *Plant Molecular Biology Reporter*, 36, 257–272. Doi: 10.1007/s11105-018-1074-2.

Gautam, K., Tripathi, J. K., Pareek, A., & Sharma, D. K. (2019). Growth and secretome analysis of possible synergistic interaction between green algae and cyanobacteria. *Journal of Bioscience and Bioengineering*, 127(2), 213–221. Doi: 10.1016/j.jbiosc.2018.07.005.

Giani, A. M., Gallo, G. R., Gianfranceschi, L., & Formenti, G. (2019). Long walk to genomics: History and current approaches to genome sequencing and assembly. *Computational and Structural Biotechnology Journal*, 18, 9–19. Doi: 10.1016/j.csbj.2019.11.002.

Gonzalez-Ballester, D., Pootakham, W., Mus, F., Yang, W., Catalanotti, C., Magneschi, L., ... Grossman, A. R. (2011). Reverse genetics in *Chlamydomonas*: a platform for isolating insertional mutants. *Plant Methods*, 7(1), 24. Doi: 10.1186/1746-4811-7-24.

Grant, M. A. A., Kazamia, E., Cicuta, P., & Smith, A. G. (2014). Direct exchange of vitamin B 12 is demonstrated by modelling the growth dynamics of algal-bacterial cocultures. *ISME Journal*, 8(7), 1418–1427. Doi: 10.1038/ismej.2014.9.

Guarnieri, M. T., & Pienkos, P. T. (2014). Algal omics: unlocking bioproduct diversity in algae cell factories. *Photosynthesis Research*, 123(3), 255–263. Doi: 10.1007/s11120-014-9989-4.

Hallmann, A. (2007). Algal transgenics and biotechnology. *Transgenic Plant Journal*, 1, 81–98.

Hao, J., Liebeke, M., Astle, W., De Iorio, M., Bundy, J. G., & Ebbels, T. M. D. (2014). Bayesian deconvolution and quantification of metabolites in complex 1D NMR spectra using BATMAN. *Nature Protocols*, 9(6), 1416–1427.

Hodges, A., Fica, Z., Wanlass, J., VanDarlin, J., & Sims, R. (2017). Nutrient and suspended solids removal from petrochemical wastewater via microalgal biofilm cultivation. *Chemosphere*, 174, 46–48. Doi: 10.1016/j.chemosphere.2017.01.107.

Hussain, F., Shah, S. Z., Ahmad, H., Abubshait, S. A., Abubshait, H. A., Laref, A., Manikandan, A., Kusuma, H. S., & Iqbal, M. (2021). Microalgae an ecofriendly and sustainable wastewater treatment option: Biomass application in biofuel and bio-fertilizer production. A review. *Renewable and Sustainable Energy Reviews*, 137, 110603. Doi: 10.1016/j.rser.2020.110603.

Jiang, L., Li, Y., & Pei, H. (2021). Algal–bacterial consortia for bioproduct generation and wastewater treatment. *Renewable and Sustainable Energy Reviews*, 149 Doi: 10.1016/j.rser.2021.111395.

Katam, K., & Bhattacharyya, D. (2019). Simultaneous treatment of domestic wastewater and bio-lipid synthesis using immobilized and suspended cultures of microalgae and activated sludge. *Journal of Industrial and Engineering Chemistry*, 69, 295–303. Doi: 10.1016/j.jiec.2018.09.031.

Kermarrec, L., Franc, A., Rimet, F., Chaumeil, P., Humbert, J. F., & Bouchez, A. (2013). Next-generation sequencing to inventory taxonomic diversity in eukaryotic communities: A test for freshwater diatoms. *Molecular Ecology Resources*, 13(4), 607–619. Doi: 10.1111/1755-0998.12105.

KrohnMolt, I., Alawi, M., Forstner, K. U., Wiegandt, A., Burkhardt, L., Indenbirken, D., Thiess, M., Grundhoff, A., Kehr, J., Tholey, A. & Streit, W. R. (2017). Insights into microalga and bacteria interactions of selected phycosphere biofilms using metagenomic, transcriptomic, and proteomic approaches. *Frontiers in Microbiology*, 8, 1941. Doi: 10.3389/Fmicb.2017.01941.

Kumar, V., & Chandra, R. (2020). Metagenomics analysis of rhizospheric bacterial communities of *Saccharum arundinaceum* growing on organometallic sludge of sugarcane molasses-based distillery. *3 Biotech* 10(7), 316. Doi: 10.1007/s13205-020-02310-5

Kumar, V., Kaushal, A., Singh, K., & Shah, M. P. (2021a). Phytoaugmentation technology for phytoremediation of environmental pollutants: opportunities, challenges and future prospects. In: Kumar, V., Saxena, G., & Shah, M. P., (Eds.), *Bioremediation for Environmental Sustainability: Approaches to Tackle Pollution for Cleaner and Greener Society*. Elsevier. Doi: 10.1016/B978-0-12-820318-7.00016-2.

Kumar, V., Shahi, S. K., & Singh, S. (2018). Bioremediation: An eco-sustainable approach for restoration of contaminated sites. In: Singh J., Sharma D., Kumar G., & Sharma N. (Eds.), *Microbial Bioprospecting for Sustainable Development*. Springer, Singapore. Doi: 10.1007/978-981-13-0053-0_6.

Kumar, V., Singh, K., & Shah, M. P. (2021c). Advanced oxidation processes for complex wastewater treatment. In: Shah, M.P. (Eds.), *Advance Oxidation Process for Industrial Effluent Treatment*. Elsevier. Doi: 10.1016/B978-0-12-821011-6.00001-3.

Kumar, V., Singh, K., Shah, M. P., Singh, A. K, Kumar, A., & Kumar, Y. (2021b). Application of omics technologies for microbial community structure and function analysis in contaminated environment. In Shah, M. P., Sarkar, A., & Mandal, S., (Ed.), *Wastewater*

Treatment: Cutting Edge Molecular Tools, Techniques & Applied Aspects in Waste Water Treatment. Elsevier. Doi: 10.1016/B978-0-12-821925-6.00013-7.

Kumar, V., & Thakur, I. S. (2020a). Extraction of lipids and production of biodiesel from secondary tannery sludge by *in situ* transesterification. *Bioresource Technology Reports*, 11, 100446. Doi: 10.1016/j.biteb.2020.100446.

Kumar, V., & Thakur, I. S. (2020b). Biodiesel production from transesterification of *Serratia* sp. ISTD04 lipids using immobilised lipase on biocomposite materials of biomineralized products of carbon dioxide sequestrating bacterium. *Bioresource Technology*, 307, 123193. Doi: 10.1016/j.biortech.2020.123193.

Kumar, V., Thakur, I. S., Singh, A. K., & Shah, M. P. (2020). Application of metagenomics in remediation of contaminated sites and environmental restoration. In: Shah, M., Rodriguez-Couto, S., & Sengor, S. S. (Eds.), *Emerging Technologies in Environmental Bioremediation.* Elsevier. Doi: 10.1016/B978-0-12-819860-5.00008-0.

Kwak, M., Park, W. K., Shin, S. E., Koh, H. G., Lee, B., Jeong, B. R., & Chang, Y. K. (2017). Improvement of biomass and lipid yield under stress conditions by using diploid strains of *Chlamydomonas reinhardtii. Algal Research*, 26, 180–189.

Lay, J. O., Liyanage, R., Borgmann, S., & Wilkins, C. L. (2006). Problems with the "omics." *TrAC Trends in Analytical Chemistry*, 25(11), 1046–1056. Doi: 10.1016/j.trac.2006.10.007.

Leng, L., Wei, L., Xiong, Q., Xu, S., Li, W., Lv, S., Lu, Q., Wan, L., Wen, Z., & Zhou, W. (2020). Use of microalgae based technology for the removal of antibiotics from wastewater: A review. *Chemosphere*, 238, 124680. Doi: 10.1016/j.chemosphere.2019.124680.

Li, K., Liu, Q., Fang, F., Luo, R., Lu, Q., Zhou, W., Huo, S., Cheng, P., Liu, J., Addy, M., Chen, P., Chen, D., & Ruan, R. (2019). Microalgae-based wastewater treatment for nutrients recovery: A review. *Bioresource Technology*, 291, 121934. Doi: 10.1016/j.biortech.2019.121934.

Liang, Z., Liu, Y., Ge, F., Xu, Y., Tao, N., Peng, F., & Wong, M. (2013). Efficiency assessment and pH effect in removing nitrogen and phosphorus by algae-bacteria combined system of *Chlorella vulgaris* and *Bacillus licheniformis. Chemosphere*, 92(10), 1383–1389.

Liu, N., Li, F., Ge, F., Tao, N., Zhou, Q., & Wong, M. (2015). Mechanisms of ammonium assimilation by chlorella vulgaris F1068: Isotope fractionation and proteomic approaches. *Bioresource Technology*, 190, 307–314. Doi: 10.1016/j.biortech.2015.04.024.

Luo, L., Lin, X., Zeng, F., Luo, S., Chen, Z., & Tian, G. (2019). Performance of a novel photobioreactor for nutrient removal from piggery biogas slurry: Operation parameters, microbial diversity and nutrient recovery potential. *Bioresource Technology*, 272, 421–432.

Mehta, S., James, D., & Reddy, M. K. (2019) Omics technologies for abiotic stress tolerance in plants: Current status and prospects. In: Wani, S. (Eds.), *Recent Approaches in Omics for Plant Resilience to Climate Change.* Springer, Cham. Doi: 10.1007/978-3-030-21687-0_1.

Mishra, A., Medhi, K., Malaviya, P., & Thakur, I. S. (2019). Omics approaches for microalgal applications: Prospects and challenges. *Bioresource Technology*, 291, 121890. Doi: 10.1016/j.biortech.2019.121890.

Mu, R., Jia, Y., Ma, G., Liu, L., Hao, K., Qi, F., & Shao, Y. (2021). Advances in the use of microalgal–bacterial consortia for wastewater treatment: Community structures, interactions, economic resource reclamation, and study techniques. *Water Environment Research*, Doi: 10.1002/wer.1496.

Muñoz Yañez, C., Rubio Andrade, M., Guangorena, J. O., Alegría Torres, J. A., García Vargas, G. G.. (2018) Las "Ómicas" como herramienta en el estudio de la Salud Ambiental. *Revista de Salud Ambiental*, 18(2), 156–165.

Olguín, E. J., Hernández, M. E., & Sánchez-Galván, G. (2021). Potential of microalgae-based biorefineries utilizing wastewater and aquatic plants. *Revista Internacional de Contaminación Ambiental*, 23(3), 139–154.

Patel, A. K., Huang, E. L., Low-Décarie, E., & Lefsrud, M. G. (2015). Comparative shotgun proteomic analysis of wastewater-cultured microalgae: Nitrogen sensing and carbon fixation for growth and nutrient removal in *Chlamydomonas reinhardtii*. *Journal of Proteome Research*, 14(8), 3051–3067. Doi: 10.1021/pr501316h.

Perera, I. A., Abinandan, S., Subashchandrabose, S. R., Venkateswarlu, K., Naidu, R., & Megharaj, M. (2019). Advances in the technologies for studying consortia of bacteria and cyanobacteria/microalgae in wastewaters. *Critical Reviews in Biotechnology*, 39 (5), 709–731. Doi: 10.1080/07388551.2019.1597828.

Piampiano, E., Pini, F., Biondi, N., Pastorelli, R., Giovannetti, L., & Viti, C. (2019). Analysis of microbiota in cultures of the green microalga *Tetraselmis suecica*. *European Journal of Phycology*, 54(3), 497–508. Doi: 10.1080/09670262.2019.1606940.

Rogato, A., Richard, H., Sarazin, A., Voss, B., Cheminant Navarro, S., Champeimont, R., Navarro, L., Carbone, A., Hess, W. R., & Falciatore, A. (2014). The diversity of small non-coding RNAs in the diatom *Phaeodactylum tricornutum*. *BMC Genomics*, 15(1), 698. Doi: 10.1186/1471-2164-15-698.

Sambles, C., Moore, K., Lux, T. M., Jones, K., Littlejohn, G. R., Gouveia, J. D., Aves, S. J., Studholme, D. J., Lee, R., & Love, J. (2017). Metagenomic analysis of the complex microbial consortium associated with cultures of the oil-rich alga *Botryococcus braunii*. *Microbiology Open*, 6, 1–9.

Schnurr, P. J., Espie, G. S., & Allen, D. G. (2013). Algae biofilm growth and the potential to stimulate lipid accumulation through nutrient starvation. *Bioresource Technology*, 136 (C), 337–344.

Sevda, S., Garlapati, V. K., Sharma, S., Bhattacharya, S., Mishra, S., Sreekrishnan, T. R., & Pant, D. (2019). Microalgae at niches of bioelectrochemical systems: A new platform for sustainable energy production coupled industrial effluent treatment. *Bioresource Technology Reports*, 7, 100290. Doi: Doi: 10.1016/j.biteb.2019.100290.

Singh, P. P., Demmitt, B. A., Nath, R. D., & Brunet, A. (2019) The genetics of aging: a vertebrate perspective. *Cell*, 177, 200– 220.

Solovchenko, A., Khozin-Goldberg, I., Selyakh, I., Semenova, L., Ismagulova, T., Lukyanov, A., & Gorelova, O. (2019). Phosphorus starvation and luxury uptake in green microalgae revisited. *Algal Research*, 43 Doi: 10.1016/j.algal.2019.101651.

Su, Y., Zhang, Y., Wang, J., Zhou, J., Lu, X., & Lu, H. (2009). Enhanced bio-decolorization of azo dyes by co-immobilized quinone-reducing consortium and anthraquinone. *Bioresource Technology*, 100(12), 2982–2987.

Teng, S. Y., Yew, G. Y., Sukačová, K., Show, P. L., Máša, V., & Chang, J.-S. (2020). Microalgae with artificial intelligence: A digitalized perspective on genetics, systems and products. *Biotechnology Advances*, 44, 107631. Doi: 10.1016/j.biotechadv.2020.107631.

Therien, J. B., Zadvornyy, O. A., Posewitz, M. C., Bryant, D. A., & Peters, J. W. (2014). Growth of *Chlamydomonas reinhardtii* in acetate-free medium when co-cultured with alginate-encapsulated, acetate-producing strains of *Synechococcus* sp. PCC 7002. *Biotechnology for Biofuels*, 7(1), 154. Doi: 10.1186/s13068-014-0154-2.

Vasseur, C., Bougaran, G., Garnier, M., Hamelin, J., Leboulanger, C., Chevanton, M. L., Mostajir, B., Sialve, B., Steyer, J.-P., & Fouilland, E. (2012). Carbon conversion efficiency and population dynamics of a marine algae–bacteria consortium growing on simplified synthetic digestate: First step in a bioprocess coupling algal production and anaerobic digestion. *Bioresource Technology*, 119, 79–87.

Vega, M. A. (2014). *caracterización y manipulación genética de microalgas marinas para la producción de compuestos de alto valor añadido*. Universidad de Huelva. Departamento de Química y Ciencia de los Materiales. http://hdl.handle.net/10272/8837.

VerBerkmoes, N. C., Denef, V. J., Hettich, R. L., & Banfield, J. F. (2009). Systems biology: Functional analysis of natural microbial consortia using community proteomics. *Nature Reviews Microbiology*, 7(3), 196–205. Doi: 10.1038/nrmicro2080.

Wan, C., Chen, B., Zhao, X., & Bai, F. (2020). Current advances in biotechnology of marine microalgae. Kim, S. K. (Ed.), *Encyclopedia of Marine Biotechnology* (1st ed., pp. 1809–1825). Wiley. Doi: 10.1002/9781119143802.ch77.

Wu, T., Li, L., Jiang, X., Yang, Y., Song, Y., Chen, L., Xu, X., Shen, Y., & Gu, Y. (2019). Sequencing and comparative analysis of three chlorella genomes provide insights into strain-specific adaptation to wastewater. *Scientific Reports*, 9(1), 9514. Doi: 10.1038/s41598-019-45511-6.

Ye, J., Song, Z., Wang, L., & Zhu, J. (2016). Metagenomic analysis of microbiota structure evolution in phytoremediation of a swine lagoon wastewater. *Bioresource Technology*, 219, 439–444. Doi: 10.1016/j.biortech.2016.08.013.

Zhang, C., Chen, X., Chou, W., & Ho, S. (2021a). Phytotoxic effect and molecular mechanism induced by nanodiamonds towards aquatic *Chlorella pyrenoidosa* by integrating regular and transcriptomic analyses. *Chemosphere*, 270 Doi: 10.1016/j.chemosphere.2020.129473.

Zhang, C., Hasunuma, T., Shiung Lam, S., Kondo, A., & Ho, S. (2021b). Salinity-induced microalgal-based mariculture wastewater treatment combined with biodiesel production. *Bioresource Technology*, 340 Doi: 10.1016/j.biortech.2021.125638.

Zocher, K., Lackmann, J.-W., Volzke, J., Steil, L., Lalk, M., Weltmann, K.-D., Wende, K., & Kolb, J. F. (2019). Profiling microalgal protein extraction by microwave burst heating in comparison to spark plasma exposures. *Algal Research*, 39, 101416. Doi: 10.1016/j.algal.2019.101416.

18 Disturbance and Stress in Coastal Ecosystems
Quantifying Responses at Multiple Levels of Biological Organization

Milton Torres-Ceron and Joshua Lerner
Texas A&M AgriLife

Antonio Leija-Tristan and Juan Antonio Vidales-Contreras
Universidad Autonoma de Nuevo Leon

CONTENTS

DOI: 10.1201/9781003247883-18

18.1   INTRODUCTION

18.1.1   ENVIRONMENTAL IMPACTS ON COASTAL ECOSYSTEMS

Coastal ecosystems are among the most valuable ecosystems in the world for many reasons, from providing critical habitat for thousands of wildlife species to supporting local economies and regulating climate change by sequestering carbon from the atmosphere and storing it in living plant biomass and the soil. Coastal ecosystems, especially estuaries and tidal wetlands, provide a variety of indispensable ecosystem services, such as commercial and recreational fisheries, nutrient cycling, erosion control, and storm and flood mitigation (Basset et al. 2013; Davis Jr. and Fitzgerald 2020). In addition, these ecosystems can also be a sink for chemical pollutants. For example, heavy metals and aromatic compounds can accumulate in plant and animal biomass or be broken down by microbes via natural bioremediation processes (Häder et al. 2020). Despite the risk of living in a changing environment exposed to increasingly intense disturbance regimes, such as tropical cyclones, half of the global population lives on the coasts due to the economic and aesthetic value that coastal ecosystems provide.

Humans continue to overpopulate coastal areas and influence their dynamic processes through land-use conversion, overfishing, overharvesting, and pollution, which result in disruptions to the ecosystem services they provide (Davis Jr. and Fitzgerald 2020). In addition to anthropogenic effects, coastal areas are shaped by weather events such as tropical cyclones, droughts, floods, heatwaves, and freezes, which in turn influence other biological properties such as biodiversity and ecosystem productivity (Ummenhofer and Meehl 2017). Anthropogenic and natural phenomena may also act synergistically to generate novel and unexpected responses. Therefore, it is a global priority to protect our coastal assets from these environmental impacts. By monitoring the physiology of organisms and the productivity of ecosystems over time, we gain an understanding of the theoretical basis needed to develop better infrastructure and policy that promotes ecosystem health, sustainability, and resilience.

18.1.2   ECOLOGICAL DISTURBANCE THEORY

In the scientific literature, the terms "stress" and "disturbance" are used inconsistently and ambiguously. For the sake of clarity, we define a disturbance as a relatively discrete pulse event that at low or intermediate levels of exposure can play a fundamental role in maintaining ecosystem function (Odum, Finn, and Franz 1979; Pickett et al. 1989; Johnson and Miyanishi 2007). When a habitat or patch is made partially or completely devoid of vegetation because of a disturbance, ecological succession begins, and new species will recolonize the cleared area and progressively regenerate species richness and biodiversity (Levin and Painet 1974; White and Jentsch 2001).

We define stress in this chapter as something that puts into action the mechanism of homeostasis (Odum 1985; Schulte 2014). In this case, a stress response is aimed at maintaining a system's functionality and stability when faced with environmental changes. Accordingly, the earliest signs of stress in an ecosystem can be detected at the organismal level. If organisms can maintain homeostasis when faced with a stressor by altering their behavior or physiology, a stress response may not be detectable at the population, community, or ecosystem levels. Therefore, stress detectable at the ecosystem level is alarming because it may indicate a breakdown in homeostasis at lower levels of biological organization (Petitjean et al. 2019).

According to our definitions, ecological disturbances are stressors that are able to alter the physical structure of an ecosystem (Figure 18.1). Disturbances in estuaries can be natural, such as tropical cyclones, droughts, freezes, floods, wildfires, and disease, or human-induced, such as oil spills, industrial waste discharges, and dredging. Stress, on the other hand, manifests itself primarily at lower hierarchical levels of biological organization as a response to stressors, while community and ecosystem response times may take longer to detect. For example, coastal vegetation stressed by high winds and flooding during tropical cyclones may lead to mass mortality of plants and subsequent habitat loss for the animals that rely upon that vegetation for food and shelter. In another example, the 2010 Deepwater Horizon oil spill in the Gulf of Mexico set off cascading effects on the entire food web, causing mass mortality of top predators such as marine mammals and seabirds, and in turn,

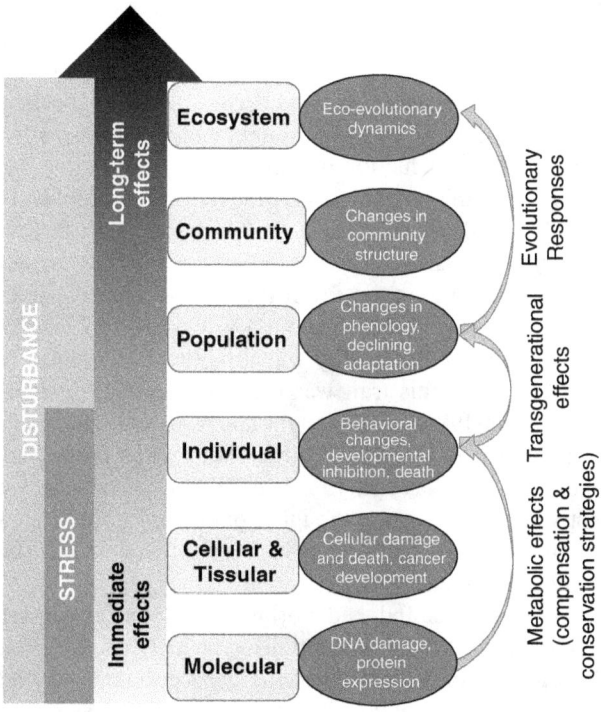

FIGURE 18.1 Hierarchical representation of bottom-up disturbance effects on ecosystems. Stress is an organism's response to a stressor, in an effort to maintain its internal homeostasis. Natural selection and genetic drift work on populations by affecting which individuals live and reproduce. Changes in community structure affect the flow of energy through an ecosystem. Responses to a disturbance can be measured at each hierarchical level.

the abundance of the prey species Gulf menhaden (*Brevoortia patronus*) increased in the absence of predation more than four standard deviations higher than its previous 39-year mean (Short et al. 2017). Therefore, disturbances can completely restructure communities and ecosystems through cascading effects up or down the food web, and also from the organismal level to the population and community levels. Moreover, short-lived disturbances can also have long-lasting effects on the abiotic environment (Borja et al. 2010). On the other hand, soil toxicity would impede plant recolonization and growth and thus reduce the ecosystem's intrinsic resilience to perturbations (El-Shatnawi and Makhadmeh 2001).

18.1.3   ECO-EVOLUTIONARY IMPLICATIONS OF DISTURBANCES IN COASTAL ECOSYSTEMS

Disturbances can be a major driver of evolutionary change over short and long timescales (Grant et al. 2017). While evolution is typically considered a gradual and incremental process, high-amplitude unpredictable disturbances, such as tropical cyclones and droughts, can deplete resources, induce stress, and cause substantial mortality over very short periods. These extreme events can be agents of natural selection if mortality is non-random concerning traits. And if the selected traits are heritable, then these events may create new selective pressures on organisms, alter the genetic composition of populations, and drive evolutionary change (Grant et al. 2017). For example, the selective force of drought on the Galapagos Island of Daphne Major in 1977 prevented reproduction in most plants, and without the production of new seeds, the soil seed bank on the islands quickly became depleted by the resident finch populations. By 1978, the only seeds that remained were large seeds that only finches with the largest beaks were able to break open, while smaller beaked birds had subsequently died of starvation. From this single disturbance event, the average beak size on the islands increased about 4% (Grant and Grant 2002, 2006). In this case, disturbance-induced mortality was directionally selective and resulted in an adapted population. Yet, this trait was only adaptive when smaller seeds were scarce. Excessive rain in 1982–1983 brought about an abundance of smaller seeds, and during a drought the following year, birds with smaller beaks had an advantage, so the direction of evolution was easily reversed. Evolution driven by extreme events can still be adaptive and long-lasting though. For example, hurricane winds resulted in natural selection for larger toe pad size in *Anolis* lizards in the Turks and Caicos archipelago (Donihue et al. 2018).

Alternatively, if disturbance-induced mortality is random or non-selective, genetic diversity is likely reduced through genetic drift. Drift alters the allele frequency of a population through random sampling of the gene pool and may lead to the fixation or loss of genotypes, especially rare variants (Star and Spencer 2013). Coastal ecosystems such as estuaries and tidal wetlands, however, are open to both the ocean and the rivers that feed into them, which facilitates rapid migration and dispersal for organisms that rely on the buoyancy and movement of water associated with tides, wind, and currents (Costanza, Kemp, and Boynton 1993). Therefore, gene flow from exogenous sources can quickly replenish a population's gene pool following a disturbance. Since functional replacements for those lost in the disturbance are nearly always available, the selective advantages of specialization are minimal, and most organisms are r-selected generalists. These organisms are highly fecund and resilient opportunists capable of filling a variety of functional roles and niches. In ecosystems that experience such frequent unpredictable disturbances, there is a lack of a stable base on which taxonomic diversity can develop, so there is no need for a specialized function. Rather, functional generalists, such as oysters, crabs, shrimp, and bass, and other dominant organisms help maintain resilience by maintaining keystone ecological processes under changing conditions.

18.1.4 Detecting Stress at Different Levels of Biological Organization

Recent technological advances, including a growing body of research using "omics" tools, have provided new insight into the mechanisms that underpin complex stress responses. Toxic waste contaminants in estuaries can cause DNA damage in fish (Almeida et al. 2021). These novel mutations, and patterns of mutation across species, can be detected using genomics approaches, which can provide insight into the genetic pathways involved in stress response. Changes in gene expression in response to stress will act on shorter timescales than DNA mutation and can be detected using transcriptomics techniques, i.e., RNA-Seq (Ferreira et al. 2021). Stress from environmental changes can induce the production of heat shock chaperone proteins, which can be detected using proteomics (Evrard et al. 2013). When a stress response is induced, organisms shift from allocating their energy toward growth and reproduction to maintaining internal homeostasis to counteract changing external conditions (Schulte 2014). Alterations in energy metabolism due to stress can be detected using metabolomics (Naour et al. 2017). In highly perturbed coastal ecosystems, these technological advances will continue to be utilized to gain a better understanding of how organisms have evolved and adapted over time to frequent and prolonged stress.

18.2 MEASUREMENT OF STRESS AND DISTURBANCE IN COASTAL ECOSYSTEMS

18.2.1 Analysis of Environmental Quality in Laguna Madre, Tamaulipas, Mexico

The Laguna Madre is one of the most important estuaries in Mexico. By its geographical location, it has been a hotspot of biodiversity, allowing the development of commercial and recreational fisheries, in addition to the ecological services provided (Leija-Tristán et al. 2000). However, resulting from human activities, this estuary has been affected by different sources of stress, such as invasive species, aquatic pollution, overfishing, and habitat modification. Because of this, researchers in the Universidad Autonoma de Nuevo Leon (UANL), Mexico, have started a research project of environmental monitoring (Torres-Ceron et al. 2014; Aguilera et al. 2019; Herrera Barroquín 2019). As part of this project, between 2009 and 2010, mollusks were used as indicators of environmental quality (Torres-Ceron et al. 2014; Aguilera et al. 2019).

In order to determine the grade of environmental impact due to pollution, an effect directed analysis (Hecker and Hollert 2009) was performed, considering three main steps: (1) chemical analysis to quantify pollutants in sediments; (2) ecological analysis of malacological community, to detect changes in communities in the environmental gradient analyzed; and (3) a battery of toxicological essays using sediments from the study area, which allows identifying the effect of the pollutants present in sediments at the physiological level.

The study area in the present case was a location called Boca de Catan, in San Fernando, Tamaulipas. This area is located in the southern part of the Laguna Madre. Three sites were chosen to sample sediments and organisms: (1) oyster reef, close to the barrier island, of which the coordinates are 24°28'58.1" N and 97°41'26" W; (2) seagrasses bed area, in the deeper zone of the estuary, 24°29'2.4" N and 97°41'52.6" W; (3) site close to the coast, in a location called Laguna de Catan, 24°29'15.5" N and 97°46'.5" W; and (4) site close to the coast, in a location called La Muela, 24°26'29.43" N and 97°46'7.71" W.

Sampling methods for sediment and mollusks are detailed in Aguilera et al. (2019) and Torres-Ceron et al. (2014). Quantification of total heavy hydrocarbons, Cd, Pb, Cu, Zn, and Fe was performed. The analytic methods are detailed in Torres-Ceron et al. (2014). Cd concentration was undetectable in the method used, while the maximum concentration of Fe, Zn, Cu, and Pb found was 91,685.25, 27.7, 6.16, and 4.02 mg/kg, respectively. All of these maximum concentrations were found in site 2, the area with seagrasses beds. According to the Mexican normativity (NOM-021-RECNA-2000), the concentration of the metals was lower for the permissible limit and, therefore, harmless. And it could be considered that these low concentrations were in the function of the low concentration of organic matter and fine particles (clay and silt). Concentrations of heavy hydrocarbons in sediment were ranging between 100 and 200 mg/kg. According to the Mexican normativity (NOM-138-SEMARNAT/SS-2003), heavy hydrocarbons concentration was under the permissible limits and, therefore, harmless. However, sediments (from both marine and freshwater environments) are not included in the Mexican legislation and normativity, so the permissible limits for pollutants in agricultural and forestall soil were considered as a threshold for the present study. A whole discussion on these variables can be found in Torres-Ceron et al. (2014).

The ecological analysis included the estimation of the Shannon diversity index and the criteria of Wilhm and Dorris to determine the degree of environmental impact in terms of diversity. In other words, the lower the diversity of the malacological community, the higher the environmental impact was used (Wilhm and Dorris 1968). In addition, adapted versions of biotic indexes according to Torres-Ceron et al. (2014) were used to determine the degree of environmental impact as a function of the number of species present in each site, and the number of species that are tolerant to harsh conditions. A detailed explanation of the application of these indexes can be found in Torres-Ceron et al. (2014) and Herrera-Barroquin (2019). This analysis showed that the study area presents a degree ranging from moderate to severe. However, given the estuarine quality paradox, a concept that explains that, as a result of the natural environmental variability of estuaries, the biota found has the characteristics of environments with high degrees of disturbance (Dauvin and Ruellet 2009; Elliott and

Quintino 2007). Therefore, this kind of ecological analysis cannot be the only way to assess environmental impacts in estuarine ecosystems, such as the Laguna Madre.

Finally, a set of toxicological assays were performed using the gastropods *Physa mexicana* and *Pomacea bridgesii* (Torres-Ceron et al. 2014). The treatments were sediment collected from the sampling sites, and the control was distilled water. Each treatment had three replicates. Additional details of experimental conditions can be found in Torres-Ceron et al. (2014). The response variables were the enzymatic activity of acetylcholinesterase (AChE) and carboxylesterase (CaE). The experimental design of the bioassays was completely randomized; an ANOVA and Tukey test ($p < 0.05$) were used to identify the differences between treatments. The results suggested a physiological response to exposure to metals and pesticides present in the sediments. In addition, individuals of Venus clam (*Chione elevata*) were collected from the sampling sites, and levels of AChE, CaE, butyrylcholinesterase (BChE), alkaline phosphatase (ALP), glutathione s-transferase (GST), and oxygen radical absorbance capacity (ORAC) were measured in the soft tissue of the clam (Aguilera et al. 2019). The findings of the authors suggest that the areas closer to the land, where the human activity was higher, are under more stress compared with the sites closer to the ocean. Additional details can be found in Aguilera et al. (2019).

The findings described above suggest an ecosystem disturbance produced by anthropogenic activities, especially fishing activity with inappropriate management of oil and its wastes. This kind of disturbance is producing stress at the individual level, which potentially could have transgenerational effects (Petitjean et al. 2019) in the aquatic organisms. However, given the estuarine quality paradox, more detailed studies are necessary for Laguna Madre, including expounding on the taxa to analyze, to include vertebrates, and increasing the study areas, to get a better idea of what is the magnitude of environmental impact in these ecosystems. Studies in crustaceans have already been done (Herrera Barroquín 2019), and ecotoxicological analyses with different taxa are expected to be started soon.

18.2.2 A Quantitative Research Framework for Analyzing Ecosystem Data in the Context of Disturbance Ecology

Ecosystem responses can be multifaceted and difficult to quantify because individual ecosystem components, e.g., nutrient concentration, organismal abundance, and productivity, may respond differently to the same disturbance. A unified research framework that captures the multitude of ecosystem responses to a disturbance is needed to facilitate the analysis and comparison of disturbance events within and across ecosystems. Disaggregating ecosystem response into individual measurable components allows for better comparisons across sites and systems (Hogan et al. 2021).

The disturbance event itself can often be disaggregated into distinct explanatory variables (e.g., wind speed, rainfall, storm surge, and storm duration) because event forcing often differs considerably over time and across heterogeneous landscapes (Turner 1989). In 2017, for example, Hurricanes Harvey and Irma both made landfall in the southern USA as category 4 storms with winds of 130 mph (209 km/h), but Harvey was slower moving and produced over 150 cm of rainfall with 1-m average storm surge, while Irma produced about 50 cm of rain with 3.5-m storm surge (Hogan et al. 2021). Therefore, a disaggregating approach to quantifying ecosystem responses will reveal the important explanatory variables that modulate response and thus enhance our ability to predict how ecosystems will respond to future disturbances by comparing them to past events.

18.2.3 MEASURING ECOSYSTEM RESPONSE: RESISTANCE AND RESILIENCE

The terms resistance and resilience were introduced into the scientific literature nearly 40 years ago by theoretical ecologist C.S. Holling to help describe an ecosystem's non-linear dynamics (Gunderson 2000). Despite this long-term recognition, the abstract nature of these terms has prevented the establishment of unified metrics that are applied to all fields of research (Ayyub 2014). The term "resistance" refers to a system's capacity to resist a forcing and remain unchanged (Hogan et al. 2021). The magnitude of change from baseline, or the effect size (E), is a measure of resistance. The term "resilience" refers to the intrinsic capacity of a system to rebound from a perturbation (Cushman and McGarigal 2019). The return time to baseline (R) is a measure of resilience (Pimm et al. 2019). Thus, ecosystems with low amplitude responses (small E) and short return times (small R) are both resistant and resilient to a forcing. Preliminary work on ecosystem responses to hurricanes shows a general pattern or trade-off between these strategies that may form the basis of an ecological law since the same trends appear to hold for different ecosystems, at different scales, and even among entirely disparate response variables. Evolutionary theory may explain this pattern, as selection toward one strategy may come at the cost of the other (Patrick et al. 2020; Hogan et al. 2021).

The concept of resilience has been gaining popularity among the general public and among coastal policymakers and managers, who are now trying to incorporate resilience into their decision-making processes (Pimm et al. 2019). Historically, coastal policymakers have not embraced strategies or infrastructures that promote resilience, since these usually demand short-term economic sacrifices. Instead, policy and infrastructure have aimed to promote stability and resistance to physical damage from disturbances. Still, management for both resistance and resilience simultaneously may not be feasible. Quantifying the trade-offs between ecosystem resistance and resilience will enhance our understanding of ecosystem vulnerability, by allowing more targeted policy and infrastructure design that addresses a given location's needs. The standardized metrics for measuring ecosystem response proposed herein will also help managers weigh the relative value of competing ecosystem services and prioritize their management actions in the face of changing climate and disturbance regimes.

18.2.3.1 Quantifying Resistance and Resilience in a Time Series: A Big-Data, Remote Sensing Approach

The quantitative resistance and resilience framework can be generalized for any response variable in a time series, but in this chapter, the utility of the framework will be illustrated using a commonly measured ecosystem response variable, gross primary productivity (GPP). GPP is a summary metric that represents the total amount of carbon fixed per unit time through the photosynthetic reduction of inorganic carbon to organic compounds by all primary producers in an ecosystem. GPP is a fundamental ecosystem process that controls the rate of return to pre-disturbance conditions since productivity generally correlates directly to growth rate and the rate of competitive displacement during succession (White and Jentsch 2001). The GPP of an ecosystem will vary as a function of disturbance size and intensity and may respond disproportionately across space and time (Hogan et al. 2021; Patrick et al. 2020). Site-specific studies that relate responses such as productivity to various explanatory variables have great utility, but limited spatial and temporal scope. Leveraging the statistical power of long-term remote sensing datasets will provide far greater insight into macroecological patterns and other large-scale phenomena.

A tidal wetland GPP dataset for the entire contiguous USA has recently been made publicly available on the Oak Ridge National Laboratory Distributed Active Archive Center for Biogeochemical Dynamics for the years 2000–2020 at 16-day return intervals and 250-m resolution (Feagin et al. 2020). Tidal wetlands in this dataset consist of salt marshes, tidal freshwater marshes, tidal forested swamps, and mangrove forests. Despite covering less than 2% of the ocean surface (Duarte, Middelburg, and Caraco 2005), these ecosystems account for about 30% of oceanic organic carbon burial (Wang et al. 2021). Disturbances can alter the GPP of an ecosystem (Ciais et al. 2005; Running 2008), yet little is known about short-term responses of tidal wetland GPP at regional to continental scales.

18.2.4   RESISTANCE AND RESILIENCE METRICS: GPP RESPONSE OF A SALT MARSH IN GALVESTON BAY, TEXAS

The first step in quantifying the resistance and resilience of responses in a time series is to empirically define the baseline antecedent ecosystem conditions. For each wetland pixel ($n=214{,}055$), average GPP and standard deviations are computed for each 16-day epoch of the year (e.g., January 1st, January 17th, February 2nd; $n=23$). Raw αGPP values vary greatly throughout the year due to seasonal changes in temperature and precipitation, so the 20-year average for the January 1st epoch will be substantially lower than the average for an epoch in July. To correct for this and remove variability among sites and dates in the time series, standardized βGPP is computed as shown in Equation 18.1 by subtracting a pixel's 20-year epoch average for a pixel from the raw αGPP value for a given epoch:

$$\beta_x = \alpha_x - \left[\left(\sum_{j=1}^{n} {}_{x,j} \right) \Big/ n \right],$$

(18.1)

where $j=1$ is the epoch of the year (23 epochs per year) and n is the total number of years (20 years). The distance from the mean standardization is advantageous because βGPP values for a given pixel can be directly compared to the pixel's standard deviations, which are essentially the average distances from the 20-year mean for each epoch. Therefore, the standard deviation is used to set the upper and lower boundaries for what is normally expected in a given epoch of the time series.

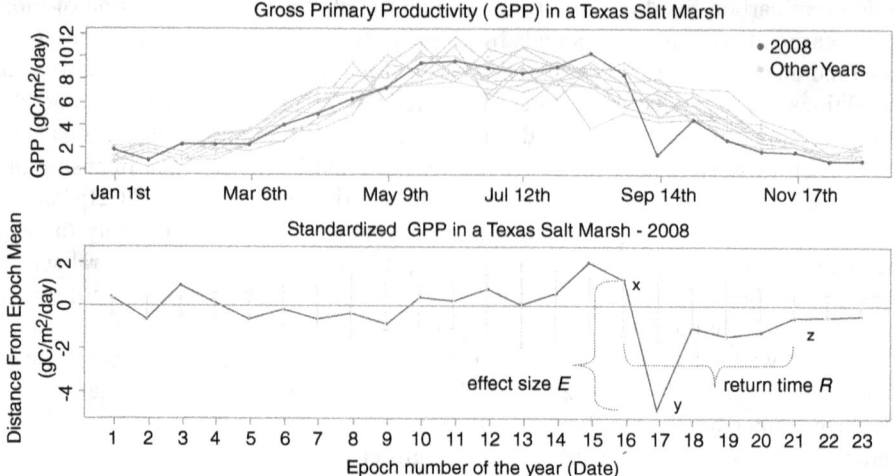

FIGURE 18.2 GPP of a single salt marsh pixel in Galveston, Texas. Hurricane Ike made landfall here on September 13, 2008. While vegetation in this wetland displayed high resilience to the hurricane, the same cannot be said about the Texas economy, which suffered an estimated $29 billion in property damages.

The effect size, E, is calculated as the maximum change in standardized response following the onset of the forcing and is a measure of resistance (Figure 18.2; equation 18.2):

$$E = \beta_y - \beta_x,\tag{18.2}$$

where β_y is the standardized GPP on date y when the change from baseline is the largest and β_x is the standardized GPP on date x before the onset of the forcing. A low amplitude response (small E) implies high resistance to the forcing. The absolute value of the differences can provide insight into response magnitude without regarding the direction of the response.

As described in Equation 18.3, the return time, R, is the amount of time between the onset of the forcing on date x and the return to the baseline condition on date z (or within a specified return threshold) and is a measure of resilience:

$$R = z - x\tag{18.3}$$

A short return time (small R) signifies relatively high resilience to the forcing. The standard deviation can be used to set an appropriate baseline threshold for return as it essentially represents the average range of variation in a given epoch. The return date z is the first epoch after date x when the following conditions are met:

$$\mu_{z,j} + \sigma_{z,j} > \beta_z > \mu_{z,j} - \sigma_{z,j},\tag{18.4}$$

where $\mu_{z,j}$ is the 20-year mean for epoch j and $\sigma_{z,j}$ is the standard deviation for epoch j. Thus, the return to baseline does not represent a return to the same pre-event GPP value (β_x), but rather, a return to the relative baseline standard deviation boundaries for a given location and time of the year.

In the following example, Hurricane Ike made landfall in Galveston, Texas, on September 13, 2008, as a category 2 hurricane with a storm surge as high as 20 ft (6.1 m). Wetland vegetation in this region was either physically destroyed and stripped from the land or became brown and wilted from prolonged saltwater inundation. In a single wetland pixel located on Smith Point in Galveston overlooking East Bay (29.53294, −94.73587), αGPP decreases from 8.44 gC/m²/day in epoch 16 (date x) to 1.22 gC/m²/day in epoch 17 (date y) due to the destructive forces of the hurricane (Figure 18.1). Once αGPP is standardized using equation (18.1), the effect size of the response, E, is calculated from equation (18.2). So, β_x is 1.27 gC/m²/day *greater* than the 20-year mean for epoch 16, and β_y is 4.79 gC/m²/day *less* than the mean for epoch 17. Therefore, the relative effect size of the response of this wetland to Hurricane Ike is $E = -4.79 - 1.27 = -6.06$ gC/m²/day. Date z is the date of the first measurement post-disturbance, where βGPP is between 1 and −1 standard deviation from the epoch's mean. In this example, date z occurs on the 21st epoch of the year, and date x, the date of the last measurement before the hurricane, is epoch 16. From equation (18.3), we can then calculate return time as: $R = 21 - 16 = 5$ epochs. So, this wetland was able to return to its dynamic baseline in 5 epochs, or approximately 80 days after the hurricane.

Until this point, the proposed framework has been free of any bias and thus can be applied to any study system and response variable in a time series. Bias is introduced when defining an appropriate baseline threshold for a given study system. For the GPP dataset, this threshold is set to one standard deviation from the epoch mean, as the entire dataset was tested for normality and determined to conform to the empirical rule; 69.73% of the 96,966,915 measurements fall between 1 and −1 standard deviations from the pixel's epoch means (expected 68.2%), 26.26% fall between ±1 and ±2 standard deviations (expected 27.2%), and 3.71% fall between ±2 and ±3 standard deviations (expected 4.2%). In addition, the αGPP values used to calculate epoch means for standardization ($n = 20$ values per epoch of the year (23 epochs) for 214,055 pixels) were normally distributed around the epoch's mean in 82.77% of epochs (Shapiro-Wilk test; $p > 0.05$). Non-normal distributions were primarily caused by outliers with extremely high or low GPP, where the extreme value drags the mean in either direction creating a skew in the data's distribution around the 20-year mean. Since these outliers are most likely extreme disturbance events, they were not removed from the dataset. When calculating resistance and resilience in non-normally distributed datasets, the data can be log-transformed or equations (18.2) and (18.3) can be multiplied by the natural log to remove skew so that statistical analyses of the data will become more valid. Of course, the baseline threshold should be adjusted to be relevant to the biological system to which it is applied to. For example, if we instead want to isolate only the most extreme and rare events in the GPP dataset, we can change the baseline departure threshold from ±1 $\sigma_{z,j}$ to ±3 $\sigma_{z,j}$ from the mean.

18.3 CONCLUSIONS AND FUTURE OUTLOOKS

Coastal ecosystems face many immediate threats, including extreme weather events, pollution, and overexploitation of resources. Among these threats is the uncertainty of how climate change, sea level rise, and increasingly intense and frequent disturbance regimes will interact with other stressors, such as drought and pollution, to alter the structure and function of ecosystems and their irreplaceable ecosystem services. Standardized metrics for measuring ecosystem response can help managers weigh the relative value of competing ecosystem services and prioritize their management actions in the face of changing climate and disturbance regimes.

Resistance and resilience are important measures of an ecosystem's vulnerability, as they provide insight into the attributes that modulate an ecosystem's response to stress or disturbance. Evolutionary theory suggests these strategies may trade off, as exploiting one strategy might come at the cost of the other. For example, resilience is primarily controlled by the species that colonize after a disturbance and their functional traits. Successional plants with short generation times and rapid growth have high resilience, but low resistance to disturbance. Quantifying ecosystem resistance and resilience will help managers and policymakers better understand the vulnerability and importance of specific areas and regions. This will allow more targeted action and policy that accounts for the unique needs of a given location and allows for more sustainable use of its natural resources. To achieve this, tools from population genetics, population ecology, and when necessary, toxicology can be integrated into a holistic approach to environmental analysis.

In this chapter, we suggest the use of ecological and toxicological methods to understand effects of stress and disturbance on an ecosystem's resident biota. Under-utilized approaches also offer promise, but come with their own sets of challenges. Effect-directed analysis is a method that can be used to identify toxins, carcinogens, and mutagens in sediment and water. It uses sophisticated fractionation procedures to identify toxicologically important pollutants from a mixture of thousands of compounds. But the biggest hindrance to using this technique, among other techniques for stress detection in coastal ecosystems, is that the flow and processing of materials and nutrients become concentrated and retained in estuaries and their adjacent wetlands, due to their spatial configuration at the interface of marine and terrestrial systems. Called the estuarine quality paradox, detecting stress effects in these ecosystems is particularly challenging because the organisms in these highly disturbed ecosystems may have significantly higher stress thresholds than expected, leading to false positives that indicate abnormal levels of stress, when the ecosystem is actually in a relatively undisturbed state. This also leads to misleading results of stress detection at the population or community levels from biodiversity indices, since biodiversity is naturally low in these unstable ecosystems due to the dominance of organisms that are highly adapted to frequent disturbance.

Therefore, assessing environmental health now and in the future should include the use of molecular tools, such as enzymatic assays, or genetic assays (i.e., comet assay), and high-throughput sequencing of genomes, transcriptomes, proteomes, and metabolomes. These techniques will provide insight into the genetic and metabolic pathways that mediate stress response. Remote sensing is another under-utilized technique for quantifying environmental stress. Time series datasets of vegetation changes can provide valuable access to long-term data from areas that are not physically accessible to researchers. Short-term vulnerabilities and long-term trends can be analyzed simultaneously to provide insight into the synergistic effects of multiple disturbances and stressors. Moreover, valuation of ecosystem service resistance and resilience to environmental impacts will help shape the direction of coastal policy and management regarding urbanization of the coasts, commercial and recreational fisheries, and tourism in the years to come.

REFERENCES

Aguilera, C., A. Leija, M. Torres, and R. Mendoza. 2019. "Assessment of environmental quality in the Tamaulipas Laguna Madre, Gulf of Mexico, by integrated biomarker response using the cross-barred venus clam *Chione elevata*." *Water, Air, and Soil Pollution* 230 (2). Doi: 10.1007/s11270-019-4078-0.

Almeida, S. F., M. R. C. Belfort, M. V. J. Cutrim, L. F. Carvalho-Costa, S. R. F. Pereira, and R. Luvizotto-Santos. 2021. "DNA damage in an estuarine fish inhabiting the vicinity of a major Brazilian port." *Anais Da Academia Brasileira de Ciencias* 93 (2). Doi: 10.1590/0001-3765202120190652.

Ayyub, B.M. 2014. "Systems resilience for multihazard environments: Definition, metrics, and valuation for decision making." *Risk Analysis* 34 (2): 340–55. Doi: 10.1111/risa.12093.

Basset, A., M. Elliott, R. J. West, and J. G. Wilson. 2013. "Estuarine and lagoon biodiversity and their natural goods and services." *Estuarine, Coastal and Shelf Science*, November 1, 2013. Doi: 10.1016/j.ecss.2013.05.018.

Borja, Á., D. M. Dauer, M. Elliott, and C. A. Simenstad. 2010. "Medium-and long-term recovery of estuarine and coastal ecosystems: Patterns, rates and restoration effectiveness." *Estuaries and Coasts* 33 (6): 1249–60. Doi: 10.1007/s12237-010-9347-5.

Ciais, Ph, M. Reichstein, N. Viovy, A. Granier, J. Ogée, V. Allard, M. Aubinet, et al. 2005. "Europe-wide reduction in primary productivity caused by the heat and drought in 2003." *Nature* 437 (7058): 529–33. Doi: 10.1038/nature03972.

Costanza, R., M. Kemp, and W. Boynton. 1993. "Predictability, scale, and biodiversity in coastal and estuarine ecosystems: Implications for management." *Ambio* 22 (2–3): 88–96. Doi: 10.1017/CBO9781139174329.006.

Cushman, S. A., and K. McGarigal. 2019. "Metrics and models for quantifying ecological resilience at landscape scales." *Frontiers in Ecology and Evolution* 7 (December). Doi: 10.3389/fevo.2019.00440.

Dauvin, J. C., and T. Ruellet. 2009. "The estuarine quality paradox: Is it possible to define an ecological quality status for specific modified and naturally stressed estuarine ecosystems?" *Marine Pollution Bulletin* 59 (1–3): 38–47. Doi: 10.1016/j.marpolbul.2008.11.008.

Davis Jr., R. A., and D. M. Fitzgerald. 2020. *Beaches and Coasts*. 2nd Ed. Hoboken, NJ: John Wiley & Sons Ltd. https://www.wiley.com/en-us/Beaches+and+Coasts%2C+2nd+Edition-p-9781119334484.

Donihue, C. M., A. Herrel, A. C. Fabre, A. Kamath, A. J. Geneva, T. W. Schoener, J. J. Kolbe, and J. B. Losos. 2018. "Hurricane-induced selection on the morphology of an island lizard." *Nature* 560 (7716): 88–91. Doi: 10.1038/s41586-018-0352-3.

Duarte, C. M., J. J. Middelburg, and N. Caraco. 2005. "Major role of marine vegetation on the oceanic carbon cycle." *Biogeosciences* 2 (1): 1–8. Doi: 10.5194/bg-2-1-2005.

El-Shatnawi, M. K. J., and I. M. Makhadmeh. 2001. "Ecophysiology of the plant-rhizosphere system." *Journal of Agronomy and Crop Science* 187 (1): 1–9. Doi: 10.1046/j.1439-037 X.2001.00498.x.

Elliott, M., and V. Quintino. 2007. "The estuarine quality paradox, environmental homeostasis and the difficulty of detecting anthropogenic stress in naturally stressed areas." *Marine Pollution Bulletin* 54 (6): 640–45. Doi: 10.1016/j.marpolbul.2007.02.003.

Evrard, A., M. Kumar, D. Lecourieux, J. Lucks, P. v. Koskull-Döring, and H. Hirt. 2013. "Regulation of the heat stress response in *Arabidopsis* by MPK6-targeted phosphorylation of the heat stress factor HsfA2." *PeerJ* 2013 (1). Doi: 10.7717/peerj.59.

Feagin, R. A., I. Forbrich, T. P. Huff, J. G. Barr, J. Ruiz-Plancarte, J. D. Fuentes, R. G. Najjar, et al. 2020. "Tidal wetland gross primary production across the continental United States, 2000–2019." *Global Biogeochemical Cycles* 34 (2). Doi: 10.1029/2019GB006349.

Ferreira, C. P., T. B. Piazza, P. Souza, D. Lima, J. J. Mattos, M. Saldaña-Serrano, R. S. Piazza, et al. 2021. "Integrated biomarker responses in oysters *Crassostrea Gasar* as an approach for assessing aquatic pollution of a Brazilian estuary." *Marine Environmental Research* 165 (March): 105252. Doi: 10.1016/j.marenvres.2021.105252.

Grant, P. R., and B. R. Grant. 2006. "Evolution of character displacement in Darwin's finches." *Science* 313 (5784): 224–26. Doi: 10.1126/science.1128374.

Grant, P. R., and B. R. Grant. 2002. "Unpredictable evolution in a 30-year study of Darwin's finches." *Science* 296 (5568): 707–11. Doi: 10.1126/science.1070315.

Grant, P. R., B. R. Grant, R. B. Huey, M. T. J. Johnson, A. H. Knoll, and J. Schmitt. 2017. "Evolution caused by extreme events." *Philosophical Transactions of the Royal Society B: Biological Sciences* 372 (1723): 20160146. Doi: 10.1098/rstb.2016.0146.

Gunderson, L. H. 2000. *Ecological Resilience-in Theory and Application.*www.annualreviews.org.

Häder, D. P., A. T. Banaszak, V. E. Villafañe, M. A. Narvarte, R. A. González, and E. W. Helbling. 2020. "Anthropogenic pollution of aquatic ecosystems: Emerging problems with global implications." *Science of the Total Environment*. Doi: 10.1016/j.scitotenv.2020.136586.

Hecker, M., and H. Hollert. 2009. "Effect-directed analysis (EDA) in aquatic ecotoxicology: State of the art and future challenges." *Environmental Science and Pollution Research International* 16 (6): 607–13. Doi: 10.1007/s11356-009-0229-y.

Herrera Barroquín, H. 2019. *Camarones Penaeidos y Carideos Como Bioindicadores de Contaminación y Su Relación Con Variaciones En La Fisicoquímica de Los Sedimentos En La Lagna Madre, Tamaulipas, México. Thesis.* Universidad Autónoma de Nuevo León. http://eprints.uanl.mx/17895/1/1080288765.pdf.

Hogan, J. A., R. A. Feagin, G. Starr, M. Ross, T. C. Lin, C. O'Connell, T. P. Huff, et al. 2021. "A research framework to integrate cross-ecosystem responses to tropical cyclones." *BioScience*. Doi: 10.1093/BIOSCI/BIAA034.

Johnson, E. A, and K. Miyanishi. 2007. "Disturbance and succession." In *Plant Disturbance Ecology*, 1–14. Doi: 10.1016/B978-012088778-1/50003-0.

Leija-Tristán, A., A. Contreras-Arquieta, M. E. Garcia-Garza, A. J. Contreras-Balderas, M. d. L. Lozano-Vilano, S. Contreras-Balderas, M. E. Garcia-Ramirez, et al. 2000. "Taxonomic, bioecological and biogeographic aspects of selected biota of the Laguna Madre, Tamaulipas, Mexico." In *Aquatic Ecosystems of Mexico: Status and Scope*, edited by M. Munawar, S.G. Lawrence, I.F. Munawar, and D.F. Malley, 399–435. Leiden: Backhuys. https://www.researchgate.net/publication/303720336.

Levin, S. A, and R. T. Painet. 1974. Disturbance, patch formation, and community structure (spatial heterogeneity/intertidal zone). *Proceedings of the National Academy of Sciences* 71 (7):2744–2747.

Naour, S., B. M. Espinoza, J. E. Aedo, R. Zuloaga, J. Maldonado, M. Bastias-Molina, H. Silva, et al. 2017. "Transcriptomic analysis of the hepatic response to stress in the red cusk-eel (*genypterus chilensis*): Insights into lipid metabolism, oxidative stress and liver steatosis." *PLoS One* 12 ssss(4). Doi: 10.1371/journal.pone.0176447.

Odum, E. P. 1985. "Trends expected in stressed ecosystems." *BioScience* 35 (7): 419–22. Doi: 10.2307/1310021.

Odum, E. P, J. T Finn, and E. H Franz. 1979. "Perturbation theory and the subsidy-stress gradient." *Deep Sea Research Part B. Oceanographic Literature Review* 26 (12): 811. Doi: 10.1016/0198-0254(79)90989-0.

Patrick, C. J., L. Yeager, A. R. Armitage, F. Carvallo, V. M. Congdon, K. H. Dunton, M. Fisher, et al. 2020. "A system level analysis of coastal ecosystem responses to hurricane impacts." *Estuaries and Coasts* 43 (5): 943–59. Doi: 10.1007/s12237-019-00690-3.

Petitjean, Q., S. Jean, A. Gandar, J. Côte, P. Laffaille, and L. Jacquin. 2019. "Stress responses in fish: From molecular to evolutionary processes." *Science of the Total Environment.* Doi: 10.1016/j.scitotenv.2019.05.357.

Pickett, S T A, J Kolasa, J J Armesto, and S L Collins. 1989. *The Ecological Concept of Disturbance and Its Expression at Various Hierarchical Levels.* Vol. 54. https://www.jstor.org/stable/3565258.

Pimm, S. L., I. Donohue, J. M. Montoya, and M. Loreau. 2019. "Measuring resilience is essential to understand it." *Nature Sustainability.* Doi: 10.1038/s41893-019-0399-7.

Running, S. W. 2008. "Climate change: Ecosystem disturbance, carbon, and climate." *Science* 321 (5889): 652–53. Doi: 10.1126/science.1159607.

Schulte, P. M. 2014. "What is environmental stress? Insights from fish living in a variable environment." *Journal of Experimental Biology* 217 (1): 23–34. Doi: 10.1242/jeb.089722.

Short, J. W., H. J. Geiger, J. C. Haney, C. M. Voss, M. L. Vozzo, V. Guillory, and C. H. Peterson. 2017. "Anomalously high recruitment of the 2010 Gulf Menhaden (*Brevoortia Patronus*) year class: Evidence of indirect effects from the Deepwater horizon blowout in the Gulf of Mexico." *Archives of Environmental Contamination and Toxicology* 73 (1): 76–92. Doi: 10.1007/s00244-017-0374-0.

Star, B., and H.G. Spencer. 2013. "Effects of genetic drift and gene flow on the selective maintenance of genetic variation." Genetics 194 (1): 235–244: Doi: 10.1534/genetics.113.149781

Torres-Ceron, M., A. Leija-Tristan, C. J. Aguilera-Gonzalez, and J. A. Vidales-Contreras. 2014. "Evaluacion de La Condicion Biologica Del Area Meridional de La Laguna Madre En San Fernando, Tamaulipas, Con Base En La Malacofauna Bentica." In *Golfo de Mexico, Contaminacion e Impacto Ambiental: Diagnostico y Tendencias*, edited by A. V Botello, J. R. Von Osten, J. A Benítez, and G. Gold-Bouchot, 901–34. UAC, UNAM-ICMYL, CINVESTAV-Unidad Mérida.

Turner, M. G. 1989. "Landscape ecology: The effect of pattern on process 1." *Annual Review of Ecology and Systematics*, 20. www.annualreviews.org.

Ummenhofer, C. C., and G. A. Meehl. 2017. "Extreme weather and climate events with ecological relevance: A review." *Philosophical Transactions of the Royal Society B: Biological Sciences.* Doi: 10.1098/rstb.2016.0135.

Wang, F., C. J. Sanders, I. R Santos, J. Tang, M. Schuerch, M. L. Kirwan, R. E Kopp, et al. 2021. "Global blue carbon accumulation in tidal wetlands increases with climate change." *National Science Review* 8 (9). Doi: 10.1093/nsr/nwaa296.

White, P. S, and A. Jentsch. 2001. "The search for generality in studies of disturbance and ecosystem dynamics." In *Progress in Botany,* Vol. 62, edited by K. Esser, U. Lüttge, J.W. Kadereit, and W Beyschlag, 399–450. Berlin, Heidelberg: Springer Berlin Heidelberg. Doi: 10.1007/978-3-642-56849-7_17.

Wilhm, J. L, and T. C Dorris. 1968. "Biological parameters for water quality criteria." *BioScience* 18 (6): 477–81. Doi: 10.2307/1294272.

19 Impact of Cadmium Toxicity on Environment and Its Remedy

Pushpa Ruwali and Niharika Pandey
M. B. Gov. P. G. College

Tanuj Kumar Ambwani
G.B. Pant University of Agriculture & Technology

Rahul Vikram Singh
Academy of Scientific and Innovative Research (AcSIR)

CONTENTS

DOI: 10.1201/9781003247883-19

19.1   INTRODUCTION

The advent of urbanization and industrialization was not solitary, and it was accompanied by a burden that is imposed on our little planet's limited natural resources and is detrimental to human and animal health. The discharge of industrial wastes, full of heavy metals and other contaminants into the water reservoirs, has amplified the environmental stress (Chandra and Kumar 2017; Kumar et al. 2021; Mishra et al. 2021). One such culprit is the heavy metal cadmium. It is a chemical element discovered in 1817 having 48 as the atomic number. It is placed between zinc and mercury in the periodic table of elements with its behavior being parallel to that of zinc. The application of cadmium in the industrial world was limited for around 50 years after its discovery, but since then, it has vastly increased. Owing to its non-corrosive property, it is mostly used in electroplating and galvanization as a cathode material in the nickel-cadmium batteries and the television sets. It is used as a coloring pigment in many paints and plastics and prominently present in tobacco smoke and many fertilizers. The combustion of fuels and industries related to zinc smelting contribute toward the release of cadmium into the environment, where it is majorly present in the form of cadmium acetate, oxide, chloride, sulfate, and carbonate (Bhattacharyya 2009; Torres-Sanchez et al. 2018; Reyes-Hinojosa et al. 2019). The rampant release of cadmium (Cd) from a variety of human activities has raised substantial concern about the toxicity of this metal. It has been categorized as one of the toxic environmental and industrial pollutants as well as a well-known animal carcinogen. The property of Cd being tasteless and odorless keeps it incognito in the environment and its prolonged exposure results in direct toxic effects with a most severe form known as 'Itai-Itai' disease in humans (Zwolak 2020). The main objective of this review is to bring forward the empirical points on the ill effects of Cd on the human body, including reproductive toxicity, cellular oxidative stress, and blockages in the DNA repair pathway. In addition, the utility of antioxidants, chelators, and biological agents as potent curatives is also discussed.

19.2   EXPOSURE AND ABSORPTION OF CADMIUM

Humans and animals are exposed to Cd through various means, the major sources being food. The poultry sector, particularly, is very vulnerable to cadmium toxicity because of the use of intoxicating ingredients of plant origin for feeding purposes and also from shell grid, fish meat, and dicalcium phosphate (Bharavi et al. 2010). Tobacco leaves absorb a high amount of Cd from the soil; therefore, smoking cigarettes is a highlighted exposure source for humans (CDC 2009; ATSDR 2012). As per the study done by Guney and Zagury in (2012), toys and jewelry can be a medium of Cd exposure to humans. When humans consume or get exposed to any of these affected substances, the cadmium gets absorbed via the gastrointestinal tract and the respiratory tract. A little amount of Cd is also absorbed through the skin (Friberg et al. 1971).

19.3   IMPACT ON HUMAN HEALTH

Cadmium is one of the most toxic elements for humans, which once absorbed, gets efficiently retained in the body accumulating for life (Bernard 2008). The intake

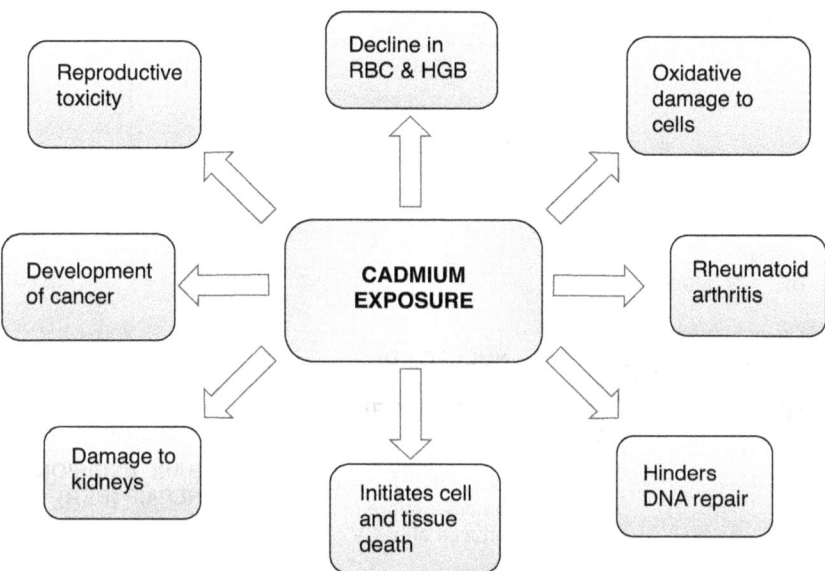

FIGURE 19.1   An overall view of the impact of cadmium exposure on the human body.

of Cd through any means can increase its concentration in the human body, which opens up a platform for various diseases (Vahter et al. 1996). Chronic Cd toxicity leads to a disease known as 'Itai-Itai', which is considered as the most severe form of Cd poisoning. Cadmium induces renal tubular osteomalacia, which is a hallmark of this disease (Aoshima 2012). Apart from being the cause of 'Itai-Itai', Cd intoxication does have other severe consequences, including hindrance in the repair mechanism within DNA; generation of oxidative stress in cells; intensive damage to various cells, tissues, and organs; rheumatoid arthritis; irregular reproductive cycles; and under grievous conditions, cell death (Figure 19.1).

19.3.1   Intervention in the DNA Repair Pathway

Cadmium has adverse effects on DNA repair mechanisms (Figure 19.2). In the context of base excision repair (BER) mechanism, the exposure of human cells to sublethal concentrations of Cd leads to a decrease in hOGGO1, which is responsible for initiating the BER of 8-oxoguanine, a mutant form of oxidized guanine (Bravard et al. 2009). In the 'nucleotide excision repair' (NER) mechanism, the repair of thymine dimers that are formed by UV irradiation gets difficult, as cadmium blocks the repair pathway at the first step itself by inhibiting the enzyme involved (Fatur et al. 2003). Cadmium intoxication leads to the inhibition of mismatch repair in human extracts, which then propagates cellular errors. Cadmium inhibits the adenosine triphosphate binding and the hydrolysis of MSH2-MSH6 protein complex of 'mismatch repair' (MMR), which are encoded by MSH2 and MSH6 genes, respectively (Jin et al. 2003).

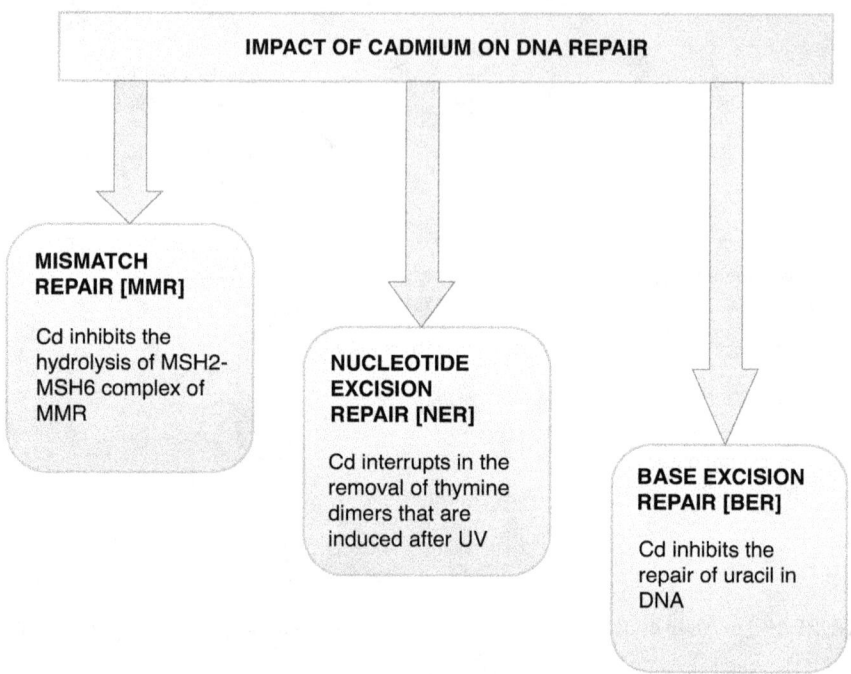

FIGURE 19.2 A representation of the impact of cadmium on the various pathways involved in the DNA repair mechanism.

19.3.2 Oxidative Damage to Cells

Cadmium enhances free radical formation, which induces injury to liver and kidney (Kumar et al. 2010). Free radicals are natural by-products of our body's metabolism and are also generated through various pollutants, including cigarette smoke, exhausts from automobiles, fertilizers and pesticides, air pollution, and radiation (Ruwali et al. 2017). These free radicals then commence chain reactions that hamper the cells to function normally leading to oxidative stress. This oxidative stress causes brutal damage to the vital biomolecules and also to various body systems and organs, including the central nervous system, kidneys, and liver (Singh et al. 2017).

19.3.3 Reproductive Toxicity

Cadmium has been included in the list of endocrine disrupting chemicals under its detrimental effect on mammalian reproduction (Chedrese et al. 2006). During human pregnancy, exposure to Cd has been known to cause a decrease in progesterone biosynthesis by the placental trophoblast, decreased birth weights, and premature delivery (Miceli et al. 2005). Additionally, it leads to the accumulation in embryos from the fourth cell stage onward, further inhibiting the progression to the blastocyst stage or causes the decompaction of the formed blastocyst, which is followed by apoptosis and disintegration of cell adhesion (Thompson and Bannigan 2008). Cadmium toxicity not only affects the pregnancy and the levels of female hormones, but also affects the male reproductive system, affecting testis, Sertoli, and Leydig cells, impairing the testicular functions, and therefore inhibiting testosterone synthesis and the process of sperm production (Taha et al. 2013). Cadmium intrudes on the functioning of the prostate gland, altering its hormonal activity and its secretion and leading to sterility in males (Sarkar et al. 2013).

19.3.4 Rheumatoid Arthritis

Rheumatoid arthritis (RA) is an autoimmune, chronic inflammatory disorder caused by genotypic and environmental factors (Turk et al. 2014). Tobacco smoke is one of the potent environmental factors that are linked with the manifestation of RA (Murphy et al. 2016). Tobacco smoke acts as a spark for those asymptomatic individuals that have a genetic predisposition to RA (Sugiyama et al. 2010). Ansari and co-workers (2015) injected a dose of 50 ppm cadmium chloride (CdCl$_2$) into RA animal model, which led to a severe inflammatory response, followed by edema, increased levels of nitric oxide, cartilage degradation, and the flow of polymorphonuclear cells. A decline in the activity of superoxide dismutase was also recorded. When 53 RA patients, including 30 smokers and 23 non-smokers, were compared to a group of 52 control individuals, including 26 smokers and 26 non-smokers, it was found that the levels of Cd were higher in the blood and hair samples of the RA patients as compared to the group of 52 controls, the difference being epochal for the patients involved in tobacco smoking (Afridi et al. 2015). Another significant study by Afridi and co-workers (2011) gave an idea that the ratio of Cd to zinc (Zn) was comparatively higher in RA patients indulged in smoking than in the other groups included in the study. In 2012, another study showed that the deficiency of Zn and higher levels of Cd (induced due to smoking) can work synergistically in the development of RA (Afridi et al. 2012).

19.3.5 Effect on Kidneys

Cadmium majorly accumulates in kidneys. About 50% of the Cd that enters the body due to chronic exposure is dumped in the kidneys (Johri et al. 2010). Exposure to cadmium oxide vapors at a dose of 5 mg/m^3 for 8 hours is stated to be lethal to humans (Agency for Toxic Substances and Disease Registry, 2012). Cadmium acts by inhibiting the reabsorption of proteins, glucose, amino acids, bicarbonates, and phosphates in the proximal tubule causing tissue damage. It also builds up in the proximal tubules, and an impaired proximal tubule becomes unable to reabsorb low molecular weight proteins such as β-2 macroglobulin and N-acetyl-β-D-glucosaminidase (Huang et al. 2019), which are then excreted in the urine, acting as a marker depicting a dysfunctional proximal tubule. Cadmium is capable of inducing apoptosis in tubular cells, becoming a major cause of oxidative stress (Thevenod 2010; Uraguchi et al. 2012). In a study conducted on a group of chain smokers exposed to Cd through any means, elevated levels of metal in the urine and serum of were observed (Taha et al. 2018).

19.3.6   DEVELOPMENT OF CANCER

Cadmium has been classified as a human carcinogen and categorized as Group 1 carcinogen according to International Agency for Research on Cancer (IARC); Group 2a carcinogen according to Environmental Protection Agency (EPA); and Group 1B carcinogen according to European Chemical Agency (ECA) (IPCS 1992; IARC 2016). Cadmium is associated with the induction and development of lung and breast cancers (IARC 2016).

19.3.7   BREAST CANCER

Cadmium intoxication and development of breast cancer have been associated with each other for a long term (Martin et al. 2003). Normal breast tissues have Cd levels of about 0.022 µg/g (Strumylaite et al. 2010). In a study, normal human breast epithelial MCF-10A cells were treated with 2.5 µg Cd for 40 weeks and characters resembling cancerous cells were observed, viz. transformation of cells that displayed increased colony formation, loss of contact inhibition, and invasiveness (Benbrahim-Tallaa et al. 2009). Cadmium is well known to mimic estrogen on a molecular basis, and therefore, it induces cell growth via estrogen receptor-mediated pro-survival signaling (Johnson et al. 2003), hence also referred to as a 'metalloestrogen'. Human mammary gland contains high Cd concentrations that mimic the effects of estradiol, which might play a putative role in the etiology of breast cancer (Brama et al. 2007).

19.3.8   LUNG CANCER

Individuals occupationally and environmentally exposed to Cd are known to develop lung cancer (Field and Withers 2012). Lei and co-workers (2008) exposed normal bronchial epithelial cells to Cd for a prolonged period *in vitro* and visualized their transformation to a malignant state; similar kinds of results were obtained in a nude mouse model. The human lung epithelial cells that are transformed due to Cd induction are called CCT-LC, which showcase the property of invasion, increased colonization, autonomous growth, and hyperproliferation (Person et al. 2013). Along with these properties, the CCT-LC suffer a loss in the expression of p16 and overexpression of the cyclin D1, both characters being associated with highly proliferating tumor cells and lung cancers (Demirhan et al. 2010).

19.3.9   IMPACT ON VARIOUS CELL TYPES

Cadmium is majorly absorbed through the gastrointestinal tract reaching the tissues via blood, where it accumulates, especially in the liver and kidneys, leading to their malfunction (Kostial 1986). Cadmium is also known to bind to the red blood cell membranes and plasma albumin, leading to a decline in the RBC and HGB count further paving a path to anemia (Haziri et al. 2012; Borane 2013).

19.3.10   CADMIUM-INDUCED CELL DEATH

Apoptosis is a kind of programmed cell death in which the affected cells in the body are deleted with the sole reason of benefiting the body. The gradual erosion of cells due to toxic stimuli has devastating effects on various organs, including the kidneys. Apoptosis occurs via two pathways either intrinsic or extrinsic. Cadmium-instigated apoptosis can make use of both the pathways that involve the concerned caspases (Eichler et al. 2006). At the time of Cd toxicity, apoptosis can also be induced by non-caspases activity or by the coupled processing of Ca^{2+}-calpain or by the translocation of apoptosis-inducing factor to the nucleus (Daugas et al. 2000; Lee et al. 2007). Along with apoptosis, necrosis is also witnessed at times (Prozialeck et al. 2009).

19.4 REMEDIATION OF CADMIUM INTOXICATION

The literature is evident with various deleterious consequences of Cd intoxication and hence the critical need for remediation. Various antioxidants and chelating agents have been reported to be effective against the ill effects of the accumulated Cd within the body. Bioremediation and phytoremediation are the two promising types of biological treatments that are being used at present to treat the Cd-contaminated soil and water resources.

19.4.1 Use of Chelating Agents

Extensive research depicts that Cd intoxication can be treated with the use of a suitable chelating agent. Chelating agents or chelators are chemical compounds that bind tightly to respective metals. Various clinically accessible chelators, including EDTA, DMPS, DMSA, and BAL are very efficient against chronic intoxication of Cd (Blanusa et al. 2005). But this efficiency is skeptical against acute poisoning, and it may contribute to the damage of the kidney tubules (Nordberg et al. 2007). These chelating agents act by increasing the Cd content in the urine (Tandon et al. 2002). Out of all other chelating agents, EDTA is known to act most effectively and is considered superior to DMSA in mobilizing intracellular cadmium. Sweat, which is excreted during sauna, does contain Cd and becomes a moderately successful method to lower the burden of cadmium on the body (Genuis et al. 2010).

19.4.2 Use of Antioxidants

Antioxidants are of critical importance as they help in inhibiting or delaying the oxidation of a substance, and hence, they also check the reactive oxygen species (ROS)-mediated oxidative damage (Kohen and Nyska 2002). Antioxidants in the body are categorized into two main classes, namely enzymatic antioxidants and non-enzymatic antioxidants. Enzymatic antioxidants include three major enzymes, catalase and glutathione dismutase that catalyze the conversion of ROS to inert species, rendering them inactive, and superoxide dismutase that degrades O_2^- (Mates 2000). Besides enzymatic antioxidants, some other compounds such as carotene, tocopherol (vitamin E), ascorbic acid (vitamin C), and lipoic acid do act as the scavengers of free radicals, healing the damage of the cells that occurred due to metal poisoning. Constant Cd toxicity augments oxidative stress and enhances the consumption of antioxidant systems (Cinar et al. 2010). $CdCl_2$ causes the attenuation of glutathione and also blocks various activities of enzymatic antioxidants in rat liver tissues (Zalups 2000). In an experiment conducted on rats, exposure of Cd in liver tissues significantly lowered the activities of catalase, glutathione peroxidase, and reduced glutathione. Supplementation of vitamins E and C was found to be extremely effective in ameliorating the ill effects. Combinational usage of vitamins E and C was much more effective in action than individual use (Layachi and Kechrid 2012). Ascorbic acid is also known to turn around the Cd-induced apoptosis and lipid peroxidation by scavenging on the ROS generated by Cd (Sen Gupta et al. 2004; Erdogan et al. 2005). Tocopherol decreases lipid peroxidation in different organs and body fluids of rats that occur due to Cd intoxication (Kara et al. 2008; Karabulut-Bulan et al. 2008). Antioxidants are therefore effective in causing the diminution of the Cd-induced ill effects.

19.4.3 Biological Methods

19.4.3.1 Bioremediation

Microbial remediation, popularly referred to as bioremediation, is an effective strategy for detoxifying the accumulated Cd present in soil and water (Kumar et al. 2018;

TABLE 19.1
Remediation of Cadmium (Cd) and Other Heavy Metals by Microorganisms

Elements	Bioremediator	Reference
Cd	*Pseudomonas aeruginosa*	Chellaiah (2018)
Cd	*Penicillium chrysogenum* XJ-1	Xu et al. (2015)
Cd	*Aspergillus versicolor, A. fumigatus, Paecilomyces* sp., *Terichoderma* sp., *Microsporum* sp., *Cladosporium* sp.	Soleimani et al. (2015), Abatenh et al. (2017)
Cd	*Bacillus safensis* (JX126862) strain (PB-5 and RSA-4)	Priyalaxmi et al. (2014), Abatenh et al. (2017)
Pb, Cr, Cd	*Aerococcus* sp., *Rhodopseudomonas palustris*	Sinha and Paul (2014), Sinha and Biswas (2014), Abatenh et al. (2017)
Cd	*Lactobacillus kefir* CIDCA 8348	Gerbino et al. (2014)
Cd	*Nostoc* sp., *Chlorella vulgaris*	Kumaran et al. (2011), Goher et al. (2016)
Cd	*Bacillus licheniformis, B. subtilis, B. cereus, B. amyloliquefaciens*	Issazadeh et al. (2011)
Cd	*Bifidobacterium longum*46	Halttunen et al. (2007)
Cd	*Bacillus licheniformis*	Shameer (2006)
Zn, Cu, Cd	*Bacillus* sp., *Pseudomonas aeruginosa*	Rajendran et al. (2003)
Pb, Au, Zn, Cd	*Chlorella vulgaris*	Rajendran et al. (2003)
Cd, Zn, Ag	*Aspergillus niger*	Rajendran et al. (2003)
Zn, Cu, Cd	*Pleurotus ostreatus*	Rajendran et al. (2003)

Agrawal et al. 2021). Several microorganisms such as bacteria, fungi, and algae have been used to bioremediate heavy metal (especially Cd) contaminant environments (Table 19.1). This environment revitalizing, cost-effective method employs bacterial strains (Yuan et al. 2017), yeast species (Hadi et al. 2002), fungi *Aspergillus* (Barros Junior et al. 2003), etc., as bioremediants for the purpose of detoxification of heavy metals. Microorganisms consume the heavy metals from soil or water via bioaccumulation, an active process, or via adsorption, a passive process (Kumar 2018). The mechanism of bioaccumulation can be intracellular or extracellular. In intracellular sequestration, the heavy metal ions are accumulated in the cell cytoplasm. *Pseudomonas putida* (Higham et al. 1986) and *Rhizobium leguminosarum* (Lima et al. 2006), display this sequestration method and accumulate the Cd ions. The extracellular sequestration involves the accumulation of metal ions in periplasm and their precipitation. *Pseudomonas aeruginosa*, under aerobic conditions, precipitates Cd very efficiently (Wang et al. 2002).

The efficiency of bioremediation primarily depends on temperature and pH. An increase in temperature increases the solubility of heavy metals, which in turn improves the bioavailability of heavy metals (Bandowe et al. 2014). The parameter of optimum climatic condition for proper growth and activity of microbes limits the use of microorganisms for detoxification (Mahajan and Kaushal 2018).

19.4.3.2 Phytoremediation

Apart from the above-mentioned microbial treatment, the biological technique of phytoremediation has been fascinating the researchers for a long time to get rid of the Cd, which is accumulated in the soil and water resources. Phytoremediation in itself is an eco-friendly, reasonable, efficient, and versatile method to reduce the toxicity of Cd-contaminated soil and water resources (Mahajan and Kaushal 2018; Chandra and Kumar 2017, 2018). Temperature, pH, amount of Cd accumulated in the soil, and the concentration of elements apart from that of Cd are significant factors that are accountable for proper remediation of Cd by plants (Yang et al. 1998). A plant is categorized as a hyperaccumulator if it accumulates 105 mg/g of Cd in shoot dry weight (Baker and Brooks 1989). *Oryza sativa, Vetiver grass, Lemna minor, Allium sativum, Thlaspi caerulescens, Solanum nigrum, Brassica juncea* (Indian mustard), and *B. napus* (Yang et al. 1998; Jiang et al. 2001; Wei et al. 2005; Quartacci et al. 2006; Sheng and Xia 2006; Murakami et al. 2007; Hou et al. 2007) are the plant species that have been reported as hyperaccumulators of cadmium. Cadmium hyperaccumulators are of high value to us because of the level of tolerance they show in captivating the heavy metals from soil. The intensity of hyperaccumulation of Cd by plants varies from species to species. *Thlaspi caerulescens* is a promising hyperaccumulator that is highly praised as a phytoremediator as it is eight times more effective than the *Brassica* species (Zhao et al. 2003; Kwinta et al. 2011). Ornamental plants, apart from having a great aesthetic value, are also very effective in Cd removal from the soil; some examples are *Calendula officinalis, Althea rosea*, and *Impatiens balsamina* (Liu et al. 2008). Table 19.2 depicts various plant species through which bioremediation has been achieved against cadmium and other two heavy metals, viz. lead and arsenic.

TABLE 19.2
Phytoremediation of Cadmium (Cd) and Other Heavy Metals

Elements	Phytoremediator	Reference
Pb, Cd	*Brassica juncea*	Raskin et al. (2005)
Cd	*Jatropha curcas*	Mangkoedihardjo et al. (2008)
Cd	*Althaea rosea*	Liu et al. (2008)
Cd	*Taraxacum mongolicum*	Wei et al. (2008)
Cd	*Calendula officinalis*	Liu et al. (2008)
Cd	*Mirabilis jalapa* L.	Yu and Zhou (2009)
Cd	*Datura innoxia*	Prabavathi et al. (2011)
Cd	*Chlorophytum comosum*	Wang et al. (2012)
Cd	*Rhytidiadelphus squarrosus*	Pipiska et al. (2013)
Cd	*Tridax procumbens*	Jameer et al. (2015)
Cd	*Pterocarpus indicus* wild	Suthep et al. (2016)
Cd	*Acorus calamus*	Jeelani et al. (2017)
Cd	*Brassica species*	Rizwan et al. (2018)
Cd	*Vicia faba* L.	Tang et al. (2019)

19.4.3.3 Detoxification through Selenium

Selenium (Se) is a chemical element with 34 as the atomic number and high tendency to limit the deleterious effects of Cd. The initial biochemical character of Se came to limelight in 1973, when it was recognized as an essential component of erythrocyte glutathione peroxidase (GPx), an enzyme with antioxidant properties. Se comes in the human body from dietary sources involving food, such as meat, seafood, and cereals (WHO 2011). It is found in the food in two forms: organic Se compounds such as selenocysteine (SeCys) and selenomethionine (SeMet) and inorganic Se compounds such as selenite (Se in +4 oxidation state) and selenate (Se in +6 oxidation state) (Rayman 2012). Se-mediated reduction of Cd content in tissues is credited to the formation of non-toxic, colloidal Cd-Se complex (Hu et al. 2018). In an experiment by Dauplais et al. (2013), $50\,\mu M$ (toxic dose) of sodium selenide (Na_2Se) was found to form insoluble aggregates with various metal ions including Cd^{+2}. This Cd-Se complex was found to be non-toxic toward *Saccharomyces cerevisiae*, and as per the hypothesis, the aggregation of Cd and Se into a complex lowered the toxicity of both the elements. Se is found to be potentially protective against Cd-induced reproductive toxicity (Wan et al. 2018), nephrotoxicity (Bao et al. 2017), and neurotoxicity (Branca et al. 2018). The renal protective effects of Se were elucidated by performing an experiment in which selenite was added to birds' feed at a supranutritional dose of 10 mg/kg of diet and observed a decline in the Cd accumulation, which led to the mitigation of $CdCl_2$-induced apoptosis of kidney cells (Liu et al. 2015). In a recent study in 2018 by Branca and his colleagues, protective effects of Se against Cd-generated neurotoxicity were investigated by pretreating undifferentiated SH-SY5Y human neuroblastoma cell line with selenite (100 nM), followed by an exposure to $CdCl_2$ ($10\,\mu M$), which resulted in mitigated levels of Cd-generated ROS and mitochondrial apoptotic pathways. Although Se at therapeutic dose is a promising agent that can minimize the Cd-related deleterious activities, the precautious use of Se is advised as there is a very minor difference between the therapeutic and toxic doses of Se compounds.

19.5 CONCLUSIONS

Scientific reports have proved the toxicity of cadmium, a metal to which humans are openly exposed via various means, including cigarette smoke, fertilizers, and nickel-cadmium batteries, and also at times via plant products. Cadmium paves the way for a bunch of diseases, for which the mechanism of toxicity is still not well understood. Therefore, for a better understanding of how cadmium leads to the occurrence of these diseases, extensive research is a need of the hour. Chelators, antioxidants, and biological treatment including microbial and phytoremediation are being used as a therapy to lower the burden of cadmium on the body. With high levels of cadmium in soil and water resources, the use of green technologies such as phytoremediation is showing great potential, as it serves the purpose without hampering the environment.

ABBREVIATIONS

BAL	British anti-Lewisite
CCT-LC	Chronic cadmium-treated lung cells
Cd	Cadmium
DMPS	2,3-Dimercapto-1-propanesulfonic acid
DMSA	Dimercaptosuccinic acid
EDTA	Ethylenediaminetetraacetic acid
Hg	Mercury
HGB	Hemoglobin
MSH	MutS homolog
nM	Nanomolar
ppm	Parts per million
RA	Rheumatoid arthritis
RBC	Red blood cell
ROS	Reactive oxygen species
UV	Ultraviolet
Zn	Zinc
μM	Micromolar

REFERENCES

Abatenh, E., Gizaw, B., Tsegaye, Z. et al. 2017. Application of microorganisms in bioremediation-review. *Environ. Microbiol.* 1(1):02–09.

Afridi, H.I., Kazi, T.G., Brabazon, D., et al. 2011. Association between essential trace and toxic elements in scalp hair samples of smokers rheumatoid arthritis subjects. *Sci. Total Environ.* 412–413:93–100.

Afridi, H.I., Kazi, T.G., Brabazon, D., et al. 2012. Interaction between zinc, cadmium, and lead in the scalp hair samples of Pakistani and Irish smokers rheumatoid arthritis subjects in relation to controls. *Biol. Trace Elem. Res.* 148(2):139–47.

Afridi, H.I., Talpur, F.N., Kazi, T.G., et al. 2015. Estimation of toxic elements in the samples of different cigarettes and their effect on the essential elemental status in the biological samples of Irish smoker rheumatoid arthritis consumers. *Environ. Monit. Assess.* 187(4):157.

Agrawal, N., Kumar, V., Shahi, S.K. 2021. Biodegradation and detoxification of phenanthrene in *in-vitro* and *in-vivo* conditions by a newly isolated ligninolytic fungus *Coriolopsis byrsina* strain APC5 and characterization of their metabolites for environmental safety. *Environ. Sci. Pollut. Res.* Doi: 10.1007/s11356-021-15271-w.

Ansari, M.M., Khan, H.A. 2015. Effect of cadmium chloride exposure during the induction of collagen induced arthritis. *Chem. Biol. Interact.* 238:55–65.

Aoshima, K. 2012. Itai-Itai disease: Cadmium-induced renal tubular osteomalacia-current situations and future perspectives. *Nippon Eiseigaku Zasshi. Japanese J. Hygiene.* 67(4):455–63.

ATSDR. 2012. *Toxicological Profile for Cadmium.* Agency for Toxic Substances and Disease Registry, U.S. Department of Health and Human Services, Atlanta, GA, pp.261–305.

Baker, A., Brooks, R. 1989. Terrestrial higher plants which hyperaccumulate metallic elements. A review of their distribution, ecology and phytochemistry. *Biorecovery.*1(2):81–126.

Bandowe, B.A.M., Bigalke, M., Boamah, L., et al. 2014. Polycyclic aromatic compounds (PAHs and oxygenated PAHs) and trace metals in fish species from Ghana (West Africa): Bioaccumulation and health risk assessment. *Environ. Int.* 65:135–46.

Bao, R.K., Zheng, S.F., Wang, X.Y. 2017. Selenium protects against cadmium-induced kidney apoptosis in chickens by activating the PI3K/AKT/Bcl-2 signaling pathway. *Environ. Sci. Pollut. Res.* 24(25):20342–353.

Barros Junior, L.M., Macedo, G.R., Duarte, M.M.L., et al. 2003. Biosorption of cadmium using the fungus *Aspergillus niger. Braz. J. Chem. Eng.* 20(3):229–39.

Benbrahim-Tallaa, L., Tokar, E.J., Diwan, B.A., et al. 2009. Cadmium malignantly transforms normal human breast epithelial cells into a basal-like phenotype. *Environ. Health Perspect.* 117(12):1847–52.

Bernard, A. 2008. Cadmium & its adverse effects on human health. *Indian J Med Res.* 128(4):557–64.

Bharavi, K., Gopala Reddy, A., Rao, G.S., et al. 2010. Reversal of cadmium-induced oxidative stress in Chicken by Herbal adaptogens *Withania somnifera* and *Ocimum sanctum. Toxicol. Int.* 17(2): 59–63.

Bhattacharyya, M.H. 2009. Cadmium osteotoxicity in experimental animals: Mechanisms and relationship to human exposures. *Toxicol. Appl. Pharmacol.* 3(238):258–65.

Blanusa, M., Varnai, V.M., Piasek, M., et al. 2005. Chelators as antidotes of metal toxicity: Therapeutic and experimental aspects. *Curr. Med. Chem.* 12(23):2771–94.

Borane, V.R. 2013. Protective role of ascorbic acid on the cadmium induced changes in hematology of the freshwater fish, *Channa orientalis* (Scheider). *Adv. Appl. Sci. Res.* 4(2):305–08.

Brama, M., Gnessi, L., Basciani, S., et al. 2007. Cadmium induces mitogenic signaling in breast cancer cell by an ERalpha-dependent mechanism. *Mol. Cell Endocrinol.* 264(1–2):102–08.

Branca, J.J.V., Morucci, G., Maresca, M., et al. 2018. Selenium and zinc: Two key players against cadmium-induces neuronal toxicity. *Toxicol in Vitro.* 48:159–69.

Bravard, A., Campalans, A., Vacher, M., et al. 2009. Inactivation by oxidation and recruitment into stress granules of hOGG1 but not APE1 in human cells exposed to sub-lethal concentrations of cadmium. *Mutat. Res.* 685:61–69.

CDC. 2009. *Fourth Report on Human Exposure to Environmental Chemicals.* Centers for Disease Control and Prevention. US Department of Health and Human Services, Atlanta, GA.

Chandra, R., Kumar, V. 2017. Phytoextraction of heavy metals by potential native plants and their microscopic observation of root growing on stabilised distillery sludge as a prospective tool for In-situ phytoremediation of industrial waste. *Environ. Sci. Pollut. Res.* 24:2605–19 Doi: 10.1007/s11356-016-8022-1.

Chandra, R., Kumar, V. 2018. Phytoremediation: A green sustainable technology for industrial waste management. In: Chandra, R., Dubey, N., Kumar, V., *Phytoremediation of Environmental Pollutants.* Boca Raton, FL: CRC Press,

Chedrese, P.J., Piasek, M., Henson, M.C. 2006. Cadmium as an endocrine disruptor in the reproductive system. *Immunol. Endocr. Metab. Agents Med. Chem.* 6:27–35.

Chellaiah, E. 2018. Cadmium (heavy metals) bioremediation by *Pseudomonas aeruginosa*: A minireview. *Appl. Water Sci.* 8:154.

Cinar, M. A., Yigit, A., Eraslan, G. 2010. Effects of vitamin C or vitamin E supplementation on cadmium induced oxidative stress and anaemia in broilers. *Revue. Med. Vet.* 161:449–54.

Daugas, E., Susin, S.A., Zamzami, N., et al. 2000. Mitochondrio-nuclear translocation of AIF in apoptosis and necrosis. *FASEB J.* 14:729–39.

Dauplais, M., Lazard, M., Blanquet, S., et al. 2013. Neutralization by metal ions of the toxicity of sodium selenide. *PLoS One.* 8(1):e54353.

Demirhan, O., Tastemir, D., Hasturk, S., et al. 2010. Alterations in p16 and p53 genes and chromosomal findings in patients with lung cancer: Fluorescence *in situ* hybridization and cytogenetic studies. *Cancer Epidemiol.* 34(4):472–77.

Eichler, T., Ma, Q., Kelly, C., et al. 2006. Single and combination toxic metal exposures induce apoptosis in cultured murine podocytes exclusively via the extrinsic caspase 8 pathway. *Toxicol. Sci.* 90:392–99.

Erdogan, Z., Erdogan, S., Celik, S., et al. 2005. Effects of ascorbic acid on cadmium-induced oxidative stress and performance of broilers. *Biol. Trace Elem. Res.* 104(1):19–32.

Fatur, T., Tusek, M., Falnoga, I., et al. 2003. Cadmium inhibits repair of UV-, methyl methane sulfonate- and N methyl-N-nitrosourea-induced DNA damage in Chinese hamster ovary cells. *Mutat. Res.* 529:109–16.

Field, R.W., Withers, B.L. 2012. Occupational and environmental causes of lung cancer. *Clin. Chest Med.* 33(4):681–03.

Friberg, L., Piscator, M., Nordberg, G. 1971. *Cadmium in the Environment.* CRC Press, Cleveland.

Genuis, S.J., Birkholz, D., Rodushkin, I., et al. 2010. Blood, urine, and sweat (BUS) study: Monitoring and elimination of bioaccumulated toxic elements. *Arch. Environ. Contam. Toxicol.* 61(2):344–57.

Gerbino, E., Carasi, P., Tymczyszyn, E.E., et al. 2014. Removal of cadmium by *Lactobacillus kefir* as a protective tool against toxicity. *J. Dairy Res.* 81:280–87.

Goher, M.E., El-Monem, A.M.A., Abdel-Satar, A.M., et al. 2016. Biosorption of some toxic metals from aqueous solution using non-living algal cells of *Chlorella vulgaris. J. Elem.* 21(3):703–14.

Guney, M., Zagury, G.J. 2012. Heavy metals in toys and low-cost jewelry: Critical review of US and Canadian legislations and recommendations for testing. *Environ. Sci. Technol.* 46:4265–74.

Hadi, B., Margaritis, A., Berruti, F., et al. 2002. Kinetics and equilibrium of cadmium biosorption by yeast cells *S. Cerevisiae* and *K. Fragilis. Int. J. Chem. Reactor Eng.* 1:1–18.

Halttunen, T., Salminen, S., Tahvonen, R. 2007. Rapid removal of lead and cadmium from water by specific lactic acid bacteria. *Int. J. Food Microbiol.* 114:30–35.

Haziri, I., Mane, B., Haziri, A., et al. 2012. Hematological effects of cadmium in hybrid isa brown. *Eur. J. Exp. Biol.* 2(6):2049–54.

Higham, D.P., Sadler, P.J., Scawen, M.D. 1986. Cadmium-binding proteins in *Pseudomonas putida*: Pseudothioneins. *Environ. Health Perspect.* 65:5–11.

Hou, W., Chen, X., Song, G., et al. 2007. Effects of copper and cadmium on heavy metal polluted water body restoration by duckweed (*Lemna minor*). *Plant Physiol. Biochem.* 45(1):62–69.

Hu, X., Chandler, J.D., Fernandes, J., et al. 2018. Selenium supplementation prevents metabolic and transcriptomic responses to cadmium in mouse lung. *BBA-Gen Subjects.* 1862(11):2417–26.

Huang, L., Liu, L., Zhang, T., et al. 2019. An interventional study of rice for reducing cadmium exposure in a Chinese industrial town. *Environ. Int.* 122:301–09.

IARC (International Agency for Research on Cancer). 2016. *Agents Classified by the IARC Monographs*. Vol. 1–116. http://monographs.iarc.fr/ENG/Classification/. Accessed 5 October 2021.

IPCS (International Programme on Chemical Safety). 1992. *Cadmium-Environmental Health Criteria 134*. World Health Organization, Geneva, 134.

Issazadeh, K., Pahlaviani, M.R.M.K., Massiha, A. 2011. Bioremediation of toxic heavy metals pollutants by *Bacillus* spp. isolated from Guilan bay sediments, north of Iran. *Conference Biotechnol. Environ. Manage. (IPCBEE).*18:67–71.

Jameer A., Sumithra, S., Senthil Kumar, P. 2015. Phytoremediation: Accumulation of cadmium by native plants. *Int. J. Current Sci. Technol.* 3(9):78–83.

Jeelani, N., Yang, W., Xu, L., et al. 2017. Phytoremediation potential of *Acorus calamus* in soils co-contaminated with cadmium and polycyclic aromatic hydrocarbons. *Sci. Rep.*7:8028.

Jiang, W., Liu, D., Hou, W. 2001. Hyperaccumulation of cadmium by roots, bulbs and shoots of garlic. *Bioresour. Technol.* 76(1):9–13.

Jin, Y.H., Clark, A.B., Slebos, R.J., et al. 2003. Cadmium is a mutagen that acts by inhibiting mismatch repair. *Nat. Genet.* 34:326–29.

Johnson, M.D., Kenney, N., Stoica, A., et al. 2003. Cadmium mimics the *in vivo* effects of estrogen in the uterus and mammary gland. *Nat. Med.* 9(8):1081–84.

Johri, N., Jacquillet, G., Unwin, R. 2010. Heavy metal poisoning: The effects of cadmium on the kidney. *Biometals.* 23(5):783–92.

Kara, H., Cevik, A., Konar, V., et al. 2008. Effects of selenium with vitamin E and melatonin on cadmium-induced oxidative damage in rat liver and kidneys. *Biol. Trace Elem. Res.* 125(3):236–44.

Karabulut-Bulan, O., Bolkent, S., Yanardag, R., et al. 2008. The role of vitamin C, vitamin E, and selenium on cadmium-induced renal toxicity of rats. *Drug Chem. Toxicol.* 31(4):413–26.

Kohen, R., Nyska A. 2002. Oxidation of biological systems: Oxidative stress phenomena, antioxidants, redox reactions and methods for their quantification. *Toxicol. Pathol.* 30:620–30.

Kostial, K. 1986. Cadmium. In: Mertz, W. (Ed.), *Trace Elements in Human and Animal Nutrition*. Academic Press, Inc., San Diego, CA, Vol. 2:319–337.

Kumar, P.V., Bricey, A.A., VeeraThamariSelvi, V., et al. 2010. Antioxidant effect of green tea extract in cadmium chloride intoxicated rats. *Adv. Appl. Sci. Res.* 1(2):9–13.

Kumar, V. 2018. Mechanism of microbial heavy metal accumulation from polluted environment and bioremediation. In: Sharma, D., Saharan, B.S., *Microbial Fuel Factories*. CRC Press, Boca Raton, FL.

Kumar, V., Ferreira, L.F.R., Sonkar, M., Singh, J. 2021. Phytoextraction of heavy metals and ultrastructural changes of *Ricinus communis* L. grown on complex organometallic sludge discharged from alcohol distillery. *Environ. Technol. Innovat.* 22:101382. Doi: 10.1016/j.eti.2021.101382.

Kumar, V., Shahi, S.K., Singh, S. 2018. Bioremediation: An eco-sustainable approach for restoration of contaminated sites. In: Singh J., Sharma D., Kumar G., Sharma N. (Eds.), *Microbial Bioprospecting for Sustainable Development*. Springer, Singapore. Doi: 10.1007/978-981-13-0053-0_6.

Kumaran, N.S., Sundaramanicam, A., Bragadeeswaran, S. 2011. Adsorption studies on heavy metals by isolated cyanobacterial strain (*Nostoc* sp.) from Uppanar estuarine water, southeast coast of India. *J. Appl. Sci. Res.* 7(11): 1609–15.

Kwinta, R.B., Bartoszek, A., Kusznierewicz, B., et al. 2011. Physiological response of plants and cadmium accumulation in heads of two cultivars of white cabbage. *J. Elem.* 16(3): 355–64.

Layachi, N., Kechrid, Z. 2012. Combined protective effect of Vitamin C and E on cadmium induced oxidative liver injury in rats. *Afr. J. Biotechnol.* 11(93):16013–020.

Lee, W.K., Torchalski, B., Thevenod, F. 2007. Cadmium-induced ceramide formation triggers calpain-dependent apoptosis in cultures kidney proximal tubule cells. *Am. J. Physiol. Cell Physiol.* 293:C839–C847.

Lei, Y.X., Wei, L., Wang, M., et al. 2008. Malignant transformation and abnormal expression of eukaryotic initiation factor in bronchial epithelial cells induced by cadmium chloride. *Biomed. Environ. Sci.* 21(4):332–38.

Lima, A.I.G., Corticeiro, S.C., de. Almeida Paula Figueira, E. M. 2006. Glutathione-mediated cadmium sequestration in *Rhizobium leguminosarum*. *Enzyme Microb. Technol.* 39(4): 763–69.

Liu, J., Zhou, Q., Sun, T., et al. 2008. Growth responses of three ornamental plants to Cd and Cd-Pb stress and their metal accumulation characteristics. *J. Hazard. Mater.* 151(1):261–67.

Liu, L., Yang, B., Cheng, Y., et al. 2015. Ameliorative effects of selenium on cadmium-induced oxidative stress and endoplasmic reticulum stress in the chicken kidney. *Biol. Trace Elem. Res.* 167(2):308–19.

Mahajan, P., Kaushal, J. 2018. Role of phytoremediation in reducing cadmium toxicity in soil and water. *J. Toxicol.* 4864365.

Mangkoedihardjo, S. 2008. *Jatropha curcas* L. for phytoremediation of lead and cadmium polluted soil. *World App. Sci. J.* 4(4):519–22.

Martin, M.B., Reiter, R., Pham, T., et al. 2003. Estrogen-like activity of metals in MCF-7 breast cancer cells. *Endocrinology.*144(6):2425–36.

Mates, J.M. 2000. Effects of antioxidant enzymes in the molecular control of reactive oxygen species toxicology. *J. Toxicol.*153:83–104.

Miceli, F., Minici, F., Tropea, A., et al. 2005. Effects of nicotine on human luteal cells in vitro: A possible role on reproductive outcome for smoking women. *Biol. Reprod.* 72:628–32.

Mishra, D., Kumari, S., Jaiswal, A., Arya, S., Yadav, B.R., Thul, S.T., Kumar, S., Pandey, R., Kumar, R., Pandey, A. 2021. Evaluation of distillery sludge as a soil amendment for improving soil quality and sugarcane (CO-265) yield. *Environ. Technol. & Innovat.* 23:101624. Doi: 10.1016/j.eti.2021.101624.

Murakami, M., Ae, N., Ishikawa, S. 2007. Phytoextraction of cadmium by rice (*Oryza sativa* L.), soybean (*Glycine max* (L.) Merr.), and maize (*Zea mays* L.). *Environ. Pollut.* 145(1):96–103.

Murphy, D., James, B., Hutchinson, D. 2016. Could the significantly increased risk of rheumatoid arthritis reported in Italian male steel workers be explained by occupational exposure to cadmium? *J. Occup. Med. Toxicol.* 11:21.

Nordberg, G.F., Nogawa, K., Nordberg, M., et al. 2007. Cadmium. In: *Handbook on the Toxicology Of Metals.* In: Nordberg, G.F., Fowler, B., Nordberg, M., Friberg, L. (Eds.), Elsevier, Amsterdam, 445–486.

Person, R.J., Tokar, E.J., Xu, Y., et al. 2013. Chronic cadmium exposure *in vitro* induces cancer cell characteristics in human lung cells. *Toxicol. Appl. Pharmacol.* 273:281–88.

Pipiska, M., Hornik, M., Remenarova, L., et al. 2013. Biosorption of cadmium, cobalt and zinc by moss *Rhytidiadelphus squarrosus* in the single and using *Pteris* sp. *Ecotoxicol. Environ. Saf.* 98(1):236–43.

Prabavathi, R., Mathivanan, V., Selvisabanayagam 2011. Analysis of concentration and accumulation of heavy metal cadmium in four selected terrestrial plants. *Int. J. Development Res.* 1(4):27–30.

Priyalaxmi, R., Murugan, A., Raja, P., et al. 2014. Bioremediation of cadmium by *Bacillus safensis* (JX126862), a marine bacterium isolated from mangrove sediments. *Int. J. Curr. Microbiol. Appl. Sci.* 3:326–35.

Prozialeck, W.C., Edwards, J.R., Lamar, P.C., et al. 2009. Expression of kidney injury molecule-1 (Kim-1) in relation to necrosis and apoptosis during the early stages of Cd-induced proximal tubule injury. *Toxicol. Appl. Pharmacol.* 238:306–14.

Quartacci, M.F., Argilla, A., Baker, A.J.M., et al. 2006. Phytoextraction of metals from a multiply contaminated soil by Indian mustard. *Chemosphere.* 63(6):918–25.

Rajendran, P., Muthukrishnan, J., Gunasekaran, P. 2003. Microbes in heavy metal remediation. *Indian J. Exp. Biol.* 41(9):935–44.

Raskin, I., Smith, R.D., Salt, D.E. 2005. Phytoremediation of metals: Using plants to remove pollutants from the environment. *Curr. Opin. Biotechnol.* 8(2):221–26.

Rayman, M.P. 2012. Selenium and human health. *Lancet.* 379(9822):1256–68.

Reyes-Hinojosa, D., Lozada-Perez, C.A., Zamudio Cuevas, Y., et al. 2019. Toxicity of cadmium in musculoskeletal diseases. *Environ. Toxicol. Pharmacol.*72:103219.

Rezvani, M., Zaefarian, F. 2011. Bioaccumlation and translocation factors of cadmium and lead in *Aeluropus littoralis. Aus. J. Agric. Eng.* 2(4): 114–19.

Rizwan, M., Ali, S., Rehman, M.Z., et al. 2018. Cadmium phytoremediation potential of Brassica crop species: A review. *Sci. Total Environ.* 631–632:1175–91.

Ruwali, P., Ambwani, T.K., Gautam, P. 2017. *In vitro* antioxidative potential of *Artemisia indica* willd. *Indian J. Anim. Sci.* 87:1326–31.

Sarkar, A., Ravindran, G., Krishnamurthy, V. 2013. A brief review on the effect of cadmium toxicity: From cellular to organ level. *Int. J. Biotech.* 3:17–36.

Sen Gupta, R., Kim, J., Gomes, C., et al. 2004. Effect of ascorbic acid supplementation on testicular steroidogenesis and germ cell death in cadmium-treated male rats. *Mol. cell Endocrinol.* 221(1–2):57–66.

Shameer, S. 2016. Biosorption of lead, copper and cadmium using the extracellular polysaccharides (EPS) of *Bacillus* sp., from solar salterns. *3 Biotech.* 6:1–10.

Sheng, X., Xia, J. 2006. Improvement of rape (*Brassica napus*) plant growth and cadmium uptake by cadmium-resistant bacteria. *Chemosphere.* 64(6):1036–42.

Singh, N., Gupta, V.K., Kumar, A., et al. 2017. Synergistic effects of heavy metals and pesticides in living systems. *Front. Chem.* 5:1–9.

Sinha, S.N., Biswas, K. 2014. Bioremediation of lead from river water through lead-resistant purple-nonsulfur bacteria. *Global J. Microbiol. Biotechnol.* 2:11–18.

Sinha, S.N., Paul, D. 2014. Heavy metal tolerance and accumulation by bacterial strains isolated from waste water. *J. Chem. Biol. Phys. Sci.* 4:812–17.

Soleimani, N., Fazli, M.M., Mehrasbi, M., et al. 2015. Highly cadmium tolerant fungi: Their tolerance and removal potential. *J. Environ. Health Sci. Eng.* 13:1–9.

Strumylaite, L., Mechonosina, K., Tamasauskas, S. 2010. Environmental factors and breast cancer. *Medicina (Kaunas).* 46(12):867–73.

Sugiyama, D., Nishimura, K., Tamaki, K., et al. 2010. Impact of smoking as a risk factor for developing rheumatoid arthritis: A meta-analysis of observational studies. *Ann. Rheum. Dis.* 69(1):70–81.

Suthep, S., Duangrat, S., Prayad, P., et al. 2016. Phytoremediation of cadmium by selected leguminous plants in hydroponics culture. *Res. Rev: J. Ecol. Environ. Sci.* 4(1): 1–7.

Taha, E.A., Sayed, S.K., Ghandour, N.M., et al. 2013. Correlation between seminal lead and cadmium and seminal parameters in idiopathic oligoasthenozoospermic males. *Cent. Eur. J. Urol.* 66:84–92.

Taha, M.M., Mahdy-Abdallah, E.M., Shahy, K.S., et al. 2018. Impact of occupational cadmium exposure on bone in sewage workers. *Int. J. Occp. Environ. Health.* 24(3–4):101–08.

Tandon, S.K., Prasad, S., Singh, S. 2002. Chelation in metal intoxication: Influence of cysteine or N-acetyl cysteine on the efficacy of 2,3-dimercaptopropane-1-sulphonate in the treatment of cadmium toxicity. *J. Appl. Toxicol.* 22(1):67–71.

Tang, L., Hamid, Y., Zehra, A., et al. 2019. Characterization of fava bean (*Vicia faba* L.) genotypes for phytoremediation of cadmium and lead co-contaminated soils coupled with agro-production. *Ecotoxicol. Environ. Saf.* 171:190–198.

Thevenod, F. 2010. Catch me if you can! Novel aspects of cadmium transport in mammalian cells. *Biometals.* 23(5):857–75.

Thompson, J., Bannigan, J. 2008. Cadmium: Toxic effects on the reproductive system and the embryo. *Reprod. Toxicol.* 25(3):304–15.

Torres-Sanchez, L., Vazquez-Salas, R.A., Vite, A., et al. 2018. Blood cadmium determinants among males over forty living in Mexico city. *Sci. Total Environ.* 1(637–638):686–94.

Turk, A.S., VanBeer-Tas, M.H., Van Schaardenberg, D. 2014. Prediction of future of rheumatoid arthritis. *Rheum. Dis. Clin. North Am.* 40:753–70.

Uraguchi, S., Fujiwara, T. 2012. Cadmium transport and tolerance in rice: Perspectives for reducing grain cadmium accumulation. *Rice (N Y).* 5(1):5.

Vahter, M., Bergland, M., Nermall, B., et al. 1996. Bioavailability of cadmium from shellfish and mixed diet in women. *Toxicol. Appl. Pharmacol.* 36:332–41.

Wan, N., Xu, Z., Liu, T., et al. 2018. Ameliorative effects of selenium on cadmium-induced injury in the chicken ovary: Mechanisms of oxidative stress and endoplasmic reticulum stress in cadmium-induced apoptosis. *Biol. Trace Elem. Res.* 184(2):463–73.

Wang, C. L., Ozuna, S.C., Clark, D.S., et al. 2002. A deep-sea hydrothermal vent isolate, *Pseudomonas aeruginosa* CW961, requires thiosulfate for Cd^{2+} tolerance and precipitation. *Biotechnol. Lett.* 24(8):637–41.

Wang, Y., Yan, A., Dai, J., et al. 2012. Accumulation and tolerance characteristics of cadmium in *Chlorophytum comosum*: A popular ornamental plant and potential Cd hyperaccumulator. *Environ. Monit. Assess.* 184(2):929–37.

Wei, S., Zhou, Q., Mathews, S. 2008. A newly found cadmium accumulator- *Taraxacum mongolicum. J. Hazard. Mater.* 159(2–3):544–47.

Wei, S., Zhou, Q., Wang, X., et al. 2005. A newly-discovered Cd-hyperaccumulator *Solanum nigrum* L. *Chin. Sci. Bull.* 50(1):33–38.

WHO. 2011. *Selenium in Drinking Water. Background Document for Development of WHO Guidelines for Drinking-Water Quality.* World Health Organization (WHO/HSE/WSH/10.01/14), Geneva.

Xu, X., Xia, L., Zhu, W., et al. 2015. Role of *Penicillium chrysogenum* XJ-1 in the detoxification and bioremediation of cadmium. *Front. Microbiol.* 6:1422. Doi: 10.3389/fmicb.2015.01422.

Yang, M.G., Lin, X.Y., Yang, X.E. 1998. Impact of Cd on growth and nutrient accumulation of different plant species. *Ying Yong Sheng Tai Xue Bao.* 9(1):89–94.

Yu, Z., Zhou, Q. 2009. Growth responses and cadmium accumulation of *Mirabilis jalapa* L. under interaction between cadmium and phosphorus. *J. Hazard. Mater.* 167(1–3):38–43.

Yuan, Z., Yi, H., Wang, T., Zhang, Y., Zhu, X., Yao, J. 2017. Application of phosphate solubilizing bacteria in immobilization of Pb and Cd in soil. *Environ. Sci. Pollut. Res.* 24(27):21877–884.

Zalups, R.K. 2000. Molecular interactions with Mercury in the kidney. *Pharmacol. Rev.* 52:113–43.

Zhao, F.J., Lombi, E., McGrath, S.P. 2003. Assessing the potential for zinc and cadmium phytoremediation with hyperaccumulator *Thlaspi caerulescens. Plant and Soil.* 249:37–43.

Zwolak, I. 2020. The role of selenium in arsenic and cadmium toxicity: An updated review of scientific literature. *Biol. Trace Elem. Res.* 193(1): 44–63.

20 Metal(loid)-Microbe Interactions

Trading on Tolerance and Transformation for Environmental Remediation

Rubina Khanam
ICAR- National Rice Research Institute

Pedda Ghouse Peera Sheikh Kulsum
C V Raman Global University

Jayanta Kumar Biswas
University of Kalyani

CONTENTS

DOI: 10.1201/9781003247883-20

20.1   INTRODUCTION

Metals and metalloids are non-biodegradable and have a high atomic weight and a density higher than 5 g/cm³. They exhibit excellent electrical conductivity, malleability, and metallic luster, and the ability to transport electrons and generate cations (Khanam et al., 2020). Essential metal(loid)s are required in small amounts in certain biological processes, where they serve as cofactors for metalloenzymes. For instance, Cu is required for mitochondrial electron transport as well as the oxidative stress enzymes cytochrome c oxidase and superoxide dismutase (Osredkar, 2011). Zinc (Zn) is involved in the catalytic activities of DNA and RNA polymerases. Ni and Cr are required for the efficient functioning of urease (Khanam et al., 2020). Non-essential metal(loid)s such as arsenic (As), mercury (Hg), cadmium (Cd), and lead (Pb) have no recognized biological roles and are hazardous even at the modest level of contamination (Khanam et al., 2020). While certain metal(loid)s are required for optimal plant development, excessive quantities are detrimental to plants and microbes. Metal(loid)s are naturally found in greater depths of soil; however, anthropogenic and geological excavations trigger them to alarming levels posing a threat to both plants and animals (Khanam et al., 2020). Unprecedented use of plant protection chemicals and continuous use of municipal solid waste and untreated sludge are the primary contributors of metal(loid)s in agricultural soils.

Phytoremediation fails beyond a certain point, as removing accumulated metal (loid) is critical for the proper functioning of plants as well as the protection of organisms that rely on them (Clemens and Ma, 2016; Chandra and Kumar, 2018). Excavation (decontamination though physical removal), stabilization (fixation by applying chemicals to change metal to an unavailable form), and soil washing are some of the methods used to clean up metal(loid) from polluted sites (reduction through physical/chemical removal). These physical processes, on the other hand, are neither efficient nor cost-effective (Khanam et al., 2020). As a result, finding cost-effective, long-lasting, and ecologically responsible metal(loid) cleaning solutions should be a top concern. Microbial technologies are active and widespread. The interaction of microorganisms with metals in natural and man-made settings has a long history. The resurgence of omics technologies for microbial development has opened up new possibilities as a remediation method. The interaction of metal microorganisms in the soil environment is extremely strict and is influenced by the type of soil, metal ions and their concentration, and diversity of microbes (Kumar and Chandra, 2018, 2020).

This chapter is not aimed to be an exhaustive review, but rather a detailed examination of recent reports on the harmful effects of metal(loid)s and leverage microbes and their metabolites to improve metal(loid) bioremediation. It also addresses factors governing metal(loid) bioremediation and their underlining mechanisms. Current research on microbial biosorption and detoxification is reviewed, and future research directions are suggested.

20.2   ENVIRONMENTAL CONTAMINATION AND (ECO)TOXICITY OF TOXIC METALS AND METALLOIDS

Soils are the primary source of metal pollution in terrestrial ecosystems. Metal concentrations in normal soil may vary from 1 to 100,000 mg/kg (Gadd, 2010).

Metal(loid)s, both essential and non-essential, become toxic when their quantities surpass certain thresholds that vary by metal. Toxic amounts of metal(loid)s may build up in bacterium cells, changing biological processes, causing structural changes, and eventually causing metal(loid)-mediated injury (Khanam et al., 2020). The formation of reactive species, which damage nucleic acids and cytoplasmic organelles, is one of the mechanisms of metal(loid) toxicity. Some of the non-essential metal(loid)s (Pb, Cd, and As) and essential metal(loid)s (Cu, Ag, and Zn) in excess are reported to cause oxidative damage in bacteria, resulting in reactive species-led nucleic acids and membrane disruption (Williams et al., 2016). Further, metal(loid)s may affect the activity of sulfhydryl group (R-SH)-containing enzymes. For instance, covalent bonding between a non-essential metal(loid) (Cd and Hg) and the sulfhydryl group alters enzyme function. Metal(loid)s may also function as competitive inhibitors, causing essential ions to displace from their target locations.

Metal(loid) toxicity is mainly expressed in terms of bioavailability of metal(loid)s and the absorbed dose and thereby detrimental effects on the microbial community. Disruption of cell membrane and damage of the DNA double helix are the very common forms of metal(loid) toxicity in microbes. This toxicity is majorly due to the relocation of metals either by ligand bonding or from common binding sites (Khanam et al., 2020). Damage of nucleic acid helix, disruption of cell membranes, altered enzymatic activity, and production of reactive species all have an impact on the morphology, metabolism, and development of microorganisms (Fashola et al., 2016) (Table 20.1).

Aluminum (Al) is known to cause acidity in soils; besides, it also has the potential to stabilize superoxide radicals. Cu species (Cu^+ and Cu^{2+}) act as soluble electron carriers and catalyze the formation of reactive species (Giner-Lamia et al., 2014).

TABLE 20.1
Metal(loid) Toxicity and Effect on Microbial Communities

Metal(loid)	Effect on Microbial Community	Reference
Ni^{2+}	Cell apoptosis, decrease in mitochondrial membrane potential of eukaryotes, oxidative stress (hypoxia)	Guo et al. (2015), Scanlon et al. (2017)
Cd^{2+} (in Zn ores)	Cd-induced protein/nucleic acid denaturation, altered gene expression, loss in mitochondrial-membrane integrity, permeability and potential	Genchi et al. (2020)
As^{3+} (arsenite) As^{5+} (arsenate)	Altered enzyme functions	Khanam et al. (2020)
CH_3-Hg^+ (organic) Hg^{2+} (inorganic)	Proteins are denatured, enzyme functions are inhibited, and cell membranes are disrupted	Bazzi et al. (2020)
Cr^{6+} and Cr^{3+}	Free radical-led lipid peroxidation, retarded growth, cellular damage by DNA abduction, and breakup of chromosomes	Tang et al. (2021), Ayele and Godeto (2021)
Pb^{2+} Pb^{4+}	Disruption in cell division, mitochondrial swelling, DNA damage	Zhai et al. (2020), Aslam et al. (2021)
Ag^+ and Ag^0	Cease of photosynthesis in algae decreased the synthesis of chloroplasts, ruptured cell wall, clumping of chromosomes at metaphase	Tripathi et al. (2017)

Species of Cr, especially Cr^{3+}, form electrostatic bonding with PO_4^{3-} in DNA and possibly affect central dogma and mutagenesis (Truu et al., 2015). By interacting with the carboxyl and thiol groups, Cr^{3+} may alter enzyme functions.

20.3 RESISTANCE AND TOLERANCE AGAINST METAL(LOID)S: COPING WITH TOXICITY

The bacterial community in metal(loid)-contaminated sites is dominated by Firmicutes, Proteobacteria, and Actinobacteria, with *Bacillus, Pseudomonas,* and *Arthrobacter* as the most common taxa (Pires et al., 2017). In fact, metal(loid) stress may make nodulation and nitrogenase activity extremely sensitive. In agricultural soils, in particular, the legume–rhizobia symbiosis is well recognized for its ability to detoxify metal(loid)s and enhances the condition of polluted soils (Khanam et al., 2020). It is evident from studies on microbial resistance against metal(loid)s that metal(loid) resistance determinants are found in almost every bacterial species. For example, isolated strains of *Acinetobacter baumannii* from agricultural soil and sediments, fuel-contaminated soil, and sewage water were found to be resistant to Hg, Ag, and As (El-Sayed, 2016; Huang et al., 2017; Khanam et al., 2020). Six mechanisms of metal(loid) resistance have been elucidated till date and found to be effective (Bazzi et al., 2020): (1) outer cell structures like a cytoplasmic membrane, cell wall rich in peptidoglycans (mucopeptides) sheath secret metal ions; (2) metal ions are extruded through efflux pumps; (3) intracellular metal ion sequestration; (4) extracellular metal sequestration; (5) toxic metal ion biotransformation/detoxification; and (6) induced microbial tolerance to metal ions. These mechanisms will be briefly discussed in the sections that follow.

20.4 BIOREMEDIATION: A PRODUCT OF MICROBIAL WISDOM AND TOLERANCE

20.4.1 BACTERIA

Microbial biomass has a wide range of biosorptive properties, which also vary greatly across microorganisms. The sorption efficiency of each microbial group is determined by its environmental conditions (Kumar 2018). When bacteria are grown in a mixed culture, they are more stable and live longer. As a result, microbial consortium is metabolically good for metal biosorption and are more suited for field use. Under in vitro condition, *Micrococcus luteus* harnessed maximum Pb (1648 mg/g). It is evident from the extensive review by Aslam et al. (2021) that *Bacillus* genera outperformed in harnessing Pb over fungi and the order is as follows: *Bacillus megaterium* > *B. subtilis* > *Aspergillus niger* > Penicillium (Table 20.2).

20.4.2 FUNGI

Fungi are often employed as biosorption tools due to their high biomass, metal adsorption, and recovery capabilities. For instance, biomass from *Rhizopus oryzae and P. chrysogenum* by bio-reduction forms Cr^{3+} from Cr^{5+} (Tang et al., 2021). Researchers also demonstrated the sorption efficiency of *Candida sphaerica* for Pb, Zn, and Fe to be 79%, 90%, and 95%, respectively (Luna et al., 2016). Because of their low toxicity, biodegradability, and variety, biosurfactants have captured a lot of attention in recent years.

20.4.3 MICROALGAE

Wider adaptability, photoautotrophic nature, inexpensive growth requirements, and instant growth cycles that complete in hours have all made microalgae a viable option among microflora for metal(loid) removal. Most commonly encountered microalgae genera for metal(loid)s removal are *Chlorella, Scenedesmus, Neochloris, Nannochloropsis*, and *Penium*. Among microalgae, *Chlorella* and *Scenedesmus* became popular because of their high adsorption capabilities. *Chlorella* is unicellular, spherical with less than 10 μm diameter (e.g., *Chlorella vulgaris)*, while *Scenedesmus* is a fresh water algae with 4–8 cells. The adsorption capacity of this group relies on cell wall architecture and composition (Spain et al., 2021). "Spectroscopic studies have shown that functional groups including carboxyl, hydroxyl, sulfate, sulfhydryl (thiol), phosphate, amino, amide, imine, thioether, phenol, carbonyl (ketone), imidazole, phosphonate, and phosphodiester have the properties to be involved in metal binding" (Sultana et al., 2020). The key players in algal metal(loid) sorption are carboxyl groups. For instance, Cd(II) can bind the carboxylic group either by forming carboxylates or by displacing H^+ in the group. *Chlamydomonas reinhardtii* is also a potent microalga in removing essential metal(loid)s such as Ca^{2+} as alginate beads (Table 20.2).

20.4.4 FACTORS INFLUENCING BIOREMEDIATION OF METAL(LOID)S

Chemical structure, concentration, and associated factors such as redox potential influence whether metal(loid)s are essential or non-essential to microbial communities. Factors such as temperature, soil reaction (pH), organic matter, and humic acids may affect the bioavailability of metal(loid)s to microbes. At acidic soil reaction, metal(loid)s form free ion species, and as the pH increases, the adsorbent surface becomes increasingly positive charged, leading to the repulsion of metal(loid)s and adsorbent and therefore triggers toxicity.

The adsorption of metal(loid)s is greatly influenced by temperature. The rate of adsorbate diffusion is directly proportional to the temperature. The solubility is temperature dependent, as the temperature rises, resulting in increased bioavailability. Microorganisms' activities, on the other hand, increase when temperature rises to an appropriate level, enhancing microbial metabolism and enzyme activity, accelerating bioremediation. Factors include microbial, substrate-related, environmental, growth substrate, and aerobic/anaerobic processes. (1) Microbial factors include prolific microbial populations, secondary metabolite synthesis, enzymes, and gene transfer. (2) Substrate factors are composition, concentration, toxicity of metal(loid)s, and solubility. (3) Environmental ones include depleting substrates and nutrient deficiencies.

TABLE 20.2

Metal(loid) Remediation by Microflora and Microfauna

Metal	Group	Microorganism	In ppm	Adsorption Capacity (g/kg)	Source
Pb	Bacteria	*Cellulosimicrobium* sp.	50–300	6.23–9.93	Bhati et al. (2019)
		Methylobacterium organophilum	-	1.8	Zhai et al. (2020)
		Micrococcus sp.	-	5.54	Aslam et al. (2021)
	Algae	*Chlorella vulgaris*	50	9.94	Igiri et al. (2018)
		Chlorella vulgaris	51	9.94	
		Nostoc	79	9.96	
		Chlamydomonas reinhardtii	62	3.8	Spain et al. (2021)
		Scenedesmus acutus	53	4.65	
Hg	Fungi	*Candida parapsilosis*	0.1	8	Aslam et al. (2021), Bazzi et al. (2020)
	Protozoa	*Tetrahymena rostrata*	0.1	4	Khanam et al. (2020)
	Algae	*Scenedesmus acutus*		1.06	Spain et al. (2021)
		Chlamydomonas reinhardtii		0.2	
Cr	Bacteria	*Bacillus cereus*	1500	9.6	Igiri et al. (2018)
		Bacillus circulans	110	7.1	
		Desulfovibrio desulfuricans	100	9.8	
		Stenotrophomonas sp.	16.59	8.13	
		Acinetobacter sp.	16	8.1	
	Fungi	*Aspergillus* sp.	100	9.2	Taştan et al. (2010)
		Aspergillus versicolor	50	9.8	Abdel-Wahab et al. (2018)
		Phanerochaete chrysosporium	100	9.8	Igiri et al. (2018)
	Algae	*Nannochloropsis oculate*	23	0.46	Spain et al. (2021)
		Scenedesmus quadricauda	31	0.37	
		Chlorella vulgaris	74	1.63	
As³⁺	Algae	*Nannochloropsis oculate*	29	0.46	Spain et al. (2021)
		Scenedesmus quadricauda	31	0.49	
		Chlorella vulgaris	49	0.5	
Cd	Bacteria	*Clostridium thermoaceticum*	110	8.2	Genchi et al. (2020)

(4) Mass transfer limitations include diffusion of gases and nutrients and solubility in water. (5) Redox potential and ease of electron acceptors.

20.5　MICROBIAL-MEDIATED MECHANISMS FOR REMEDIATION OF METAL(LOID)S

The most common bacterial mechanisms that impart metal(loid) resistance are either hydrolysis of ATP or an electrochemical potential gradient of H^+ or both (Bazzi et al., 2020). Microorganisms have five main efflux system families: (1) the ABC

(ATP-binding cassette), (2) RND (resistance-nodulation-cell division), (3) SMR (small multi-metal resistance), (4) MATE transporters (multi-metal and toxic metals efflux), and (5) the MFS (major facilitator superfamily). Energy requirements, base sequence, and substrates all vary for each of these transporters.

For instance, ABCs efflux mechanism operates through ATP hydrolysis, while chemiosmosis is a key player in SMRs and RNDs. In most of the bacterial species, basal levels of efflux are insufficient to impart metal(loid) resistance. Mutations in regulator and promotor regions or, in some cases, inactivated repressor leads to overexpression of transporters. Biosorption, bio-reduction, ion exchange, redox processes, and electrostatic bonding are some of the methods; microorganisms take advantage to counteract metal toxicity. The five important mechanisms highlighted commonly by several researchers are discussed in the following sections.

20.5.1 BIOSORPTION

Metal(loid) uptake by microbes through this mechanism may be either metabolism dependent or independent. It is cell surface resistance mechanism and includes redox reactions, bio-conversion/reduction, and sequestration steps. It is also a passive adsorption process led by dead biomass via surface complexation onto the plasma membrane, capsule, and cell wall layers (Gadd, 2010) (Figure 20.1).

20.5.2 BIOSEQUESTRATION

20.5.2.1 Intracellular Sequestration

Intracellular metal ions are sequestrated with the help of metal ion-binding proteins such as metallochaperones, metallothioneins (MTs), and glutathione (GSH). For

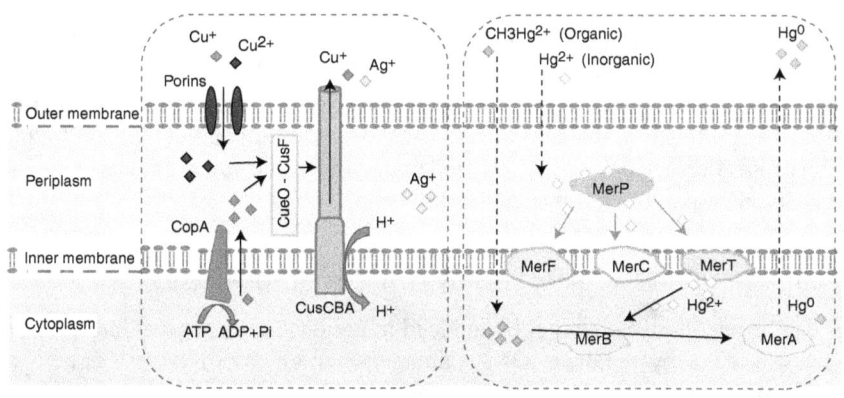

Resistance to Copper via the Cue and Cus systems operon of clusters CueRAO and CusRCFBA

Bacterial resistance to mercury. The mer operon consists of a cluster up to 8 genes merTPCAGBDE.

CopA - Cu1C exporting P-type ATPase
CueO - Periplasmic multi-Cu Oxidase
CusF - Periplasmic metallo-chaperone
CusCBA - multi-Cu/Ag efflux pump

MerP,F,C,T,B - Chain of binding proteins
MerA-Mercury (II) reductase (key detoxifier)

FIGURE 20.1 Bacterial resistance against Cu and Hg toxicity.

example, in *Pseudomonas* and *Anabaena* genera, non-essential metal(loid)s such as Cd, Pb, and Hg are captured on cysteine-rich MT-polypeptides. GSH is used by certain bacteria as an alternative chelator to sequester metal ions. Through its R-SH group, GSH scavenges and detoxifies metals. For example, *Rhizobium* spp. utilizes GSH in regulating Cd tolerance. MTs are popular transporters/pumps for essential metal(loid)s; for example, Cu plays a vital role in taking Cu to metalloenzymes, thereby protecting cytoplasmic organelles.

20.5.2.2   Extracellular Sequestration

Extracellular sequestration in microbial cells is achieved by retention of metal(loid)s in the periplasm. The retention is facilitated by chelating-proteins such as sidero-phores, PO_4^{3-}, S^{2-}, and oxalates. This mechanism is selective and requires consistent metal(loid) concentrations to work efficiently. For example, *Clostridium thermoace-ticum* follows this mechanism to sequester Cd via sulfides and *Streptomyces acidis-cabies* sequesters Ni through siderophores (Table 20.2).

20.5.3   OUTER CYTOPLASMIC STRUCTURES PREVENTING METAL ENTRY INTO MICROBIAL CELL

The outer cytoplasmic structure in the order: plasma membrane-cell wall-capsule/sheath act as a frontline defense to prevent metal ions from entering into microbial cells. It is possible via different functional groups ($-COOH$, $-NH_2$, and PO_4^{3-}) present in cell wall. For example, bacteria belonging to *Brevibacterium, Bacillus,* and *Pseudomonas genera* actively exclude the metal(loid)s such as Pb and Cu entry into the cell.

20.5.4   BIOTRANSFORMATION

Enzymes are the key players in the reduction of metal toxicity either by catalyzing redox reactions or by methylation mechanisms (Igiri et al., 2018). For instance, Hg^{2+} reductase (governed by MerA gene) converts Hg^{2+} to Hg^0 (Bazzi et al., 2020); reduction of Cr (Cr^{6+} to Cr^{3+}); oxidation of As (As^{3+} to As^{5+}) (Khanam et al., 2020); methylation of Pb (Fernandez et al., 2021); and under anaerobic environment, methylation of Hg (Spain et al., 2021) (Table 20.3).

20.5.5   REDUCING BACTERIAL SENSITIVITY TO METAL IONS

This can be accomplished either by induced mutations in the gene of interest to trigger resistance, or by repairing DNA damage when an SOS response is activating alternate-target on plasmids.

20.6   FUTURE PERSPECTIVES

Many studies have emphasized the value of whole genome sequencing (WGS) as a technique for detecting genome-wide changes and the development of metal(loid) resistance genes. It is now possible to analyze the complete bacterial genome at a

TABLE 20.3

Metal(loid)-Wise Microbial Resistance Mechanisms

Metal (loid)	Forms	Mechanisms	Transporters and Enzymes Involved	Microorganisms	Reference
As	As^{3+} (arsenite) As^{5+} (arsenate) As^0 (elemental As) As^3 (arsenide)	a. Oxidation and reduction b. Methylation c. Efflux via ars operon d. Intracellular sequestration	Oxidases-reductases GSH	*Enterobacter* sp. *Klebsiella pneumonia*	Bazzi et al. (2020)
Hg	CH_3-Hg^+ (organic) Hg^{2+} (inorganic)	Mer operon: oxidation and methylation	Hg (II) reductase	*Acinetobacter seohaensis* *Klebsiella pneumonia FY2*	Hobman and Crossman (2015), Pushkar et al. (2019)
Cu	Cu^+ Cu^{2+}	1. Cue system (at low concentration, aerobic and cytoplasmic Cu) 2. Cus system (at high Cu under anaerobic and both periplasmic and cytoplasmic Cu) 3. Pco system 4. System of Cops	P-type ATPase Periplasmic-multi-Cu-oxidase Metallochaperones	*E. coli* *Pseudomonas* spp. *Cupriavidus metallidurans*	Khanam et al. (2020), Aslam et al. (2021)
Cr	Cr^{6+} Cr^{3+}	Enzyme-mediated Cr reduction from VI to III (less toxic form)	Reductases	*Escherichia coli* *Aeromonas* spp. *Pantoea* spp. *Acinetobacter* spp.	Khanam et al. (2020), Aslam et al. (2021)
Pb	Pb^{2+} Pb^{4+}	1. Intracellular sequestration 2. Prevention of entry by forming insoluble complexes 3. Methylation 4. Extrusion of Pb through efflux pumps	Metallothioneins (MTs)	*Bacillus* RPB5-3	Jaroslawiecka and Piotrowska-Seget (2014), Fernandez et al. (2021)

cheap cost and in a rapid way. It provides a realistic option for evaluating genomes and identifying resistance genes for chemicals that aren't tested very often. In addition, this method enables scientists to uncover novel resistance pathways and offers useful information to researchers.

20.7   CONCLUSIONS

The present level of metal(loid) bioremediation reviewed in this chapter holds a lot of potential for metal biosorption and detoxification, particularly from biofilm and genetically engineered microorganisms. The function of microbial cells, biofilms, and their metabolites in metal(loid) remediation and environmental studies is explored in this book chapter. The emphasis of gene transfer inside biofilms for metal(loid) cleanup needs to be expanded further. These would make it easier to create better metal(loid) bioremediation methods in the environment.

REFERENCES

Abdel-Wahab, N. M., Scharf, S., Özkaya, F. C., Kurtán, T., Mándi, A., Fouad, M. A., & Proksch, P. (2019). Induction of secondary metabolites from the marine-derived fungus Aspergillus versicolor through co-cultivation with *Bacillus subtilis*. Planta Medica, 85 (06), 503–512.

Aslam, M., Aslam, A., Sheraz, M., Ali, B., Ulhassan, Z., Najeeb, U., Zhou, W., & Gill, R. A. (2021). Lead toxicity in cereals: Mechanistic insight into toxicity, mode of action, and management. *Frontiers in Plant Science, 11*, 2248. Doi: 10.3389/fpls.2020.587785.

Ayele, A., & Godeto, Y. G. (2021). Bioremediation of chromium by microorganisms and its mechanisms related to functional groups. *Journal of Chemistry, 2021*, 7694157. Doi: 10.1155/2021/7694157.

Bazzi, W., Abou Fayad, A. G., Nasser, A., Haraoui, L.-P., Dewachi, O., Abou-Sitta, G., Nguyen, V.-K., Abara, A., Karah, N., Landecker, H., Knapp, C., McEvoy, M. M., Zaman, M. H., Higgins, P. G., & Matar, G. M. (2020). Heavy metal toxicity in armed conflicts potentiates amr in *A. baumannii* by selecting for antibiotic and heavy metal co-resistance mechanisms. *Frontiers in Microbiology, 11*, 68. Doi: 10.3389/fmicb.2020.00068.

Bhati, T., Gupta, R., Yadav, N., Singh, R., Fuloria, A., Waziri, A., & Purty, R. S. (2019). Assessment of bioremediation potential of *Cellulosimicrobium* sp. for treatment of multiple heavy metals. *Microbiology and Biotechnology Letters, 47*(2), 269–277.

Clemens, S., & Ma, J. F. (2016). Toxic heavy metal and metalloid accumulation in crop plants and foods. *Annual Review of Plant Biology, 67*, 489–512. Doi: 10.1146/annurev-arplan t-043015-112301.

Chandra, R., & Kumar, V. 2018. Phytoremediation: A green sustainable technology for industrial waste management. In: Chandra, R., Dubey, N., & Kumar, V. (Eds.), *Phytoremediation of Environmental Pollutants*. Boca Raton, FL: CRC Press, Doi: 10.1201/9781315161549-1.

El-Sayed, M. H. (2016). Multiple heavy metal and antibiotic resistance of *Acinetobacter baumannii* Strain HAF – 13 Isolated from industrial effluents. *American Journal of Microbiological Research, 4*(1), 26–36. Doi: 10.12691/ajmr-4-1-3.

Fashola, M. O., Ngole-Jeme, V. M., & Babalola, O. O. (2016). Heavy metal pollution from gold mines: Environmental effects and bacterial strategies for resistance. *International Journal of Environmental Research and Public Health, 13*(11), 1047. Doi: 10.3390/ijerph13111047.

Fernandez, V., Abdelaziz, S., Yousaf, B., Ali Gill, R., Aslam, M., Aslam, A., Sheraz, M., Ali, B., Ulhassan, Z., Najeeb, U., & Zhou, W. (2021). Lead toxicity in cereals: Mechanistic insight into toxicity, mode of action, and management. *Front. Plant Sci, 11*, 587785. Doi: 10.3389/fpls.2020.587785.

Gadd, G. M. (2010). Metals, minerals and microbes: Geomicrobiology and bioremediation. *Microbiology (Reading, England), 156*(Pt 3), 609–643. Doi: 10.1099/mic.0.037143-0.

Genchi, G., Sinicropi, M. S., Lauria, G., Carocci, A., & Catalano, A. (2020). The effects of cadmium toxicity. *International Journal of Environmental Research and Public Health, 17*(11). Doi: 10.3390/ijerph17113782.

Giner-Lamia, J., López-Maury, L., & Florencio, F. J. (2014). Global transcriptional profiles of the copper responses in the cyanobacterium *Synechocystis* sp. PCC 6803. *PLoS One, 9*(9), e108912. Doi: 10.1371/journal.pone.0108912.

Guo, H., Chen, L., Cui, H., Peng, X., Fang, J., Zuo, Z., Deng, J., Wang, X., & Wu, B. (2015). Research advances on pathways of nickel-induced apoptosis. Doi: 10.3390/ijms17010010.

Hobman, J. L., & Crossman, L. C. (2015). Bacterial antimicrobial metal ion resistance. *Journal of Medical Microbiology, 64*, 471–497. Doi: 10.1099/jmm.0.023036-0.

Huang, Z., Lu, Q., Wang, J., Chen, X., Mao, X., & He, Z. (2017). Inhibition of the bioavailability of heavy metals in sewage sludge biochar by adding two stabilizers. *PLoS One.* Doi: 10.1371/journal.pone.0183617.

Igiri, B. E., Okoduwa, S. I., Idoko, G. O., Akabuogu, E. P., Adeyi, A. O., & Ejiogu, I. K. (2018). Toxicity and bioremediation of heavy metals contaminated ecosystem from tannery wastewater: A review. *Journal of Toxicology*, Doi: 10.1155/2018/2568038

Jarosławiecka, A., & Piotrowska-Seget, Z. (2014). Lead resistance in micro-organisms. *Microbiology (Reading, England), 160*(Pt 1), 12–25. Doi: 10.1099/mic.0.070284-0.

Kumar, V. (2018). Mechanism of microbial heavy metal accumulation from polluted environment and bioremediation. In: Sharma, D., & Saharan, B.S., *Microbial Fuel Factories*. Boca Raton, FL: CRC Press.

Kumar, V., & Chandra, R. (2018). Bacterial assisted phytoremediation of industrial waste pollutants and eco-restoration. In Chandra, R., Dubey, N.K., & Kumar, V. (Eds.), *Phytoremediation of Environmental Pollutants*. Boca Raton, FL: CRC Press.

Kumar, V., & Chandra, R. (2020). Metagenomics analysis of rhizospheric bacterial communities of *Saccharum arundinaceum* growing on organometallic sludge of sugarcane molasses-based distillery. *3 Biotech 10*(7), 316. Doi: 10.1007/s13205-020-02310-5.

Khanam, R., Kumar, A., Nayak, A. K., Shahid, M., Tripathi, R., Vijayakumar, S., Bhaduri, D., Kumar, U., Mohanty, S., Panneerselvam, P., Chatterjee, D., Satapathy, B. S., & Pathak, H. (2020). Metal(loid)s (As, Hg, Se, Pb and Cd) in paddy soil: Bioavailability and potential risk to human health. *Science of the Total Environment, 699*, 134330. Elsevier B.V. Doi: 10.1016/j.scitotenv.2019.134330.

Luna, J. M., Rufino, R. D., & Sarubbo, L. A. (2016). Biosurfactant from *Candida sphaerica* UCP0995 exhibiting heavy metal remediation properties. *Process Safety and Environmental Protection, 102*, 558–566. Doi: 10.1016/j.psep.2016.05.010.

Osredkar, J. (2011). Copper and zinc, biological role and significance of copper/zinc imbalance. *Journal of Clinical Toxicology, s3*(01), 1–18. Doi: 10.4172/2161-0495.s3-001.

Pires, C., Franco, A. R., Pereira, S. I. A., Henriques, I., Correia, A., Magan, N., & Castro, P. M. L. (2017). Metal(loid)-contaminated soils as a source of culturable heterotrophic aerobic bacteria for remediation applications. *Geomicrobiology Journal, 34*(9), 760–768. Doi: 10.1080/01490451.2016.1261968.

Pushkar, B., Sevak, P., & Singh, A. (2019). Bioremediation treatment process through mercury-resistant bacteria isolated from Mithi river. *Applied Water Science, 9*, 117. Doi: 10.1007/s13201-019-0998-5.

Scanlon, S. E., Scanlon, C. D., Hegan, D. C., Sulkowski, P. L., & Glazer, P. M. (2017). Nickel induces transcriptional down-regulation of DNA repair pathways in tumorigenic and non-tumorigenic lung cells. *Carcinogenesis, 38*(6), 627–637. Doi: 10.1093/carcin/bgx038.

Spain, O., Plöhn, M., Funk, C., & Jensen, P.-E. (2021). The cell wall of green microalgae and its role in heavy metal removal. *Physiologia Plantarum, 173*(2), 526–535. Doi: 10.1111/ppl.13405.

Sultana, N., Hossain, S. M. Z., Mohammed, M. E., Irfan, M. F., Haq, B., Faruque, M. O., Razzak, S. A., & Hossain, M. M. (2020). Experimental study and parameters optimization of microalgae based heavy metals removal process using a hybrid response surface methodology-crow search algorithm. *Scientific Reports, 10*(1), 15068. Doi: 10.1038/s41598-020-72236-8.

Tang, X., Huang, Y., Li, Y., Wang, L., Pei, X., Zhou, D., He, P., & Hughes, S. S. (2021). Study on detoxification and removal mechanisms of hexavalent chromium by microorganisms. *Ecotoxicology and Environmental Safety, 208*, 111699. Doi: 10.1016/j.ecoenv.2020.111699.

Taştan, B. E., Ertuğrul, S., & Dönmez, G. (2010). Effective bioremoval of reactive dye and heavy metals by Aspergillus versicolor. Bioresource *Technology*, 101(3), 870–876.

Tripathi, D. K., Tripathi, A., Shweta, Singh, S., Singh, Y., Vishwakarma, K., Yadav, G., Sharma, S., Singh, V. K., Mishra, R. K., Upadhyay, R. G., Dubey, N. K., Lee, Y., & Chauhan, D. K. (2017). Uptake, accumulation and toxicity of silver nanoparticle in autotrophic plants, and heterotrophic microbes: A Concentric Review. *Frontiers in Microbiology, 8*, 7. Doi: 10.3389/fmicb.2017.00007.

Truu, J., Truu, M., Espenberg, M., Nõlvak, H., & Juhanson, J. (2015). Phytoremediation and plant-assisted bioremediation in soil and treatment wetlands: A review. *The Open Biotechnology Journal, 9*, 85–92.

Williams, C. L., Neu, H. M., Gilbreath, J. J., Michel, S. L. J., Zurawski, D. V, & Merrell, D. S. (2016). Copper resistance of the emerging pathogen *Acinetobacter baumannii*. *Applied and Environmental Microbiology, 82*(20), 6174–6188. Doi :10.1128/AEM.01813-16.

Zhai, Q., Qu, D., Feng, S., Yu, Y., Yu, L., Tian, F., Zhao, J., Zhang, H., & Chen, W. (2020). Oral supplementation of lead-intolerant intestinal microbes protects against lead (Pb) toxicity in mice. *Frontiers in Microbiology, 10*, 3161. Doi: 10.3389/fmicb.2019.03161.

21 Insights into Pathways of Biodegradation of Endocrine Disrupting Chemicals by Microbes

Kulal Deekshitha and Shetty K. Vidya
National Institute of Technology

CONTENTS

21.1 INTRODUCTION

Endocrine disrupting chemicals (EDCs) are exogenous compounds that are known to alter the endocrine system by interfering with synthesis, secretion, storage, metabolism, receptor binding, or elimination of endogenous hormones (Diamanti-Kandarakis et al. 2009; Lauretta et al. 2019). Endocrine system through the secretion of hormones is responsible for regulating various sets of body functions and maintaining the relative stability of the internal environment. EDCs act like hormones and exert action through hormone receptors altering the hormonal and homeostatic system, potentially affecting the development, metabolism, reproduction, and growth (La Merrill et al. 2020). EDCs disturb hormonal regulation even at low concentration and pose a threat to humans and animals. They are heterogeneous and represent a broad class of molecules such as industrial solvents/lubricants and their by-products (polychlorinated biphenyls, polybrominated biphenyls, and dioxins), pesticides (dich lorodiphenyltrichloroethane (DDT)), fungicides (vinclozolin), plastics (bisphenol A (BPA)), plasticizers (phthalates), and pharmaceuticals (Lauretta et al. 2019). Many of these EDCs have widely been used; for example, BPA is used as a plasticizer in

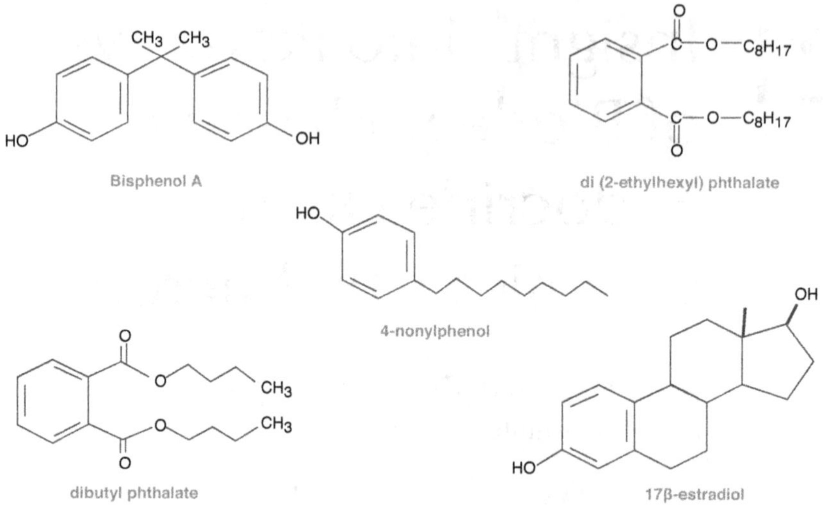

FIGURE 21.1 Structure of few endocrine disrupting compounds.

polycarbonate plastic and epoxy resin production; DEHP in food containers, flexible vinyl plastics, toys, packaging film, etc.; diethyl phthalate as a fragrance component in personal care products, housekeeping products, etc.; and nonylphenol as a degradation product of alkylphenol ethoxylates, which are used in detergents, paints, etc. (Jacksona and Sutton 2008). Figure 21.1 presents the structure of few endocrine disrupting compounds.

There are various ways of exposure to EDCs, including air, water, food, consumer products, or direct contact with EDCs. Aquatic organisms are adversely affected due to their continuous exposure to these EDCs in contaminated surface water. Further, these may also enter the food chain and accumulate in animal tissues up to humans (Lauretta et al. 2019). Due to their biologically active and toxic nature, it is extremely important to eliminate or reduce their concentration in the environment. Time of exposure of individuals to EDCs may have different consequences and might not be immediate, but may manifest later. Exposure to EDCs during sensitive developmental stages can cause detrimental effects on an individual (Diamanti-Kandarakis et al. 2009). Long-term harmful effects in glucose metabolism may be seen later in life in mothers exposed to BPA during pregnancy (Alonso-Magdalena et al. 2015). BPA exposure in pregnant mice affected pancreatic β-cell function, altered gene expression, and presented symptoms of diabetes associated with obesity in male offspring (Garcia-Arevalo et al. 2014). Feminization of native minnow in Oldman River, Canada, was seen due to the presence of EDCs that showed estrogen-like activity. EDCs caused altered gene regulation, which led to the female-biased ratio (Evans et al. 2012).

A study conducted by Jacksona and Sutton (2008) in urban wastewater, Oakland, showed the presence of phthalates, BPA, triclosan, and tris(2-chloroethyl) phosphate (TCEP) with phthalates being present in majority of the samples. Manickum and

John (2014) detected the presence of steroid hormones such as 17β-estradiol (E2), estrone (E1), and (17α-ethinylestradiol (EE2) in the effluent of activated sludge wastewater treatment plant in Africa, which indicates that sewage treatment plants are significant point source of EDCs. EE2 is found to be highly resistant to biodegradation when compared to E2, but the presence of both in surface water causes adverse effects to humans and aquatic life. Seasonal variation was also known to influence the level of steroid estrogens, particularly for E1 in treated water, which showed maximum removal in winter and the least in spring. Steroid estrogens are not completely removed during wastewater treatment due to the presence of benzene ring, and hence, they are largely found in the surface waters downstream of wastewater works (Manickum and John 2014). Due to their widespread presence in the environment, various treatment methods such as adsorption, membrane, coagulation/flocculation, and biological treatment were employed for the removal of EDCs. However, these treatment methods have some limitations such as EDCs being transferred from one phase to another without being degraded/mineralized as in adsorption and fouling, and sludge disposal issues associated with reverse osmosis/nanofiltration membranes. Treatment methods such as coagulation/flocculation exhibited low removal efficiency for BPA and nonylphenol (Choi et al. 2006). Advanced oxidation processes such as ozonation and photocatalytic degradation were known to show better removal efficiency compared to conventional treatment methods. Microbial-mediated degradation of the EDCs is a promising technology to eliminate these toxic pollutants from the environment (Chandra and Kumar 2017a, b). Moreover, these methods are cost-effective and environment-friendly.

Microbes have the ability to utilize these EDCs and metabolize it. Researchers have found the involvement of certain enzymes responsible for the degradation of EDCs. Enzymes such as 7α-hydroxysteroid dehydrogenase, laccase, peroxidase, monooxygenase, and dioxygenase were known to play a crucial role in the degradation of EDCs (Chen et al. 2017; Yu et al. 2007; Kresinova et al. 2012; Qiu et al. 2019). Various studies showed that bacteria (*Pseudomonas* sp., *Rhodococcus sp., Novosphingobium* sp., etc.) and fungi (*Gliocephalotrichum simplex, Pleurotus ostreatus*, etc.) can degrade the EDCs effectively. A few algal species were also studied for their ability to degrade EDCs. The mechanism of degradation of EDCs exhibits different pathways depending on the compound. In aquatic ecosystems, estrogen biodegradation occurs through 4,5-seco pathway with the formation of pyridinestrone acid (Chen et al. 2017). Degradation of phthalates may occur through hydrolysis of ester linkage (Lamraoui et al. 2020). Likewise, numerous authors have proposed different microbial degradation pathways for the breakdown of EDCs. In this chapter, a review on the microbes involved in the degradation of EDCs and their biotransformation/biodegradation pathways is presented.

21.2 MICROBIAL REMOVAL OF EDCS

Many research groups have identified the ability of microbes to transform harmful compounds to less toxic compounds (Chandra and Kumar 2015). Figure 21.2 shows the schematic representation of degradation of EDCs by microbes. Microbes either degrade these EDCs completely, or reduce their concentration in the environment.

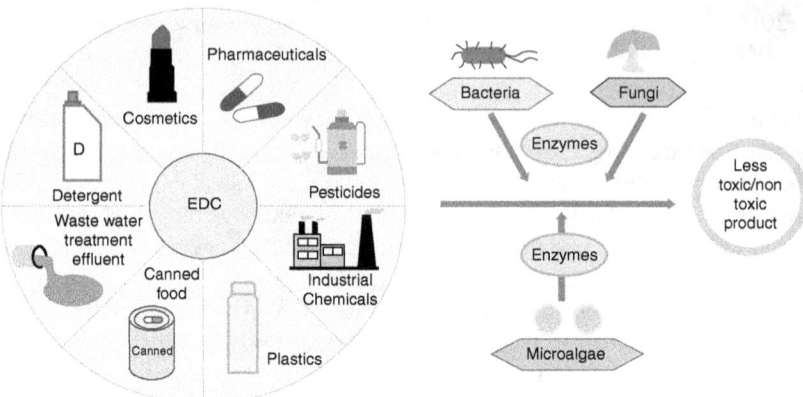

FIGURE 21.2   Schematic representation of microbial degradation of endocrine disrupting chemicals (EDCs).

21.2.1   BACTERIAL DEGRADATION OF EDCs

Bacteria have the ability to utilize these compounds and use them as the carbon source and detoxify them; hence, bacterial degradation has been an advanced technique in removing these compounds from the environment (Kumar and Chandra 2018). Lee et al. (2008) investigated the microbial degradation of endocrine disruptor vinclozolin and its toxic metabolite 3,5-dichloroaniline. Vinclozolin is a chlorinated fungicide used in the treatment of fungal infections. 3,5-Dichloroaniline, a degradation product of vinclozolin, is found to be more toxic and persistent than the parent compound. It was found that *Rhodococcus* sp. T1-1 strain isolated from pesticide-polluted agriculture soil was able to biodegrade both vinclozolin and metabolite 3,5-dichloroaniline with degradation ratios of 90% and 84.1%, respectively. Ruiz et al. (2013) isolated five bacterial strains capable of utilizing polyethoxylated nonylphenol as its carbon source, among which *Pseudomonas fluorescens* (strain Yas2) and *Klebsiella pneumoniae* (strain Yas1) were found to be efficient degraders under optimum conditions of pH 8 and temperature of 27°C. After 24 hours of incubation, COD values were reduced by 95% and 85% by *Klebsiella pneumoniae* and *Pseudomonas fluorescens*, respectively. Qhanya et al. (2017) reported the isolation and enrichment of six nonylphenol-degrading bacterial strains from different soil samples, which can utilize nonylphenol as its carbon source. This is the first report of endocrine disruptor nonylphenol-degrading bacteria from South Africa. From phylogenetic analysis, it was found that out of six bacterial isolates, four belonged to *Pseudomonas*, while the other two belonged to *Enterobacteria* and *Stenotrophomonas*. All six bacterial isolates showed around 41%–46% degradation of nonylphenol 12 hours after addition in broth cultures.

Toyama et al. (2013) found that *Phragmites australis* plant in association with its rhizobacteria is able to degrade phenolic endocrine disrupting chemicals in polluted water. *Sphingobium fuliginis* TIK1 and *Sphingobium* sp. IT4 bacterial strain present in the rhizosphere of *Phragmites australis* utilized root extract as the carbon source and were capable of degrading BPA, bisphenol B, bisphenol E, bisphenol F, bisphenol P, and bisphenol S (BPS). Further, batch reactor experiments showed that the *Phragmites* in association with TIK1 or IT4 were able to treat polluted secondary effluent water containing BPA, BPS, 4-tert-butylphenol, 4-tert-octylphenol, and 4-nonylphenol. It was found that the organic compounds exuded from the roots supported bacterial growth, which sustainably colonized the root surface and removed phenolic EDCs from polluted water. Suyamud et al. (2018) isolated a novel bacterium strain *Bacillus megaterium* ISO-2 from polycarbonate industrial wastewater and tested it against BPA. The bacteria not only degraded the BPA in synthetic water, but also were able to remove it from the wastewater. BPA degradation was studied using *Sphingobium* sp. YC-JY1 under optimum conditions (30°C temperature and 6.5 pH) by Jia and group (2020), where the strain was able to utilize the BPA as its carbon and energy source. Their study also showed significant inhibition in BPA degradation when NaCl at a concentration of 0.6%–1% was used.

Eltoukhy et al. (2020) isolated a strain of *Pseudomonas putida* YC-AE1 (gram-negative bacterium) from polluted soil, capable of degrading an endocrine disruptor BPA. The YC-AE1 strain showed complete degradation of 500 mg/L concentration of BPA within 72 hours. There are various factors affecting the biodegradation process, and optimization of these parameters is necessary to enhance the biodegradation process. Temperature is one of the most important parameters as it affects the growth rate of microbes, which in turn affects the enzyme activity. Maximum degradation of BPA was achieved at a temperature of 30°C and pH of 7.2. Further, inoculum size was also optimized to achieve better degradation. The optimum value of 2.5% inoculum size exhibited maximum degradation, and at values lower than the optimum, the degradation efficiency decreased due to lesser number of bacterial cells. Reduction in degradation was also found in inoculum size higher than the optimum value (2.5%), which could be due to the increased competition for nutrients and reduced dissolved oxygen. Furthermore, this bacterial strain could also degrade other BPA-related pollutants such as bisphenol B, bisphenol F, bisphenol S, dibutyl phthalate, diethylhexyl phthalate, and diethyl phthalate with greater degradation of bisphenol B and F when compared to bisphenol S, dibutyl phthalate and diethylhexyl phthalate. *Pseudomonas putida* YC-AE1 could survive both low and high concentrations of BPA and degrade it in a short time (200 mg/L in 20 hours), making it a promising bacterium compared with other reported strains.

Biodegradation of dibutyl phthalate (DBP) by a novel strain, *Methylobacillus* sp. V29b, was reported by Kumar and Maitra (2016). Their study reports the degradation of DBP in both minimal media and contaminated samples from landfill site, which are known to contain PAEs. Recently, Lamraoui et al. (2020) have reported a novel *Enterobacter* spp. YC-IL1 strain isolated from contaminated soil in Algeria, which has the ability to degrade di-(2-ethylhexyl) phthalate (DEHP) and a wide range of phthalic acid esters (PAEs). They have found that YC-IL1 strain showed significant degradation of DEHP within 7 days with optimum activity at neutral pH and temperature of 30°C. The effect of salinity stress was assessed, which was shown to negatively influence the degradation percentage of DEHP as the concentration of NaCl was increased up to 4%, which might be due to cell lysis. Furthermore, the strain also showed the ability to degrade other PAEs such as butyl benzyl phthalate and dicyclohexyl phthalate.

Several authors have studied the ability of bacteria to degrade the estrogens effectively. Fourteen strains of bacteria belonging to genera *Aminobacter* (strains KC6 and KC7), *Brevundimonas* (strain KC12), *Escherichia* (strain KC13), *Flavobacterium* (strain KC1), *Microbacterium* (strain KC5), *Nocardioides* (strain KC3), *Rhodococcus* (strain KC4), and *Sphingomonas* (strains KC8-KC11 and KC14) were isolated from activated sludge (Yu et al. 2007). All these strains were capable of converting 17β-estradiol to estrone and showed three degradation patterns. Strains KC6, KC7, and KC8 were capable of degrading estrone, of which only KC8 strain was able to utilize 17β-estradiol as the sole carbon source. KC8 strain exhibited non-specific monooxygenase activity and degraded 17β-estradiol and estrone within 3 days (Yu et al. 2007). Pauwels et al. (2008) for the first time studied the co-metabolism of 17α-ethinylestradiol using six bacterial strains when metabolizing estrone (E1), 17β-estradiol (E2), and estriol (E3). All the six strains of bacteria belonging to α-, β-, and γ-*Proteobacteria* were able to co-metabolize 17α-ethinylestradiol in the presence of 17β-estradiol, estrone, and estriol. The degradation of EE2 begins only after the degradation of E2. The initial concentration of E2 was known to be an important parameter in EE2 co-metabolic removal. The higher the ratio of E2/EE2, the higher is the removal of EE2. It was found that the degradation begins with the most accessible functional groups such as 17-hydroxylgroup of E2 or the 17-ketogroup of E1. Due to the steric hindrance by the ethinyl group, 17-hydroxylgroup is not accessible in EE2. The study also suggests the production of ATP during the breakdown of E1 and its metabolites (Pauwels et al. 2008). A swine (*Sus scrofa*) manure-borne bacteria was examined by Yang et al. (2010) for the degradation of testosterone, 17β-estradiol (E2), and progesterone. Testosterone, 17β-estradiol, and progesterone were degraded in 25 hours of incubation under aerobic condition. Breakdown of testosterone yields dehydrotestosterone, androstenedione, and androstadienedione with dehydrotestosterone being the major degradation product. Villemur et al. (2013) for the first time described the use of solid-liquid two-phase partitioning systems in the degradation of endocrine disruptors by enrichment culture. Three endocrine disruptor-degrading microbial enrichment cultures were used in solid-liquid two-phase partitioning systems that are loaded with estrone, estradiol, estriol, ethinylestradiol, nonylphenol, and BPA. These enrichment cultures were able to degrade all estrogen disruptors with the exception of ethinylestradiol. *Rhodococcus* isolate was found to significantly degrade estrone, estradiol, and estriol. Chen et al. (2017) for the first time studied the identification and characterization of catabolic genes and enzymes involved in the degradation of estrogen. They identified three catabolic gene clusters in *Sphingomonas* sp. strain KC8 and characterized 17β-estradiol dehydrogenase and 4-hydroxyestrone-4,5-dioxygenase involved in estrogen degradation. *oecB* in gene cluster I and *oecC* in gene cluster II were expressed during the aerobic growth on E2, and cluster III genes were expressed during the aerobic growth of strain KC8 on testosterone and E2. Flavin-dependent monooxygenase (estrone 4-hydroxylase) and extradiol dioxygenase (4-hydroxyestrone-4,5-dioxygenase) might be encoded by *oecB* and *oecC*, respectively. The product of *oecA* exhibited dehydrogenase activity to oxidize the C-17 hydroxyl group of both E2 and testosterone.

Another study conducted by Chen and group (2018) showed that *Novosphingobium* sp. strain SLCC isolated from the activated sludge of a sewage treatment plant in

Taiwan was the major estrogen-degrading Alphaproteobacteria. The strain showed aerobic growth in minimal medium with androst-4-ene-3,17-dione, E1, E2, cholic acid, progesterone, or testosterone as sole source of carbon and energy under the optimal growth condition of 28°C and pH 7.0. However, it couldn't utilize androsta-1,4-diene-3,17-dione, cholesterol, or 17α-ethinylestradiol (EE2) as its carbon and energy source. The degradation pathway followed 4,5-seco pathway with the production of pyridinestrone and the A/B-ring cleavage products. Li et al. (2018) demonstrated significant removal of 17β-estradiol (E2) using bacterial co-culture than incubating them individually. Two strains of bacteria, *Acinetobacter* and *Pseudomonas*, isolated from animal wastes removed ~98% of 17β-estradiol (5 mg/L) in 7 days of co-culture incubation, whereas around 77% and 68% of E2 degradation were observed with *Acinetobacter* and *Pseudomonas*, respectively, when incubated alone. Various factors such as C/N ratio, pH, carbon coexistence, and heavy metals were known to influence the bacterial growth and 17β-estradiol degradation. C/N ratio of 1:35 and pH range of 7–9 were found to be the optimum for bacterial growth and 17β-estradiol degradation. The coexistence of sodium acetate, sodium citrate, and glucose reduced E2 degradation in the initial 4 days, but more degradation of E2 was found on their depletion. Heavy metals such as Ni, Pb, Cd, or Cu at the concentrations of 0.8, 1.2, 1.6, or 0.8 mg/L, respectively, did not significantly decrease the growth of bacterial culture. Therefore, co-culture of bacteria may be more useful in the rapid removal of these contaminants (Li et al. 2018). An estrogen-degrading bacteria, *Acinetobacter* sp. DSSKY-A-001 isolated from soil samples showed good degradation ability in the removal of estrogen (Qiu et al. 2019). The degradation rate of estrogen was 90% after 6 days of incubation with three metabolic intermediates detected using HPLC and mass spectrometry. It was found that the enzymes catechol 1,2-dioxygenase, dioxygenase, and 7α-hydroxysteroid dehydrogenase were expressed in the strain and hence these enzymes along with six other enzymes play a crucial role in the degradation of estrogen. Khattab and group (2019) worked on the degradation of 17β-estradiol using different bacterial isolates, namely *Bacillus* sp., *Enterobacter* sp. I, *Enterobacter* sp. II, *Klebsiella* sp., *Aeromonas veronii*, and *Aeromonas punctata*. The degradation experiments were studied under aerobic and anaerobic conditions, of which aerobic condition favored faster degradation. The degradation was 100% with *Klebsiella* sp. under aerobic condition, which indicates the complete conversion of 17β-estradiol into carbon dioxide and water. Under anaerobic condition, a maximum degradation of 91.56% was obtained for *Bacillus* sp. The CO_2 production was known to increase with time with the highest CO_2 production generated using *Bacillus* sp. at 48 hours.

Studies on *Novosphingobium tardaugens* NBRC 16725 showed that the strain is capable of utilizing 17β-estradiol and testosterone. 3β/17β-Hydroxysteroid dehydrogenase is involved in the initial step of testosterone degradation. Genes of *edc* cluster encoding different enzymes take part in the different steps of degradation (Ibero et al. 2019, 2020).

Menashe et al. (2020) developed an innovative approach for the degradation of 17α-ethinylestradiol using the SBP (Small Bioreactor Platform) macro-encapsulation method. Two bacterial cultures, *Rhodococcus zopfii* (*R. zopfii*) and *Pseudomonas putida* F1 (*P. putida*), were encapsulated and studied for the degradation of

17α-ethinylestradiol. *R. zopfii* showed better degradation efficiency (86.5%) compared to *P. putida* (73.8%) after 24 hours when grown on minimal media. The degradation efficiency of 17α-ethinylestradiol decreased in the presence of additional carbon sources. However, this effect was not much seen in domestic secondary effluents and the degradation efficiency of 17α-ethinylestradiol was almost similar in both cases. Tian et al. (2020) in their study of estrogen degradation by *Rhodococcus equi* DSSKP-R-001 found that not only estrogen-degrading enzymes, but also several transporters and metabolism-related enzymes were responsible for the biodegradation of estrogen. Transcriptome analysis indicated the involvement of 720 (estrone), 983 (17β-estradiol), and 845 (17α-ethinylestradiol) genes, which showed significant expression along with differentially expressed genes.

21.2.2 FUNGAL DEGRADATION OF EDCs

Many fungal strains including ligninolytic fungi have been employed in the degradation of EDCs. White-rot fungi belong to basidiomyctes and are mostly associated with woody materials playing a role in lignin decomposition. Non-ligninolytic fungi colonize rapidly in the environment and can play a significant role in metabolizing chemicals in the environment. Cajthaml and group (2009) investigated the ability of eight ligninolytic fungal strains, namely *Irpex lacteus* 617/93, *Bjerkandera adusta* 606/93, *Phanerochaete chrysosporium* ME 446, *Phanerochaete magnoliae* CCBAS 134/I, *Pleurotus ostreatus* 3004 CCBAS 278, *Trametes versicolor* 167/93, *Pycnoporus cinnabarinus* CCBAS 595, and *Dichomitus squalens* CCBAS 750, in the degradation of 4-n-nonylphenol, technical 4-nonylphenol, BPA, 17a-ethinylestradiol, and triclosan. It was found that almost all strains of fungi were able to degrade EDCs with *I. lacteus* and *P. ostreatus* being the most efficient EDC degraders exhibiting degradation in 7 days. Another study on the degradation of 4-n-nonylphenol by a non-ligninolytic filamentous fungus was reported by Rozalska et al. (2010). The filamentous fungus *Gliocephalotrichum simplex* showed almost 88% degradation of 50 mg/L of 4-n-nonylphenol at the end of 24 hours (Rozalska et al. 2010). Krupinski et al. (2014) for the first time showed that the non-ligninolytic filamentous fungus *Aspergillus versicolor* is able to degrade nonylphenol co-metabolically and as the only growth substrate in the culture medium. The fungus was capable of utilizing 4-n-nonylphenol as the sole carbon and energy source and mineralized the compound completely after 3 days of incubation.

Kresinova et al. (2012) investigated the participation of intracellular and extracellular enzymes involved in the degradation of 17α-ethinylestradiol using the fungus *Pleurotus ostreatus*. Extracellular degradation may take place through manganese-dependent peroxidase and extracellular laccase activity, whereas mycelium-associated laccase-like activity and microsomal enzymes might be responsible for the intracellular degradation of 17α-ethinylestradiol. Further studies by Kresinova and group (2018) showed that the strain *Pleurotus ostreatus* HK 35 exhibited more than 76% EDCs (BPA, estrone, 17β-estradiol, estriol, 17α-ethinylestradiol, triclosan, and 4-n-nonylphenol) removal after 10 days in a trickle-bed reactor in a wastewater treatment plant. Muszynska et al. (2018) used the mycelium of an edible mushroom *Lentinula edodes* to degrade testosterone and 17α-ethinylestradiol.

17α-Ethinylestradiol was not detected in any samples indicating that the in vitro cultures of *L. edodes* cultivated in Oddoux liquid medium were able to degrade 17α-ethinylestradiol, whereas a small amount of undegraded testosterone was detected in one of the samples.

21.2.3 Degradation of EDCs by Microalgae

Sami et al. (2020) studied the role of catabolic enzymes in the degradation of estrone by cyanobacteria *Spirulina* CPCC-695. The strain utilized estrone as a sole carbon source and found that esterase, peroxidase, and laccase play a crucial role in degrading estrone and form pyruvate as a major product along with poly-β-hydroxybutyrate as a by-product. These enzymes help to bring about a cleavage in the ring structure of estrone, leading to its degradation. The maximum activity of these enzymes was at 20 mg/L estrone with laccase (55.4 U/L), peroxidase (24.0 U/L), and esterase (8.8 U/L). A higher concentration of estrone caused reduction in growth and also decrease in enzyme activity. Four marine microalgae were evaluated for their ability to degrade 4-nonylphenol (NP) in culture medium. The four microalgae, namely *Phaeocystis globosa*, *Nannochloropsis oculata*, *Dunaliella salina*, and *Platymonas subcordiformis*, were able to remove nonylphenol through biosorption, biodegradation, or biotransformation. *P. subcordiformis* was found to be highly tolerant to NP and found to be an excellent candidate for the treatment of NP (Wang et al. 2019). Freshwater microalgae such as *Chlamydomonas mexicana* and *Chlorella vulgaris* were able to bioaccumulate and biodegrade BPA with degradation rates of 24% and 23%, respectively, at 1 mg/L BPA. An increase in BPA concentrations led to a decrease in the degradation efficiency of the microalgae. *C. mexicana* exhibited better tolerance to BPA and could be used in bioremediation of BPA-polluted aqueous systems (Ji et al. 2014). Table 21.1 summarizes the degradation of EDCs by microorganisms.

21.3 BIODEGRADATION PATHWAYS OF EDCS BY MICROBES

Due to the toxicity and persistence of EDCs in the environment, it is important to study their mechanism of degradation by microbes. Several authors have proposed the degradation pathways based on the intermediates produced in the degradation process. Figure 21.3 presents the degradation mechanism of di-(2-ethylhexyl) phthalate by YC-IL1 strain (Lamraoui et al. 2020). Lamraoui et al. (2020) observed two intermediates during the breakdown of DEHP by *Enterobacter* spp. YC-IL1, i.e. mono-(2-ehtylhexyl) phthalate (MEHP) (m/z, 277) with a retention time of 2.60 minutes and phthalic acid (PA) (m/z, 164.9 and 121.0) with a retention time of 1.443 minutes. Hydrolysis of ester linkage occurs in DEHP between each alkyl chain and its aromatic ring, which leads to the appearance of MEHP at the fourth day of incubation period. It was further hydrolyzed to PA at ester bond, which appeared on fifth day and then disappeared due to its conversion into benzoic acid, finally utilizing it for cell growth. Four metabolic intermediates, dibutyl phthalate (DBP), monobutyl phthalate (MBP), phthalic acid (PA), and pyrocatechol (PC), were identified with the breakdown of DBP by *Methylobacillus* sp. (Kumar and Maitra 2016). The DBP levels decreased with time, whereas the MBP and PA levels increased with the highest

TABLE 21.1

Degradation of Endocrine Disrupting Chemicals by Microorganisms

Endocrine Disrupting Chemical (EDC)	Microorganism	References
Di-(2-ethylhexyl) phthalate (DEHP)	*Enterobacter* spp. YC-IL1	Lamraoui et al. (2020)
Bisphenol A (BPA)	*Sphingobium fuliginis* TIK1	Toyama et al. (2013)
Bisphenol B	*Sphingobium* sp. IT4	
Bisphenol E		
Bisphenol F		
Bisphenol P		
Bisphenol S		
Bisphenol A (BPA)	*Pseudomonas putida* YC-AE1	Eltoukhy et al. (2020)
Dibutyl phthalate (DBP)	*Methylobacillus* sp. V29b	Kumar and Maitra (2016)
Estrogen	*Novosphingobium* spp.	Chen et al. (2018)
Nonylphenol	*Pseudomonas, Enterobacteria, Stenotrophomonas*	Qhanya et al. (2017)
Vinclozolin and its metabolite 3,5-dichloroaniline	*Rhodococcus*	Lee et al (2008)
Estrogens	*Rhodococcus equi* DSSKP-R-001	Tian et al. (2020)
4-n-Nonylphenol	*Aspergillus versicolor*	Krupinski et al. (2014)
Bisphenol A	*Sphingobium* sp. YC-JY1	Jia et al. (2020)
Bisphenol A	*Bacillus megaterium ISO-2*	Suyamud et al. (2018)
17β-Estradiol	*Acinetobacter and Pseudomonas* co-culture	Li et al. (2018)
Polyethoxylated nonylphenol	*Pseudomonas fluorescens* (strain Yas2), *Klebsiella pneumoniae* (strain Yas1)	Ruiz et al. (2013)
17β-Estradiol	*Bacillus* sp., *Enterobacter* sp. I, *Enterobacter* sp. II, *Klebsiella* sp., *Aeromonas veronii, Aeromonas punctate*	Khattab et al. (2019)
17β-Estradiol	*Aminobacter* (strains KC6 and KC7), *Brevundimonas* (strain KC12), *Escherichia* (strain KC13), *Flavobacterium* (strain KC1), *Microbacterium* (strain KC5), *Nocardioides* (strain KC3), *Rhodococcus* (strain KC4), *Sphingomonas* (strains KC8-KC11 and KC14)	Yu et al. (2007)
Testosterone and 17α-ethinylestradiol	*Lentinula edodes*	Muszynska et al. (2018)
17α-Ethinylestradiol	*α-, β-,* and *γ-Proteobacteria*	Pauwels et al. (2008)
Estrogen	*Acinetobacter* sp. DSSKY-A-001	Qiu et al. (2019)
17α-Ethinylestradiol	*Rhodococcus zopfii* and *Pseudomonas putida*	Menashe et al. (2020)

(Continued)

TABLE 21.1 (*Continued*)
Degradation of Endocrine Disrupting Chemicals by Microorganisms

Endocrine Disrupting Chemical (EDC)	Microorganism	References
Estrone, estradiol, estriol, ethinylestradiol, nonylphenol, and bisphenol A	*Sphingomonadales* and *Rhodococcus*	Villemur et al. (2013)
17β-Estradiol	*Sphingomonas* sp. strain KC8	Chen et al. (2017)
17β-Estradiol	*Novosphingobium tardaugens* NBRC 16725	Ibero et al. (2020)
Testosterone	*Novosphingobium tardaugens* NBRC 16725	Ibero et al. (2019)
Testosterone	Manure-borne bacteria	Yang et al. (2010)
4-n-Nonylphenol, technically 4-nonylphenol, bisphenol A, 17a-ethinylestradiol, triclosan	*Irpex lacteus* 617/93, *Bjerkandera adusta* 606/93, *Phanerochaete chrysosporium* ME 446, *Phanerochaete magnoliae* CCBAS 134/I, *Pleurotus ostreatus* 3004 CCBAS 278, *Trametes versicolor* 167/93, *Pycnoporus cinnabarinus* CCBAS 595, *Dichomitus squalens* CCBAS 750)	Cajthaml et al. (2009)
4-n-Nonylphenol	*Gliocephalotrichum simplex*	Rozalska et al. (2010)
17α-Ethinylestradiol	*Pleurotus ostreatus*	Kresinova et al. (2012)
Bisphenol A, estrone, 17β-estradiol, estriol, 17α-ethinylestradiol, triclosan, and 4-n-nonylphenol	*Pleurotus ostreatus* HK 35	Kresinova et al. (2018)
Estrone	*Spirulina* CPCC-695	Sami et al. (2020)
Nonylphenol	*Phaeocystis globosa, Nannochloropsis oculata, Dunaliella salina*, and *Platymonas subcordiformis*	Wang et al. (2019)
Bisphenol A	*Chlamydomonas mexicana* and *Chlorella vulgaris*	Ji et al. (2014)

level of PC being present on the eighth day of incubation. The pathway suggests the conversion of DBP to MBP, which gets converted to PA. PA finally gets converted to carbon dioxide through the formation of PC (Kumar and Maitra 2016).

Eltoukhy et al. (2020) proposed two degradation pathways based on the compounds detected during the biodegradation of BPA by *Pseudomonas putida* YC-AE1 strain. They identified eight compounds, namely BPA (m/z 227.1), 4,4-dihydroxy-alpha-methylstilbene (m/z 225), p-hydroxybenzaldehyde (p-HBAL) (m/z 122), p-hydroxyacetophenone (p-HAP) (m/z 136), 4-hydroxyphenylacetate (HPA) and 4-hydroxyphenacylalcohol (same m/z, 152), 2,2-bis(4-hydroxyphenyl)-1-propanol and 1,2-bis(4-hydroxyphenyl)-2-propanol (same m/z, 244) and 2,2-bis(4-hydroxyphe

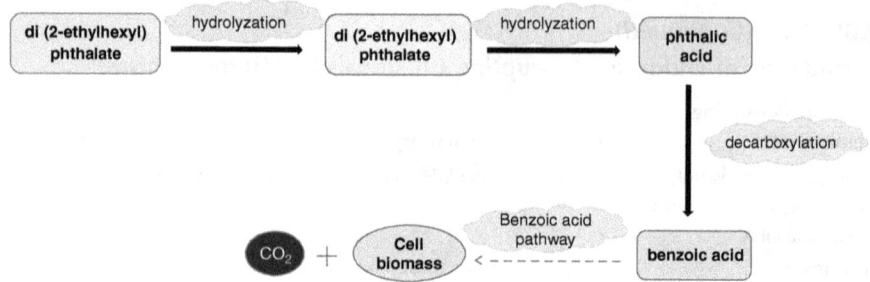

FIGURE 21.3   Degradation pathway proposed by Lamraoui et al. (2020) for di-(2-ethylhexyl) phthalate.

nyl)propanoate (m/z 258). Both the pathways show the same initial step of hydroxyl-ation of BPA to form 1,2-bis(4-hydroxyphenyl)-2-propanol and 2,2-bis(4-hydroxyph enyl)-1-propanol in pathways (I) and (II), respectively. The intermediate 1,2-bis(4-h ydroxyphenyl)-2-propanol produced in pathway (I) is dehydrated to 4,4-dihydroxy-a lpha-methylstilbene. This further gets oxidized to p-HBAL and p-HAP. p-HAP was metabolized to HPA and then to HQ. Both hydroxybenzoic acid (HBA) and HQ were assumed to be mineralized to carbon dioxide (CO_2) and bacterial biomass through benzoate degradation pathway. HBA and HQ were not detected and presumed in the pathway. In pathway (II), the compound 2,2-bis(4-hydroxyphenyl)-1-propanol is metabolized to form two products, 2,2-bis (4-hydroxyphenyl)propanoate and 2,3-bis(4-hydroxyphenyl)-1,2-propanediol. The latter compound (2,3-bis(4-hydroxyphenyl)-1,2-propanediol) finally forms HBA after mineralization through many steps, which further gets converted to CO_2 and bacterial biomass through benzoate degradation pathway.

The degradation of BPA by *Sphingobium* sp. YC-JY1 was known to take place in two routes (Jia et al. 2020). In one route, BPA was converted into 1,2-bis(4-hydr oxyphenyl)-2-propanol (1-BP), which is later converted to 4-hydroxybenzaldehyde (4-HBD) and 4′-hydroxyacetophenone (4-HAP) through the formation of 4,4′-dihydroxy-α-methylstilbene (4-DM). Both the metabolites 4-HBD and 4-HAP were utilized by the strain as the sole carbon source. In another route, degrada-tion occurs through the conversion of BPA to 2,2-bis(4-hydroxyphenyl)-1-propan ol (2-BP) and then to 2,3-bis(4-hydroxyphenyl)-1,2-propanediol (3,4-BP), with the latter being accumulated in the medium (Jia et al. 2020). Based on the interme-diates, 4-(2-hydroxypropan-2-yl) phenol, 4-isopropylphenol, 4-isopropenylphenol, benzoic acid, butanoic acid, propanoic acid, benzeneacetic acid, phenylethyl alcohol, 4-hydroxy-3-methoxybenzaldehyde, and phenolic compounds, two pathways were proposed by Suyamud et al. (2018). In pathway (I), the breakdown of BPA results in the formation of benzoic acid through the formation of 4-isopropylphenol and/or 4-isopropenylphenol obtained from carbocationic intermediates. Further, the benzoic acid gets converted into propanoic acid and butanoic acid. 4-(2-Hydroxypropan-2-yl) phenol was produced in pathway (II), which is assumed to be converted to 4-hydroquinone to give carbon dioxide and water (Suyamud et al. 2018).

Pauwels et al. (2008) suggested that conversion of E2 to E1 occurs as a consequence of oxidation of hydroxyl group at C-17β position of E2 to a ketone group, which is further converted to 4-(3a-methyl-3,7-dodecane-6H-cyclopentadiene[a] naphthalene-6-subunit)-2-methoxy-3-butenoic acid (R2) by oxygenase that cleaves the ring A of E1. R2 is unstable, and therefore, it is converted to (Z)-8-(7a-methyl-1n-5-dioxo-octahydro-1H-inden-4-yl)-2o-6-dioxy-4-butenoic acid (R3) by opening the B ring of R2, which is mediated by oxygenase enzyme. Yu et al. (2007) observed three degradation patterns (A–C) in the degradation of 17β-estradiol by the fourteen strains, namely Aminobacter (strains KC6 and KC7), Brevundimonas (strain KC12), Escherichia (strain KC13), Flavobacterium (strain KC1), Microbacterium (strain KC5), Nocardioides (strain KC3), Rhodococcus (strain KC4), and Sphingomonas (strains KC8-KC11 and KC14). Eleven strains showed degradation pattern A, where 17β-estradiol was converted to estrone. Degradation pattern B was exhibited by KC6 and KC7 strains, which showed the degradation of both 17β-estradiol and estrone with slower degradation rate in 17β-estradiol when compared to degradation pattern A. Further, only KC8 strain followed degradation pattern C with rapid degradation of both 17β-estradiol and estrone (Yu et al. 2007).

Chen et al. (2017) proposed that the degradation of E1 begins with the oxygenolytic breakdown of the A ring through the 4-hydroxylation and subsequent meta-cleavage reactions. However, the meta-cleavage product formed is unstable and undergoes abiotic recyclization to form pyridinestrone acid. Conversion of E2 to estrone is known to be mediated by 17β-estradiol dehydrogenase, which is converted to 4-hydroxyestrone by estrone 4-hydroxylase. 4-Hydroxyestrone-4,5-dioxygenase might be responsible for the opening of A ring to form ring-cleaved products, which on further reactions forms pyridinestrone acid. Pyridinestrone acid is known to have negligible effects on estrogenic activity. β-Oxidation enzymes might be responsible for the further opening of C, D rings. Figure 21.4 presents the degradation

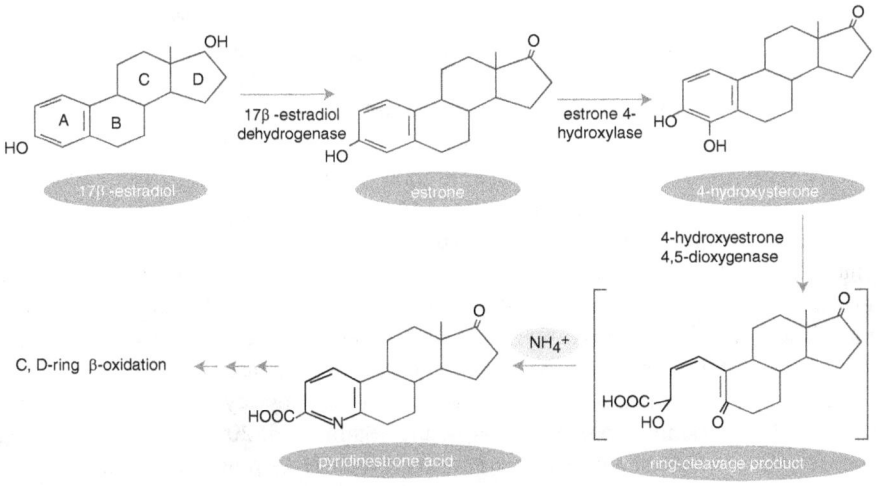

FIGURE 21.4 Degradation pathway of 17β-estradiol by *Sphingomonas* sp. proposed by Chen et al (2017).

pathway of 17β-estradiol by *Sphingomonas* sp. (Chen et al. 2017). Ibero and group (2020) suggested the transformation of 17β-estradiol to estrone is performed by 17β-hydroxysteroid dehydrogenase, which is followed by 4-hydroxylation of estrone by cytochrome P450 hydroxylase CYP450 encoded by *edcA* gene. The cleavage of A ring occurs after 4-hydroxylation by 4-hydroxyestrone-4,5-dioxygenase, which is encoded by *edcB* gene. Further, several reactions occur to form 3aα-H-4α(3-propanoate)7a-β-methylhexahydro-1,5-indanedione by the activity of β-oxidation-like enzymes. The study also showed the involvement of TonB-dependent receptor protein EdcT in estrogen uptake for the first time.

In the fungal degradation of 4-n-nonylphenol *by Gliocephalotrichum simplex*, Rozalska et al. (2010) proposed two pathways: One shows the detachment of carbon to yield 4-hydroxyphenylheptanoic acid in 6 hours of incubation. Further, the carbons were detached subsequently from the aliphatic chain to form 3-(4-hydroxyphenyl)propanoic acid at the 12th hour of cultivation. This metabolite is further converted to 4-(1-hydroxyvinyl)phenol, on a side route and on the main route to 2-(4-hydroxyphenyl)acetic acid and 4-hydroxybenzoic acid. In the second pathway, hydroxylation in the ninth (next to aromatic ring) position and carboxylation in the first (distal carbon) position of the nonyl-moiety occurs to form 9-hydroxy-9-(4-hydroxyphenyl)nonanoic acid. 4-Hydroxybenzoic acid was found as one of the main metabolites at the 6th hour of incubation, which confirms the detachment of linear chain. 9-Hydroxy-9-(4-hydroxyphenyl)nonanoic acid was converted to 8-hydroxy-8-(4-hydroxyphenyl)octanoic acid in a similar manner as in the first route. The removal of 4-n-NP in *Aspergillus versicolor* IM 2161 cultures yielded the following metabolites: 9-(4-hydroxyphenyl)nonanoic acid, 8-(4-hydroxyphenyl)octanoic acid, 7-(4-hydroxyphenyl)heptanoic acid, 6-(4-hydroxyphenyl)hexanoic acid, 5-(4-hydroxyphenyl)pentanoic acid, 3-(4hydroxyphenyl)propanoic acid, 4-hydroxybenzoic acid, and 3,4-dihydroxybenzoic acid. [The degradation of 4-n-NP proceeds through the formation of hydroxyalkyl phenols as a result of monohydroxylation of terminal carbon of the alkyl moiety, oxidation of the alcohols to carboxylic acids followed by the detachment of single-carbon fragments leading to the formation of 4-hydroxybenzoic acid (Krupinski et al. 2014). The degradation of testosterone and 17α-ethinylestradiol by *Lentinula edodes* occurs by the oxidation and cleavage of C and D rings of the steroids. Further oxidation of B ring was also observed, and to a lesser extent, ring A was also affected (Krupinski et al. 2014). Kresinova and group (2012) suggested the involvement of several enzymes to take part in degradation mechanisms, such as 17β-hydroxysteroid dehydrogenase in the formation of dehydrogenated EE2 intermediates and laccase in the oxidation of phenolic functional group of dioxo-E2. Methoxylation on estrone benzene ring might be due to the consequent methylation of hydroxyl derivatives, which is sequential, requiring hydroxylation by cytochrome P450 followed by the action of methyl transferases.

The degradation pathway in degrading estrone by cyanobacteria *Spirulina* CPCC-695 was found to be aerobic in nature (Sami et al. 2020) and meta-cleavage type, where the breakdown of estrone forms 4-hydroxyestrone. This product finally gets converted into pyruvate, which gets metabolized to carbon dioxide and water through TCA cycle. The other product acetate forms acetyl-CoA, which on condensation forms poly-β-hydroxybutyrate. The initial step of estrone degradation includes hydroxylation of A ring at the C-4 position followed by a cleavage in the ring structure, further undergoing keto-enol conversion to form 2-oxopent-4-enoate. This compound finally forms pyruvate and acetate (Sami et al. 2020).

21.4 CONCLUSIONS

EDCs are a great concern due to their widespread occurrence and persistence. This review suggests that microorganisms have great potential in the removal of EDCs in the environment. A variety of bacteria and fungi have been investigated for its role in the biodegradation of EDCs. Bacterial degradation is environment-friendly and cost-effective, among which *Acinetobacter, Pseudomonas,* and *Rhodococcus* genera are widely studied. Microbial enzymes are known to play a significant role in metabolizing and transforming these EDCs. Enzymes such as 17β-estradiol dehydrogenase and 4-hydroxyestrone-4,5-dioxygenase are known to cause the breakdown of estrogens by cleaving the ring structures. This review summarizes the mechanisms involved in the transformation of EDCs by bacteria and fungi. The knowledge about the biodegradation pathway of EDCs is very important to ensure that the degradation intermediates or the products are not more toxic than the parent pollutant. This knowledge would serve as one of the major decision factors in choosing a particular microorganism for the biodegradation of a target EDC and also in determining whether bioremediation is a favorable strategy of treatment. Most of the degradation processes of EDCs by microorganisms have produced non-toxic or less toxic compounds, hence proving it as an effective approach in the bioremediation of EDCs.

REFERENCES

Alonso-Magdalena, P., Garcia-Arevalo, M., Quesada, I. and Nadal, A. (2015) 'Bisphenol-A treatment during pregnancy in mice: A new window of susceptibility for the development of diabetes in mothers later in life', *Endocrinology* 156(5), pp. 1659–70 Doi: 10.1210/en.2014-1952.

Cajthaml, T., Kresinova, Z., Svobodova, K. and Moder, M. (2009) 'Biodegradation of endocrine-disrupting compounds and suppression of estrogenic activity by ligninolytic fungi', *Chemosphere* 75, pp. 745–50 Doi: 10.1016/j.chemosphere.2009.01.034.

Chandra, R. and Kumar, V. (2015). Biotransformation and biodegradation of organophosphates and organohalides. In: Chandra, R. (Ed.) *Environmental Waste Management.* Boca Raton, FL: CRC Press Doi:10.1201/b19243-17.

Chandra, R. and Kumar, V. (2017b). Detection of androgenic-mutagenic compounds and potential autochthonous bacterial communities during in-situ bioremediation of post methanated distillery sludge. *Frontiers in Microbiology* 8, p. 87 Doi: 10.3389/fmicb.2017.00887.

Chandra, R. and Kumar, V. (2017a). Detection of *Bacillus* and *Stenotrophomonas* species growing in an organic acid and endocrine-disrupting chemicals rich environment of distillery spent wash and its phytotoxicity. *Environmental Monitoring and Assessment* 189, p. 26 Doi: 10.1007/s10661-016-5746-9.

Chen, Y.-L., Fu, H-Y., Lee, T.-H., Shih, C.-J., Huang, L., Wang, Y.-S., Ismail, W. and Chiang, Y.-R. (2018) 'Estrogen degraders and estrogen degradation pathway identified in an activated sludge', *Applied and Environmental Microbiology* 84, pp. e00001–18 Doi: 10.1128/AEM.00001-18.

Chen, Y.-L., Yu, C.-P., Lee, T.-H., Goh, K.-S., Chu, K.-H., Wang, P.-H., Ismail, W., Shih, C.-J. and Chiang, Y.-R. (2017) 'Biochemical mechanisms and catabolic enzymes involved in bacterial estrogen degradation pathways', *Cell Chemical Biology* 24, pp. 712–24 Doi: 10.1016/j.chembiol.2017.05.012.

Choi, K. J., Kim, S.G., Kim, C.W. and Park, J.K. (2006) 'Removal efficiencies of endocrine disrupting chemicals by coagulation/flocculation, ozonation, powdered/granular activated carbon adsorption, and chlorination', *The Korean Journal of Chemical Engineering* 23(3), pp. 399–408.

Diamanti-Kandarakis, E., Bourguignon, J-P., Giudice, L. C., Hauser, R., Prins, G. S., Soto, A. M. and Zoeller, R. T. (2009) 'Endocrine-disrupting chemicals: An endocrine society scientific statement', *Endocrine Reviews* 30(4), pp. 293–342 Doi: 10.1210/er.2009-0002.

Eltoukhy, A., Jia, Y., Nahuriral, R., Abo-Kadoum, M.A., Khokhar, I., Wang, J. and Yan, Y. (2020) 'Biodegradation of endocrine disruptor Bisphenol A by *Pseudomonas putida* strain YC-AE1 isolated from polluted soil, Guangdong, China', *BMC Microbiology* 20(-11) Doi: 10.1186/s12866-020-1699-9.

Evans, J.S., Jackson, L.J., Habibi, H.R. and Ikonomou, M.G. (2012) 'Feminization of longnose dace (*Rhinichthys cataractae*) in the Oldman River, Alberta, (Canada) provides evidence of widespread endocrine disruption in an agricultural basin', *Scientifica* Article ID 521931 Doi: 10.6064/2012/521931.

Garcia-Arevalo, M., Alonso-Magdalena, P., Santos, J. R. D., Quesada, I., Carneiro, E.M. and Nadal, A. (2014) 'Exposure to bisphenol-A during pregnancy partially mimics the effects of a high-fat diet altering glucose homeostasis and gene expression in adult male mice', *PLoS One* 9(6), p. e100214 Doi: 10.1371/journal.pone.0100214.

Ibero, J., Galan, B., Diaz, E. and Garcia, J. L. (2019) 'Testosterone degradative pathway of Novosphingobium tardaugens', *Genes* 10 (871) Doi:10.3390/genes10110871.

Ibero, J., Galán, B., Rivero-Buceta, V. and Garcia, J.L. (2020) 'Unraveling the 17b-estradiol degradation pathway in *Novosphingobium tardaugens* NBRC 16725', *Frontiers in Microbiology* 11, p. 588300 Doi: 10.3389/fmicb.2020.588300.

Jacksona, J. and Sutton, R. (2008) 'Sources of endocrine-disrupting chemicals in urban wastewater, Oakland, CA', *Science of the Total Environment* 405, pp. 153–60 Doi: 10.1016/j.scitotenv.2008.06.033.

Ji, M-K., Kabra, A. N., Choi, J., Hwang, J-H., Kim, J. R., Aou-Shanab, R. A.I., Oh, Y-K. and Jeon, B-H. (2014) 'Biodegradation of bisphenol A by the freshwater microalgae *Chlamydomonas mexicana* and *Chlorella vulgaris*', *Ecological Engineering* 73, pp. 260–9 Doi: 10.1016/j.ecoleng.2014.09.070.

Jia, Y., Eltoukhy, A., Wang, J., Li, X., Hlaing, T.S., Aung, M.M., Nwe, M.T., Lamraoui, I. and Yan, Y. (2020) 'Biodegradation of bisphenol A by *Sphingobium* sp. YC-JY1 and the essential role of cytochrome P450 monooxygenase', *International Journal of Molecular Science* 21, 3588 Doi: 10.3390/ijms21103588.

Khattab, R. A., Elnwishy, N., Hannora, A., Mattiasson, B., Omran, H., Alharbi, O. M. L. and Ali, I. (2019) 'Biodegradation of 17-β-estradiol in water', *International Journal of Environmental Science and Technology* 16, pp. 4935–44 Doi: 10.1007/s13762-018-1929-y.

Kresinova, Z., Moeder, M., Ezechias, M., Svobodova, K. and Cajthaml, T. (2012) 'Mechanistic study of 17α-ethinylestradiol biodegradation by *Pleurotus ostreatus*: Tracking of extracelullar and intracelullar degradation mechanisms', *Environmental Science & Technology* 46, pp. 13377–85 Doi: 10.1021/es3029507.

Kresinova, Z., Linhartova, L., Filipova, A., Ezechias, M., Masin, P. and Cajthaml, T. (2018) 'Biodegradation of endocrine disruptors in urban wastewater using *Pleurotus ostreatus* bioreactor', *New Biotechnology* 43, pp. 53–61 Doi: 10.1016/j.nbt.2017.05.004.

Krupinski, M., Janicki, T., Palecz, B. and Dlugonski, J. (2014) 'Biodegradation and utilization of 4-n-nonylphenol by *Aspergillus versicolor* as a sole carbon and energy source', *Journal of Hazardous Materials* 280, pp. 678–84 Doi: 10.1016/j.jhazmat.2014.08.060.

Kumar, V. and Chandra, R. (2018). 'Characterisation of manganese peroxidase and laccase producing bacteria capable for degradation of sucrose glutamic acid-Maillard reaction products at different nutritional and environmental conditions', *World Journal of Microbiology and Biotechnology* 34, p. 32.

Kumar, V. and Maitra, S. S. (2016) 'Biodegradation of endocrine disruptor dibutyl phthalate (DBP) by a newly isolated *Methylobacillus* sp. V29b and the DBP degradation pathway', *3 Biotech* 6, p. 200 Doi: 10.1007/s13205-016-0524-5.

La Merrill, M.A., Vandenberg, L.N., Smith, M.T., Goodson, W., Browne, P., Patisaul, H.B., Guyton, K.Z., Kortenkamp, A., Cogliano, V.J., Woodruff, T.J., Rieswijk, L., Sone, H., Korach, K.S., Gore, A.C., Zeise, L. and Zoeller, R.T. (2020) 'Consensus on the key characteristics of endocrine- disrupting chemicals as a basis for hazard identification', *Nature Reviews Endocrinology* 16 Doi: 10.1038/s41574-019-0273-8.

Lamraoui, I., Eltoukhy, A., Wang, J., Lamraoui, M., Ahmed, A., Jia, Y., Lu, T. and Yan, Y. (2020) 'Biodegradation of Di (2-Ethylhexyl) phthalate by a novel *Enterobacter* spp. Strain YC-IL1 isolated from polluted soil, Mila, algeria', *International Journal of Environmental Research and Public Health* 17, p. 7501 Doi:10.3390/ijerph17207501

Lauretta, R., Sansone, A., Sansone, M., Romanelli, F. and Appetecchia, M. (2019) 'Endocrine disrupting Chemicals: Effects on endocrine glands', *Frontiers in Endocrinology* 10 (178) Doi: 10.3389/fendo.2019.00178.

Lee, J-B., Sohn, H-Y., Shin, K-S., Kim, J-S., Jo, M-S., Jeon, C-P., Jang, J-O., Kim, J-E. and Kwon, G-S. (2008) 'Microbial biodegradation and toxicity of vinclozolin and its toxic metabolite 3,5-dichloroaniline,' *The Journal of Microbiology Biotechnology* 18(2), pp. 343–9.

Li, M., Zhao, X., Zhang, X., Wu, D. and Leng, S. (2018) 'Biodegradation of 17β-estradiol by bacterial co-culture isolated from manure', *Scientific Reports* 8, p. 3787 Doi:10.1038 /s41598-018-22169-0.

Manickum, T. and John, W. (2014) 'Occurrence, fate and environmental risk assessment of endocrine disrupting compounds at the wastewater treatment works in Pietermaritzburg (South Africa)', *Science of the Total Environment* 468–469, pp. 584–97 Doi: 10.1016/j. scitotenv.2013.08.041.

Menashe, O., Raizner, Y., Kuc, M.E., Cohen-Yaniv, V., Kaplan, A., Mamane, H., Avisar, D. and Kurzbaum, E. (2020) 'Biodegradation of the endocrine-disrupting chemical 17α-ethynylestradiol (EE2) by *Rhodococcus zopfii* and *Pseudomonas putida* encapsulated in small bioreactor platform (SBP) capsules', *Applied Science* 10 Doi: 10.3390/ app10010336.

Muszynska, B., Zmudzki, P., Lazur, J., Kala, K., Sulkowska-Ziaja, K. and Opoka, W. (2018) 'Analysis of the biodegradation of synthetic testosterone and 17α-ethynylestradiol using the edible mushroom *Lentinulaedodes*', *3 Biotech* 8, p. 42 Doi: 10.1007/s13205-018-1458-x.

Pauwels, B., Wille, K., Noppe, H., De Brabander, H., de Wiele, T.V., Verstraete, W. and Boon, N. (2008) '17α-ethinylestradiol cometabolism by bacteria degrading estrone, 17β-estradiol and estriol', *Biodegradation* 19, pp. 683–93 Doi: 10.1007/s10532-007-9173-z.

Qhanya, L.B., Mthakathi, N.T., Boucher, C.E., Mashele, S.S., Theron, C.W. and Syed, K. (2017) 'Isolation and characterisation of endocrine disruptor nonylphenol-using bacteria from South Africa', *South African Journal of Science* 113(5/6), pp. 1–7.

Qiu, Q., Wang, P., Kang, H., Wang, Y., Tian, K. and Huo, H. (2019) 'Genomic analysis of a new estrogen-degrading bacterial strain, *Acinetobacter* sp. DSSKY-A-001', *International Journal of Genomics*, Article ID 2804134 Doi: 10.1155/2019/2804134.

Rozalska, S., Szewczyk, R. and Długonski, J. (2010) 'Biodegradation of 4-n-nonylphenol by the non-ligninolytic filamentous fungus *Gliocephalotrichum simplex*: A proposal of a metabolic pathway', *Journal of Hazardous Materials* 180, pp. 323–31 Doi: 10.1016/j. jhazmat.2010.04.034.

Ruiz, Y., Medina, L., Borusiak, M., Ramos, N., Pinto, G. and Valbuena, O. (2013) 'Biodegradation of polyethoxylated nonylphenols', *ISRN Microbiology* Article ID 284950 Doi: 10.1155/2013/284950.

Sami, N., Ansari, S., Yasin, D. and Fatma, T. (2020) 'Estrone degrading enzymes of Spirulina CPCC-695 and synthesis of bioplastic precursor as a by-product', *Biotechnology Reports* 26, p. e00464 Doi: 10.1016/j.btre.2020.e00464.

Suyamud, B., Inthorn, D., Panyapinyopol, B. and Thiravetyan, P. (2018) 'Biodegradation of bisphenol A by a newly isolated *Bacillus megaterium* strain ISO-2 from a polycarbonate industrial wastewater', *Water, Air, & Soil Pollution* 229, p. 348 Doi: 10.1007/s11270-018-3983-y.

Tian, K., Meng, F., Meng, Q., Gao, Y., Zhang, L., Wang, L., Wang, Y., Li, X. and Huo, H. (2020) 'The analysis of estrogen-degrading and functional metabolism genes in *Rhodococcus equi* DSSKP-R-001', *International Journal of Genomics*, Article ID 9369182 Doi: 10.1155/2020/9369182.

Toyama, T., Ojima, T., Tanaka, Y, Mori, K. and Morikawa, M. (2013) 'Sustainable biodegradation of phenolic endocrine-disrupting chemicals by *Phragmites australis*–rhizosphere bacteria association', *Water Science & Technology* 68, p. 3 Doi: 10.2166/wst.2013.234.

Villemur, R., dos Santos, S. C. C., Ouellette, J., Juteau, P., Lepine, F. and Deziel, E. (2013) 'Biodegradation of endocrine disruptors in solid-liquid two-phase partitioning systems by enrichment cultures' *Applied and Environmental Microbiology* 79(15), pp. 4701–11 Doi: 10.1128/AEM.01239-13.

Wang, L., Xiao, H., He, N., Sun, D. and Duan, S. (2019) 'Biosorption and biodegradation of the environmental hormone nonylphenol by four marine microalgae', *Scientific Reports* 9(5277) Doi: 10.1038/s41598-019-41808-8.

Yang, Y-Y., Borch, T., Young, R.B., Goodridge, L.D. and Davis, J.G. (2010) 'Degradation kinetics of testosterone by manure-borne bacteria: Influence of temperature, pH, glucose amendments, and dissolved oxygen', *Journal of Environmental Quality* 39, pp. 1153–60 Doi: 10.2134/jeq2009.0112.

Yu, C-P., Roh, H. and Chu, K-H. (2007) '17β-estradiol-degrading bacteria isolated from activated sludge', *Environmental Science & Technology* 41, pp. 486–49.

22 New Insights into Horizontal Gene Transfer among Bacterial Pathogens to Acquire Antibiotic Resistance and Culture-Independent Techniques to Study ARG Dissemination

Sanchita Das
CSIR-National Environmental Engineering
Research Institute (CSIR-NEERI)

Sukdeb Pal
CSIR-National Environmental Engineering
Research Institute (CSIR-NEERI)
Academy of Scientific and Innovative Research (AcSIR)

CONTENTS

DOI: 10.1201/9781003247883-22

22.1   INTRODUCTION

The consistently expanding of anti-microbial (AMR) obstruction experienced in pathogenic microbes are of colossal for public health sectors around the world, restricting treatment choices for bacterial diseases and subsequently decreasing clinical viability while expanding treatment expenses and mortality (Ben et al., 2019). With the absence of improvement of new anti-microbials, and expanding opposition even to final retreat anti-toxins (Nordmann et al., 2012), there is a need to ration the ones accessible. Protection from anti-toxins can happen either by changes or by obtaining obstruction giving qualities by means of flat quality exchange (HGT), of which the last is viewed as the main factor in the current ascent of AMR among bacterial microorganisms. The HGT of anti-resistance genes like-wise originates far before the creation and utilization of anti-toxins by people. For instance, OXA-type β-lactamase genes were observed to be plasmid-borne and mobile among bacterial species (Yang et al., 2018). Nonetheless, while anti-toxin obstruction and its spread by HGT are old instruments, the rate at which these cycles happen and the quantity of safe strains have expanded colossally in the course of recent many years due to particular tension through human anti-toxin use (Zaman et al., 2017).

To comprehend the scattering of antibiotic obstruction, it is important to map the resistome of different ecological niches and unwind how various niches affect as a repository for the dispersal of ARGs to bacterial microbes (Savoldi et al., 2018). Lately, there has been an increasing interest in this, as many experiments have uti-lized different strategies to test the resistome of conditions, for example, however not restricted to the human gut microbiota, soil, and wastewater (Tang et al., 2017). It has thus become evident that ARGs, including clinically important ones, are far and wide in such conditions. That commensals and the climate are significant repositories for obstruction is upheld by a few instances of ARGs on MGEs in human microorganisms that seem to have started from those supplies. A notable model of the bla CTXM qualities has turned into the most common reason for expanded range β-lactamases (ESBLs) in Enterobacteriaceae worldwide and a significant reason for scientific treatment issues. OXA-48-type carbapenem-hydrolysing β-lactamase qualities, which are progressively announced in enterobacterial species around the world, were likewise found to begin from the chromosomes of natural *Shewanella species*. Similarly, as with these couple of models, numerous clinically applicable opposition qualities are accepted to have begun from non-pathogenic microscopic organisms, featuring the tremendous capability of HGT for these microorganisms in defeating human utilization of anti-toxins. The resistome is separated into 'natural resistome or intrinsic' and 'extraneous resistome or obtained'. As Galán et al. referenced, in the European Committee on Antimicrobial Susceptibility Testing (EUCAST) master rules on anti-microbial helplessness testing, the 'natural resistome' is characterized as the arrangement of chromosomic qualities that partake in the intrinsic opposition, its essence in strains of a bacterial animal varieties is free to the past openness to anti-toxins, and it isn't identified with HGT. In addition, the 'extraneous resistome' is characterized as the arrangement of qualities procured with straightforward or different changes in the genome, which can be acquired in a steady way from one age to another or by means of HGT (Liu et al., 2019). This load of qualities from the outward resistome can be prepared by HGT components to pathogenic and non-pathogenic microorganisms from various conditions, including people and creatures. The assortment of these opposition qualities that exist in nature is known as resistome.

Local area episodes of irresistible sicknesses driven by HGT further outline the likely effects of HGT occasions in clinical settings. For instance, Shigella flare-ups arose in the UK because of the exchange of a plasmid-encoded ARG. The scattering of a plasmid-encoding azithromycin obstruction permitted beforehand low recurrence microorganisms to spread on the grounds that conventional anti-toxins were as of now not successful. At last, various flare-ups of various strains happened because of free obtaining of a similar plasmid (Lerminiaux et al., 2019). Comprehend the essential components and reasons of ARG spread in the climate. Numerous bacterial species have inborn components to battle anti-microbial build-ups present in their general climate (Zhen et al., 2019). It is important to decipher the genuine reason for ARG scattering and furthermore to consider the various potential occasions occurring being developed of ARGs and sequential flat quality exchange.

22.2 HORIZONTAL GENE TRANSFER (HGT) MECHANISMS

22.2.1 METHODS OF HGT

Various modes of mobile genetic element transfer among bacterial species are pictorially depicted in Figure 22.1.

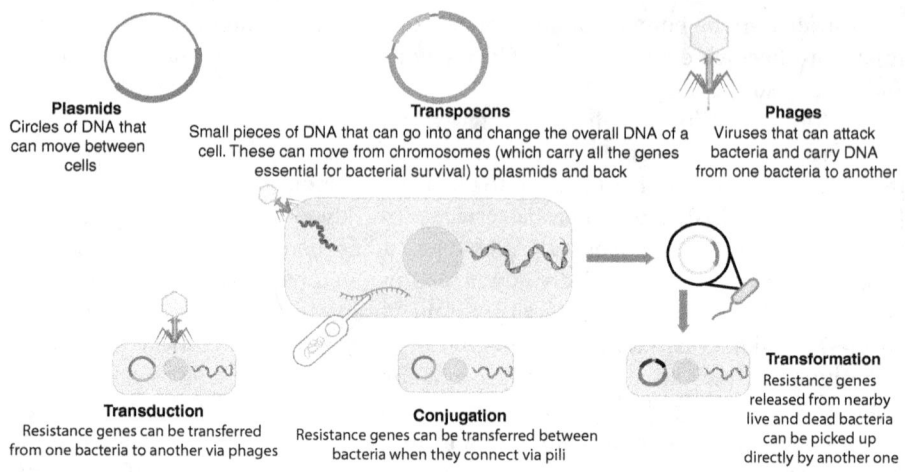

FIGURE 22.1   Various modes of mobile genetic element transfer among bacterial species.

22.2.1.1   Transformation

During transformation, free DNA from surrounding environment is absorbed by a competent bacterium consecutively inserted in its genome. The process of transformation relies on the bacteria being typically competent or transformable (Von Wintersdorff et al., 2016). Significant pathogens, comprising *Staphylococcus aureus, Pseudomonas aeruginosa*, and *Streptococcus pneumoniae* are among environmentally occurring effective bacteria that acquire antibiotic resistance via transformation (Sultan et al., 2018). Experiments revealed *Escherichia coli* could possibly be transmuted through extrachromosomal DNA in ingenuous conditions, proving *E. coli* can carry the plasmid DNA into the human gut through transformation, without contributing to the dissemination of ARGs (Yan et al., 2018).

22.2.1.2   Membrane Vesicles

Of late, a second method of development has been portrayed to clarify the varieties of ARGs isolated by single duplicates of IS26 generally found in opposition. Membrane vesicles are normally around 20–250 nm globular formed designs, for the most part created by gram-negative microorganisms when the external layer swells from the cell and is then delivered through narrowing. Film vesicles combine with their objective cells, in this manner conveying their freight. MVs are delivered in vitro through the human gut of commensal Bacteroides containing β-lactamases. These vesicles secure objective cells against β-lactam anti-microbials. With regard to HGT, MVs delivered through gut microorganisms could likewise contain cytoplasmic substance, such as DNA. DNA-holding MVs are believed to be framed through distension by both external and inward film, which motivates the consideration of cytoplasmic parts into the cysts. Without a doubt, film vesicles confined from microorganisms in the variety Acinetobacter can move anti-microbial obstruction plasmids in vitro. Essentially, vesicle-intervened move of DNA has been accounted for

E. coli, while MVs that are delivered in gut and might possibly impact have safe reactions (Yaffe et al., 2020).

22.2.1.3 Transduction

Transduction is the transfer of genetic material among bacteria through and intermediate virus such as a bacteriophage. Transduction mechanisms' types are generalized and specialized transductions. When in lytic cycle, a bacteriophage incorporates segments of host bacterial genome during capsid synthesis and it is called generalized transduction. In contrast, specialized transduction is characterized by excision and packaging of nearby regions flanking the integration site of a lysogenic phage in a capsid. In lateral transduction, prophages commence DNA replication being integrated with the host genome. The human gut is a breeding spot for a vast community of bacteriophages carrying significant amount of ARGs. The bacteriophage-harbouring ARGs are also present in other natural environments. The overflow of the phages carrying ARGs in human gut surges upon antibiotic consumption. Experiments suggest that transduction might contribute to the development of drug resistance in gut-colonizing *E. coli* strains and other gut bacteria (Ross et al., 2015).

22.2.1.4 Conjugation

MGEs such as plasmids and transposons are transferred during conjugation from one bacteria to another. Plasmids are proven to be the most pertinent for the dissemination of ARGs among MGEs class, as they possess multiple resistance genes because of their considerable size (~ 90 kbp). In addition, plasmids often carry other genes that contribute to bacterial fitness (resistance genes to disinfectants or heavy metals). ARGs henceforth are often co-selected under environmental pressure, since they inhabit genetically with other fitness determinants. Nevertheless, plasmids may provide mechanisms permitting the mobilization of DNA, thus highly boosting the potential for HGT of resistance genes (Bickhart et al., 2019).

22.3 MODES OF ANTIBIOTIC RESISTANCE DISSEMINATION IN THE ENVIRONMENT

Mobile genetic elements that promote dissemination of ARGs among intercellular genomic DNA are transposons (Tn), integrons (In), and insertion sequences (IS). IS and Tn are individual DNA segments that move randomly to different positions within the same or different genome inside a bacterial cell. On the other hand, In elements use site-specific recombination in order to shift ARGs between defined sites. The interplay between the MGEs reinforce the swift evolution of distinct multidrug-resistant pathogens in the environment (Podell et al., 2007). Mobile genetic elements bearing ARGs and their individual characteristics are presented in Table 22.1.

22.3.1 Insertion Sequences

Among various IS elements, IS6, IS26, and IS*1216 family elements* have played a very important role in the spreading of ARGs among gram-positive and gram-negative

TABLE 22.1

Mobile Genetic Elements Bearing ARGs and Their Individual Characteristics

Mobile Genetic Element	Characteristic	Example of Resistance Determinants	Reference
Plasmid	Self-transmissible or mobilizable. Variable range (0.5 to >110 kb)	Plasmid pP2G1 harbours ARGs conferring resistance to distinct antibiotics	Marti and Balcázar (2012)
Insertion sequence	Consists of inverted repeats at the terminal and encodes a transposase. Small (size less than 2.5Kb)	IS18 transmits gene overexpressing blaOXA-type genes	Higgins et al. (2013)
ISCR elements	DNA replication through rolling circle transposition through transposase enzyme activity	ISCR1 mediates the mobility of blaCTX-M genes	Sun et al. (2018)
Transposon	Encodes transposase along with other functional genes; large (more than 10 Kb) flanked by IS or inverted repeats	Tn1 and Tn3 render resistance to beta-lactam resistance genes	Alekshun et al. (2007)
Integron	Encodes for integrase gene and other transcriptional elements. Mediates the capture and expression of gene cassettes	Class 1 integrons harbour various gene cassettes rendering multidrug resistance	Marti and Balcázar (2012)
Genomic island	Very large mobile regions in genomic DNA encoding for proteins used in complex biological functions	SGI1 harbours ARGs of spectinomycin, streptomycin, beta-lactams, and macrolides	Levings et al. (2005)
Integrating conjugative elements	MGEs are integrated in the genomic DNA and travel through conjugation method	ICEVchHai1 harbour various clinically relevant ARGs	Karkey et al. (2018)
Bacteriophages	Bacterial viruses that infect and their spread lets the DNA to transfer from one bacterium to another	Many beta-lactams, aminoglycosides, and quinolone resistance genes have been found among isolated environmental phages	Colomer-Lluch et al. (2011)

bacteria. Positioning the mentioned IS was initially shown to take place by replicative transposition (Press et al., 2017). Recently, a second method of development has been portrayed to clarify the varieties of ARGs isolated by duplicates of IS26 generally found in plasmids. The unit of versatility comprises 1 duplicate of IS26 and adjoining locale (which could be up to the following IS26 intersection) and termed as 'translocatable unit' (TU). TU specially embeds close to a current duplicate of IS26 in a beneficiary atom through a moderate interaction (no replication of IS26 and production of TSD; however, any TSD previously flanking the objective IS26 is saved), creating the equivalent cointegrate structure made by replicative interpretation.

Significantly, this cycle, which is reliant upon the IS26 transposase (Tnp26) and is recA-free, has exhibited a recurrence greater than multiple times of untargeted replicative rendering. It signifies that once a chromosome/plasmid has a duplicate of IS26, it was inclined to procure further adjoining IS26 TU (Blokesch et al., 2016).

22.3.2 TRANSPOSONS

Transposons are thought as materials bigger than insertion sequences, which include transposase genes and an internal 'passenger' gene encoding ARGs. ARGs are frequently found overlapping with Tn3 family transposons, and Tn3 family members mainly consist of 38-bp terminal IR, along with IRR and IRL sequences towards the transcription direction of transposase gene tnpA. Transposition takes place through a mechanism where tnpA catalyses a sequence of cointegrate structures with direct repeat copies of transposon splitting the original donor molecules from the recipient ones.

Another transposon family Tn7 consists of members bearing ARGs, and examples are Tn402 and Tn7 elements in various Tn552 and gram-negative bacteria (GNB) in *Staphylococcus* species (Liu et al., 2015). The elements share the same features, but bear distinct transposition mechanisms. Also, these transposons usually target a particular site (Peterson et al., 2018).

22.3.3 INTEGRONS

Transposon Tn402 has resulted from the acquisition of intI1, which is frequently found on the chromosomes of Betaproteobacteria. As the reports suggests, class 1 integrons, which are a part of tni region, have been replaced by three conserved segments. INTEGRALL (123; http://integrall.bio.ua.pt/) presently keeps vault of Intergron (In) numbers; however, these truly compare just to various cassettes. Purported 'complex' class 1 integrons, normally with fractional duplications of three CS, made by the addition of circles comprise ISCR1 and a related opposition quality by recombination into three CS or a current ISCR1 component. Class 2 integrons, related to variations and Tn7, frequently have a non-functional IntI2 quality because of an interior stop codon and, likely as an outcome, house a restricted assortment of tapes. Class 3 integrons are more like class 1 integrons and, furthermore, have all the earmarks of being related to Tn402-like transposons. A couple of models have been distinguished, generally conveying tapes that encode beta-lactamases. Class 4 has recently been used to allude with an integron found inside *Vibrio cholerae* chromosome. Despite the fact that tapes containing opposition qualities make up a minority of those in SCI, they seem, by all account, to be the wellspring of tapes found in 'portable' integrons. 'Versatile' integron types, presently assigned class 4 and class 5 integrons, have all the earmarks of being uncommon and have not been recognized (Qian et al., 2016).

A diverse range of gene cassettes bearing ARGs have been identified. The ones considered to be most clinically relevant are the ones encoding aminoglycoside or beta-lactamases resistance genes. The genes that are cassette-borne also encode

ARGs that are ESBL or carbapenemases and Bla$_{OXA}$ genes. Variants of aacA4/aac (6=)-Ib usually confer hindrance to amikacin, gentamicin, and tobramycin and are also sometimes found to confer resistance to fluoroquinolones. Various gene cassettes (dfrA12, aadA5, and dfrA17) conferring hindrance to trimethoprim, streptomycin, and spectinomycin are frequently found in class 1 integrons (Lester et al., 2006).

22.3.4 PLASMIDS

ARGs in *Enterococci* sp. are mainly encoded by theta replicating plasmids. These plasmids are mostly subdivided into Rep3, Inc18, and RepA families. In *Enterococci* spp., plasmids belonging to Inc18 class often confer hindrance towards MLS antibiotics (ermB), but can potentially confer hindrance to other antibiotics as well. These plasmids widely contribute resistance towards vancomycin (vanA) (Stalder et al., 2017). Furthermore, Inc18 plasmids were the first plasmids that were identified to harbour the cfr gene, which gives resistance towards antibiotics belonging to classes such as phenicols, lincosamides, and oxazolidinones. Many studies have reported the occurrence of ARGs on resistance islands in *Enterobacteriaceae* and *P. aeruginosa* species. A PCR-based replicon typing (PBRT) scheme 'AB-PBRT', based on 18 plasmid sequences has been proposed.

22.4 BACTERIAL EFFLUX PUMPS AND THEIR ROLE IN ANTIBIOTIC RESISTANCE

Horizontally acquired ARGs are among the major elements involved in the development of multidrug-resistant bacterial species. However, it has recently been discovered that the upgraded expression of efflux system genes plays a pivotal role in multidrug resistance. Antibiotics liberated through the cell membrane led to a reduction in its build-up and an increase in its MICs (minimum inhibitory concentrations). Henceforth, efflux pumps result in the ejection of drugs from cells, which decreases antibiotic accumulation and surges the MICs. All efflux pumps contain three important sections: the inner membrane transporter, the periplasmic lipoprotein, and the outer membrane channel (Podnecky et al., 2015). Classes of efflux pumps related to antibiotic resistance include resistance-nodulation-cell division (RND), major facilitator superfamily (MFS), multidrug and toxic compound extrusion (MATE), and small multidrug resistance (SMR) families. Among these classes, RND-type and AdeABC efflux pumps are only related to aminoglycoside resistance, but also render resistance to other such antibiotic classes, for example chloramphenicol, tigecycline, tetracycline, and macrolides. Efflux pumps outstandingly remove commonly applied antibiotics on the cell during the therapy of infections caused by bacterial pathogens. Several studies suggest the efflux pumps often confer lesser susceptibility towards antibiotics. Efflux pumps are usually specific for a single substrate and sometimes for a range of different chemical compounds or antibiotics. These pumps that transport varied elements may possibly be related to MDR. Several efflux pumps are clinically significant since they possibly lessen the bacterial infection. Since these pumps are controlled by other nearby genes, any type of variations in the expression profile of those genes might possibly disrupt the efflux pumps activity. Many of these genes are generally carried by the MGEs and are acquired by nearby bacterial cells.

22.5 CO-SELECTION OF ARGS AND MRGS IN BACTERIAL PATHOGENS

It has been suggested that antibiotic sequestration could be a potential resistance strategy for acquiring antibiotic resistance. Nevertheless, evidences suggest that functional and structural attributes of antibiotic resistance are familiar to those of metallic resistance (Bengtsson-Palme & Larsson, 2016). Bacterial vulnerability to heavy metals predates is nothing new, and heavy metal (HM) contamination by anthropogenic activity is responsible for the contamination of the environment (Pal et al., 2017). A number of researchers propose that HM pollution in environmental niches might have a pivotal role in the proliferation and dissemination of antibiotic resistance. It is of particular concern since the anthropogenic extent of HM is currently several folds higher than the antibiotic levels. Additionally, other toxic contaminants are also co-selected along with ARGs, including detergents and quaternary ammonium compounds (Pal et al., 2017). Co-selection happens through three variant mechanisms: co-regulation, cross-resistance, and co-resistance. The co-selection of metal-associated antibiotic resistance in bacteria happens either by co-resistance or by cross-resistance. Metal-accompanied ARGs found in bacteria that are present on the same MGEs are horizontally transferred to remotely related human bacterial pathogens (Li et al., 2015). ARGs and MRGs identified in various microbes are listed in Table 22.2.

Co-resistance takes place when the ARGs and MRGs are positioned alongside on a single MGE such as an integron, a transposon, or a plasmid. The physical gene linkage thus results from the co-selection of other genes present on the same MGE. Summers et al. demonstrated the genetic linkage between ARGs and mercury when resistance genes for mercury were found to co-transfer in subsets of Enterobacteriaceae matings with other bacterial species. Studies also reported traits of genes encoding beta-lactamases (blaCTX-M) and copper and silver resistance genes linked on plasmids. An *S. aureus* plasmid named pAFS11 bearing methicillin resistance genes (MRSA) also featured co-occurrence with various other ARGs (macrolides, tetracycline, and trimethoprim) and MRGs for copper and cadmium. Resistance genes for cadmium and zinc were found to be linked and present on the same plasmid along with ARGs for macrolides and aminoglycosides (Pal et al., 2017).

Another possible mechanism involved in the co-selection of antibiotic resistance is cross-resistance, which takes place when different antibiotic classes attack a similar target, initiating a common cell death pathway. For example, an MDR pump in *Listeria monocytogenes* exports metals along with antibiotics. Recently, it has been found that a membrane-oriented DsbA–DsbB system in *Burkholderia cepacia* showed that the system is associated with the formation of a multidrug resistance system along with a metal efflux pump. Experiments with mutants lacking a DsbA–DsbB system were found to show lower resistance to several metals, and antibiotics, including zinc, cadmium, β-lactams, erythromycin, ofloxacin, novobiocin, sodium dodecyl sulphate, and kanamycin.

TABLE 22.2

Antibiotic Resistance Genes (ARGs) and Metal Resistance Genes (MRGs) Identified among Various Microbes

Host Organisms	ARG	MRG	MGE	References
Salmonella species	Beta-lactams	Copper resistance gene	Plasmid	Ahmed et al. (2018)
Enterococcus faecalis	Norfloxacin	Zinc resistance gene	Plasmid	Garbisu et al. (2018)
Escherichia coli	Quinolones	Zinc resistance genes	Plasmid	Bengtsson-Palme et al. (2016)
Acinetobacter baumannii	Beta-lactams	Mercury, arsenic resistance genes	Plasmid	Baker-Austin et al. (2006)
Pseudomonas aeruginosa	Aminoglycosides	Nickel, iron resistance genes	Plasmid	Pal et al. (2017)
	Tetracycline, chloramphenicol	Copper resistance genes	Plasmid	Pal et al. (2017)
Salmonella maltophilia	Erythromycin	Cadmium resistance genes	Plasmid	Gillings et al. (2017)
Salmonella spp.	Streptomycin	Cadmium resistance genes	Plasmid	Zhang et al. (2017)
Proteobacteria	Ampicillin	-	Tn1	Zhang et al. (2016)
Escherichia coli	Kanamycin, Neomycin	Mercury	Tn903, Tn5, Tn6	Martinez et al. (2015)
Proteobacteria	Tetracycline	Copper	Tn10	Staehlin et al. (2016)
Escherichia coli	Chloramphenicol	-	Tn9	Bengtsson-Palme (2016)

22.6 MOLECULAR TECHNIQUES USED TO STUDY HORIZONTAL GENE TRANSFER IN BACTERIAL COMMUNITY

22.6.1 MOLECULAR EXPERIMENTAL TECHNIQUES

22.6.1.1 Flow Cytometry Using gfp-Encoding Protein

Sorensen et al. demonstrated a novel culture-independent technique for examining collateral gene transfer among bacterial species. The technique was established on the basis of a direct detection method and transconjugant bacterial cell count via flow cytometry. Recognition of transconjugants was acquired through plasmid with *gfp* tag, which is a reporter gene that encodes for a green fluorescent protein. GFP repressed expression among donors was achieved by a repressor encoded chromosomally (*lacI*^{q1}). Donor transconjugant cells were enumerated after GFP expression through adding isopropyl-thio-B-D-galactoside inducer (IPTG). This method helped to achieve easy and accurate quantification of HGT among *Pseudomonas putida* and *Escherichia coli* strains (Sørensen et al., 2003).

22.6.1.2 Population Genetic Model Prediction Method

Surveillance techniques have failed to monitor HGT events that occur between bacterial population present in soil and transgenic plants. Nevertheless, a population-based genetic model helps predict whether bacterial transformants acquire transgenes that require a long duration of growth to mimic wild bacteria. Population genetic approach has been developed for interpreting the essential sample size and numbers of sampling for monitoring HGT events among huge bacterial populations. Some alterations in present monitoring techniques are required, the bacterial generation time, consideration of the population size of exposed bacteria, the sample size necessary to verify or falsify the HGT hypotheses tested, and the strength of selection acting on the transgene-carrying bacteria (Nielsen et al., 2004).

22.6.1.3 MetaCHIP

Metagenomic datasets impart a chance to study HGT at microbial community level. MetaCHIP refers to a pipeline that provides reference-independent HGT study opportunity at the community level. On investigation, the MetaCHIP revealed that it can forecast HGTs even with various genetic divergence through metagenomic datasets. Outcomes of experiments performed by Song et al. showed that the identification of recent gene transfers from metagenomics datasets is greatly influenced by the read assembly step (McInnes et al., 2020). A comparison of MetaCHIP with a former analysis on bacteria present in soil indicated a higher level of prediction for recent HGT events. An analysis of various environmental metagenomic data with MetaCHIP confirmed the pivotal role of HGT in the spread ARGs in the various environment (McInnes et al., 2020). Additionally, tests reveal that functions subjected to energy production, conversion along with carbohydrate transport, and metabolism are often transferred among free-living microbes (Song et al., 2019).

22.6.1.4 MetaCherchant

MetaCherchant is an algorithm that is built for the extraction of genomic data from environmental metagenomic samples in order to identify and locate ARGs and produce data in the form of a graph. Olekhnovich et al. validated this algorithm through various simulated and published datasets; its application to new 'shotgun' metagenomes also helped reveal its efficiency (Olekhnovich et al., 2018). Also, the genomic context was restructured for many resistance genes. Taxonomic annotations reveal that inside a metagenome, the obstruction qualities can be contained in genomes of numerous species. MetaCherchant permits the reproduction of portable hereditary components with obstruction qualities inside the genomes of microbes utilizing metagenomic information. Utilization of MetaCherchant in differential mode created explicit chart structures proposing the proof of conceivable obstruction quality transmission inside a portable component that happened because of the anti-toxin treatment (Bridier et al., 2019). MetaCherchant is a promising apparatus offering specialists a chance to get an understanding into elements of obstruction transmission in vivo basis on metagenomic information (Olekhnovich et al., 2018).

22.6.1.5 Computational Binning Using DNA Methylation

Shotgun metagenomics techniques empower portrayal of microbial networks in human microbiome and natural environments. Get together of metagenome groupings doesn't yield entire genomes, so computational hiving strategies have been formed to bunch successions into genome 'receptacles'. Following techniques exploit species wealth, grouping creation, or chromosome association, yet can't completely recognize firmly related species and strains. Beaulaurier et al. introduced a binning technique that consolidates bacterial DNA methylation marks, which are distinguished utilizing single-atom ongoing sequencing. This strategy exploits these endogenous epigenetic standardized identifications to determine individual peruses and gathered contigs into species- and strain-level containers. Approval of this technique was finished utilizing manufactured and genuine microbiome arrangements. Notwithstanding genome binning, they showed that the strategy joins plasmids and other portable hereditary components into their host species in a genuine microbiome test. Fusing DNA methylation data into shotgun metagenomics investigations will supplement existing techniques to empower more precise grouping binning (Beaulaurier et al., 2018).

22.6.1.6 In vivo Experimental Studies

Transient colonization by resistant enterococci species to vancomycin beginning has been reported in the digestion tracts of people. Nonetheless, little is thought concerning whether the mobility of the vanA quality happens in the human digestive tract. Lester et al. included six isolates of vancomycin resistant *Enterococcus faecium* belonging to chicken gut, along with a vancomycin non-resistant *E. faecium* from of human gut. Transconjugants were recuperated in three of six strains. In one strain, not exclusively was vancomycin opposition moved, yet additionally quinupristin-dalfopristin obstruction. This review shows that the exchange of the vanA genes from an *E. faecium* strain beginning to an *E. faecium* residing in human gut can happen in the digestive organs of people (Jahan et al., 2015). It proposes that transient digestive colonization by enterococci conveying portable components with opposition qualities addresses a danger for the spread of obstruction qualities to other enterococci that are essential for the human native verdure, which can be liable for diseases in specific gatherings of patients (Lester et al., 2006).

22.6.1.7 Culturomics Using MALDI-TOF

Culturomics is a culturing method where multiple culture conditions are used. Culturomics using MALDI-TOF mass spectrometry is one of them where along with MALDI-TOF, 16S rRNA sequencing for species identification is used. Gut microorganisms have a pivotal role in maintaining the health and immune system of humans. This was understood after the development of omics tools and its applications to improve our understanding about the gut microbial diversity. Nevertheless, metagenomics has unravelled the microbial diversity of the microorganisms present in gut and has also highlighted that most of the bacteria in the gut remain uncultured. Hence, culturomics was created to culture and recognize obscure microscopic organisms that occupy the human gut as a piece of resurrection for culture methods in microbial science. Comprising different culture conditions joined with fast-distinguishing proof of microscopic organisms, the culturomics approach has empowered the way of life for many new microbiota that are related to people, giving invigorating new viewpoints on having microbes connections (Lagier et al., 2018).

22.6.1.8   Fluorescent Reporter Systems

Conjugated plasmids can give organisms full supplements of new qualities and establish powerful vehicles for even quality exchange. These plasmids are considered responsible for the quick spread of ARGs opposition among microorganisms. While wide host range plasmids are known to move to assorted hosts in unadulterated culture, the degree of their capacity to move in the complex bacterial networks present in many living spaces has not been exhaustively considered. Klumper et al. segregated and described transconjugants with a level of affectability not recently acknowledged to research the exchange scope of IncP- and IncPromA-type expansive host range plasmids from three proteobacterial donors to a microbial contaminated area. They recognized exchange to a wide range of beneficiaries having a place with 11 unique bacterial phyla. The predominance of transconjugants having a place with assorted gram-positive *Firmicutes* and *Actinobacteria* recommends that plasmid movement between gram-positive strains containing IncP1- and IncPromA-type plasmids is a successive marvel. The plasmid getting parts of the local genetic area were both plasmid and benefactor subordinate, it was observed that a centre portion highly resistant could take up various plasmids from assorted donor strains. This division, involving 80% of the distinguished transconjugants, along these lines can possibly overwhelm IncP- and IncPromA-type plasmid transfer in soil. Their outcomes exhibit that these wide host range plasmids have an unnoticed potential to move promptly to extremely assorted microbes and can, subsequently, straightforwardly interface huge extents of the dirt bacterial genetic stock. This finding builds up the developmental and clinical meanings of these plasmids (Klümper et al., 2015).

22.6.1.9   epicPCR

Spencer et al. displayed epicPCR by sequencing 16S rRNA genes from cells bearing the dissimilar sulphate reductase gene *dsrB* through detection of sulphate-reducing cell types among the microbial community of a lake. They affirmed that the noticed phylogenetic appropriation of dsrB qualities matches forecasts dependent on the noticed lithospheric chemistry, while additionally uncovering beforehand undetected putative sulphate reducers. The effectiveness of microbial cell rupture can be estimated by contrasting untargeted epicPCR and mass 16S rRNA quality information. Their mass emulsion configuration can enquire a huge number of cells in correspondence with costs similar to one genomic library prep, expanding outturn and decreasing cost contrasted and existing strategies. This versatile strategy can interpret hereditary relationship from any example into a sequencing library that answers designated environmental inquiries, another procedure that joins useful qualities and phylogenetic markers in crude single cells, giving an output of a huge number of cells with costs tantamount to one genomic library planning (Spencer et al., 2016).

22.6.1.10   meta3C Approach

Marbouty et al. demonstrated a computational and experimental method that uses the genetic linkage among DNAs to know their propinquity. In an analysis, a couple of phage genomes and bacteria living in a rodent gut biota were collected and scaffolded all over again. The bacteria/phage genetic materials were then doled out into separate hosts as per the actual linkage among the distinctive DNA atoms, opening a novel viewpoint for a thorough picture of the genomic design of the gut verdure. Along these lines, this work holds broad ramifications for environmental studies expecting to connect the virome to the microbiome (Marbouty et al., 2017). This finding surpasses certain limitations such as:

    i. Ability to characterize complete bacterial and viral genomes from a complex mix of species.
    ii. Difficulty assigning phage sequences to their bacterial hosts.

22.6.1.11 Hi-C Sequencing Technique

Many environmental habitats have been identified as ARGs hotspots, yet we miss the understanding of the origins of antibiotic resistance and its dissemination from various ecological niches to hospital facilities. This problem is partly associated with the inability to identify the bacterial host and MGE interactions. Stalder et al. showed that the *in vivo* proximity ligation technique Hi-C reconstructs an identical plasmid-host union from an effluent microbial population, and spot *in situ* plasmids among hosts, also integrons and ARGs by physically relating them to host chromosomes. Hi-C distinguished both recently known and novel relationship among ARGs, portable hereditary components, and host genomes, consequently approving this strategy. The study suggested that, among other MGEs, class 1 integron and IncQ plasmids have the broadest host range in the wastewater, and recognized microorganisms having such as, *Moraxellaceae, Bacteroides,* and *Prevotella,* particularly *Aeromonadaceae* as the most probable repositories of ARGs locally (Stalder et al., 2019).

22.6.1.12 CRISPR Recorder

A study conducted by Munck et al. demonstrated a unique recorder highlighting HGT events, which is based on the integration of fragments of acquired genomes through CRISPR-Cas9 systems. As it is known, CRISPR-Cas systems are adaptive immune response systems used by several bacterial strains against invading foreign DNA. The invader gene is integrated in the bacterial chromosome adjacent to the CRISPR locus in a process called spacer incorporation. Thus, the sequences are transcribed to target the invading DNA through Cas9 activation. In order to identify HGT events, Munck et al. developed an *E. coli* strain bearing a plasmid containing *cas1-cas2* controlled by a promoter to trap foreign DNA sequences entering the cell. Subsequently, the CRISPR-spacer regions were sequenced in order to identify HGT events and their respective occurrence order. This technique gives a better understanding of HGT events and an unprecedented resolution of mobilomes associated with antibiotic resistance spread (Munck et al., 2020).

22.7 CONCLUSIONS

It is known that even quality exchange is interceded by a wide range of hereditary components and different instruments. It has been seen that bacterial species have their own guard frameworks (efflux siphons and CRISPR-Cas9 pathway) to battle ecological pressure and poisonous specialists such as anti-infection agents and substantial metals. The current sequencing-based and exploratory advances are revealing the degree by which opposition qualities spread among strains and species. Bacterial genes that present resistance against clinically pertinent anti-microbials and are carried on portable hereditary components that recreate in microorganisms are viewed as a quick danger to the effective treatment of clinical contaminations. The conjugative spread of plasmids is still broadly viewed as the main way obstruction qualities can be moved among microscopic organisms.

Improved comprehension of HGT in the environment opens up roads for advancement of novel strategies to limit the spread of ARGs, for instance, by the conjugative conveyance of CRISPR-Cas to specifically exhaust strains conveying resistant qualities (Rodrigues et al., 2019). We conceive that the original strategies audited here will support further explaining the mobilomes involved in transfer of ARGs among bacterial microorganisms. The data gathered from these investigations would then be able to help in the improvement of designated ways to deal with controlling or decreasing the quantity of AMR bacterial species in the environment.

ACKNOWLEDGEMENTS

The Director, CSIR-NEERI, is thankfully acknowledged for giving the opportunity to pursue the work in CSIR-NEERI, Nagpur, India. This chapter is checked for plagiarism using the iThenticate software and recorded in the Knowledge Resource Center, CSIR-NEERI, Nagpur, for anti-plagiarism (KRC No.: CSIR-NEERI/KRC/2021/OCT/WWTD/2).

REFERENCES

Ahmed, W., Zhang, Q., Lobos, A., Senkbeil, J., Sadowsky, M. J., Harwood, V. J., … Ishii, S. (2018). Precipitation influences pathogenic bacteria and antibiotic resistance gene abundance in storm drain outfalls in coastal sub-tropical waters. *Environment Internassstional*, *116*, 308–318. Doi: 10.1016/j.envint.2018.04.005.

Alekshun, M. N., & Levy, S. B. (2007). Molecular mechanisms of antibacterial multidrug resistance. *Cell*, *128*(6), 1037–1050. Doi: 10.1016/j.cell.2007.03.04.

Baker-Austin, C., Wright, M. S., Stepanauskas, R., & McArthur, J. V. (2006). Co-selection of antibiotic and metal resistance. *Trends in Microbiology*, *14*(4), 176–182. Doi: 10.1016/j.tim.2006.02.006.

Beaulaurier, J., Zhu, S., Deikus, G., Mogno, I., Zhang, X. S., Davis-Richardson, A., … Schadt, E. E. (2018). Metagenomic binning and association of plasmids with bacterial host genomes using DNA methylation. *Nature Biotechnology*, *36*(1), 61. Doi: 10.1038/nbt.4037.

Ben, Y., Fu, C., Hu, M., Liu, L., Wong, M. H., & Zheng, C. (2019). Human health risk assessment of antibiotic resistance associated with antibiotic residues in the environment: A review. *Environmental Research*, *169*, 483–493. Doi: 10.1016/j.envres.2018.11.040

Bengtsson-Palme, J., Hammaren, R., Pal, C., Östman, M., Björlenius, B., Flach, C. F., … Larsson, D. J. (2016). Elucidating selection processes for antibiotic resistance in sewage treatment plants using metagenomics. *Science of the Total Environment*, *572*, 697–712. Doi: 10.1016/j.scitotenv.2016.06.228.

Bengtsson-Palme, J., & Larsson, D. J. (2016). Concentrations of antibiotics predicted to select for resistant bacteria: Proposed limits for environmental regulation. *Environment International*, *86*, 140–149. Doi: 10.1016/j.envint.2015.10.015

Bickhart, D. M., Watson, M., Koren, S., Panke-Buisse, K., Cersosimo, L. M., Press, M. O., … Heiner, C. (2019). Assignment of virus and antimicrobial resistance genes to microbial hosts in a complex microbial community by combined long-read assembly and proximity ligation. *Genome Biology*, *20*(1), 1–18. Doi: 10.1186/s13059-019-1760-x.

Blokesch, M. (2016). Natural competence for transformation. *Current Biology*, *26*(21), R1126-R1130. Doi: 10.1016/j.cub.2016.08.058.

Bridier, A., Le Grandois, P., Moreau, M. H., Prénom, C., Le Roux, A., Feurer, C., & Soumet, C. (2019). Impact of cleaning and disinfection procedures on microbial ecology and *Salmonella* antimicrobial resistance in a pig slaughterhouse. *Scientific Reports*, *9*(1), 1–13. Doi: 10.1038/s41598-019-49464-8

Colomer-Lluch, M., Jofre, J., & Muniesa, M. (2011). Antibiotic resistance genes in the bacteriophage DNA fraction of environmental samples. *PloS One*, *6*(3), e17549. Doi: 10.1371/journal.pone.0017549.

Garbisu, C., Garaiyurrebaso, O., Lanzén, A., Álvarez-Rodríguez, I., Arana, L., Blanco, F., … Alkorta, I. (2018). Mobile genetic elements and antibiotic resistance in mine soil amended with organic wastes. *Science of the Total Environment*, *621*, 725–733. Doi: 10.1016/j.scitotenv.2017.11.221.

Gillings, M. R. (2017). Class 1 integrons as invasive species. *Current Opinion in Microbiology*, *38*, 10–15. Doi: 10.1016/j.mib.2017.03.002.

Higgins, P. G., Pérez-Llarena, F. J., Zander, E., Fernández, A., Bou, G., & Seifert, H. (2013). OXA-235, a novel class D β-lactamase involved in resistance to carbapenems in *Acinetobacter baumannii*. *Antimicrobial Agents and Chemotherapy*, *57*(5), 2121–2126. Doi: 10.1128/AAC.02413-12.

Karkey, A., Thwaites, G. E., & Baker, S. (2018). The evolution of antimicrobial resistance in *Salmonella typhi*. *Current Opinion in Gastroenterology*, *34*(1), 25–30. Doi: 10.1097/MOG.0000000000000406.

Klümper, U., Riber, L., Dechesne, A., Sannazzarro, A., Hansen, L. H., Sørensen, S. J., & Smets, B. F. (2015). Broad host range plasmids can invade an unexpectedly diverse fraction of a soil bacterial community. *The ISME Journal*, *9*(4), 934–945. Doi: 10.1038/ismej.2014.191.

Lagier, J. C., Dubourg, G., Million, M., Cadoret, F., Bilen, M., Fenollar, F., … Raoult, D. (2018). Culturing the human microbiota and culturomics. *Nature Reviews Microbiology*, *16*(9), 540–550. Doi: 10.1038/s41579-018-0041-0.

Lerminiaux, N. A., & Cameron, A. D. (2019). Horizontal transfer of antibiotic resistance genes in clinical environments. *Canadian Journal of Microbiology*, *65*(1), 34–44. Doi: 10.1139/cjm-2018-0275.

Lester, C. H., Frimodt-Møller, N., Sørensen, T. L., Monnet, D. L., & Hammerum, A. M. (2006). *In vivo* transfer of the vanA resistance gene from an *Enterococcus faecium* isolate of animal origin to an *E. faecium* isolate of human origin in the intestines of human volunteers. *Antimicrobial Agents and Chemotherapy*, *50*(2), 596–599. Doi: 10.1128/AAC.50.2.596-599.2006.

Levings, R. S., Lightfoot, D., Partridge, S. R., Hall, R. M., & Djordjevic, S. P. (2005). The genomic island SGI1, containing the multiple antibiotic resistance region of *Salmonella enterica* serovar *Typhimurium* DT104 or variants of it, is widely distributed in other *S. enterica* serovars. *Journal of Bacteriology*, *187*(13), 4401–4409. Doi: 10.1128/JB.187.13.4401-4409.2005.

Li, B., Yang, Y., Ma, L., Ju, F., Guo, F., Tiedje, J. M., & Zhang, T. (2015). Metagenomic and network analysis reveal wide distribution and co-occurrence of environmental antibiotic resistance genes. The ISME *Journal*, *9*(11), 2490–2502. Doi: 10.1038/ismej.2015.59

Liu, X., Guo, X., Liu, Y., Lu, S., Xi, B., Zhang, J., … Bi, B. (2019). A review on removing antibiotics and antibiotic resistance genes from wastewater by constructed wetlands: Performance and microbial response. *Environmental Pollution*, *254*, 112996. Doi: 10.1016/j.envpol.2019.112996.

Marbouty, M., Baudry, L., Cournac, A., & Koszul, R. (2017). Scaffolding bacterial genomes and probing host-virus interactions in gut microbiome by proximity ligation (chromosome capture) assay. *Science Advances*, *3*(2), e1602105. Doi: 10.1126/sciadv.1602105.

Marti, E., & Balcázar, J. L. (2012). Multidrug resistance-encoding plasmid from *Aeromonas* sp. strain P2G1. *Clinical Microbiology and Infection*, *18*(9), E366-E368. Doi: 10.1111/j.1469-0691.2012.03935.x.

Martinez, J. L., Coque, T. M., & Baquero, F. (2015). Prioritizing risks of antibiotic resistance genes in all metagenomes. *Nature Reviews Microbiology*, *13*(6), 396-396. Doi: 10.1038/nrmicro3399-c2.

McInnes, R. S., McCallum, G. E., Lamberte, L. E., & van Schaik, W. (2020). Horizontal transfer of antibiotic resistance genes in the human gut microbiome. *Current Opinion in Microbiology*, *53*, 35–43. Doi: 10.1016/j.mib.2020.02.002.

Munck, C., Sheth, R. U., Freedberg, D. E., & Wang, H. H. (2020). Recording mobile DNA in the gut microbiota using an *Escherichia coli* CRISPR-Cas spacer acquisition platform. *Nature Communications*, *11*(1), 1–11. Doi: 10.1038/s41467-019-14012-5.

Nordmann, P., Dortet, L., & Poirel, L. (2012). Carbapenem resistance in Enterobacteriaceae: Here is the storm! *Trends in Molecular Medicine*, *18*(5), 263–272. Doi: 10.1016/j.molmed.2012.03.003.

Nielsen, K. M., & Townsend, J. P. (2004). Monitoring and modeling horizontal gene transfer. *Nature Biotechnology*, *22*(9), 1110–1114. Doi: 10.1038/nbt1006.

Olekhnovich, E. I., Vasilyev, A. T., Ulyantsev, V. I., Kostryukova, E. S., & Tyakht, A. V. (2018). MetaCherchant: Analyzing genomic context of antibiotic resistance genes in gut microbiota. *Bioinformatics*, *34*(3), 434–444. Doi: 10.1093/bioinformatics/btx681.

Pal, C., Asiani, K., Arya, S., Rensing, C., Stekel, D. J., Larsson, D. J., & Hobman, J. L. (2017). Metal resistance and its association with antibiotic resistance. *Advances in Microbial Physiology*, *70*, 261–313. Doi: 10.1016/bs.ampbs.2017.02.001.

Peterson, E., & Kaur, P. (2018). Antibiotic resistance mechanisms in bacteria: Relationships between resistance determinants of antibiotic producers, environmental bacteria, and clinical pathogens. *Frontiers in Microbiology*, *9*, 2928. Doi: 10.3389/fmicb.2018.02928.

Podell, S., & Gaasterland, T. (2007). DarkHorse: A method for genome-wide prediction of horizontal gene transfer. *Genome Biology*, *8*(2), R16. Doi: 10.1186/gb-2007-8-2-r16.

Podnecky, N. L., Rhodes, K. A., & Schweizer, H. P. (2015). Efflux pump-mediated drug resistance in Burkholderia. *Frontiers in Microbiology*, 6, 305. Doi: 10.3389/fmicb.2015.00305

Press, M. O., Wiser, A. H., Kronenberg, Z. N., Langford, K. W., Shakya, M., Lo, C. C., … Liachko, I. (2017). Hi-C deconvolution of a human gut microbiome yields high-quality draft genomes and reveals plasmid-genome interactions. *Biorxiv*, 198713. Doi: 10.1101/198713.

Qian, X., Sun, W., Gu, J., Wang, X. J., Zhang, Y. J., Duan, M. L., … Zhang, R. R. (2016). Reducing antibiotic resistance genes, integrons, and pathogens in dairy manure by continuous thermophilic composting. *Bioresource Technology*, *220*, 425–432. Doi: 10.1016/j.biortech.2016.08.101.

Ross, J., & Topp, E. (2015). Abundance of antibiotic resistance genes in bacteriophage following soil fertilization with dairy manure or municipal biosolids, and evidence for potential transduction. *Applied and Environmental Microbiology*, *81*(22), 7905–7913. Doi: 10.1128/AEM.02363-15.

Savoldi, A., Carrara, E., Graham, D. Y., Conti, M., & Tacconelli, E. (2018). Prevalence of antibiotic resistance in *Helicobacter pylori*: A systematic review and meta-analysis in World Health Organization regions. *Gastroenterology*, *155*(5), 1372–1382. Doi: 10.1053/j.gastro.2018.07.007.

Song, W., Wemheuer, B., Zhang, S., Steensen, K., & Thomas, T. (2019). MetaCHIP: Community-level horizontal gene transfer identification through the combination of best-match and phylogenetic approaches. *Microbiome, 7*(1), 36. Doi: 10.1186/s40168-019-0649-y.

Sørensen, S. J., Sørensen, A. H., Hansen, L. H., Oregaard, G., & Veal, D. (2003). Direct detection and quantification of horizontal gene transfer by using flow cytometry and gfp as a reporter gene. *Current Microbiology, 47*(2), 0129–0133. Doi: 10.1007/s00284-002-3978-0.

Spencer, S. J., Tamminen, M. V., Preheim, S. P., Guo, M. T., Briggs, A. W., Brito, I. L., … Alm, E. J. (2016). Massively parallel sequencing of single cells by epicPCR links functional genes with phylogenetic markers. *The ISME journal, 10*(2), 427–436. Doi: 10.1038/ismej.2015.124.

Staehlin, B. M., Gibbons, J. G., Rokas, A., O'Halloran, T. V., & Slot, J. C. (2016). Evolution of a heavy metal homeostasis/resistance island reflects increasing copper stress in enterobacteria. *Genome Biology and Evolution, 8*(3), 811–826. Doi: 10.1093/gbe/evw031.

Stalder, T., Press, M. O., Sullivan, S., Liachko, I., & Top, E. M. (2019). Linking the resistome and plasmidome to the microbiome. *The ISME Journal, 13*(10), 2437–2446. Doi: 10.1038/s41396-019-0446-4.

Stalder, T., Rogers, L. M., Renfrow, C., Yano, H., Smith, Z., & Top, E. M. (2017). Emerging patterns of plasmid-host coevolution that stabilize antibiotic resistance. *Scientific Reports, 7*(1), 1–10. Doi: 10.1038/s41598-017-04662-0

Sultan, I., Rahman, S., Jan, A. T., Siddiqui, M. T., Mondal, A. H., & Haq, Q. M. R. (2018). Antibiotics, resistome and resistance mechanisms: A bacterial perspective. *Frontiers in Microbiology, 9*, 2066. Doi: 10.3389/fmicb.2018.02066.

Sun, Z., Li, Y., Wang, M., Wang, X., Pan, Y., & Dong, F. (2019). How does vertical integration promote innovation corporate social responsibility (ICSR) in the coal industry? A multiple-step multiple mediator model. *Plos One, 14*(6), e0217250. Doi: 10.1371/journal.pone.0217250.

Tang, K. L., Caffrey, N. P., Nóbrega, D. B., Cork, S. C., Ronksley, P. E., Barkema, H. W., … Ghali, W. A. (2017). Restricting the use of antibiotics in food-producing animals and its associations with antibiotic resistance in food-producing animals and human beings: A systematic review and meta-analysis. *The Lancet Planetary Health, 1*(8), e316-e327. Doi: 10.1016/S2542-5196(17)30141-9.

Von Wintersdorff, C. J., Penders, J., Van Niekerk, J. M., Mills, N. D., Majumder, S., Van Alphen, L. B., … Wolffs, P. F. (2016). Dissemination of antimicrobial resistance in microbial ecosystems through horizontal gene transfer. *Frontiers in Microbiology, 7*, 173. Doi: 10.3389/fmicb.2016.00173.

Yaffe, E., & Relman, D. A. (2020). Tracking microbial evolution in the human gut using Hi-C reveals extensive horizontal gene transfer, persistence and adaptation. *Nature Microbiology, 5*(2), 343–353. Doi: 10.1038/s41564-019-0625-0.

Yan, W., Guo, Y., Xiao, Y., Wang, S., Ding, R., Jiang, J., … Zhao, F. (2018). The changes of bacterial communities and antibiotic resistance genes in microbial fuel cells during long-term oxytetracycline processing. *Water Research, 142*, 105–114. Doi: 10.1016/j.watres.2018.05.047.

Yang, Y., Song, W., Lin, H., Wang, W., Du, L., & Xing, W. (2018). Antibiotics and antibiotic resistance genes in global lakes: A review and meta-analysis. *Environment International, 116*, 60–73. Doi: 10.1016/j.envint.2018.04.011.

Zaman, S. B., Hussain, M. A., Nye, R., Mehta, V., Mamun, K. T., & Hossain, N. (2017). A review on antibiotic resistance: Alarm bells are ringing. *Cureus, 9*(6). Doi: 10.7759/cureus.1403.

Zhen, X., Stålsby Lundborg, C., Sun, X., Hu, X., & Dong, H. (2019). The clinical and economic impact of antibiotic resistance in China: A systematic review and meta-analysis. *Antibiotics, 8*(3), 115. Doi: 10.3390/antibiotics8030115.

Zhang, Y., Li, H., Gu, J., Qian, X., Yin, Y., Li, Y., ... Wang, X. (2016). Effects of adding different surfactants on antibiotic resistance genes and intI1 during chicken manure composting. *Bioresource Technology*, *219*, 545–551. Doi: 10.1016/j.biortech.2016.06.117.

Zhang, Y. J., Hu, H. W., Gou, M., Wang, J. T., Chen, D., & He, J. Z. (2017). Temporal succession of soil antibiotic resistance genes following application of swine, cattle and poultry manures spiked with or without antibiotics. *Environmental Pollution*, *231*, 1621–1632. Doi: 10.1016/j.envpol.2017.09.074.

23 Genetically Engineered Microorganisms
A Promising Approach for Bioremediation

Debabrata Nath, Vandana Kumari, and Ranjan Laik
Dr. Rajendra Prasad Central Agricultural University

Raj Mukhopadhyay
ICAR-Central Soil Salinity Research Institute

CONTENTS

23.1 INTRODUCTION

The overuse of inorganic (heavy metals and metalloids) and organic chemicals (polycyclic aromatic hydrocarbons (PAHs), antibiotics, pesticides, and dyes) in industries and intensive agriculture causes severe harmful effects on plants, animals, and humans (Kang, 2014). Thus, the elimination of these geogenic and anthropogenic contaminants is the need of the hour to achieve environmental sustainability. In the fields of environmental remediation, the term "bioremediation" is commonly used for contaminants' removal (Marinescu et al., 2009; Joutey et al., 2013). Bioremediation using microorganisms has come up a simple, efficient, and environment-friendly approach to remediate contaminants (Kumar et al., 2018; Bhat et al., 2018). However,

DOI: 10.1201/9781003247883-23

a one-step procedure is uncommon; typically, a series of biological changes are required to remove a specific component. One solution to this issue is to use genetic engineering. There are three phases of the bioremediation processes: (1) biostimulation: by addition of nutrients and oxygen to the system, system efficacy is improved and biodegradation process is sped up; (2) bioaugmentation: microorganisms are added into the systems to facilitate remediation; (3) natural attenuation: contaminants are reduced by local microorganisms without the need for human intervention. These additional organisms exhibit better efficiency than native flora in degradation of the target pollutant (Diez, 2010). Microorganisms capable of fast adaptation and effective utilization of pollutants of interest in a given environment within an acceptable period of time are required for a successful restoration technique (Seo et al., 2009). Microorganisms' capacity to utilize pollutants as substrates or metabolize them is influenced by multiple variables, including biotic and abiotic components that govern microorganismal capacity as a bioremediator, such as temperature, pH, and nutrient supplier. These factors affect the rate and amount of degradation of the contaminants (Fritsche and Hofrichter, 2008). Genetically engineered microorganisms (GEMs) improve the overall efficiency and effectiveness of bioremediation approaches. There are many instances of transgenic bacteria being used to remediate inorganic and organic pollutants from the environment (Ruiz et al., 2011; Dash et al., 2014). An engineered microorganism accumulates a high concentration of contaminants and may remove the pollutant from polluted environment through different processes such as volatilization, adsorption, and degradation. However, it is feasible to introduce specific functional genes into the genome to be applied in field conditions using genetic engineering technologies (Tozzini, 2000). As a result, there has been a lot of interest in using GEMs in bioremediation. These GEMs have high degradation capacity and have been proven to efficiently degrade a variety of pollutants under controlled conditions. However, environmental and ecological concerns, as well as regulatory restrictions, are substantial roadblocks for GEMs to apply in field conditions (Menn et al., 2008; Joutey et al., 2013). In this chapter, we discuss the sustainable and efficient removal process of inorganic and organic contaminants using GEMs with their associated challenges in order to achieve United Nations (UN's) Sustainable Development Goals (SDGs): "Clean Water and Sanitation", "Life on Land", and "Life Below Water", within 2030 (UN, 2016). Further, we also focus on future research opportunities on bioremediation using GEMs.

23.2 GEMS FOR ENVIRONMENTAL RESCUE AGAINST POLLUTANTS

23.2.1 Bioremediation of Inorganic Pollutants by GEMs

Heavy metal contamination in the environment has become a significant danger to the ecosystem-living organisms (Deepa and Suresha, 2014; Siddiquee et al., 2015; Okolo et al., 2016; Kumar, 2018). Their non-biodegradability, bioaccumulation, and toxicity properties are major environmental concerns (Wai et al., 2012; Gautam et al., 2015, 2017). Various metals, including manganese (Mn^{2+}), calcium (Ca^{2+}), magnesium (Mg^{2+}), sodium (Na^+), and zinc (Zn^{2+}), are essential for metabolic and redox activities in small quantities. However, cadmium (Cd^{2+}), lead (Pb^{2+}), mercury (Hg^{2+}), and copper (Cu^{2+}) are heavy metals that are harmful to living organisms (Turpeinen et al., 2002; Lakherwal, 2014; Siddiquee et al., 2015). The use of GEMs (*Pseudomonas aeruginosa, Neurospora crassa, Escherichia coli*) is an innovative approach for the remediation of contaminants from contaminated water and soils (Shukla et al., 2010; Liu et al., 2011). The capacity to withstand metal stress, the overexpression of metal-chelating proteins and peptides, and the accumulation of metal are desirable properties of GEMs that have made them successful for the effective bioremediation of a wide range of pollutants. For example, Frederick et al. (2013) developed trehalose bacteria to degrade chromium (Cr^{6+}) to chromium (Cr^{3+}). Similarly, the context of genes manipulated directly by *Chlamydomonas reinhardtii* led to a major rise in Cd^{2+} tolerance (Ibuot et al., 2017). The *Escherichia coli* (ArsR ELP153AR) degraded arsenic (As(III)) (Kostal et al., 2004), while *Saccharomyces cerevisiae* (CP2 HP3) degraded (Cd^{2+}) and (Zn^{2+}) from contaminated environment (Vinopal et al., 2007). Moreover, the genetically modified *Corynebacterium glutamicum* with the use of ars operons overexpression (ars1 and ars2) remediated As-polluted areas (Mateos et al., 2017). Table 23.1 lists the various GEMs, their expressing genes, and functions for heavy metal remediation from the contaminated environment. The environment-friendly bioremediation approach using GEMs is only successful in laboratory condition. However, its use in the decontamination of real wastewater or contaminated land is urgently needed to promote its field scale applicability and efficiency for improving its wider acceptability to the end-users.

23.2.2 BIOREMEDIATION OF ORGANIC POLLUTANTS BY GEMS

Organic pollutants include phenols; phenols of chlorine; phthalic esters; azo dyes; pesticides; hydrocarbons; explosives such as 2,4,6-trinitrotoluene (TNT), hexahydro-1,3,5-trinitro-1,3,5-triazine or hexogen (RDX), and octahydro-1,3,5,7-tetrazocine or octogen (HMX); and persistent organic pollutants (POPs) are also harmful to living organisms. Some of the POPs (e.g. poly- and perfluoroalkyl substances) are difficult to remove completely from environment. GEMs are being developed to overcome the conventional bioremediation constraints for the organic xenobiotics transformation. Researches have shown that *Pseudomonas* sp. (B13 strain genetically engineered) inoculation enhanced 3-chlorotoluene and 4-chlorotoluene degradation (Brinkmann and Reineke, 1992) over the control. In contrast, *Deinococcus radiodurans* was genetically modified for the degradation of toluene, but it was failed to be marketed as bioremediation applications, owing to its expected dangers and regulatory difficulties (Ezezika and Singer, 2010). The progression of the newest biotechnological techniques such as recombinant DNA or natural gene transfer may enable the production of particular enzymes that enhance the degradation of organic hazardous compounds (rhizospheric or endophytic) (Chakraborty and Das, 2016; Pandotra et al., 2018). This technique contributes to improve the breakdown of the hazardous organic pollutants in polluted areas via plant-based endophytic and rhizospheric bacteria (Fasani et al., 2018). Various investigations have shown that *Pseudomonas* strains are capable of biodegradation of total petroleum hydrocarbons (TPH) with significant boost of dehydrogenase activity in the soil, which increased their capacity

TABLE 23.1

Detoxification and Degradation of Heavy Metals by GEMs

Name of the Inorganic Pollutants	GEM Used	Gene Involved/Modified	Gene Function	Reference
Arsenic (As)	*Escherichia coli*	Cloned *E. coli* with *ars*M (arsenite S-adenosylmethionine methyltransferase gene), which is retrieved from *Rhodopseudomonas palustris* Also AtACR2 gene from *A. thaliana* L. and *ars*M taken from *Sphingomonas desiccabilis and Bacillus idriensis*	Arsenic volatilization (conversion of methylated inorganic arsenic to non-toxic volatile form trimethylarsine (TMA)	Qin et al. (2006) Nahar et al. (2017) Chen et al. (2013) Yang et al. (2010)
Lead (Pb)	*Pseudomonas aeruginosa* *Salmonella choleraesuis* *Proteus penneri*	Strain 4EA, strain 4A, and strain GM10, respectively, cloned with *smt*A and *smt*AB bacterial metallothioneins	Rhizosphere degradation of Pb^{2+} rather than uptake	Naik et al. (2012).
Copper (Cu)	*Neurospora crassa*	A strong affinity copper (Cu) transporter gene (tcu-1) from the fungus *Neurospora crassa* was introduced into *Nicotiana tabacum*	Copper (Cu) acquisition in plants body reduces heavy metal toxicity in the soil readily	Singh et al. (2011)
Zinc (Zn)	*Neurospora crassa*	*Nicotiana tabacum* L. with Zn transporter gene (tzn1) from the fungus *Neurospora crassa*	A potential fungal gene aids transgenic plants in increasing Zn absorption while avoiding Cd co-transport	Dixit et al. (2010)
Nickel (Ni)	*E. coli* SE5000 strain *Pseudomonas fluorescens* 4F39	Nickel transporting expression (nix-A gene products) Phytochelatin synthase (enzyme secreted by *P. fluorescens*)	Degradation of nickel (aqueous system)	Farnham and Dube (2015) Lopez et al. (2002)

(Continued)

TABLE 23.1 (*Continued*)
Detoxification and Degradation of Heavy Metals by GEMs

Name of the Inorganic Pollutants	GEM Used	Gene Involved/Modified	Gene Function	Reference
Chromium (Cr)	*Methylococcus capsulatus* *Escherichia coli*	Gene name of M capsulatus. nix-A, an oxygen-insensitive nitro-reductase (flavoprotein) of *E. coli*	Potential Cr^{6+}reductase activity leads to bioremediation of Cr^{6+} from a wide range of concentrations in soil (1.4–1000 mg L^{-1} of Cr^{6+}). Reduce chromate to less soluble and less toxic form Cr^{3+}	Al Hasin et al. (2010) Ackerley et al. (2004)
Mercury (Hg)	*Deinococcus radiodurans* *Acidithiobacillus ferrooxidans* *Rhodopseudomonas palustris* *Pseudomonas* K-62	MerH (a new ion transporter gene) incorporation from *marinum* strain containing merC (mercuric ion transporter gene) mercuric ion (Hg) transporter gene	Degrade ionic mercury Faster degradation of mercury Hg^{+2} removal from wastewater	Gupta and Walther (2016) Ouyang et al. (2013) Ye et al. (2012) Chang et al. (2015)
Cadmium (Cd)	*Mesorhizobium huakuii*	Transformed CIPK24 (CBL-interacting protein kinase 24) genes from *Arabidopsis thaliana*	Degrade cadmium	Porter et al. (2017)

for oil breakdown by 100 times over wild isolates (natural) in extra-chromosome plasmids (Wang, 2014; Gao et al., 2015; Chebbi et al., 2017). Table 23.2 lists various GEMs and their functions for organic pollutant remediation in the environment.

The potentiality of the transgenic microorganisms should be better explored for herbicide-tolerant transgenic plants at large-scale field applications for the remediation of organic pollutants. However, for many years, agriculture used highly developed herbicide-tolerant transgenic plants. Glutathione S-transferase and cytochrome P450 are two major enzymes that are key factors for an herbicidal breakdown among various major enzymatic groups (Duhoux et al., 2015). Due to an adapted metabolism, transgenic rice plants with transgenes overexpression such as CYP2C19, CYP1A1, and CYP2B6 were more herbicide-tolerant than non-transgenic rice plants

TABLE 23.2

Detoxification and Degradation of Organic Pollutants by GEMs

Name of the Organic Pollutants	GEM Used	Gene Involved/Modified	Gene Function	Reference
Toluene/benzoate	*Pseudomonas putida* KT2442	Modified pathway	Degradation of toluene/benzoate	Khan et al. (2016); Wang et al. (2010)
Trichloroethylene (TCE), benzene, toluene, and fluorobiphenyl	*Pseudomonas (pseudoalcaligenes)* (KF707-D2))	Modified substrate specificity	Degradation of trichloroethane, benzene, toluene, and fluorobiphenyl	Chen et al. (2016)
Naphthalene	*Pseudomonas putida* VM1441 (pNAH7)	Several self-transmissible plasmids' conjugal transfer mechanisms	Naphthalene degradation	Germaine et al. (2009) Pant et al. (2020)
Phenol and cyanide	*Pseudomonas putida* and *P. stutzeri*	Specificity of substrate in modified form	Cyanide and phenol degradation	Singh et al. (2018) Pant et al. (2020)
Organophosphorus pesticides	*Escherichia coli* *Pseudomonas diminuta*	Insertion of Pds plasmid and pL-DsRed–pL-OPH Recombinant expression	Degradation of organophosphorus pesticides Phosphotriesterase enzyme activation for organophosphorus degradation	Li and Wu (2014) Bigley and Raushel (2019)
Organophosphorus pesticides and *p*-nitrophenol (PNP)	*Moraxella* sp.	Organophosphorus hydrolase is surface expressed	Degradation of organophosphorus and *p*-nitrophenol	Schüürmann et al. (2014)
Methyl parathion and carbofuran	*Sphingomonas* sp. CDS1	The methyl parathion hydrolase gene (mpd) was inserted into chromosome	Degrade carbofuran and methyl parathion	Jiang et al. (2007) Pant et al. (2020)
Methyl parathion	*Pseudomonas putida* X3 strain	Contains methyl parathion (MP)-degrading gene and enhanced green fluorescent protein (EGFP) gene	Degrade soil microcosms contaminated with cadmium and methyl parathion	Zhang et al. (2016)
Xenobiotics	*Pandoraea* sp.	Lactase gene expression	Bioremediation of xenobiotics by production of various primary metabolites (halogenated aromatic compounds, DDT, etc.)	Peeters et al. (2019)
Chlorobenzoate	*Rhodococcus* RHA1	Contains pPC3 with *fcb* operon from *Arthrobacter globiformis*	Degrade 4-chlorobenzoate	Li et al. (2016)

(Kawahigashi et al., 2006). Tobacco (*Nicotiana tabacum*) was the first genetically modified plant used to develop resistance to xenobiotics (organic) such as halogenated organics and explosives (French et al., 1999; Doty 2008; Doty et al., 2000). In hydroponic solutions, reduced levels of atrazine and simazine were found in transgenic paddy containing CYP1A1 genes (Kawahigashi et al., 2007a, b). A well-researched cytochrome, *i.e.* CYP2E1, can be used to remediate pollutants such as halogenated organic compounds and explosives, in transgenic organisms (Abhilash et al., 2009, 2013; Zhang et al., 2013).

23.3 SUICIDAL GENETICALLY ENGINEERED MICROORGANISMS (S-GEMS)

The GEMs can have two possible outcomes: (1) the desired outcome is the GEM to perform the specified job and then be completely removed from the ecosystems; (2) the alternative process is typically unpopular, involving in the proliferation of microorganism rather than being killed. The first approach is the way forward because if GEMs with recombinant technology are allowed to persist in the environment, they may have negative consequences on ecosystems (Paul et al., 2005). As a result, the strains' capacity to survive should be reduced by the development of specific containment devices in order to reduce the negative effects on the environment.

The drawback of biological confinement systems is that a significant proportion of cells survive during death function owing to random mutagenesis. The main disadvantage appears to be the negative regulation of suicidal functions, which means that the killer gene is turned off by default in the presence of the pollutant(s), and the killer protein is produced only when the pollutant is depleted, allowing random mutations in the control elements by the time the pollutant is depleted (Paul et al., 2005). Antidote production may be regulated by the presence or absence of the pollutant(s), which would then inactivate killer gene product, reducing these issues. A unique method for the development of S-GEMs is based on catabolic regulatory operons and plasmid addiction systems of bacteria (S-GEMs). In the proposed construct, the killer gene would be continuously generated, while the antidote would be expressed under the strict control of a promoter that would be inducible by the pollutant of concern. Antidote synthesis starts as soon as the pollutant is recognized by the S-GEM (inducible promoter) in this killer system, neutralizing the toxin's killing effect. However, as soon as the pollutant is depleted (or the concentration falls below the detection thresholds of the S-GEM), antidote synthesis ceases, resulting in cell death.

23.4 VARIOUS APPROACHES FOR THE DEVELOPMENT OF GEMS

The discovery and subsequent modification of specific genomic sequences have resulted in the development of GEMs. GEMs are designed using knowledge on microbe-xenobiotic interaction, the genetic basis of interaction, biochemical processes, operon structure, molecular biology, and ecological application (Kulshreshtha, 2013). The four approaches for the development of GEMs are discussed as follows:

a. The identification of specific microorganisms with specific genes is the main step in the development of GEMs. The use of aquatic microbes has been suggested for identifying specific genes that can be modified to provide bioremediation properties. Scientists have developed two new algal species: *Anabaena* sp. and *Ellipsosporum*, which can reduce the pollutant concentration from water sources by degrading hexachlorocyclobenzoates.

b. The development and expansion of metabolic engineering of pathways by combining capabilities of two or more microorganisms is the second strategy. GEMs can improve or expand existing pathways to degrade compounds that are resistant to degradation by wild strain. A combination of bacteria that execute one or more of the bioremediation stages encode the whole catabolic pathway. Constructed GEMs can recognize different microbial communities and improve their degradation capabilities.

c. Modification of the enzymes' specificity and affinity is the third technique. Enzymes that catalyse the transcription and translation of specific genes start the metabolic processes (Kulshreshtha, 2013). Various gene clusters have been developed to improve the enzyme-substrate specificity of toluene dioxygenase enzyme, which was expressed in *E. coli* strains to boost the degradation of trichloroethylene.

d. Bioprocess development, monitoring, and control, as well as affinity, bioreporter, and sensor applications for chemical sensing, and toxicity reduction are the fourth method. A lux gene-based method for monitoring bioremediation processes has been created, which has numerous benefits. Bioluminescence is easily detectable and does not necessitate the use of costly instruments or the inclusion of external chemicals or co-factors. Furthermore, GEMs are equipped with chemical sensors that allow the monitoring of contaminant bioavailability rather than just the presence of contaminants. From bioremediation's point of view, those GEMs produce bioluminescence and can also help to proliferate microorganisms in a polluted area (Figure 23.1).

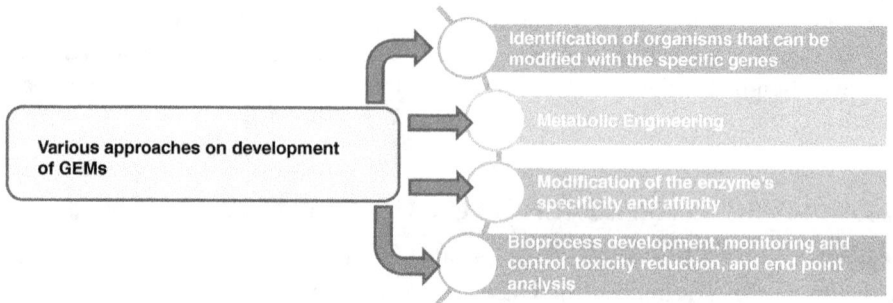

FIGURE 23.1 Approaches for the development of GEMs.

23.5 USE OF GEMS FOR BIOREMEDIATION: FROM LABORATORY TO FIELD

Using GEMs, determining the overall effectiveness of an *in situ* bioremediation strategy is a tough task (Ripp et al., 2000). These issues derive mostly from determining how much the microbe contributed to the degradation process, given that mechanisms such as volatilization and chemical transformations occur simultaneously in the system. It might be difficult to distinguish between GEM-specific breakdown and biodegradation due to the presence of indigenous soil microbial biota (Sayler et al., 1999). The environmental release of *Pseudomonas fluorescens* (strain HK-44) in highly contaminated soil with PAHs is an excellent example of *in situ* bioremediation. After introducing the pUTK21 (plasmid that breaks down naphthalene) into the strain of *P. fluorescens* (HK-44) (King et al., 1990) along with inherent lux gene (based on transposon produce bioluminescence) fused within a naphthalene catabolic promoter genes, *P. fluorescens* (HK-44) is capable of sensing contaminants such as PAHs through its easily detectable signal, i.e. bioluminescence (Chatterjee and Meighen, 1995). In this way, GEMs can serve as a living tool for the biodegradation of pollutants through bioluminescence for *in situ* field monitoring of bioremediation. In future, more pilot-scale studies focusing on using GEMs for the bioremediation of large contaminated land need to be promoted to gain popularity and effectiveness of this technique.

23.6 ADVANTAGES AND OBSTACLES OF GEMS

23.6.1 ADVANTAGES OF GEMs APPLICATION

The use of GEMs involves strains that truly address the issues through the execution of bioremediation procedures for the removal of recalcitrant chemicals of environmental concern. Evaluating the overall effectiveness of an *in situ* bioremediation program, whether employing genetically engineered or naturally occurring microorganisms, is a challenging task. These challenges come largely from determining how much the microbe itself contributes to the degradation process, even though mechanisms such as volatilization and chemical transition occur inside the system at the same time. The GEMs are not natural, but natural is not always desirable. For bioremediation GEMs have been developed after research and doing artificial DNA implantation from one species to the next.. The objective may be achieved instantly using GEMs instead of waiting for conventional process outcomes. The key advantages of GEMs approach of bioremediation are the following:

1. More efficient and effective than the conventional method of bioremediation.
2. Easy and eco-friendly method of cleaning environmental contaminants.
3. Multiple pollutants can be degraded by GEMs without producing toxic substances in the environment.

23.6.2 OBSTACLES IN THE GEMs APPLICATION

While genetic engineering has produced a number of modified microbial strains capable of breaking down resistant pollutants in a petri dish or bioreactor, the practical application of this research into actual *in situ* bioremediation techniques has been lacking (Sayler and Ripp, 2000). The significant challenge in this regard is the strains, and bacterial species used most often in conventional enrichment approaches do not conduct biodegradation in natural environments and may not even be acceptable as bioremediation mediators (Joutey et al., 2013). Hence, the main issue in successful bioremediation is adverse field conditions for the specific microorganisms. Furthermore, molecular applications are primarily limited to a few well-characterized bacteria such as *E. coli*, *Pseudomonas putida*, and *Bacillus subtilis*. Other bacterial strains must be explored in order to produce the modified microorganisms. To address the increased difficulties, the particular characteristics of application of biotechnology have demanded the modified bacterial strains development. The main concerns of using GEMs are as follows: (1) the development of GE bacteria with a high level of environmental assurance for field release in bioremediation. The performance of modified bacteria in terms of survival and the potential of horizontal gene transfer, which might influence the indigenous microflora in a complex environmental scenario, should be investigated; (2) in the majority of cases, bacteria developed for the bioremediation procedures were created in the laboratory for a specific purpose, neglecting field requirements and other complicated scenarios (Garbisu et al., 2017). However, there is no proof that releasing GE bacteria for bioremediation has had a detectable detrimental effect on the native microbial population. At the very least, the oversimplified notion of risk assessment has inspired a lot of discussions and a lot of studies, all of which have contributed significantly to the area of environmental microbiology. However, the survival of GE bacteria in complicated environmental conditions remains an open topic that must be addressed in light of recent results (Singh et al., 2011, 2016).

23.7   CONCLUSIONS AND FUTURE CHALLENGES

A large variety of synthetic compounds and other chemicals with ecotoxicological effects can be destroyed or converted by GEMs in order to achieve environmental sustainability. In most cases, however, this statement refers to prospective degradation evaluation in the laboratory using specific cultures and ideal growth conditions. In natural conditions, biodegradation is hampered by a number of factors, including competition with microorganisms, a lack of essential substrate, unfavourable environmental conditions (aeration, moisture, pH, and temperature), and low contaminant bioavailability. As a result, the goal of environmental biotechnology is to address and solve these issues to allow the use of microorganisms in bioremediation methods more effectively and timely manner. Due to eco-friendly and human-friendly strategies, the use of GEMs-based xenobiotic remediation is at the forefront. However, information on the relevant genes is scarce, limiting the development of GEMs. The regulatory problems and dangers associated with GEMs are the second major impediment to the use of GEMs. This issue, however, may be addressed by the development and deployment of S-GEMs. If in-depth information on bioremediation tools (e.g. microorganisms), their genomes, biochemical pathways and functions, and remediation mechanisms is discovered, this will be the most efficient technology in the future enabling the use of GEMs for large-scale bioremediation approaches.

REFERENCES

Abhilash, P. C., Jamil, S., & Singh, N. (2009). Transgenic plants for enhanced biodegradation and phytoremediation of organic xenobiotics. *Biotechnology Advances*, *27*(4), 474–488.

Abhilash, P. C., Singh, B., Srivastava, P., Schaeffer, A., & Singh, N. (2013). Remediation of lindane by *Jatropha curcas* L: Utilization of multipurpose species for rhizoremediation. *Biomass and Bioenergy*, *51*, 189–193.

Ackerley, D. F., Gonzalez, C. F., Keyhan, M., Blake, R., & Matin, A. (2004). Mechanism of chromate reduction by the *Escherichia coli* protein, NfsA, and the role of different chromate reductases in minimizing oxidative stress during chromate reduction. *Environmental Microbiology*, *6*(8), 851–860.

Al Hasin, A., Gurman, S. J., Murphy, L. M., Perry, A., Smith, T. J., & Gardiner, P. H. (2010). Remediation of chromium (VI) by a methane-oxidizing bacterium. *Environmental Science & Technology*, *44*(1), 400–405.

Bigley, A. N., & Raushel, F. M. (2019). The evolution of phosphotriesterase for decontamination and detoxification of organophosphorus chemical warfare agents. *Chemico-Biological Interactions*, *308*, 80–88.

Brinkmann, U., & Reineke, W. (1992). Degradation of chlorotoluenes by *in vivo* constructed hybrid strains: Problems of enzyme specificity, induction and prevention of meta-pathway. *FEMS Microbiology Letters*, *96*(1), 81–87.

Bhat, S.A., Singh, S., Singh, J., Kumar, S., & Vig, A.P., 2018. Bioremediation and detoxification of industrial wastes by earthworms: Vermicompost as powerful crop nutrient in sustainable agriculture. *Bioresource Technology*, 252, 172–179.

Chakraborty, J., & Das, S. (2014). Characterization and cadmium-resistant gene expression of biofilm-forming marine bacterium *Pseudomonas aeruginosa* JP-11. *Environmental Science and Pollution Research*, *21*(24), 14188–14201.

Chang, S., Wei, F., Yang, Y., Wang, A., Jin, Z., Li, J., He, Y., & Shu, H. (2015). Engineering tobacco to remove mercury from polluted soil. *Applied Biochemistry and Biotechnology*, *175*(8), 3813–3827.

Chatterjee, J., & Meighen, E. A. (1995). Biotechnological applications of bacterial bioluminescence (lux) genes. *Photochemistry and Photobiology*, *62*(4), 641–650.

Chebbi, A., Hentati, D., Zaghden, H., Baccar, N., Rezgui, F., Chalbi, M., … & Chamkha, M. (2017). Polycyclic aromatic hydrocarbon degradation and biosurfactant production by a newly isolated *Pseudomonas* sp. strain from used motor oil-contaminated soil. International Biodeterioration & Biodegradation, 122, 128–140.

Chen, D., Wang, H., & Yang, K. (2016). Effective biodegradation of nitrate, Cr (VI) and p-fluoronitrobenzene by a novel three dimensional bioelectrochemical system. *Bioresource Technology*, *203*, 370–373.

Chen, J., Qin, J., Zhu, Y. G., de Lorenzo, V., & Rosen, B. P. (2013). Engineering the soil bacterium *Pseudomonas putida* for arsenic methylation. *Applied and Environmental Microbiology*, *79*(14), 4493–4495.

Dash, H. R., Mangwani, N., & Das, S. (2014). Characterization and potential application in mercury bioremediation of highly mercury-resistant marine bacterium *Bacillus thuringiensis* PW-05. *Environmental Science and Pollution Research*, *21*(4), 2642–2653.

Deepa, C. N., & Suresha, S. (2014). Biosorption of lead (II) from aqueous solution and industrial effluent by using leaves of *Araucaria cookii*: Application of response surface methodology. *IOSR Journal of Environmental Science, Toxicology and Food Technology*, *8*(7), 67–79.

Diez, M. C. (2010). Biological aspects involved in the degradation of organic pollutants. *Journal of Soil Science and Plant Nutrition*, *10* (3), 244–267.

Dixit, P., Singh, S., Vancheeswaran, R., Patnala, K., & Eapen, S. (2010). Expression of a *Neurospora crassa* zinc transporter gene in transgenic *Nicotiana tabacum* enhances plant zinc accumulation without co-transport of cadmium. *Plant, Cell & Environment, 33*(10), 1697–1707.

Doty, S. L. (2008). Enhancing phytoremediation through the use of transgenics and endophytes. *New Phytologist, 179*(2), 318–333.

Doty, S. L., Shang, T. Q., Wilson, A. M., Tangen, J., Westergreen, A. D., Newman, L. A., ... & Gordon, M. P. (2000). Enhanced metabolism of halogenated hydrocarbons in transgenic plants containing mammalian cytochrome P450 2E1. *Proceedings of the National Academy of Sciences, 97*(12), 6287–6291.

Duhoux, A., Carrère, S., Gouzy, J., Bonin, L., & Délye, C. (2015). RNA-Seq analysis of ryegrass transcriptomic response to an herbicide inhibiting acetolactate-synthase identifies transcripts linked to non-target-site-based resistance. *Plant Molecular Biology, 87*(4–5), 473–487.

Ezezika, O. C., & Singer, P. A. (2010). Genetically engineered oil-eating microbes for bioremediation: Prospects and regulatory challenges. *Technology in Society, 32*(4), 331–335.

Farnham, K. R., & Dube, D. H. (2015). A semester-long project-oriented biochemistry laboratory based on *Helicobacter pylori* urease. *Biochemistry and Molecular Biology Education, 43*(5), 333–340.

Fasani, E., Manara, A., Martini, F., Furini, A., & Dal Corso, G. (2018). The potential of genetic engineering of plants for the remediation of soils contaminated with heavy metals. *Plant, Cell & Environment, 41*(5), 1201–1232.

Frederick, T. M., Taylor, E. A., Willis, J. L., Shultz, M. S., & Woodruff, P. J. (2013). Chromate reduction is expedited by bacteria engineered to produce the compatible solute trehalose. *Biotechnology Letters, 35*(8), 1291–1296.

French, C. E., Rosser, S. J., Davies, G. J., Nicklin, S., & Bruce, N. C. (1999). Biodegradation of explosives by transgenic plants expressing pentaerythritol tetranitrate reductase. *Nature Biotechnology, 17*(5), 491–494.

Fritsche, W., & Hofrichter, M. (2001). Aerobic degradation by microorganisms. In: Rehm, H.-J., & Reed, G., (2nd Ed., Eds.), *Biotechnology Set*. Wiley-VCH Verlag GmbH, Weinheim, Germany, 144–167. Doi: 10.1002/9783527620999.ch6m.

Gao, C., Jin, X., Ren, J., Fang, H., & Yu, Y. (2015). Bioaugmentation of DDT-contaminated soil by dissemination of the catabolic plasmid pDOD. *Journal of Environmental Sciences, 27*, 42–50.

Garbisu, C., Garaiyurrebaso, O., Epelde, L., Grohmann, E., & Alkorta, I. (2017). Plasmid-mediated bioaugmentation for the bioremediation of contaminated soils. *Frontiers in Microbiology, 8*, 1966.

Gautam, R. K., Soni, S., & Chattopadhyaya, M. C. (2015). Functionalized magnetic nanoparticles for environmental remediation. In: Soni, S., Salhotra, A., & Suar, M. (Eds.) *Handbook of Research on Diverse Applications of Nanotechnology in Biomedicine, Chemistry, and Engineering*. IGI Global, New Delhi. 518–551.

Gautam, S., Kaithwas, G., Bhaeagava, R.N. & Saxena, G. (2017). Pollutants in tannery wastewater: Pharmacological effects, and bioremediation approaches for human health protection and environmental safety. In: Bharagava, R.N. (Ed.), *Environmental Pollutants and Their Bioremediation Approaches*. CRC Press, Boca Raton, Florida 369–396.

Germaine, K. J., Keogh, E., Ryan, D., & Dowling, D. N. (2009). Bacterial endophyte-mediated naphthalene phytoprotection and phytoremediation. *FEMS Microbiology Letters, 296*(2), 226–234.

Gupta, D. K., Chatterjee, S., Datta, S., Voronina, A. V., & Walther, C. (2016). Radionuclides: Accumulation and transport in plants. *Reviews of Environmental Contamination and Toxicology, 241*, 139–160.

Ibuot, A., Dean, A. P., McIntosh, O. A., & Pittman, J. K. (2017). Metal bioremediation by CrMTP4 over-expressing *Chlamydomonas reinhardtii* in comparison to natural wastewater-tolerant microalgae strains. *Algal Research, 24*, 89–96.

Jiang, J., Zhang, R., Li, R., Gu, J. D., & Li, S. (2007). Simultaneous biodegradation of methyl parathion and carbofuran by a genetically engineered microorganism constructed by mini-Tn5 transposon. *Biodegradation, 18*(4), 403.

Joutey, N. T., Bahafid, W., Sayel, H., & Ghachtouli, N. E. (2013) Biodegradation: Involved microorganisms and genetically engineered microorganisms. Doi: 10.5772/56194.

Kang, J. W. (2014). Removing environmental organic pollutants with bioremediation and phytoremediation. *Biotechnology Letters, 36*(6), 1129–1139.

Kawahigashi, H., Hirose, S., Iwai, T., Ohashi, Y., Sakamoto, W., Maekawa, M., & Ohkawa, Y. (2007a). Chemically induced expression of rice OSB2 under the control of the OsPR1. 1 promoter confers increased anthocyanin accumulation in transgenic rice. *Journal of Agricultural and Food Chemistry, 55*(4), 1241–1247.

Kawahigashi, H., Hirose, S., Ohkawa, H., & Ohkawa, Y. (2006). Phytoremediation of the herbicides atrazine and metolachlor by transgenic rice plants expressing human CYP1A1, CYP2B6, and CYP2C19. *Journal of Agricultural and Food Chemistry, 54*(-8), 2985–2991.

Kawahigashi, H., Hirose, S., Ohkawa, H., & Ohkawa, Y. (2007b). Herbicide resistance of transgenic rice plants expressing human CYP1A1. *Biotechnology Advances, 25*, 75–85.

King, J. M. H., DiGrazia, P. M., Applegate, B., Burlage, R., Sanseverino, J., Dunbar, P., ... Sayler, G. A. (1990). Rapid, sensitive bioluminescent reporter technology for naphthalene exposure and biodegradation. *Science, 249*(4970), 778–781.

Kostal, J., Yang, R., Wu, C. H., Mulchandani, A., & Chen, W. (2004). Enhanced arsenic accumulation in engineered bacterial cells expressing ArsR. *Applied and Environmental Microbiology, 70*(8), 4582–4587.

Kulshreshtha, S. (2013). Genetically engineered microorganisms: A problem solving approach for bioremediation. *J BioremedBiodegr, 4*(4), 1–2.

Kumar, V. (2018). Mechanism of microbial heavy metal accumulation from polluted environment and bioremediation. In: Sharma, D., & Saharan, B. S. (Eds.), *Microbial Fuel Factories.* CRC Press, Boca Raton, FL.

Kumar, V., Shahi, S. K., & Singh, S. (2018). Bioremediation: An eco-sustainable approach for restoration of contaminated sites. In: Singh, J., Sharma, D., Kumar, G., & Sharma, N. (Eds.), *Microbial Bioprospecting for Sustainable Development.* Springer, Singapore. Doi: 10.1007/978-981-13-0053-0_6.

Lakherwal, D. (2014). Adsorption of heavy metals: A review. *International Journal of Environmental Research and Development, 4*(1), 41–48.

Li, C., Zhang, C., Song, G., Liu, H., Sheng, G., Ding, Z., Wang, Z., Sun, Y., Xu, Y., & Chen, J. (2016). Characterization of a protocatechuate catabolic gene cluster in *Rhodococcusruber* OA1 involved in naphthalene degradation. *Annals of Microbiology, 66*(1), 469–478.

Li, Q., & Wu, Y. J. (2014). A safety type genetically engineered bacterium with red fluorescence which can be used to degrade organophosphorus pesticides. *International Journal of Environmental Science and Technology, 11*(4), 891–898.

Liu, S., Zhang, F., Chen, J., & Sun, G. (2011). Arsenic removal from contaminated soil via biovolatilization by genetically engineered bacteria under laboratory conditions. *Journal of Environmental Sciences, 23*(9), 1544–1550.

López, A., Lázaro, N., Morales, S., & Marqués, A. M. (2002). Nickel biosorption by free and immobilized cells of *Pseudomonas fluorescens* 4F39: A comparative study. *Water, Air, and Soil Pollution, 135*(1), 157–172.

Marinescu, M., Dumitru, M., & Lăcătuşu, A. R. (2009). Biodegradation of petroleum hydrocarbons in an artificial polluted soil. *Research Journal of Agricultural Science, 41*(2), 157–162.

Mateos, L. M., Villadangos, A. F., Alfonso, G., Mourenza, A., Marcos-Pascual, L., Letek, M., Pedre, B., Messens, J., & Gil, J. A. (2017). The arsenic detoxification system in corynebacteria: Basis and application for bioremediation and redox control. *Advances in Applied Microbiology, 99*, 103–137.

Menn, F. M., Easter, J. P., & Sayler, G. S. (2008). Genetically engineered microorganisms and bioremediation. In: H.-J. Rehm, & G. Reed, (2nd Ed., Eds.), *Biotechnology: Environmental Processes II 11b*. Wiley-VCH Verlag GmbH, Weinheim, Germany. Doi: 10.1002/9783527620951.ch21.

Naik, M. M., Shamim, K., & Dubey, S. K. (2012). Biological characterization of lead-resistant bacteria to explore role of bacterial metallothionein in lead resistance. *Current Science, 103*(4), 426–429.

Okolo, N. V., Olowolafe, E. A., Akawu, I., & Okoduwa, S. I. R. (2016). Effects of industrial effluents on soil resources in Challawa industrial area, Kano, Nigeria. *Journal of Global Ecology and Environment, 5*(1), 1–10.

Ouyang, J., Guo, W., Li, B., Gu, L., Zhang, H., & Chen, X. (2013). Proteomic analysis of differential protein expression in *Acidithiobacillus ferrooxidans* cultivated in high potassium concentration. *Microbiological Research, 168*(7), 455–460.

Pandotra, P., Raina, M., Salgotra, R. K., Ali, S., Mir, Z. A., Bhat, J. A., Tyagi, A., & Upadhahy, D. (2018). Plant-bacterial partnership: A major pollutants remediation approach. In: Oves, M., Zain Khan, M., & M.I. Ismail, I. (Eds.), *Modern Age Environmental Problems and their Remediation*. Springer, Cham, 169–200.

Pant, G., Garlapati, D., Agrawal, U., Prasuna, R. G., Mathimani, T., & Pugazhendhi, A. (2020). Biological approaches practised using genetically engineered microbes for a sustainable environment: A review. *Journal of Hazardous Materials, 405*, 124631.

Paul, D., Pandey, G., & Jain, R. K. (2005). Suicidal genetically engineered microorganisms for bioremediation: Need and perspectives. *Bioessays, 27*(5), 563–573.

Peeters, C., De Canck, E., Cnockaert, M., Brandt, E. D., Snauwaert, C., Verheyde, B., Depoorter, E., Spilker, T., LiPuma, J. J., & Vandamme, P. (2019). Comparative genomics of pandoraea, a genus enriched in xenobiotic biodegradation and metabolism. *Frontiers in Microbiology*, 10: 2556.

Porter, S. S., Chang, P. L., Conow, C. A., Dunham, J. P., & Friesen, M. L. (2017). Association mapping reveals novel serpentine adaptation gene clusters in a population of symbiotic Mesorhizobium. *The ISME Journal, 11*(1), 248–262.

Qin, J., Rosen, B. P., Zhang, Y., Wang, G., Franke, S., & Rensing, C. (2006). Arsenic detoxification and evolution of trimethylarsine gas by a microbial arsenite S-adenosylmethionine methyltransferase. *Proceedings of the National Academy of Sciences, 103*(7), 2075–2080.

Ripp, S., Nivens, D. E., Ahn, Y., Werner, C., Jarrell, J., Easter, J. P., … Sayler, G. S. (2000). Controlled field release of a bioluminescent genetically engineered microorganism for bioremediation process monitoring and control. *Environmental Science & Technology, 34*(5), 846–853.

Ruiz, O. N., Alvarez, D., Gonzalez-Ruiz, G., & Torres, C. (2011). Characterization of mercury bioremediation by transgenic bacteria expressing metallothionein and polyphosphate kinase. *BMC Biotechnology, 11*(1), 1–8.

Sayler, G. S., Cox, C. D., Burlage, R., Ripp, S., Nivens, D. E., Werner, C., … Matrubutham, U. (1999). Field application of a genetically engineered microorganism for polycyclic aromatic hydrocarbon bioremediation process monitoring and control. In: Fass, R., Flashner, Y., & Reuveny, S. (Eds.), *Novel Approaches for Bioremediation of Organic Pollution*. Springer, Boston, MA, 241–254.

Sayler, G. S., & Ripp, S. (2000). Field applications of genetically modified bacteria for bioremediation processes. *Current Opinion in Biotechnology, 11*, 286–289.

Schüürmann, J., Quehl, P., Festel, G., & Jose, J. (2014). Bacterial whole-cell biocatalysts by surface display of enzymes: Toward industrial application. *Applied Microbiology and Biotechnology, 98*(19), 8031–8046.

Seo, J. S., Keum, Y. S., & Li, Q. X. (2009). Bacterial degradation of aromatic compounds. *International Journal of Environmental Research and Public Health, 6*(1), 278–309.

Shukla, K. P., Singh, N. K., & Sharma, S. (2010). Bioremediation: Developments, current practices and perspectives. *Journal of Genetic Engineering and Biotechnology, 3,* 1–20.

Siddiquee, S., Rovina, K., Azad, S. A., Naher, L., Suryani, S., & Chaikaew, P. (2015). Heavy metal contaminants removal from wastewater using the potential filamentous fungi biomass: A review. *Journal of Microbial and Biochemical Technology, 7*(6), 384–393.

Singh, J. S., Abhilash, P. C., Singh, H. B., Singh, R. P., & Singh, D. P. (2011). Genetically engineered bacteria: An emerging tool for environmental remediation and future research perspectives. *Gene, 480*(1–2), 1–9.

Singh, U., Arora, N. K., & Sachan, P. (2018). Simultaneous biodegradation of phenol and cyanide present in coke-oven effluent using immobilized *Pseudomonas putida* and *Pseudomonas stutzeri. Brazilian Journal of Microbiology, 49,* 38–44.

Singh, V., Singh, P., & Singh, N. (2016). Synergistic influence of Vetiveriazizanioides and selected rhizospheric microbial strains on remediation of endosulfan contaminated soil. *Ecotoxicology, 25*(7), 1327–1337.

Tozzini, A. C., Martínez, M., Lucca, M. F., Vázquez Rovere, C., Distéfano, A. J., del Vas, M., & Hopp, E. (2000). Semi-quantitative detection of genetically modified grains based on CaMV 35S promoter amplification. *Electronic Journal of Biotechnology, 3* (2), 25–34.

Turpeinen, R., Kairesalo, T., & Haggblom, M. (2002). Microbial activity community structure in arsenic, chromium and copper contaminated soils. *Journal of Environmental Microbiology, 35*(6), 998–1002.

United Nations. (2016). *The Sustainable Development Goals 2016.* Working papers, eSocialSciences.

Vinopal, S., Ruml, T., & Kotrba, P. (2007). Biosorption of Cd^{2+} and Zn^{2+} by cell surface-engineered *Saccharomyces cerevisiae. International Biodeterioration & Biodegradation, 60*(2), 96–102.

Wai, W.L., Kyaw, N. A. K., & Nway, N. H. N. (2012) Biosorption of lead (Pb2+) by using *Chlorella vulgaris.* In *Proceedings of the International Conference on Chemical engineering and Its applications (ICCEA),* Bangkok, Thailand.

Wang, Y., Jiang, Q., Zhou, C., Chen, B., Zhao, W., Song, J., Fang, R., Chen, J., & Xiao, M. (2014). *In-situ* remediation of contaminated farmland by horizontal transfer of degradative plasmids among rhizosphere bacteria. *Soil Use and Management, 30*(2), 303–309.

Wang, Y., Li, H., Zhao, W., He, X., Chen, J., Geng, X., & Xiao, M. (2010). Induction of toluene degradation and growth promotion in corn and wheat by horizontal gene transfer within endophytic bacteria. *Soil Biology and Biochemistry, 42*(7), 1051–1057.

Yang, C., Xu, L., Yan, L., & Xu, Y. (2010). Construction of a genetically engineered microorganism with high tolerance to arsenite and strong arsenite oxidative ability. *Journal of Environmental Science and Health Part A, 45*(6), 740–745.

Ye, J., Rensing, C., Rosen, B. P., & Zhu, Y. G. (2012). Arsenic biomethylation by photosynthetic organisms. *Trends in Plant Science, 17*(3), 155–162.

Zhang, R., Xu, X., Chen, W., & Huang, Q. (2016). Genetically engineered *Pseudomonas putida* X3 strain and its potential ability to bioremediate soil microcosms contaminated with methyl parathion and cadmium. *Applied Microbiology and Biotechnology, 100*(4), 1987–1997.

Zhang, Y., Liu, J., Zhou, Y., Gong, T., Wang, J., & Ge, Y. (2013). Enhanced phytoremediation of mixed heavy metal (mercury)–organic pollutants (trichloroethylene) with transgenic alfalfa co-expressing glutathione S-transferase and human P450 2E1. *Journal of Hazardous Materials, 260,* 1100–1107.

24 Microbial Biofilm in Remediation of Environmental Contaminants from Wastewater

Mechanisms, Opportunities, Challenges, and Future Perspectives

Pallavi Singh and Akshita Maheshwari
CSIR-National Botanical Research Institute

Varsha Dharmesh and Vandana Anand
Academy of Scientific and Innovative Research
CSIR-National Botanical Research Institute

Jasvinder Kaur
CSIR-National Botanical Research Institute

Sonal Srivastava
Academy of Scientific and Innovative Research
CSIR-National Botanical Research Institute

Satish Kumar Verma
Banaras Hindu University (BHU)

Suchi Srivastava
Academy of Scientific and Innovative Research
CSIR-National Botanical Research Institute

DOI: 10.1201/9781003247883-24

CONTENTS

24.1 INTRODUCTION

Wastewater comprises effluents discharged from the household sewage, industries, and agriculture. It includes both the organic and biological materials. Major organic contaminants in effluents include pesticides, detergents, oils, and fats along with various microbes, *viz.* viruses, bacteria, fungus, and protozoa. Additionally, wastewater is rich in inorganic materials, metals, and basic nutrients, such as hydrogen sulphide, mercury, cadmium, ammonia, nitrogen, and phosphorous. Releasing inadequately treated wastewater in water bodies has adverse effects on human health, aquatic life, and the environment (Kumar et al., 2020, 2021a, b). Loss of biodiversity, eutrophication, and aquatic ecosystem imbalance are the major causes of the discharge of such pollutants in the environment (Deyerling et al., 2013). Along with that, the process of breakdown of organic materials releases greenhouse gases such as methane and nitrous oxide. Thus, there is an immediate need to clean the wastewater and transform it to an attractive resource majorly to protect the population health by formulating novel scientific and engineering solutions.

In the past few decades, biofilms, a process of attached growth, have been used for the biological treatment of wastewater by attaching them to biocarriers and delivering substrates such as nitrogen (ammonia and nitrates), BOD (biological oxygen demand), and dissolved oxygen from bulk solution to the interface. A new generation of microbes is synthesized, and their metabolic consumption helps to remove wastewater contaminants by exploiting these delivered nutrients. However, during the lifecycle, wastewater biofilm has been summarized by only a few studies; thus, a knowledge gap is there on the characterization and regulation of various stages of biofilm. Thus, this chapter fills the gap and presents an effective assistance for innovative applications of process of biofilm by providing a complete idea of the process, reason and factors affecting the formation and development of biofilm, along with various other microbial, physiological, and ecological activities, especially initial formation and characterization processes. An outline of the entire life cycle of biofilm has been summarized with the concept of ageing. Moreover, future prospects of challenges and opportunities in biofilm-based technologies aim to offer guidelines on the control of biofilm in wastewater treatment.

24.2 SOURCES OF INDUSTRIAL WASTEWATER

Industrial revolution has brought in new technologies with enormous power, which constituted the misuse and deterioration of natural resources. Instantaneous development of various industries has led to the utilization of hefty amounts of fresh water for cooling purposes and fabrication. Effluents thus formed are discharged directly into the ocean or other water bodies, which wrecks the aquatic life, is generally venomous, and can consequentially cause health issues. These effluents also display physical properties forging the water aesthetically obnoxious. Along with heavy metals content, the key physical attributes of the effluents discharged are colour and odour. Solid effluents perceive a dual nature as suspended and insoluble form. Some parameters of effluents, *viz.* pH, biological and chemical oxygen demand (BOD and COD), total dissolved solids (TDS), and total organic carbon (TOC), should be controlled before discharge (Chandra and Kumar, 2017). A heavy chunk of organic as well as inorganic dyes is discharged by textile industries into the marine environment. Different types of industries well known as a source of industrial wastewater pollutants are as follows:

a. **Textile industries** obsessively use water in fabric processing. Voluminous numbers of chemicals are used for cleaning and dyeing procedures; thus, the end products from these industries consist of dangerous contaminants, heavy metals, aqueous waste and organic dyes. These contaminants affect the photosynthetic rate of aquatic species and reduce the sunlight penetration through the water and are highly poisonous to the aquatic life. Increased BOD levels of organic and inorganic pollutants are detrimental to the environment. The dye-containing effluent is carcinogenic and mutagenic in living vital force. It leads to severe health issues such as respiratory disorders, hypersensitive reaction in the eyes, allergies and skin irritation due to contact dermatitis. The release of raw wastewater severely influences the groundwater quality and soil productivity.

b. **Power plants and nuclear industries**: The growth and development of power plants is directly affiliated with energy requirements in various agricultural and industrial sectors along with the human population. Each step in the process of power generation is guided by the utilization of tremendous amount of water from the generation of steam to auxiliary cooling; thus, the discharged wastewater is suspended with various solids and heavymetals such as chlorine, phosphates, zinc, and chromium. It affects the pH, suspended solids, BOD, COD, heavymetals and TDS in water and land ecosystems.

c. **Iron and steel industry** is the major site of production of crude steel in India. Wastewater released for 1 million tonnes of ingot steels is 250–500 m³/h, and thus, water is reused to meet this huge demand.

d. **Pulp and paper industry** is directly related to both human and industrial development requirements (Kumar et al., 2020). A high amount of water (20, 000 and 60, 000 gallons) is used for the production of a ton of product. Different paper, pharmaceutical, bleaching, and tannery industries discharge various dyes of organic and inorganic nature. In terms of usage of the amount of water, the chemical industry utilizes much less than the textile industry; thus, high demand and less availability of resources call for recycling.

e. **Domestic sewage wastes** are produced from the human consumption, utilization and excretion. These are natural products quickly broken down by natural decomposers in the aquatic environment to liberate nutrients such as nitrogen and phosphorus.. In lakes with excessive concentrations of nitrates and phosphates, with an availability of sufficient temperature and adequate sunlight, lay the groundwork for a rapid algal bloom, causing vandalism to water and intensifying the turbidity of the water bodies. Algal blooms generated due to sewage discharges destruct the lake ecosystem by accelerating the natural ageing process of lakes known as eutrophication. Kroiss et al. (2011) stated that sewage sludge also contains other hazardous substances such as heavy metals, micropollutants, and pathogens.

TABLE 24.1

Types of Pollutant Present in Industrial Sectors

Sectors	Pollutants
Iron and steel	Metals, oil, phenols, and cyanide
Paper and pulp	Chlorinated hydrocarbons
Leather and textiles	Sulphates and organic compounds
Petrochemicals and refineries	Minerals, oil, and phenols
Chemicals	Organic chemicals, heavy metals, and cyanide
Non-ferrous metals	Fluorine
Mining	Metals, acids, and salts

24.2.1 INDUSTRIAL EFFLUENTS

Enormous amount of industrial wastewater discharged into rivers, lakes, and ponds results in the contamination of water ecosystem and thus causes a threat to human life. Compositions of wastewater released from different industries are described in Table 24.1

24.2.2 INORGANIC SOURCE

Steel coal and other metallic industries use certain basic facilities such as floor cleaner, which pollute water and produce inorganic wastewater. In coal industries, acids and scrubbers are being used to separate coal from dead rocks, which are further transported through water. These particles of coal and rock contaminate the water. Industries are another wastewater pollutant source as mineral oil is present in the rolling mills, which is further removed by flocculation method. Metal processing industries discharge highly toxic wastes such as non-ferrous metals, chromates, and cyanides in acidic and alkaline solutions along with fluorides that are produced from aluminium works into the wastewater.

24.2.3 ORGANIC SOURCE

Organic industrial wastewater is potentially produced from the organic chemicals used in chemical industries. Organic waste is contributed in large amounts from various different sources such as vegetable, fruit, and meat packaging; dairy and poultry processing; and paper, wood, oil and tanning industries (Kumar et al., 2021c; Kumar and Thakur, 2020). Organic wastes with low biodegradability are produced by textile industries, pharmaceuticals, paper and cellulose industries, brewery, etc. Along with unpleasant odour, organic waste is one of the primary causes of oxygen depletion in receiving wastewater streams. Such streams have high pH, have bad colour, and require long-term biological treatment and pre-biological treatment processes. Ammonium is present in large quantities in the wastewater produced by solid waste disposal sites of urban pharmaceutical, petrochemical, food, and fertilizer industries.

Free ammonia diluted in the water is one of the most hazardous contaminants depleting aquatic life (Foroughi et al., 2013). Types of pollutants present in wastewater vary with their sources; most of the wastewater generated from industries has high heavy metals and less nitrogen and phosphorous compared to others. Agro-industrial wastewater has high nitrogen and phosphorous due to obligatory anaerobic treatment. Synthetic detergents mainly constitute phosphorous, which also inevitably reaches the wastewater. Effluents from industries contain total phosphorous in the range of 1–5 mg/L, out of which 5–9 mg/L is inorganic and the rest is organic. The two major forms of phosphorous in the environment and in the aqueous solutions are polyphosphates and orthophosphates. Secondary treatments are only able to remove a very little amount of phosphorous, and thus, the rest gets discharged into surface water causing eutrophication.

24.3 BIOFILM

The most common mode of existence of microorganisms is surface attachment, but the word "biofilm" was described as "microbial slimes" until the year 1960s and 1970s. Biofilm is the communities of microorganisms encapsulated in the self-formed extracellular polymeric substance (EPS) matrices that are attached to the living or non-living surfaces. The biofilm matrix produces EPSs that make the cells aggregate together and form the structure of the matrix. Biofilm is very common, and it is encountered in day-to-day life; it is found on the surface of stones in any water source, in the water pipes, in showers as slimy coating, on the surface of boats, on teeth as dental plaque, etc. The biofilm bacterial cells show some properties that are different from planktonic properties, *viz.* tolerance to antimicrobial substances, ability to act as a physical barrier to the environmental stress, and degradation of toxic chemicals. The structural biofilm matrix is also overstated by several other conditions, such as attachment to surface; surface and interface properties; nutrient availability; composition type of microbial community; and hydrodynamics.

24.3.1 Extracellular Polymeric Substances

Extracellular polymeric substances are complex polymers of high molecular weight substances that are produced as a result of hydrolysis as well as adsorption of the organic pollutants from the wastewater. They include polysaccharides, proteins, phospholipids, and humic acids (Flemming and Wingender, 2010). Apart from this, they serve as the scaffolds for the adherence of other carbohydrates, DNA oligomers, and lipids on the surface. EPS matrices have widely been studied in *A. baumannii, E. faecalis, K. pneumonia, P. aeruginosa,* and *S. aureus* biofilms having most abundant carbohydrates such as glucose, galactose and mannose, followed by galacturonic acid, n-acetyl glucosamine, arabinose, xylose, rhamnose, and fucose (Bales et al., 2013). Alginic acid and colanic acids are extracellular polymeric heteropolysaccharides that are produced mostly under stress conditions. Alginates associated with the biofilms of *P. aeruginosa* are composed of interspersed L-glucouronic acid residues in d-mannuronic acid residues and act as a critical virulence factor for causing chronic infections (May et al., 1991; Friedman and Kolter, 2004; Jackson et al.,

2004). EPS matrices can integrate large amounts of water into their structure with the help of hydrogen bonding. Microbial cells linked with hydrophobic interactions with multivalent cations form an intact gel-like matrix. Along with this, EPS provides a strategy to adhere to the surfaces in biofilm, granulation and flocculation to protect the bacteria from toxic conditions in the environment and improves the soil fertility by retaining the soil moisture (Costa et al., 2018). It enables the bacteria to aggregate soil particles and benefits the plant by maintaining soil moisture and nutrient capturing from the surroundings. Various biofilms produce different amounts of EPSs, and there is an increase in amounts with the maturation of biofilm.

24.3.2　Biofilm Formation Process

A major prerequisite of microbial assemblage in the formation of biofilm is that the microbial population present on the surface should get close enough. Various forces, i.e. repulsive and attractive forces, act upon as the bacteria approach the surface. The repulsion occurs at an approximate distance of 10–20 nm, if the negative charges present on the surface of the bacteria face the negative charges present on the environmental surfaces (Palmer et al., 2007). However, these repulsive forces can be subdued by attractive forces (van der Waals) in between bacteria and the surface environment along with that mechanical attachment can also be provided to the surface by flagella and fimbriae (Palmer et al., 2007). Generally, the biofilm development can be described in five major stages (Figure 24.1): (1) primary attachment of planktonic microbial community to the aqueous medium surface; (2) irreversible adhesion of the microorganism *via* microorganism-mediated EPSs since the EPSs have polyhydroxyl groups that lead to surface colonization of microorganisms with the help of hydrogen

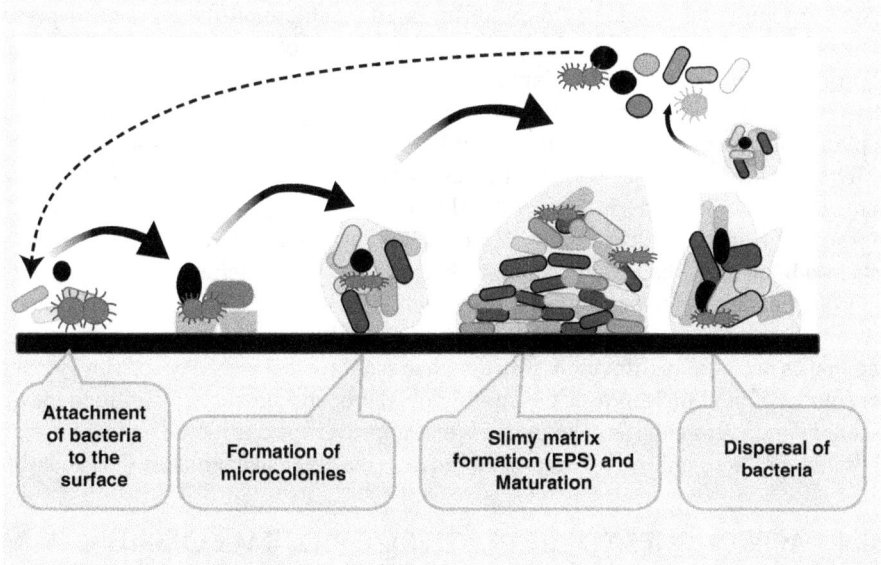

FIGURE 24.1　Stages of biofilm development.

bonding (Judd, 2008). The irreversible adhesion of biofilm is able to tolerate stronger chemical or physical shear forces. (3) Formation of monolayer microcolonies; (4) maturation or ageing of biofilm as debris present in adjacent environment attaches to it and utilization by novel planktonic microorganism; and (5) dispersion where matrix-enclosed biofilm cells are converted to free swimming planktonic microbes *via* quorum sensing (QS) mechanism or a mechanism of cell-to-cell signalling (Webb, 2007). The attachment requires type IV pili-mediated motilities and flagella, which play a significant role in initial interactions between the surface and cells. Twitching motilities enable microbes for aggregation and microcolony formation. Surface components of the microbes recognize the adhesion-dependent adhesive matrix molecules and get covalently linked to the peptidoglycan present on the cell wall. Initial attachment of biofilms is also contributed by non-covalent adhesions such as autolysin-mediated adhesion (Heilmann et al., 1997). Alginate, the complex polysaccharide, and EPSs that are produced by the microbes attached to the surface are present in manifold quantities as compared to planktonic cells. Alternative (sigma) σ factor *alg* T, is important for the production of alginates, also downregulates flagella genes (Garrett et al., 1999). The biofilm grows in mushroom shape structure from a monolayer biofilm (just after establishment). Microbial cells in the biofilm may be the same, or cells from other species from the bulk fluid can also get associated with the biofilm. A thick biofilm is formed in accordance with the aerotolerance and metabolism. The crosstalk in the biofilm communities gets developed and involves specialized functions. With the maturation of biofilm, more scaffolds of biofilms such as polysaccharides, proteins and DNA get secreted by the entrapped microbes into the biofilm. After maturation of the biofilm, it gets dispersed, which is a significant step for the life cycle of the biofilm. Various factors are responsible for the dispersal of biofilms, such as passionate competition, unavailability of nutrition and overpopulation. A part of the biofilm or the whole biofilms can get dispersed depending upon these factors. The process of initiation of new biofilm at another site is promoted by the release of planktonic bacterial cells. Usually, the dispersal of a few parts of the biofilm takes place as a continuous process, which is the final step (a planktonic form from a mature biofilm) in the development process of a biofilm.

The planktonic state of bacteria possesses a relatively high growth and reproduction rate. However, bacteria grow naturally and profoundly in the biofilm state. The presence of bacteria switches between biofilm and planktonic state, where biofilm state acts as a boon to the bacteria residing in this state and the bacteria present in the harsh environment develop tolerance due to their biofilm state. Second, it enhances their surface attachment in running water flow as the slimy layer gets attached to the surface and makes bacterial biofilm more resistant than planktonic form. EPS, the fundamental constituent of biofilm, improves the antimicrobial property probably by limiting the diffusion of antimicrobial agents in the cells and improves the nutritional and water availability by promoting the persistent metabolism in the atypical conditions (Stone, 2018).

24.4   APPROACHES TO CHARACTERIZE BIOFILM COMMUNITY

The upcoming sections constitute the discussion on characterization approaches of various biofilm community.

24.4.1 TRADITIONAL METHODS

24.4.1.1 Weight Determination

The weight of the biofilm can be determined using a digital weighing balance in terms of both wet and dry weight. The wet weight can be measured after rinsing the biofilm with distilled water. On the other hand, the dry weight is predicted by drying it in aseptic conditions (laminar flow) until a constant weight of polypropylene and polystyrene filter media is attained (Naz et al., 2013; Khatoon et al., 2014). However, oven-drying of natural filter media, such as rocks, stones, or granite, should be done at 60°C temperature. Then the biofilm weight is calculated by subtracting the weight of medium with and without biofilm.

24.4.1.2 Optical Density Determination

The absorbance method can also be used to measure the biofilm. The biofilm supported by filter media is first washed with sterilized water in order to remove the contaminants present on the surface. Then extraction of the biofilm from the filter media takes place by sonication in 0.9% saline for 15 min. Lastly, the absorbance is recorded by a spectrophotometer at 550 nm wavelength of the biofilm taking saline as blank (Naz et al., 2015).

24.4.1.3 Microscopic Biofilm Analysis

Microscopic technique is a non-invasive and more accurate way to visualize the biofilm without interrupting its structural integrity. However, the traditional microscopic techniques for biofilm sample analysis by imaging involve electron and light (EMand LM). But scanning electron microscopy (SEM) has proven as an ingrained elemental technique to examine the bacterial morphology and to measure the cell attachment, with the capability to demonstrate the surface-biofilm relationship. In addition to this, various advanced and novel techniques such as simple and confocal laser scanning microscopy (LSM and CLSM) and magnetic resonance imaging (MRI) are established. Another technique called scanning transmission X-ray microscopy (STXM) is also being used in biofilm analysis. *In situ* analysis of biofilm is achieved by these new techniques on the basis of composition, structure, dynamics, and process of microbial communities. Thus, mixed microbial communities can also be examined using these powerful tools and techniques, which are usually present in aggregate and biofilm forms (Palmer and Sternberg, 1999).

24.4.1.4 Biofilm Activity Determination

Metabolic activities of the biofilm constituting microbial communities can be determined by calculating the conversion rate of a specific substrate after inoculating it into the seed of biomass. A common example of *Nitrosomonas* spp., its physiological activity can be anticipated by analysing the strength of nitrites (NO_2-N)produced from a known concentration of nitrogen source ((NH_4)$_2SO_4$) in the growth media after a specific time interval (Naz et al., 2015). In a similar manner, the estimation of the rate of removal of nitrogenous (NH_4-N) and carbonaceous (COD and BOD) contaminants by using biofilms can be done.

24.4.2 Advanced Methods

24.4.2.1 Clone Library Technique

Sequencing of the 16S rRNA gene and cloning have been successfully employed for the study of microbial biofilms. The cloning technology includes the following steps: (1) extraction of the nucleic acid from the biofilm sample; (2) amplification of the 16S rRNA gene by polymerase chain reaction (PCR), usually using universal primers for bacteria or archaea, for obtaining a mixture of rDNA copies of the microorganisms; (3) cloning of the PCR products into an appropriately high number of copies of plasmid and then transformation of competent *Escherichia coli* cells with this vector; (4) selection of the transformed clones based on a plasmid containing an indicator; (5) plasmid DNA extraction from the clones; (6) cloned gene sequencing and creation of a cone library; and ultimately (7) sequence identification of isolated clone, using various phylogenetic software and computer programs such as PHYLIP, ARB, PAUP, and Seqlab. In general, advanced techniques in combination with construction of rRNA gene library and cloning are applied to explore biofilm communities for the treatment of wastewater (Sanz et al., 2007).

24.4.2.2 DNA Microarray Technology

DNA microarray technology is used to identify hundreds or even thousands of differentially expressed genes in the formation of biofilms (Shemesh et al., 2008). This process involves (1) extraction of DNA from the sample biofilms; (2) amplification of DNA by PCR; (3) PCR-amplified product hybridization from the total DNA to microarrays attached with known molecular probes; and (4) positive signal scoring using CLSM after hybridizing PCR amplicons to the probes labelled fluorescently and, as a result, a hybridization signal intensity in direct proportion to the abundance of target organism. This technique has a major limitation, cross-hybridization, which creates difficulty in identifying and detecting newly discovered microorganisms. In addition to this, if the genus is not having a consequent probe attached on the microarray, then it would be difficult to signify the genus. However, there is nominal application of this technique in the field of biofilm-mediated wastewater management.

24.4.2.3 Next-Generation Sequencing (NGS) Technology

Next-generation sequencing is a revolutionized DNA sequencing technology, where sequencing by synthesis principle is applied. In NGS, sequencing of millions of small fragments of DNA occurs. This technology provides an insight into microbial ecology with exploration of deeper layers of communities of microbes and gives impartial outlook of diversities and composition of communities. The steps involved in NGS are (1) DNA extraction from the sample biofilms; (2) checking the extracted DNA's purity and quantity using NanoDrop spectrophotometer; (3) PCR amplification of the samples using 16SrRNA gene along with universal primers, i.e. 28F and 519R, with the different barcodes incorporated between the forward primer and 454 adaptor; (4) PCR products which are purified are further used for pyrosequencing and then short adaptors are ligated to both ends for sequence segregation; (5) modified products are attached to the DNA beads; (6) clonal amplification; (7) pyrosequencing for 16S rRNA gene sequence, pre-processing at Ribosomal Database Project (RDP) for trimming of barcodes and primers are removed from the partial ribo tags along with discarding short and low-quality sequences; (8) generation of the FASTA file data sets; (9) analysis of these sequence through analysis pipeline (MOTHUR) and R-Scripts. The NGS technique has the potential utility in confirming the sequencing and removing the conventional technique of characterizing of microbes because it has the advantages of flexibility, accuracy, and easy automation (Ronaghi, 2001).

24.5 BIOFILM IN WASTEWATER TREATMENT

Wastewater is a comprehensive term including discharge or effluent from the households and industrial and agricultural sewage. Biofilm technology is a well-executed technology, where the process of biofilm development on varied filter media provides an attachment surface and increases the microbial concentration. The wastewater microbial cells have prominent nature of attachment to the surfaces as well as various substrates such as BOD, nitrogen, ammonia, nitrates and dissolved oxygen, which are delivered to the interface from bulk liquid. The increased microbial concentration enhances the biosorption of heavy metals and organic and inorganic pollutants. The biofilm system is advantageous over the suspended growth system in the removal of contaminants through the process of biosorption, bioaccumulation, biomineralization, and biodegradation. Besides, the biofilm-associated microbial communities breakdown different nutrients, such as carbonaceous materials, nitrogen-containing compounds and trapped pathogens from the wastewater. The efficient biosorption of heavy metals and pollutants by biofilm matrix has been found. Various organic and inorganic pollutants such as pesticides, chlorophenols, polycyclic aromatic hydrocarbons and heavy metals are adsorbed and accumulated by the EPSs produced by the microbial communities.

Biofilm-mediated wastewater treatment has emerged as an advantageous process due to several reasons such as operational flexibility, low-cost requirement, highly active biomass concentration, amplified biomass residence time, elasticity to the environmental changes, improved degradation of recalcitrant compounds, and lesser amount of sludge production. The treated water obtained from the pollutant removal can be used for recreational activities such as agricultural and other household activities. Biofilm system has become a hotspot in advanced wastewater treatment process due its high rate of removal of nitrogen and other organics, strong adaptability, convenient operational management, low sludge production, etc., thus making it ideal for application in wastewater treatment plants. Biofilm based on autotrophic denitrification, ammonium oxidation (anammox), steadily improves wastewater standards in many countries of the world (Augusto et al., 2018).In the biofilm system, microbial growth takes place on natural and man-made surfaces (*viz.* stone or filter media), enhancing the surface area of wastewater facilitating the passage and increasing the volume of the substrate that can be absorbed from the influent. Biofilms developed provide diverse habitats for the transformation and mineralization of constituents of wastewater such as carbon, nitrogen and phosphorus. An individual biofilm performs aerobic, anaerobic and anoxic processes and limiting substrate may amend the thickness of the biofilm. Thus, there is complexity in designing the fixed film-based technology (Eaton, 2005). Conventional wastewater treatment has been improved with a wide variety of biofilm reactors. Development and application of biofilm reactors to act on industrial wastewater and domestic sewage is done to treat the wastewater. These bioreactors can be of various types such as biological aerated filter (BAF), biological rotating disc, biological contact oxidation tank, biological fluidized bed, integrated fixed film activated sludge reactor and moving bed biofilm reactor (IFAS and MBBR) (Odegaard, 2006; Andreottola et al., 2002; Zhang et al., 2015; Boltz et al., 2017).

24.5.1 Types of Reactors Used in Secondary Treatment of Wastewater

24.5.1.1 Rotating Biological Contactor (RBC)

The rotating biological contactor (RBC) has same basic mechanism similar to the process of trickling filtration system. There has been overuse of the RBC in the 1960s–1970s; however, due to some pitfalls in meeting the expected performance and proliferation of unwanted microorganisms, their use has decreased. Nonetheless, various problems existed and this framework is again generally applied. The design involves a bunch of discs commonly produced using plastic on a similar shaft, which is circulated across the progression of influents. The shaft moves at the rate of 1–2 rpm that quickly permits the microbial growth thickness at around the moist surface of the shaft. The pivot continues and directs wastewater towards the air, moving oxygen to the layer of microbial development. The inordinate wastewater substances are taken out as sloughing by moving it with the progression of profluent. Nowadays, MRBC, a novel approach, is used, which replaces the conventional RBC in addition to membrane filtration. The performance of MBRC is approximately 92.5% higher than the conventional RBC practices (Waqas et al., 2021) (Figure 24.2).

24.5.1.2 Membrane Bioreactor (MBR)

Membrane bioreactor for wastewater treatment is a combination of low-pressure microfiltration equipment or ultrafiltration equipments and suspended growth bioreactors which are usually activated sludge. Membrane bioreactors used perform on

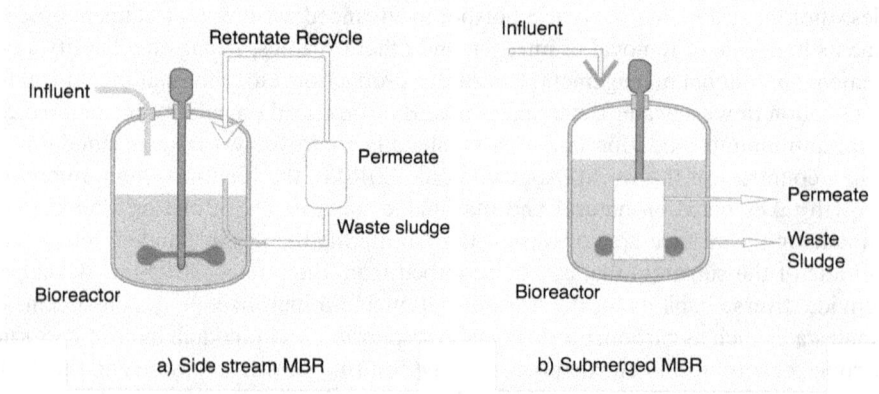

FIGURE 24.2 Side stream and submerged MBR configuration.

the principle of solid-liquid separation. The membrane bioreactor uses polymeric flat sheet membranes with a porosity of 0.003 and 0.001 µm (Judd, 2010). Nowadays, submerged MBR is more commonly employed, particularly for the domestic effluents due to their low energy demands.

24.5.1.3 Membrane Biofilm Reactor

A membrane biofilm reactor (MBfR) is the integration of separation membrane biofilm that regularly utilizes a gas-penetrable membrane that conveys the vaporous substrate (e.g., nitrogen, oxygen, hydrogen and methane) on the external surface of the membrane-shaped biofilm (Rittmann, 2006). Looking at the regular biofilms, the MBfR acts distinctively because of the counter-inclination dispersion of electron benefactor and acceptor substrates. The dispersion property of the two substrates resembles one of the substrates dispersed from the foundation of the biofilm, while the other substrate diffuses from the fluid stage (Nerenberg, 2016). Another type is membrane-aerated biofilm reactor (MABR), where the membrane utilizes the air or oxygen. Unlike the traditional activated sludge-based wastewater treatment process, the MABR with closed-end operation allows 100% oxygen transfer efficiency.

The MBfR technology is responsible for the elimination of organic carbon and nitrogen of upto 95%–98% and 80%–85%, respectively. High-strength industrial wastewater and other organic and inorganic pollutants have been successfully treated with MBfR technology. Findings from the study of Wu et al.(2019) showed reduction of per-chlorate in the presence and absence of oxygen in laboratory-scale methane MBfR system in around 1100–1150 days. While the system showed per-chlorate removal at the rate of 4 mg-Cl/L/day, the removal rate increased by about four times when limited oxygen (10 mg/L/day) was externally supplied. The biofilm framework accomplished a high ammonium and nitrate evacuation rate of 0.48 and 0.55 kg-N/m³/day, respectively, with no expansion of organic carbon. Lai et al. (2018) utilized methane-based MBfR with the functioning volume of 65 mL for concurrent evacuation of nitrate and bromate. When treating influent enriched with bromate concentration of 800 µg/L and having 40 hours hydraulic maintenance time, MBfR eliminated the bromate at a maximum rate of 40.7 µg/L/min.

24.5.1.4 Nutrient Removal Using Biofilm Process

Wastewater is the amalgamation of high levels of nitrogen and phosphorus components impacting the environment, like reduced oxygen level ecological turmoil and loss of aesthetic values of water system. Consequently, expulsion of harmful constituents from the environment is vital to diminish their damage to the environment (Wang et al., 2006). Biological wastewater treatment depends on the expulsion of suspended biomass as they are proficient in the evacuation of natural constituents through sedimentation tanks and reuse of biomass.. Application of biofilm reactors has ended up being more significant for the removal of nutrient, and there are fewer issues associated with membrane biofilm reactors.

24.5.1.4.1 Nitrogen Removal

The process of nitrification and denitrification are normally applied for the removal of nitrogenous components present in domestic as well as industrial effluents in the form of ammonia, particulate matter, and dissolved organic nitrogen. There are two aerobic stages related to nitrification; the initial one is oxidation of ammonia where it gets converted to nitrite *via* autotrophs. In the subsequent stage, nitrite is further converted to nitrate through oxidation by another group of autotroph (Vymazal, 2007). Microbial reduction of nitrates and nitrites to gaseous form of nitrogen, particularly nitrous oxide (N_2O) and nitrogen gas (N_2), in the absence of oxygen is known as denitrification. Denitrification can be accomplished in anaerobic conditions, utilizing heterotrophy, where nitrogen gas is created as eventual outcome.

24.5.1.4.2   Phosphorus Removal

Wastewater from household and industrial influents generally has phosphorus concentration between 3 and 15 mg/L. Above all, the important phosphorus additives in wastewater are as orthophosphate linked with a small quantity of organic phosphorus. The other common chemical means of removing phosphorus are now modified by removing organic phosphorus. The biological phosphorus removal is a widely used microbial method of eliminating phosphorus from wastewater before it is dumped into water bodies, preventing eutrophication. It is an economically flexible way to treat wastewater before it is actually disposed in water bodies. The biological treatment of phosphorus was suggested by Kim et al. (2011). Their group examined the intermittent variable aeration system biofilm reactor having mobile media and examined the performance of artificial wastewater purification devices. As the device was designed for laboratory scale, it was fitted with a 5.5 L reactor and removed 73.1%, 9.4%, and 97.7% of total nitrogen (TN), total phosphorus (TP), and total organic matter with the given cycle operation of 4 hours. Liu et al. (1996) studied a system that integrates the use of biofilms and activated sludge for removing organic matter, phosphorus, and nitrogen. The device uses fibre-containing carriers mounted in an anaerobic pond where the sludge is funnelled to stimulate the growth of denitrifying bacteria in order to evacuate phosphate. The preliminary wastewater contains a high amount of COD, ammonium nitrogen, and total phosphorus. Following several hours of process at a high temperature, it was confirmed that the total phosphorus and ammonium nitrogen were reduced.

24.5.1.5   Moving Bed Biofilm Reactor (MBBR)

The MBBR integrates the suspended as well as the attached growth system where microorganisms grow on bed using carriers such as plastic as a biofilm. The aerobic agitation and mechanical stirrers (anaerobic process) support the movement in the biofilm reactors (Ødegaard, 2006). The reactor includes two units, a submerged unit (biofilm reactor) and a separation unit for liquid and solid. The MBBR system allows for the growth of biomass throughout the entire tank as compared to biofilm reactors. Thus, only a single reactor can be used for nutrient removal that can reduce the burden of ground space requirements for wastewater treatment plants. The MBBR framework allows the upgrading of traditional methods for wastewater treatment to reduce energy consumption and enhance pollutant removal efficiency (Leyva-Díaz et al. 2017). Furthermore, various evidences have suggested that this framework is a high-strength wastewater treatment technology under extreme environmental conditions that can be applied for domestic, agricultural and municipal wastewater treatment.

24.5.1.6 Nanotechnology Applications of Biofilm

For the treatment of industrial wastewater, Kriklavova and Lederer (2010) employed nanofibre carriers in biofilm reactors. The term nanomaterial refers to naturally or chemically synthesized materials produced in a size ranging between 1 and 100 nm. Nanofibres offer various distinguishing qualities, including increased strength, reactivity of chemicals and growth rate. A layer that is very minute in size will not sustain microbial growth, while a layer that is too large would make it expensive to manufacture so that the optimum amount of nanolayer fibre filling is utilized. Due to their recalcitrant activity of the nanofibers, they are being utilized in the wastewater treatment technology. Utilization of electrospun nanofibres improves the capability of water filtration membrane (Botes and Eugene Cloete, 2010). A comparative study between commercial Anox Kaldnes carriers and carriers with the biofilm showed that bacterial biofilm captured very slowly on the commercial Anox Kaldnes carriers and their rapid colonization results in better technology than the commercial Anox Kaldnes technology.

24.6 REMEDIATION OF CHEMICALS AND HEAVY METALS THROUGH BIOFILM

Several different types of pollutants [such as nitrogen-based ones (cyanide, nitrate and nitrite) and inorganic anionic pollutants (phosphate, chlorite, perchlorate, bromate and fluoride)] persist in water bodies as a result of industrialization, which has caused an increase in health and environmental problems (Chen et al., 2017). There are a few techniques available for the removal of persistent anionic pollutants, such as photo-oxidation process, electro-coagulation and ion exchange method. Meanwhile, a novel approach of hydrotalcite-based Getters are nowadays used for the removal of anionic pollutants and all of these are high capital-intensive methods that require high scale and regeneration of surfaces (Bo et al., 2016). The United States Environmental Protection Agency (EPA) has estimated that the maximum contamination level in drinking water is 10 mg/L for nitrate and 1 mg/L for nitrite. Similarly, the level of sulphate concentration is 250 mg/L as the second maximum contaminant level. Moreover, 0.025 mg/L phosphate concentration in water reservoirs, *viz.* lakes and ponds, can avoid the eutrophication caused due to high concentration of phosphate ions. Apart from anionic pollutants, several heavy metals have been considered as the environmental contaminants risking the human health and ecosystem. Although heavy metals are naturally obtained metals from the earth crusts, the source of contamination is from anthropogenic activities such as mining, smelting, domestic and industrial production, untreated sewage discharge, erosion of soil and natural weathering of rocks.

24.7 CHALLENGES OF MICROBIAL BIOFILM IN WASTEWATER REMEDIATION

The immobilization of biofilms on permeable membranes for the remediation of environment pollutants is of interest for applications, where in traditional treatment

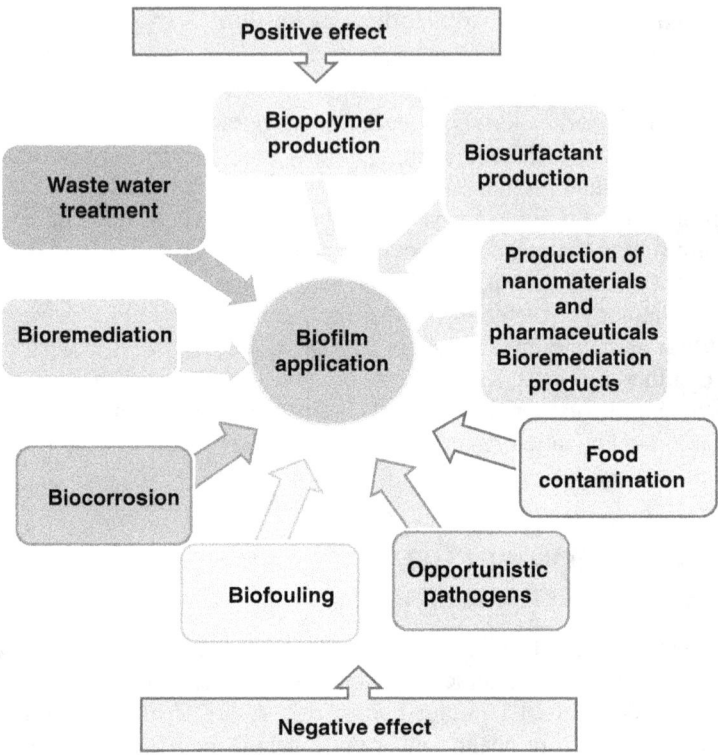

FIGURE 24.3   Positive and undesirable impacts of biofilm development in environment.

technologies are unsuitable. These bioreactors are more efficient as compared to the fixed microorganisms-based method (Lewandowski and Boltz, 2011). Primarily, biofilm-based systems include structures that contain living biomass that might be influenced by the existing situations such as a change in pH, osmolarity, temperature and microbial metabolisms. Consequently, it is complex to explain and difficult for practical use (Asri et al., 2017). Despite various advantages of these biofilm-based bioreactors, there are challenges and limitations that influence the overall performance and in near future need to be addressed (Figure 24.3).

24.7.1   Challenges in Biofilm-Based Processes

Microorganisms in biofilm reactors require support with a variety of materials. The performance of depollution is significantly influenced by the properties and the nature of those supports, and the ability of biofilm to link with this particular support (Asrietal, 2017). Granular activated carbon (GAC) is utilized as a support material in microbe-based biofilm wastewater treatment plants (Quintelas et al., 2010; Muhamad et al., 2013). However, its use has been severely limited because of its excessive price and the difficult conditions related to its regeneration have strongly restricted its operation.

24.7.2 CHALLENGES FACED IN AEROBIC/ANAEROBIC TREATMENT

Treatment of organic wastewaters is done in aerobic biological processes. These processes are more effective in the removal of soluble biodegradable organic material than anaerobic treatment. The biomass obtained is simply settled out from the liquid media providing a higher flocculation property. This property can be used to settle out the produced biomass from the liquid media and a reduced suspended material concentration in the effluent, resulting in a higher-quality effluent. Anaerobic treatments outperform aerobic treatments in treating higher COD concentration influents. Despite its promising approach, anaerobic treatment still faces a number of obstacles in the quest for practices for pollution management (Seghezzo et al., 1998; Chan et al., 2009). The key obstacle is that because of the high COD content, full stabilization of organic matter is difficult, resulting in a final effluent comprising solubilized organic matter.

24.7.3 CHALLENGES FACED IN DIFFERENT BIOREACTORS

24.7.3.1 Challenges in Membrane Biofilm Reactors

Membrane biofilm reactors (MBRs) are a potential biotechnology for eliminating pollutants from aqueous solutions and restoring the useful constituents from them. It combines biological waste degradation through biofilm-based framework with substantial separation by utilizing a membranous unit that replaces the separators. Membrane fouling and clogging layers are a key obstacle for this helpful approach, which has ramifications in terms of operational expenses and reactor maintenance. Aside from that, a limiting issue is the high energy requirement associated with air ultrafiltration and the high expenses of membranes are restricting factors. Clogging and fouling of membranes are the frequent problems in membrane biofilm reactors along with MBR systems (Ngo et al., 2006). However, the fraction for the most part adding to this issue stays indistinct; the reason for this could be colloidal and soluble natural substances such as EPSs and suspended solids (Ivanovic and Leiknes, 2012).

24.7.3.2 Challenges in Moving-Bed Biofilm Reactors

Moving-bed biofilm reactors highlight the utilization of both biofilm and activated sludge, deliver good effluent quality, and reduce sludge generation. It is used in the management of different industrial and municipal wastewaters, demonstrating a compelling evacuation ability of natural, organic and inorganic materials (Di Fabio et al., 2013). It additionally gives an incredible adaptability towards influent biofilm-based systems for industrial wastewater treatment variability, high volumetric loading and great sanitization proficiency. This kind of reactor might be utilized for high-impact anaerobic cycles. The transporters that are utilized for biofilm development in this framework move totally in the tank volume. In aerobic cases their development is brought about by the agitation of fresh air, whereas in anaerobic cycles, the transporters are kept in development by a power-driven blending (Ødegaard, 2006).

The MBBR is an intense powerful reactor for wastewater treatment that is targeted for oxidation of carbon as well as for nitrification and denitrification as a single framework or in consolidated frameworks (Gilbert et al., 2014). The working outcomes are fulfilling in both laboratory scale and higher scopes. The limitation of these frameworks is that the carriers or transporters ought to be eliminated so that it will help the reactor parts (Butler and Boltz, 2014).

24.7.3.3 Challenges in Trickling Filter in Biofilm Reactors

Trickling filter in a biofilm reactor comprises influent recirculation pump station, the filter media (rock beds and plastic sheets), and an underdrain collection unit that stores treated water. Treatments with a trickling filter require a liquid-solid separation to eliminate suspended solids. Circular or rectangular secondary clarifiers are involved in this step. While designing these trickling filters, major focus should be given to filter media. Stones and gravel have been preferred for a long time. The major issue it faces is these materials restrict the air circulation in the filter as well as available oxygen quantity for the growth of microbial biofilm. These problems affect the performance of trickling filters in wastewater treatment. During the treatment of high organic loads, stone bed trickling filters' void spaces get clogged due to excessive microbial cell growth. Also trickling filter is not a volume effective system.

24.7.3.4 Challenges in Microbial Fuel Cells

A recently emerged biofilm-based system is the microbial fuel cell (MFC).Two current world challenges can be resolved by using MFCs, which provide solutions to decontamination of wastewater and green energy production. MFCs are microbial-based devices capable of producing electric current by transferring of electrons with an electrode. Electroactive biofilms, formed by microbes on electrodes, have long been known as promising technologies in pollution sequestration and electricity production, mainly because they are efficient in energy production and produces lower biomass. However, significant obstacles still stand in the way of their practical application. An increase in reactor volume was proposed to improve MFC performance; nevertheless, larger MFCs yielded poor results. When laboratory-sized experiments are scaled up to pilot scale, these limits become most apparent. In reality, in larger systems, sustaining the generated power density is difficult. The volumetric power density of MFCs has tended to decrease as their size has increased, although a minimum threshold volumetric power density of 1 kW/m^3 should be maintained (Zhang et al., 2013). Similar to the voltage loss generated by the substrate cross-conduction effect, this loss could significantly limit full-scale operation (Asri et al., 2017).

Other than this, some frontline challenges are faced. While biofilm formation includes extracting microscopic and macroscopic measurements within millions of biofilm cells, the challenge it faces is in extracting high-quality single cell measurements from the population of assorted biofilm cells. It requires particular attention while reproducing desired readouts and then extracting its cellular dynamics and heterogeneity.

Bacterial cells during exponential phase are best suited for synthetic biology work. But it's not easy to maintain bacteria in a particular phase outside of laboratory condition. The change in cellular fates and the cellular decisions influences the metabolism and transcriptional process; as a result, cell fate has an impact on numerous genes. Other cell fate pathways, such as sporulation, mortality, and wrinkle formation due to localized cell death, can muddle the cellular commitment to form a biofilm when creating biofilms. Therefore, the lack of molecular genetics tools planned to operate in non-exponential growth phases exacerbates the problem (Tran and Prindle, 2021). Within biofilm cells, non-uniform behaviour in terms of matrix production and motility is difficult, so to engineer cells in stationary phase, a proper attention is still needed. Although some stationary phase promoters are in line and characterized, due to low numbers, they are not much effective.

24.8 OPPORTUNITIES OF BIOFILMS IN WASTEWATER TREATMENT TECHNOLOGY

Due to self-renewing properties, biofilms are far superior to other wastewater treatment methods, such as activated carbon filters and reverse osmosis. Biofilms require only the seeding of the right bacteria on the plastic scaffolds. Depending on nutrient balance and moisture, biofilm can form on any surface. Biofilms are reasonable to advancements in synthetic biology such as medicine, genetic and metabolic research, biofuels, and other sectors due to enhanced understanding of their biology and the creation of new synthetic biological tools (Tran and Prindle, 2021). Water treatment plants include fixed-bed biofilm reactors (SFBBRs) to filter water using biofilm water filtration. SFBBR biofilm water filtration is applied in large-scale water purification systems. SFBBR biofilm's ability to degrade solids that can clog filtration is its salient feature. Researchers at Sam Houston State University have tested and demonstrated that SFBBR systems have the ability to degrade solids that can clog filtration. Not only does this system clean wastewater as well as the old system, but it does so more efficiently as well. This system is leaving behind merely 10% "sludge", a vexatious by-product of wastewater treatment. Typical septic systems take a longer duration than SFBR and leave behind up to 50% sludge (Cheung, 2019).

Biofilms are effective for the development of portable water purification systems. There are a wide range of applications for biofilm-based water filtration; if well optimized, they can be employed during both natural disasters and floods as a backup water supply during contamination of the main supply. In one standard 20-foot shipping container, all the components necessary to set up a mini water treatment centre can be assembled. It is also a cost-effective source of water supply in developing countries throughout the world (Cheung, 2019).On-site remediation of pollutants and pollution sequestration upstream is well suited to biofilms as environmental remedies. Researchers have studied the ability of biofilms to trap or sequester pollutants from their environment. In addition to the chelation of heavy metals such as mercury, biofilms can also immobilize mercury compounds. *E. coli* biofilms are genetically engineered, which can immobilize mercury compounds in the fibrils. *Pseudomonas putida* biofilms were engineered to express haloalkane dehalogenases, which have further catalytic activity after being subjected to tuneable control that led to the degradation of toxic halogenated compounds (Tayet al., 2017; Benedetti et al., 2021).

24.9 CONCLUSIONS

The wastewater biotreatment process is an unstable process utilized by wastewater industries. Despite that, microbial biofilm-based technology provides suitable alternatives for the management of rigorously generated waste from different industries because it's cheap compared to chemical and physical methods. The remarkable designing of the biofilm reactors provide options for better stability and tougher surface attachment of microbial biofilm. The viability of microbial biofilm bioremediation has been demonstrated in several bioreactors, such as biofilm-based filters, MBfR, and RBCs that give promising results. Thus, the growing biofilm can not only remove the pollutants from wastewater, but also recover valuable resources. A sound knowledge about the various constituents of filter media and development of biofilm on them is very important. Although there is a great deal of literature available that discusses the beneficial role of biofilm and EPS for removing various pollutants, the industrial application of such systems is still unknown.

Traditional and advanced techniques are used for the study of biofilms, such as analysing through spectroscopic absorbance, viable plate count, substrate utilization, and microscopic techniques. Advanced techniques are employed for complete biofilm community profiling, such as microbial sequencing, DNA microarray, NGS, nanotechnology, and DNA fingerprinting to increase the feat, stability, and robustness of biofilm reactors. There are still challenges for creating a robust biofilm-mediated remediation technology for large-scale applications. Furthermore, biofilms are associated with membrane fouling and clogging, raising some concerns about bio-containment. To improve treatment efficiency, it is therefore imperative to understand biofilm formation and its esoteric mechanisms. Consequently, further research is necessary to trounce these issues with the purpose of developing microbial biofilm-based treatment system with high efficacy and lower management outlay.

24.10   FUTURE PERSPECTIVES

Majority of studies have focused on wastewater treatment plants in municipal and industrial systems, but wastewater generated in rural regions has received little attention. Sewer model in rural regions should be upgraded to vacuum sewers, or more sustainable models with less time consumption and stable flow. As flow significantly influences the biofilm formation. As a result, different operational modes should be explored in order to construct exceptional wastewater treatment facilities in rural locations with minimal requirements of labour cost and high maintenance machineries.

In situ experiments should be conducted as the results vary in actual conditions. There is a need to build a better perspective about the management of greenhouse gases such as methane and hydrogen sulphide produced in traditional biofilm treatment of wastewater along with more efficient removal of nitrogen and emerging micropollutants. On the other hand, the successful application of dynamic biological reactors can be explored to achieve a more realistic and detailed stimulation biofilm wastewater treatment plants.

Biofilm reactors can not only be used as a treatment model, but should also be explored as a recovery model to recover valuable metals and other resources. A better understanding of applicability of biofilms is required. More research is needed to overcome the problem of solids suspended in influents interfering with substrate and oxygen diffusion processes reducing the biofilm activity. Many nanoparticles are undergoing active research and development due to their particular activity against refractory pollutants and application versatility. However, there are reported contradictions among experimental researches with nanoparticles with the success of the treatment. Thus, a multidisciplinary approach to overcome the scientific, technical, and environmental constraints should be adopted. The overall future research should be based on efficient and sustainable plants with safe, cost-effective, and feasible applications with more focus on water recovery and reusability.

REFERENCES

Andreottola, G., Foladori, P., Ragazzi, M., and Villa, R. 2002. Dairy wastewater treatment in a moving bed biofilm reactor. *Water Science and Technology* 45, no. 12, 321–328.

Asri, M., Elabed, S., Koraichi, S.I., and Ghachtouli, N.E. 2018. Biofilm-based systems for industrial wastewater treatment. *Handbook of Environmental Materials Management*, pp.1–21.

Augusto, M.R., Camiloti, P.R., and de Souza, T.S.O. 2018. Fast start-up of the single-stage nitrogen removal using anammox and partial nitritation (SNAP) from conventional activated sludge in a membrane-aerated biofilm reactor. *Bioresource Technology* 266, 151–157.

Bales, P.M., Emilija, M.R., Sarah, L.M., Yang, S., and Daniel, C.N. 2013. Purification and characterization of biofilm-associated EPS exopolysaccharides from ESKAPE organisms and other pathogens. *PloS One* 8, no. 6: e67950.

Benedetti, I., De, L.V., and Nikel, P.I. 2016. Genetic programming of catalytic *Pseudomonas putida* biofilms for boosting biodegradation of haloalkanes. *Metabolic Engineering* 33: 109–118.

Botes, M., and Eugene Cloete, T. 2010. The potential of nanofibers and nanobiocides in water purification. *Critical Reviews in Microbiology* 36, no. 1: 68–81.

Bo, A., Sarina, S., Liu, H., Zheng, Z., Xiao, Q., Gu, Y., Ayoko, G.A., and Zhu, H. 2016. Efficient removal of cationic and anionic radioactive pollutants from water using hydrotalcite-based getters. ACS Applied Materials & Interfaces 8, no. 25: 16503–16510.

Boltz, J.P., Smets, B.F., Rittmann, B.E., Van Loosdrecht, M.C., Morgenroth, E., and Daigger, G.T. 2017. From biofilm ecology to reactors: a focused review. *Water Science and Technology* 75, no. 8, 1753–1760.

Butler, C.S., and Boltz, J.P. 2014. Biofilm processes and control in water and wastewater treatment. In: Satinder Ahuja (ed.), *Comprehensive Water Quality and Purification*, pp.90–107. Elsevier.

Chan, Y.J., Chong, M.F., and Law, C.L. 2009. A review on anaerobic–aerobic treatment of industrial and municipal wastewater. *Chemical Engineering Journal* 155: 1–18.

Chandra, R., and Kumar, V. 2017. Detection of *Bacillus* and *Stenotrophomonas* species growing in an organic acid and endocrine-disrupting chemicals rich environment of distillery spent wash and its phytotoxicity. *Environmental Monitoring and Assessment* 189: 1–19. Doi: 10.1007/s10661-016-5746-9.

Chen, M., Wu, Y., and Jafvert, C.T. 2017. Synthesis of cross-linked cationic surfactant nanoparticles for removing anions from water. *Environmental Science: Nano* 4, no. 7: 1534–1543.

Cheung, E. 2019. *Waste Not, Want Not: Harnessing the Power of Microbes in Wastewater Recycling*. Science in the news. Springer International Publishing, Morocco.

Costa, O.Y.A., Raaijmakers, J.M., and Kuramae, E.E. 2018. Microbial extracellular polymeric substances: ecological function and impact on soil aggregation. *Frontiers in Microbiology* 9: 1636.

Deyerling, D., Wang, J., Bi, Y., Peng, C., Pfister, G., Henkelmann, B., and Schramm, K.W. 2016. Depth profile of persistent and emerging organic pollutants upstream of the three gorges dam gathered in 2012/2013. *Environmental Science and Pollution Research* 23, no. 6: 5782–5794.

Di Fabio, S., Malamis, S., Katsou, E., Vecchiato, G., Cecchi, F., and Fatone, F. 2013. Are centralized MBRs coping with the current transition of large petrochemical areas? A pilot study in Porto-Marghera (Venice). Chemical *Engineering Journal* 214: 68–77.

Di, F.S., Lampis, S., and Zanetti, L. 2013. Role and characteristics of problematic biofilms within the removal and mobility of trace metals in a pilot-scale membrane bioreactor. *Process Biochemistry* 48, no. 11: 1757–1766.

Drinking Water Contaminants–Standards and Regulations. United States Environmental Protection Agency website, 2017. https://www.epa.gov/dwstandardsregulations (accessed 20 April 2019).

Flemming, H.-C., and Wingender, J. 2010. The biofilm matrix. *Nature Reviews Microbiology* 8, no. 9: 623–633.

Foroughi, M., Najafi, P., Toghiani, S., Toghiani, A., and Honarjoo, N. 2013. Nitrogen removals by *Ceratophyllum demersum* from wastewater. *Journal of Residuals Science and Technology* 10:63–68.

Friedman, L., and Kolter, R. 2004. Two genetic loci produce distinct carbohydrate-rich structural components of the *Pseudomonas aeruginosa* biofilm matrix. *Journal of Bacteriology* 186, no. 14: 4457–4465.

Garrett, E.S., Perlegas, D., and Wozniak, D.J. 1999. Negative control of flagellum synthesis in *Pseudomonas aeruginosa* is modulated by the alternative sigma factor AlgT (AlgU). *Journal of Bacteriology* 181, no. 23: 7401–7404.

Gilbert, E.M., Agrawal, S., and Karst, S.M. 2014. Low temperature partial nitritation/anammox in a moving bed biofilm reactor treating low strength wastewater. *Environmental Science & Technology* 48, no. 15, 8784–8792.

Heilmann, C., Hussain, M., Peters, G., and Götz, F. 1997. Evidence for autolysin-mediated primary attachment of *Staphylococcus epidermidis* to a polystyrene surface. *Molecular Microbiology* 24, no. 5: 1013–1024.

Ivanovic, I., and Leiknes, T.O. 2012. The biofilm membrane bioreactor (BF-MBR) – a review. *Desalination and Water Treatment* 37, no. 1–3: 288–295.

Jackson, K.D., Starkey, M., Kremer, S., Parsek, M.R., and Wozniak, D.J. 2004. Identification of PSL, a locus encoding a potential exopolysaccharide that is essential for *Pseudomonas aeruginosa* PAO1 biofilm formation. Journal of Bacteriology: 4466–4475.

Judd, S. 2008. The status of membrane bioreactor technology. *Trends in Biotechnology* 26, no. 2: 109–116.

Judd, S. 2010. MBR Book: *Principles and Applications of Membrane Bioreactors for Water and Wastewater Treatment.* Elsevier.

Khatoon, N., Naz, I., Ali, M.I., Ali, N., Jamal, A., Hameed, A., and Ahmed, S. 2014. Bacterial succession and degradative changes by biofilm on plastic medium for wastewater treatment. *Journal of* Basic Microbiology 7, no. 54: 739–749.

Kikuchi, T., and Tanaka, S. 2012. Biological removal and recovery of toxic heavy metals in water environment. *Critical Reviews in Environmental Science and Technology* 42, no. 10: 1007–1057.

Kim, B.K., Chang, D., Son, D.J., Kim, D.W., Choi, J.K., Yeon, H.J., Yoon, C.Y., Fan, Y., Lim, S.Y., and Hong, K. H. 2011. Wastewater treatment in moving-bed biofilm reactor operated by flow reversal intermittent aeration system. *International Journal of Environmental and Ecological Engineering* 5, no. 12: 867–870.

Kriklavova, L. and Lederer, T. 2010. The use of nanofiber carriers in biofilm reactor for the treatment of industrial wastewaters. Nanocon, Olomouc, Ceska Republika. http://konsys-t.tanger.cz/files/proceedings/nanocon_10/lists/papers/393.pdf.

Kroiss, H., Rechberger, H., and Egle, L. 2011. Phosphorus in water quality and waste management. In: Sunil Kumar (ed.), *Integrated Waste Management-Volume II*. DOI: 10.5772/18482 (pp. 181–214). IntechOpen.

Kumar, V., Shahi, S.K., Ferreira, L.F.R., Bilal, M., Biswas, J.K., and Bulgariu, L. 2021b. Detection and characterization of refractory organic and inorganic pollutants discharged in biomethanated distillery effluent and their phytotoxicity, cytotoxicity, and genotoxicity assessment using *Phaseolus aureus* L. and *Allium cepa* L. *Environmental Research* 201: 111551. Doi: 10.1016/j.envres.2021.111551.

Kumar, V., Singh, K., and Shah, M.P., 2021a. Advanced oxidation processes for complex wastewater treatment. In: Shah, M.P. (Eds.), *Advance Oxidation Process for Industrial Effluent Treatment*. Elsevier. Doi: 10.1016/B978-0-12-821011-6.00001-3.

Kumar, V., Srivastava, S., and Thakur, I.S., 2021c. Enhanced recovery of polyhydroxyalkanoates from secondary wastewater sludge of sewage treatment plant: Analysis and process parameters optimization. *Bioresource Technology Reports* 15: 100783.

Kumar, V., and Thakur, I.S., 2020. Extraction of lipids and production of biodiesel from secondary tannery sludge by *in situ* transesterification. *Bioresource Technology Reports* 11: 100446. Doi: 10.1016/j.biteb.2020.100446.

Kumar, V., Thakur, I.S., and Shah, M.P. 2020. Bioremediation approaches for pulp and paper industry wastewater treatment: Recent advances and challenges. In: Shah, M.P. (Ed.), *Microbial Bioremediation & Biodegradation*. Springer, Singapore. Doi: 10.1007/978-981-15-1812-6_1.

Lai, C.-Y., Lv, P.-L. Dong, Q.-Y., Yeo, S.L., Rittmann, B.E., and Zhao, H.-P. 2018. Bromate and nitrate bioreduction coupled with poly-β-hydroxybutyrate production in a methane-based membrane biofilm reactor. *Environmental Science &Technology* 52, no. 12: 7024–7031.

Lewandowski, Z. and Boltz, J.P. 2011. Biofilms in water and wastewater treatment. In Water-Quality Engineering (pp. 529–570). Elsevier.

Le, C.P., Chen, V., and Fane, TAG. 2006. Fouling in membrane bioreactors used in wastewater treatment. *Journal of Membrane Science* 284, no. 1–2: 17–53.

Leyva-Díaz, J.C., Martín-Pascual, J., and Poyatos, J.M. 2017. Moving bed biofilm reactor to treat wastewater. *International Journal of Environmental Science and Technology* 14, no. 4: 881–910.

Liu, J.X., Van Groenestijn, J.W., Doddema, H.J., and Wang, B.Z. 1996. Removal of nitrogen and phosphorus using a new biofilm-activated-sludge system. *Water Science and Technology* 34, no. 1–2: 315–322.

May, T.B., Shinabarger, D., Maharaj, R.A., Kato, J., Chu, L., DeVault, J.D., Roychoudhury, S.A., Zielinski, N.A., Berry, A., and Rothmel, R.K. 1991. Alginate synthesis by *Pseudomonas aeruginosa*: A key pathogenic factor in chronic pulmonary infections of cystic fibrosis patients. *Clinical Microbiology Reviews* 4, no. 2: 191–206.

Muhamad, M.H., Sheikh, A.S.R., and Mohamad, A.B. 2013. Application of response surface methodology (RSM) for optimisation of COD, NH_3-N and 2, 4-DCP removal from recycled paper wastewater in a pilot-scale granular activated carbon sequencing batch biofilm reactor (GAC-SBBR). *Journal of Environmental Management* 121: 179–190.

Naz, I., Batool, S.A.-u., Ali, N., Khatoon, N., Atiq, N., Hameed, A., and Ahmed, S. 2013. Monitoring of growth and physiological activities of biofilm during succession on polystyrene from activated sludge under aerobic and anaerobic conditions. *Environmental Monitoring and Assessment* 185, no. 8: 6881–6892.

Naz, I., Seher, S., Perveen, I., Saroj, D.P., and Ahmed, S. 2015. Physiological activities associated with biofilm growth in attached and suspended growth bioreactors under aerobic and anaerobic conditions. *Environmental Technology* 36, no. 13: 1657–1671.

Nerenberg, R. 2016. The membrane-biofilm reactor (MBfR) as a counter-diffusional biofilm process. *Current Opinion in Biotechnology* 38: 131–136.

Ngo, H.H., Nguyen, M.C., Sangvikar, N.G., Hoang, T.T.L., and Guo, W.S. 2006. Simple approaches towards the design of an attached-growth sponge bioreactor (AGSB) for wastewater treatment and reuse. *Water Science and Technology* 54, no. 11–12: 191–197.

Ødegaard, H. 2006. Innovations in wastewater treatment: The moving bed biofilm process. *Water Science and Technology* 53, no. 9: 17–33.

PalmerJr, R.J., and Sternberg, C. 1999. Modern microscopy in biofilm research: Confocal microscopy and other approaches. *Current Opinion in Biotechnology* 10, no. 3: 263–268.

Palmer, J., Flint, S., and Brooks, J. 2007. Bacterial cell attachment, the beginning of a biofilm. *Journal of Industrial Microbiology and Biotechnology* 34, no. 9: 577–588.

Quintelas, C., Silva, B., Figueiredo, H., and Tavares, T. 2010. Removal of organic compounds by a biofilm supported on GAC: modelling of batch and column data. *Biodegradation* 21, no. 3: 379–392.

Rittmann, B.E. 2006. The membrane biofilm reactor: The natural partnership of membranes and biofilm. *Water Science and Technology* 53, no. 3: 219–225.

Ronaghi, M. 2001. Pyrosequencing sheds light on DNA sequencing. *Genome Research* 11, no. 1: 3–11.

Sanz, J.L., and Köchling, T. 2007. Molecular biology techniques used in wastewater treatment: an overview. *Process Biochemistry* 42, no. 2: 119–133.

Seghezzo, L., Zeeman, G., van Lier, J.B., Hamelers, H.V.M., and Lettinga, G. 1998. A review: The anaerobic treatment of sewage in UASB and EGSB reactors. *Bioresource Technology* 65, no. 3: 175–190.

Shemesh, M., Tam, A., Kott-Gutkowski, M., Feldman, M., and Steinberg, D. 2008. DNA-microarrays identification of *Streptococcus mutans* genes associated with biofilm thickness. *BMC Microbiology* 8, no. 1: 1–12.

Stone, W. and Wolfaardt, G. 2018. Measuring microbial metabolism in atypical environments. *Methods in Microbiology* 45: 123–144. Academic Press.

Tay, P.K.R., Nguyen, P.Q., and Joshi, N.S. 2017. A synthetic circuit for mercury bioremediation using self-assembling functional amyloids. *ACS Synthetic Biology* 6, no. 10: 1841–1850.

Tran, P., and Prindle, A. 2021. Synthetic biology in biofilms: Tools, challenges, and opportunities. *Biotechnology Progress* 37: e3123.

Vymazal, J. 2007. Removal of nutrients in various types of constructed wetlands. *Science of the Total Environment* 380, no. 1–3: 48–65.

Waqas, S., Bilad, M.R., Man, Z.B., Suleman, H., Nordin, N.A.H., Jaafar, J., Othman, M.H.D., and Elma, M. 2021. An energy-efficient membrane rotating biological contactor for wastewater treatment. *Journal of Cleaner Production* 282: 124544.

Wang, Y.M., Yang, J.B., and Xu, D.L. 2006. Environmental impact assessment using the evidential reasoning approach. *European Journal of Operational Research* 174, no. 3: 1885–1913.

Webb, J.S. 2007. Differentiation and dispersal in biofilms. In *The Biofilm Mode of Life: Mechanisms and Adaptations* (pp. 167–178). Horizon Bioscience.

Wu, M., Luo, J.-H., Hu, S., Yuan, Z., and Guo, J. 2019. Perchlorate bio-reduction in a methane-based membrane biofilm reactor in the presence and absence of oxygen. *Water Research* 157: 572–578.

Zhang, L., Zhang, S., Peng, Y., Han, X., & Gan, Y. 2015. Nitrogen removal performance and microbial distribution in pilot-and full-scale integrated fixed-biofilm activated sludge reactors based on nitritation-anammox process. *Bioresource Technology* 196: 448–453.

Zhang, L., Li, J., Zhu, X., Ye, D., and Liao, Q. 2013. Anodic current distribution in a liter-scale microbial fuel cell with electrode arrays. *Chemical Engineering Journal* 223: 623–631.

25 Artificial Intelligence in Waste Management/ Wastewater Treatment

Rao Muhammad Mahtab Mahboob
University of Agriculture Faisalabad (UAF)
Institute of Management and Applied Sciences

Kiran Mustafa
The Women University Multan
Govt. Graduate College for Women Khanewal

Mahrukh Khan, Fakhr-un-Nisa, and Sara Musaddiq
The Women University Multan

Rao Muhammad Shahbaz Mahboob
University of Agriculture Faisalabad (UAF)

CONTENTS

25.1 INTRODUCTION

Because of increased human activities and industrialization, toxic effluents are being released into freshwater systems and clean water is not readily available to all (Xie et al., 2014, Dubey et al., 2015). Inorganic as well as organic pollutants, including fertilizers, persistent organic pollutants (POPs), and heavy metals, cause water contamination (Cheung et al., 2003, Hoh and Hites, 2005, Luo et al., 2013, Li et al., 2016b). Both the general public and environmentalists have recently given far more attention to water contamination issues than in the past (Ma et al., 2011). In the wastewater treatment, solvent extraction, adsorption, chemical or biological oxidation, coagulation,

DOI: 10.1201/9781003247883-25

reduction, flocculation, and membrane filtration have all been thoroughly explored (Kumar et al., 2008a, 2008b). Some vital operations in pollutant removal systems are optimization and modelling that help to boost efficiency without increasing expenses (Shojaeimehr et al., 2014). In batch studies, identification of a dependent variable is necessary for each and every set of independent factors in traditional approaches for the simulation and analysis of processes involve in pollutants, with only one variable varying at a time while the others remain constant (Singh et al., 2010). Such procedures, of course, necessitate a wide range of investigation, which would be costly and time-consuming (Tak et al., 2015). Furthermore, these methodologies would be unable to reveal the impact of interdependencies between the independent variables (Sahu et al., 2009). Traditional modelling methods may be hampered by the complexity of water treatment appendages such as reduction, adsorption, and oxidation (Nandi et al., 2010). Artificial intelligence (AI) has recently made significant progress in a variety of areas, including intelligent search, big data, autonomous driving, pattern recognition, automatic programming, image understanding, human-computer games, and robotics and will have a major influence on human civilization. Genetic algorithms (GAs), random forests (RFs), support vector machines (SVMs), ant colony algorithm (ACA), artificial neural networks (ANNs), simulated annealing (SA), boosted regression trees (BRTs), imperialist competitive algorithms (ICAs), particle swarm optimization (PSO), Monte Carlo simulation (MCS), immune algorithms (IAs), and decision trees (DTs) are some of the most prevalent AI tools. To increase the accuracy of optimal solution prediction, AI technologies have been integrated using research set-ups such as uniform design and response surface approach. Both ANN and RSM models take full advantage of experimental data due to their inherent capacity to extract information from operation parameters, allowing them to represent complicated connections in environmental engineering without having to comprehend the underlying mechanisms of removal (Nandi et al., 2010, Fan et al., 2017).

Traditional wastewater treatment methods rely on calculations of design factors and complicated mathematical formulas, process parameters, operation parameters, as well as costs, regulatory, social, governmental, and technical constraints (Kumar et al., 2021). Previously, the removal efficiency of contaminants and uncertainty estimations were assessed using multivariate and statistical data analysis approaches. Without the need of complicated mathematical formulas or extensive information about the relationship between the output and input variables, AI technologies, on the other hand, enable lower average error and more accuracy in the prediction of R2 (regression coefficients).

Artificial intelligence (AI) has demonstrated its ability to handle complicated, dynamic, and interactive wastewater treatment problems. They can be used to estimate the effluent quality of the treatment method and to take prompt action to prevent quality degradation. Automated sampling aids the operator in recognizing patterns and trends, as well as preventing disasters through quick intervention. Regular monitoring of systems using sensors aids in optimizing energy usage, enhancing operational efficiency and lowering costs. In the days ahead, AI is likely to play a key role in wastewater treatment.

The manuscript outlines the research that has been conducted in this area, as well as the many AI-based technologies that have been employed to date. The most crucial stage in reducing aqueous pollution and improving water quality is wastewater treatment. The composition of wastewater is extremely diverse, with pollutant concentrations, influent qualities, and treated effluent differing dramatically among water treatment plants. Wastewater treatment is a multi-step process influenced by a variety of chemical, microbiological, and physical variables. Furthermore, the stochastic disturbances and influent variability necessitate operators performing adequate system operating controls.

Many uncertainties exist in wastewater treatment systems due to the natural processes' complexity, wastewater treatment process, and anthropogenic activities. Furthermore, given the amount, removal efficiency, and quality of wastewater, these uncertainties fluctuate at random (Long et al., 2019). Modern water treatment plants must comply with more severe emission restrictions, and resource recycling guidelines and new energy efficiency. To address these issues, researchers have attempted to apply AI technology to WWTPs.

In the last few decades, the use of AI models in many parts of wastewater treatment has increased dramatically. It's used to predict water treatment performance, estimate effluent quality, optimize operating parameters and treatment unit design, design sensors for key component estimation, optimize micropollutants as well as other emerging pollutants, perform operating instructions based on performance, predict maintenance strategies based on fault identification, optimize energy use in the treatment, and automate the system.

Each intelligent control approach has benefits and drawbacks and should be carefully chosen based on the treatment system's mechanism and the purpose for which it is being utilized to achieve the greatest outcomes (Gupta et al., 2021, Wang and Man, 2021, Zhao et al., 2020).

Because of the aforementioned benefits such as the usage of reduced reagent and experimental strategies, this technology has been adopted for pollutants removal with the strengthening of the global sustainability idea (Acharya et al., 2006). However, because of their emphasis on water purification, AI technologies already have limited uses in environmental protection (Mandal et al., 2015, Mendoza-Castillo et al., 2015, Reynel-Avila et al., 2015).

The goal of this analysis is to objectively examine the basics and benefits of tools used in AI (for example, PSO, GA, and ANNs) as well as coupled techniques (e.g. uniform design, orthogonal design, and response surface methodology). This paper also reviews recent research on employing self-organizing map (SOM) networks, fuzzy neural networks, radial basis function (RBF) networks, and multilayer perceptron (MLP) networks to maximize and boost contaminant removal procedures in water purification. Furthermore, this evidence found that ANN's hybrid models combined with PSO or GA (ANN-PSO or ANN-GA) can be used successfully and with good accuracy in water purification. Ultimately modelling and improving pollutants removal processes, we describe the current AI tools' limits as well as their new enhancements (e.g. SA, SVM, MCS, and deep learning ANNs).

25.2 NEURAL NETWORKS

The most used AI techniques formed from biological procedures are ANNs (Zhang and Pan, 2014, López et al., 2017), which can tackle multivariate non-linear problems given enough data and a sufficient training algorithm (Wang and Deng, 2016, Li et al., 2016a). Each ANN is made up of artificial neurons that have been grouped into layers and connected via connections. Because of their ability to learn highly complex functions, such as data mining tools and non-linear statistical data modelling, ANNs are popular machine learning tools (Hu et al., 2010). In removal procedures, a framework for mapping relationships between input and output parameters is provided by ANNs (Chu, 2003, Aleboyeh et al., 2008). The ability of ANNs to self-learn and self-adapt can be successfully used to forecast output while taking into account a number of operational characteristics. ANNs have the advantage of not requiring mathematical explanations for the phenomena involved in the procedure; hence, the model development takes less time than standard mathematical modelling (Assefi et al., 2014). As a result, traditional empirical modelling based on linear regressions and polynomial can be replaced with ANNs. (Kose, 2008, Luo et al., 2015a, 2015b). ANNs such as the SOM network, RBF network, MLP network, convolutional neural network, fuzzy neural network (FNN), chaotic neural network (CNN), deep belief network (DBN), and recurrent neural network (RNN) can be used to simulate water treatment processes and establish models (Ouarda and Shu, 2009).

An output layer, one or more hidden layers, and an input layer formulate MLP-ANN. The RBF-ANN and MLP-ANN are feed-forward ANNs inside which the neurons in each layer are connected from one layer to the next without any lateral or feedback links. RBF-ANN has a lot in common with MLP-ANN; however, it has different activation functions. The function of Gaussian density is the generic activation function of RBF-ANN, which is specified through a "central" location and a parameter, i.e. "width" (Deshmukh et al., 2012). The FNN model is a combination of systems that combines fuzzy logic theories with neural network theories, allowing it to make efficient use of fuzzy logic's easy interpretability as well as neural network's superior learning and adaptive characteristics (Mingzhi et al., 2009). In a competitive layer, SOM networks learn to aggregate set of related input patterns from a strong input space onto a weak (most commonly two-dimensional) discontinuous lattice of neurons in a non-linear fashion (Kalteh et al., 2008).

The RNN can only handle one sample at a time, unlike feed-forward neural networks, keeping the context and recollection; on the other hand, other networks have long been difficult to train due to a variety of factors. Deep learning has recently evolved from research on ANNs with numerous hidden layers, supervised, reinforcement learning, and unsupervised; for example, a deep neural network is a supervised learning method that decreases the number of parameters in order to improve generalization ability by using the relative relation of space. DBN, on the other hand, is an uncontrolled learning model that can solve the optimization challenges of deep learning frameworks.

In water treatment, the ESN and CNN models can be used to predict removal of contaminants. Huang and Fang (2010) employed the model of CNN to undertake a complete assessment of water resource effects of allocation, which enhances the effectiveness of the algorithm, reinforces the objectivity of the evaluated outcomes, and is highly adaptable and replicable. This model of CNN takes benefits of chaotic motion's ergodicity as well as neural networks' ability to solve sophisticated mapping that is non-linear. Sacchi et al. (2007) also designed the echo-state network (ESN), which is an RNN model for reservoir water prediction influx at hydropower plants. When compared to the SOM, RBF, and FNN models, the ESN model performs well in forecasting water inflows. Furthermore, as an emerging method, the use of SVM in the field of environment safety is currently limited. Actually, the main idea behind SVM is as follows: Use a kernel function "K" for mapping the input variables into a higher-dimensional feature space and then use this space to do a linear regression. SVM has been shown to be useful in a variety of situations, including small samples, non-linear correlations, and high-dimensional pattern identification. Furthermore, because it is a convex optimization issue, SVM may determine the global optimal solution.

An input layer, an output layer, and one or more hidden layers make up MLP-ANN. The RBF-ANN and MLP-ANN are feed-forward ANNs in which each layer's neurons are interlinked from the previous layer to the coming layer without any lateral or feedback connections. RBF-ANN has a structure that is similar to that of MLP-ANN, but its artificial neurons are different. RBF-general ANN's activation function is the Gaussian density function, which is specified by a "centre" position and a "width" parameter (Deshmukh et al., 2012). The FNN model is a combination of systems that combine fuzzy logic theories with neural network theories, allowing it to make efficient use of fuzzy logic's easy interpretability as well as neural network's superior learning and adaptive capabilities (Mingzhi et al., 2009). In a competitive layer, SOM systems learn to aggregate set of related input patterns from a high-dimensional input space onto a low-dimensional (most commonly two-dimensional) discontinuous lattice of neurons in a non-linear fashion (Kalteh et al., 2008).

RNN is totally different than feed-forward neutral network; it can only handle one parameter at a time by keeping its memory as well as state intact, whereas other networks require a very long training session because they work on different sorts of parameters. Progress in the deep learning is a result of the number of modern researches on ANNs with various hidden folders with unsupervised learning, reinforcement learning, and supervised learning. For example, convolutional neural networks are supervised learning models that use the relative relationship of space to reduce the number of parameters and increase training performance, whereas unsupervised learning model, e.g. DBN, can tackle deep learning structure's optimization difficulties.

CNN and ESN models are useful in water treatment. These models can be used to predict the removal of contaminants from water. CNN model was employed by Huang and Fang (2010) for the complete evaluation of allocation effects of water resource that increases algorithm's efficiency, reinforces the objectivity of the evaluated outcomes, and is highly adaptable and replicable. Chaotic motion's ergodicity and neural networks' ability are the plus points of CNN in order to handle sophisticated mapping, especially non-linear mapping. RNN model is an ESN model developed by Sacchi et al. (2007), and it is useful in hydropower plants for forecasting reservoir water influx. The ESN model performs reasonably well in forecasting water influx when compared to SOM, RBF, and FNN models. Although SVM is a newly emerging method, its uses in the environment protection are limited. SVM uses a kernel function, K, to translate the input variables into a higher-dimensional feature space and then perform a linear regression in that space, which is the main principle of small sample problems; some relationships, i.e. non-linear, and recognition of high-dimensional pattern have all shown the benefits of SVM. Furthermore, because it is a convex optimization issue, SVM may determine the global optimal solution.

25.3 HEURISTIC ALGORITHMS

It's widely known that shallow local optima can trap a plain gradient descent algorithm (Mohanraj et al., 2012). The simplex method, which is characterized by a line search derivative-free method, is another similar approach. This method is based on the main concept of comparing objective function values in N-dimensional space at the N1 vertices of a polytope and shifting the vertex, as an optimization is in progress towards the minimal point (El-Wakeel, 2014). This limitation related to ANN has prompted researchers to consider combining or hybridizing ANN with other heuristic techniques in order to improve performance. As a result, tools of AI (e.g. MCS GA, PSO, IA, ICA, SA, and ACA) for fast and efficient searching of global optima have been shown to be good alternatives to conventional algorithms (e.g. gradient search techniques) and that tools were inspired by natural phenomena or social behaviour (Arhami et al., 2013, Dehghani et al., 2014, Mohebbi et al., 2008).

The principles of genetics and natural selection are similar with GAs (Marseguerra et al., 2005). GAs can be used to adjust fixed set of connections' output layers and the weights of hidden layers once an ANN-based process model has been constructed and has generalization ability and good prediction accuracy (Chang and Hou, 2006, Ghaedi et al., 2015). The GA-based optimization technique can be regarded as a global optimization procedure with the advantage of achieving convergence regardless of the initial value. The GA method is created through steps that occur in series that include encoding solutions, computing fitness on the basis of objective function, and selecting the best chromosomes, genetic mutation, and crossover (Jiang et al., 2014). To make new and improved chromosome populations, both mutation and crossover are used. The process is repeated multiple times until a chromosome which has the best fitness is obtained and it is considered the optimal solution. In addition, immune operator is introduced by IA, which is based on the GA that is original and consists of two phases (immune selection and vaccination). The first is to prevent population decrease, while the second is to improve the solution's fitness. Because its search goal has some dispersion and independence, IA is considered good at multi-target search, but the search target of GA is exclusive and single global optimum solution can be found by using GA (Luo et al., 2015a, Wu et al., 2015).

The behaviour of a flock of birds inspired the well-known heuristic approach particle swarm optimization (PSO). It is a developmental algorithm suggested by Kennedy and Eberhart that, due to the fact that it does not use the gradient descent algorithm, can prevent trapping in a local minimum. Steps to start the procedure are as follows: (1) production of an initial population with irregular velocities and placements; (2) for each particle, evaluation of the fitness function (when a new position with a higher fitness value is obtained, the prior value will be substituted); (3) calculation of new velocity of particles; (4) by progressing towards the maximal objective function, updation of the position of the particles (Khajeh et al., 2013). PSO has the advantage of relying just on velocity of particle to accomplish the search process (no mutation or crossover or operations), resulting in a higher convergence rate than GA. PSO can also remember things; this enables it to remember a particle's best location and pass it on to subsequent particles. Furthermore, for PSO real number coding (which is decided directly by a problem's solution) is used. One of the main disadvantages of PSO is that it is quite easy to fall into a local optimum and also it lacks dynamic velocity adjustment, which results in low convergence accuracy and difficult convergence.

The SA methods are different from traditional descent algorithms in that it allows not just downhill, but also moves upwards when searching for a neighbourhood solution.

Because of its high robustness, the ACA is better than other algorithms; nonetheless, it requires a long search time, and it is really easy to fall into a local optimum and has a slower convergence rate. ICA, unlike ACA, GA, and PSO, was influenced by social behaviour, and it simulates the empire's competition mechanism and the colonial assimilation mechanism. ICA has a greater rate of convergence and accuracy than GA and PSO. Beyond this, ideas from statistical mechanics (which is a probabilistic optimization technique) are the basis of SA, which are capable of providing a global optimal solution for large-scale practical concerns. The representativeness and accuracy of the predictions' input parameters are linked to input parameter uncertainty. MCS is a technique for statistical sampling used to provide a probabilistic approximation to the solution of a mathematical model. The forms and magnitudes of the probability density functions (PDFs) of individual inputs' uncertainties determine the level of uncertainty in model predictions as a result of input parameter uncertainties (Hanna et al., 1998). This method works by repeatedly generating from the probability distributions random parameters and then computing resulting statistics. For modelling uncertainty analysis, the MCS process is frequently used, and it allows measurement of model output uncertainty caused by uncertain model parameters, model structure, or input data (Shrestha et al., 2009). MCS can thus be utilized to deal with the input parameters uncertainty to be able to properly utilize ANNs. MCS uses two random sampling procedures, Latin hypercube sampling (LHS) and simple random sampling (SRS). Because MCS is simple to use, its accuracy and robustness are not affected by the complexity or dimension of the problem, except that for obtaining a good estimate, it necessitates a significant number of simulation runs. Because it converges with small sample sizes, the LHS technique can be used to overcome this issue.

25.4 APPLICATIONS OF AI TOOLS ON OPTIMIZATION AND MODELLING OF POLLUTANT REMOVAL PROCESS

In water treatment, many ANN approaches (hybrid methods and single methods) are used for modelling, optimizing, and predicting pollutant removal. By analysing variables such as COD, total suspended solids (TSS), and BOD, ANN can be used for the evaluation of wastewater treatment performance. Esquerre et al. (2004) used a link networking and multilayer network analysis method to determine BOD for wastewater biological treatment units (lagoon). In separate studies, Akratos and co-workers, and Hamed and researchers established a BOD elimination model based on ANNs (Hamed et al., 2004, Akratos et al., 2008). For variables such as COD, TSS, and BOD, Nourani and fellow researchers compared three different models for assessing the efficiency of a wastewater treatment facility. They found that the ensemble neural network model was more reliable in predicting BOD (a 24% increase) and COD, as well as total nitrogen in the effluent (both about 5%) (2018). A computational modelling technique was also used to forecast BOD concentrations in a highly eutrophic river. The findings revealed that using 16-4-1 modelling with phosphate as one of the key input variables, the parameter could be predicted ahead of time (Hasan, 2009). Emad et al. looked at how artificial intelligence may be used to model the degradation of antibiotics such as penicillin and other of this type (amoxicillin, cloxacillin, and ampicillin) in aqueous medium using the Fenton process, which is interpreted as COD elimination. The model's suitability was confirmed by the results generated by ANN, which exhibited a correlation coefficient of 0.997. Time, concentration, pH, and molar ratio of hydrogen peroxide, as well as ferrous ions, all had an impact on COD elimination (Elmolla et al., 2010). ANN investigated the prediction of COD removal effectiveness in sewage water (Tarke et al., 2016). The performance of the Khorramabad wastewater treatment plant (WWTP) was analysed using AI in one of Samaneh et al.'s studies. Temperature, dissolved oxygen, pH, BOD, dissolved solids, and COD were all taken into account, and the output performance was measured in terms of BOD, total solids, and COD. With a maximum contaminant removal effectiveness of 87.68%, good correlation coefficients were found (Khademikia et al., 2016).

A lot of scholars have looked at BP-ANN model, and it was suggested for optimization and modelling of water treatment (Akratos et al., 2009, Kundu et al., 2013, Santin et al., 2016). The BP-ANN model was used by Kundu et al. (2013) at the laboratory scale, to estimate the efficiency of a sequencing batch reactor. This model has the unique ability to understand non-linear functional correlations without relying on the complex process's underpinning mechanism. The ANN model offers a lot of potential for predicting nutrient removal efficiency in biological systems, according to this study. Akratos et al. developed the BP-ANN model to evaluate phosphorus removal in horizontal subsurface flow constructed wetlands (HSFCWs). The modelling results revealed a great match between experimental and predicted data, providing theoretical directions by employing a physical technique to remove nutrients from wastewater. Santin et al. (2016) suggested a BP-ANN model to help in riparian buffer strip (RBS) design based on the intended nitrogen filtering efficacy. The BP-ANN model can accurately represent the relationships between the variables relevant to the mitigation potential of nutrient pollution based on this result.

For phosphorous (P) and nitrogen (N2) removal, Gernaey and fellow researchers employed mathematical models that took into account the amount of influent, effluent quality, and biomass activity (Gernaey et al., 2004). ANN was used to analyse substances' adsorption characteristics such as resorcinol and phenol, and the results were compared to batch experimental results. *Carbonaceous* adsorbent amount, adsorbate concentrations, pH, and time were all input factors. The removal efficiency of the pollutants was determined by the neural network's output, which matched the experimental results (Aghav et al., 2011). Pinto et al. used Clementine 11.1 software to estimate the presence of manganese and turbidity in a similar attempt. A 1-year sample collection from water resources was used as the input data (2009). The removal effectiveness of reactive dye over a seed gum produced from Cassia species was determined using single-layer network analysis. To establish the extraction tendency, an error analysis was performed and the correlation coefficient value was calculated. For predicted and practical values, there was a strong correlation (Bui et al., 2016). Manu and Thalla used AI models such as adaptive neuro-fuzzy interference system (ANFIS) and support vector machine (SVM) to investigate the true removal efficiency of nitrogen from a wastewater treatment plant in Mangalore, India. Input variables such as influent pH, COD, total solids, NH3, and influent nitrogen were used in the study. Forecasting was found to be more efficient when SVM was used rather than ANFIS (2017). Mustafa Yasmeen and co-researchers used artificial intelligence to examine and simulate the oil-degrading capabilities of photo-Fenton's process. The ANN was trained using the backpropagation technique, and it was found that variables such as hydrogen peroxide, impact time, and ferrous ions have a considerable impact on oil destruction ability (Mustafa et al., 2018). The use of artificial intelligence (AI) has been successful in predicting zinc uptake onto defatted oil cake. The data were tested for the ANN-based model, and by utilizing parameters such as pH, temperature, initial concentration, flow rate, and adsorbent bed height, both batch and continuous operations were investigated. According to the correlation coefficient and error analysis value, the ANN has a good prediction capacity (Shanmugaprakash et al., 2018). Greywater and rainwater systems have also been studied using AI-based modelling. The Kosko fuzzy cognition mapping and MATLAB (matrix laboratory) software were used to create this modelling system. The importance of instrumentation, automation, and control in water systems and treatment units has been underlined by Yuan et al. (2019).

Using a BP-ANN model, Daneshvar et al. (2006) studied the solution of methine dye's decolourization, which contains C. I. Basic Yellow 28. This ANN model accurately represented the electrocoagulation process, allowing it to predict the system's behaviour under various situations. Asfaram et al. (2016a, b) used BP-ANN and RSM to optimize and model dye adsorption onto several materials simultaneously. On the basis of R^2, mean absolute error (MAE), root-mean-square error (RMSE), and absolute average deviation (AAD), the BP-ANN model was found to be more predictive than the RSM model. The combination of BP-ANN and RSM was used by Bagheri et al. (2016) to model and optimize the entire adsorption process after derivative spectrophotometry was used to determine the concentration of ternary dyes. The values predicted by the experimental data and the ANN model were found to agree to a satisfactory degree. Dastkhoon et al. (2017) used, under various conditions, BP-ANN and RSM to model the removal process of dye and predict the removal capacity of nanowires for the dye indigo carmine (IC) and Safranin O (SO). The BP-ANN model's sensitivity analysis revealed that the duration of sonication was the inessential and essential factors for IC and SO removal, with 12.60% and 36.62% relative importance, respectively. Furthermore, using an ANN-GA model, to remove an azo metal complex dye, Çelekli et al. (2012) used the lentil straw. The findings revealed that the initial concentration of dye has the greatest influence on dye adsorption, followed by pH. Computational technology was also used to automate wastewater treatment in the industrial and municipal sectors for the US army's water supply. pH monitoring, sludge pumping, alum dose, chlorination, and lime addition were all successfully automated. In the aerated activated sludge process, experiments were conducted to eliminate phosphorus and nitrogen. An anaerobic index was developed using a three-layered neuron network to build a quantitative relationship between organic compounds, which was determined to be suitable for the majority of organics. The ANN (Kohonen self-organizing feature map, KSOFM) was effectively used to evaluate the treatment plants and to comprehend the process' dependencies on factors such as dissolved oxygen, concentration, and pH level. The ANN clustered the generated map for about 400 days (Cinar, 2005).

25.5   CONCLUSIONS

AI systems are popular in a variety of fields due to their accuracy, precision, and continuous error-free operation. AI has been utilized successfully in fields such as wastewater treatment monitoring. Areas such as contaminant detection and elimination, detection of levels for process initiation, release of chemicals for the treatment of wastewater, identification of contaminants, and subsequent selection of various treatment processes still need to be studied in the context of the mentioned technology and are still of interest to researchers. Because of the following advantages, artificial neural networks (ANNs) are useful AI techniques for optimizing removal processes: (1) ANNs, compared to other methods, can be used to model a multivariable system and, by data training, extract complex non-linear correlations between variables. (2) Traditional mathematical models have limitations that ANNs can overcome by the extraction of the essential information using the data of training, and this does not demand a prior specification of fitting functions. The following are some of the disadvantages of ANNs: (1) Because the bias and weight between neurons are provided randomly, ANNs have bad reproducibility; (2) ANNs cannot produce an optimal condition for the treatment of water because it is possible for them to fall into a local optimal solution, for the engineering application. ANNs combined with other AI tools (e.g. SA, PSO, MCS, IA, ICA, ACA, and DT) have the potential to create optimal operational variables for removal procedures owing to their ability to search for global optima.

REFERENCES

Acharya, C., Mohanty, S., Sukla, L. & Misra, V. 2006. Prediction of sulphur removal with *Acidithiobacillus* sp. using artificial neural networks. *Ecological Modelling,* 190, 223–230.
Aghav, R. M., Kumar, S. & Mukherjee, S. N. 2011. Artificial neural network modeling in competitive adsorption of phenol and resorcinol from water environment using some carbonaceous adsorbents. *Journal of Hazardous Materials,* 188, 67–77.
Akratos, C., Papaspyros, I. & Tsihrintzis, V. 2008. An artificial neural network model and design equations for BOD and COD removal prediction in horizontal subsurface flow constructed wetlands. *Chemical Engineering Journal,* 143, 96–110.
Akratos, C. S., Papaspyros, J. N. E. & Tsihrintzis, V. A. 2009. Artificial neural network use in ortho-phosphate and total phosphorus removal prediction in horizontal subsurface flow constructed wetlands. *Biosystems Engineering,* 102, 190–201.

Aleboyeh, A., Kasiri, M., Olya, M. & Aleboyeh, H. 2008. Prediction of azo dye decolorization by UV/H_2O_2 using artificial neural networks. *Dyes and Pigments,* 77, 288–294.

Arhami, M., Kamali, N. & Rajabi, M. M. 2013. Predicting hourly air pollutant levels using artificial neural networks coupled with uncertainty analysis by Monte Carlo simulations. *Environmental Science and Pollution Research,* 20, 4777–4789.

Asfaram, A., Ghaedi, M., Azqhandi, M. H. A., Goudarzi, A. & Dastkhoon, M. 2016a. Statistical experimental design, least squares-support vector machine (LS-SVM) and artificial neural network (ANN) methods for modeling the facilitated adsorption of methylene blue dye. *RSC Advances,* 6, 40502–40516.

Asfaram, A., Ghaedi, M., Hajati, S. & Goudarzi, A. 2016b. Synthesis of magnetic γ-Fe2O3-based nanomaterial for ultrasonic assisted dyes adsorption: Modeling and optimization. *Ultrasonics Sonochemistry,* 32, 418–431.

Assefi, P., Ghaedi, M., Ansari, A., Habibi, M. & Momeni, M. 2014. Artificial neural network optimization for removal of hazardous dye Eosin Y from aqueous solution using Co_2O_3-NP-AC: Isotherm and kinetics study. *Journal of Industrial and Engineering Chemistry,* 20, 2905–2913.

Bagheri, A. R., Ghaedi, M., Asfaram, A., Hajati, S., Ghaedi, A. M., Bazrafshan, A. & Rahimi, M. R. 2016. Modeling and optimization of simultaneous removal of ternary dyes onto copper sulfide nanoparticles loaded on activated carbon using second-derivative spectrophotometry. *Journal of the Taiwan Institute of Chemical Engineers,* 65, 212–224.

Bui, H., Perng, Y.-S. & Duong, G. 2016. The use of artificial neural network for modeling coagulation of reactive dye wastewater using *Cassia fistula* Linn. (CF) gum. *Journal of Environmental Science and Management,* 19, 9.

Çelekli, A., Bozkurt, H. & Geyik, F. 2012. Use of artificial neural networks and genetic algorithms for prediction of sorption of an azo-metal complex dye onto lentil straw. *Bioresource Technology,* 129C, 396–401.

Chang, H. & Hou, W.-C. 2006. Optimization of membrane gas separation systems using genetic algorithm. *Chemical Engineering Science,* 61, 5355–5368.

Cheung, K., Poon, B., Lan, C. & Wong, M. H. 2003. Assessment of metal and nutrient concentrations in river water and sediment collected from the cities in the Pearl River Delta, South China. *Chemosphere,* 52, 1431–1440.

Chu, K. 2003. Prediction of two-metal biosorption equilibria using a neural network. *(European Journal of Mineral Processing and Environmental Protection),* 3, 119–127.

Cinar, O. 2005. New tool for evaluation of performance of wastewater treatment plant: Artificial neural network. *Process Biochemistry,* 40, 2980–2984.

Daneshvar, N., Khataee, A. R. & Djafarzadeh, N. 2006. The use of artificial neural networks (ANN) for modeling of decolorization of textile dye solution containing C. I. Basic Yellow 28 by electrocoagulation process. *Journal of Hazardous Materials,* 137, 1788–1795.

Dastkhoon, M., Ghaedi, M., Asfaram, A., Ahmadi Azqhandi, M. H. & Purkait, M. K. 2017. Simultaneous removal of dyes onto nanowires adsorbent use of ultrasound assisted adsorption to clean waste water: Chemometrics for modeling and optimization, multicomponent adsorption and kinetic study. *Chemical Engineering Research and Design,* 124, 222–237.

Dehghani, M., Saghafian, B., Nasiri Saleh, F., Farokhnia, A. & Noori, R. 2014. Uncertainty analysis of streamflow drought forecast using artificial neural networks and Monte-Carlo simulation. *International Journal of Climatology,* 34, 1169–1180.

Deshmukh, S. C., Senthilnath, J., Dixit, R. M., Malik, S. N., Pandey, R. A., Vaidya, A. N., Omkar, S. N. & Mudliar, S. N. 2012. Comparison of radial basis function neural network and response surface methodology for predicting performance of biofilter treating toluene. *Journal of Software Engineering and Applications,* 5(8), 595–603.

Dubey, R., Bajpai, J. & Bajpai, A. 2015. Green synthesis of graphene sand composite (GSC) as novel adsorbent for efficient removal of Cr (VI) ions from aqueous solution. *Journal of Water Process Engineering,* 5, 83–94.

El-Wakeel, A. S. 2014. Design optimization of PM couplings using hybrid particle swarm optimization-simplex method (PSO-SM) algorithm. *Electric Power Systems Research,* 116, 29–35.

Elmolla, E. S., Chaudhuri, M. & Eltoukhy, M. M. 2010. The use of artificial neural network (ANN) for modeling of COD removal from antibiotic aqueous solution by the Fenton process. *Journal of Hazardous Materials,* 179, 127–34.

Esquerre, K., Seborg, D., Mori, M. & Bruns, R. 2004. Application of steady-state and dynamic modeling for the prediction of the BOD of an aerated lagoon at a pulp and paper mill Part II. Nonlinear approaches. *Chemical Engineering Journal,* 105, 61–69.

Fan, M., Li, T., Hu, J., Cao, R., Wei, X., Shi, X. & Ruan, W. 2017. Artificial neural network modeling and genetic algorithm optimization for cadmium removal from aqueous solutions by reduced graphene oxide-supported nanoscale zero-valent iron (nZVI/rGO) composites. *Materials,* 10, 544.

Gernaey, K. V., Van Loosdrecht, M. C. M., Henze, M., Lind, M. & Jørgensen, S. B. 2004. Activated sludge wastewater treatment plant modelling and simulation: State of the art. *Environmental Modelling & Software,* 19, 763–783.

Ghaedi, M., Shojaeipour, E., Ghaedi, A. M. & Sahraei, R. 2015. Isotherm and kinetics study of malachite green adsorption onto copper nanowires loaded on activated carbon: Artificial neural network modeling and genetic algorithm optimization. *Spectrochimica Acta Part A: Molecular and Biomolecular Spectroscopy,* 142, 135–149.

Gupta, P. K., Yadav, B., Kumar, A. & Himanshu, S. K. 2021. Machine learning and artificial intelligence application in constructed wetlands for industrial effluent treatment: Advances and challenges in assessment and bioremediation modeling. In: Saxena, G., Kumar, V., Shah, M.P. (Eds.), *Bioremediation for Environmental Sustainability: Toxicity, Mechanisms of Contaminants Degradation, Detoxification and Challenges.* Elsevier, pp. 403–414.

Hamed, M. M., Khalafallah, M. G. & Hassanien, E. A. 2004. Prediction of wastewater treatment plant performance using artificial neural networks. *Environmental Modelling and Software,* 19, 919–928.

Hanna, S. R., Chang, J. C. & Fernau, M. E. 1998. Monte carlo estimates of uncertainties in predictions by a photochemical grid model (UAM-IV) due to uncertainties in input variables. *Atmospheric Environment,* 32, 3619–3628.

Hasan, Y. A. 2009. Predicting biochemical oxygen demand as indicator of river pollution using artificial neural networks. In *18th World IMACS / MODSIM Congress,* Cairns, Australia.

Hoh, E. & Hites, R. A. 2005. Brominated flame retardants in the atmosphere of the east-central United States. *Environmental Science & Technology,* 39, 7794–7802.

Hu, J., Zhang, X. & Wang, Z. 2010. A review on progress in QSPR studies for surfactants. *International Journal of Molecular Sciences,* 11, 1020–1047.

Huang, X. & Fang, G. 2010. Water resources allocation effect evaluation based on chaotic neural network model. *Journal of Computers,* 5(8), 1169–1176.

Jiang, B., Zhang, F., Sun, Y., Zhou, X., Dong, J. & Zhang, L. 2014. Modeling and optimization for curing of polymer flooding using an artificial neural network and a genetic algorithm. *Journal of the Taiwan Institute of Chemical Engineers,* 45, 2217–2224.

Kalteh, A. M., Hjorth, P. & Berndtsson, R. 2008. Review of the self-organizing map (SOM) approach in water resources: Analysis, modelling and application. *Environmental Modelling & Software,* 23, 835–845.

Khademikia, S., Haghizadeh, A., Godini, H. & Shams Khorramabadi, G. 2016. The performance evaluation of Khorramabad wastewater treatment plant by using artificial intelligence network. *Scientific Magazine Yafte,* 18, 12–23.

Khajeh, M., Kaykhaii, M. & Sharafi, A. 2013. Application of PSO-artificial neural network and response surface methodology for removal of methylene blue using silver nanoparticles from water samples. *Journal of Industrial and Engineering Chemistry*, 19, 1624–1630.

Kose, E. 2008. Modelling of colour perception of different age groups using artificial neural networks. *Expert Systems with Applications*, 34, 2129–2139.

Kumar, J. R., Lee, H.-I., Lee, J.-Y., Kim, J.-S. & Sohn, J.-S. 2008a. Comparison of liquid–liquid extraction studies on platinum (IV) from acidic solutions using bis (2, 4, 4-tri-methylpentyl) monothiophosphinic acid. *Separation and Purification Technology*, 63, 184–190.

Kumar, K. V., Porkodi, K., Rondon, R. A. & Rocha, F. 2008b. Neural network modeling and simulation of the solid/liquid activated carbon adsorption process. *Industrial & Engineering Chemistry Research*, 47, 486–490.

Kumar, V., Singh, K., & Shah, M.P. 2021. Advanced oxidation processes for complex wastewater treatment. In: Shah, M.P. (Eds.), *Advance Oxidation Process for Industrial Effluent Treatment*. Elsevier. Doi: 10.1016/B978-0-12-821011-6.00001-3.

Kundu, P., Debsarkar, A. & Mukherjee, S. 2013. Artificial neural network modeling for biological removal of organic carbon and nitrogen from slaughterhouse wastewater in a sequencing batch reactor. *Advances in Artificial Neural Systems*, 2013, 268064.

Li, L., Hu, J., Shi, X., Fan, M., Luo, J. & Wei, X. 2016a. Nanoscale zero-valent metals: A review of synthesis, characterization, and applications to environmental remediation. *Environmental Science and Pollution Research*, 23, 17880–17900.

Li, L., Hu, J., Shi, X., Ruan, W., Luo, J. & Wei, X. 2016b. Theoretical studies on structures, properties and dominant debromination pathways for selected polybrominated diphenyl ethers. *International Journal of Molecular Sciences*, 17, 927.

Long, Z., Pan, Z., Wang, W., Ren, J., Yu, X., Lin, L., Lin, H., Chen, H. & Jin, X. 2019. Microplastic abundance, characteristics, and removal in wastewater treatment plants in a coastal city of China. *Water Research*, 155, 255–265.

López, M. E., Rene, E. R., Boger, Z., Veiga, M. C. & Kennes, C. 2017. Modelling the removal of volatile pollutants under transient conditions in a two-stage bioreactor using artificial neural networks. *Journal of Hazardous Materials*, 324, 100–109.

Luo, J., Chen, C. & Xie, J. 2015a. Multi-objective immune algorithm with preference-based selection for reservoir flood control operation. *Water Resources Management*, 29, 1447–1466.

Luo, J., Hu, J., Wei, X., Fu, L. & Li, L. 2015b. Dehalogenation of persistent halogenated organic compounds: A review of computational studies and quantitative structure–property relationships. *Chemosphere*, 131, 17–33.

Luo, J., Hu, J., Zhuang, Y., Wei, X. & Huang, X. 2013. Electron-induced reductive debromination of 2, 3, 4-tribromodiphenyl ether: A computational study. *Journal of Molecular Modeling*, 19, 3333–3338.

Ma, X., Li, Y., Li, X., Yang, L. & Wang, X. 2011. Preparation of novel polysulfone capsules containing zirconium phosphate and their properties for Pb^{2+} removal from aqueous solution. *Journal of Hazardous Materials*, 188, 296–303.

Mandal, S., Mahapatra, S., Sahu, M. & Patel, R. 2015. Artificial neural network modelling of As (III) removal from water by novel hybrid material. *Process Safety and Environmental Protection*, 93, 249–264.

Manu, D. S. & Thalla, A. K. 2017. Artificial intelligence models for predicting the performance of biological wastewater treatment plant in the removal of Kjeldahl nitrogen from wastewater. *Applied Water Science*, 7, 3783–3791.

Marseguerra, M., Zio, E. & Podofillini, L. 2005. Multiobjective spare part allocation by means of genetic algorithms and Monte Carlo simulation. *Reliability Engineering & System Safety*, 87, 325–335.

Mendoza-Castillo, D., Villalobos-Ortega, N., Bonilla-Petriciolet, A. & Tapia-Picazo, J. 2015. Neural network modeling of heavy metal sorption on lignocellulosic biomasses: Effect of metallic ion properties and sorbent characteristics. *Industrial & Engineering Chemistry Research*, 54, 443–453.

Mingzhi, H., Ma, Y., Jinquan, W. & Yan, W. 2009. Simulation of a paper mill wastewater treatment using a fuzzy neural network. *Expert systems with Applications*, 36, 5064–5070.

Mohanraj, M., Jayaraj, S. & Muraleedharan, C. 2012. Applications of artificial neural networks for refrigeration, air-conditioning and heat pump systems—A review. *Renewable and Sustainable Energy Reviews*, 16, 1340–1358.

Mohebbi, A., Taheri, M. & Soltani, A. 2008. A neural network for predicting saturated liquid density using genetic algorithm for pure and mixed refrigerants. *International Journal of Refrigeration*, 31, 1317–1327.

Mustafa, Y., Alwared, A. & Majeed, G. 2018. Environmental science and pollution research the use of artificial neural network (ANN) for the prediction and simulation of oil degradation in wastewater by AOP The use of artificial neural network (ANN) for the prediction and simulation of oil degradation in wastewater by AOP. *Environmental Science and Pollution Research*, 21, 7530–7537.

Nandi, B., Moparthi, A., Uppaluri, R. & Purkait, M. 2010. Treatment of oily wastewater using low cost ceramic membrane: Comparative assessment of pore blocking and artificial neural network models. *Chemical Engineering Research and Design*, 88, 881–892.

Nourani, V., Elkiran, G. & Abba, S. I. 2018. Wastewater treatment plant performance analysis using artificial intelligence - an ensemble approach. *Water Science and Technology*, 78, 2064–2076.

Ouarda, T. B. & Shu, C. 2009. Regional low-flow frequency analysis using single and ensemble artificial neural networks. *Water Resources Research*, 45. Doi: 10.1029/2008WR007196

Pinto, A., Fernandes, A., Vicente, H. & Neves, J. 2009. Optimizing water treatment systems using artificial intelligence based tools. *WIT Transactions on Ecology and the Environment*, 125, 185–194.

Reynel-Avila, H. E., Bonilla-Petriciolet, A. & De La Rosa, G. 2015. Analysis and modeling of multicomponent sorption of heavy metals on chicken feathers using Taguchi's experimental designs and artificial neural networks. *Desalination and Water Treatment*, 55, 1885–1899.

Sacchi, R., Ozturk, M. C., Príncipe, J. C., Carneiro, A. A. F. M. & da Silva, I.N. 2007. Water Inflow Forecasting using the Echo State Network: a Brazilian Case Study. In *International Joint Conference on Neural Networks*, Orlando, FL, IEEE. Doi: 10.1109/IJCNN.2007.4371334.

Sahu, J., Acharya, J. & Meikap, B. 2009. Response surface modeling and optimization of chromium (VI) removal from aqueous solution using tamarind wood activated carbon in batch process. *Journal of Hazardous Materials*, 172, 818–825.

Santin, F. M., Da Silva, R. V. & Grzybowski, J. M. V. 2016. Artificial neural network ensembles and the design of performance-oriented riparian buffer strips for the filtering of nitrogen in agricultural catchments. *Ecological Engineering*, 94, 493–502.

Shanmugaprakash, M., Venkatachalam, S., Rajendran, K. & Pugazhendhi, A. 2018. Biosorptive removal of Zn(II) ions by *Pongamia* oil cake (*Pongamia pinnata*) in batch and fixed-bed column studies using response surface methodology and artificial neural network. *Journal of Environmental Management*, 227, 216–228.

Shojaeimehr, T., Rahimpour, F., Khadivi, M. A. & Sadeghi, M. 2014. A modeling study by response surface methodology (RSM) and artificial neural network (ANN) on Cu^{2+} adsorption optimization using light expended clay aggregate (LECA). *Journal of Industrial and Engineering Chemistry*, 20, 870–880.

Shrestha, D. L., Kayastha, N. & Solomatine, D. P. 2009. A novel approach to parameter uncertainty analysis of hydrological models using neural networks. *Hydrology and Earth System Sciences*, 13, 1235–1248.

Singh, K. P., Gupta, S., Singh, A. K. & Sinha, S. 2010. Experimental design and response surface modeling for optimization of Rhodamine B removal from water by magnetic nanocomposite. *Chemical Engineering Journal*, 165, 151–160.

Tak, B.-Y., Tak, B.-S., Kim, Y.-J., Park, Y.-J., Yoon, Y.-H. & Min, G.-H. 2015. Optimization of color and COD removal from livestock wastewater by electrocoagulation process: Application of Box–Behnken design (BBD). *Journal of Industrial and Engineering Chemistry*, 28, 307–315.

Tarke, P., Sarda, D. P. & Sadgir, P. 2016. Performance of ANNs for prediction of TDS of Godavari River, India. *International Journal of Engineering Research*, ISSN, 2319-68902347.

Wang, J. & Deng, Z. 2016. Modeling and prediction of oyster norovirus outbreaks along Gulf of Mexico coast. *Environmental Health Perspectives*, 124, 627–633.

Wang, Z. & Man, Y. 2021. Artificial intelligence algorithm application in wastewater treatment plants: Case study for COD load prediction. In: Ren, J., Shen, W., Man, Y. & Dong, L. (Eds.), *Applications of Artificial Intelligence in Process Systems Engineering*. Elsevier.

Wu, J., Peng, D., Li, Z., Zhao, L. & Ling, H. 2015. Network intrusion detection based on a general regression neural network optimized by an improved artificial immune algorithm. *Plos One*, 10, e0120976.

Xie, H., Li, J., Zhang, C., Tian, Z., Liu, X., Tang, C., Han, Y. & Liu, W. 2014. Assessment of heavy metal contents in surface soil in the Lhasa–Shigatse–Nam Co area of the Tibetan Plateau, China. *Bulletin of Environmental Contamination and Toxicology*, 93, 192–198.

Yuan, Z., Olsson, G., Cardell-Oliver, R., Van Schagen, K., Marchi, A., Deletic, A., Urich, C., Rauch, W., Liu, Y. & Jiang, G. 2019. Sweating the assets - The role of instrumentation, control and automation in urban water systems. *Water Research*, 155, 381–402.

Zhang, Y. & Pan, B. 2014. Modeling batch and column phosphate removal by hydrated ferric oxide-based nanocomposite using response surface methodology and artificial neural network. *Chemical Engineering Journal*, 249, 111–120.

Zhao, L., Dai, T., Qiao, Z., Sun, P., Hao, J. & Yang, Y. 2020. Application of artificial intelligence to wastewater treatment: A bibliometric analysis and systematic review of technology, economy, management, and wastewater reuse. *Process Safety and Environmental Protection*, 133, 169–182.

Index

Note: **Bold** page numbers refer to tables; *italic* page numbers refer to figures.

For Product Safety Concerns and Information please contact our EU
representative GPSR@taylorandfrancis.com
Taylor & Francis Verlag GmbH, Kaufingerstraße 24, 80331 München, Germany

www.ingramcontent.com/pod-product-compliance
Lightning Source LLC
Chambersburg PA
CBHW060442240326
41598CB00087B/2244